T0145095

Advances in Intelligent Systems and Computing

Volume 274

Series editor

Janusz Kacprzyk, Polish Academy of Sciences, Warsaw, Poland
e-mail: kacprzyk@ibspan.waw.pl

For further volumes:
http://www.springer.com/series/11156

About this Series

The series "Advances in Intelligent Systems and Computing" contains publications on theory, applications, and design methods of Intelligent Systems and Intelligent Computing. Virtually all disciplines such as engineering, natural sciences, computer and information science, ICT, economics, business, e-commerce, environment, healthcare, life science are covered. The list of topics spans all the areas of modern intelligent systems and computing.

The publications within "Advances in Intelligent Systems and Computing" are primarily textbooks and proceedings of important conferences, symposia and congresses. They cover significant recent developments in the field, both of a foundational and applicable character. An important characteristic feature of the series is the short publication time and world-wide distribution. This permits a rapid and broad dissemination of research results.

Advisory Board

Jong-Hwan Kim · Eric T. Matson
Hyun Myung · Peter Xu · Fakhri Karray
Editors

Robot Intelligence Technology and Applications 2

Results from the 2nd International Conference
on Robot Intelligence Technology
and Applications

 Springer

Editors

Jong-Hwan Kim
Electrical Engineering
KAIST
Daejeon
Korea

Eric T. Matson
Computer and Information Technology
Purdue University
Indiana
USA

Hyun Myung
Civil and Environmental Engineering
KAIST
Daejeon
Korea

Peter Xu
Department of Mechanical Engineering
University of Auckland
Auckland
New Zealand

Fakhri Karray
University of Waterloo
Ontario
Canada

ISSN 2194-5357   ISSN 2194-5365   (electronic)
ISBN 978-3-319-05581-7   ISBN 978-3-319-05582-4   (eBook)
DOI 10.1007/978-3-319-05582-4
Springer Cham Heidelberg New York Dordrecht London

Library of Congress Control Number: 2014934673

Preface

Since the industrial revolution, technological evolution has been the defining aspect of the societal progress. Among all the technologies that emerged over the years, information technology (IT) is the one that truly revolutionized the modern lifestyle. IT, first coined in 1958 and booming since the early 1990s, is the application of computers and telecommunication equipments to store, retrieve, transmit and manipulate data. In 1990s, the concept of Information Superhighway was developed to realize the goals of IT across the globe.

Now we are facing a new technological challenge on how to store and retrieve knowledge and manipulate intelligence, in addition to the management of information and data, for autonomous services by intelligent systems. By utilizing the technology, the systems should be capable of carrying out real world tasks autonomously. To address this issue, robot researchers have been developing intelligence technology (InT) for "robots that think." InT is intelligence operating technology, which is the application of machines and agents to perceive and process data and information for knowledge-based reasoning and utilize their own reasoning to execute an appropriate action. InT covers all aspects of intelligence from perception at sensor level and reasoning at cognitive level to behavior planning at execution level for each low level segment of the machine. It is equipped with technologies for cognitive reasoning, social interaction with humans, behavior generation, ability to cooperate with other robots, ambience awareness, and an artificial genome that can be passed on to other robots. These technologies are to materialize cognitive intelligence, social intelligence, behavioral intelligence, collective intelligence, ambient intelligence and genetic intelligence.

As IT has been flourished through information superhighway for the last two decades, InT also needs a central and ubiquitous medium with the ability to intelligently process and autonomously act on data by operating the technology. The medium should be interconnected with the information superhighway. Utilizing the revolutionary advances in information technology and combining them with the potential prospects of InT, the medium would be intelligence super agent (iSA) that can enhance human capability in perception, reasoning and action and provide intelligent supervision to the less intelligent agents and robots within its domain. iSA derives some of its inspiration from similar intelligent agents in popular science fictions, such as VIKI (Virtual Interactive

Kinetic Intelligence) and JARVIS (Just A Rather Very Intelligent System). VIKI is a supercomputer based intelligent agent responsible for managing a commercial building in the famous science fiction movie I,Robot (2004). JARVIS is a super intelligent and multifunctional software agent that manages its owner's lab and mansion in the famous science fiction series called Iron Man (2008). The key properties of these agents are intelligence, autonomy and ubiquity. In this sense, the future research direction of InT would be the development of iSA with such challenging properties.

Following the first edition, this second edition also aims at serving the researchers and practitioners in related fields with a timely dissemination of the recent progress on robot intelligence technology and its applications, based on a collection of papers presented at the 2nd International Conference on Robot Intelligence Technology and Applications (RiTA), held in Denver, USA, December 18–20, 2013. For better readability, this edition has the total 84 papers grouped into 3 chapters: Part I: Ambient, Collective, Cognitive and Social Intelligence, Part II: Behavioral, Collective and Genetic Intelligence, Part III: Emerging Applications of Robot Intelligence Technology, where individual chapters, edited respectively by Eric T. Matson, Hyun Myung, Peter Xu, Fakhri Karray along with Jong-Hwan Kim, begin with a brief introduction written by the respective chapter editors.

I do hope that readers find the second edition of Robot Intelligence Technology and Applications 2, RiTA 2, stimulating, enjoyable and helpful for their further research.

Jong-Hwan Kim
General Chair
RiTA 2013

Contents

Part III: Emerging Applications of Robot Intelligence Technology

Part I

Ambient, Collective, Cognitive and Social Intelligence

Eric T. Matson

Ambient, collective, cognitive and social intelligence technology gives artificial entities the capabilities normally only represented in human beings. The abilities to quickly gather, discern and manage knowledge about the human ambient environment. The ability to work together as a team, organization, society or collective is another high level human-like capability. Cognition is an application of higher order mental use. Social intelligence imparts the trait of working together and in many ways, enables collective capability. The theory, development and realization of technology to create a more human-like robot depends on the at least rudimentary mastery of ambient, cognitive, collective and social intelligence. This section demonstrates new efforts to achieve this technological development with a total of 27 papers.

Ambient and Collective Intelligence represents a group of 13 papers covering various research areas. Several papers focus on localization of robots using sensors, complex cameras or magnetic field interpretation. Another area complementary to localization is mapping and there are entries that use laser range finders to map using glass walls. A new area of research is using directional antennas as sensors for robotic following and communication is also represented. Larger scale systems, such as swarm systems and those using an inverted robotic space for agriculture are also within this section.

1) Feature-based 6-DoF Camera Localization using Prior Point Cloud and Images.
2) Programming an E-Puck Robot to Create Maps of Virtual and Physical Environments.
3) A method to localize transparent glass obstacle using laser range finder in mobile robot indoor navigation.
4) Robotic Follower System using Bearing-only Tracking with Directional Antennas.
5) Oscillator aggregation in redundant robotic systems for emergence of homeostasis.
6) The ChIRP Robot: a Versatile Swarm Robot Platform.
7) Aeroponic Greenhouse as an Autonomous System using Intelligent Space for Agriculture Robotics.
8) A Convex Fuzzy Range-Free Sensor Network Localization Method.
9) Multi-Directional Weighted Interpolation for Wi-Fi Localisation.

10) Visual Loop-Closure Detection Method Using Average Feature Descriptors.
11) Mobile Robot Localization using Multiple Geomagnetic Field Sensors.
12) GA-based Optimal Waypoint Design for Improved Path Following of Mobile Robot.
13) Indoor Mobile Robot Localization Using Ambient Magnetic Fields and Range Measurements.

In the collection of papers representing **Cognitive Intelligence,** there are 9 papers in the areas of path planning with high-dimension lattices, artificial creature composite behavior organization, brain wave identification for robotic arm control, word recognition using learning, action verification in manufacturing, intention recognition for human performance, knowledge representation for orthopaedic surgery and dynamic environments and kit building using sensor networks.

1) Combined Trajectory Generation and Path Planning for Mobile Robots Using Lattices with Hybrid Dimensionality.
2) Organization and Selection Methods of Composite Behaviors for Artificial Creatures. Using the Degree of Consideration-based Mechanism of Thought.
3) Brainwave Variability Identification in Robotic Arm Control Strategy.
4) Acquisition of Context-based Active Word Recognition by Q-Learning Using a Recurrent Neural Network.
5) An Ontology Based Approach to Action Verification for Agile Manufacturing.
6) Evaluating State-Based Intention Recognition Algorithms Against Human Performance.
7) Knowledge and Data Representation for Motion Planning in Dynamic Environments.
8) A Simulated Sensor-based Approach for Kit Building Applications.
9) A Survey on Biomedical Knowledge Representation for Robotic Orthopaedic Surgery.

Social Intelligence has a total of 5 papers approaches social robotic intelligence in the applications of human to robotic interfaces for fire-fighting robots and behavior recognition. Also, a direct connection using direct brain machine interface is contained. Using configurable modular robotics for learning and explorations of using robotics for artistic creation are also demonstrated.

1) Consideration about the Application of Dynamic Time Warping to Human Hands Behavior Recognition for Human-Robot Interaction.
2) Lessons Learned in Designing User-configurable Modular Robotics.
3) Playware Explorations in Robot Art.
4) Navigation Control of a Robot from a Remote Location via the Internet using Brain-Machine Interface.
5) Design of Knowledge-based Communication between Human and Robot Using Ontological Semantic Technology in Firefighting Domain.

Feature-Based 6-DoF Camera Localization Using Prior Point Cloud and Images

Hyongjin Kim[1], Donghwa Lee[1], Taekjun Oh[1], Sang Won Lee[1],
Yungeun Choe[2], and Hyun Myung[1]

[1] Urban Robotics Lab, KAIST(Korea Advanced Insititute of Science and Technology),
291 Daehak-ro, Yuseong-gu, Daejeon 305-701, Korea
{hjkim86,leedonghwa,buljaga,lsw618,hmyung}@kaist.ac.kr
[2] Robot Research Lab, KAIST(Korea Advanced Insititute of Science and Technology),
291 Daehak-ro, Yuseong-gu, Daejeon 305-701, Korea
yungeun@kaist.ac.kr

Abstract. In this paper, we present a new localization algorithm to estimate the localization of a robot based on prior data. Over the past decade, the emergence of numerous ways to utilize various prior data has opened up possibilities for their applications in robotics technologies. However, challenges still remain in estimating a robot's 6-DoF position by simply analyzing the limited information provided by images from a robot. This paper describes a method of overcoming this technical hurdle by calculating the robot's 6-DoF location. It only utilizes a current 2D image and prior data, which consists of its corresponding 3D point cloud and images, to calculate the 6-DoF position. Furthermore, we employed the SURF algorithm to find the robot's position by using the image's features and the 3D projection method. Experiments were conducted by the loop of 510m trajectory, which is included the prior data. It is expected that our method can be applied to broad areas by using enormous data such as point clouds and street views in the near future.

Keywords: Localization, Feature Matching, Point Cloud, 3D to 2D projection.

1 Introduction

For the past few decades, the topic of autonomous mobile robot and UGVs(Unmanned Ground Vehicles) has been widely investigated [1, 2]. The most essential element of an autonomous robot navigation system is to identify the physical location of the robot, which in technical terms, is the localization problem.

Currently, there exist numerous methods proposed to solve the localization problem. However, there still remain several challenges. Although GPS(Global Position System) is generally used as a choice system to resolve the localization problem, there are some technical difficulties, mainly the occurrence of poor results under conditions of non-line-of-sight or in radio shadow areas of GPS. In order to overcome this complication, an integrated sensor that is a fusion of GPS and INS(Inertial Navigation System) is widely used [3]. However, the state of INS technology is imperfect as well, as there are difficulties in estimating global positions due to accumulative errors.

J.-H. Kim et al. (eds.), *Robot Intelligence Technology and Applications 2*,
Advances in Intelligent Systems and Computing 274,
DOI: 10.1007/978-3-319-05582-4_1, © Springer International Publishing Switzerland 2014

Recently, as the use of prior information became remarkably easier through such programs as Google Street View [4], the global research endeavor to solve the localization problem using prior information has gained momentum. One notable study that has exploited prior images information employed an integrated sensor that was comprised of a fusion of GPS, INS, and positions of prior images [5]. Another research team reported the use of LADARs sensors to construct a prior point cloud, which constituted the foundation on which localization was conducted with LADARs sensors [6]. However, the high cost of LADAR served as a bottleneck for practical applications. Thus, another study dealt with the development of a system with a cheaper sensor, which utilized a pre-constituted prior point cloud, a low-cost camera, and odometry information to perform localization [7].

Therefore, this study proposes the use of existing prior data as a novel localization method that only necessitates a camera. The precedent study of this system is introduced [8] in the work-in-progress secsion. In this system, the features are extracted from prior images, and 3D coordinates are obtained from the prior point cloud. Similarly, features are extracted from the current image. Matching is performed with the features of the current image and those of prior images. Finally, a 3D projection algorithm is applied to estimate the robot's location.

2   Proposed Localization System

In this study, our approach utilizes only the current 2D image data and prior data. The processing steps of our proposed localization system are illustrated in figure 1. For preprocessing, the 2D features of prior images are extracted using the SURF(Speeded Up Robust Features) algorithm [9]. Furthermore, the 3D coordinates of the 2D features are also extracted using a ray tracing algorithm in the prior point cloud. After performing preprocessing, the 2D features of the current image are extracted on the fly. The features of the current image are matched to those of prior images using a RANSAC(RANdom SAmple Consensus) algorithm based on 2D homography. Then, the 3D coordinates of the matched features from preprocessing are projected to the 2D features of the current image to estimate candidate positions. Finally, the final position of the robot is estimated by median and mean filter among the candidate positions.

2.1   Preprocessing

In the preprocessing step, it is possible to reduce the operating time by performing SURF and ray tracing algorithm in advance using prior data. First, a survey vehicle that is equipped with costly devices such as LIDARs and camera collects prior data. The prior data consists of a point cloud and images. Each image contains 3D coordinates of a camera center in the point cloud at the time when the image was taken. Each image utilizes a SURF algorithm to extract its own feature descriptors. The 2D coordinates of the extracted features in the image are expressed as lines in the point cloud, using a pin hole camera model. The first meeting 3D point is found on the line formed by connecting the camera center, the feature point, and point cloud by utilizing a ray tracing algorithm. The concept of the ray tracing algorithm has been visually depicted in

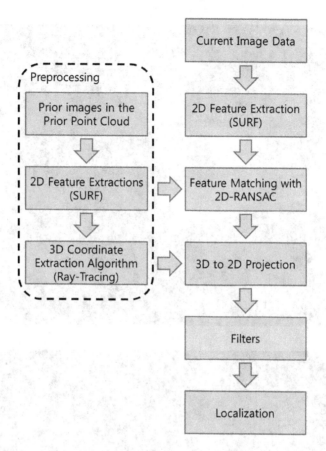

Fig. 1. Overview of the proposed localization system

figure 2(a), and figure 2(b) shows the ray tracing results for all images on the point cloud map. All feature descriptors of all prior images and its corresponding feature coordinates are saved for the main localization algorithm.

2.2 Feature Extraction, Matching, and 2D-RANSAC

In order to accomplish feature matching of the current image with prior images, the SURF algorithm is utilized. The intensities of brightness between the current image and prior images may show remarkable differences, even if the photos were taken from similar locations. This contrast can be attributed to the fact that the data were obtained from different cameras and time frames. Due to this discrepancy of light intensities, poor results are drawn in feature matching, even in similar places, as shown in figure 3(a). In order to solve this problem, all images udergo histogram equalization method before feature extraction. In figure 3(b), it is clear that the matching result is more successful than that without pretreatment. However, there is still much mismatching, due to the fact that the images were taken from different cameras and time frames.

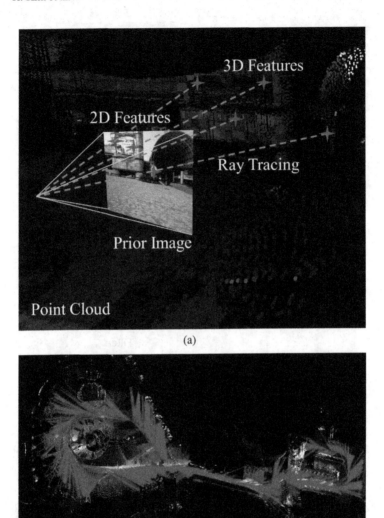

Fig. 2. (a) The green dashed lines show a concept of the ray tracing algorithm in the point cloud. The red stars indicate 2D and 3D features (b) The red lines indicate the ray tracing result of each 2D feature for a full path.

In order to eliminate the mismatching results, the RANSAC algorithm based on 2D homography was used. The final matching result is shown in figure 3(c). It is obvious that the outliers have been successfully removed. After carrying out feature matching of the current image and the prior images, the reference images are selected among the prior images by selecting the images that project points more than a predetermined threshold value.

Fig. 3. (a) Feature matching between a prior image(left) and a current image(right) without any pretreatment. (b) Feature matching after a histogram equlization method. (c) The result of the RANSAC algorithm based on 2D homography.

2.3   3D Projection Method

As the 3D projection method is essentially a mapping of 3D points to a 2D plane, the location of the camera center is estimated by mapping the 3D coordinates in the point cloud to a 2D image plane. The 3D coordinates in the point cloud are the values obtained by conducting the ray tracing algorithm of the feature points in the reference image. On the other hand, the 2D coordinates are the feature points that are matched with the current image. A 6-DoF(Degree of Freedom) position is estimated from a perspective projection method utilizing the parameters of a pin hole camera as follows:

$$[u, v, 1]' = K[R|T][x, y, z, 1]',$$

where K is the parameter of the pin hole camera model of the current image, R is 3-DoF rotation matrix and T is 3-DoF translation matrix. The 3D coordinates features in the point cloud from the reference image are expressed by homogeneous coordinates as $[x, y, z, 1]'$. The 2D coordinates features of the current images are also expressed by homogeneous coordinates as $[u, v, 1]'$. Between the reference image and the current image, one 6-DoF position is estimated. This is defined as the candidate position. As such, several candidate positions are estimated from the reference images that exhibit notable similarity to the current image.

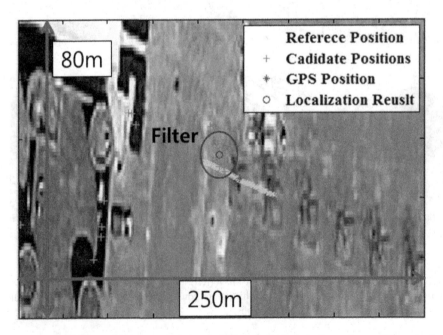

Fig. 4. The yellow x, green crosses, red star, and blue small circle indicate the positions of reference images, the candidate positions, the GPS positions as ground truth, and the final localziation result. The blue large circle indicates a boundary of filters.

2.4 Localization

There exist numerous false candidate positions which occur due to mismatching of features and ray tracing errors. To remove such incorrect results, median and mean filters are employed.

In this system, the median value of the geometric distances of candidate positions is calculated. This value is then used to determine outliers of the candidate positions by comparing a distance. If the number of inlier candidate positions is lower than a specific threshold, the localization is considered as a failure. Then, the final 6-DoF localization result is calculated by determining the mean value of inlier candidate positions. The result of a particular position is shown in figure 4.

3 Experiments

We have conducted experiments at the National Science Museum in Daejeon, South Korea. A survey vehicle was equipped with three SICK LMS-291 LIDARs, a camera, a Huace B20 GPS receiver, an Xsens MTi IMU, and a wheel odometry sensor. The camera had 1280×960 resolution and was mounted in the front. EKF(Extended Kalman Filter) sensor fusion of GPS, IMU, and odometry was used for localization in the survey phase. The prior data were obtained from one closed loop of 510m in the science museum. The prior data consisted of 1.0 million point cloud(x, y, z, r, g, b) and 3466 images. Each image had a 6-DoF position in the prior point cloud. The RMSE(Root Mean Squared Error) of the prior point cloud was lower than 1.5m.

To obtain the current data, a second vehicle was equipped with a Bumblebee XB3 stereo vision and a NovAtel OEM-Star GPS receiver. The current image was obtained from a left image of the stereo vision, which also has 1280×960 resolution. The GPS

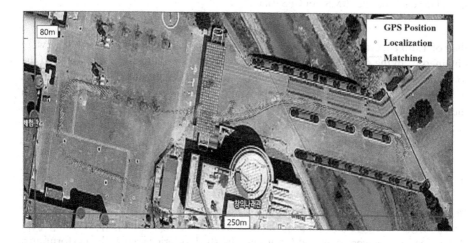

Fig. 5. The red crosses and blue circles indicate GPS position and the result of localization for full path. The yellow lines show matched result between them.

Table 1. The result of accuracy

	X (m)	Y (m)
Standard deviation	3.22	2.18
RMSE	3.64	2.29

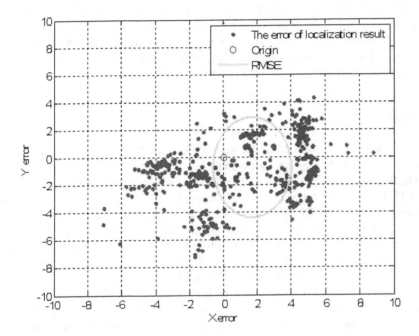

Fig. 6. The blue dots denote error of each localization result and the red circle denotes origin, and the yellow circle denotes RMSE of GPS data

data of 1.5m accuracy was used as a ground truth in this paper. The current data were obtained from the same closed loop used in the survey vehicle, by rotating around the loop three times. The result of 1,082 current images for all paths is shown in figure 5. The localization succeeded with a probability of 56%, a value which depended on the number of features on the images. The average Euclidean distance error of 606 successful localization points was 4.0m. The result of accuracy is shown in table 1. The error of each localization result is presented in the figure 6.

4   Conclusion and Future Works

In this paper, a novel localization method using a current image and prior data was proposed. The 3D features of the prior images were extracted in the preprocessing step. Reference images were searched for among the stack of prior images, by performing a correlative comparison of 2D features of the current image with the features of prior images. Furthermore, the candidate positions from reference images were obtained using a 3D projection method. Subsequently, the final positions were estimated by filtering of the candidate positions.

In this experiment, the RSME was comparably high than expected. This may be attributed to the erroneous 3D features due to inaccurate positions of the prior image in the point cloud. To resolve this problem, future studies will necessitate a closer investigation into the matching procedure between images and point clouds.

Acknowledgment. This work was supported by the R&D program of the Korea Ministry of Trade, Industry & Energy (MOTIE) and the Korea Evaluation Institute of Industrial Technology (KEIT). (The Development of Low-cost Autonomous Navigation Systems for a Robot Vehicle in Urban Environment, 10035354) The students are supported by Korea Ministry of Land, Infrastructure and Transport (MOLIT) as U-City Master and Doctor Course Grant Program.

References

1. Kanayama, Y., Kimura, Y., Miyazaki, F., Noguchi, T.: A stable tracking control method for an autonomous mobile robot. In: Proc. of IEEE Int. Conf. on Robotics and Automation, pp. 384–389 (1990)
2. Gage, D.W.: UGV history 101: A brief history of unmanned ground vehicle (UGV) development efforts. Unmanned Systems Magazine 13(3) (1995)
3. Obst, M., Bauer, S., Reisdorf, P., Wanielik, G.: Multipath detection with 3D digital maps for robust multi-constellation GNSS/INS vehicle localization in urban areas. In: Proc. of IEEE Intelligent Vehicles Symposium (IV), pp. 184–190 (2012)
4. Anguelov, D., Dulong, C., Filip, D., Frueh, C., Lafon, S., Lyon, R., Ogale, A., Vincent, L., Weaver, J.: Google street view: Capturing the world at street level. Computer 43(6), 32–38 (2010)
5. Saito, T., Kuroda, Y.: Mobile robot localization using multiple observations based on place recognition and gps. In: Proc. of IEEE Int. Conf. on Robotics and Automation, pp. 1540–1545 (2013)
6. Baldwin, I., Newman, P.: Road vehicle localization with 2D push-broom lidar and 3D priors. In: Proc. of IEEE Int. Conf. on Robotics and Automation, pp. 2611–2617 (2012)
7. Stewart, A.D., Newman, P.: Laps-localisation using appearance of prior structure: 6-DOF monocular camera localisation using prior pointclouds. In: Proc. of IEEE Int. Conf. on Robotics and Automation, pp. 2625–2632 (2012)
8. Kim, H., Oh, T., Lee, D., Choe, Y., Chung, M.J., Myung, H.: Mobile Robot Localization by Matching 2D Image Features to 3D Point Cloud. In: Proc. of IEEE Int. Conf. on Ubiquitous Robots and Ambient Intelligence, pp. 266–267 (2013)
9. Bay, H., Tuytelaars, T., Van Gool, L.: SURF: Speeded up robust features. In: Leonardis, A., Bischof, H., Pinz, A. (eds.) ECCV 2006, Part I. LNCS, vol. 3951, pp. 404–417. Springer, Heidelberg (2006)

Programming an E-Puck Robot to Create Maps of Virtual and Physical Environments

Pablo Tarquino and Kevin Nickels

Trinity University Engineering Department, San Antonio, Texas
{ptarquin,knickels}@trinity.edu

Abstract. This project is a first step towards research on implementing Simultaneous Localization and Mapping (SLAM) techniques in robots. The paper provides theoretical background for SLAM and occupancy grids, which are used in the project to create maps. The paper also describes the software, Python and Player/Stage, and the hardware, the E-Puck robot, used in the project. This project successfully programmed an E-Puck robot to map an unknown virtual and physical environment. The virtual environment has perfect localization conditions, while the physical environment has error in its localization. A comparison of these maps shows that the map of the virtual environment is highly accurate, while the map of the physical environment is less accurate due to odometry errors.

Keywords: Occupancy Grid, Simultaneous Localization and Mapping, E-Puck Robot, Table-Top Robot, Robot Programming.

1 Introduction

Imagine the task set before the next-generation vacuum-cleaning robot - to vacuum around an unknown environment, clean the floor as it goes, and make a map not only of the rooms but of the dirty spots in the environment. In order to do this, it will need to create a map of the house, and to figure out where it is within this (incomplete) map at the same time.

This project focused on programming a simulated and physical E-Puck robot to create a map of an unknown environment. The project used the popular Player/Stage software package. Player [1] is a robot device interface used to communicate with the E-Puck robots. Stage [2] is a simulator used for testing of the code. The computer language Python was used to implement the mapping behaviors of the robot. In addition, the graphic Python library Matplotlib was used to visually represent the map the robot constructed.

This project is an initial step towards researching Simultaneous Localization and Mapping (SLAM). This project explored only the mapping aspect a robot might encounter within SLAM. For this work, perfect conditions were assumed, meaning that the robot had a perfect knowledge of its position at every point within the map. With this assumption, the robot observed its surroundings via infrared proximity sensors and constructed a map of the surroundings of the robot.

J.-H. Kim et al. (eds.), *Robot Intelligence Technology and Applications 2*,
Advances in Intelligent Systems and Computing 274,
DOI: 10.1007/978-3-319-05582-4_2, © Springer International Publishing Switzerland 2014

This paper is organized as follows. Section 2 provides a general overview of SLAM theory. Section 3 explains the software used in the project: the programming language Python and the software package Player/Stage. Section 4 describes the components of the E-Puck robot. Section 5 gives the basic theory of occupancy grids and shows how they were used within the project. Section 6 provides the experimental setup of the project, which includes an explanation of the mapping technique used and the virtual and physical environment created in the lab. Section 7 makes a comparison between the virtual and physical maps created. Finally, Section 8 discusses the results and conclusions obtained from the project.

2   Simultaneous Localization and Mapping

Localization is the process of estimating the position of a robot within a known environment, while mapping is the process of creating a model of the robot's environment [3].

The Simultaneous Localization and Mapping (SLAM) problem asks if it is possible for a mobile robot to be placed at an unknown location in an unknown environment and for the robot to incrementally build a consistent map of this environment while simultaneously determining its location within this map [4]. In this manner, the problem of SLAM can be seen as a "chicken and egg" problem: A robot needs an accurate map to know its location within the environment, yet a precise knowledge of the location within an environment is needed to create an accurate map [5].

SLAM is a process by which a mobile robot can build a map of an environment and at the same time use this map to deduce its location. In SLAM both the trajectory of the platform and the location of all landmarks are estimated on-line without the need for any *a priori* knowledge of location [4].

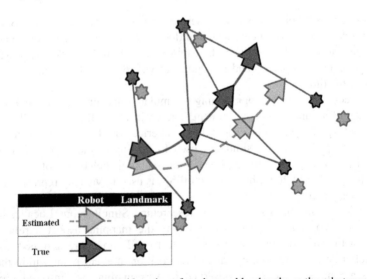

Fig. 1. The true and estimated location of a robot and landmarks as the robot moves

As the robot moves, the estimated robot and landmark locations become distinctly different from the true robot and landmark locations. This is because the sensors are subject to error. The primary source of this error occurs within the odometer readings, often called the odometer noise. The continual discrepancy within the estimated location of a robot and the true location of a robot affects the interpretation of all other sensor readings. In Figure 1 the robot's true location becomes increasingly different than its estimated location, which in turn affects the estimated landmark locations. A secondary source of error, measurement noise, comes from sensors that perceive the landmarks. The measurement and process noise are correlated, which means that the error accumulates over time and affects all future measurements. Figure 10 in Section 7 depicts a map made with accumulated error [6].

It is the task of SLAM to obtain more accurate estimates of the true location of the robot and the landmarks. This is accomplished by using probabilistic techniques to filter out the noise found in the measurements, primarily by re-observing features seen previously (with lower uncertainty). Probabilistic algorithms approach the problem by explicitly modeling different sources of noise and their effects on the measurements [4][6].

SLAM also aids in solving the data association problem. The data association problem is the problem of determining if sensor measurements taken at different points in time correspond to the same physical object in the world. For example, a robot may complete a cycle inside an environment while mapping. When closing the cycle, the robot has to find out where it is relative to its previously built map. A robot that successfully localizes itself while closing the cycle reduces the uncertainty associated with that landmark, helping to solve the data association problem. Once this is accomplished, the robot can map multiple cycles around the environment. Though the sensor data accumulates error, using probabilistic methods allows the data to be filtered, creating accurate estimates of the environment [6].

The following discusses three of the major approaches taken towards developing SLAM in robotics, namely the Extended Kalman Filter (EKF) method, graph based SLAM, and particle filtering. An in depth discussion of these three SLAM approaches can be found in [7].

2.1   Extended Kalman Filters

The EKF approach assumes the position of the robot to be a Gaussian distribution. The robot goes through a process of movement and observation; when it moves it takes time updates of the position of the robot through odometer readings and then the robot uses proximity sensors to make observation updates of the landmarks surrounding the robot.

The odometer readings will have error that will affect future readings, so that the robot's position is represented as a Gaussian probability distribution showing where the robot is likely located. In turn, these readings affect the landmark observations, so these too will be represented by a probability distribution. Figure 2 illustrates the EKF approach.

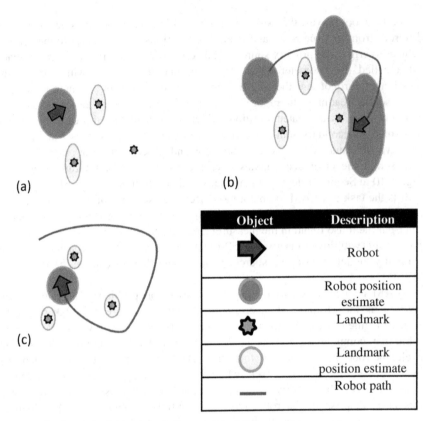

Fig. 2. Three basic stages of the EKF approach

Figure 2(a) shows a robot surrounded by its position estimate and three landmarks in the environment. The robot only senses two landmarks, which in turn are surrounded by the landmark position estimate the robot observes. In Figure 2(b) the robot travels and observes the landmarks. The more the robot travels, the greater the increase in uncertainty of both the robot's position and the landmark's position, represented by a larger ellipses surrounding the robot and the landmarks. In Figure 2(c) the robot circle back near its original starting point and observes the original two landmarks. Recognizing its position, the robot reduces the uncertainty of both its position and that of the landmark.

In this manner, the EKF SLAM approach addresses the data association problem previously discussed. This approach is strongest when a robot can recognize and associate landmarks that had been previously observed. In this case, a robot would "ask" which of the landmarks in the map most likely corresponds to the landmark just observed.

There are some drawbacks to the EKF approach. First, the size of the measurements needed to keep the map updated grows quadratically. This means that high-speed computing and large memory storage is required for this approach to have a practical application. Second, this approach generally fails against the "kidnapping problem." If the robot were to be picked up and taken to a different area of a map (essentially the robot is "kidnapped") then all calculations would be corrupted. The robot would not know it had been moved, so it would try to associate completely different landmarks, thereby leading to an erroneous map [4][7].

2.2   Graph Based SLAM

Often the graph based SLAM approach is likened to a mass-spring system. In this case the mass is represented by landmarks and springs signify correlations between these landmarks. As the robot travels through a map, it makes more correlations between landmarks (there are more springs between the masses). Theoretically, at one point all springs would be stiff, as all masses would be tightly connected. Putting together the knowledge of how all "masses" are connected with "springs" a map can be created. Figure 3 helps illustrate Graph Based SLAM.

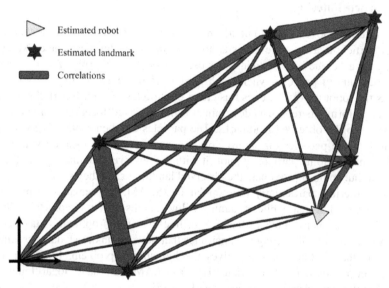

Fig. 3. Spring network analogy for the graph based SLAM (taken from [4])

In the figure, as the robot moves several times through the environment the correlations (red links) become increasingly thicker. Since landmarks are fixed, the correlations between landmarks are the thickest. By making the robot run multiple times through the environment landmarks are observed multiple times and estimated locations are corrected. It should be noted that the robot itself is also correlated to the map.

One way to apply Graph Based SLAM is to use sub-maps over very large environments. The point of the sub-maps is to gather data for only small sections of an overall environment. In this manner, the sub-maps are accurate, since the smaller space means that the odometer drift error is relatively low. Sub-maps are taken over the same area multiple times and then are compared to each other. Maps that overlay the same area are placed side by side and the correspondence between them is observed. This is an iterative process, where propagated correspondence is maintained and those with no correspondence are dissolved.

Graph based SLAM has the advantage that it scales to much higher-dimensional maps than EKF SLAM. It was mentioned that EKF SLAM grew quadratically; the time updates of Graph Based SLAM remains constant and the amount of memory required for a map is linear. There is a drawback. Ideal correlations between robot and landmarks allow a map to be created. However, obtaining the optimal data for these ideal correlations is difficult. In fact, it is believed that gathering the optimal data association is an NP-hard problem. This means that in practicality less than accurate data is used. Graph based SLAM can build large maps, but this needs to be done offline [4][7].

2.3   Particle Filtering

Particle filtering moves the simulation rather than the analysis of the location of a robot. A particle is a concrete guess as to where a robot may be located. Numerous particles, or a "cloud" of particles, are used in particle filtering. As an example, consider a cloud of particles where each particle could represent the location of the robot. As the robot moves, so does this cloud of particles. For each particle, a "filter" process is perform whereby it is determined if the expected location of a particle fits with respect to the observed landmarks. The particles are then replaced, giving more importance to the particles that more closely approximated the location with respect to the landmarks. Some particles do not fit with the observed landmarks, and they are filtered out, but some approximate the observed landmarks, so they survive.

Particle filtering, particularly the popular FastSLAM, has remarkable properties. First, it creates a full map while remaining online. As discussed, Graph Based SLAM can create large maps, but it does so offline by using optimal data. Second, FastSLAM is capable of dealing with particles that have an unknown data association. This means that particle filtering solves the kidnapping problem. Finally, particle filtering can be implemented efficiently; FastSLAM can have logarithmic time updates and the size of the map grows linearly [4][7].

3   Software

This section explains the software used throughout the project. The robot was programmed using the user-friendly Python. When simulating, the python controller interfaced with the Player/Stage software package, and when running on the E-Puck, the same controller was interfaced with the Player package.

3.1   Python

Python was used as the primary programming language throughout the project. Python is an interpreted, object-oriented, high-level programming language with dynamic semantics [8][9].

Two characteristics made Python ideal for the project. In the first place, Python is a language that can be supported with the freeware Player, which was extensively used throughout the project. Second, Python is easy to learn due to its simple syntax and extensive online support, furthering the project objective of improving access to robotics research for undergraduate students.

3.2   The Player/Stage Project

The *Player/Stage Project* provides open-source tools that simplify controller development, particularly for multiple-robot and sensor network systems. The project offers a combination of flexibility and speed that makes it one of the most useful and popular robot development environments available [5].

Player is a socket-based device server that allows control of a wide variety of robotic sensors and actuators. Player executes on a machine that is physically connected to a collection of such devices and offers a TCP socket interface to clients that wish to control them. Clients connect to Player and communicate with the devices by exchanging messages with Player over a TCP socket. Because Player's external interface is simply a TCP socket, client programs can be written in any programming language that provides socket support, such as C, C++, Tcl, Python, Java, and Common LISP [1].

Stage can simulate a population of mobile robots, sensors and environmental objects. Its purposes are to enable rapid development of controllers that will eventually drive real robots, enable robot experiments without access to the real hardware and environments and to enable experiments with devices that do not exist yet. Stage enables experiments with large populations of robots that would be prohibitively expensive to buy and maintain [1].

4   E-Puck Robot

The project used an E-Puck robot due to the affordability and ease of use that this table-top robot provides. The E-Puck robot was developed by the Swiss Federal Institute of Technology in Lausanne (EPFL) for teaching purposes and is currently a commercial product available from Cyberbotics Ltd [10][11].

E-Puck robots are small, having a diameter of 70 mm, height of 55 mm, and a weight of 150 g. Apart from its small size, the robot's dual motors, proximity sensors, wireless communication, and programmability made it ideal for this project. The robot has two stepper motors with a 50:1 reduction gear. Each of these motors can be separately programmed to turn and each motor keeps track of the displacement of the robot with integrated odometers. The robot has eight infrared sensors that measure ambient light and proximity of objects up to 6 cm. The robot supports computer-robot or robot-robot wireless communication via Bluetooth [11]. Additionally, the robot is easy to program and compatible with popular robotic software, such as the commercial Webots or the open-license Player/Stage project.

A Player Driver for the E-Puck [11] was modified for the current version of Player (3.0.2), and integrated with the demonstration programs distributed with the E-Puck [13]. This player driver runs on the E-Puck robot, taking Bluetooth commands from the Player server and returning sensor data to the Player server.

The E-Puck's major components can be seen in Figure 4. A full list of the E-Puck's specifications can be found in [11].

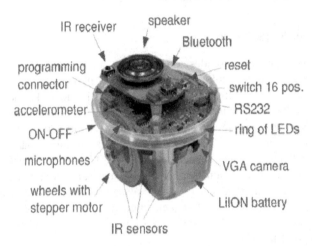

Fig. 4. An E-Puck robot with numerous components labeled (from [11])

5   Occupancy Grid

A popular and useful way that robots can map a physical area is by using occupancy grids. An occupancy grid representation employs a multidimensional tessellation of space into cells, where each cell stores a probabilistic estimate of its state [14]. In this project a two-dimensional area was correlated to cells in a two-dimensional matrix, each cell being filled with a value ranging from 0 (meaning that cell was certain to be empty) to 1 (meaning that cell was certain to be filled).

The top of Figure 3 shows a two-dimensional environment containing a triangle, a square, and a circle. The environment is segmented into a grid. This grid correlates to the image on the bottom of Figure 4, which shows a two-dimensional matrix filled with numerical values ranging from 0 to 1. Areas in the environment that are completely filled, such as the center of the circle, have a corresponding value in the matrix of 1, while areas in the environment that are empty, such as the upper area of the environment, have a corresponding value in the matrix of 0. Cells that are not completely filled have a value between 0 and 1.

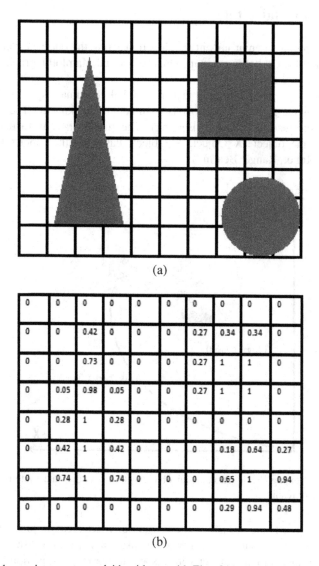

(a)

0	0	0	0	0	0	0	0	0	0
0	0	0.42	0	0	0	0.27	0.34	0.34	0
0	0	0.73	0	0	0	0.27	1	1	0
0	0.05	0.98	0.05	0	0	0.27	1	1	0
0	0.28	1	0.28	0	0	0	0	0	0
0	0.42	1	0.42	0	0	0	0.18	0.64	0.27
0	0.74	1	0.74	0	0	0	0.65	1	0.94
0	0	0	0	0	0	0	0.29	0.94	0.48

(b)

Fig. 5. Simple environment overlaid with a grid Fig. 3(a) and the corresponding two-dimensional matrix representing the occupancy grid of the environment Fig. 3(b)

These values represent the probability that the cell is occupied, so that a value of 0 shows there is a 0% probability that the cell is occupied, and a value of 1 shows there is a 100% probability that the cell is occupied.

6   Experimental Setup

The task of mapping an environment requires translating the odometer and proximity sensor readings of the robot into a map. The Numpy and Matplotlib graphical libraries helped to visualize the map.

Figure 4 shows a robot and an object in a global coordinate system. The robot's current position (X_Robot, Y_Robot) is assumed to be known. The robot takes proximity sensor readings (Range, Bearing). The robot's task is to determine the coordinates of the object (X_Object, Y_Object) based on the robot's known data (X_Robot, Y_Robot, Range, Bearing).

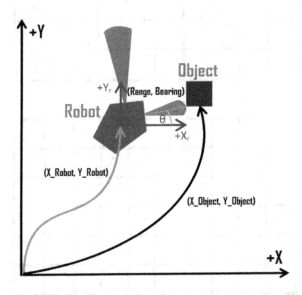

Fig. 6. A robot sensing the location of an object within a coordinate system based on its position and sensor readings

Rigid body coordinate transformations are fully covered in [15]. The project simplified these transformations for a two-dimensional space as follows. Equation 1 shows the coordinates of the object with respect to the robot. Equation 2 shows the coordinates of the object with respect to the global coordinate system. Both equations have been condensed into matrix form.

$$\begin{bmatrix} x\_object_{Relative} \\ y\_object_{Relative} \end{bmatrix} = Range \times \begin{bmatrix} \cos\theta \\ \sin\theta \end{bmatrix} \tag{1}$$

$$\begin{bmatrix} X\_Object \\ Y\_Object \end{bmatrix} = \begin{bmatrix} x\_object_{Relative} + X\_Robot \\ y\_object_{Relative} + Y\_Robot \end{bmatrix} \tag{2}$$

In these equations, x_object$_{Relative}$ and y_object$_{Relative}$ describe the position of the object *relative to the robot*. Range refers to the numerical reading obtained from a proximity sensor on the robot. The value θ is the angle of the proximity sensor within the global coordinate system. X_Object and Y_Object describe the position of the object within the global coordinate system, and X_Robot and Y_Robot describe the position of the robot within the global coordinate system.

To determine the coordinates of an object within a coordinate system first the position of the object with respect to the robot is determined. For the object in Figure 4, this means using the numerical reading of the proximity sensor that registers the object and the value θ of the angle of this proximity sensor within the global coordinate system and applying these values to Equation 1. This then gives the coordinates of the object with respect to the robot.

The next step is to localize the object within the global coordinate system. This is a simple matter of adding the coordinates of the object relative to the robot (x_object$_{Relative}$, y_object$_{Relative}$) to the coordinates of the robot (X_Robot, Y_Robot).

For the project, the E-Puck robot could determine its position and bearing by means of the odometers in both of its wheels. The E-Puck robot's eight infrared sensors gave the range of a detected object up to 6 cm away. The bearing of each proximity sensor reading was determined by two factors. First, the bearing of the E-Puck robot itself was determined, once again by the odometers in the wheels. Second, each of the eight sensors was positioned at a specific angle along the E-Puck robot. The angle on the E-Puck robot of a specific sensor was added as an offset to the bearing of the E-Puck robot. Figure 5 shows the angles of separation of each of the infrared sensors on the E-Puck.

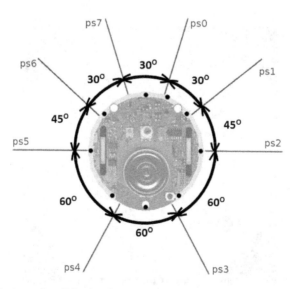

Fig. 7. Overhead view of an E-Puck robot with the angles between proximity sensors labeled (from [16])

In the lab, a physical environment 4 feet long by 4 feet wide (1.22m by 1.22m) was arranged. Figure 6 illustrates the physical environment. Likewise, a virtual environment was created in Stage that approximated the physical environment in both the dimensions specified and the arrangement of the shapes. Figure 7 illustrates the virtual environment.

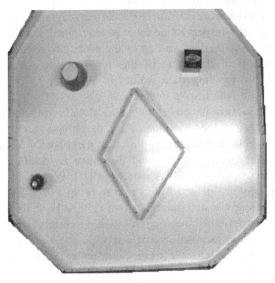

Fig. 8. Picture of the physical environment used for mapping. The E-Puck robot is seen in the bottom left.

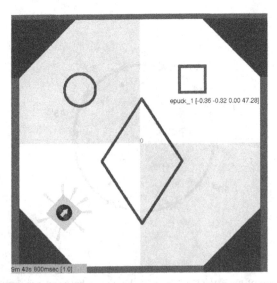

Fig. 9. Picture of the virtual environment used for mapping. The E-Puck robot is seen in the bottom left with eight proximity sensors "rays."

7   Results of Mapping

The project used both a virtual and a physical E-Puck robot to create maps of the environments. In the virtual environment, the E-Puck robot had perfect localization conditions. For the physical environment, the E-Puck computed its coordinates from the odometer readings. These coordinate readings were subject to error, meaning the localization conditions were inaccurate.

7.1   Map for the Virtual Environment

The best maps were obtained within the simulated Stage environment. Figure 8 shows the map of the virtual Stage environment. This map closely replicates the virtual environment seen in Figure 7. During simulation, the location of the robot is exact. That is, Stage allows the simulated robot to have perfect odometer readings and eliminates the error caused due to the slippage of the wheels. The majority of the error can be seen around curved surfaces or sharp edges. This is due mainly to three factors. First, there are some inaccuracies in the readings of the range of the proximity sensors. Second, the size of the grid may not be large enough to give a good resolution of the map. Third, the wide angles of the proximity sensors make it hard to accurately depict small objects.

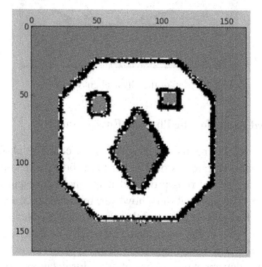

Fig. 10. Map of the virtual environment

7.2   Map for the Physical Environment

The results of the real E-Puck were less accurate than those of the virtual E-Puck. Figure 9 shows the map of the physical environment. As can be seen, this is an inaccurate representation of the environment shown in Figure 6. This inaccuracy is caused by error due mainly to well-known odometer drift [3]. The inaccuracies are greatly noticeable along surfaces that required the robot to turn sharply, causing wheel slippage, but they are less noticeable along flat surfaces. In this manner, surfaces that required numerous turns for the E-Puck robot to fully map them are greatly distorted, as is the case with the square and the circle found within the environment. Flat surfaces, such as the diamond or the borders of the environment, were mapped as fairly straight (albeit skewed) lines.

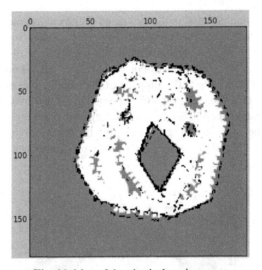

Fig. 11. Map of the physical environment

7.3   Accumulated Error for the Physical Environment

The E-Puck robot could be manually controlled, or it could be allowed to "wander" around the environment. Regardless of whether the robot was manually controlled or allowed to wander, the virtual runs produced maps that consistently looked like the map seen in Figure 8. The physical runs, however, required strict manual control. The most efficient routes had to be selected so that the E-Puck traveled along straight lines as much as possible. This is far from ideal, especially considering that the goal of SLAM is to allow a robot to make maps without any *a priori* knowledge.

To illustrate the detrimental effects of the localization error, the physical E-Puck robot was allowed to wander for some time; the results can be seen in Figure 10.

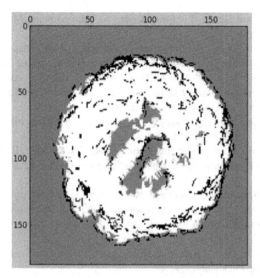

Fig. 12. Map of the physical environment when the robot was allowed to wander

8   Conclusions

The goal of this paper was to explore the mapping aspect of Simultaneous Localization and Mapping. An overview of SLAM and the theory used throughout the project was provided. This project ultimately produced very accurate maps when localization was exact, yet it also showed how inaccurate maps can result if localization has error. This illustrates the well-known effect of odometer drift on mapping, motivating further the SLAM problem. The drift is accumulated over time and affects future landmark readings. In this manner, it was shown that odometer drift increases the uncertainty of the location of a robot, and SLAM techniques such as the ones discussed in Section 2 must be used to decrease and correct this uncertainty.

This project was an excellent first step into SLAM research. It acquainted the student with the Player/Stage software and it allowed for hands-on programming of a robot.

Future research can focus on implementing SLAM techniques in order to integrate the localization aspect that was not covered throughout this project. This would allow a robot to accurately map an environment even with localization error, such as the physical scenario presented in this project.

References

1. Vaughan, R., Howard, A., Gerkey, B.: The Player/Stage Project: Tools for Multi-Robot and Distributed Sensor Systems. In: Proceedings of the 11th International Conference on Advanced Robotics (ICAR 2003), Coimbra, Portugal, pp. 317–323 (2003)
2. Vaughan, R.: Massively Multiple Robot Simulations in Stage. Swarm Intelligence 2(2-4), 189–208 (2008)

 3. Siegwart, R., Nourbakhsh, I.: Introduction to Autonomous Mobile Robots. MIT, Massachusetts (2004)
 4. Bailey, T., Durrant-Whyte, H.: Simultaneous Localisation and Mapping (SLAM): Part I The Essential Algorithms. IEEE Robotics and Automation Magazine 2 (2006)
 5. Bennet, S., Nieto-Wire, C., Peche, J., Timotheu, M., Vasili-Acevedo, D.: Mobile Robot Simultaneous Localization and Mapping in Static Environments. Technical Report, The City College of New York of the City University of New York
 6. Thrun, S.: Robotic mapping: A survey. In: Exploring Artificial Intelligence in the new Millennium, pp. 1–35 (2002)
 7. Siciliano, B., Khatib, O.: Springer Handbook of Robotics. Springer, Berlin (2008)
 8. Zelle, J.: Python Programming: An Introduction to Computer Science. Franklin, Beedle (2004)
 9. Official Python Programming Language Website, http://www.python.org/
10. Mondada, F., et al.: The e-puck, a robot designed for education in engineering. In: Proceedings of the 9th Conference on Autonomous Robot Systems and Competitions, vol. 1(1) (2009)
11. E-Puck Brochure, http://www.cyberbotics.com/e-puck/e-puck.pdf
12. Player Driver for E-Puck Robots,
    https://code.google.com/p/epuck-player-driver/
13. Basic E-Puck Robot Demos,
    http://www.gctronic.com/doc/index.php/E-Puck#Examples
14. Elfes, A.: Using occupancy grids for mobile robot perception and navigation. IEEE Computer, 46–57 (1989)
15. Wolovich, W.: Robotics: Basic Analysis and Design. Holt, Rinehart and Winston, New York (1987)
16. Webots User Guide,
    http://www.cyberbotics.com/dvd/common/doc/webots/guide/
    section8.1.html

A Method to Localize Transparent Glass Obstacle Using Laser Range Finder in Mobile Robot Indoor Navigation

Jungsoo Park[1], Eric T. Matson[2], and Jin-Woo Jung[1,*]

[1] Dept. of Computer Science and Engineering, Dongguk University, Seoul, Korea
jshostkit@gmail.com, jwjung@dongguk.edu
[2] Dept. of Computer and Information Technology, Purdue University, USA
ematson@purdue.edu

Abstract. The problem to localize transparent glass obstacles using laser range finder is very difficult and still open problem. Most of applications use additional sensor device such as sonar sensor to cope with the transparent glass obstacle environment. This paper deals with that problem only using laser range finder. By the insight from human sensing mechanism which uses the fusion of more data with different view directions or different measurement locations, a novel method to localize transparent glass obstacles is addressed. And, the effectiveness of the proposed algorithm is evaluated by the real robot experiments with three case studies.

Keywords: transparent obstacle, laser range finder, localization, mobile robot.

1 Introduction

To recognize the nearby obstacles rapidly and localize their exact positions accurately is one of the most important problems in the mobile robot navigation. For this problem, there have been many approaches so far such as using sonar sensor [1,2], using robot vision [3,4] and using laser range finder [5,6,7]. Laser range finder is one of the most promising methods in the sense of high accuracy, high resolution and fast measurement speed even though it is very expensive [8].

But laser range finder is not complete except for the high cost. When only laser range finder is used for mobile robot indoor navigation, many applications are usually assumed that there are no transparent obstacles since lights such as laser may be easily penetrated or refracted on the surface of transparent materials [9].

Sometimes, it is not easy to recognize the existence or exact location of a transparent glass obstacle even by human eyes. But, we, the human beings, can estimate the existence and location of a transparent glass obstacle by gathering and fusion of more data with different view directions or different measurement locations.

---

[*] Corresponding author.

J.-H. Kim et al. (eds.), *Robot Intelligence Technology and Applications 2*,
Advances in Intelligent Systems and Computing 274,
DOI: 10.1007/978-3-319-05582-4_3, © Springer International Publishing Switzerland 2014

In this paper, a method is addressed to deal with the problem to localize the transparent glass obstacles only by using laser range finder for mobile robot indoor navigation.

2   Problem Specification

When a laser range finder is used to detect a transparent glass obstacle, reflected noise can be easily founded such as Fig.1. In Fig.1, blue line means the real location of transparent glass obstacle but the detected signal may be distorted by the reflected noise or missing by the penetration or refraction. (See Fig.2)

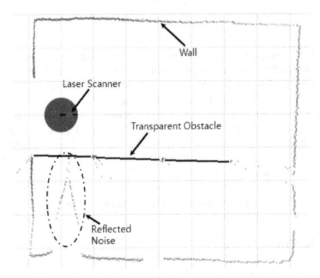

Fig. 1. Noisy data for transparent obstacle from laser scanner (laser range finder)

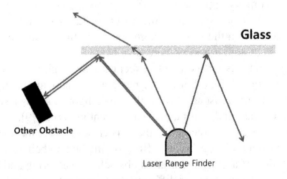

Fig. 2. Distortion and missing of laser when a transparent glass obstacle exists

3   Proposed Algorithm

It is not easy to recognize the transparent glass obstacles only with a Fig.1-like data but if there are more data with different view directions or different measurement locations, we can estimate the location of the glass with more confidence.

To make the problem in a simple form, several assumptions are used as follows:

- The transparent glass obstacles have polygonal structure and every type of curvilinear surfaces is excluded.
- There is no non-transparent obstacle near the outside of transparent glass obstacle.
- The locations of all obstacles are fixed during the experiment.

Based on these assumptions, the proposed algorithm uses an idea that the measurement by laser range finder will be always made not before the glass surface but after the glass surface. After all, what we should do is just to find what is the nearest meaningful obstacle data using the various sensor data distribution. But, one more thing we should consider is that the algorithm should be effective not only for the transparent glass obstacles but also for the non-transparent normal obstacles. If we try to recognize just the nearest measurement data and disregard the remaining, another problem may be occurred by unintended 'erosion' of detected data for the non-transparent glass obstacles.

The specific algorithm based on the grid map is as Fig.3. Here, θ_n is the biased angle of n-th sensor, , $[\theta_{min} + \theta_{step} \times (n - 1)]$ (n: 1~1081, θ_{min} = -45(deg), θ_{step}=0.25(deg)). $d_n(t)$ is the distance of measurement by n-th sensor at time t, and $D_{min} \leq d_n(t) \leq D_{max}$. D_{min} and D_{max} are the minimum and maximum sensing distance of the sensor, respectively. $P_R(t)$ and $\theta_R(t)$ are the position and orientation angle of the robot at time t. $cell_n(t) := f(d_n(t), \theta_n, P_R(t), \theta_R(t))$ is the cell number of the grid map which is pointed by the n-th sensor. $cell_R(t)$ is the cell number of the grid map which includes the robot position $P_R(t)$. $DCMap(cell)$ is a grid map with gray values, 0~255, whose element value is the detected counting number from various sensor data distribution.

In the proposed algorithm, the first step is to check the robot motion. If there is no change in the robot motion, nothing will be updated by the algorithm. But, if robot moves, new sensor data is analyzed by making a set L whose elements are the cells on the virtual line between the robot cell and sensor data cell. And the most nearest obstacle cell with a certain confidence, bigger than the threshold value $Th_{obstacle}$, is emphasized by the detected counting map $DCMap(cell)$. In addition, the sensor data with no ambiguity is directly emphasized by $DCMap(cell)$ to cope with the non-transparent obstacles. Here, $Th_{obstacle}$ is actually a variable dependent with the robot moving speed but we assume that as a constant during our experiments because the resulting map was not very sensitive to the robot moving speed among the normal speed range of Pioneer P3-DX robot. And, $DCMap_{max}$ is simply set to 255 for the direct conversion into gray-scale colors.

```
INPUT: sensor data $d_n(t), \theta_n$ and robot posture $P_R(t), \theta_R(t)$ at
       the current time
OUTPUT: a grid map with the detected counting numbers,
       $DCMap(cell)$

PROCEDURE:
  IF $P_R(t) \neq P_R(t-1)$ THEN
     Calculate the robot cell $cell_R(t)$ which includes the
     robot position $P_R(t)$ in the grid map

     FOR each $d_n(t)$
        IF $d_n(t) \in [D_{min}, D_{max}]$ THEN
           Calculate the cell $cell_n(t) := f(d_n(t), \theta_n, P_R(t), \theta_R(t))$
           which was pointed by $n$-th sensor data $d_n(t)$ in the
           grid map

           Make a set $L$ whose elements are the cells on the
           virtual line segment between $cell_R(t)$ and $cell_n(t)$
           excluding $cell_R(t)$ and $cell_n(t)$

           LOOP
              Find the nearest cell $cell_{L1}$ in $L$ from the robot cell $cell_R(t)$
              IF $DCMap_{max} > DCMap(cell_{L1}) \geq Th_{obstacle}$ THEN
                 $DCMap(cell_{L1})$ += 1
                 EXIT LOOP
              END IF
              Find the next nearest cell $cell_{L2}$ in $L$ from $cell_R(t)$
              $cell_{L1} := cell_{L2}$
           END LOOP

           IF $DCMap$ values for all cells in $L < Th_{obstacle}$ THEN
              $DCMap(cell_n(t))$ += 1
           END IF
        END IF
     END FOR
  END IF
```

Fig. 3. Proposed algorithm to detect the surface of transparent glass obstacle

4   Experimental Results

Three types of different environments are considered for the evaluation of effectiveness of the proposed algorithm. The first experiment was designed with the straight glass wall such as Fig.4. And Pioneer P3-DX and SICK LMS-100 laser range finder was used for the experiment. The size of environment was 4m×3.53m and there was no transparent wall on the outskirts. And the size of cell was 15cm×15cm. The second experiment was basically same with the first case but additional static obstacles, diagonally located A4 box case, is also included. The last experiment was designed with the right-angled glass wall.

Fig. 4. Experimental environment with Pioneer P3-DX robot and SICK laser range finder

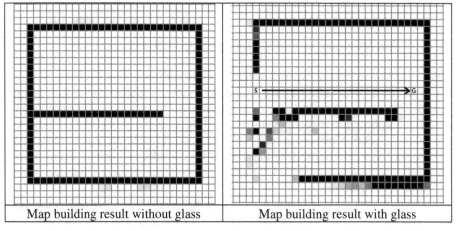

Map building result without glass	Map building result with glass

Fig. 5. Experimental results following with the straight glass wall

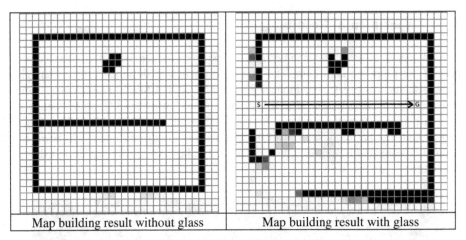

| Map building result without glass | Map building result with glass |

Fig. 6. Experimental results following with the straight glass wall including a static obstacle on the top middle side

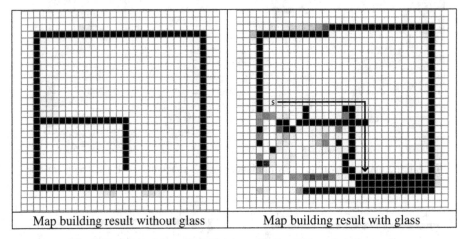

| Map building result without glass | Map building result with glass |

Fig. 7. Experimental results following with the right-angled glass wall

Fig.5 to Fig.7 shows the experimental results. Here, the robot moves from 'S' position to 'G' position as following the wall and continually gathers the sensor data to recognize the nearby obstacles and localize them. Even though there was some distortions in the starting phase (Fig.5 and Fig.6) and the corner region (Fig.7) but the results after the starting phase can be considered good. In addition, a non-transparent obstacle was also detected well in Fig.6 when we consider the visible region. Here, one pixel difference can be regarded as the inside of error boundary because of the characteristics of grid map.

The main reason of distortion in the starting phase could be understood as the lack of information compared with the other region since the proposed algorithm does not update the information during the stop phase to prevent with the data overflow by excessive update. In addition, the main reason of distortion in the corner may be understood as the inexact synchronization between robot localization and sensor data during the turning phase.

5   Concluding Remarks

In this paper, a novel algorithm to recognize the transparent glass obstacles is proposed only by using the laser range finder based on the data with different view directions and different measurement locations. Even though the results in the starting phase and corner region were not very good, the effectiveness can be confirmed by three different types of experiments with real robot, Pioneer P3-DX and SICK LMS-100 laser range finder. In addition, the proposed algorithm can deal with not only the transparent glass obstacles but also non-transparent normal obstacles together.

Acknowledgement. This work was partially supported by Basic Science Research Program (2010-0025247) and the fostering project of LINC through the National Research Foundation of Korea (NRF) funded by the Ministry of Education, Science and Technology (MEST).

References

1. Lee, Y.C., Lim, J.H., Cho, D.W., Chung, W.K.: SonarMap Construction for Autonomous Mobile Robots Using a Data Association Filter. Advanced Robotics 23(1-2), 185–201 (2009)
2. Fazli, S., Kleeman, L.: Wall Following and Obstacle Avoidance Results from a Multi-DSP Sonar Ring on a Mobile Robot. In: Proceedings of the IEEE International Conference on Mechatronics & Automation, pp. 432–437 (2005)
3. Wang, Y.T., Feng, Y.C., Hung, D.Y.: Detection and Tracking of Moving Objects in SLAM using Vision Sensors. IEEE Trans. Instrumentation and Measurement Technology Conference, 1–5 (2011)
4. Ma, Y., Soatto, S., Kosecka, J., Sastry, S.S.: An Invitation to 3-D Vision. Springer (2003)
5. Rudan, J., Tuza, Z., Szederkenyi, G.: Using LMS-100 Laser Range finder for Indoor Metric Map Building. In: IEEE International Symposium on Industrial Electronics, pp. 525–530 (2010)
6. Aboshosha, A., All, A.: Robust Mapping and Path Planning for Indoor Robots based on Sensor Integration of Sonar and a 2D Laser Range Finder. In: IEEE 7th International Conference on Intelligent Engineering Systems (2003)
7. Diosi, A., Kleeman, L.: Advanced Sonar and Laser Range Finder Fusion for Simultaneous Localization and Mapping. In: Proc. of IEEE/RSJ International Conference on Intelligent Robots and Systems, vol. 2, pp. 1854–1859 (2004)
8. Lumelsky, V., Skewis, T.: Incorporation Range Sensing in The Robot. IEEE Transactions on System, Man, and Cybernetics 20(5), 1058–1068 (1990)
9. LMS100 Laser Measurement System Operating Instructions, http://www.sick.com

Robotic Follower System Using Bearing-Only Tracking with Directional Antennas

Byung-Cheol Min[1] and Eric T. Matson[1,2]

[1] Machine-to-Machine (M2M) Lab, Department of Computer and Information Technology, Purdue University, West Lafayette, IN 47907 USA
[2] Department of Computer Engineering, Dongguk University, Seoul 100-715, Republic of Korea
{minb,ematson}@purdue.edu

Abstract. This paper presents the development of a robotic follower system with the eventual goal of autonomous convoying to create end-to-end communication. The core of the system is a bearing-only tracking with directional antennas and an obstacle avoidance algorithm with sonar sensors. For bearing estimation with directional antennas, we employ a Weighted Centroid Algorithm (WCA), which is a method for active antenna tracking and Direction Of Arrival (DOA) estimation. We also discuss the use of sonar sensors that can detect objects, which could improve our robotic follower system in mobile robot navigation. Through extensive field experiments in different environments, we show feasibilities of our proposed system, allowing a follower robot to track a leader robot effectively in convoying fashion. We expect that our system can be applied in a variety of applications that need autonomous convoying.

1 Introduction

In a disaster area, where previously established networks are destroyed, autonomous mobile robots carrying wireless devices can be deployed to create end-to-end communication. In the event of an earthquake disaster, like Fukushima, Japan, rapid establishment of a wireless backbone is useful, because it would allow rescuers and first responders to communicate and to coordinate evacuation and search-and-rescue missions effectively.

Typically, there are two means to build an end-to-end communication link with mobile robots. The first way is realized by planning robots' final positions prior to deployment of robots [1]−[5]. This planning should be designed for optimizing the communication link, and thus this approach is suitable for static environment rather than dynamic environments. Also, this is useful for cases where a rapid establishment of the network is required, because this way does not require a search task.

The second way is realized by deploying a team of leader-follower robots in convoying fashion [6]−[8]. The overview of this way is depicted in Fig. 1. In this way, multiple robots can be used, and only the leader requires navigation capabilities to create the network while followers do not require any planning.

J.-H. Kim et al. (eds.), *Robot Intelligence Technology and Applications 2,*
Advances in Intelligent Systems and Computing 274,
DOI: 10.1007/978-3-319-05582-4_4, © Springer International Publishing Switzerland 2014

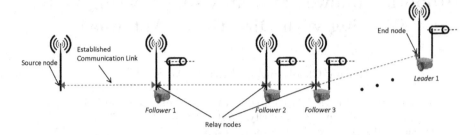

Fig. 1. The overview of a team of leader-follower robots to create end-to-end communication link

Alternatively, they need to follow the leader or the precedent robot. Therefore, this approach is suitable more for dynamic environments because this way is based on reactive approaches, not pre-planning approaches.

In this paper, we deal with the later way and propose a robotic follower system to create long-distance end-to-end communication. Leader-follower robots have been studied for a long time. Most of the work in this area focuses on control and algorithm aspects related to steering a follower robot so that it follows a leader robot. Other works study on sensors to enable follower behavior [11]. Typically, followers use one of the sensors such as a laser range finder, computer vision, GPS, Inertial Measurement Units (IMUs) to implement leader following strategies.

A vision based leader-follower system is the most common approach in convoying [9]. By estimating the leader's position form the sequence of video image, the follower can follow the leader. Although this approach shows powerful performances, there is a prerequisite condition that the sight of its leader should be guaranteed all the times in their system. Since the sight of its leaser is often lost (e.g., when the follower moves down the slope, the leader turns sharply, or objects lie between them), alternative ways to compensate it have to be incorporated. Also, it has been shown that identifying its leader among other possible objects is not easy such as obstacles and robots.

Other ways employ GPS or IMU to estimate the leader's position and heading [10][11]. During motion, the leader produces absolute way points, which are then followed by the follower. Because of the absolute position provided by GPS, position errors can be bounded and their leader-follower team can travel together by sharing the information. As this approach does not require line-of-light at all, it is known to be a good alternative way to use a vision-based follower system. However, outage of GPS signal due to limited places may result in position errors, finally causing the system to fail. Once the failure in communication takes place, the follower may not be able to track the leader.

In our system, we introduce directional antennas as a directional finder for the robotic follower system. For example, if there are two robots – one is a leader and the other is a follower – they are all equipped with a network device with antennas and wirelessly connected to each other. More specifically, the leader is equipped with an omni-directional antenna and the follower has a directional antenna (this configuration will be explained later in detail). The follower estimates a bearing from the transmitter at the leader using the directional antenna and follows it with the estimated bearing.

Directional antennas have several advantages. First, they are easy to obtain and are affordable as compared to other sensory devices such a laser range finder and a vision system, they can be used in both indoor and outdoor environments as directional sensors, and they can also detect some objects [12]−[14]. Moreover, in contrast to the vision based leader-follower system, a directional antennas-based follower system can cope with cases where the sight of its leader is lost.

Although directional antennas have many advantages, it remains a challenge to increase their accuracy enough to use them as typical sensory devices, similar to laser or ultrasonic sensors in the field of mobile robotics. One common type of directional antenna (the type used in this study) has a beam width that is conical in shape [15]. This broad beam width allows directional antennas to measure a wide area; however, it also yields a coarser measurement resolution than that of the non-expanding beamwidth generated by a laser. In addition, because of the presence of walls and other objects that act as reflectors or scatters, the signals received by a directional antenna can consist of multiple copies of the transmitted signal that arrive via different paths. This effect gives rise to varying levels of received power, represented as sensor noise or uncertainty. Because the magnitude of this sensor noise is much larger than typical noise in other types of sensors, it may hinder directional antennas from being used as sensory devices. This necessitates some filtering of the received signal to remove such interference.

In this paper, we use a Direction Of Arrival (DOA) estimation technique for bearing estimation that is called the Weighted Centroid Algorithm (WCA) and is a type of weighted centroid approach. Weighted centroid approaches have been adopted by several research groups [16]−[19]. The previous studies used the "distance" as the weighting factor, through power measured from multiple anchor nodes. In [20], we examined the directionality of the radiation pattern with a stand-alone directional antenna for DOA estimation. As the basic concept of using weights to obtain the centroid of a data set is similar to the previous studies, we recommend referring to the papers referenced above for a more detailed explanation of the concept of weighted centroid approaches.

2 Follower Robotic System

In this paper, we adopt the strategy for deploying mobile relay nodes that was introduced in [6], in order to create long-distance end-to-end communication.

The overview of the strategy was depicted in Fig. 1. Multiple robots can be used in this system with the following rule − one robot should be a leader to act as an end node, and others should be followers to act as relay nodes connecting the wireless source node to the end node. They are linked sequentially with a wireless device. Then, all followers follow the leader in convoying fashion.

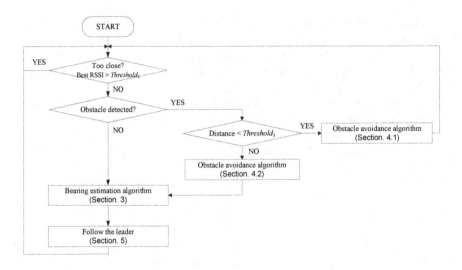

Fig. 2. A flow chart of the follower robotic system

When the link between the source node and the last follower (e.g., *Follower* 1 in Fig. 1) in the convoy is about to disconnect, that follower stops, and the rest of the convoy continues on. After that, when the link between the now stationary relay (*Follower* 1) and the last follower (*Follower* 2) in the remaining convoy is about to disconnect, that follower also stops and becomes a stationary relay. The process continues until all followers have been deployed.

The leader can be programmed to reach the goal position autonomously, but we control it manually through a remote control, in order to focus on the follower robotic system in this research. Alternatively, follower robots are autonomous to follow the leader. For follower robots to be autonomous, the follower robotic system runs with the steps, depicted in Fig. 2.

As shown in Fig. 2, the follower robot is composed mainly of two algorithms – a bearing estimation algorithm and an obstacle avoidance algorithm. The bearing estimation algorithm allows a follower robot to track the leader. We will detail this algorithm in Section 3. The obstacle avoidance algorithm allows the follower robot to avoid the obstacle between itself and the leader. For the obstacle avoidance algorithm, we employ two different ways, depending on a distance between the object to the follower robot. First, if the distance is too close (i.e., when measured distance is smaller than the pre-defined threshold), obstacle avoidance becomes a top priority, i.e., the robot stops following the leader and avoids the obstacle and tries to be in a safe place from any crash. For this way, we develop a simple obstacle avoidance algorithm and will detail it in Section 4.1. Second, if the distance is not close, but an object on the path is detected, we employ both bearing estimation algorithm and obstacle avoidance algorithm at the same time, i.e., the robot keeps following the leader while avoiding the object. For this later way, we introduce a penalty function and will detail it in Section 4.2.

3   Weighted Centroid Algorithm

In this section, we briefly describe the WCA, introduced in [20], for bearing estimation of the follower robot[1]. In Fig. 3, the receiver for bearing estimation attached to the follower is a directional antenna that can rotate horizontally by a servo motor, and the transmitter attached to the leader is a stationary omni-directional antenna, which is installed perpendicular to the ground. The omni-directional antenna on both sides is the real one for data transmission in end-to-end communication (not covered in this paper), so it is wirelessly paired together. Fig. 4 shows defined parameters for bearing estimation with a directional antenna, and detailed parameters are described in Table 1. For the interval angle θ^t, it is assumed that this angle can be computed by dividing the interesting range by the total number of measurements N_t.

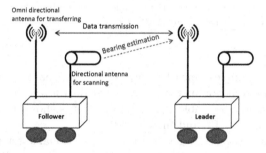

Fig. 3. The directional sensing model for leader-follower robotic system

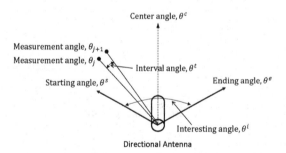

Fig. 4. Defined parameters for bearing estimation with a directional antenna, when scanning clockwise

---

[1] In [20], we used the term "DOA estimation" to represent bearing estimation that we use in this paper. Compared to the previous paper, this paper develops a considerably more comprehensive analysis on WCA.

Table 1. Setting of parameters

Variables	Description
θ^i	Interesting range where a scanning task performs, defined at the center of the antenna's body
θ^s	Starting angle where to start the range
θ^e	Ending angle where to end the range, going either clockwise or counter-clockwise from the starting angle in turn
θ^c	Center angle between the staring angle and the ending angle (at the beginning of scanning, the center angle is the front of the device)
θ^t	Interval angle of measurement
θ^j	Measurement angle, where j is the index of the measurement such that $j \in \{1, 2, \ldots, N_t\}$
N_t	Total number of measurements while scanning from the starting angle to the ending angle
$RSSI_j$	The measured RSSI at the jth measurement

Fig. 5 shows an example of measured RSSI (Radio Signal Strength Index) from an experiment that was conducted indoors, with a rotary directional antenna, showing the parameters in Table 1. In this figure, it is shown that $\theta^i = 180°$, $\theta^s = -90°$, $\theta^e = 90°$, $N_t = 13$, and therefore $\theta^t = 10°$.

In the first step of the WCA, a single rotary directional antenna measures the signal strength by rotating from θ^s to θ^e and produces a set of $RSSI_j$, as shown in Fig. 5. Here, θ^s and θ^e are determined by the center angle θ^c that is set to be aligned exactly on the previously calculated bearing to prevent the estimation from approaching the end where an actual bearing dwells [20].

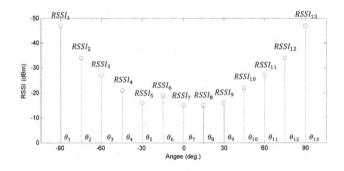

Fig. 5. An example of measured signal strength with a rotary directional antenna. The horizontal axis is the measurement angle and the vertical is the measured signal strength.

In the second step, a weight is computed by the measured signal strengths at θ_j using the following expression

$$w_j = 10^{\left(\frac{RSSI_j}{\gamma_1}\right)}. \tag{1}$$

where γ_1 is a positive gain that should be appropriately determined in every application scenario so that stronger signal strengths are more weighted than weaker signal strengths. Then, the bearing can be estimated by means of weighted centroid approaches as follows,

$$\tilde{\Theta} = \frac{\sum_{j=1}^{N_t} w_j \theta_j}{\sum_{j=1}^{N_t} w_j} \tag{2}$$

If we use the measured RSSI shown in Fig. 5 again and depict all variables used in Eq. (2) in polar coordinates, it should look like Fig. 6. Here, γ_1 was set to 10, the estimated bearing $\tilde{\Theta}$ using the WCA was depicted with a symbol "★" (See nearby $0°$ on angle-axis between -20 dBm and -30 dBm) in a polar coordinate, and the actual angle was depicted with a symbol "■".

The variance of the estimated bearing $\tilde{\Theta}$ can be derived as follows. First, let the uncorrelated random variable X_j be a function of $\overline{p_{Rj}}$ multiplied by θ_j, where $\overline{p_{Rj}}$ is a normalized signal strength

$$\overline{p_{Rj}} = p_{Rj} \frac{N_t}{\sum_{j=1}^{N_t} p_{Rj}}. \tag{3}$$

In [19], the authors showed that random variables follow a Gaussian (Normal) distribution, with a mean μ_j and variation $\sigma_j{}^2$ as follows,

$$X_j \sim N\left(\mu_j, \sigma_j{}^2\right), \tag{4}$$

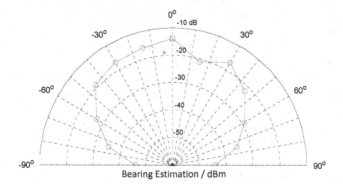

Fig. 6. Weighted Centroid Algorithm(WCA) in a polar coordinate frame

where $\sigma_j{}^2$ is the variation of received signal strength, represented as sensor noise or uncertainty, as mentioned in Section 1. Then, the mean of the random variables can be obtained as follows,

$$\overline{X} = \frac{1}{N_t} \sum_{j=1}^{N_t} \mu_j = \frac{1}{N_t} \sum_{j=1}^{N_t} \overline{p_{R_j}} \theta_j \qquad (5)$$

and the variation is computed as

$$V\left(\overline{X}\right) = \frac{1}{N_t{}^2} \sum_{j=1}^{N_t} \sigma_j{}^2. \qquad (6)$$

Consequently, when the variations of all observations are almost equal, $\sigma_j{}^2 \approx d$, we have $V\left(\overline{X}\right) \approx d/N_t$, which is related to the central limit theorem [21]. It is clearly shown that, as the sample number N_t increases, the variance $V\left(\overline{X}\right)$ decreases. This also verifies that, as we calculate the bearing based on multiple sampled data, the variation of bearing can be smaller in proportion to the sample number. In fact, the authors in [22]−[24] conducted multiple measurements from multiple devices, or from a long period, to have more sample data, and could produce a better estimation. However, because of the multiple measurements, it was very slow and computationally expensive to process their estimation methods. As shown in the equations derived in this section, the WCA only requires the summation and multiplication of data. Not only is the WCA computationally effective, but it estimates the bearing very accurately.

4   Obstacle Avoidance Algorithm

With the bearing estimation algorithm presented in the previous section, the follower can track the leader. However, if there is a static or dynamic obstacle between the two robots, the follower should avoid it first and continue following the leader. Therefore, a decent obstacle avoidance algorithm is necessary for a successful follower.

For this research, we use P3AT which is a mobile robot research platform available from Adept MobileRobots, Inc. [7]. There are eight sonar sensors attached to the front of the P3AT that can detect an object in front of the robot and measure the distance between the object and the robot. For an obstacle avoidance algorithm, we utilize these sensors.

We define in this paper that the follower may face with three different situations in terms of an obstacle − one is when the robot is free to obstacles so it can keep following the leader, the second is when the obstacle is too close (e.g., the distance between the robot and the obstacle is less than 1 meter), and the last is when the obstacle is detected, but not close.

4.1   When an Obstacle Is Too Close

In the second situation, since there is a high chance that a collision takes place, the robot should stop following the leader, and avoid the obstacle first with a set of sonar sensors. Therefore, we have developed a simple obstacle avoidance algorithm that is also based on weighted centroid approach, calculating a direction guiding the robot to a safe region. This algorithm is mainly designed to avoid an obstacle, so it should be activated only when there is an object detected by a sonar sensor and its measured distance is shorter than a pre-determined threshold, $Threshold_1$.

In the first step of the algorithm, we compute a weight by the measured sonar distances using the following expression

$$w_k = 10^{\left(\frac{-Distance_k}{\gamma_2}\right)}, \tag{7}$$

where,

$$\gamma_2 = \text{a positive gain}$$
$$Distance_k = \text{a measured sonar distance in centimeters at } \phi_k,$$
$$\text{where } k \in \{1, 2, \ldots, N_s\}$$
$$N_s = \text{the total number of sonar measurements.}$$

As we defined in the previous section for bearing estimation, the center angle is the front of the device. P3AT has eight sonar sensors so each sensor has a field of view of approximately $25.7°$, and the horizontal range of all sensors is from $-90°$ to $+90°$, implying that measurement angle $\phi_k \in \{-90°, -64.3°, \ldots, +90°\}$. Then, the direction $\tilde{\Phi}$, guiding the robot to a safe region, can be estimated by means of weighted centroid approaches as follows,

$$\tilde{\Phi} = \begin{cases} \hat{\Phi} + \left|\phi_1 + \phi_{N_s/2}\right| & \text{if } \hat{\Phi} \leq 0° \\ \hat{\Phi} - \left|\phi_1 + \phi_{N_s/2}\right| & \text{else} \end{cases} \tag{8}$$

where,

$$\hat{\Phi} = \frac{\sum_{k=1}^{N_s} w_k \phi_k}{\sum_{k=1}^{N_s} w_k}. \tag{9}$$

With Eqs. (7) - (9), the measured data with long distances to the object are depicted further from the center in the polar coordinate frame, and their angle values become more important to determine the weighted centroid. Conversely, shorter distances are rarely weighted because of the log scale. Therefore, the measured data with shorter distances are depicted closer to the center, and their angle values become less important. As a result, it can be said that Eqs. (8) and (9) calculates a reasonable direction by averaging the measured data with appropriate weighting.

4.2   When an Obstacle Is Detected, But Not Close

The algorithm for the second situation was developed to prevent the follower robot from colliding with objects. With this algorithm, we can prevent most of the crashes into obstacles. However, if we take only this situation into consideration, motions of the robot would become too large as approaching the obstacle.

For this reason, we have developed another algorithm for dealing with situations where an obstacle is detected, but not close. We believe that implementing this third algorithm remarkably helps reducing the chances that the robot faces with a dangerous situation from close objects like the second situation. Since the robot changes its heading in advance approaching the close objects, it will follow the leader more effectively and safely.

We utilize sonar sensors again for this algorithm in form of a penalty function. The basic concept of the penalty function is to integrate a sonar sensor measurement into an antenna measurement (i.e., if an object on the path is detected, then the function generates a pseudo RSSI measurement that is levied into a real RSSI measurement). This penalty function is activated only when there is an object detected by a sonar sensor, and its measured distance is longer than the pre-determined threshold for the second situation.

The steps for constructing and using the penalty function are as follows:

1. A function, sonar(ϕ), includes data of measured distances to an object, using the eight sonar sensors. This function can be expressed as

$$\text{sonar}(\phi_k), \text{where } k \in \{1, 2, \ldots, N_s\} \tag{10}$$

where ϕ_k is a measurement angle, N_s is the total number of sonar measurements (eight for P3AT).

2. The sonar sensors and the directional antenna have different data set lengths ($N_s \neq N_t$). For instance, there would be eight sonar sensor measurements from the robot and 13 measurements from the directional antenna. When two measured data sets are combined into one data set, they must have the same length. Therefore, we expand the measured sonar sensor data, which is coarser, to have the same length as the data obtained from the antenna. This expansion is accomplished by linear interpolation that is obtained by passing a straight line between two adjacent data points, as follows,

$$\widetilde{\text{sonar}}(\phi) = \sum_{k=1}^{N_s} \text{sonar}(\phi_k) L_k(\phi) \tag{11}$$

where for each $k = 1, 2, ..., N_s$,

$$L_k(\phi) = \prod_{i=1, i \neq k}^{N_s} \frac{(\phi - \phi_i)}{(\phi_k - \phi_i)} (1 \leq k \leq N_s). \tag{12}$$

sonar(ϕ) is the exact function for which values are known only at a discrete set of data points, and the function $\widetilde{\text{sonar}}(\phi)$ is the interpolated approximation to sonar(ϕ).

Using Eq. (11), the interpolated approximation of sonar(ϕ) can have the expanded range of $\widetilde{\text{sonar}}(\phi_j)$, where $j \in \{1, 2, \ldots, N_T\}$, meaning the sonar measurements could have the same data length as the antenna measurements. An example of this measurement expansion, using the linear interpolation, is depicted in Fig. 7. There are some gaps between the exact data and the approximated data, but they are negligible.

3. Using the expanded sonar sensor measurement, we generate penalty values (pseudo RSSI measurements), as follows,

$$\text{penalty}(\phi) = \alpha\exp(-\beta\widetilde{\text{sonar}}(\phi)),\tag{13}$$

where α and β are constants for regulating the level of the penalty function. The effects of different α values are depicted in Fig. 8. As shown in this figure, when α is higher, more penalty is levied. Here, as an example, we set $\alpha = 60$ and $\beta = 0.1$. Note that these two parameters should be carefully determined, depending on the material of the object that is detected by the sonar sensors. For example, if the material of the object is impenetrable with wireless signal, α could be set to a lower value. However, if the material of the object is penetrable, α should be set high enough for the obstacle to be recognized.

4. Using the penalty function in Eq. (13), The pseudo RSSI measurements $PRSSI_j$ is formed, and the real RSSI measurement using the antenna becomes,

$$NRSSI_j = RSSI_j + PRSSI_j\tag{14}$$

5. Returning to the original algorithm, $NRSSI_j$ is used instead of $RSSI_j$ as an input variable for WCA in Eq. (1).

The following two figures (Figs. 9 and 10) show the effectiveness of the penalty function. In these simulations, the leader was placed at the center of the map as a stationary transmitter. The first set was run without the penalty function (i.e. it depicts a case where there are only real RSSI measurements with a directional antenna). As shown in Fig. 9, the bearing was estimated to around 0° from the follower. Considering the follower depends on this estimated bearing for navigation, the follower would crash into the wall that lies between the transmitter and the follower's current position.

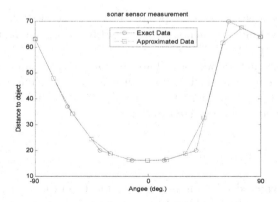

Fig. 7. Sonar measurements and their interpolation. The horizontal axis is the measurement angle and the vertical is the measured distance.

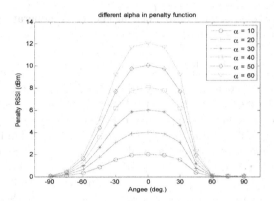

Fig. 8. Different α values in the penalty function. The horizontal axis is the measurement angle and the vertical is the pseudo signal strength, *PRSSI*.

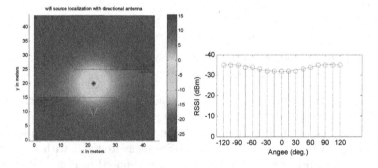

Fig. 9. Failure of bearing estimation with only real RSSI measurement. The left figure shows that the bearing was estimated to around $0°$(See the black arrow), and the right figure shows the measured RSSI.

The second set was run with the penalty function (i.e. it depicts a case where there are real RSSI measurements, with pseudo RSSI measurements levied as a penalty). As shown in Fig. 10, the bearing was estimated to over $+90°$, pointing toward a roadway that allows the follower to come out to an open space. With the penalty function, even when the follower was located behind walls that could not be detected by a directional antenna, the bearing could be estimated to a safe and open region for robot navigation.

5   Mobile Robots Control

P3AT is a four-wheeled robot; however, two wheels on the same side are physically interconnected with a rubber belt. For the simple control of this robot for the follower robotic system, differential-drive mobile robots with characteristics of non-slipping and pure rolling are considered. The robot can be then controlled

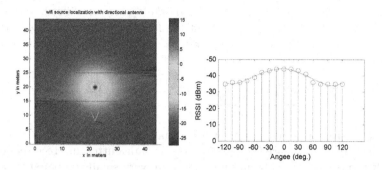

Fig. 10. Success in bearing estimation with real RSSI measurement, with penalty levied. The left figure shows that the bearing was estimated to over $+90°$ (see the black arrow), and the right figure shows the measured RSSI with pseudo RSSI added.

to move to any posture by adjusting the velocity of the left wheel V_L and the velocity of the right wheel V_R. V_L and V_R are calculated with either Eqs. (15) or (16), depending on the current situation of the robot.

$$V_L = v_1 + k_{p1}\widetilde{\Theta} + k_{d1}(\widetilde{\Theta} - \widetilde{\Theta}^{t-1})$$
$$V_R = v_1 - k_{p1}\widetilde{\Theta} - k_{d1}(\widetilde{\Theta} - \widetilde{\Theta}^{t-1}) \quad (15)$$

$$V_L = v_2 + k_{p2}\widetilde{\Phi} + k_{d2}(\widetilde{\Phi} - \widetilde{\Phi}^{t-1})$$
$$V_R = v_2 - k_{p2}\widetilde{\Phi} - k_{d2}(\widetilde{\Phi} - \widetilde{\Phi}^{t-1}) \quad (16)$$

If the follower lies in the first or third situation that we defined in the previous section, it runs with bearing estimation algorithm, activating Eq. (15) for the velocity control. If the follower lies in the second situation, it runs with the obstacle avoidance algorithm, activating Eq. (16) for the control. Therefore, in Eq. (15), $\widetilde{\Theta}$ is the current estimated bearing obtained by Eq. (2), $\widetilde{\Theta}^{t-1}$ is the old estimated bearing, k_{p1} and k_{d1} are positive gains, and v_1 is the background velocity of the robot, set to change according to a value of the best RSSI measurement from one scanning, i.e., v_1 is calculated by

$$v_1 = -\omega_1 RSSI^* - \omega_2, \quad (17)$$

where $RSSI^*$ indicates the best RSSI measurement in one scanning, ω_1 and ω_2 are should be set to a positive value and $w_2 \le |w_1 \cdot RSSI^*|$ for v_1 to be a positive value.

In the same way, in Eq. (16), $\widetilde{\Phi}$ is the current estimated direction by Eq. (8), v_2 is a constant of the background velocity of the follower that we set to be low to avoid any dangerous situations (e.g., here we set v_2 to be 100, meaning 0.1 m/sec in a P3AT library), and k_{p2} and k_{d2} are positive gains.

For the robot stopping criteria, we use the following condition,

$$\begin{cases} V_L \text{ and } V_R = 0 & \text{if } RSSI^* \ge Threshold_2 \\ V_L \text{ and } V_R \text{ from Eqs.(15) or (16)} & \text{else.} \end{cases} \quad (18)$$

In Eq. (18), depending on a value of $Threshold_2$, we can differ how close the follower can get to the leader or prevent the follower from getting too close to the leader. Actually, the received power at the follower from the transmitter at the leader can be given by [25]

$$P_{dBm} = \underbrace{L_0 - 10n \cdot \log\left(\left\|x^t - x\right\|\right)}_{Fading} - \underbrace{f\left(x^t - x\right)}_{Shadowing} - \underbrace{\varepsilon}_{multipath} , \qquad (19)$$

where L_0 is the measured power at 1m from the transmitter, n is the decay exponent, and x^t and x are the positions of the transmitter and receiver respectively. If terms of shadowing and multipath are very small compared to a term of fading, they can be negligible. Then, we can roughly calculate P_{dBm} by pre-obtaining L_0 and n with experiments. Therefore, we can select a proper value of $Threshold_2$ with Eq. (19) for a desired motion of our follower system. For example, we identified through experiments that -15 dBm of $Threshold_2$ keeps the follower away from the leader at intervals of 1 m in indoor environments and -20 dBm for outdoor environments.

6   Experiments

6.1   Preparation for Experiments

To test the proposed methods, we have developed a prototype of the leader-follower robotic system as shown in Fig. 11. The complete system mainly consists of a leader system and a follower system. The both systems can equip the same components, but we simplified the leader system for this research in order to focus on the follower system. The follower robotic system is made up of the P3AT mobile robot, a laptop, a yagi antenna, Wi-Fi USB adapter, and a pan-tilt servo device. And, for the ultimate goal of this research on end-to-end communication, we have installed two access points (AP) and a network switch. Later, by using a network switch in the communication system, we will be able to easily add additional network devices or laptops to the established communication link between the robots. The leader robotic system is equipped with almost same components as the follower has, but it does not have the yagi antenna and Wi-Fi USB adapter for this test.

For bearing estimation, we installed a small and light yagi antenna, manufactured by PCTEL. This device looks like a can and can be seen on the bottom of the system on the right side of Fig. 11 (a). This device has 10 dBi of gain, uses 2.4 GHz frequency range, and has $55°$ horizontal and vertical beamwidth at $1/2$ power. For the transmitter at the leader, requiring an omnidirectional antenna, we use a state of the art, low cost, high performance, and small wireless AP, Pi-coStation M2-HP, manufactured by Ubiquiti Networks Inc. This AP is equipped with a 5 dBi omnidirectional antenna, and supports passive Power over Ethernet (PoE), so it does not require an additional power code. Also, it runs with IEEE 802.11g protocol having an operating frequency of 2.4 GHz, and produces up to 28 dBm output power. As this device was designed to be deployed either indoor

(a) leader-follower robotic system   (b) field test

Fig. 11. Leader-follower robotic system composed of the follower system (left side on (a)) and the leader system (right side on (a))

or outdoor environments, it is ideal for applications requiring medium-range performance and a minimal installation footprint.

The laptop is connected by a serial connection to the P3AT, the pan-tilt device, and Alfa USB adapter. A pan-tilt device allows the directional antenna to be oriented in specific angle autonomously. In this paper, we employ a pan angle only as the directional antenna we chose for this project has about 55° beamwidth vertically, and therefore there are few cases that our robot is deployed out of the range. However, it should be noted that vertical beamwidth would also affect wireless communication in some cases.

For parameters needed in WCA and the obstacle avoidance algorithm, we set them as shown in Table 1. Due to the physical limitation of servo motors in our pan-tilt system, we set θ^i to be 180°. This setting results in the initial scan performed at $\theta^s = -90°$, $\theta^e = 90°$. N_t was approximately 25 for most of the tests. These settings were applied to all of the tests.

Table 2. Setting of parameters

Parameter	Value
θ^i	180°
$Threshold_1$	800 cm
$Threshold_2$	−25 dBm
γ_1, γ_2	10, 10
k_{p1}, k_{d1}	1.0, 0.3
k_{p2}, k_{d2}	1.2, 0.6
α, β	60, 0.1
w_1, w_2	10, 150

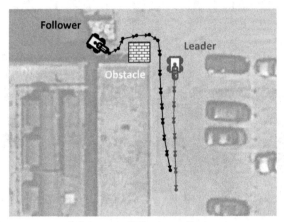

(a) traces of the two robots

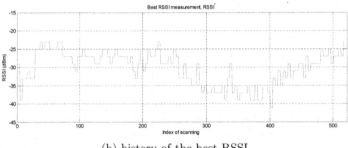

(b) history of the best RSSI

Fig. 12. Field test 1 (a video of this test is available at http://web.ics.purdue.edu/
~minb/rita2013.html)

6.2   Experiments

In order to validate the proposed system, we conducted three different field tests.
We chose ENAD parking lot at Purdue University for these tests, as shown in
Figs. 12 to 14.

The first test was designed to analyze the performance of the obstacle avoidance algorithm. The leader was manually controlled so that it moves straight
to about 15 m with a constant velocity at 0.2 m/sec. The follower was initially
placed behind the building and the squared obstacle about the size of 0.5x0.5 m
when viewed from above. In this planned situation, the follower should avoid the
obstacle and the side of the building in order to follow the leader successfully.
Otherwise, the follower fails to achieve its goal.

In Fig. 12 (a), the red lines indicate moved paths by the leader. The black
lines indicate moved paths by the follower. These lines were drawn by referring
to videos recorded during the test and odometer information from the robots. As
shown in this figure, the follower could avoid the obstacle and the side of building
without any contacts and follow the leader in the long run. Fig. 12 (b) shows

that a history of the measured best RSSI denoted with $RSSI^*$. As shown in the horizontal axis, approximately 520 times of scanning were performed during this test. During the first half of scanning, there were few decreases in measured RSSI as the leader and the follower were close to each other. While the leader bore off gradually and the follower focused on escaping from obstacles, measured RSSI became decreased up to about -40 dBm. However, as soon as the follower avoided the obstacles and became free, it resumed following the leader. After that, as shown in the end of the history in Fig. 12 (b), the measured RSSI reached to the pre-defined threshold, -25 dBm, making the follower stop with a close distance to the leader.

The second and third tests were designed to run for about 10 minutes each, in order to validate a robustness of the proposed system, including bearing estimation and the obstacle avoidance algorithm. The leader was manually controlled so that it moves along with pre-planned paths. The follower was initially placed just behind the leader.

Figure 13 shows the results of the second test. In Fig. 13 (a), the red lines show moved paths by the leader. The black lines show moved paths by the follower. As shown in this figure, the follower tracked way points that the leader produced relatively well during the entire test. It is shown that there are some noticeable gaps in the paths that two robots moved, but it results from the fact that the leader always moved ahead, resulting the follower changed its heading at a corner before it reaches the path that the leader moved. Figure 13 (b) shows a history of the best RSSI measurements. As shown in the horizontal axis, approximately 1500 times of scanning were performed during this test. Around the 1200th scanning, the leader was stopped intentionally, so the two robots got very closer to each other, resulting in about -20 dBm of the best RSSI. Since this value was lower than the threshold, the follower also stopped for a while. Because of this stop, the follower had to keep its pose pointing to a west direction. As the leader resumed moving to a south direction, the bearing between the two robots became almost a right angle, resulting in $-90°$ of bearing estimation from the follower. (see around the 1200th scanning in Figure 13 (c) that shows a history of the estimated bearing).

Figure 14 shows the results of the third test. This test includes multiple stops performed by the leader and very sharp paths requiring almost $180°$ turning for the follower to successfully follow the leader. Figure 14 (a) shows the tracked paths by the leader and the follower. As shown in this figure, the follower followed the leader well during the entire test. As the paths produced by the leader were relatively smooth until it rotated at the right top of the map, the difference in the paths the two robots moved was not shown unlike the previous test. There were two times of stop taken by the leader during the first half of the test. As shown in Fig. 14 (b) around the 300th and 500th scannings, the best RSSI reached the threshold accordingly, and therefore the follower stopped as well with a close distance to the leader. These behaviors validate that two robots in convoying performs successfully with the proposed system.

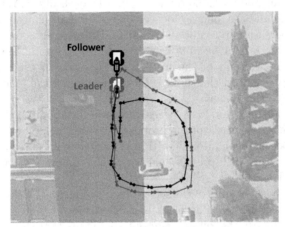

(a) traces of the two robots

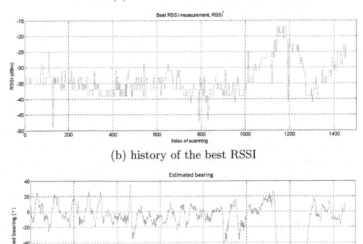

(b) history of the best RSSI

(c) history of the estimated bearing

Fig. 13. Field test 2 (a video of this test is available at http://web.ics.purdue.edu/
~minb/rita2013.html)

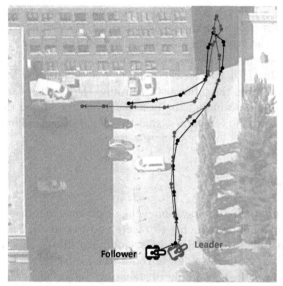

(a) traces of the two robots

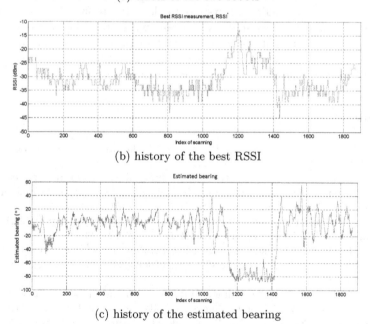

(b) history of the best RSSI

(c) history of the estimated bearing

Fig. 14. Field test 3 (a video of this test is available at http://web.ics.purdue.edu/
~minb/rita2013.html)

From the right top of the map in Fig. 14 (a), one can see that the leader wheeled about to the other extreme. This was intended motion controlled by a human to see if the follower could follow the leader or not. As the leader turned to the opposite direction that the follower headed, the estimated bearing by the reader showed all the way to the left, meaning that the follower needs to change its heading to the left as well. However, since the measured RSSIs were lower than the threshold at that time, the follower had to stop for a while until the leader moved away from the follower (see around the 1200th scanning in Fig. 15 (b)). After the 1300th scanning, the best RSSI became higher than the threshold, and finally the follower turned sharply and resumed following the leader again.

7   Conclusion

We proposed a robotic follower system using directional antennas with the eventual goal of autonomous convoying to create end-to-end communication. With DOA estimation called the Weighted Centroid Algorithm, directional antennas could be utilized for guiding a follower robot to a leader robot. For mobile robot navigation, we developed a simple obstacle avoidance algorithm that is also based on weighted centroid approach, calculating a direction guiding the robot to a safe region. As a result, our system yielded very robust direction estimations in a constrained environment, specifically when the follower was placed in a region out of line of sight with the leader. Our future works will be devoted to increase the number of robots in convoying test, and discuss the use of this system in building long-distance end-to-end communication.

References

1. Bezzo, N., Fierro, R.: Tethering of mobile router networks. In: American Control Conference (ACC), pp. 6828–6833 (2010)
2. Pei, Y., Mutka, M.W.: Steiner traveler: Relay deployment for remote sensing in heterogeneous multi-robot exploration. In: 2012 IEEE International Conference on Robotics and Automation (ICRA), pp. 1551–1556 (2012)
3. Yan, Y., Mostofi, Y.: Robotic Router Formation in Realistic Communication Environments. IEEE Transactions on Robotics 28, 810–827 (2012)
4. Tekdas, O., Kumar, Y., Isler, V., Janardan, R.: Building a Communication Bridge With Mobile Hubs. IEEE Transactions on Automation Science and Engineering 9, 171–176 (2012)
5. Dixon, C., Frew, E.W.: Maintaining Optimal Communication Chains in Robotic Sensor Networks using Mobility Control. Mobile. Netw. Appl. 14, 281–291 (2009)
6. Nguyen, H.G., Pezeshkian, N., Raymond, M., Gupta, A., Spector, J.M.: Autonomous Communication Relays for Tactical Robots. In: Proceedings of the International Conference on Advanced Robotics (ICAR) (2003)

7. Nguyen, C.Q., Min, B.-C., Matson, E.T., Smith, A.H., Dietz, J.E., Kim, D.: Using Mobile Robots to Establish Mobile Wireless Mesh Networks and Increase Network Throughput. International Journal of Distributed Sensor Networks 2012, Article ID 614532, 1–13 (2012)
8. Tuna, G., Gungor, V.C., Gulez, K.: An autonomous wireless sensor network deployment system using mobile robots for human existence detection in case of disasters. Ad Hoc Networks (2012)
9. Giesbrecht, J.L., Goi, H.K., Barfoot, T.D., Francis, B.A.: A vision-based robotic follower vehicle. In: Proc. of the SPIE Defence, Security and Sensing, vol. 7332, pp. 14–17 (2009)
10. Hogg, R., Rankin, A.L., McHenry, M.C., Helmick, D., Bergh, C., Roumeliotis, S.I., Matthies, L.H.: Sensors and algorithms for small robot leader/follower behavior. In: Proc. of the 15th SPIE AeroSense Symposium (2001)
11. Borenstein, J., Thomas, D., Sights, B., Ojeda, L., Bankole, P., Fellars, D.: Human leader and robot follower team: correcting leader's position from follower's heading. In: Proc. of the SPIE Defence, Security and Sensing, vol. 7692 (2010)
12. Tokekar, P., Vander Hook, J., Isler, V.: Active target localization for bearing based robotic telemetry. In: 2011 IEEE/RSJ International Conference on Intelligent Robots and Systems (IROS), pp. 488–493 (2011)
13. Kim, M., Chong, N.Y.: RFID-based mobile robot guidance to a stationary target. Mechatronics 17, 217–229 (2007)
14. Kim, M., Chong, N.Y.: Direction Sensing RFID Reader for Mobile Robot Navigation. IEEE Transactions on Automation Science and Engineering 6, 44–54 (2008)
15. Graefenstein, J., Albert, A., Biber, P., Schilling, A.: Wireless node localization based on RSSI using a rotating antenna on a mobile robot. In: 6th Workshop on Positioning, Navigation and Communication (WPNC 2009), pp. 253–259 (2009)
16. Blumenthal, J., Grossmann, R., Golatowski, F., Timmermann, D.: Weighted Centroid Localization in Zigbee-based Sensor Networks. In: IEEE International Symposium on Intelligent Signal Processing (WISP 2007), pp. 1–6 (2007)
17. Behnke, R., Salzmann, J., Grossmann, R., Lieckfeldt, D., Timmermann, D., Thurow, K.: Strategies to overcome border area effects of coarse grained localization. In: 6th Workshop on Positioning, Navigation and Communication (WPNC 2009), pp. 95–102 (2009)
18. Pivato, P., Palopoli, L., Petri, D.: Accuracy of RSS-Based Centroid Localization Algorithms in an Indoor Environment. IEEE Transactions on Instrumentation and Measurement 60, 3451–3460 (2011)
19. Wang, J., Urriza, P., Han, Y., Cabric, D.: Weighted Centroid Localization Algorithm: Theoretical Analysis and Distributed Implementation. IEEE Transactions on Wireless Communications 10, 3403–3413 (2011)
20. Min, B.-C., Matson, E.T., Khaday, B.: Design of a Networked Robotic System Capable of Enhancing Wireless Communication Capabilities. In: 11th IEEE International Symposium on Safety, Security, and Rescue Robotics (SSRR 2013), Sweden, October 21-26 (2013)
21. Montgomery, D.C., Runger, G.C., Hubele, N.F.: Engineering Statistics, Student Study edn. John Wiley & Sons (2009)

22. Yang, C.-L., Bagchi, S., Chappell, W.J.: Topology Insensitive Location Determination Using Independent Estimates Through Semi-Directional Antennas. IEEE Transactions on Antennas and Propagation 54, 3458–3472 (2006)
23. Malajner, M., Planinsic, P., Gleich, D.: Angle of Arrival Estimation Using RSSI and Omnidirectional Rotatable Antennas. IEEE Sensors Journal 12, 1950–1957 (2011)
24. Sun, Y., Xiao, J., Li, X., Cabrera-Mora, F.: Adaptive Source Localization by a Mobile Robot Using Signal Power Gradient in Sensor Networks. In: IEEE Global Telecommunications Conference (IEEE GLOBECOM 2008), pp. 1–5 (2008)
25. Fink, J., Kumar, V.: Online methods for radio signal mapping with mobile robots. In: 2010 IEEE International Conference on Robotics and Automation (ICRA), pp. 1940–1945 (2010)

Oscillator Aggregation in Redundant Robotic Systems for Emergence of Homeostasis

Sho Yamauchi, Hidenori Kawamura, and Keiji Suzuki

Graduate School of Information Science and Technology, Hokkaido University
{sho-yamauchi,kawamura,suzuki}@complex.ist.hokudai.ac.jp

Abstract. The main feature that keeps states and structures stable can be seen in living organisms. This adjusting and adaptive feature is called homeostasis. This integrated adaptive feature is achieved by the cooperation of organs in living organisms. Living organisms in nature act dynamically due to this feature. Highly adaptive behavior caused by this feature is also observed in simple living organisms that have no neural circuits such as amoebas. Based on these facts, a method of control to generate homeostasis in robotic systems is proposed by assuming a robot system is an aggregation of oscillators in this paper and each parameter in a robot system is allocated to an oscillator. Such oscillators interact so that the whole system can adapt to the environment. Also, a redundant robot arm is made to confirm the effect of this control method to generate homeostatic behaviors in robotic systems.

1 Introduction

The main feature that keeps states and structures stable can be seen in living organisms. This adjusting and adaptive feature is called homeostasis. This integrated adaptive feature is achieved by the cooperation of organs in living organisms. Living organisms in nature act dynamically due to this feature.

Robot systems that use neural oscillators and central pattern generators(CPGs) have been developed to enable such features to emerge in robotic systems. Some have been developed to generate rhythmic arm motion [1], while others have been developed to generate stable quadruped locomotion [2] and bipedal locomotion[3].

However, we focused on a flocking algorithm in multi-agent systems to generate homeostasis in a robotic system. We applied an autonomous and distributed algorithm for agent control mechanisms to a single autonomous system by assuming the robot was a set of agents to stabilize its state [4]. In addition, we experimented on these agents and confirmed that they could be assumed to be oscillators and their locomotion had aspects of synchronization phenomena. We therefore concluded that it might be possible for highly adaptive behavior to emerges[5] by constructing a robot system that was the aggregation of oscillators.

However, predictive behaviors for periodic stimuli have been observed in living organisms that have no neural circuits such as amoebas. Synchronization is considered to have occurred [6] in these phenomena when each cell acts as an oscillator and an amoeba acts as an aggregation of oscillators.

J.-H. Kim et al. (eds.), *Robot Intelligence Technology and Applications 2*,
Advances in Intelligent Systems and Computing 274,
DOI: 10.1007/978-3-319-05582-4_5, © Springer International Publishing Switzerland 2014

We consider a method of control to generate homeostasis in robotic systems by assuming a robot system is a set of oscillators in this paper. Such oscillators interact so that the whole system can adapt to the environment.

2   Concept

The earth and living organisms are open non-equilibrium systems with flows of energy and materials [7] that are closely related. For example, although climate change affects living organisms [8] many of them have mechanisms to deal with periodic changes on earth such as circadian rhythms [9]. However, these adaptations in living organisms are not only for periodic phenomena. Amoebas, for instance, can expect further stimuli even after the periodic stimuli have stopped[6].

We modeled these relationships with the environment and autonomous systems such as living organisms and robots by assuming that the environment and autonomous systems were both oscillators. These oscillators were also the aggregation of oscillators with some interactions. In addition, we assumed that each oscillator had a condition and there was a term that caused the state to become unstable if the condition was not satisfied. When we considered the aggregation of oscillators that interacted, its state was multistable and it transitsed to other states due to the effect of the initial state and disturbances [10]. We expected that these oscillators would transit to other available states because of such features when they were unstable until they became stable. After they reached a stable state, they synchronized and maintained their relationships. They constructed relationships where the autonomous system was more stable this time because the speed at which the state changed in an autonomous system appeared to be faster locally. We considered such phenomena and their processes as homeostasis and achieved these in robotic systems.

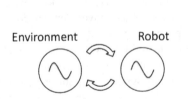

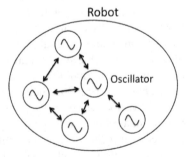

Fig. 1. Environment and robot as oscillators   **Fig. 2.** Robot as oscillator set

3   Method of Control with the Interactions of Oscillators

We assumed parameters of an autonomous system to be oscillators after the example of the method of control with a flocking algorithm[4]. We constructed the oscillators to correspond to all parameters. We considered the set of n oscillators. Let $x^{(k)}$ denote the state of oscillator k. We define the dynamics of oscillator k as

$$\frac{d^2 x^{(k)}}{dt^2} = f_o^{(k)} + f_p^{(k)} + f_s^{(k)} + f_{prob}^{(k)}. \tag{1}$$

The four terms on the right are determined as follows.

3.1 f_o

Each oscillator is designed based on a van der Pol oscillator. Term f_o is designed to express the behavior of a van der Pol oscillator. A van der Pol oscillator has dynamics of

$$\frac{d^2 x}{dt^2} + \lambda (x^2 - 1) \frac{dx}{dt} + x = 0. \tag{2}$$

We extend this oscillation from around the origin to around x_d and multiply its amplitude by $\frac{1}{c}$.

$$\frac{d^2 x}{dt^2} + \lambda \{ c^2 (x - x_d)^2 - 1 \} \frac{dx}{dt} + (x - x_d) = 0. \tag{3}$$

Therefore, we assume that oscillator k oscillate around $x_d^{(k)}$ and then

$$f_o^{(k)} = -\lambda \{ c^2 (x^{(k)} - x_d^{(k)})^2 - 1 \} \frac{dx^{(k)}}{dt} - (x^{(k)} - x_d^{(k)}). \tag{4}$$

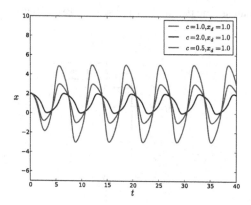

Fig. 3. Example of extended van der Pol oscillator

3.2 f_p

To keep the speed at which the state changes in each oscillator, we determine f_p as

$$f_p^{(k)} = c_p \sum_i^n \left(\frac{dx^{(i)}}{dt} - \frac{dx^{(k)}}{dt} \right).$$
(5)

where c_p is a positive constant.

3.3 f_s

Interaction is needed to synchronize oscillators. We define simple interaction at this time as

$$f_s^{(k)} = c_s \sum_i^n \left(x_s^{(i)} - x_s^{(k)} \right).$$
(6)

where c_s is a positive constant.

3.4 f_{prob}

We define the instability of whole aggregation $\varepsilon(t)$ at time t. This indicates how much the current state is unperturbed and $0 \le \varepsilon(t) \le 1$. Term f_{prob} provides probabilistic fluctuations proportional to its instability $\varepsilon(t)$. f_{prob} is determined as

$$f_{prob}^{(k)} = c_{prob}\varepsilon(t)r.$$
(7)

where c_{prob} is a positive constant and r is a random number in $(-0.5, 0.5)$. Because of this definition, the state transition probability of the aggregation of oscillators becomes larger in proportion to instability[11].This definition is typical compared to the example in nature [6]. We allocate all the parameters and sensors to oscillators. These allocations are shown in Fig. 4.

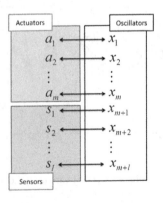

Fig. 4. Allocation of actuators and sensors

4 Experimental Environment

It is difficult to separate the state of oscillators into synchronized and asynchronous states under noisy environments[11]. It is also difficult to evaluate the states of robots only from the states of oscillators for this reason. Therefore, we defined the task of evaluating the effect of this mechanism in robots by converting it to an observable phenomenon. Let us consider a robot with 20 joints. Each joint is a servo and this robot has a link mechanism that moves in 2-Dimensional space. Each joint can move at angles between $-15°$ to $15°$. The main task is to move the head of the robot closer to the object (Fig. 5).

4.1 Convergence of Mechanism to Evaluate Oscillation and Locomotion

We introduced two additional mechanisms to this robotic system. First, we introduced a mechanism to converge oscillation according to closeness to the object. Coefficient c in Eq. (4) at time t is represented by using instability $\varepsilon(t)$ at time t as

$$c(t) = \frac{m_b}{\left(\frac{m_b - m_s}{m_s}\right)\varepsilon(t) + 1} \tag{8}$$

where $m_b > m_s > 0$. We also introduced a mechanism to evaluate locomotion following the example of the method of hill climbing. The amount of change in $x_d^{(k)}$ at time t is determined as

$$\Delta x_d^{(k)}(t) = \left(1 - \frac{\varepsilon(t)}{\varepsilon(t-1)}\right)\{x^{(k)}(t) - x^{(k)}(t-1)\} \tag{9}$$

We expected that oscillators would maintain specific phase relationships by interaction at time t by introducing these mechanisms. This is expressed as the specific locomotion by a robot in the real world and its head moves on a specific limit cycle. Then, Eq. (9) is used to evaluate locomotion and it expresses which direction to go to get closer to

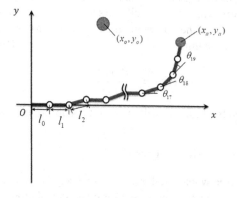

Fig. 5. 20-Dof robot and object

the object with the specific relationships among oscillators at time t. The distance from the object is expressed as potential (Fig. 6) and the potential is radially and monotonically increasing so it is possible to reach the object with the method of hill climbing. However, if the oscillators do not maintain a specific relationship, such an effect is not expected by hill climbing and the number of possible angle combinations is 30^{20}. Therefore, it is difficult to reach the object in an executable period of time.

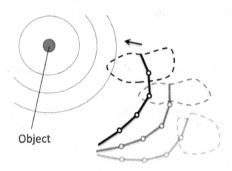

Object

Fig. 6. Cyclic arm motion and potential

4.2 Allocation of Servos and Sensors

We allocate the state of oscillator k to the servo angle to apply the behaviors of oscillators to robot control. Let θ_{min}, θ_{max} denote the minimum and maximum angles. Then, servo angle θ_k is represented by

$$\theta_k = \frac{x^{(k)} - \theta_{min}}{(\theta_{max} - \theta_{min})} \tag{10}$$

There are 20 servos and we prepare the same number of oscillators. We also make the oscillators act as sensors. The state of oscillators acting as sensors is denoted as $x_s(t)$ at time t. The distance between the object and the head of the robot is denoted as $d(t)$ and we let d_{min}, d_{max} denote the minimum and maximum distances. Then, x_s is represented as

$$x_s(t) = \frac{d(t) - d_{min}}{(d_{max} - d_{min})} \tag{11}$$

The robot is controlled by these 21 oscillators and instability $\varepsilon(t) = x_s(t)$ in this case.

4.3 Phase Calculations Using Isochron

Cases must be considered where oscillators are not on the limit cycle when calculating the phase of oscillators. An isochron is used for phase calculations[12][13] to deal with this, where the isochron is preliminarily calculated for a fixed area and the actual phases of oscillators are obtained by using a reference. Fig. 7 shows the example of calculated isochron. When the phases of oscillators are obtained, x' is defined as

$$x' = c(x - x_d) \tag{12}$$

and the phase of x' is obtained by using a reference.

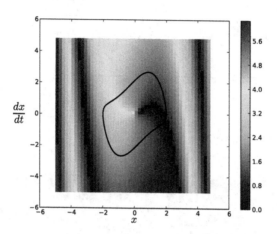

Fig. 7. Example of isochron

5   Experimental Result

Fig. 8 shows the distance between the object and the head of a robot and Fig. 10 shows the robot we used in this experiment. The positions of robot head and the object are illustrated in Fig. 11. First, the robot occasionally gets away from the object, but after a

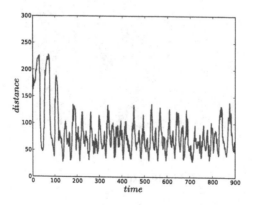

Fig. 8. Distance between object and arm

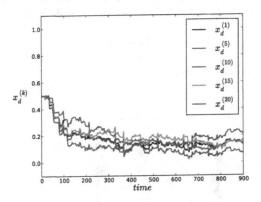

Fig. 9. x_d transition

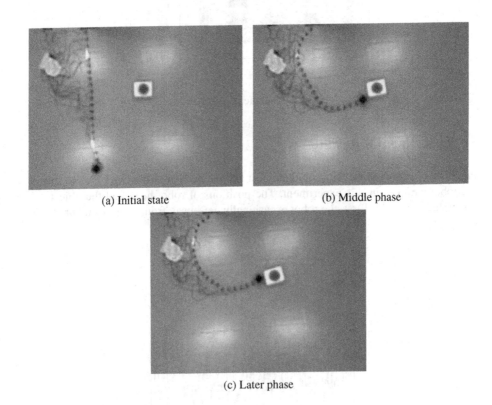

(a) Initial state   (b) Middle phase

(c) Later phase

Fig. 10. Experiment

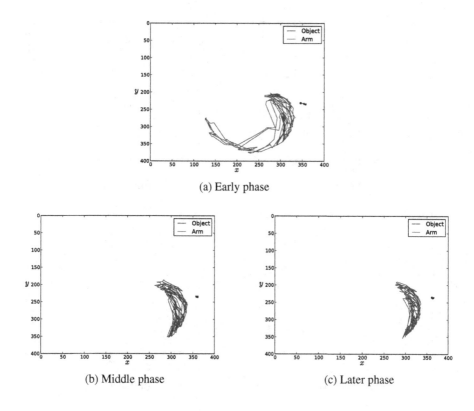

(a) Early phase

(b) Middle phase   (c) Later phase

Fig. 11. The object and the arm head positions

while, the robot starts to get closer to it and repeats approximately the same oscillating locomotion. Some oscillator behaviors are presented in Figs. 12 and 9. Fig. 9 has the $x_d^{(k)}(t)$ of all oscillators and their states are shown in Fig. 12. The specific state of the robot emerged from these results and the oscillators tried changing relationships frequently in the early phase. The oscillators kept their fluctuations smaller and adjusted their states repeatedly after their states had partially stabilized.

Synchronization was also observed to occur among oscillators. We used return maps [14] to observe synchronization. $\Delta\theta_i$ denotes the phase of an oscillator in a return map when the phase of an other oscillator becomes 0 at the ith time and this is plotted them as $\Delta\theta_i$ on the x-axis and $\Delta\theta_{i+1}$ on the y-axis. Here, 21 oscillators repeated synchronous and asynchronous states and it was difficult to split the states into synchronous and asynchronous states as has already been stated. However, as the examples in Fig. 13 indicate, a number of synchronized states were observed. Also, most of these phase relationships were not in-phase synchronization. We observed that complex phase relationships occurred such as $5.5 \sim 6.0(rad)$ and $0.0 \sim 1.0(rad)$.

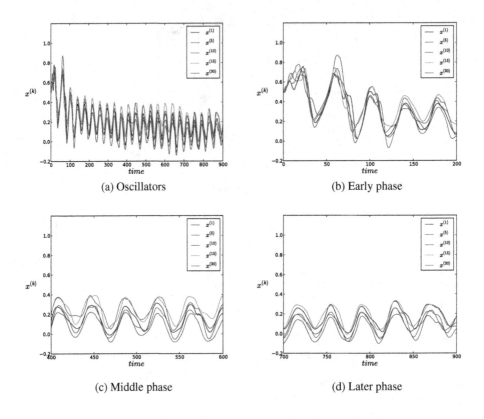

(a) Oscillators

(b) Early phase

(c) Middle phase

(d) Later phase

Fig. 12. Examples of oscillator states

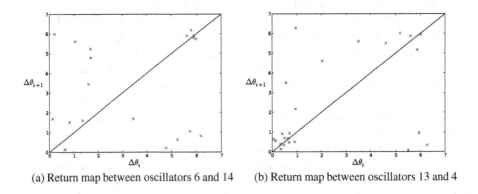

(a) Return map between oscillators 6 and 14

(b) Return map between oscillators 13 and 4

Fig. 13. Return map examples

6 Conclusion

We proposed a method of controlling robots as an aggregation of oscillators to generate homeostasis in robotic systems. We also confirmed the effect of this control by applying tasks that made the effect observable.

References

1. Williamson, M.M.: Neural control of rhythmic arm movements. Neural Networks 11(7), 1379–1394 (1998)
2. Ito, S., Yuasa, H., Luo, Z.-W., Ito, M., Yanagihara, D.: Quadrupedal robot system adapting to environmental changes. Journal of the Robotics Society of Japan 17(4), 595–603 (1999)
3. Aoi, S., Tsuchiya, K.: Locomotion control of a biped robot using nonlinear oscillators. Autonomous Robots 19(3), 219–232 (2005)
4. Yamauchi, S., Kawamura, H., Suzuki, K.: Extended flocking algorithm for self-parameter tuning. IEEJ Transactions on Electronics, Information and Systems 133(6), Sec. C (2013)
5. Yamauchi, S., Kawamura, H., Suzuki, K.: Observation of synchronization phenomena in structured flocking behavior. Journal of Advanced Computational Intelligence and Intelligent Informatics
6. Saigusa, T., Tero, A., Nakagaki, T., Kuramoto, Y.: Amoebae anticipate periodic events. Physical Review Letters 100(1), 018101 (2008)
7. Strogatz, S.H.: SYNC: The Emerging Science of Spontaneous Order, 1st edn. Hyperion (March 2003)
8. Kleidon, A., Lorenz, R.: 1 Entropy production by earth system processes. In: Non-Equilibrium Thermodynamics and the Production of Entropy, pp. 1–20. Springer (2005)
9. Sassone-Corsi, P.: Molecular clocks: mastering time by gene regulation. Nature 392, 871–874 (1998)
10. Collins, J.J., Stewart, I.N.: Coupled nonlinear oscillators and the symmetries of animal gaits. Journal of Nonlinear Science 3(1), 349–392 (1993)
11. Pikovsky, A., Rosenblum, M., Kurths, J.: Synchronization: A Universal Concept in Nonlinear Sciences. Cambridge Nonlinear Science Series. Cambridge University Press (January 2002)
12. Winfree, A.T.: Biological rhythms and the behavior of populations of coupled oscillators. Journal of Theoretical Biology 16(1), 15–42 (1967)
13. Guckenheimer, J.: Isochrons and phaseless sets. Journal of Mathematical Biology 1(3), 259–273 (1975)
14. Nakata, S., Miyata, T., Ojima, N., Yoshikawa, K.: Self-synchronization in coupled salt-water oscillators. Physica D: Nonlinear Phenomena 115(3), 313–320 (1998)

The ChIRP Robot: A Versatile Swarm Robot Platform

Christian Skjetne, Pauline C. Haddow*, Anders Rye,
Håvard Schei, and Jean-Marc Montanier

CRAB Lab, Gemini Centre of Applied Artificial Intelligence,
The Norwegian University of Science and Technology (NTNU), Trondheim, Norway
pauline@idi.ntnu.no

Abstract. Swarm Robotic experiments are ideally performed on real robots. However, a cost versus versatility trade-off exists between simpler and more advanced swarm robots. The simpler swarm robots provide limited features and thus, although suitable for simpler swarm tasks, lack versatility in the types of tasks that may be approached. On the other hand, advanced swarm robots provide a broader range of features, enabling a wide range of tasks to be approached, from simple to advanced. To address this trade-off, an available and versatile robotic platform is proposed: *the Cheap, Interchangeable Robotic Platform — the ChIRP robot*. The basic platform implements mandatory features required for most swarm experiments, providing a cheap simple and extendible platform. Further, extensions (including both electronic and mechanical features) enable an advanced specialised swarm robot tailored to the needs of a given research agenda. The design considerations and implementation details are presented herein. Further, an example swarm task for the basic ChIRP robot is presented together with an example task illustrating an extension of the ChIRP robot.

Keywords: Robotic Platform, Swarm Robots, Robot Design, Swarm Intelligence, IR Sensors.

1 Introduction

Advanced mobile robots today provide for the achievement of a variety of complex tasks, such as warehouse management (Kiva company), space exploration (NASA missions) or car manufacturing (industrial robots). The Spirit and Opportunity robots sent by NASA to Mars are such robots, evaluated at a staggering $625 million. Although an extreme case, with respect to cost, advanced mobile robots may be said to be, not only expensive, but also limited by a single point of failure i.e. one single robot. The swarm robotic approach focuses on solving varied and complex tasks through collective behaviours emerging from the interaction between autonomous simple (and cheap) robots [1,2].

Despite the reduced cost per robot, Swarm Robotic experiments are, most often, conducted in simulation alone [3,4]. However, solutions optimised in simulation can not, in general, be applied directly to real robots. Modelling the interactions between robots and the real world e.g. movement, collisions, communication and sensing; are challenging due to noise [5]. Approaches which help to bridge the gap between simulation and

---

* The author is currently on sabbatical at the University of Adelaide, Australia.

J.-H. Kim et al. (eds.), *Robot Intelligence Technology and Applications 2*,
Advances in Intelligent Systems and Computing 274,
DOI: 10.1007/978-3-319-05582-4_6, © Springer International Publishing Switzerland 2014

the real world provide various forms of hybrid simulated/embodied regimes. Values resulting from sampling the real robot in its environment may be applied so to fine-tune the simulator for further experimentation in simulation [6] and/or the simulator can be dynamically tuned through regular transfers between the real robots and the simulator [7,8]. However, these approaches are still using approximations of a real world environment and as such will be subject to bias and a reduction of the search space. Moreover, the realistic simulation of a larger number of robots requires increased computational power and thus further limits the swarms that may be realistically simulated.

Real swarm robot experiments are cost prohibitive — bearing in mind the large number of robots required. Some labs resort to experimenting with swarms of only a few robots [9], limiting the complexity of the functionality that may be tested and the true potential of swarm robots. Other labs resort to simpler swarm robots which due to the limited functionality of the robots themselves, also limit the actions that may be tested [10].

In this paper, a comparison of existing swarm robots is made so as to identify the strengths and weaknesses of the underlying designs. Further, a new swarm robot — the ChIRP robot is presented, which provides a minimum yet customisable swarm robot. The remainder of the article is presented as follows. In section 2 related work in Swarm Robot design is presented and compared. In section 3, an overview of the ChIRP robot is provided together with a more detailed design description. Section 4 provides two example tasks, applying the ChIRP robot in its base and an extended form, respectively. Finally, section 5 concludes the work.

2   Motivation

Swarm Robotics experiments often require multiple types of interactions with the environment e.g. battery loading [4], object pushing [11], object grasping [12] and aggregation [13]. In the design of any robot, the specification depends on the task(s) that the robot is being designed to solve. In recent years multiple robot platforms targeting swarm robotics have been developed. Such platforms range from simpler robots with few features able to solve simpler swarm tasks, to those that have features enabling solution of a range of simpler to more advanced swarm tasks. Table 1 provides a comparison of a number of existing swarm robot designs. In selecting the robots for comparison, care has been given to the inclusion of robots covering simpler to the more advanced swarm robots. Further, the robots selected provide a range of costs — a further important design issue.

The most basic features found in all robots developed for Swarm Robotics are the capacity to move and to sense their environment. Table 1 shows that differential wheels (diff.wheels) are the most common steering mechanism. The Kilobot [10], on the other hand, has vibrating rigid legs. However, such a solution is limited due to the lack of precision in their control. Further, although various sensors may be incorporated, IR sensors (IR) are implemented in all the robots. In addition, although the selected robots vary in size, they are mostly hand-sized or smaller. At 170mm, the MarXbot, the largest of the swarm robots investigated, challenges the practical feasibility of larger swarms due to the need for a larger experimental arena. Building on the lessons learnt to date,

Table 1. A Comparison of Swarm Robots

	Size (mm) HxW	Actuators / Sensors	IR location	Battery (hours)	re-charge	Available p:parts r:retail	open source
Jasmine	30mm³	diff.wheels , IR, compass, colour light	open	1-2	off-line	p: €100	yes
Khepera III	70x130	diff.wheels, IR, ground, ultra sonic, ambient light, WiFi	open	8	off-line	r: €2600	no
R-One	n.a.x82	diff.wheels, IR, radio, ultrasonic, accelerometer, cliff detector, temperature	open	4	off-line	r: €320	no
Kilobot	34x33	vibration rigid legs, IR, ambient light	open	3-24	on-line	p: €10	yes
E-puck	50x70	diff.wheels, IR, camera, microphone, accelerometer	open	1-2	off-line	r: €800	no
S-bot	150x120	diff.wheels, IR, position, humidity, accelerometer, temperature, light, accelerometer, camera, microphones	open	1	off-line	no	no
MarXbot	170x170	diff.wheels, IR, ground, camera, accelerometer, gyroscope, microphones, axis force	open	8	off-line	n.a.	no

the common features of differential wheels, IR sensors and hand-sized or smaller provide a minimum set of proven features for a simple swarm robot.

Looking further at the range of available features, it is clear that the task-oriented specification of these robots varies considerably. On one end of the spectrum, providing fewer features, are the Kilobot [10] and Jasmine [14] robots. These robots are particularly aimed at achieving lower costs and smaller size with a limited set of features. The Kilobot is the smallest and cheapest currently available swarm robot. However, its movement can not be controlled precisely. The Jasmine robot, on the other hand, suffers from a limited battery life, similar to that of the E-puck robot — see table 1. Further, although small size is a sought after feature, such small sizes limit the possibility for a flexible design and thus the potential to extend the robots, specialised to advanced tasks.

On the other end of the spectrum are advanced robots, such as, MarXbot[15], S-bot[16], E-Puck [17]; Khepera III [18] and R-one [19]. Rather than specialised, such robots may be considered to be more generalised task solvers, involving a range of features for the solution of simpler to more advanced tasks. Further, some robots also propose extensions to these features, thus providing task specialisation e.g. the E-puck and S-bot. A further advantage of these robots, with the exception of the S-bot, is that they are obtainable, enabling comparison of results. However, the cost of such robots is a hindrance to experiments with larger swarms i.e. over 30 robots.

Advanced robot features enable a broader range of tasks to be conducted. However, most experiments only apply a sub-set of the standard features available and thus much expense is lost on generality. The higher cost of the generalised base platform often restricts experimentation to a relatively small number of robots e.g. between 10 and 20 robots.

The practical application of such robots highlights some further issues. Battery life is, of course, important. Demonstrating with robots or running longer swarm experiments, requires sufficient battery life and/or the potential to re-charge batteries. As stated, the E-puck and the Jasmine robots have limited battery life and although the E-puck robot allows re-charging off-line — see table 1, an on-line solution is more desirable.

A further issue is with the IR sensors. Most commonly IR sensors are placed "open" such that any ambient light (external IR light e.g. natural light from a window) will affect the sensor readings. The affect can range from total failure due to blindness, to reduced sensitivity and/or reduced contrasts between objects and the background. As such, the issue of ambient light is a critical issue in the use of IR sensors [14]. A further issue, is availability. Some robots have limited availability, such as those limited to a particular project e.g. the s-bot.However, most robots investigated are available and the Kilobot and Jasmine robots may be said to be highly available, providing both open source material and robots kits.

There are inherent advantages and disadvantages with all swarm robots that are not tailored to meet a specific task and specific cost constraints for a given research agenda. However, focusing on the advantages of each of the different solutions discussed, it is clear that there is a need for a swarm robot that can provide generality at a low cost as well as tailoring to a specific task so as to reap many of the benefits of both the simpler and the more advanced robots presented.

3   The ChIRP Robot

The key issues driving the design of the ChIRP robot include availability and versatility as well as ease of use in an experimental setting. The term availability is interpreted as physically available as well as available to research labs with limited resources e.g. arena limitations, robot purchase limitations. Further, a versatile robot needs to be tunable to the research agenda without conflicting with the availability requirements.

The ChIRP robot design consists of an extendible simplified platform offering the basic swarm features highlighted in section 2 i.e. differential wheels, IR sensors and smaller size, as well as providing extension capabilities. The core features of the robot in its simplified form are summarised in table 2.

The ChIRP robot clearly offers a versatile platform, being designed to enable extensions — see section 3.1. Two features in table 2 highlight the availability of the platform. Firstly, the low cost due to the simplified platform, provides for a comparable cost with the low-end simplified robots described in table 1. Secondly, the robot is fully open-source, including mechanical and electrical design and all software libraries. Further,

Table 2. Features of the ChIRP robot

	Size (mm) HxW	Actuators / Sensors	IR location	Battery (hours)	re-charge	Available p:parts r:retail	open source
ChIRP	55x85	diff.wheels, IR	hidden	4	on-line	r: €100	yes

the smaller size (comparable with the E-puck) together with the larger battery capacity (comparable with the R-one) together with the hidden placement of the IR sensors and the on-line battery re-charging capability, make for a practical solution.

3.1 Hardware Design

The ChIRP robot without extensions is illustrated in figure 1. As shown, the ChIRP robot is a cylindrical robot with a 3D printed chassis and two 3D printed wheels attached to the motors.

Two ATtiny microcontrollers (**A**) control the IR sensors (**B**). Eight pairs of IR emitters and receivers are available for distance measurement, IR light sensing and inter-robot communication. As stated, the receivers are positioned under the board so as to avoid sensitivity to ambient light. Such a solution provides a unique but simple hardware solution, avoiding the need for existing software solutions that reduce the affect of ambient light. Further, the robot is equipped with two motors and a motor driver board (not shown) located under the sensor board. The motors are simple geared stepper motors. The stepper motors have been chosen due to their accuracy and low cost. The motor driver board is used to control the movement of the motors. This board is composed of one Darlington chip and one ATtiny microcontroller to drive it. Further,

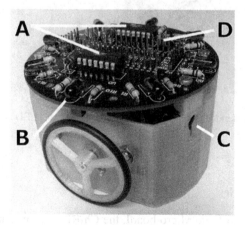

Fig. 1. The ChIRP base robot. A) The two ATtinys that control the sensors, B) An IR emitter (top of board) and receiver (bottom), C) slot for attaching expansions to the chassis and D) The row of pins along the middle are the exposed pins of the Arduino, the I^2C bus and the TTL port, for attaching expansions.

a battery charger board, enabling on-line battery re-charging, and the battery itself are also hidden from view. The battery chosen has 2600mAh capacity and provides approximately four hours of runtime (measured with motors at full speed).

The main processor board — an Arduino Micro; is fixed under the sensor board and manages the activity of the sensor board and the motor driver board. The main processor runs the robot program and is, therefore, the only processor the user accesses. It communicates with the other microcontrollers over an I^2C bus. This design offloads the main processor from recurrent tasks such as moving the motors and polling of sensors. The Arduino Micro offers sufficient capacities to run swarm robotic algorithms. The robot is programmed in C/C++ using the freely available Arduino IDE and numerous libraries.

Features (C) and (D) are designed to provide extension capabilities for the robot. (C) shows one of the four slots in the chassis, enabling physical extensions to the robot to be added e.g. a gripper (see figure 2). In (D), the exposed Input/Output pins of the main processor are shown, enabling extension of the robot upwards with further boards that can access the main processor. The I^2C bus and TTL serial port are also accessible. Therefore any I^2C and TTL enabled device can be used when adding an extension to the ChIRP robot. Examples of such extensions are: LED light shield, Bluetooth communication, extra sensors and additional actuators such as gripper arms.

Fig. 2. ChiRP robot extended with a gripper arm, bluetooth, LED shield and ultrasonic rangefinder

3.2 Software Libraries

By incorporating the Arduino Micro board, the ChIRP robot benefits from numerous available libraries supporting both software development and hardware control. Further, two software libraries have been created: one for the sensor board (InfraRed library) and one for the motor board (Motors library). The advantage of these libraries are that the user does not need to concern themselves about the underlying communication between the main and the other processors. For example, the reading of distance sensors is achieved by simply calling the function *getDistSensors* of the `InfraRed` library and setting of motor speeds is achieved by calling the function *moveAtSpeeds* of the `Motors` library. Consequently, the creation of a program for the task of wall avoidance behaviour, is a matter of manipulating these two functions.

3.3 Availability

To ensure availability to as broad a research base as possible the ChIRP robots are fully open source, including mechanical, electrical and software. Such resources are already available on the ChIRP website [20], enabling those in the community that are interested in such robots to build their own. Further, it is intended to make kits and complete robots available to the community at as low a cost as possible. The estimated price is €100 for a complete robot or, of course, if the open source option is chosen, the robots can be built at a cheaper cost.[7] No price is currently available for the robot kits. Currently the robots are undergoing a thorough testing process and the estimated availability of the completed robots is spring 2014.

4 ChIRP Robot Demonstrations

To demonstrate the base ChIRP robot capabilities in a swarm setting, a simple swarm task was performed — box pushing (with no internal/external communication). The second demonstration focuses on an extended ChIRP robot. Instead of a swarm task, a single robot wall avoidance task was performed through remote control. This demonstration serves to highlight the ease in which the hardware and software may be extended.

4.1 Box Pushing Demonstration

Box Pushing is a task that is suitable for a simplified robot or in this case the basic ChIRP robot i.e. it requires the ability to move and sense the environment. The box pushing experiment is inspired by the food foraging behaviour of ants. The objective of the experiment is for the swarm of robots (ants) to collectively push the box (food) to a wall. No specific designation on the walls is given. Further, the smell of food is represented by an IR-light emitted from the box that the robots can detect and, therefore, distinguish the box from other obstacles. Further, the box is designed to be too heavy for a single robot to push alone, requiring cooperating robots pushing together. The robots have no notion of where they are situated in the environment or where the box is located and are not provided with any communication capabilities with each other. However, the robots are looking for the IR transmission from the box (food smell) and will react to such a transmission. Further, they will avoid any other obstacles that they come across.

The controller for this task is inspired by [21]. However, no communication is enabled between the robots. If the box is detected, the robot starts pushing for a predefined time. If a robot detects that it cannot move the box, it tries to realign itself and finally

Fig. 3. ChiRP robot performing the box pushing experiment

reposition itself by moving away and tracking the box again. Finally when sufficient robots are pushing the box in a similar direction the box will move and this movement will provide a positive feedback to the robots, supporting further pushing by the robots.

The box measures 25x25cm and the environment is 121.5x121.5cm. Each run of the experiment is set up with the box in the middle and the robots randomly placed in the environment. A run is successful when the robots are able to collectively move the box to one of the walls of the environment, as depicted in figure 3. Averaged over 10 runs, 5 robots completed the task in 167 seconds. The maximum time used to successfully complete the task was 367 seconds. The varied time used reflects the random movement of the robots (ants) due to lack of communication between the robots. However, it shows that communication is not needed for the robots to solve the task.

Fig. 4. Box pushing stages: A) Starting position B) Stagnation C) Repositioning and D) Success

The behaviour of the robots is illustrated in figure 4. As stated, the ChIRP robots are placed randomly in the environment (**A**). In (**B**), several robots have found the box but their placement results in the box being pushed from two sides and thus no movement.

(C) highlights the realignment process where the robots move so as to find the box again. It should be noted that this is not a group movement and thus the robots individually will pull away and reposition themselves. As such, stagnation can again happen if robots find themselves on different sides of the box again. (D), shows the success of the swarm in moving the box to the wall.

This experiment highlights the capabilities of autonomous navigation, object avoidance, object detection through IR-light emission and the achievement of a simple swarm task. A video of this demonstration and the source codes are available [20].

4.2 Expansion Demonstration

An Android application was implemented to provide an example extension of the ChIRP robot. The Android application controls the movement of a ChIRP robot for an object avoidance task. The robot's sensor data are displayed in real time on the device screen and the LED's on the robot itself illustrate the current status of the robot i.e. in which direction the robot is moving. This experiment requires two extension shields for the robot, as illustrated in figure 5: one for bluetooth communication (A) and one for LEDs (B). Both of these shields require access to the I^2C bus and the TTL port of the Arduino processor, thus exploiting the exposed pins available on the sensor board.

Fig. 5. ChIRP with bluetooth board (A) and LEDshield (B) expansions

The TTL serial port's role is to provide high speed communication between the Arduino processor and the external hardware i.e. the Android device in this case, through the bluetooth expansion board. The LED shield, as shown, has eight blue LEDs and is placed on top of the robot. The board is populated by two microcontrollers, each controlling four LEDs on either side of the robot. Android commands are transmitted from the Android device, interpreted by the Arduino processor and result in the commands to the motors, sensors or LEDs, sent over the I^2C bus.

Figure 6 provides a snapshot of the Android application. The button on the top is used to connect and disconnect to/from the robot. The 8 IR sensor receivers status are

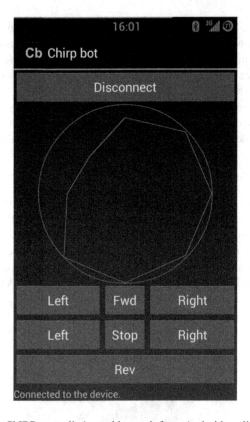

Fig. 6. ChIRP controlled over bluetooth from Android application

displayed in real time inside a circle on the screen. The 8 points in the octagon represent the 8 sensor readings. As the robot approaches an object, the sensor reading is interpreted and shown to be closing in to the centre of the circle. As shown, neighbouring sensors may also be affected. The buttons at the bottom of the screen provide manual control of the movement of the robot. The following orders are implemented (from the top left to bottom right): turn left, move forward, turn right, rotate left, stop and rotate right. The button at the bottom is used for backward motion. A video of this demonstration and the source code for the Android application is available [20].

5   Conclusion

The comparison of existing swarm robots presented illustrates that existing swarm robots have advantages and disadvantages and for a given research agenda, a trade-off is needed so as to select the robot design most suitable for the task(s) and budget sought. The ChIRP robot, seeks to provide a further solution, filling the gap identified where cheap simple extendible swarm robot platforms are needed that, further, may be tuned to the needs of particular research agenda. Thus the ChIRP robot combines features from simple to advanced robots in a unique package with some extra twists to improve practical application of the robots. These features include:

- a simple cheap platform
- suitable swarm size
- good battery life with on-line re-charge capabilities
- multitude of extension possibilities
- avoidance of external IR interference
- numerous available libraries
- fully open source [20]
- robot kits and completed robots (limited availability currently)

Two distinctly different tasks have been demonstrated illustrating both the application of the basic ChIRP robot for a typical simpler swarm task — box pushing; as well as a remote control task exploiting an extended ChIRP robot, tuned to the needs of the task. Further work will include testing of both the base robot and extended versions, involving several research groups and their associated research agendas.

References

1. Şahin, E.: Swarm robotics: From sources of inspiration to domains of application. In: Şahin, E., Spears, W.M. (eds.) Swarm Robotics 2004. LNCS, vol. 3342, pp. 10–20. Springer, Heidelberg (2005)
2. Sahin, E., Girgin, S., Bayindir, L., Turgut, A.E.: Swarm robotics. In: Blum, C., Merkle, D. (eds.) Swarm Intelligence. Natural Computing Series, pp. 87–100. Springer, Heidelberg (2008)
3. Karafotias, G., Haasdijk, E., Eiben, A.E.: An algorithm for distributed on-line, on-board evolutionary robotics. In: Proceedings of the 13th Annual Conference on Genetic and Evolutionary Computation, GECCO 2011, pp. 171–178. ACM, New York (2011)
4. Montanier, J.M., Bredeche, N.: Surviving the tragedy of commons: Emergence of altruism in a population of evolving autonomous agents. In: Proceedings of the 11th European Conference on Artificial Life, ECAL 2011, pp. 550–557 (2011)
5. Jakobi, N., Husband, P., Harvey, I.: Noise and the reality gap: The use of simulation in evolutionary robotics. In: Morán, F., Merelo, J.J., Moreno, A., Chacon, P. (eds.) ECAL 1995. LNCS, vol. 929, pp. 704–720. Springer, Heidelberg (1995)
6. Nolfi, S., Floreano, D., Miglino, O., Mondada, F.: How to Evolve Autonomous Robots: Different Approaches in Evolutionary Robotics. In: Brooks, R.A., Maes, P. (eds.) Artificial life IV: Proceedings of the 4th International Workshop on Artificial Life, pp. 190–197. MIT Press, MA (1994)
7. Bongard, J., Lipson, H.: Nonlinear system identification using coevolution of models and tests. IEEE Transactions on Evolutionary Computation 9, 361–384 (2005)
8. Koos, S., Mouret, J.B., Doncieux, S.: Automatic system identification based on coevolution of models and tests. In: Proceedings of the Eleventh conference on Congress on Evolutionary Computation, CEC 2009, pp. 560–567. IEEE Press, Piscataway (2009)
9. Prieto, A., Becerra, J.A., Bellas, F., Duro, R.J.: Open-ended evolution as a means to self-organize heterogeneous multi-robot systems in real time. Robot. Auton. Syst. 58, 1282–1291 (2010)

10. Rubenstein, M., Ahler, C., Nagpal, R.: Kilobot: A low cost scalable robot system for collective behaviors. In: 2012 IEEE International Conference on Robotics and Automation, ICRA, pp. 3293–3298 (2012)
11. Maris, M., Boeckhorst, R.: Exploiting physical constraints: heap formation through behavioral error in a group of robots. In: Proceedings of the 1996 IEEE/RSJ International Conference on Intelligent Robots and Systems, IROS 1996, vol. 3, pp. 1655–1660. IEEE (1996)
12. Mouret, J.-B., Doncieux, S.: Encouraging behavioral diversity in evolutionary robotics: an empirical study. Evolutionary Computation 20(1), 91–133 (2011)
13. Soysal, O., Sahin, E.: Probabilistic aggregation strategies in swarm robotic systems. In: Proceedings 2005 IEEE Swarm Intelligence Symposium, SIS 2005, pp. 325–332. IEEE (2005)
14. Kornienko, S.: IR-based communication and perception in microrobotic swarms. arXiv preprint arXiv:1109.3617 (2011)
15. Bonani, M., Longchamp, V., Magnenat, S., Retornaz, P., Burnier, D., Roulet, G., Vaussard, F., Bleuler, H., Mondada, F.: The marXbot, a miniature mobile robot opening new perspectives for the collective-robotic research. In: 2010 IEEE/RSJ International Conference on Intelligent Robots and Systems, IROS, pp. 4187–4193 (2010)
16. Dorigo, M., Floreano, D., Gambardella, L.M., Mondada, F., Nolfi, S., Baaboura, T., Birattari, M., Bonani, M., Brambilla, M., Brutschy, A., et al.: Swarmanoid: a novel concept for the study of heterogeneous robotic swarms. IEEE Robotics & Automation Magazine (2013) (in press)
17. Mondada, F., Bonani, M., Raemy, X., Pugh, J., Cianci, C., Klaptocz, A., Magnenat, S., Christophe Zufferey, J., Floreano, D., Martinoli, A.: The e-puck, a robot designed for education in engineering. In: Proceedings of the 9th Conference on Autonomous Robot Systems and Competitions, pp. 59–65 (2009)
18. K-Team: Khepera iii homepage (2013),
http://www.k-team.com/mobile-robotics-products/khepera-iii/
19. McLurkin, J., Lynch, A.J., Rixner, S., Barr, T.W., Chou, A., Foster, K., Bilstein, S.: A low-cost multi-robot system for research, teaching, and outreach. In: Martinoli, A., Mondada, F., Correll, N., Mermoud, G., Egerstedt, M., Hsieh, M.A., Parker, L.E., Støy, K. (eds.) Distributed Autonomous Robotic Systems. STAR, vol. 83, pp. 597–609. Springer, Heidelberg (2013)
20. CRABlab-Team: Chirp homepage (2013), https://chirp.idi.ntnu.no
21. Berg, J., Karud, C.H.: Swarm intelligence in bio-inspired robotics. Master's thesis, Department of Computer and Information Science, The Norwegian University of Science and Technology (2011)

Aeroponic Greenhouse as an Autonomous System Using Intelligent Space for Agriculture Robotics

Martin Pala[1], Ladislav Mizenko[1], Marian Mach[1], and Tyler Reed[2]

[1] Department of Cybernetics and Artificial Intelligence, Technical University of Košice
{martin.pala,ladislav.mizenko,marian.mach}@tuke.sk
[2] Independent Consultant, Maya Culpa, LLC, Galloway, Ohio
tyler@mayaculpa.com

Abstract. This paper describes a novel approach to aeroponic and hydroponic system monitoring, fault detection and automation. The first part of this paper is dedicated to brief literature preview about hydroponics and aeroponics, its common and distinctive features and the description of the needs for its automation. The second part of this paper deals with aeroponic greenhouse control scheme proposal. We consider a greenhouse covered by a sensor network, actuators and hydroponic or aeroponic platforms to be a robotic system in so called intelligent space. The aeroponic platform design is described besides the conclusions and future work ideas in the last part of this paper.

Keywords: Agriculture Robotics, Intelligent Space, Aeroponics, Hydroponics, Greenhouse.

1 Introduction

The current world population of 7.2 billion is projected to increase by almost one billion people within the next twelve years, reaching 8.1 billion in 2025 and 9.6 billion in 2050, according to a new United Nations report [1]. As the world population continues to grow, the rising demand for agricultural production is significant. More than half of global population growth between now and 2050 is expected to occur in Africa. According to the UN's medium-variant projection, the population of Africa could more than double by mid-century, increasing from 1.1 billion today to 2.4 billion in 2050, and potentially reaching 4.2 billion by 2100. High population numbers are putting further strain on natural resources, fuel supplies, employment, housing and food supplies. In addition, an increased demand for biofuels could further increase pressure on inputs, prices of agricultural produce, land, and water and endanger a global food security.

The main motivation of this paper is to provide overview information about progressive techniques and methods for producing green food with consideration for environmental factors and energy efficiency. The main idea behind the aeroponic greenhouse in intelligent space is full automation, scalability, anytime-anyplace access monitoring and fault diagnostics for home or enterprise farming. In the first

J.-H. Kim et al. (eds.), *Robot Intelligence Technology and Applications 2*,
Advances in Intelligent Systems and Computing 274,
DOI: 10.1007/978-3-319-05582-4_7, © Springer International Publishing Switzerland 2014

part of this paper the hydroponics and aeroponics methods of growing plants are introduced and the connection between greenhouse as a robotic system and intelligent space is explained. The second part is dedicated to aeroponics control system architecture proposal and its main features.

2   Hydroponics, Aeroponics, and Current Trends in Controlled Environment Agriculture

In order to understand hydroponics and aeroponics methods of growing plants, some of the basic terms and concepts need to be clarified. The concepts and terms provided in the following words appear in the well-known publications related to hydroponics or aeroponics [2, 3, 4, 5, 6, and 7].

Horticulture: Horticulture is the science, technology, and business involved in intensive plant cultivation for human use. It is practiced from the individual level in a garden up to the activities of a multinational corporation. Horticulture is very often described as both science and art [2, 3].

Hydroculture: Hydroculture is the growing of plants in soilless medium, or an aquatic based environment. All the plant nutrients are distributed via water.

Hydroponics: Hydroponics is a subset of hydroculture and is a method of growing plants using mineral nutrient solutions, in water, without soil. Terrestrial plants may be grown with their roots in the mineral nutrient solution only (liquid hydroponic systems) or in an inert medium, such as perlite, mineral wool, gravel, expanded clay pebbles or coconut husk (aggregate hydroponic systems). Hydroponics is a subset of soilless culture, but many types of soilless culture do not use the mineral nutrient solutions required for hydroponics.

Aeroponics: Aeroponics on the other hand is considered to be another form of hydroponic technique as water is used to transmit nutrients as well. Aeroponics is the process of growing plants in an air or mist environment without the use of soil or an aggregate medium [4]. According to AgriHouse (product outcome of NASA research program), growers choosing to employ the aeroponics method can reduce water usage by 98 percent, fertilizer usage by 60 percent, and pesticide usage by 100 percent, all while maximizing their crop yields by 45 to 75 percent [4].

2.1   Hydroponics vs. Aeroponics

Hydroponics and aeroponics are both highly efficient methods of growing plants without them ever touching any soil. Both perform very well indoors or outdoors, are easy to maintain and relatively easy to automate, so during the plant growing process less attention from humans is needed. In addition the following advantages are being considered for both aeroponics and hydroponics systems [4, 5, 6, 7, and 8]:

- No soil is needed, no nutrition pollution
- The water stays in the system, so can be reused
- Lower nutrition requirements, nutrition control possibility
- Stable and high yields, healthier plants as no pests are used
- Energy efficient, easy harvesting

Despite their many similarities (and the fact that aeroponics is actually a type of hydroponics), aeroponics and hydroponics techniques also have some important differences.

The most distinctive characteristic is that aeroponics uses no growing medium at all, while hydroponics uses growing medium. Also the distribution of nutrients is different. Using hydroponics systems, plants may be suspended in the water full-time or the nutrients can be distributed by a continuous or even an intermittent flow. Aeroponic plants are never placed into water, even for a minute. Instead, aeroponic plants receive nutrients from a nutrient-rich water solution that is sprayed onto their dangling roots and lower stem several times an hour. The main advantage of nutrient delivery using aeroponics systems is that the plants are kept in a relatively closed environment, so diseases are not spread rapidly, while in traditional hydroponic methods plant diseases can be spread via nutrient distribution system from plant to plant. Another advantage of aeroponics is that suspended aeroponic plants receive 100% of the available oxygen and carbon dioxide to the roots zone, stems, and leaves, thus accelerating growth and reducing rooting times [stoner]. According to the NASA research aeroponics requires 65% less water than hydroponics approach and aeroponically grown plants requires ¼ the nutrient input compared to hydroponically grown plants [4, 9].

The main problem of both hydroponics and especially aeroponics is that without soil as some kind of water and nutrient buffer, any failure of the aeroponic or hydroponic system leads to rapid death of grown plant.

For this reason, more sophisticated methods have to be used for fault detection, real-time monitoring, control and automation of such systems. Utilization of artificial intelligence in hydroponic and aeroponic systems may lead not only to early fault detection, thus avoiding damage to grown plants, but may also help to fully automate all the processes required in aeroponics and hydroponics and adapt to current needs of grown plants in real-time without any or small interventions of human operators, help to lower costs and make the whole process more efficient and likely more profitable. In the following section we provide a preview of some advanced methods and techniques currently applied in real life solutions or currently being developed or considered for real application.

2.2 Current Trends and Related Work

As previously mentioned, hydroponic and aeroponic systems have substantial vulnerability which can increase the complexity of their successful unattended application in real production. Problems like the failure of water pumps, nutrient distribution and preparation, nozzle clogging and many others require special attention to avoid damage or rapid death of growing plants. In this chapter we provide a basic preview of current trends and selected related work to hydroponics and aeroponics done so far.

For example, the optimization of long-term plant growth in hydroponics, a hierarchical intelligent control system consisting of an expert system and a hybrid system based on genetic algorithms and neural networks was proposed in [10]. These two control systems were used appropriately, depending on the plant growth. Using this approach, the plant growth is controlled by the nutrient concentration of the solution. The expert system was used for determining the appropriate set-points of nutrient concentration through the whole of the growth stages, and the hybrid system for determining the optimal set-points of nutrient concentration which maximize total leaf length (TLL) and stem diameter (SD) during the initial growth (seedling) stage. In the hybrid system, TLL/SD as affected by nutrient concentration was first identified using neural networks and then the optimal value was determined through simulation of the identified model using genetic algorithms. The set-points from the expert system were similar to those used by a skilled grower. Further details and experimental results can be found in [10].

Early disease detection is crucial especially when using hydroponic configuration as the disease may spread quickly via the nutrient distribution system. JAPIEST, an integral intelligent system for the diagnosis and control of tomatoes diseases and pests in hydroponic greenhouses was developed [11]. JAPIEST is focused on the prevention, diagnosis and control of diseases that affect tomatoes and is capable of early detection of candidate diseases and suitable control treatment proposals.

Another recent work deals with fault detection using artificial intelligence methods. Neural-network systems capable of detecting of mechanical and biological faults in deep-trough hydroponics are based on utilization of two separate fault detection systems. The first fault detection system is based on detection of faulty operation using sensor information, as inputs for neural network (pH, EC, air temperature, nutrient solution temperature, humidity, light intensity). The second fault detection system is focused on the detection of a category of biological faults (transpiration rate). The proposed neural-network model was able to detect mechanical problems using the first approach in most of the cases within 20 to 40 minutes, which may be sufficient for both hydroponics and aeroponics configurations. In the case of the biological fault detection, neural network model generally detected faults within 2-3 hours, where the output was either 0 – normal, or 1 – faulty operation [12, 13].

All the previously described growth optimization methods or disease and fault detection systems require input information which can be easily acquired by installing suitable sensors. The recent publication "Monitoring of an Aeroponic Greenhouse with Sensor Network" describes a possible design and implementation of a wireless sensor network for greenhouse environment monitoring [14]. The real-time information obtained from the sensor nodes may be utilized to optimize temperature, ventilation, artificial lighting and nutrient solution properties control. More recent work about the real time greenhouse monitoring can be found in [15, 16 and 17].

The following subchapter is dedicated to a robust system architecture that aims to utilize all the available methods to achieve rapid power optimization, plant growth, early fault detection and the highest possible automation level.

2.3   HAPI

The Hydroponic Automation Platform Initiative (HAPI) develops and distributes control modules for automated food production systems, including and especially hydroponic and aeroponic systems. Through the initiative, a steady line of monitoring and control options are being evolved and optimized. Control options include control of lighting, water pumps, nutrient and pH dosing pumps, automated shades and louvers, heaters, air coolers, water chillers, humidifiers and dehumidifiers and security mechanisms. Support is also being developed for a wide array of sensors including temperature, humidity, moisture of growth mediums, water levels, light intensity and color, pH, electrical conductivity and weather-related information. Sensing can be performed in plant, section, zone and site-wide formats.

Artifacts from this aspect of the project include: hardware design, schematic drawings, firmware, management software, reporting functions, structural designs and process documentation.

HAPI is not solely concerned with automation. The initiative includes several programs that together will serve to achieve the primary goal of increasing high-yield hydroponic production of clean food in urban settings. A core value of the initiative is "clean" food. In this case, clean means as close to organic as possible and without genetic modifications. One challenge in this regard is the diminishing supply of non-gmo seeds. To this end, HAPI will initiate a worldwide network of "clean seed" production and distribution points. The primary goal of this program is to establish a global distribution web, by collaborating with similar programs with local focus, so that local access to a worldwide supply of diverse clean seeds is provided to as many urban farmers as possible. By providing automation modules and structural designs, a clean seed network and a best practice application, HAPI will dramatically lower the barriers to high-yield, sustainable food production to individuals and communities across the planet. HAPI is an open source project with many collaborators.

The research project of aeroponic greenhouse proposed in this paper will share a lot of features and is planned to be fully compatible with all HAPI elements.

3   Intelligent Space

Intelligent space (iSpace) is an area or room that monitors what is happening in it, can make decisions and react according to these decisions [18]. Monitoring is performed by using sensors distributed in intelligent space, which may include cameras, microphones, temperature and humidity sensors, ultrasonic and other range finding sensors. Actuators on the other hand provide information for inhabitants of iSpace using monitors or speakers, or can execute actions through robotic devices to manipulate with the real world objects.

Since Hashimoto Laboratory proposed Intelligent Space in 1996, research has started in many different fields from robotics and mobile robot positioning and navigation [19] to health care [20], intelligent vehicles, transportation spaces and traffic control [21], intelligent homes and rooms [22] or service robotics in large-scale dynamic environments [23]. Various research disciplines are involving into intelligent space, while it incorporates wide scope of issues needed to be solved. Most of the research is done in following scope: smart cameras and devices, flexibility and scalability of system architecture and networking, signal processing and inferences, and real-time execution of actions. Typical high-level architecture of intelligent space consists of a communication layer connecting all devices, sensing, an acting layer for collecting data and providing feedback by performing actions to physical world, and a processing layer responsible for handling and storing data and making decisions.

In an effort to achieve the intelligent exchange of information, it is necessary to build up pervasive communication infrastructure. Substantial research in this area has focused on various aspects of network communication. Recent studies focus on wireless networking with a goal of creating a communications layer for the fast and reliable transfer of information and knowledge.

ISpace is characterized by specific properties. One important aspect is modularity. It is necessary to have the ability to add and remove components in real-time without system downtime. In order to use iSpace in a wide variety of environments, all of the component systems must be scalable. As mentioned, iSpace consists of devices like sensors, actuators and other components spatially distributed in the environment. All these devices are connected into the network, which creates a space for the emergence of difficulties to solve.

Greenhouses provide ideal environments for the application of intelligent space methods. Covering a greenhouse complex with networked sensors and actuators can have a strong impact on the entire automation process and fundamentally transforms the greenhouse to a "robot ready" environment. This opens paths to not only monitoring and fault diagnostics, but also to robotic crop collection and many other robotic applications.

4   Aeroponics Control System Architecture Proposal

We propose the control architecture for remote operation, monitoring, diagnostics and automation of the processes realized during the standard operation of aeroponic or hydroponic greenhouse configuration. Our research is specialized for aeroponic configuration, but is fully compatible with the approach adopted by the previously described HAPI project. The main reason of this compatibility is further cooperation and acceleration of technology application.

The following picture describes the main architecture of control and communication mechanisms between a remote human or virtual operator (e.g. expert system for autonomous plant growing) and an aeroponic or hydroponic greenhouse.

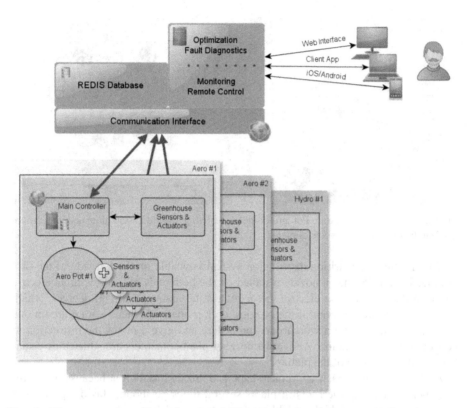

Fig. 1. The proposed architecture of control and communication mechanism between human/virtual operator and aeroponic/hydroponic system

The greenhouse is equipped with its own sensor network and actuator network for climate monitoring and adjustment and basically creates the intelligent space and robot-ready environment. Every module is connected to the main controller which provides the communication with "Aero Pots" and remote modules (top of the picture). Main controller is capable of independent control or might be dependent on control signals from a remote human or virtual operator. A virtual operator consists of three important modules: Communication module, module for monitoring, fault diagnostics remote control and optimization and database module. Interaction with a human operator is provided using a human machine interface via web interface, client application or iOS/Android application on a smartphone. During the first stage of the project optimization will be aimed to lower power consumption of aeroponic system preserving maximum possible growth of the plants using various methods of artificial intelligence (e.g. genetic algorithms, adaptive fuzzy cognitive maps, etc...).

Aero Pot is working codename for a highly scalable aeroponic system designed in our department. The Aero Pot prototype is visualized on the following picture.

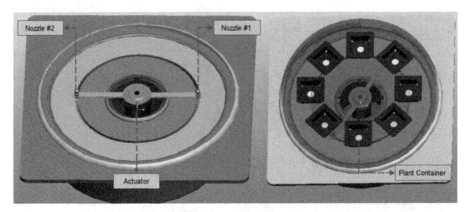

Fig. 2. CAD design of Aero Pot prototype, visualizing nozzle installation and process of nutrient distribution to the roots of plants

The finishing of nutrition distribution system is visible on the left side of the picture. Each Aero Pot has a nozzle installed on the both ends of a rod with an electric motor in the middle. The electric motor rotates the main rod in 180 degrees, which gives us a full circle if both of the nozzles work properly. If there is a failure on one nozzle, we can perform a workaround procedure and give a motor the instruction to perform full circle rotation. Above the nozzles, plant containers are situated as visualized on the right side of the picture.

Aero Pot has this specific circular shape and nozzle installation because of the requirement of misting the roots of plants several times per hour. This design is very energy efficient, cheap, high scalable and easy to control.

The future work will be focused on sensor installation and piping of the nutrient distribution system predesign so Aero Pot can work stand-alone or can be connected to other Aero Pots, thus sharing some sensors and actuators which will significantly lower the initial costs per one grown plant.

4.1   Experiments and Simulations

For the first experimental and simulation work we designed and created software based on genetic algorithm to optimize power consumption of hydroponic or aeroponic system. Using this software user is able to define various properties and virtually configure hydroponic or aeroponic system. This software allows user to add and remove lights, pumps and define consumption of added devices. In addition user is later asked to specify details about system configuration. Both hydroponic and aeroponic systems may be running in different modes. Different plants require different light and water conditions. All these conditions are specified by the user using intuitive graphical user interface. Genetic algorithm designed primarily for power consumption optimization then optimizes and create a power schedule for 24 hour window for all the devices in hydroponic or aeroponic system. First experimental work was performed on simple hydroponic system created in our laboratory as Aero Pot creation is in progress. The following picture shows example of power optimization framework GUI.

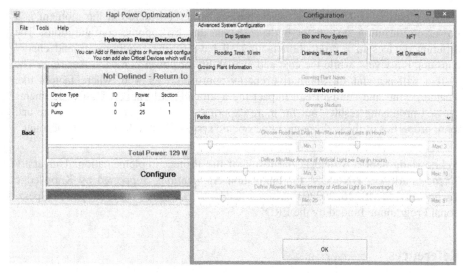

Fig. 3. Detailed system configuration and growing plant conditions and requirements

Final fitness function of genetic algorithm at this stage of the project consists of 3 elements. These are partial fitness of lights plus partial fitness of water pumps and penalties defined by user. Both, light and water pumps configuration have to meet the conditions defined by user for every growing plant. Penalties are also defined by user. User can choose to optimize system for plant growth, eliminate peak consumptions or optimize power consumption for specific time of the day. The output of genetic algorithm is the power schedule of lights, light intensities and water pumps cycles during 24 hour time window. Initial experiments showed that plants are doing well while power consumption of the system is dropped down. This is achieved with minimum human operator effort.

5   Conclusions

This work is aimed to explore some possibilities for improvement of the hydroponic and aeroponic systems by utilization of methods of artificial intelligence. We showed that the demand for clean food becomes increasingly alarming as the world population rises. In this paper we provided a brief literature preview, described the basics of aeroponics and hydroponics, their common and distinctive features. In addition some previous work done in this field of research and current trends has been presented. We provided a brief literature preview of intelligent space problem and presented the idea how it can be utilized in greenhouse environments. In the second part of this paper we proposed the control architecture for the aeroponic greenhouse system being developed in our department, which is fully compatible with HAPI. Also the Aero Pot sketches and some of its features were uncovered in this paper.

Future work will be focused on finishing the Aero Pot design, construction of an aeroponic greenhouse as proposed in chapter 4 and the commencement of work on control algorithms. The first software project being considered at the moment is a power optimization system for both aeroponic and hydroponic configurations. The different devices in a production facility all have timing requirements and energy consumption levels, while the cost and availability of power can vary over time. The system will account for alternative energy approaches as well, where factors like weather, wind and sunlight can impact the availability of energy. First experiments showed promising results, but far more experiments need to be done and more and more properties of the hydroponic or aeroponic system need to be taken into account.

Acknowledgments. Article is the result of the Project implementation: University Science Park TECHNICOM for Innovation Applications Supported by Knowledge Technology, ITMS: 26220220182, supported by the Research & Development Operational Programme funded by the ERDF.

References

[1] World Population Prospects: The 2012 Revision. UN Press Release, New York (June 13, 2013)

[2] Pittinger, D.R.: Introduction to Horticulture. In: California Master Gardener Handbook (Publication 3382), ch. 2, The Regents of the University of California, Division of Agriculture and Natural Resources (2002)

[3] Doyle, O., Aldous, D., Barrett-Mold, H., Bijzet, Z., Darnell, R., Martin, B., McEvilly, G., Stephenson, R.: 2012 Defining Horticulture, Horticulturist and Horticultural Scientist. In: Doyle, O. (ed.) Ad Hoc Committee for Global Horticulture Advocacy. University College Dublin, Ireland (2012)

[4] NASA Spinoff 2006, Innovative Partnership Program, Publications and Graphics Department NASA Center for Aerospace Information (CASI) (2006)

[5] Thiyagarajan, G., Umadevi, R., Ramesh, K.: Hydroponics. Science Tech Entrepreneur (January 2007)

[6] Sholto, D.J.: Advanced guide to hydroponics, No. new edn. Pelham Books (1985)

[7] Nir, I.: Growing Plants in Aeroponics Growth System. Acta Hort (ISHS) 126, 435–448 (1982)

[8] Amos, J., et al.: Final Report for the Robotic Construction of a Permanently Manned Mars Base. Mars Investment for a New Generation (M.I.N.G.). NASA (1989)

[9] Ritter, E., Angulo, B., Riga, P., Herran, C., Relloso, J., San Jose, M.: Comparison of hydroponic and aeroponic cultivation systems for the production of potato minitubers. Potato Research 44, 127–135 (2001)

[10] Morimoto, T., Hatou, K., Hashimoto, Y.: Intelligent Control for a Plant Production System, Control Eng. Practice 4(6), 773–784 (1996)

[11] Lopez-Morales, V., Lopez-Ortega, O., Ramos-Fernandez, J., Munoz, L.B.: JAPIEST: An integral intelligent system for the diagnosis and control of tomatoes diseases and pests in hydroponic greenhouses. Expert Systems with Applications 35, 1506–1512 (2008)

[12] Ferentinos, K.P., Albright, L.D., Selman, B.: Neural network-based detection of mechanical, sensor and biological faults in deep-trough hydroponics. Computers and Electronics in Agriculture 40, 65–85 (2003)

[13] Ferentinos, K.P., Albright, L.D.: Fault Detection and Diagnosis in Deep-trough Hydroponics using Intelligent Computational Tools. Biosystems Engineering 84(1), 13–30 (2003), doi:10.1016/S1537-5110(02)00232-5

[14] Tik, L.B., Khuan, C.T., Palaniappan, S.: Monitoring of an Aeroponic Greenhouse with a Sensor Network. IJCSNS International Journal of Computer Science and Network Security 9(3) (March 2009)

[15] Sahu, K., Mazumdar, S.G.: Digitally Greenhouse Monitoring and Controlling of System based on Embedded System. International Journal of Scientific & Engineering Research 3(1) (January 2012) ISSN 2229-5518

[16] Song, Y., Ma, J., Zhang, X., Feng, Y.: Design of Wireless Sensor Network-Based Greenhouse Environment Monitoring and Automatic Control System. Journal of Networks 7(5), 838–844 (2012), doi:10.4304/jnw.7.5.838-844.

[17] Ahonen, T., Virrankoski, R., Elmusrati, M.: Greenhouse Monitoring with Wireless Sensor Network. In: IEEE/ASME International Conference on Mechtronic and Embedded Systems and Applications, MESA 2008, October 12-15, pp. 403–408 (2008), doi:10.1109/MESA.2008.4735744

[18] Liu, B., Wang, F.-Y., Geng, J., Yao, Q., Gao, H., Zhang, B.: Intelligent spaces: An overview. In: IEEE International Conference on Vehicular Electronics and Safety, ICVES, December 13-15, pp. 1–6 (2007)

[19] Sasaki, T., Brscic, D., Hashimoto, H.: Human-Observation-Based Extraction of Path Patterns for Mobile Robot Navigation. IEEE Transactions on Industrial Electronics 57(4), 1401–1410 (2010)

[20] Tivatansakul, S., Tanupaprungsun, S., Areekijseree, K., Achalakul, T., Hirasawa, K., Sawada, S., Saitoh, A., Ohkura, M.: The intelligent space for the elderly — Implementation of fall detection algorithm. In: 2012 Proceedings of SICE Annual Conference, SICE, August 20-23, pp. 1944–1949 (2012)

[21] Qu, F., Wang, F.-Y., Yang, L.: Intelligent transportation spaces: vehicles, traffic, communications, and beyond. IEEE Communications Magazine 48(11), 136–142 (2010)

[22] Lu, F., Tian, G., Zhou, F., Xue, Y., Song, B.: Building an Intelligent Home Space for Service Robot Based on Multi-Pattern Information Model and Wireless Sensor Networks. Intelligent Control and Automation 3, 90–97 (2012)

[23] Lu, S., Qi, W.: Navigation and positioning research of service robot based on intelligent space. In: Automation and Logistics, ICAL 2009, August 5-7, pp. 2015–2017 (2009)

A Convex Fuzzy Range-Free Sensor Network Localization Method

Fatma Kiraz, Barış Fidan, and Fakhri Karray

School of Engineering, University of Waterloo, Waterloo, ON, N2L 3G1, Canada

Abstract. We propose a new fuzzy range-free sensor localization methodology, using certain Euclidean notions to build convex fuzzy sets and circumvent multiple stable local minima issues, encountered in some recent range-free localization approaches. A range-free localization algorithm is developed based on the proposed methodology. Performance of the developed algorithm is tested via a set of simulations, comparatively to some recent fuzzy range-free localization algorithms in the literature.

Keywords: range-free localization, radical axis, convexity, fuzzy logic.

1 Introduction

Wireless sensor network (WSN) localization algorithms are classified based on various criteria [1–3]. One such categorization, which has motivated this study as well, is absolute range-based vs. range-free localization. Absolute range-based techniques employ absolute distance or angle measurements between anchors and sensor nodes. They utilize inter-sensor measurement techniques such as received signal strength (RSS), time of arrival (TOA), time difference of arrival (TDoA), or angle of arrival (AoA) [1–3].

Range-based methods can produce more accurate location estimates with the help of additional hardware equipment. However, they bear higher cost and power consumption, and generally require higher traffic in network communication [1]. They are more sensitive to measurement noises as well. Furthermore, certain WSNs might not have hardware devices needed to implement absolute range based algorithms. In such cases, range free methods [4–6] are preferred because they use only standard features found in most radio modules. There is no single perfect algorithm working very well for all localization problems. Thus both absolute range-based and range-free algorithms are used in practice, choice depending on the specific application and requirements [7]. Nevertheless, the focus of this work will be settings with inaccurate measurements, where range-free algorithms are more applicable.

There have been various studies (see, e.g. [4–6, 8, 9]) engaging fuzzy logic in range-free localization. He et al. [9] introduced a range-free algorithm based on regions inside and outside the triangles defined by anchor triples. Liu et al. [5] proposed ROCRSSI (ring-overlapping based on comparison of RSS indicator) algorithm, where intersecting overlapping rings centered at anchors constrain

J.-H. Kim et al. (eds.), *Robot Intelligence Technology and Applications 2*,
Advances in Intelligent Systems and Computing 274,
DOI: 10.1007/978-3-319-05582-4_8, © Springer International Publishing Switzerland 2014

the location of a sensor node. RSS between anchors define radii of the rings. The RSS of the sensor node at anchors are also employed to localize the sensor. However, there are some issues in the proposed method preventing the algorithm functioning properly [6]. As a remedy for ROCRSSI, fuzzy ring overlapping range free (FRORF) localization technique was proposed by Velimirovic et. al [4]. The FRORF algorithm represents localization areas with overlapping rings similar to ROCRSSI. Although FRORF overcomes some of the RSS uncertainty issues, it is affected by another issue: Non-convexity or multiple stable local minima.

Non-convexity is a common issue in most of the localization problems [1–3]. This issue is also encountered in FRORF, since FRORF employs ring intervals to represent fuzzy sets [10]. To work with the fuzzy logic based framework of [4] more efficiently, there is a need for an alternative approach which can handle the non-convexity problem. In this paper, we propose such an alternative approach using certain overlapping rectangles instead of rings. Absolute distances between anchor pairs and a geometric feature, called *radical axis*, are used to build these regions.

We develop our methodology combining the fuzzy framework of (FRORF) and a new convex geometric technique proposed in [8]. The general structure and main components of our algorithm are introduced in Section 2. The mentioned convex geometric technique is briefly introduced in Section 3. Detailed design of the proposed algorithm is presented in Section 4. Simulation test results are presented in Section 5. Concluding remarks of the paper are given in Section 6.

2   General Structure and Main Components

The localization problem of interest can be formally defined as follows: Consider a WSN with N anchors $A_1, \ldots, A_N$ ($N > 2$) and a sensor/source node S with unknown position $y^* \in \Re^2$. The goal is to estimate the coordinates of S, from the distance estimates/measurements $d_i \approx \|y^* - A_i\|$.

The main motivator of our new design is the non-convexity (multiple stable local minima) issue mentioned in Section 1. This issue stems from the non-convex nature of range measurement: The set of points having a fixed distance x from a sensor A is non-convex; it is a circle for planar settings. The fuzzy logic based studies in the literature, including FRORF [4], consider fuzzy sets defined based on such natural non-convex sets. Instead of fixed distances, ranges at distances are used, leading to rings, which are also non-convex.

To overcome the non- convexity issue we propose using the notions of *radical axes* and *radical axis powers* [8,11]. As detailed in Section 3, these notions help to compose convex fuzzy regions, still based on distance information. Our proposed algorithm consists of two main components: (i) a fuzzy logic framework similar to that of FRORF, and (ii) a new methodology for generating convex fuzzy sets based on the *radical axis* notion, which is explained in detail in the following section.

3   Constructing Convex Fuzzy Regions

One of the main components of our methodology is construction of an alternative set of convex regions in place of the non-convex rings used by FRORF. In a slightly different context, in [8], a geometric methodology is proposed to construct such convex regions. The methodology is based on the following Euclidean geometric property:

Theorem 1. *(Fact 45 of [11]) Given two non-concentric circles with centres C_1, C_2 and radii r_1, r_2, respectively, there is a unique line consisting of points S holding equal powers with regard to these circles, i.e., satisfying*

$$\|S - C_1\|^2 - r_1^2 = \|S - C_2\|^2 - r_2^2 \tag{1}$$

This line is perpendicular to the line connecting C_1 and C_2, and if the two circles intersect, passes through the intersection points.

The unique line mentioned in Theorem 1 is called the *radical axis* of $C(c_1, r_1)$ and $C(c_2, r_2)$ [11]. In this work, we call the difference $\|S - C_1\|^2 - \|S - C_2\|^2 = r_1^2 - r_2^2$ the *radical axis power* of S with respect to the pair (C_1, C_2) and denote by $P(S, (C_1, C_2))$.

Lemma 1. *[8] In 2 dimensions, if $d_i = \|y^* - A_i\|$, the intersection of the radical axes of any $N - 1$ distinct circle pairs $C(A_i, d_i)$, $C(A_j, d_j)$ $(i \neq j)$ is y^*.*

Proof. The result straightforwardly follows from the the problem the last statement of Theorem 1.

Using Theorem 1 for a WSN with N anchors $A_1, \ldots, A_N$, one can consider the $N - 1$ node-distance measurement (estimate) pairs $(A_i, d_i), (A_{i+1}, d_{i+1})$ and the corresponding radical axes line l_i perpendicularly intersecting $x_i x_{i+1}$ at y_i (refer to Fig. 1). Hence [8], any point y on l_i satisfies

$$(y - y_i)^T e_i = 0, \forall i \in \{1, ..., N - 1\} \tag{2}$$

where

$$e_i = x_{i+1} - x_i. \tag{3}$$

y_i can be computed as

$$y_i = x_i + a_i \frac{e_i}{\|e_i\|}, \tag{4}$$

as well as $d_i^2 - a_i^2 = d_{i+1}^2 - a_{i+1}^2 = d_{i+1}^2 - (\|e_i\| - a_i)^2$, from which a_i can be calculated as

$$a_i = \frac{\|e_i\|^2 + d_i^2 - d_{i+1}^2}{2\|e_i\|}. \tag{5}$$

Remark 1. Lemma 1, further implies that the intersection of $l_1, \ldots, l_{N-1}$ is S, i.e., S is the unique point satisfying (2).

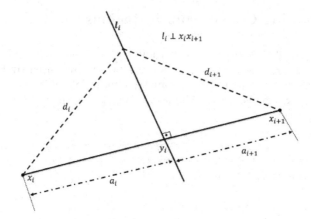

Fig. 1. Representation of node-distance measurement pairs (x_i, d_i), (x_{i+1}, d_{i+1}) and the corresponding radical axis l_i [8]

4   The Algorithm

In this section, based on the two main components mentioned in Section 2, we introduce our proposed algorithm, *radical axis based fuzzy range-free (RAFRF)* localization algorithm. The working scheme of RAFRF is as follows: Consider an anchor pair (A_i, A_j) and, based on estimate distance measurement/estimate of the sensor node S to A_i and A_j the radical axis power $P(S, (A_i, A_j))$. Order all anchors $A_1, \ldots, A_N$ according to radical axis power $P_{(i,j)}(A_k) := P(A_k, (A_i, A_j))$. Define the permutation set $I_{(i,j)} = \{\rho_k^{(i,j)}\}$ of $\{1, \ldots, N\}$ based on this ordering, so that

$$P_{(i,j)}(A_{\rho_k^{(i,j)}}) \leq P_{(i,j)}(A_{\rho_{k+1}^{(i,j)}}) \text{ for all } k. \tag{1}$$

With respect to the anchor ordering, as illustrated in Fig. 2, define the *power line intervals*

$$PI_k^{(i,j)} = \{S | \alpha_k \leq P_{(i,j)}(S) \leq \beta_k\}, for \quad k = 0, 1, \ldots, N \tag{2}$$

where $\alpha_k^{(i,j)} = \beta_{k-1}^{(i,j)} = P_{(i,j)}(A_{\rho_k^{(i,j)}})$ for $k = 1, \ldots, N$, $\alpha_0^{(i,j)} = -\infty$, and $\beta_N^{(i,j)} = \infty$.

Each such line interval PI_k (dropping the superscript (i,j) for brevity) can be considered as the intersection

$$PI_k = LT_k \cap GT_k \tag{3}$$

of the region $LT_k = \{S | P_{(i,j)}(S) \leq \beta_k\}$ with maximum radical axis power β_k and the region $GT_k = \{S | P_{(i,j)}(S) \geq \alpha_k\}$ with minimum radical axis power α_k.

For the sensor S with crisp value x, representing the radical axis power estimate with respect to (A_i, A_j), LT_k and GT_k crisp intervals are represented by fuzzy sets $\widetilde{LT_k} = \{(x, \mu_{LT_k}(x)) | x \in R^+\}$ and $\widetilde{GT_k} = \{(x, \mu_{GT_k}(x)) | x \in R^+\}$.

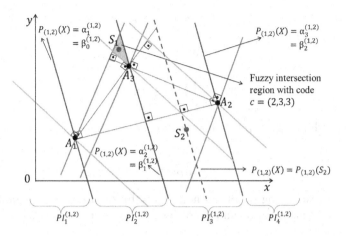

Fig. 2. RAFRF interval parameter and area code definitions for a 3-anchor setting

Here, $\mu_{LT_k}(x)$ and $\mu_{GT_k}(x)$ are membership functions are formally defined as follows for a pre-defined fuzzification parameter $0 \leq p \leq 1$:

$$\mu_{LT_k}(x) = \begin{cases} 1 & \text{if} \quad x \leq sgn(\beta_k)(1-p)\beta_k \\ 0 & \text{if} \quad x \geq sgn(\beta_k)(1+p)\beta_k \\ \frac{(1+p)\beta_k - x}{2p\beta_k} & \text{if} \quad else \end{cases}$$

$$\mu_{GT_k}(x) = \begin{cases} 0 & \text{if} \quad x \leq sgn(\alpha_k)(1-p)\alpha_k \\ 1 & \text{if} \quad x \geq sgn(\alpha_k)(1+p)\alpha_k \\ \frac{x-(1-p)\alpha_k}{2p\alpha_k} & \text{if} \quad else \end{cases}$$

$$\mu_{PI_k}(x) = \mu_{LT_k}(x) + \mu_{GT_k}(x) - 1. \tag{4}$$

Similar to FRORF, fuzzification parameter $p \in [0,1]$ enables the system or the user decide to the width of the region in the neighbourhood of power line boundaries corresponding α_k and β_k. Note that the number of power line intervals corresponding to each anchor pair $A_i A_j$ is equal to $N+1$ for $k = 1, \ldots, N$. Also, the number of anchor pairs $A_i A_j$ for $i, j = 1, \ldots, N$ in the network is $M = \binom{N}{2}$.

Next step is to build fuzzy inference and fuzzy sets based on the fuzzy intervals and membership functions defined for each anchor pair (A_i, A_j). The fuzzy set for (A_i, A_j) is defined as

$$\widetilde{PS}_{(i,j)} = \{(k, \mu_{PI_k^{(i,j)}}) | \mu_{PI_k^{(i,j)}} > 0, k \in \{0, 1, ..., N\}\}.$$

Overall fuzzy regional map $\widetilde{PM}$ is constructed by taking Cartesian product over the fuzzy sets of the M anchor pairs:

$$\widetilde{PM} = \widetilde{PS}_{(1,2)} \times \ldots \times \widetilde{PS}_{(1,N)}$$
$$\times \widetilde{PS}_{(2,3)} \times \ldots \times \widetilde{PS}_{(N-1,N)} = \{(c, \mu_{PM}(c))\}$$

where $c = (k_{(1,2)}, k_{(1,3)}, ..., k_{(N-1,N)})$ denotes the intersection region of $PI_{k_{(1,2)}}^{(1,2)}$, $PI_{k_{(1,3)}}^{(1,3)}, ..., PI_{k_{(N-1,N)}}^{(N-1,N)}$, and

$$\mu_{PM}(c) = \prod_{i=1}^{N-1} \prod_{j=i+1}^{N} \mu_{PI_{k_{(i,j)}}^{(i,j)}}$$

is the membership degree of this intersection region. The feasible area code set is defined as

$$\widetilde{P} = \{c | (c, \mu_P(c)) \in \widetilde{PM}\}$$

i.e, the set corresponding to all the non-empty fuzzy intersection regions. Consequently, total (maximum) number of intersection (fuzzy) regions is equal to $n_F = (N+1)^M$.

As an example of the region area coding, examine the system of 3-anchors and 1-sensor node depicted in Fig. 2. Let the sensor node located in the shaded area. Considering the power line of anchor pair (A_1, A_2), this region is inside the fuzzy interval $PI_2^{(1,2)}$. From the aspects of (A_1, A_3) and (A_2, A_3) anchor pairs, respectively, it is inside $PI_3^{(1,3)}$ and $PI_3^{(2,3)}$. Thus, the corresponding area code is $c = (2, 3, 3)$.

In the defuzzification step, similiar to FRORF, the location estimate is calculated using membership degrees of the sensor node to regions in $\widetilde{PM}$ and the center of gravity (CoG) values associated with those regions. Thus, the location estimate is given by

$$\hat{S} = Estimated\ location = \frac{\sum_{c \in \widetilde{P}} g_c(c) \mu_{PM}(c)}{\sum_{c \in \widetilde{P}} \mu_{PM}(c)} \tag{5}$$

where $g_c(c)$ is COG of the region with area code c.

5   Simulation Tests

In this section, we present the results of two sets of simulations performed to test performance of the proposed RAFRF algorithm in general as well as comparative to the FRORF algorithm of [4], noting that the results of a significant number of other simulations are skipped here due to space limitations.

In the first set of simulations we present, we consider a network of 4 anchors located at $A_1 = [1, 1]^T$, $A_2 = [3, -0.5]^T$, $A_3 = [2, -2]^T$, $A_4 = [0.5, 0.8]^T$ (noting that the location assignments are done randomly). The fuzzification parameter is taken as $p = 0.1$. Figures 5 and 4 show the estimation errors $e_S = \|\hat{S} - S\|$ vs. the actual sensor position $S = [S_x, S_y]^T$ to be estimated. For easier comparison of performances the difference of these plots, $e_{S,FRORF} - e_{S,RAFRF}$, is plotted in Fig. 5.

As can be seen in these figures, RAFRF performed superior to FRORF in terms of localization accuracy, in general. Average and maximum values of e_S in

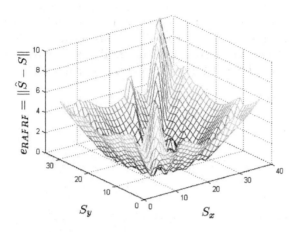

Fig. 3. Estimation errors in the first simulation set using RAFRF

Fig. 5 with RAFRF are 2.4626 and 10.9628, respectively; while the corresponding values for Figure 4(FRORF) are 6.0977 and 12.1231.

In the second set of simulations we present next, we consider 10 WSNs of 4 anchors, each anchor having a randomly generated location $(x, y) \in [-3, 6] \times [-6, 6]$. The localization performances for all $S = [S_x, S_y]^T$ locations, $S_x \in \{-6, -5.5, -5, \ldots, +9\}$, $S_y \in \{-9, -8.5, -8, \ldots, +9\}$ using FRORF and RAFRF are summarized in Table 1. It can be inferred from this table that RAFRF works more accurately than FRORF, in general.

Table 1. Average and maximum error values using FRORF and RAFRF

	$ave(e_S)$		$max(e_S)$	
Set No	FRORF	RAFRF	FRORF	RAFRF
1	5.1400	4.6540	12.7279	12.6989
2	4.4296	4.2332	12.9292	11.4641
3	4.4969	4.0387	12.5084	10.1078
4	5.0208	3.4789	12.3840	12.0416
5	4.5949	3.9338	12.7279	10.8167
6	4.4684	3.4756	12.7279	10.5178
7	4.4228	2.9798	12.7279	10.8167
8	4.9771	3.9922	12.6352	10.6381
9	4.4297	3.9224	15.9784	12.6945
10	5.0646	3.4247	15.8777	10.7072

Previous results were obtained assuming noiseless/perfect distance information. To test the effect of distance measurement/estimation noises, the third
set of simulations presented next consider a white Gaussian distance noise with
$\sigma = 0.1$. Results are summarized in Table 2. The results demonstrate that superior performance of RAFRF is valid in noisy measuremen/estimation cases as
well.

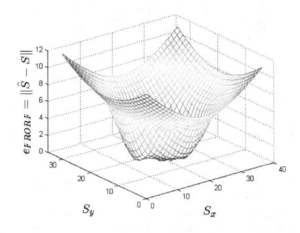

Fig. 4. Estimation errors obtained in the first simulation set using FRORF

Table 2. Average and maximum error values using FRORF and RAFRF with noise
($\sigma = 0.1$)

Set No	$ave(e_S)$		$\max(e_S)$	
	FRORF	RAFRF	FRORF	RAFRF
1	5.1427	4.6475	12.7279	12.3794
2	4.4310	4.2344	12.9255	11.9586
3	∞	4.0516	12.5588	9.9138
4	5.0233	3.6158	12.3840	12.3794
5	4.5948	3.9434	12.7279	10.8167
6	4.4710	3.4899	12.7279	10.5085
7	4.4242	3.0331	12.7279	11.0114
8	4.9788	4.0992	12.6405	10.5867
9	4.4344	3.9466	15.9201	12.7279
10	5.0707	3.4823	15.8777	10.5649

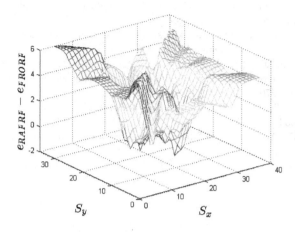

Fig. 5. Differences of error values between RAFRF and FRORF for 4 anchors when fuzzification parameter $p = 0.1$

6   Concluding Remarks

We have proposed a new methodology for fuzzy range-free sensor localization, using the Euclidean geometric notion of "radical axis" for building convex fuzzy sets. A range-free localization algorithm has been developed based on the proposed methodology. The algorithm is demonstrated to have superior performance, especially in settings where non-convexity issue of overlapping ring based algorithms is effective. The authors currently work on application of the methodology for networks with large number of sensors.

References

1. Han, G., Xu, H., Duong, T., Jiang, J., Hara, T.: Localization algorithms of wireless sensor networks: a survey. Telecommunication Systems 47(3-4), 1–18 (2011)
2. Zekavat, S., Buehrer, R. (eds.): Handbook of Position Location. IEEE Press/Wiley (2012)
3. Mao, G., Fidan, B.: Localization algorithms and strategies for wireless sensor networks. Information Science Reference. IGI Publishing (2009)
4. Velimirovic, S., Djordjevic, G., Velimirovic, M., Jovanovic, M.: Fuzzy ring-overlapping range-free (FRORF) localization method for wireless sensor network. Computer Communications 35(13), 1590–1600 (2012)
5. Liu, C., Scott, T., Wu, K., Hoffman, D.: Range-free sensor localization with ring overlapping based on comparison of received signal strength indicator. International Journal of Sensor Networks (IJSNet) 2(5), 399–413 (2007)
6. Velimirovic, S., Djordjevic, G., Velimirovic, M., Jovanovic, M.: A fuzzy set-based approach to range-free localization in wireless sensor networks. Facta Univ. Ser.: Elec. Energ. (2010)

7. Sanford, J., Slijepcevic, S., Potkonjak, M.: Localization in wireless networks: foundations and applications. Springer (2012)

8. Fidan, B., Kiraz, F.: On convexification of range measurement based sensor and source localization problems. ArXiv, 1310.7042 (2013)

9. He, T., Krishnamurthy, S., Stankovic, J., Abdelzaher, T., Luo, L., Stoleru, R., Yan, T., Gu, L.: Energy-efficient surveillance system using wireless sensor networks. In: Mobisys, pp. 270–283. ACM Press (2004)

10. Karray, F., Silva, C.: Soft Computing and Intelligent Systems Design. Addison Wesley (2004)

11. Johnson, R.: Advanced Euclidean Geometry. Dover Publications (2007)

Multi-Directional Weighted Interpolation for Wi-Fi Localisation

Dale Bowie, Jolon Faichney, and Michael Blumenstein

School of Information and Communication Technology
Griffith University, Gold Coast, Australia
Dale.Bowie@griffithuni.edu.au,
{j.faichney,m.blumenstein}@griffith.edu.au

Abstract. The rise in popularity of unmanned autonomous vehicles (UAV) has created a need for accurate positioning systems. Due to the indoor limitations of the Global Positioning System (GPS), research has focused on other technologies which could be used in this landscape with Wi-Fi localisation emerging as a popular option. When implementing such a system, it is necessary to find an equilibrium between the desired level of final precision, and the time and money spent training the system. We propose Multi-Directional Weighted Interpolation (MDWI), a probabilistic-based weighting mechanism to predict unseen locations. Our results show that MDWI uses half the number of training points whilst increasing accuracy by up to 24%.

1 Introduction

Methods of human and robot localisation are fast becoming important areas of research. The Global Positioning System (GPS) has widely been accepted as a means of determining a device's location outdoors. However, there are a number of limitations with GPS that make it undesirable for indoor use: the inability to detect levels, poor location quality because it requires direct contact with satellites, and poor location accuracy because indoor localisation requires a higher level of precision.

As a result, various methods of indoor localisation have been evaluated in literature. The methods differ by the technology as well as their applications; papers have proposed the use of infrared, radio frequency identifier (RFID), Bluetooth, ZigBee and Wi-Fi [2,6,10,12], for various applications of human or robot tracking or navigation [7,9]. Wi-Fi is seen as a desirable option due to its low setup costs and high availability [1,7,9].

Traditionally Wi-Fi localisation works by training a system on signal strength readings obtained from wireless networks inside a building. Readings from a single location are collated together to form a fingerprint. When deployed, the system will take a device's current fingerprint and compare it to all existing fingerprints to find the best-matching location [1]. Systems using other technologies have followed similar training and positioning phases [12].

J.-H. Kim et al. (eds.), *Robot Intelligence Technology and Applications 2*, 105
Advances in Intelligent Systems and Computing 274,
DOI: 10.1007/978-3-319-05582-4_9, © Springer International Publishing Switzerland 2014

As one of the main limitations of indoor localisation schemes is the physical training time required, efforts are being made to attempt to reduce this [3,8]. Of course, reducing the training space may seem like a simple and effective measure to reduce costs, however that generally results in lower accuracy rates [5]. In this paper we propose Multi-Directional Weighted Interpolation (MDWI) which utilises the probabilities produced by a localisation algorithm to recognise a location outside of the training space.

2   Related Work

The RADAR system, published by Microsoft Research in 2000 [1], is generally recognised as the founding paper in the field of Wi-Fi localisation, with 5,790 citations to date [4]. The authors gathered Wi-Fi data in one of two ways. In the first approach, samples were manually taken at a number of locations facing different directions, totalling a training space of at least 5,600 readings. The second approach used a signal propagation model to predict signal strengths at those same locations, based on the known locations of access points. The system had consistent results with a median error distance of between 2 and 3 metres, with the first method performing better overall.

In 2002, Youssef et al. [11] proposed the application of probabilistic techniques, such as Bayesian Inference, to the problem of indoor localisation. Based on prior training records as well as a sample reading obtained from a device, they calculated the probability for each possible location. The output was the location with the highest probability, above a pre-defined confidence threshold.

As part of a wider study in 2004, Elnahrawy et al. [3] proposed the Interpolated Map Grid (IMG). IMG utilised a small sample of training fingerprints to interpolate the expected signal strength at various intermediary locations. IMG essentially combines the two methods proposed in the RADAR paper.

In 2006, Li et al. [8] evaluated the use of inverse distance weighting (IDW) and universal kriging (UK). As the name suggests, IDW applies weights inversely proportionate to the distance between the known points. UK generates a polynomial model representative of the known points, and uses this to interpolate intermediary points. The researchers' results reduced the error distance in all cases, with UK consistently outperforming both the standard and IDW methods. When 16 access points were used, UK reduced the original error distance of approximately 2.5 metres down to 1.5 metres.

In their 2007 Signal Strength Difference (SSD) paper Hossain et al. [5] adopted Li et al.'s approach, however, they used a weighted linear regression technique to develop their training set. Equations were created for each pair of access points and intermediary points. Hossain et al.'s test results show the usage of these intermediary points consistently improves the average error distance, with a maximum improvement of approximately 3.5 metres.

The techniques mentioned above are compared in table 1 with our technique, MDWI. The authors of prior works propose various methods of interpolation which predict new locations prior to runtime. One of the main disadvantages of

Table 1. Comparison of techniques

Technique	Signal strength	Inverse distance	Probability	Infinite locations
RADAR [1]	No	No	No	No
Prob. Cluster [11]	No	No	Yes	No
IMG [3]	Yes	No	No	No
IDW [8]	No	Yes	No	No
SSD [5]	No	Yes	No	No
MDWI	No	No	Yes	Yes

performing interpolation prior to runtime is it still enforces a limited state space when deployed. Some propose the use of averaging the n best-matching locations, however, this disregards the inherent probabilities and potentially results in unlikely locations being included in the final result.

3   Multi-Directional Weighted Interpolation

In this paper we propose Multi-Directional Weighted Interpolation which provides greater precision in localisation without an increase in the training time or associated costs. We utilise the probabilities output from an algorithm, such as a Bayesian Network, as weights between the n best-matching, known fingerprints. In general this will return an intermediary location not included in the original training set.

The basic procedure can be expressed as a series of formulas which were derived from a weighted centre of gravity algorithm. Equations 1 and 2 are equivalent to calculating the dot product of the individual coordinate vectors and the probability vector. The result is the estimated x and y coordinates for the current location when given the coordinates of n best matches (i.e. x_i and y_i) as well as their probabilities (p_i).

$$x = \sum_{i=1}^{n} x_i p_i \tag{1}$$

$$y = \sum_{i=1}^{n} y_i p_i \tag{2}$$

These formulas can be used for any number of matches. Traditional interpolation would be where $n = 2$. Examples follow which show $n = 3$ and $n = 4$.

Take the example shown in figure 1. The three outer nodes are known fingerprints while the central node is the actual location of a user. Some sample probabilities from a Bayesian Network are shown inside the outer nodes. In traditional approaches, either of the two nodes with 45% probability would be chosen. In our approach these probabilities will be used as weights determining

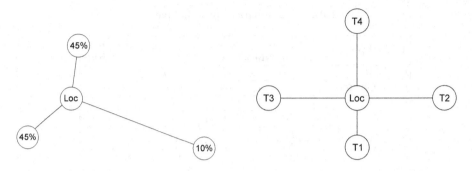

Fig. 1. MDWI where $n = 3$ **Fig. 2.** MDWI where $n = 4$

a location in the centre of the three fingerprints, much closer to the user's actual location.

Using figure 2 as an example, we shall now explicitly utilise the formulas. Suppose we have 4 training locations which are labelled T1 through to T4. Their coordinates are (3, 4), (5, 3), (1, 3) and (3, 1), respectively. The coordinates of the actual location are (3, 3). In an ideal scenario, where the algorithm accurately calculates these probabilities, we will have a probability of 40% for T1, and 20% for each of the remaining nodes. From this information we can calculate the predicted location:

$$
\begin{aligned}
x = \sum_{i=1}^{n} x_i p_i \\
= (3 \times 0.4) + (5 \times 0.2) \\
+ (1 \times 0.2) + (3 \times 0.2) \\
= 1.2 + 1 + 0.2 + 0.6 \\
= 3
\end{aligned}
\qquad
\begin{aligned}
y = \sum_{i=1}^{n} y_i p_i \\
= (4 \times 0.4) + (3 \times 0.2) \\
+ (3 \times 0.2) + (1 \times 0.2) \\
= 1.6 + 0.6 + 0.6 + 0.2 \\
= 3
\end{aligned}
$$

Therefore the algorithms correctly predicted the location as (3, 3).

4   Experimental Analysis

4.1   Setup

For our tests, we utilised a corridor in Griffith University's G09 building. Locations for Wi-Fi scans were chosen at three-metre intervals along the corridor. These are signified by the coloured dots in figure 4. During the tests, a total of 21 access points were discovered. We later filtered this list to access points with a specific network name (SSID), resulting in a collection of 4 access points. The locations of access points in the environment were not known.

To prepare out dataset for processing we created a subset containing all of the data from every second location. This dataset, as represented by the dark

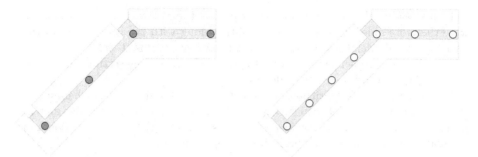

Fig. 3. G09 building with the training locations denoted by dark dots

Fig. 4. G09 building with the testing locations denoted by light dots

dots in figure 3, contains four locations and was used as a training set. This dataset would be representative of one taking Wi-Fi readings along the corridor at intervals of 6 metres. For testing the accuracy of our algorithm, we used all locations depicted in figure 4.

A test set containing unseen samples, such as that used in our experiments, is representative of a real-world situation in which a user may be standing in the middle of two training locations. Traditional approaches would pick one of the two locations, and could provide vastly different accuracy rates. An example where two training locations are at opposite ends of the corridor is shown in figure 5. Traditionally, a device in the intermediary area (as denoted by the X in the figure) would be positioned at the closest training location (the right-hand dark dot), not its actual location. The proposed MDWI approach allows for a location in the middle to be selected, much closer to the device's actual location.

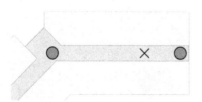

Fig. 5. A corridor with two training locations and an arbitrary device location, as depicted by the dark dots and X, respectively

4.2 Algorithms

We ran our real world datasets through both a Bayesian Network (BN) and a K-Nearest Neighbour (KNN) algorithm. These two machine learning algorithms are commonly used in Wi-Fi localisation literature.[7] For both of the algorithms, we apply MDWI with n set to 5.

BNs already calculate probabilities as part of their classification process. We were therefore able to use these probabilities along with the coordinates of the locations in our weighted interpolation calculations. We used the University of Waikato's Weka implementation of a Bayesian Network. The settings in version 3.6.9 of the software were left as defaults.

Unlike a BN, KNN algorithms do not calculate probabilities but rather output the sum of differences between two instances. As a smaller sum is a closer match, we took the inverse of the lowest-cost (or best-matching) for each location and calculated probabilities. This procedure is represented in equation 3.

$$P = \frac{1}{minCost} \qquad (3)$$

4.3   Results

The results from our tests are summarised in figure 6 and table 2. The figure shows the perfect accuracy rates of various algorithms, both before and after the application of MDWI. We define a perfectly accurate case as one which returns the same location as where the samples were originally taken. When MDWI is applied, we round the x and y coordinates to the nearest three metres, to match the granularity of the original readings. The figure shows perfect accuracy rates increase in all cases when MDWI is applied.

The percentages in table 2 are calculated by taking the actual location and comparing it to the algorithm's prediction. Therefore a 0m error distance indicates the algorithm correctly predicted the location, 3m indicates an adjacent

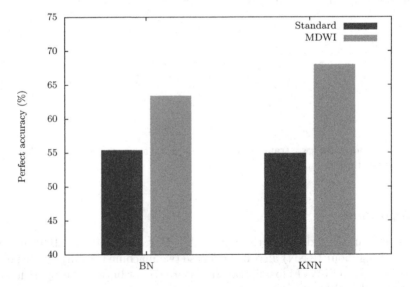

Fig. 6. Perfect accuracy rate comparison

Table 2. Error distribution

Settings	Rounded Error Distance					
	0m	3m	6m	9m	12m	15m
BN	55.4%	42.9%	1.7%	–	–	–
BN+MDWI	63.4%	32.0%	4.6%	–	–	–
KNN	54.9%	42.3%	2.3%	0.6%	–	–
KNN+MDWI	68.0%	26.3%	3.4%	0.6%	1.1%	0.6%

location, and so on. Again, the coordinates output from MDWI have been rounded to the nearest three metres.

The tests performed on the BN showed positive results. In these tests, MDWI shifted the majority of the error distribution towards smaller error distances, capping all errors at 6m. The BN with MDWI correctly predicted perfect matches 63.4% of the time, an increase of 8 percentage points. The number of 6m errors did slightly increase, however, we feel the 14% increase in perfect predictions offsets this.

The KNN tests also provided positive results. When KNN was used along with MDWI perfect accuracy was at 68%, an increase of 13.1 percentage points, providing the best result in all of our tests. However, MDWI's application to KNN also had the unintended result of stretching the error distribution out to 15m. The main reason why this happened with KNN and not with the BN to the same extent was the distribution of the probabilities calculated for KNN were more evenly distributed than that of the BN.

The application of MDWI clearly provides a number of benefits. Accuracy rates can be increased when unseen intermediary locations are input as the algorithm will not immediately pick the closest-matching location. Additionally, if one is limited in the time spent and/or in the associated costs in training a system, MDWI can provide greater precision.

5   Conclusions and Future Work

In this paper we propose Multi-Directional Weighted Interpolation (MDWI) for any Wi-Fi localisation system, whether it be designed for human or robot navigation. MDWI utilises probabilistic-based weighting mechanism to output locations which have not been included in a training set. Our results show that, as with any localisation system, it is necessary to find a balance between the number of training locations and the desired level of final precision. The number of training locations are proportional to distance-based accuracy rates, however, more training locations also come at a higher cost. Our results show MDWI facilitates the ability for training costs to be halved without affecting accuracy rates to the same degree.

Future work will focus on utilising the theoretical concepts proposed in this paper to evaluate its performance in larger areas and areas which feature more dense training locations, as well as how signal strength variations between multiple devices are handled by our algorithm.

References

1. Bahl, P., Padmanabhan, V.N.: RADAR: an in-building RF-based user location and tracking system. In: Nineteenth Annual Joint Conference of the IEEE Computer and Communications Societies, vol. 2, pp. 775–784 (2000)
2. Dimitrova, D.C., Alyafawi, I., Braun, T.: Experimental Comparison of Bluetooth and WiFi Signal Propagation for Indoor Localisation. In: Koucheryavy, Y., Mamatas, L., Matta, I., Tsaoussidis, V. (eds.) WWIC 2012. LNCS, vol. 7277, pp. 126–137. Springer, Heidelberg (2012)
3. Elnahrawy, E., Li, X., Martin, R.P.: The limits of localization using signal strength: a comparative study. In: First Annual IEEE Communications Society Conference on Sensor and Ad Hoc Communications and Networks, pp. 406–414 (2004)
4. Google Scholar. Bahl: RADAR: an in-building RF-based user location and tracking system (October 2013),
   http://scholar.google.com/scholar?cites=5301100132960192733
   (accessed October 15, 2013)
5. Hossain, A.K.M.M., Van, H.N., Jin, Y., Soh, W.-S.: Indoor localization using multiple wireless technologies. In: IEEE Internatonal Conference on Mobile Adhoc and Sensor Systems, pp. 1–8 (2007)
6. Kaemarungsi, K., Ranron, R., Pongsoon, P.: Study of received signal strength indication in ZigBee location cluster for indoor localization. In: 10th International Conference on Electrical Engineering/Electronics, Computer, Telecommunications and Information Technology, pp. 1–6 (2013)
7. Kjærgaard, M.B.: A taxonomy for radio location fingerprinting. In: Hightower, J., Schiele, B., Strang, T. (eds.) LoCA 2007. LNCS, vol. 4718, pp. 139–156. Springer, Heidelberg (2007)
8. Li, B., Salter, J., Dempster, A.G., Rizos, C.: Indoor positioning techniques based on wireless lan. In: First IEEE International Conference on Wireless Broadband and Ultra Wideband Communications (2006)
9. Ocana, M., Bergasa, L.M., Sotelo, M.A., Nuevo, J., Flores, R.: Indoor robot localization system using wifi signal measure and minimizing calibration effort. In: Proceedings of the IEEE International Symposium on Industrial Electronics, vol. 4, pp. 1545–1550 (2005)
10. Want, R., Hopper, A., Falcão, V., Gibbons, J.: The Active Badge Location System. ACM Transactions on Information Systems 10(1), 91–102 (1992)
11. Youssef, M.A., Agrawala, A., Shankar, A.U., Noh, S.H.: A probabilistic clustering-based indoor location determination system. Technical report, Department of Computer Science and UMIACS, University of Maryland, College Park, MD 20742 (March 2002)
12. Zhou, J., Shi, J.: RFID localization algorithms and applications–a review. Journal of Intelligent Manufacturing 20(6), 695–707 (2009)

Visual Loop-Closure Detection Method Using Average Feature Descriptors

Deok-Hwa Kim and Jong-Hwan Kim

Department of Electrical Engineering, KAIST
335 Gwahangno, Yuseong-gu, Daejeon 305-701, Republic of Korea
{dhkim,johkim}@rit.kaist.ac.kr

Abstract. This paper proposes a novel visual loop-closure method using average feature descriptors. The average feature descriptors are computed by averaging the descriptors of feature points at each frame. Through GPGPU (General-Purpose computing on Graphics Processing Units) technique, the proposed method selects a frame having a minimum distance with the current average feature descriptor from the average feature descriptor history. After the minimum distance calculation, loop-closure is determined by matching feature points between the selected frame and current frame. Experiments results demonstrate that the proposed method successfully detects the visual loop-closure and is much faster than the conventional visual loop-closure method in detecting the visual loop-closure.

1 Introduction

Visual simultaneous localization and mapping (SLAM) [1][2][3][4] systems are widely used by robots and autonomous vehicles to build up a map within an unknown environment using a vision sensor. In applications of the visual SLAM, loop-closure detection is a issue that require the capacity to recognize a previously visited place from current vision sensor measurements. In a graph-based visual SLAM system, once a loop closure is detected, an additional constraint between data frames is added to optimize poses and the map.

In the field of computer vision, many researchers have conducted research to detect loop-closure using vision sensors. In 2006, Newman et al. presented outdoor SLAM system using a visual appearance and laser ranging [5]. This research detected the loop-closure in an image classification scheme. In 2008, Angeli et al. proposed a real-time visual loop-closure method using a concept of visual words to recognize a previously visited place. This research relied on the Bayesian filter to estimate the loop-closure probability [6]. In 2012, Henry et al. presented a keyframe-based loop-closure method [3]. They defined the keyframes that are a subset of the aligned frames to save unnecessary time. Once the loop closure is detected, the new correspondence between data frames can be used as an additional constraint in the graph-based SLAM [7] consisting a pose-graph. Because of the new correspondence, the pose-graph is dynamically optimized by graph optimization techniques [8][9][10].

J.-H. Kim et al. (eds.), *Robot Intelligence Technology and Applications 2*,
Advances in Intelligent Systems and Computing 274,
DOI: 10.1007/978-3-319-05582-4_10, © Springer International Publishing Switzerland 2014

Although the previous research have been applied to detect loop-closure and achived successful results, they have not taken computation time increasing linearly as time goes by. To detect the loop-closure, measurements of a current frame are matched with all of measurements of previously visited places. As the map size is increased, the computation time to recognize a previously visited place is increased.

In this paper, a visual loop-closure detection method using average feature descriptors is proposed. The average feature descriptors are computed by averaging the descriptors of feature points at each frame. Through the concept of average feature descriptors, parallel processing for the GPGPU is maximized and a memory efficiency is increased.

This paper is organized as follows. Section 2 proposes the novel visual loop-closure detection method using the average feature descriptors. In Section 3, experimental environment and scenarios are presented and the experimental results to detect the loop-closure in a real-time are discussed. Finally, concluding remarks follow in Section 4.

2   Visual Loop-Closure Detection Method Using Average Feature Descriptors

The proposed visual loop-closure detection method is shown in Fig. 1. This proposed method is based on descriptors of feature points. The descriptors of

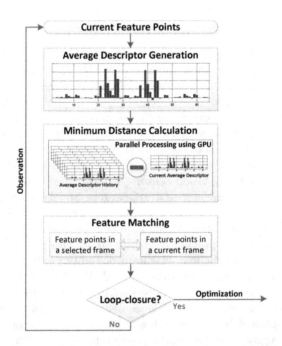

Fig. 1. A diagram of the proposed visual loop-closure detection method

the feature points are obtained by various feature point detection algorithm, such as SIFT, SURF and ORB [11][12][13]. The average feature descriptors are generated using the feature descriptors at each frame. Through the GPGPU, the proposed method selects the frame having a minimum distance with the current average feature descriptor from the average feature descriptor history. After the minimum distance calculation, the loop-closure is determined by matching the feature points between the selected frame and current frame.

2.1 Average Feature Descriptor

The usage of the feature points in all previous image frames is too slow and too large data to recognize the previously visited place. Therefore, the proposed visual loop-closure method uses the average feature descriptors to maximize the parallel processing and increase the memory efficiency. The average feature descriptors at the *i-th* is calculated from

$$m_{i,j} = \frac{1}{N} \sum_{k=1}^{N} f_{i,j,k} \qquad (1)$$

where j is an index of the feature descriptor element; k is an index of the feature points; N is a number of the feature points; $m_{i,j}$ is an average feature descriptor of *j-th* element at *i-th* frame; $f_{i,j,k}$ is a *k-th* feature descriptor of *j-th* element at *i-th* frame. In this paper, the average feature descriptor is used as a frame identification at each frame. Fig. 2 shows an example of the average feature descriptor.

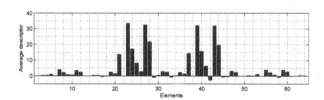

Fig. 2. An average feature descriptor of each image frame

2.2 Minimum Distance Calculation Using GPGPU

The GPGPU is the technique of using a GPU to perform computations that are usually handled by the CPU. Usually, the GPGPU is used to processing parallel tasks at high speed in the field of computer vision. Therefore, in this paper, parallel processing to detect the visual loop-closure is maximized through the concept of the average feature descriptors.

A distance of the average feature descriptors between the *i-th* frame and current frame is calculated by the Euclidean distance. A frame having the minimum distance of the average feature descriptor among the previous frames is decided by

$$i_{min,cur} = \arg \min_i \left\{ \sqrt{\sum_{j=1}^{O}(m_{i,j} - m_{cur,j})^2} \right\} \qquad (2)$$

where $i_{min,cur}$ is a frame index having the minimum distance of the average feature descriptors; O is a size of the feature descriptor elements; $m_{cur,j}$ is an average feature descriptor of j-th element at the the current frame.

After the minimum distance calculation, the proposed method conducts matching the feature points between the selected frame and the current frame. The matching process of the feature points is performed by the Brute-force matcher. If the number of matched feature points exceeds a threshold, which is assigned heuristically, then the proposed method considers that the loop-closure is detected.

3   Experiment

The proposed visual loop-closure method was tested using an RGB-D sensor, known as the Kinect sensor. The experimental setup was composed of Gentoo OS, Intel i5 3.3GHz dual-core processor, NVIDIA GTX 560 GPU and 6GB RAM. The proposed visual odometry algorithm was tested in the Robot Intelligence Technology laboratory at KAIST. To test the proposed visual loop-closure method, Euclidean error RANSAC (EE-RANSAC) algorithm was used to estimate visual odometry of the RGB-D sensor [3]. The EE-RANSAC algorithm computed the visual odometry recursively through minimizing the Euclidean error between the two consecutive frames. Once a loop closure was detected, a pose-graph of the RGB-D sensor was optimized by iSAM algorithm [8]. Keyframe-based visual loop-closure method [3] was used for comparison purpose. This paper defined the keyframes based on visual overlap.

The experiment results of the computation time are shown in Fig. 3. As a result, the keyframe-based visual loop-closure method was much slow than the

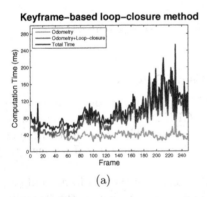 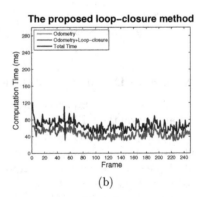

(a)   (b)

Fig. 3. Computation time results of the visual SLAM. (a) Using the keyframe-based visual loop-closure method. (b) Using the proposed visual loop-closure method.

Before the Loop–Closure Optimization   **After the Loop–Closure Optimization**

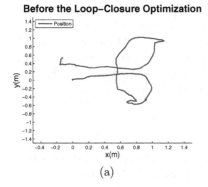 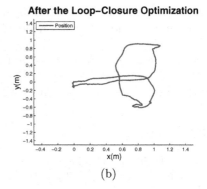

(a)   (b)

Fig. 4. The loop-closure optimization results with the proposed visual loop-closure method. (a) Before the loop-closure optimization. (b) After the loop-closure optimization.

proposed visual loop-closure method. As time went by, while the computation time of the keyframe-based method linearly increased, the computation time of the proposed method was steady.

After the loop-closure was detected, the pose-graph of the RGB-D sensor was optimized by the iSAM algorithm as shown in Fig. 4. Through the loop-closure optimization technique, an accurate pose of the RGB-D sensor could be obtained.

4   Conclusion

In this paper, a novel visual loop-closure method using average feature descriptors was proposed. The usage of feature points in all previous image frames was too slow to recognize the previously visited place. Therefore, the proposed visual loop-closure method used the average feature descriptors to maximize parallel processing and increase memory efficiency. The average feature descriptors were computed by averaging the descriptors of the feature points at each frame. Through the GPGPU technique, the proposed method selected a frame having a minimum distance with the current average feature descriptor from the average feature descriptor history. Finally, after the minimum distance calculation, loop-closure was determined by matching feature points between the selected frame and current frame. To verify the effectiveness of the proposed method, a visual SLAM experiment was conducted. Through the comparison with the conventional visual loop-closure method, our proposed method was verified to be much faster in detecting the visual loop-closure.

Acknowledgment. This work was supported by the Technology Innovation Program, 10045252, Development of robot task intelligence technology that can perform task more than 80% in inexperience situation through autonomous knowledge acquisition and adaptational knowledge application, funded By the Ministry of Trade, industry & Energy (MOTIE, Korea).

This research was also supported by the MOTIE (The Ministry of Trade, Industry and Energy), Korea, under the Human Resources Development Program for Convergence Robot Specialists support program supervised by the NIPA (National IT Industry Promotion Agency)(H1502-13-1001).

References

[1] Wang, J., Zha, H., Cipolla, R.: Coarse-to-fine vision-based localization by indexing scale-invariant features. IEEE Transactions on Systems, Man, and Cybernetics, Part B: Cybernetics 36(2), 413–422 (2006)

[2] Davison, A.J., Reid, I.D., Molton, N.D., Stasse, O.: Monoslam: Real-time single camera slam. IEEE Transactions on Pattern Analysis and Machine Intelligence 29(6), 1052–1067 (2007)

[3] Henry, P., Krainin, M., Herbst, E., Ren, X., Fox, D.: Rgb-d mapping: Using kinect-style depth cameras for dense 3d modeling of indoor environments. The International Journal of Robotics Research 31(5), 647–663 (2012)

[4] Han, S., Kim, J., Myung, H., et al.: Landmark-based particle localization algorithm for mobile robots with a fish-eye vision system. IEEE/ASME Transactions on Mechatronics PP(99), 1–12 (2012)

[5] Newman, P., Cole, D., Ho, K.: Outdoor slam using visual appearance and laser ranging. In: Proceedings of the 2006 IEEE International Conference on Robotics and Automation, ICRA 2006, pp. 1180–1187. IEEE (2006)

[6] Angeli, A., Doncieux, S., Meyer, J.A., Filliat, D.: Real-time visual loop-closure detection. In: IEEE International Conference on Robotics and Automation, ICRA 2008, pp. 1842–1847. IEEE (2008)

[7] Thrun, S., Montemerlo, M.: The graph slam algorithm with applications to large-scale mapping of urban structures. The International Journal of Robotics Research 25(5-6), 403–429 (2006)

[8] Kaess, M., Ranganathan, A., Dellaert, F.: isam: Incremental smoothing and mapping. IEEE Transactions on Robotics 24(6), 1365–1378 (2008)

[9] Kuemmerle, R., Grisetti, G., Strasdat, H., Konolige, K., Burgard, W.: g2o: A general framework for graph optimization. In: Proc. of the IEEE Int. Conf. on Robotics and Automation (ICRA) (2011)

[10] Kaess, M., Johannsson, H., Roberts, R., Ila, V., Leonard, J., Dellaert, F.: isam2: Incremental smoothing and mapping with fluid relinearization and incremental variable reordering. In: 2011 IEEE International Conference on Robotics and Automation (ICRA), pp. 3281–3288. IEEE (2011)

[11] Lowe, D.G.: Object recognition from local scale-invariant features. In: The Proceedings of the Seventh IEEE International Conference on Computer Vision, vol. 2, pp. 1150–1157. IEEE (1999)

[12] Bay, H., Tuytelaars, T., Van Gool, L.: SURF: Speeded up robust features. In: Leonardis, A., Bischof, H., Pinz, A. (eds.) ECCV 2006, Part I. LNCS, vol. 3951, pp. 404–417. Springer, Heidelberg (2006)

[13] Rublee, E., Rabaud, V., Konolige, K., Bradski, G.: Orb: an efficient alternative to sift or surf. In: 2011 IEEE International Conference on Computer Vision (ICCV), pp. 2564–2571. IEEE (2011)

Mobile Robot Localization Using Multiple Geomagnetic Field Sensors

Seung-Mok Lee, Jongdae Jung, and Hyun Myung

Urban Robotics Laboratory, KAIST
291 Daehak-ro, Yuseong-gu, Daejeon 305-701, Korea
hmyung@kaist.ac.kr

Abstract. This paper proposes a novel approach to substantially improve the performance of the conventional vector field SLAM (simultaneous localization and mapping) by using multiple geomagnetic field sensors. The main problem of the conventional vector field SLAM is the assumption of known data association. If a robot has a high uncertainty of the pose estimate, the probability of data association failure increases when the robot's pose is located in a wrong cell. To deal with this problem, we propose a multi-sensor approach utilizing multiple geomagnetic field sensors. As the multi-sensor approach updates nodes of one or more cells simultaneously, the probability of data association failure significantly decreases. The proposed multi-sensor-based localization is implemented based on a Rao-Blackwellized particle filter (RBPF) with geomagnetic field sensors. Simulation results demonstrate that the proposed approach greatly improves the performance of the vector field SLAM compared to the conventional approach with a single sensor.

Keywords: Simultaneous localization and mapping (SLAM), Vector field SLAM, Rao-Blackwellized particle filter (RBPF).

1 Introduction

The aim of simultaneous localization and mapping (SLAM) technology is to provide estimates of both the robot pose and map in unknown environments where an accurate map is unavailable. The SLAM technology has received a great deal of attention in the field of autonomous robot navigation as it has numerous potential applications such as navigation, exploration, reconnaissance, coverage, and 3D reconstruction. For practical application of the SLAM technology, real-time operation with low-priced sensors is essential. This paper therefore focuses on the development of an efficient SLAM algorithm requiring low computational cost and low-priced sensors.

The use of signal strength to localize mobile nodes or robots has gained great attention because it can be implemented with low-priced sensors. Range-only SLAM [1] is performed by ranging from the received signal strength (RSS) of wireless sensor nodes, through inversion of a path loss model. However, the range estimate from the path loss model is not sufficiently accurate due to the stochastic nature of the radio propagation, especially in indoor environments. A sequence-based localization [2] is

J.-H. Kim et al. (eds.), *Robot Intelligence Technology and Applications 2,*
Advances in Intelligent Systems and Computing 274,
DOI: 10.1007/978-3-319-05582-4_11, © Springer International Publishing Switzerland 2014

proposed to localize a mobile node based on a sequence that represents signal strength ranks of multiple nodes with known locations. In [3],[4], WiFi-based SLAMs are proposed using Gaussian processes in order to estimate the location of a pedestrian. Recently, Gutmann et al. [5] proposed another signal strength-based SLAM, called vector field SLAM, that can be applied to commercialized cleaning robots with low-priced sensors such as NorthStar system [6]. The vector field SLAM divides a working space into a number of regular cells, and then each node of the cells is considered to be a virtual landmark. The robot estimates both its pose and the signal strength of the surrounding nodes by bilinear interpolation and linear extrapolation while moving through the working space.

In this paper, a novel approach is proposed to substantially improve the performance of the conventional vector field SLAM by using a mobile robot equipped with multiple geomagnetic field sensors. By deploying multiple geomagnetic field sensors on different places on the robot, the proposed multi-sensor approach updates four or more nodes simultaneously. The selection of nodes to be updated depends on the placement of the sensors. By using the multi-sensor approach, the vector field SLAM reduces the probability of data association failure significantly, and it provides robust performance with respect to the varying cell size. To evaluate the effectiveness of the proposed approach, we performed simulations with multiple sensors. The proposed multi-sensor-based vector field SLAM is implemented by a Rao-Blackwellized particle filter (RBPF) [7],[8]. In simulations, the performance of the conventional approach using one sensor is compared to that of the proposed approach with multiple sensors used under conditions where the probability of data association failure is high.

2   Conventional Vector Field SLAM

Before introducing the vector field SLAM, a SLAM problem is briefly reviewed from a probabilistic perspective. We denote a robot state sequence from time 1 up to time t by $\mathbf{x}_{1:t}=\mathbf{x}_1,...,\mathbf{x}_t$, a measurement sequence by $\mathbf{z}_{1:t}=\mathbf{z}_1,...,\mathbf{z}_t$, a control input sequence by $\mathbf{u}_{1:t}=\mathbf{u}_1,...,\mathbf{u}_t$, and map features by $\mathbf{m}=\{\mathbf{m}_1,...,\mathbf{m}_N\}$. The SLAM algorithm provides an estimate for a posterior over the entire robot trajectory along with the map features, given all available sensor data:

$$p(\mathbf{x}_{1:t}, \mathbf{m} \mid \mathbf{z}_{1:t}, \mathbf{u}_{1:t}). \tag{1}$$

The motion of the robot through an environment is modeled as follows:

$$\mathbf{x}_t = f(\mathbf{x}_{t-1}, \mathbf{u}_t) + \mathbf{e}_u \tag{2}$$

where e_u is normally distributed process noise with zero mean and covariance $\mathbf{P}_t$. An observation given the current robot pose and map features can be modeled as follows:

$$\mathbf{z}_t = h(\mathbf{x}_t, \mathbf{m}_1,...,\mathbf{m}_N) + \mathbf{e}_v \tag{3}$$

where e_v is normally distributed measurement noise with zero mean and covariance $\mathbf{R}_t$.

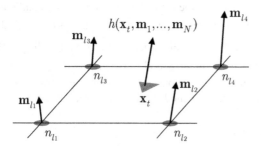

Fig. 1. Measurement model by bilinear interpolation. $\mathbf{m}_{lj}$ denotes the vector of signal values of node l_j, n_{lj} the position of node l_j, and $\mathbf{x}_t$ robot's pose at time t.

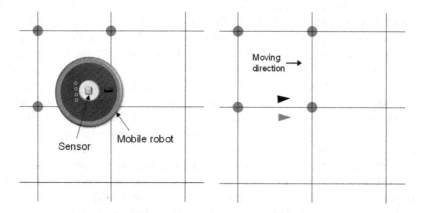

Fig. 2. Conventional approach using a single sensor. The gray circles on a regular grid denote the updated nodes determined from the estimate of the current robot pose (*left*). If the pose estimate (gray triangle) is in a different cell from the cell of the true pose (black triangle), the vector field SLAM might diverge (*right*).

The vector field SLAM, introduced by Guttmann et al. [5], is a SLAM algorithm that estimates both the robot pose and vector field signals simultaneously, in a space with vector field signals such as radio frequency (RF) signal strength, magnetic field strength, and so on. The main concept of the vector field SLAM is as follows: The work space is divided into a number of grid cells, and then four nodes of each cell are considered to be virtual landmarks. Moving through the working space, the robot approximates the vector field signals of four nodes in the corners of the cell where the robot is located as a piecewise linear function. It should be noted that the main difference between the conventional landmark-based SLAM and the vector field SLAM is the dimension and type of mapping variables. That is, the previous SLAM algorithms estimate physical poses of landmarks or map features, whereas the vector field SLAM estimates the signal values of nodes located at fixed ground positions.

For an arbitrary robot pose, a measurement can be modeled by bilinear interpolation from the four nodes of the cell where the robot is located with the assumption that the sensor to measure the vector field is installed on the robot. Let $\mathcal{N}_t=\{l_1, l_2, l_3, l_4\}$ be an index set of four nodes of the cell in which the robot is located at time t, $\mathbf{m}_{l_j}$ be the vector of signal values of the corresponding nodes for $l_j \in \mathcal{N}_t$, and $n_{l_j}=[n_{l_j,x}, n_{l_j,y}]^T$ be the position of l_j node, as shown in Fig. 1. If the vector field is variable depending on the orientation of the sensor, the sensor orientation θ_t should be considered in the measurement model. The measurement considering the orientation can be computed as

$$h(\mathbf{x}_t,\mathbf{m}_1,...,\mathbf{m}_N) = \mathbf{T}(\theta_t)\sum_{j=1}^{4}\omega_j\mathbf{m}_{l_j} \qquad (4)$$

where $\mathbf{T}(\theta_t)$ is a transformation matrix from a global frame to a robot fixed frame considering the robot's orientation θ_t and ω_j is the weight of node l_j. The weight can be computed by bilinear interpolation [5] from the current robot pose $\mathbf{x}_t=[x_t, y_t, \theta_t]^T$.

The most serious problem of the vector field SLAM is that the performance depends on the cell size, as pointed out in [5]. This is caused by the assumption of known data association. The conventional vector field SLAM updates only four nodes of a cell corresponding to the current robot pose estimate, as shown in Fig. 2(*left*). This is problematic when the robot moves near and along the boundary of cells with a high uncertainty of the robot pose, as illustrated in Fig. 2(*right*). In a situation when the robot moves along the boundary lines, the likelihood of choosing a wrong cell can be large due to the robot pose uncertainty. If incorrect nodes that were not initialized before are chosen, then the observed nodes are registered as new nodes and the vector field SLAM diverges. For this reason, the robot's moving trajectory can affect the performance of localization in the conventional vector field SLAM approach. In other words, the performance of the conventional approach depends on the cell size. The smaller cell size improves the accuracy of the measurement model by bilinear interpolation, but if the cell size is too small, the estimate becomes inaccurate since the probability of choosing a wrong cell increases. Finally, if the data association failure continues while the robot moves, the robot's trajectory estimated by the conventional vector field SLAM mostly shifts to wrong cells and fails to close loops.

3   Vector Field SLAM Using Multiple Geomagnetic Sensors

3.1   Multi-sensor Approach on Mobile Robot

In order to overcome this limitation, a multi-sensor approach is proposed. The concept of the proposed multi-sensor approach is described in Fig. 3. Two or more sensors to measure vector field signals are mounted on different positions of a mobile robot, and then the nodes of a cell corresponding to each sensor are updated simultaneously using the measurements from multiple sensors. As four or more nodes can be updated simultaneously according to the sensors' placement on the robot, this approach greatly reduces the likelihood of data association failure even though the pose estimate has a high uncertainty.

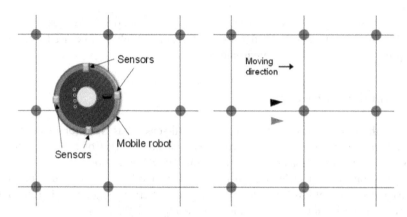

Fig. 3. Proposed approach using multiple measurement sensors. The gray circles on grid denote the updated nodes determined from the estimate of the current robot pose (*left*). Even though the pose estimate (gray triangle) is in a cell different from the cell of the true pose (black triangle), the vector field SLAM can converge into the cell of the true pose (*right*).

3.2   Rao-Blackwellized Particle Filter

In the RBPF approach [7],[8], the posterior is estimated by factoring the SLAM problem into separate problems, i.e., a robot localization problem and a feature estimation problem conditioned on the robot pose estimate. The SLAM posterior (1) can be rewritten in a factored form as follows:

$$p(\mathbf{x}_{1:t},\mathbf{m} \mid \mathbf{z}_{1:t},\mathbf{u}_{1:t}) = p(\mathbf{x}_{1:t} \mid \mathbf{z}_{1:t},\mathbf{u}_{1:t})\prod_{n=1}^{N} p(\mathbf{m}_{n} \mid \mathbf{x}_{1:t},\mathbf{z}_{1:t},\mathbf{u}_{1:t}). \tag{5}$$

According to the typical RBPF framework, the RBPF for the vector field SLAM uses a particle filter to estimate the robot pose, and an EKF to estimate signal values of nodes, i.e., the vector field. Each particle in the vector field SLAM with RBPF represents a potential pose estimate of a robot and a map associated with the pose estimate.

3.3   Implementation with Geomagnetic Field

The proposed multi-sensor-based vector field SLAM with RBPF is implemented using geomagnetic field. In the implementation of the vector field SLAM, geomagnetic field not only satisfies the properties of continuity, large spatial variation, and small temporal variation, but also has the following advantages. First, the magnetic field signal is robust against moving nonmetallic obstacles or passersby. Second, the signal is very helpful to compensate the heading angle of a mobile robot, since geomagnetic field changes with the orientation of the robot and the orientation can be incorporated in the measurement models described by (4). Third, because geomagnetic field can be detected almost everywhere, it is not necessary to set up beacons throughout the environment. Last, as most building structures produce distortions in the magnetic field, the distorted magnetic field can be used as a good fingerprint for SLAM in indoor environments.

4   Simulation Results

To verify the effectiveness of the proposed multi-sensor-based vector field SLAM, the performance of the proposed approach is compared to that of the conventional approach where a single sensor is used under the RBPF framework.

In the simulations, a differential drive wheeled mobile robot is used. The three dimensional vector field $\mathbf{m}=[\mathbf{m}_X, \mathbf{m}_Y, \mathbf{m}_Z]^T$ formed by Earth's magnetic field (unit: μT) over a two dimensional position (x, y) is simulated by

$$
\begin{aligned}
m_X &= 3x^3 - 22(x-0.1y)^2 + 42x + 50 \\
m_Y &= 9(x-0.1y)^3 - 66(x-0.1y)^2 + 126x + 30 \\
m_Z &= 0.9(x-y)^3 - 6.6(x+0.1y)^2 + 12.6x + 150.
\end{aligned}
\tag{7}
$$

The sensor measurement data are generated by bilinear interpolation of the vector field. The cell size is set to 0.5 m. The process and measurement noises are assumed to be sampled from the following normal distributions:

$$
\mathbf{e}_u \sim N(0, \sigma_u^2), \quad \mathbf{e}_v \sim N(0, \sigma_v^2)
\tag{8}
$$

where $\sigma_u = 0.02$ m/s and $\sigma_v = 5$ μT. The initial error covariance of each node is set to 50 μT.

The robot is simulated to move along a path that starts from the lower left corner, and continues to move along the rows, and changes rows on the left and right sides. To observe the cases in which the likelihood of data association failure increases, the distance between rows is set to be slightly different from the cell size. The simulation is run for two cases with the numbers of sensors 1 and 2, with fixed number of particles $M=30$. The sensors are installed on the mobile robot 0.2 m apart from the center of the robot, while a single sensor is installed at the center of the mobile robot for the conventional approach.

Fig. 4(a) shows the trajectory computed from odometry, and Fig. 4(b) and (c) show the trajectories computed from the conventional approach in which a single sensor is used and the proposed approach with two sensors, respectively. The cells are represented by dotted lines, and the crossing points of the lines are considered to be nodes. As shown in Fig. 4(b), the trajectory by the conventional approach has large errors, because data association failure occurs frequently while moving along the boundary lines. However, as shown in Fig. 4(c), the trajectory by the proposed approach matches the true trajectory closely. Fig. 5(a) shows the true vector field produced by (7). Fig. 5(b) and (c) show the vector fields estimated by the conventional and proposed approaches, respectively. Comparing these results, it is found that the proposed approach produces more accurate signal maps than the conventional approach.

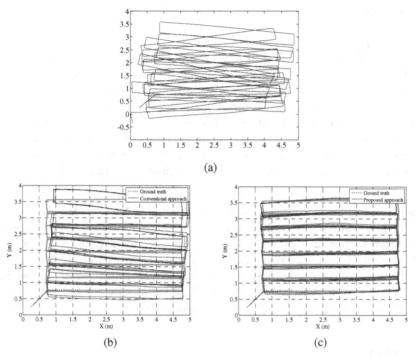

(a)

(b)   (c)

Fig. 4. (a) Trajectory computed from odometry. (b) Results of the conventional approach with one sensor. (c) Results of the proposed approach with two sensors.

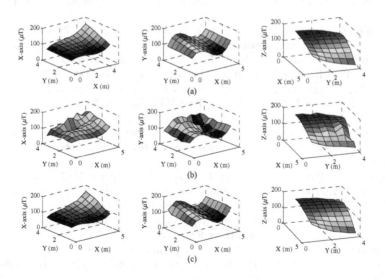

Fig. 5. (a) True three-axis geomagnetic field produced by (7). (b) Three-axis geomagnetic field estimated by the conventional approach. (c) Three-axis geomagnetic field estimated by the proposed approach.

5   Conclusion

In this paper, we have presented a novel idea to substantially improve the performance of the vector field SLAM by using a mobile robot equipped with multiple geomagnetic field sensors. The proposed multi-sensor-based vector field SLAM greatly reduces the probability of data association failure by updating four or more nodes of multiple cells depending on the sensors' placement. The effectiveness of the proposed multi-sensor-based vector field SLAM is verified through simulations with Earth's magnetic field. Consequently, the proposed approach greatly improves the accuracy of localization and mapping, and also it enhances the ability of the system to close loops compared to the conventional approach with a single sensor.

Acknowledgements. This research was supported by the MOTIE (The Ministry of Trade, Industry and Energy), Korea, under the Human Resources Development Program for Convergence Robot Specialists support program supervised by the NIPA(National IT Industry Promotion Agency) (NIPA-2013-H1502-13-1001). This research was also supported by Basic Science Research Program through the National Research Foundation of Korea (NRF) funded by the Ministry of Science, ICT & Future Planning (grant number NRF-2013R1A1A1A05011746).

References

1. Menegatti, E., Zanella, A., Zilli, S., Zorzi, F., Pagello, E.: Range-only SLAM with a mobile robot and a wireless sensor networks. In: Proc. IEEE Int. Conf. Robot. Autom., pp. 8–14 (2009)
2. Yedavalli, K., Krishnamachari, B.: Sequence-based localization in wireless sensor networks. IEEE Trans. Mob. Comput. 7(1), 81–94 (2008)
3. Ferris, B., Fox, D., Lawrence, N.: WiFi-SLAM using Gaussian process latent variable models. In: Proc. Int. Joint Conf. Artificial Intelligence, pp. 2480–2485 (2007)
4. Huang, J., Millman, D., Quigley, M., Stavens, D., Thrun, S., Aggarwal, A.: Efficient, generalized indoor WiFi GraphSLAM. In: Proc. IEEE Int. Conf. Robot. Autom., pp. 1038–1043 (2011)
5. Gutmann, J.-S., Eade, E., Fong, P., Munich, M.E.: Vector field SLAM-localization by learning the spatial variation of continuous signals. IEEE Trans. Robot. 28(3), 650–667 (2012)
6. Yamamoto, Y., Pirjanian, P., Brown, J., Munich, M., DiBernardo, E., Goncalves, L., Ostrowski, J., Karlsson, N.: Optical sensing for robot perception and localization. In: Proc. IEEE Workshop Adv. Robot. Social Impacts, pp. 14–17 (2005)
7. Grisetti, G., Stachniss, C., Burgard, W.: Improved techniques for grid mapping with Rao-Blackwellized particle filters. IEEE Trans. Robot. 23(1), 34–46 (2007)
8. Montemerlo, M., Thrun, S.: Simultaneous localization and mapping with unknown data association using FastSLAM. In: Proc. IEEE Int. Conf. Robot. Autom., pp. 1985–1991 (2003)

GA-Based Optimal Waypoint Design for Improved Path Following of Mobile Robot

Jae-Seok Yoon[1,*], Byung-Cheol Min[2], Seong-Og Shin[1],
Won-Se Jo[1], and Dong-Han Kim[1]

[1] Department of Electronics and Radio Engineering, Kyung-Hee University
Yongin, KS, 009, Republic of Korea
{blackyjs,wonsu0513}@khu.ac.kr,
tmfb@naver.com, donghani@gmail.com
[2] Department of Computer and Information Technology, Purdue University
West Lafayette, IN, 47907, USA
minb@purdue.edu

Abstract. Mobile robot can follow the planned path using a waypoint following guidance scheme. As this type of guidance scheme only uses the position of waypoints to navigate the path, the waypoint following is relatively simple and efficient to implement. However, it is non-trivial to determine the number and size of waypoints, which heavily affect the performance of robot. Thus, we tackle the problem of finding the optimal number and size of waypoints in this paper. For this optimization problem, we use genetic algorithm, where the effectiveness of the proposed method is verified in MATLAB simulation. The proposed method shows that mobile robot effectively navigates the planned path and successfully reaches the destination with the minimum path following error and travel time.

Keywords: Waypoints, mobile robot, navigation, limit-cycle, genetic algorithm.

1   Introduction

Mobile robot navigation can be divided into two methods: deliberative method [1] and reactive method [2]. The first method generates and follows the path to the target point with accurate information about the environment. However, if the surrounding environment suddenly changes, many problems occur due to the difference form the existing information. The latter method practically does not require the environment model because it overcomes the uncertainties of environment by continuously reflecting on the changing state of environment with low computation. However, it is a less goal-oriented approach. In order to overcome these limitations, hybrid navigation method, which combines both deliberative method and reactive method, has been introduced for mobile robot [3].

---

* Corresponding author.

J.-H. Kim et al. (eds.), *Robot Intelligence Technology and Applications 2*,
Advances in Intelligent Systems and Computing 274,
DOI: 10.1007/978-3-319-05582-4_12, © Springer International Publishing Switzerland 2014

For a mobile robot to arrive the desired target with obstacles avoidance, path planning and path following are needed to generate a path and to follow the generated path, respectively. The most common methods of path planning are Dijkstra's algorithm, A* search algorithm, limit-cycle algorithm, potential field algorithm, etc., where the path is generated by connecting the number of waypoints in series. However, these methods have a tendency to reduce the system performance as the number of waypoints increases due to the increased amount of computation. Mobile robot can use the waypoint to generate a path and to follow to reach the target point. The accuracy of path following generally depends on the number of waypoint, where Shair et al. have also showed that the tracking performance depends on the size of waypoint [4]. Boucher et al. have introduced an approach to follow the path smoothly using waypoints [5]. These waypoint following methods have the advantage of being easy to design, but there is a big downside to following error [6].

Few researches have been studied showing that two parameters, the number and size of waypoints, could affect the performance of the waypoint following guidance scheme. Moreover, any methods considering the two parameters in the guidance problem have not yet been proposed. It is for this reason that we propose an improved waypoint following method capable of adjusting the number and size of waypoints. This will be used easily for user because it is easy to design. Genetic algorithm (GA), one of the evolutionary algorithms, is used to find the optimized number and optimized size of waypoints. GA is widely used because it does not require a mathematical model of the system and it can solve most of the problems [7]. The objective function consists of the weighted sum of the time and following error for robot to reach the target point from the starting point. Note that the form of this function is multi-objective optimization problem [8]. Most of practical engineering problems have multi-objectives either to minimize the cost or to maximize the performance. These problems are faced frequently, which are difficult to obtain the right answer. Since GA is known as a popular meta-heuristic for these problems, the proposed method uses GA to solve the multi-objective optimization problem.

This paper is organized as follows: Section 2 describes the path following methods, limit-cycle, and GA. Section 3 presents the problem of following waypoints and proposes the method to design the optimized waypoint, which fixes the presented problem. Section 4 presents the simulation method, three different types of simulation environment and the corresponding results for each case. Lastly, concluding remarks follow in Section 5.

2   Background

2.1   Path Following Methods

Three basic path following schemes are path following method, cross-track following method, and waypoint following method [6]. Path following method adjusts towards the planned path by minimizing the distance between the robot's position and the path. The path following method effectively follows along the given path, but it is much more complex than two other methods.

Cross-track following method follows the path, which consists of the track of connected waypoints. Note that the track is a path drawn by a line between two consecutive waypoints. The cross-track following method is a little difficult compared to waypoint following method, but it follows along the path with better accuracy.

Waypoint following method sets the number of waypoints on the path, where it changes to face the next waypoint once the circle of waypoint is within the desired range. The waypoint following method is simple and easy to implement, but following error is large because it does not follow the exact path. However, this error can be minimized by adding more number of waypoints to the generated path. In addition, the following error can be decreased by adjusting the size of circle when robot is about to head to the next waypoint. Thus, the performance of waypoint following method can be improved if both the number and size of waypoints are adjusted according to situation and purpose.

2.2   Limit-Cycle Navigation

Limit-cycle is a method for determining the stability of second-order linear/non-linear system. When a phase portrait is drawn, the stability can be determined by analyzing the characteristics and shape of circle and its trajectory, which converges to the circle. Consider the following second-order nonlinear system [9],

$$
\begin{aligned}
\dot{x}_1 &= x_2 + x_1(r^2 - x_1^2 - x_2^2) \\
\dot{x}_2 &= -x_1 + x_2(r^2 - x_1^2 - x_2^2)
\end{aligned}
\tag{1}
$$

where r is the radius of circle, which is generated when using the limit-cycle. By setting the proper value of radius, the size of circle can be defined and the trajectory started from all points on the coordinates can be converged. The direction of convergence depends on the location of negative sign. Note that equation (1) converges to the clockwise, whereas its direction changes to counter-clockwise when there is a negative sign in front of x_2 in $\dot{x}_1$. The characteristics of limit-cycle can be applied to mobile robot navigation, which is known as limit-cycle navigation [10]. Since the limit-cycle navigation does not require high computational power, its greatest advantage is to apply in unknown environment in real-time for mobile robot navigation. However, the performance is highly influenced by the location and size of obstacles. To overcome these problems of early limit-cycle navigation, improved limit-cycle navigation has been studied [11]. This paper uses the limit-cycle navigation as the path planning method and deals with the path following method to follow the generated path.

3   Waypoint Following

3.1   Problem Description

If mobile robot attempts to precisely follow the planned path, the path following can be accomplished by setting many number of waypoints on the path. However, in this case, it takes a long time to reach the target point because the increased number of waypoints leads to the frequent acceleration/deceleration of robot speed. It is important to follow the path with precision, but it is also important to get to the target point in the fastest time. This problem can be solved by setting the number and size of waypoints according to each case because the performance of path following depends on these two parameters.

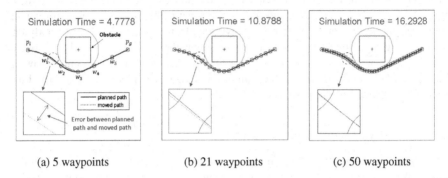

(a) 5 waypoints   (b) 21 waypoints   (c) 50 waypoints

Fig. 1. Travel time and error to reach target point by varying the number of waypoints

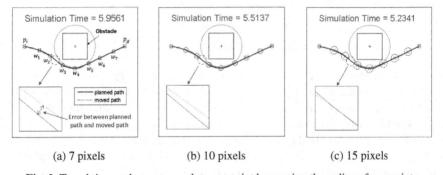

(a) 7 pixels   (b) 10 pixels   (c) 15 pixels

Fig. 2. Travel time and error to reach target point by varying the radius of waypoints

Fig. 2 and Fig. 3 show the total travel time to precisely follow the path and to quickly reach the target point according to the different number and size of waypoint, respectively. Note that the robot is moved along the planned waypoint from left to right. The total travel time is shown on the top, whereas the error between the planned path and moved path is shown at the bottom.

As shown in Fig. 2, the robot precisely follows the planned path as the number of waypoint increases, but it takes a longer time to get to the target point. Similarly, as shown in Fig. 3, the travel time depends on the size of waypoint even if the number of waypoint is identical in all three cases. Thus, an optimized number and size of waypoint are needed for mobile robot to quickly reach to the target while minimizing the path error.

3.2   Designing Optimal Waypoint

As stated above, there is a difference in the results of path following of mobile robot according to the number and size of waypoints. In order to follow the path with the minimum error, mobile robot can reach to the target by passing many waypoints, but it takes a longer time due to the increased number of waypoints. Thus, the tradeoff between the travel time and following error must be considered. This paper proposes an improved path following method by optimizing the number and size of waypoints such that mobile robot can reach to the target efficiently.

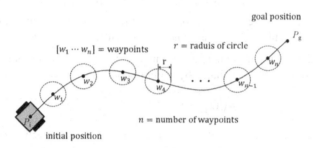

Fig. 3. Path following method of mobile robot using waypoints

Fig. 3 shows the path tacking method of mobile robot using waypoints in the given path. The mobile robot starts from p_i and arrives at p_g after passing through n number of waypoints, { $w_1, w_2, ..., w_n$ }, sequentially. In all waypoints, there is the same size of circle with the radius, r. If the robot reaches within the circle of waypoint, it updates and moves toward the next waypoint. As a mobile robot approaches to each waypoint, the speed of the mobile robot is decreased; however, it does not decelerate under the pre-defined speed limit in order to prevent frequent acceleration/deceleration of robot speed.

Table 1. Parameter settings of GA

Variable	Bits	Lower Bound	Upper Bound	Resolution
n	6	1	64	1
r	5	1	32	1

GA is used to find the optimal number and size of waypoints, where the parameter settings of GA are summarized in Table 1. Even though there are many methods to represent the chromosome in GA, the binary string is used in the proposed method. All parameters in Table 1 consist in equation (2) as follows [13]:

$$x_i^R = \frac{x_i^U - x_i^L}{(2^{b_i} - 1)} \qquad (2)$$

where x_i^R represents the resolution between the discretized values of x_i , l_i is the number of bits to code x_i , x_i^U and x_i^L are the upper bound and lower bound on variable, respectively. The resolution value is set to 1 because the number and size of waypoint must be increased by integer. Since the number of possible solution and size are 64 and 32, respectively, the total number of possible solution becomes 2048. The proposed method finds the most optimized one among these possible solutions.

The main objective is to find n^* and r^* until the value of objective function, $f(n,r)$, is minimized. The fitness criteria are needed to find these optimal values. In this paper, the fitness criteria consist of the summation of two factors: the error between the planned path, (x_p, y_p), and the moved path, (x_m, y_m), and the total travel time from the starting point to the target point. The distance error between the planned path and the moved path is calculated by using Euclidean distance as follows:

$$d_i = \sqrt{(x_{p_i} - x_{m_i})^2 + (y_{p_i} - y_{m_i})^2} \tag{3}$$

Since the objective function of proposed method is multi-objective optimization problem, weighted sum approach – one of classical approaches – is used in the objective function as follows:

$$f(n,r) = \alpha f_1 + \beta f_2 \quad \text{with} \quad f_2 = \sum d_i \tag{4}$$

where f_1 and f_2 represent the objective function of simulation time and error between the planned path and moved path, respectively. Note that the value of each objective function is normalized. α and α correspond to the weights of f_1 and f_2, respectively, where the summation of α and α equals 1. Different weight values can be assigned depending on the situation. As an example, the value of α can be increased if the objective is to get to the target point in less time rather than following the path with precision, or vice versa to increase the value of α. In Section 4, the simulation is conducted by adjusting these weights depending on the situation.

4   Simulation

4.1   Simulation Method

The proposed method was verified through the simulation environment in MATLAB. TOC function was used to measure the total travel time from the starting point to the target point. For accurate performance verification, all simulations were conducted on the same computer.

In the simulation, the waypoints were initially designed by the proposed GA and then the mobile robot followed the designed path. The simulation environment was consisted of two cases. The first case was when there is one obstacle, which describes the simple path following situation for robot. In the first case, higher importance should be given to the total travel time to the target compared to the path following accuracy. Thus, the higher value of a was set in the simulation to provide higher importance to the arrival time.

The second case was when there are three obstacles, which describes the accurate path following situation for robot. If the importance of a is lower than a like the first case, the robot is likely to collide with one of obstacles. Thus, in the second case, higher importance should be given to the path following by increasing the value of a. In each case, the location of target point and obstacles were fixed, but the simulation was conducted by changing the starting point to three different locations.

4.2   Simulation Result

In each case, the results showed the average of optimized values, which were found 10 times separately using GA. Note that the global solution was manually found and used for performance analysis.

Case 1: The presence of one obstacle

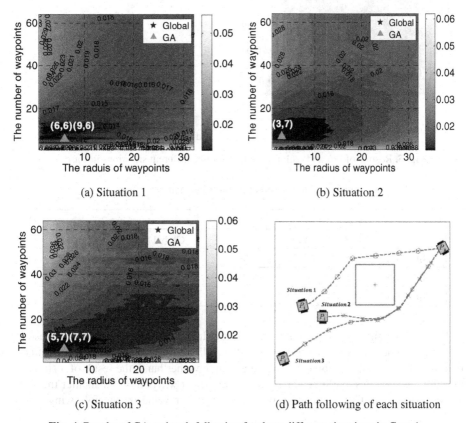

(a) Situation 1   (b) Situation 2

(c) Situation 3   (d) Path following of each situation

Fig. 4. Results of GA and path following for three different situations in Case 1

Case 2: The presence of three obstacles

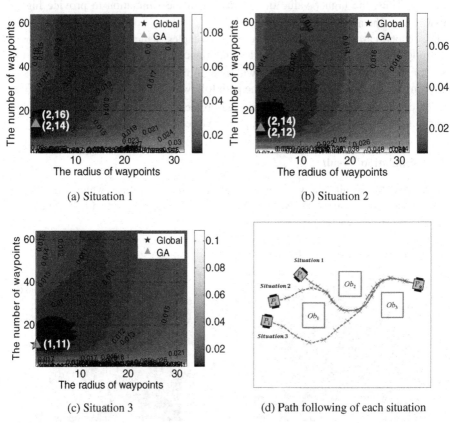

(a) Situation 1

(b) Situation 2

(c) Situation 3

(d) Path following of each situation

Fig. 5. Results of GA and path following for three different situations in Case 2

Table 2. Comparison of global value and GA value

	Case 1		Case 2	
	Global *(n,r)*	GA *(n,r)*	Global *(n,r)*	GA *(n,r)*
Situation 1	(6,9)	(6,6)	(16,2)	(14,2)
Situation 2	(7,3)	(7,3)	(14,2)	(12,2)
Situation 3	(7,7)	(7,5)	(11,1)	(11,1)

From the result of a simulation of Case 1, it didn't need many waypoints because it sets more higher value on the travel time than following error. Therefore, it is effective to follow path using about 7 waypoints. On the other hand, the result of a simulation of Case 2 shows that it needs more the number of waypoints and smaller the size of waypoints than Case 1 because it sets more higher value on the following error than travel time. In this case, it needed accurate following since the robot is likely to collide with obstacles than Case 1. We can see that the mobile robot effectively reach the target point depending on the situation through simulation result.

5   Conclusion

This paper showed that the performance of path following of mobile robot depends on the number and size of waypoints. In order to optimize the performance, this paper proposed the GA-based path following method, where the effectiveness and applicability of the proposed method was verified through simulation. The advantage of the proposed method is that the minimum time and path error to the target point are guaranteed if robot has prior knowledge of the path information. Since the proposed method is simple and easy to implement, it will be a useful tool for engineers who need their robot to precisely follow the path and to quickly arrive at the target point.

Acknowledgements. This research was supported by Technology Innovation Program of the Knowledge economy (No. 10041834, 10045351) funded by the Ministry of Knowledge Economy (MKE, Korea) and Basic Science Research Program through the National Research Foundation of Korea (NRF) funded by the Ministry of Education, Science and Technology (No. 2012R1A1A2043822)

References

1. Ibrahim, M.T.S., Ragavan, S.V., Ponnambalam, S.G.: Way point based deliberative path planner for navigation. In: IEEE/ASME International Conference on Advanced Intelligent Mechatronics, AIM 2009, pp. 881–886 (2009)
2. Pradhan, S.K., Parhi, D.R., Panda, A.K., Behera, R.K.: Potential field method to navigate several mobile robots. Applied Intelligence 25(3), 321–333 (2006)
3. Mulvaney, D., Wang, Y., Sillitoe, I.: Waypoint-based Mobile Robot Navigation. In: The 6th World Congress on Intelligent Control and Automation, vol. 2, pp. 9063–9067 (2006)
4. Shair, S., Chandler, J.H., Gonzalez-Villela, V.J., Parkin, R.M., Jackson, M.R.: The use of aerial images and GPS for mobile robot waypoint navigation. IEEE/ASME Trans. Mechatronics 13(6), 692–699 (2008)
5. Boucher, P., Cohen, P.: A smoothness-preserving waypoints follower for mobile platforms. In: IEEE/ASME International Conference on Advanced Intelligent Mechatronics, pp. 471–476 (2010)
6. Shima, T., Rasmussen, S. (eds.): UAV Cooperative Decision and Control: Challenges and Practical Approaches, vol. 18. SIAM (2009)
7. Gen, M., Cheng, R.: Genetic algorithms and engineering optimization, vol. 7. John Wiley & Sons (2000)
8. Konak, A., Coit, D.W., Smith, A.E.: Multi-objective optimization using genetic algorithms: A tutorial. Reliability Engineering & System Safety 91(9), 992–1007 (2006)
9. Khalil, H.K.: Nonlinear System, 2nd edn. Prentice Hall (1996)

10. Kim, D.H., Kim, J.H.: A real-time limit-cycle navigation method for fast mobile robots and its application to robot soccer. Robotics and Autonomous Systems 42, 17–30 (2003)
11. Lim, Y.W., Kim, J.W., Nam, S.Y., Kim, D.H.: Local-path planning using the limit-cycle navigation method with the edge detection method. In: ICEIC: International Conference on Electronics, Informations and Communications, pp. 232–234 (2010)
12. Davidor, Y.: Genetic Algorithms and Robotics: a heuristic strategy for optimization. World Scientific Publishing Singapore (1991)
13. Min, B.C., Lewis, J., Matson, E.T., Smith, A.H.: Heuristic optimization techniques for self-orientation of directional antennas in long-distance point-to-point broadband networks. Ad Hoc Networks (2013)

Indoor Mobile Robot Localization Using Ambient Magnetic Fields and Range Measurements

Jongdae Jung, Seung-Mok Lee, and Hyun Myung

Dept. of Civil and Environ. Engg., KAIST,
291 Daehak-ro, Yuseong-gu, Daejeon 305-701, Korea
hmyung@kaist.ac.kr

Abstract. In this paper we present a method for solving an indoor SLAM and relocation problem using ambient magnetic fields and radio sources. Specifically, we exploit the magnetic field anomalies and noisy radio ranges in indoor environments. A robot with two magnetometers and one active radio range sensor first explores the unknown environment based on the simultaneous localization and mapping (SLAM) technique, gathering its path and some useful multisensory observations. The gathered data are then applied to the Monte Carlo localization (MCL)-based relocation algorithms. The performance of the proposed methods is validated by simulation using the real-world data.

Keywords: SLAM, relocation, ambient magnetic field, radio ranges.

1 Introduction

Localization is a key component in autonomous navigation systems for mobile robots. Given a prior map, a robot can locate itself rather easily by evaluating the likelihood of observing the current sensory inputs. In real world application, however, obtaining a precise map in advance is often not practical and sometimes impossible - e.g., in disaster areas.

The simultaneous localization and mapping (SLAM) aims to solve a localization problem concurrently with a mapping problem. During the recent decades, the SLAM problem was intensively researched and many novel and high-performing SLAM methodologies were proposed [1, 2].

Since the online SLAM is usually recursive, maintaining a certain error bound on current position is a critical issue. For examples, when the robot is kidnapped or suffers from locomotive failures (due to large slip or falling, etc.), it is inevitable that the robot loses its current position [3, 4]. In that case, an immediate recovery of the robot position is required for the robot's seamless operations.

In this paper, both SLAM and relocation issues are addressed by exploiting the geo-magnetic field anomalies [5–8]. During the SLAM procedure, only magnetic fields are used as external landmark sources while in the relocation phase

J.-H. Kim et al. (eds.), *Robot Intelligence Technology and Applications 2*,
Advances in Intelligent Systems and Computing 274,
DOI: 10.1007/978-3-319-05582-4_13, © Springer International Publishing Switzerland 2014

additional radio source is employed to address the ambiguity issues related to the magnetic-only representations of the map. We employed the Rao-Blackwellized particle filter (RBPF) [9] and bilinear interpolation [10] in our SLAM frameworks and also employed the Monte Carlo localization (MCL) method to solve the relocation problem.

2   Magnetic SLAM

2.1   State Representation

It is assumed that a time series of the robot state $\{\mathbf{x}_t\}$ is defined in a $\mathbf{SE}(2)$ pose set as

$$\mathbf{x}_t = \left[x_t, y_t, \theta_t\right]^T, \quad t = 0, \ldots, t_f \tag{1}$$

with an initial pose $\mathbf{x}_0 = [0, 0, 0]^T$. In the proposed magnetic SLAM formulation, the planar space of the robot's movement is divided into finite number of cells and each cell is assigned with a cell index k. The map state $\{\mathbf{m}^l\}$ then can be defined at the fixed landmark points $\{n_l\}_k = \{n_{k_1}, \ldots, n_{k_4}\}$ as a set of vectors describing the ambient magnetic field in cell k:

$$\mathbf{m}^l = \left[m_x, m_y, m_z\right]^T, \quad l = 1, \ldots, M. \tag{2}$$

where the three magnetic field elements are represented in a earth-fixed reference frame defined at $\mathbf{x}_0$ and M denotes the number of the landmark points.

The measurement $\mathbf{z}_t$ also consists of three orthogonal components of the magnetic field but these are represented in a sensor-fixed reference frame:

$$\mathbf{z}_t = \left[m_x^b, m_y^b, m_z\right]^T, \quad t = 0, \ldots, t_f \tag{3}$$

where we assumed that a robot rotate on a horizontal plane defined by geographic north and east, which makes the vertical component of the measurement vector same in both earth- and sensor-fixed reference frames.

Finally, the augmented state variable $\mathbf{y}_t$ is defined as follows:

$$\mathbf{y}_t = \left[\mathbf{x}_t, \mathbf{m}^1, \ldots, \mathbf{m}^M\right]^T. \tag{4}$$

2.2   SLAM Formulation

In the RBPF framework, the particles $\{\mathbf{y}_t^i\}$ containing all the state variables and the relevant weight value β^i are generated first. After proper initialization, two key steps for the SLAM update procedure – prediction and measurement – are performed in a recursive manner.

In the prediction step, the state transition is predicted based on the state transition model. Usually, robot's kinematic model is used to describe this model.

In the measurement step, both the robot state and the map states are updated according to the measurement model $\mathbf{h}(\mathbf{y}_t)$. As defined in the previous section, the measurement $\mathbf{z}_t$ is described in a sensor-fixed reference frame and is related to the sensor's orientation as well as the map states. This can be modeled as follows:

$$\hat{\mathbf{z}}_t = \mathbf{h}(\mathbf{y}_t)$$
$$= \mathbf{h}_R(\mathbf{h}_0(x, y, \mathbf{m}^1, \ldots, \mathbf{m}^M), \theta) \tag{5}$$

where $\hat{\mathbf{z}}_t$ is a predicted measurement, $\mathbf{h}_0$ is a measurement model for the pre-defined sensor orientation (e.g., $\theta = 0$), and $\mathbf{h}_R$ is a model for the measurement transformation from the predefined orientation to the current orientation. In the implementation, $\mathbf{h}_0$ is calculated by bilinear interpolation of the map states [10].

In the RBPF framework, the measurement update of the map states are done by the Kalman update using the Jacobian of the measurement model (5).

The robot state is then updated by updating the weight of each particle:

$$\beta_{t+1}^i = \beta_t^i \frac{1}{\sqrt{2\pi|\mathbf{S}_{t+1}^i|}} \exp[-\frac{1}{2}\nu_i^T(\mathbf{S}_{t+1}^i)^{-1}\nu_i] \tag{6}$$

where $\nu_i = \mathbf{z}^i - \hat{\mathbf{z}}^i$ is the measurement residual and $\mathbf{S}^i$ is the innovation covariance for the i-th particle.

3   Relocation

3.1   Design of Fingerprints

For a successful relocation, the sensor signature values called 'fingerprints' should be carefully designed. The designed fingerprints are encoded during the SLAM procedure and will be used in the relocation phase later.

For the magnetic fingerprints, three types of raw sensor readings $[m_x^b, m_y^b, m_z]$ are available. The norm of the magnetic vector, the total strength of the magnetic field, can also be a valid fingerprint which is easy to compute. Since the true magnetic north is unknown, however, the available independent measurements falls into two elements: the total strength and vertical component (or total strength and inclination). Therefore we employed additional radio sources to enhance the location-uniqueness of the fingerprints. In this paper we used chirp spread spectrum (CSS) ranging measurements [11] as additional radio fingerprints. To supplement the deficiency in magnetic fingerprint, we also added another magnetometer in our robot system. This also enhanced the observability of the measurement model.

3.2   MCL-Based Relocation

For the MCL-based relocation, the fingerprints are constructed in cell-level. During the SLAM procedure several statistics related to the field characteristics are encoded for each cell k. The encoded statistics are mean and standard deviation of the sensor readings from two magnetometers and a number of radio ranging beacons:

$$\Lambda^k = \begin{bmatrix} \mu_m^k & \mu_r^k \\ \sigma_m^k & \sigma_r^k \end{bmatrix} \tag{7}$$

where m and r are the subscripts for the magnetic field and radio range data, $\mu_{(\cdot)}$ and $\sigma_{(\cdot)}$ are the mean and standard deviation value of the sensor data, respectively.

In the relocation phase, each particle pose is updated by the robot's odometric sensor data. The particle weight is then updated by the likelihood of the current measurement ζ. The vector ζ is composed as

$$\zeta = \begin{bmatrix} m_L^1 \\ m_L^2 \\ r \end{bmatrix} \tag{8}$$

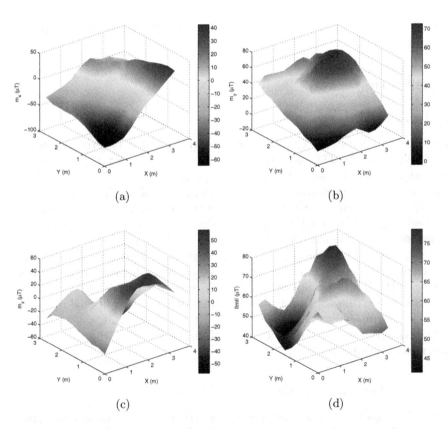

(a)

(b)

(c)

(d)

Fig. 1. Indoor geo-magnetic fields constructed by the magnetic SLAM. The surface plot was generated by the bilinear interpolation of the estimated map states.

where $m_L^j = [m_z, ||m||]^T$ is the j-th magnetometer's readings and $r = [r_1, \ldots, r_q]^T$ is the low-pass filtered radio ranges from q beacons. The weight update equation is then described as:

$$\gamma_{t+1}^{(i)} = \gamma_t^{(i)} \frac{1}{\sqrt{2\pi |\mathbf{R_f}^{(i)}|}} \exp[-\frac{1}{2}\nu_{(i)}^T (\mathbf{R_f}^{(i)})^{-1}\nu_{(i)}] \tag{9}$$

where γ_t is a particle weight, $\nu = \zeta_t - \hat{\zeta}_t$ is the measurement residual, $\mathbf{R_f} = \mathrm{diag}([\sigma_m, \sigma_r])$ is the error covariance matrix.

The estimated robot position is then finally calculated as a weighted sum of the particles.

4   Simulation Results

Using the real-world sensor data gathered online, we evaluate the proposed algorithm in off-line simulations. For the simulation data, the robot took several loops along the window-shaped trajectory in a 4 m x 3 m area and gathered all the sensor data. The robot was equipped with two magnetometers, and one active radio range sensor. Three radio beacons were installed around the area where the robot moves.

Fig. 1 shows the surface plot of the indoor magnetic fields constructed by the magnetic SLAM. We can see that the norm value $||m||$ has more power to discriminate the location ambiguities than other magnetic fingerprints.

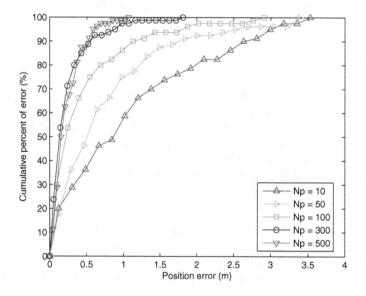

Fig. 2. MCL relocation results with different particle size

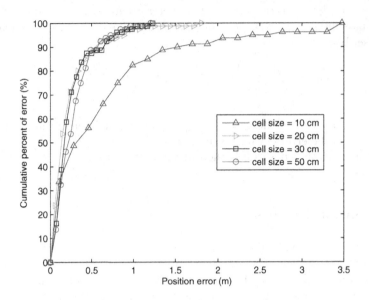

Fig. 3. MCL relocation results with different cell size

Figs. 2 and 3 show the MCL relocation results with different parameters. Obviously, the MCL performance is better with a lager particle size. In Fig. 3, we can also see that the performance is rapidly degraded with a smaller cell size less than 10 *cm*. This can be explained by the fact that with a smaller cell it is more difficult to gather the enough samples to describe the statistical values for the fingerprints.

5   Conclusion

In this paper we proposed a method for solving both the SLAM and the relocation problem. We exploited magnetic field anomalies for the SLAM landmark and relocation fingerprints. The magnetic SLAM was realized in the RBPF and bilinear interpolation-based SLAM framework. For the relocation, additional radio source was also applied to solve the ambiguity problem related to the magnetic only fingerprints. Simulation results with real-world data validated the performance of the MCL-based relocation algorithms.

Acknowledgment. This research was supported by the MKE(The Ministry of Knowledge Economy), Korea, under the Human Resources Development Program for Convergence Robot Specialists support program supervised by the NIPA(National IT Industry Promotion Agency) (NIPA-2012-H1502-12-1002).

References

1. Thrun, S., Burgard, W., Fox, D.: Probabilistic Robotics. The MIT Press, Cambridge (2005)
2. Fernandez-Madrigeal, J., Claraco, J.L.: Simultaneous Localization and Mapping for Mobile Robots: Introduction and Methods. Information Science Reference. Hershey, PA (2013)
3. Gutmann, J., Fong, P., Munich, M.E.: Localization in a Vector Field Map. In: IEEE International Conference on Intelligent Robots and Systems, pp. 3144–3151 (2012)
4. Lee, H., Jung, J., Choi, K., Park, J., Myung, H.: Fuzzy-logic-assisted interacting multiple model (FLAIMM) for mobile robot localization. Robot. Auton. Syst. 60(12), 1592–1606 (2012)
5. Gozick, B., Subbu, K., Dantu, R., Maeshiro, T.: Magnetic Maps for Indoor Navigation. IEEE Trans. Instrum. Meas. 60(12), 3883–3891 (2011)
6. Chung, J., Donahoe, M., Schmandt, C., Kim, I., Razavai, P., Wiseman, M.: Indoor Location Sensing Using Geo-magnetism. In: The 9th International Conference on Mobile Systems, Applicatins, and Services, pp. 141–154 (2011)
7. Li, B., Gallagher, T., Dempster, A., Rizos, C.: How Feasible is the Use of Magnetic Field Alone for Indoor Positioning? In: Proc. IEEE International Conference on Indoor Positioning and Indoor Navigation, pp. 1–9 (2012)
8. Bilke, A., Sieck, J.: Using the Magnetic Field for Indoor Localisation on a Mobile Phone. In: Krisp, J. (ed.) Lecture Notes in Geoinformation and Cartography, pp. 195–208. Springer, Heidelberg (2013)
9. Montemerlo, M., Thrun, S.: A Scalable Method for the Simultaneous Localization and Mapping Problem in Robotics. Springer, New York (2007)
10. Gutmann, J., Eade, E., Fong, P., Munich, M.: Vector Field SLAM–Localization by Learning the Spatial Variation of Continuous Signals. IEEE Trans. Robot. 28(3), 650–667 (2012)
11. Jung, J., Myung, H.: Indoor localization using particle filter and map-based NLOS ranging model. In: Proc. IEEE International Conference on Robotics and Automation, pp. 5185–5190 (2011)

Combined Trajectory Generation and Path Planning for Mobile Robots Using Lattices with Hybrid Dimensionality

Janko Petereit, Thomas Emter, and Christian W. Frey

Fraunhofer Institute of Optronics, System Technologies
and Image Exploitation IOSB, Karlsruhe, Germany
{janko.petereit,thomas.emter,christian.frey}@iosb.fraunhofer.de

Abstract. Safe navigation for mobile robots in unstructured and dynamic environments is still a challenging research topic. Most approaches use separate algorithms for global path planning and local obstacle avoidance. However, this generally results in globally sub-optimal navigation strategies. In this paper, we present an algorithm which combines these two navigation tasks in a single integrated approach. For this purpose, we introduce a novel search space, namely, a state × time lattice with hybrid dimensionality. We describe a procedure for generating high-quality motion primitives for a mobile robot with four-wheel steering to define the motion in this lattice. Our algorithm computes a hybrid solution for the path planning problem consisting of a trajectory (i.e., a path with time component) in the imminent future, a dynamically feasible path in the near future, and a kinematically feasible path for the remaining time to the goal. Finally, we provide some results of our algorithm in action to prove its high solution quality and real-time capability.

Keywords: mobile robot motion planning, hybrid-dimensional planning, state lattice planner.

1 Introduction

A common usage scenario for autonomous mobile robots is the support of rescue teams after natural disasters or industrial accidents. Robots can assist by exploring and mapping the disaster area, acquiring important environmental data, searching for victims or simply carrying heavy loads. This task is characterized by a mostly unknown and unstructured environment, which is highly dynamic due to human rescue workers and other rescue vehicles operating in the close vicinity of the robot.

In order to efficiently accomplish the mission while at the same time moving safely between the (possibly dynamic) obstacles in such an environment, the robot needs fast and high-quality path planning as well as a strategy for reliable obstacle avoidance. In the last few years mainly two approaches for global path planning that consider the kinematic constraints of car-like robots have emerged for planning in unstructured environments. The first one combines continuous

J.-H. Kim et al. (eds.), *Robot Intelligence Technology and Applications 2,*
Advances in Intelligent Systems and Computing 274,
DOI: 10.1007/978-3-319-05582-4_14, © Springer International Publishing Switzerland 2014

motion primitives with a discrete search space by spanning a tree of continuous states, thus constituting a hybrid search space (Hybrid A* [1]). The second approach utilizes specifically constructed motion primitives which cause the reachable set to form a lattice structure in the robot's state space [2]. Therefore, the graph search, although using motion primitives which represent *continuous* motion, operates in a *discrete* search space.

In their original versions these global path planners focused merely on the search of *kinematically* feasible paths. However, for the application to autonomous vehicles, it is favorable to guarantee that the found solutions are also *dynamically* feasible. This naturally results in a search space with increased dimensionality, which makes the path planning even more challenging and complex. In the last two decades a lot of effort went into the development of algorithms to tackle this increased complexity. Two examples are the well-known Probabilistic Roadmaps [3] and Rapidly-exploring Random Trees (RRTs) [4]. Although both algorithms are probabilistically complete and have been applied to mobile robots in the past, they have some disadvantages for this particular application area. Probabilistic Roadmaps lose their benefit of precomputed roadmaps of complex configuration spaces in a rapidly changing environment, which is the case in the presence of dynamic obstacles. RRTs generally provide non-optimal solutions and it is very hard to incorporate additional information (like terrain quality or traversal risk) into the search. To overcome these limitations, particularly with regard to mobile robot applications, state lattices have been successfully extended to search spaces of higher dimensionality to allow for a planning of dynamically feasible maneuvers [5]. However, the algorithm proposed in [5] plans a completely dynamically feasible maneuver from the start to the goal, which is generally not necessary and thus may waste valuable computation time. For this purpose, we propose a novel algorithm that relaxes the accuracy requirements of the motion plan with increasing time while still guaranteeing dynamically feasible motions in the close future.

1.1   Problem Statement and Proposed Solution

Especially in a highly dynamic environment (e.g., in the presence of human rescue workers), it is important to include these dynamic obstacles already in the global path planning as an unaware global path planner followed by a subsequent local obstacle avoidance would generally lead to sub-optimal driving strategies.

In this paper, we present a novel algorithm, which combines the trajectory generation (i.e., planning in state × time space) with global path planning. It exploits the fact that the requirements imposed on the accuracy of the planned robot motion decrease the more it extends into the future. For this purpose, the dimensionality of the search space is successively reduced: The search starts in the full-dimensional state × time space for the immediate future (thus generating time-parametrized trajectories), then continues through still dynamically feasible maneuvers, and finally considers merely kinematically feasible paths in the far future. For distant regions an even further reduction could be made by also dropping the kinematic feasibility and reducing to a simple grid search.

However, we deliberately refrain from this option as we impose some minimum requirements on the resulting path (namely, kinematic feasibility).

1.2   Related Work

There are some recent research results closely related to our approach. For example, Ziegler and Stiller [6] used spatiotemporal lattices for planning on-road driving maneuvers. However, this algorithm is not suitable for planning in unstructured environments. In [7] a hybrid approach which considers time during planning for a specific time horizon is proposed, but after reaching this point in time, it reduces to a simple 2D grid search not considering dynamic or kinematic constraints any more. In past research, we have extended this approach to at least guarantee the kinematic feasibility of the resulting motion plan [8], and proposed a multi-resolution concept to speed-up the hybrid-dimensional planning algorithm [9]. However, all these previous research results share the drawback of an abrupt transition from full-dimensional state × time states to only kinematically feasible states. The contribution of this paper is to fill this gap by introducing a concept of successive dimensionality reduction in order to lower the fidelity of the robot's state representation in a more gradual way. This will allow for a better trade-off between planning quality and computational performance.

The general idea of path planning in a search space with adaptive dimensionality has been addressed in [10], however, it is restricted to mere *path* planning – *trajectories* are not considered.

2   Algorithm

The presented algorithm consists mainly of the following three steps, of which steps 1 and 2 can be precomputed off-line.

1. Construction of a state × time lattice L_0 with full dimensionality.
2. Repeated projection of the state × time lattice L_0 to state lattices with lower dimensionality: $L_1, L_2, \ldots, L_{\max}$.
3. Graph search in the generated state lattices, starting in L_0 and weaving through the state lattices with decreasing dimensionality. Special edges connect lattices of subsequent dimensionality to allow for transitions from L_k to L_{k+1}.

Throughout this paper, we will explain the algorithm using the example of a mobile robot, whose dynamics can be described using a general nonlinear system model

$$\dot{\boldsymbol{x}} = \boldsymbol{f}(\boldsymbol{x}, \boldsymbol{u}) \tag{1}$$

where the state $\boldsymbol{x}$ consists of the robot's position x and y, its orientation θ, and its translational velocity v. The input $\boldsymbol{u}$ contains the commanded acceleration u_a and the steering angle u_β.

However, our approach is not restricted to this specific system model, which is why we do not go into detail here. In fact, it can be applied to a wide range of robotic systems which shall be empowered to act in a dynamic environment.

2.1   Motion Primitive Generation

State lattice planners are commonly based on a set of motion primitives which span the search space. A motion primitive is a short time driving strategy which connects a state s to a succeeding state s' in the robot's state space. By storing the associated inputs that are needed to drive the system from s to s' along with each motion primitive, an overall driving strategy can be reconstructed from the motion primitives that form the solution of the trajectory/path planning problem.

There is a vast variety of approaches for the construction of these motion primitives. Bicchi et al. have shown in [11] which conditions a system and its inputs must satisfy so that its reachable set forms a lattice. However, for mobile robot applications this generally has the disadvantage of non-uniform heading discretization. Rufli and Siegwart overcome this limitation by bending the state lattice towards an a priori known path [12] but this approach is not well suited for heavily unstructured environments. Pivtoraiko et al. [2] use the approach presented by Kelly and Nagy [13], which is based on curvature polynomials, which approximate the vehicle motion.

In [8] we have shown that in the case of a simplified system model of a mobile robot with four-wheel steering the integration of the system of differential equations (1) can be done analytically for constant inputs u_a and u_β. This allows for a very efficient simulation of robot trajectories which are the basis for acquiring the needed motion primitives. In this section we will recap from [8] the procedure for constructing a high dimensional state × time lattice by sampling high-dimensional motion primitives, however, we will put it on a more formal basis in order to be able to conveniently derive the gradual dimensionality reduction in the following sections and to enable the smooth integration in the multi-resolution concept that we have proposed in [9].

First, the desired quantization of each dimension of the full-dimensional state × time lattice

$$L_0 = X \times Y \times \Theta \times V \times T \qquad (2)$$

has to be set. Although, in principle, these could be chosen arbitrarily, the final choice has a huge impact on the outdegree of the lattice points. Furthermore, as we require only "translational invariance" of a motion primitive with respect to the x, y, and t dimensions, the remaining θ and v dimensions may be discretized in a non-equidistant way, which is especially useful for the v dimension. The consequence of the fact that motion primitives are only "translationally invariant" with respect to the x, y, and t dimensions is that it is necessary to sample individual motion primitive sets for states that start at different θ and v configurations. Because of its shape, we call the set $B(\theta, v)$ of all motion primitives that originate from an identical start configuration a *bunch*.

For each $\theta \in \Theta$ and $v \in V$ the system model (1) is now used to run a large number of simulations of the robot's motion for randomly sampled inputs u_a and u_β in order to generate the motion primitives that constitute each $B(\theta, v)$. After each time interval Δt, which corresponds to the quantization of the t dimension, we assess the quantization error e_Q of the end point of the motion primitive. The quantization error e_Q is simply calculated as the normalized distance to the closest state in L_0. If e_Q is larger than a given threshold $e_{Q,max}$ (e.g. 5 %), we continue simulating the robot's motion until a simulation horizon t_{max}. If by then e_Q is still too large, we drop the motion primitive. If e_Q is less then $e_{Q,max}$, we check whether another motion primitive exists in the bunch ending at an identical state. If this is the case, we score both motion primitives by a weighted sum of their length and e_Q in order to decide which one to keep.

The union of all bunches that are generated using the above-described procedure constitutes the high-dimensional motion primitive set M_0 that defines the valid state transitions for L_0.

$$M_0 = \bigcup_{\substack{\theta \in \Theta \\ v \in V}} B_0(\theta, v) \tag{3}$$

The index "0" indicates that the bunch/motion primitive set has not been projected to a lower dimensionality so far.

The presented approach is capable to generate and test several million motion primitives per second, which enables the algorithm to quickly generate a set of high-quality motion primitives with small quantization errors e_Q. Furthermore, additional constraints on the state variables (like a larger turning radius for higher velocities) can be easily integrated in our approach. However, for more complex systems a numerical integration of the system of differential equations might be necessary, but as all this is done off-line, computation time is not an issue. Furthermore, a simple stop criterion for this probabilistic sampling approach can be employed by using a performance measure like the RMS of the quantization error e_Q.

2.2   Repeated Projection of High-Dimensional Lattice

The high-dimensional state lattice L_0 can guarantee a good planning quality of the robot motion, however, due to its high number of dimensions it is prone to the curse of dimensionality, which may result in a very poor planning speed. Therefore, we propose a novel concept of successive dimensionality reduction of the search space in order to gradually lower the planning fidelity with increasing time. For this purpose, we start by projecting the full-dimensional state × time lattice L_0 onto its $X \times Y \times \Theta \times V$ subspace in order to define a new state lattice

$$L_1 = X \times Y \times \Theta \times V . \tag{4}$$

This essentially means that the corresponding set of – still dynamically feasible – motion primitives M_1 is generated by projecting the set M_0 of motion primitives corresponding to L_0 onto L_1. This is done for each motion primitive in each bunch separately. Although an arbitrarily chosen projection rule would be possible, we perform the projections by simply dropping the t dimension.

The projection process generally results in multiple motion primitives ending in an identical state. For these cases, we assign a cost to each of them using the same cost function as the subsequent graph search will use, and drop all motion primitives except the one with the least cost. Furthermore, the motion primitive which results from the "wait" action in M_0 (i.e, the robot does not move at all) is removed from M_1 because of its identical start and end state. Although the motion primitives of

$$M_1 = \bigcup_{\substack{\theta \in \Theta \\ v \in V}} B_1(\theta, v) \tag{5}$$

do not contain an explicit time component any more, the duration for executing a motion primitive is still stored to allow the graph search to incorporate this information into the cost function.

Transitions from L_0 to L_1 are defined by connecting all states $(x, y, \theta, v, t) \in L_0$, $t > t_0$ with the states reachable by the motion primitive $m \in M_1$ that starts at the corresponding projected state $(x, y, \theta, v) \in L_1$. The parameter t_0 determines for which time horizon dynamic obstacles should be considered during the planning. This threshold can be either a fixed value or computed adaptively depending on the robot motion and the prediction of the motion of the dynamic obstacles (cf. [7]).

In the following steps, the state lattices are repeatedly projected onto a subspace to successively reduce the order of the system of differential equations associated with the state lattice until it finally contains no more dynamic components but still maintains kinematic feasibility of the motion. For our given exemplary motion model this is already the case for one further projection of L_1, which contains (x, y, θ, v) states, onto the $X \times Y \times \Theta$ space, thus generating the state lattice L_2. The corresponding set of motion primitives M_2 is constructed in a similar manner. However, as the v dimension is dropped during the projection, the new bunches that constitute M_2 are only dependent on the starting θ.

$$M_2 = \bigcup_{\theta \in \Theta} B_2(\theta) \tag{6}$$

Again, appropriate transitions from L_1 to L_2 have to be defined, which are generated as needed during the graph search. For this purpose, each node stores the time of it being first visited during the search. If this time exceeds a given threshold t_1, a transition from L_1 to L_2 is inserted defined by the motion primitive $m \in M_2$ which starts at the projection of the corresponding state $(x, y, \theta, v) \in L_1$ onto $(x, y, \theta) \in L_2$.

At this stage, the maximum depth of projections is reached as a further projection (e.g. onto the $X \times Y$ sub-space) would discard the kinematic feasibility of the representable motion. Thus, for our example, $L_2 = L_{\max}$ follows.

Due to the repeated projections of the state ($\times$ time) spaces, it is obvious that the goal region for the subsequent graph search must be defined using the

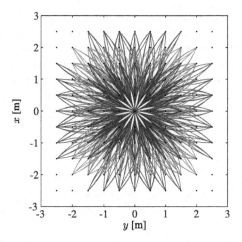

Fig. 1. Set of motion primitives $M_2 = M_{\max}$, $|M_2| = 812$, average outdegree: 17.3, average length: 2.02 m

farthest projected state lattice $L_{\max}$. Consequently, the graph search can easily check an expanded node that belongs to any lattice $L_0, L_1, \ldots, L_{\max}$ by simply projecting its corresponding state onto $L_{\max}$ and testing if it lies in the specified goal region.

3   Graph Search

The search space is completely defined by the procedure described in the previous section. The corresponding graph is constructed during the search as needed using the motion primitive sets $M_0, M_1, \ldots M_{\max}$.

For finding the optimal trajectory/path through the sequence of lattices, standard algorithms for finding shortest paths in graphs can be employed. As on the one hand especially the first seconds of the search space are relatively high-dimensional but on the other hand a fast generation of (possibly sub-optimal) trajectories is required to allow a safe navigation in the presence of dynamic obstacles, anytime graph search algorithms are particularly well suited for this type of problem.

For our implementation of the presented algorithm, we chose to use the ARA* algorithm (Anytime Repairing A*, [14]). It first starts a weighted A* search to find a valid, but possibly ϵ-sub-optimal solution. For this purpose, it uses a heuristic, which is inflated by a factor ϵ, to determine the order of the node expansion. If a solution is found and there is still enough computation time left, the search starts again using a decreased inflation factor ϵ. This step may be repeated several times. To speed up planning, ARA* exploits intermediate results from the previous iteration. The integration of such an anytime graph search algorithm with our sequence of lattices is straightforward.

As especially in a highly dynamic environment a fast replanning is very important, it would be interesting to explore the potential of employing an explicit replanning algorithm like D* Lite [15]. However, this is a rather challenging task because all these replanning algorithms generally search backwards from the goal to the robot's current position to make use of their inherent advantages. This conflicts with our approach, which uses the accumulated time of a node since the start to determine the transition between two lattices. Of course, this time is not available when searching backwards.

4 Results

We implemented the individual components of the presented algorithm in C++ and evaluated them on an Intel® Xeon® E3-1270 CPU using both simulated and real data.

4.1 Construction of State (× Time) Lattices

The choice of an optimal quantization of the state × time lattice turned out to be a challenging task. On the one hand, a relatively fine discretization is desirable to enable planning of near optimal paths and to guarantee completeness of the search. On the other hand, the outdegree of each node increases with higher resolutions, which slows down the graph search because of the higher branching factor.

After a thorough look at the tradeoffs, we finally chose the following quantization. For the discretization of x and y we chose $\Delta x = \Delta y = 0.5$ m. To enable sufficiently smooth paths, we allow 16 discrete orientations, which corresponds to a heading resolution $\Delta\theta = 22.5°$. For the discretization of the velocity, we exploited the fact that the quantization may be done in a non-equidistant way, thus, we chose the set $\{-2\frac{m}{s}, 0\frac{m}{s}, 3\frac{m}{s}\}$ to represent the admissible discrete velocities. The duration of the time increment Δt is set to 0.5 s and the maximum duration t_{max} of a motion primitive is limited to $2\Delta t = 1$ s.

With this quantization setup, we are able to generate and test roughly 8 million motion primitives per second. This part of our algorithm profits enormously from parallelization as each motion primitive can be generated independently. In order to obtain a high-quality set of motion primitives, it is sufficient to run the generation process for one minute. The generated set M_0 consists of 1064 motion primitives with an average length of 1.96 m and an average outdegree of 22.6. M_1, which results from the projection of M_0 onto L_1, contains the same number of motion primitives as M_0 does. This implies that no end states of any two motion primitives from M_0 share the same (x, y, θ, v) components. This is due to the chosen quantization in conjunction with the short maximum motion primitive duration of $2\Delta t = 1$ s.

The subsequent projection of M_1 onto L_2 results in the expected reduction of the motion primitive count. A total of 814 motion primitives remains in $M_2 = M_{max}$ (see Fig. 1). Consequently, the average outdegree decreases to 17.3, the average length increases slightly to 2.02 m.

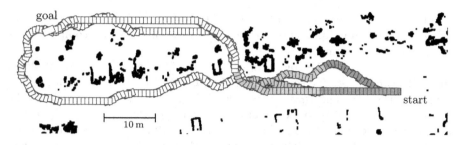

Fig. 2. Three iterations of the ARA* algorithm. First pass (blue) with $\epsilon = 2.0$ (computation time 6 ms), second pass (red) with $\epsilon = 1.3$ (computation time 47 ms), and third pass (green) with $\epsilon = 1.0$ (computation time 156 ms). The robots shape is shown every 0.25 s. The dark parts of the hybrid trajectory/path consist of states from L_0, i.e., they were planned in the $X \times Y \times \Theta \times V \times T$ space. The medium-dark patches belong to L_1 (i.e., the $X \times Y \times \Theta \times V$ space), and, finally, the light patches result from planning in $L_2 = L_{\max}$, i.e., the $X \times Y \times \Theta$ space.

4.2 ARA* Graph Search

The integration of the sequence of lattices with the ARA* algorithm is straightforward. As expected, the anytime nature of this algorithm is beneficial for this particular application. It has the ability to quickly find a hybrid trajectory/path to initiate an obstacle avoidance maneuver if it encounters a critical situation like an upcoming collision with a dynamic obstacle in the vicinity of the robot. However, this might come at the expense of the optimality of the solution. If there is some computation time left, the path can be successively improved. Fig. 2 shows three ARA* iterations for planning in an unstructured environment which has been mapped by laser scanners.

From Fig. 2 also our concept for planning in sequences of lattices with variable dimensionality becomes clear. The dark tiles of each solution represent the *trajectory* part of the solution, i.e., they consist of states from L_0, which explicitly contain a time dimension. This full-dimensional trajectory extends to the point $t_0 = 5\,\text{s}$ in time for the shown example. The subsequent medium-dark tiles originate from planning in L_1, thus, still representing a dynamically feasible motion in the $X \times Y \times \Theta \times V$ space. At the time $t_1 = 10\,\text{s}$ the solution transitions to a pure *path*, which comprises only (x, y, θ) states; however, it is still kinematically feasible (white tiles).

Overall, the dynamically feasible part of the solution extends to the time t_1, which results in planning a smooth trajectory in a relatively large vicinity of the robot. However, due to the quantization of the heading dimension, it might nonetheless be advisable to employ an additional trajectory smoothing for the very first seconds of the path.

To focus the graph search towards the goal, we used the following very simple heuristic.

$$h(x, y) = \|(x_{\text{goal}}, y_{\text{goal}}) - (x, y)\| \left(1 + \frac{\lambda_t}{v_{\max}}\right) \tag{7}$$

It is the sum of the Euclidean distance and the estimated time to reach the goal using the maximum admissible velocity. The ratio of these two components is controlled by the parameter λ_t.

To gain a sense for the computational effort of our algorithm, we simulated the path planning for the scenario depicted in Fig. 2 for different values of the transition time t_0. The time parameter t_1, i.e., the transition time from the dynamically feasible to the kinematically feasible phase, was set to $t_1 = 2t_0$. Fig. 3 shows the results. It can be seen, that according to the measured times, an optimal ($\epsilon = 1$) motion planning with 10 Hz is possible for transition times $t_0 \leq 3$ s and $t_1 = 2t_0$.

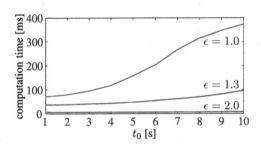

Fig. 3. Computation time for the trajectory/path planning of the scenario depicted in Fig. 2. The transition time t_1 is set to $2t_0$.

4.3   Dynamic Obstacles

The most interesting result is the behavior of our algorithm in the presence of dynamic obstacles because, after all, the safe operation in dynamic environments was our main motivation for the development of the presented algorithm on the basis of a search space with variable dimensionality. In the following, we will describe two exemplary scenarios for such a dynamic environment. The prediction of the dynamic obstacles is done using the methods described in [8], which model obstacles in a probabilistic way in order to incorporate a measure for the collision risk into the cost function during the graph search.

In the first scenario (see Fig. 4), the robot arrives at an intersection to its left, but it cannot turn left immediately because of an oncoming dynamic obstacle. Therefore, the robot plans an avoidance maneuver and positions itself right above the intersection to enter it as soon as the obstacle has passed. All this is possible because the robot explicitly considers time (and, thus, also the predicted position of dynamic obstacles) during the planning for the first t_0 seconds. In our implementation of the algorithm, we set this transition time to a fixed value of $t_0 = 5$ s, however, as already mentioned, it would be also possible to choose t_0 depending on the predicted motion of the dynamic obstacles by using an approach similar to the one described in [7]. From this scenario, it is particularly apparent, that the local obstacle avoidance must not be decoupled from the global path planning. If this had been the case, the robot would probably have

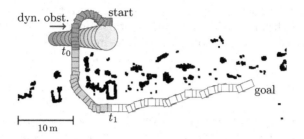

Fig. 4. Planning of a turning maneuver (green) with oncoming traffic (red). Again the dark-green tiles represent the *trajectory* part of the solution, the medium-dark tiles the dynamically feasible part, and the light tiles the kinematically feasible path. The color gradient of the dynamic obstacle directly encodes the time (lightness increases with time).

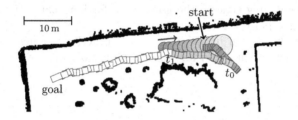

Fig. 5. Robot (green) with oncoming traffic (red). The robot backs off, lets the dynamic obstacle pass, and proceeds. Again the dark-green tiles represent the *trajectory* part of the solution, the medium-dark tiles the dynamically feasible part, and the light tiles the kinematically feasible path. The color gradient of the dynamic obstacle directly encodes the time (lightness increases with time).

missed the intersection leading to an unnecessary detour and thus sub-optimal solution of the path planning problem. Also note, how – in addition to the initial full-dimensional trajectory till t_0 – planning a still dynamically feasible motion for the interval $[t_0, t_1]$ leads to high smoothness of the planned motion till t_1.

The second scenario (see Fig. 5) shows a more complex maneuver. Initially, the robot is oriented towards the goal on the left when it encounters an oncoming dynamic obstacle. As the passage is too narrow for both vehicles, the robot backs off, lets the obstacle pass, and then proceeds towards the goal. The whole maneuver is planned consistently only using the presented algorithm; no additional rules or logic are necessary. Nevertheless, it takes only 43 ms to compute this relatively complex maneuver (even the optimal solution with $\epsilon = 1$, and including the probabilistic prediction of the dynamic obstacle). This is only moderately more than planning a mere path (i.e., $t_0 = t_1 = 0$), which takes 12 ms; however, of course, the mere path would not be able to cope with the dynamic obstacle. On the other hand, planning a full-dimensional trajectory from the start to the goal, would take about 1500 ms, which is clearly infeasible for real-time applications. These benchmarks demonstrate the capability of our algorithm to significantly reduce the computation times while still guaranteeing high-quality solutions for the close future of the motion plan.

5   Conclusions

In this paper, we have presented a novel algorithm for the integrated global path planning and local dynamic obstacle avoidance. For this purpose, we used a state × time lattice with hybrid dimensionality. The algorithm can efficiently generate the required motion primitives using a probabilistic sampling strategy. We applied the ARA* algorithm to quickly find an initial solution for the planning problems in order to safely avoid collisions with moving objects. Our algorithm finds hybrid solutions consisting of a trajectory in the imminent future, a dynamically feasible path in the near future, and a kinematically feasible path for the remaining time to the goal. Even for long paths the computation time of our algorithm is quite moderate.

In future work, we will have a look at the practical application to systems with higher dimensionality and investigate the use of improved heuristics.

References

1. Dolgov, D., Thrun, S., Montemerlo, M., Diebel, J.: Path planning for autonomous vehicles in unknown semi-structured environments. The International Journal of Robotics Research 29(5), 485–501 (2010)
2. Pivtoraiko, M., Knepper, R.A., Kelly, A.: Differentially constrained mobile robot motion planning in state lattices. Journal of Field Robotics 26(3), 308–333 (2009)
3. Kavraki, L.E., Švestka, P., Latombe, J.C., Overmars, M.H.: Probabilistic roadmaps for path planning in high-dimensional configuration spaces. IEEE Transactions on Robotics and Automation 12(4), 566–580 (1996)
4. LaValle, S.M., Kuffner Jr., J.J.: Randomized kinodynamic planning. The International Journal of Robotics Research 20(5), 387–400 (2001)
5. Likhachev, M., Ferguson, D.: Planning long dynamically feasible maneuvers for autonomous vehicles. The International Journal of Robotics Research 28(8), 933–945 (2009)
6. Ziegler, J., Stiller, C.: Spatiotemporal state lattices for fast trajectory planning in dynamic on-road driving scenarios. In: Proceedings of the IEEE/RSJ International Conference on Intelligent Robots and Systems (2009)
7. Kushleyev, A., Likhachev, M.: Time-bounded lattice for efficient planning in dynamic environments. In: Proceedings of the IEEE International Conference on Robotics and Automation (2009)
8. Petereit, J., Emter, T., Frey, C.W.: Safe mobile robot motion planning for waypoint sequences in a dynamic environment. In: Proceedings of the IEEE International Conference on Information Technology (2013)
9. Petereit, J., Emter, T., Frey, C.W.: Mobile robot motion planning in multi-resolution lattices with hybrid dimensionality. In: Proceedings of the IFAC Intelligent Autonomous Vehicles Symposium (2013)

10. Gochev, K., Cohen, B., Butzke, J., Safonova, A., Likhachev, M.: Path planning with adaptive dimensionality. In: Proceedings of the Symposium on Combinatorial Search (2011)
11. Bicchi, A., Marigo, A., Piccoli, B.: On the reachability of quantized control systems. IEEE Transactions on Automatic Control 47(4), 546–563 (2002)
12. Rufli, M., Siegwart, R.: On the design of deformable input- / state-lattice graphs. In: Proceedings of the IEEE International Conference on Robotics and Automation (2010)
13. Kelly, A., Nagy, B.: Reactive nonholonomic trajectory generation via parametric optimal control. The International Journal of Robotics Research 22(7-8), 583–601 (2003)
14. Likhachev, M., Gordon, G., Thrun, S.: ARA*: Anytime A* search with provable bounds on sub-optimality. In: Proceedings of the Conference on Neural Information Processing Systems (2003)
15. Koenig, S., Likhachev, M.: Fast replanning for navigation in unknown terrain. IEEE Transactions on Robotics and Automation 21(3), 354–363 (2005)

Organization and Selection Methods of Composite Behaviors for Artificial Creatures Using the Degree of Consideration-Based Mechanism of Thought

Woo-Ri Ko and Jong-Hwan Kim

Department of Electrical Engineering, KAIST, 291 Daehak-ro, Yuseong-gu, Daejeon, 305-701,
Republic of Korea
{wrko,johkim}@rit.kaist.ac.kr

Abstract. This paper proposes organization and selection methods of composite behaviors for artificial creatures. Using the degree of consideration-based mechanism of thought (DoC-MoT), each pre-defined atom behavior is evaluated by the fuzzy integral of the partial evaluation values of atom behaviors over the artificial creature's wills and external contexts, with respect to the fuzzy measure values representing its degrees of consideration (DoCs). Based on these evaluation values of atom behaviors, a composite behavior is organized as a set of atom and composite behaviors which are connected by the relationships of 'parallel,' 'choice' and 'sequence.' However, in the organized composite behavior, the behaviors connected by 'choice' relationship can not be generated at the same time, and therefore, only one atom or composite behavior is randomly remained in each set of atom or composite behaviors connected by 'choice' relationship. The effectiveness of the proposed scheme is demonstrated by simulations carried out with an artificial creature, "DD" in the 3D virtual environment. The results show that the diversity of the generated behaviors is increased fourfold compared to the behavior selection without the organization process of composite behaviors. Moreover, the generated composite behaviors satisfy the artificial creature's wills more and the logical connectivity of them is increased compared to the method without the process.

Keywords: Composite behavior, artificial creature, behavior selection, fuzzy integral, fuzzy measure.

1 Introduction

An artificial creature needs an intelligent behavior selection method to be used as an entertainment robot or an intermediate interface for interaction with users [1]. In other words, the generated behaviors from the artificial creature should be the ones provided from the human-like mechanism of thought considering both its internal wills and external contexts. Besides, they should show various series of behaviors even in the same situation to get and hold a user's attention. For this purpose, there has been much research on the behavior selection method for artificial creatures. The architectures consisting of perception, motivation, behavior and actuator modules were proposed for behavior selection [2]-[4]. A behavior selection method for entertainment robots was proposed

J.-H. Kim et al. (eds.), *Robot Intelligence Technology and Applications 2*,
Advances in Intelligent Systems and Computing 274,
DOI: 10.1007/978-3-319-05582-4_15, © Springer International Publishing Switzerland 2014

using intelligence operating architecture [5], [6]. However, in the previous research, there was a limit to the variety of generated behaviors, since a behavior was selected among pre-defined list of behaviors.

In this paper, organization and selection methods of composite behaviors are proposed for artificial creatures. Using the degree of consideration-based mechanism of thought (DoC-MoT), each pre-defined atom behavior is evaluated by the fuzzy integral of the partial evaluation values of atom behaviors over the artificial creature's wills and external contexts, with respect to the fuzzy measure values representing the degrees of consideration (DoCs) or preference defined by a user [7]. Based on these evaluation values of atom behaviors, a composite behavior is organized as a set of atom and composite behaviors which are connected by the relationships of 'parallel,' 'choice' and 'sequence.' However, in the organized composite behavior, the behaviors connected by 'choice' relationship can not be generated at the same time, and therefore, only one atom or composite behavior is randomly remained in each set of atom or composite behaviors connected by 'choice' relationship. The finalized composite behavior is called a definite composite behavior. To show the effectiveness of the proposed scheme, simulations are carried out with an artificial creature, "DD" in the 3D virtual environment.

This paper is organized as follows. Section 2 presents the degree of consideration-based mechanism of thought (DoC-MoT), which is a well-modeled mechanism of human thought. Section 3 proposes the organization and selection methods of composite behaviors using the DoC-MoT. Section 4 presents the simulation results to demonstrate the effectiveness of the proposed scheme. The concluding remarks follow in Section 5.

2 Degree of Consideration-Based Mechanism of Thought (DoC-MoT)

The DoC-MoT is inspired by the basic concept of 'confabulation theory' that explains how a human brain functions [8]. The human brain is composed of approximately 100 billion neurons and the sets of adjacent neurons represent input or target symbols, such as "red," "round," as input symbols and "apple" as a target symbol. The link between the input and target symbols represents a knowledge about the perceived entities, e.g. the "likeliness" of an "apple" to be "red" or the "likeliness" of an "apple" to be "round." Therefore, a link between the input and target symbols is called knowledge link and its strength describes the degree of belief about the target symbol. The cognitive information processing is accomplished by selecting a target symbol with the highest degree of belief when some input symbols are perceived.

Fig. 1 shows the cognitive information processing in the DoC-MoT. In the DoC-MoT, the degree of belief $I(z)$ of target symbol z over the input symbols $X = \{x_1, x_2, \ldots, x_n\}$, where n is the number of input symbols, is calculated by the Choquet fuzzy integral as follows [9]:

$$I(z) = \sum_{i=1}^{n} g(A_i)\{h(x_i) - h(x_{i-1})\},$$
(1)

where the input symbol set X is sorted so that $h(x_i) \geq h(x_{i+1}), i = \{1, \ldots, n-1\}$ and $h(x_0) = 0$, $g(A_i)$ is the fuzzy measure value of A_i, $A_i = \{x_i, x_{i+1} \ldots, x_n\}$ is the subset

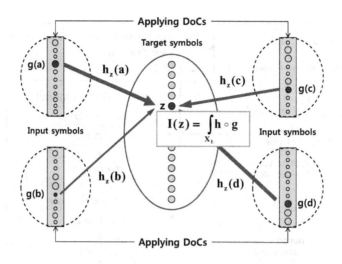

Fig. 1. Cognitive information processing in the DoC-MoT

of X, and $h(x_i)$ is the partial evaluation value of the target symbol z over the ith input symbol x_i. Note that the value of $I(z)$ also represents the evaluation value of z to be a conclusion of the information-processing.

The fuzzy measure value of a subset of input symbols is calculated by the Sugeno λ-fuzzy measure as follows [10]:

$$g(A \cup B) = g(A) + g(B) + \lambda g(A)g(B), \tag{2}$$

where $g(A)$ and $g(B)$, $A, B \subset X$ represent the degrees of consideration of the subsets A and B, respectively, and $\lambda \in [-1, +\infty]$ denotes an interacting degree index. If the two subsets A and B have negative (positive) correlation, (2) becomes a plausible (belief) measure and λ is a positive (negative) value so that $g(A \cup B) < g(A) + g(B)$ ($g(A \cup B) > g(A) + g(B)$). If the two subsets are independent, (2) becomes a probability measure and the value of λ is zero so that $g(A \cup B) = g(A) + g(B)$. Another interaction degree index ξ, which is scaled to be in $[0, 1]$, is employed to efficiently calculate the fuzzy measure values.

3   Organization and Selection Methods of Composite Behaviors

In this section, organization and selection methods of composite behaviors using the DoC-MoT is described. Fig. 2 shows the overall architecture of the proposed scheme. In the internal state and context modules, the strengths of input symbols on will and context are updated, respectively. The memory module stores all the necessary memory contents including the normalized weights of input symbols and the partial evaluation values of atom behaviors over wills and contexts. In the behavior selection module, each atom behavior is evaluated by the fuzzy integral of the partial evaluation values over wills and contexts, with respect to the fuzzy measure values. Based on their evaluation

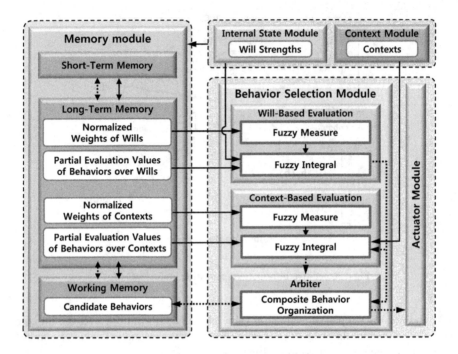

Fig. 2. Overall architecture. The solid arrows denote the movement of data related to wills and contexts, and the dotted arrows denote the behavior recommendation.

values, a proper composite behavior is organized and selected as a set of atom and composite behaviors which are connected by the relationships of 'parallel,' 'choice' and 'sequence.' In the actuator module, the selected behavior is generated through actuators. In the following, the key modules for behavior selection, namely internal state, context and behavior selection modules are described.

3.1   Internal State Module

The internal state module deals with the input symbols on will and updates the strengths of them. Since the human needs are categorized into five levels, i.e. physiological, safety, love and belonging, esteem and self-actualization needs, 15 input symbols on will are defined, as shown in Table 1 [11]. The will strength at time t, $\Omega_j(t)$ of the jth will w_j, $j = 1, 2, \ldots, n$, where n is the number of wills, is updated by

$$\Omega_j(t+1) = \Omega_j(t) + \alpha_j(\overline{\Omega_j} - \Omega_j(t)) + S^T \cdot W_j(t) + \sum_i \delta_{ij}(t), \qquad (3)$$

where α_j is the difference gain, $\overline{\Omega_j}$ is the steady-state value, S is the stimulus vector, W_j is the strength vector between stimulus and the jth will and $\delta_{ij}(t)$ is the amount of change of the jth will strength caused by the ith atom behavior of the previous composite behavior. If the ith atom behavior affects positively on the jth will, $\delta_{ij}(t)$ is a

Table 1. Maslow's five human needs and input symbols on will

Maslow's human need	Input symbol
Physiological needs	Fatigue (w_1), Hunger (w_2), Thirst (w_3), Excretion (w_4)
Safety needs	Safety (w_5), Dirt (w_6), Exercise (w_7)
Love and belonging needs	Comrade (w_8), Express (w_9), Play (w_{10})
Esteem needs	Decorate (w_{11}), Pride (w_{12}), Study (w_{13}), Wealth (w_{14})
Self-actualization needs	Self-actualization (w_{15})

positive value. If the ith atom behavior affects negatively on the jth will, $\delta_{ij}(t)$ is a negative value. Note that the calculated strengths of wills are scaled to be in $[0, 1]$ and they are used in the behavior selection module.

3.2 Context Module

In the context module, 35 input symbols for "time", "place" and "object" are defined, as shown in Table 2. In each category, only one input symbol is perceived as context. It means that three input symbols are perceived as context at each time.

Table 2. Input symbols on context

Classification	Input symbol
Time	Morning (c_1), Afternoon (c_2), Evening (c_3), Night (c_4)
Place	Livingroom (c_5), Bedroom (c_6), Kitchen (c_7), Restroom (c_8)
Object	Sofa (c_9), TV - Document (c_{10}), TV - Entertainment (c_{11}), Livingroom table (c_{12}), Chess table (c_{13}), Chess chair 1 (c_{14}), Chess chair 2 (c_{15}), Phone (c_{16}), Door (c_{17}), Flower (c_{18}), Bed (c_{19}), Dress (c_{20}), Seasonal clothes (c_{21}), Book (c_{22}), Diary (c_{23}), Bedroom chair (c_{24}), Radio (c_{25}), Mirror (c_{26}), Food (c_{27}), Kitchen chair 1 (c_{28}), Kitchen chair 2 (c_{29}), Water (c_{30}), Close stool (c_{31}), Basin (c_{32}), Comrade 1 (c_{33}), Comrade 2 (c_{34}), Comrade 3 (c_{35})

3.3 Behavior Selection Module

In the behavior selection module, a proper determinate composite behavior considering both internal wills and external contexts is selected by the following procedure. First of all, each atom behavior in Table 3 is evaluated by the fuzzy integral of the partial evaluation values over wills and contexts, with respect to the fuzzy measure values. Then, based on their evaluation values, a composite behavior is organized as a set of atom and composite behaviors which are connected by the relationships of 'parallel,' 'choice' and 'sequence.' Finally, a determinate composite behavior is selected by randomly remaining a behavior in each set of behaviors connected by 'choice' relationships.

Table 3. A list of atom behaviors

Classification	Body part	Atom behavior
Facial movement	Face (head)	Watch TV (b_1^f), Listen radio (b_2^f), Talk (b_3^f), Shout (b_4^f), Sing loudly (b_5^f), Sing softly (b_6^f), Look at (b_7^f), Observe (b_8^f), Mumble (b_9^f), Sermon loudly (b_{10}^f), Sermon softly (b_{11}^f), Look around (b_{12}^f)
Gesture	Upper body	Eat cheerfully (b_1^g), Eat slowly (b_2^g), Drink (b_3^g), Sleep (b_4^g), Nap (b_5^g), Scratch fast (b_6^g), Scratch slowly (b_7^g), Wash quickly (b_8^g), Wash slowly (b_9^g), Close door (b_{10}^g), Put on (b_{11}^g), Chess hard (b_{12}^g), Chess roughly (b_{13}^g), Call (b_{14}^g), Exercise cheerfully (b_{15}^g), Exercise normal (b_{16}^g), Exercise slowly (b_{17}^g), Read hard (b_{18}^g), Read roughly (b_{19}^g), Write hard (b_{20}^g), Write roughly (b_{21}^g), Work hard (b_{22}^g), Work roughly (b_{23}^g), Raise flowers (b_{24}^g), Clean hard (b_{25}^g), Clean normal (b_{26}^g), Clean roughly (b_{27}^g), Make up (b_{28}^g), Wave hand strongly (b_{29}^g), Wave hand softly (b_{30}^g), Shake hand strongly (b_{31}^g), Shake hand softly (b_{32}^g), Hit strongly (b_{33}^g), Hit softly (b_{34}^g), Meditate (b_{35}^g), Pray cheerfully (b_{36}^g), Pray normal (b_{37}^g)
Movement	Lower body	Sit (b_1^m), Walk (b_2^m), Stand (b_3^m), Lie (b_4^m), Kneel (b_5^m), Bow (b_6^m), Follow closely (b_7^m), Follow slowly (b_8^m), Wander cheerfully (b_9^m), Wander normal (b_{10}^m), Wander slowly (b_{11}^m), Kick strongly (b_{12}^m), Kick softly (b_{13}^m), Excrete (b_{14}^m), Urine (b_{15}^m)

Evaluation of Atom Behaviors. The evaluation value $E(b_i^a)$ of the ith atom behavior $b_i^a, i = 1, 2, \ldots, l$ over wills and contexts, where l is the number of atom behaviors, is calculated as follows:

$$E(b_i^a) = E_w(b_i^a) \cdot E_c(b_i^a), i = 1, 2, \ldots, l, \tag{4}$$

where $E_w(b_i^a)$ is the evaluation value of b_i^a over wills and $E_c(b_i^a)$ is the evaluation value of b_i^a over contexts. The evaluation values $E_w(b_i^a)$ and $E_c(b_i^a)$ are calculated by the following Choquet fuzzy integrals:

$$E_w(b_i^a) = \int_{X_w} h_w \circ g_w = \sum_{j=1}^{n} \{h_{ij}^w \cdot \Omega_j(t) - h_{i(j-1)}^w \cdot \Omega_{j-1}(t)\} g(A_w), \tag{5}$$

$$E_c(b_i^a) = \int_{X_c} h_c \circ g_c = \sum_{j=1}^{m} \{h_{ij}^c - h_{i(j-1)}^c\} g(A_c), \tag{6}$$

where $A_w \subset X_w$ and $A_c \subset X_c$ are the subset of will and context symbols, respectively, h_{ij}^w and h_{ij}^c are the partial evaluation values of b_i^a over the jth will and the perceived context, respectively, and $g(A_w)$ and $g(A_c)$ are the fuzzy measure values of A_w and A_c, respectively.

To measure the fuzzy measure value $g(A)$, ϕ_s transformation method is employed [7], [12]. In this method, the fuzzy measure values are calculated using a hierarchy diagram of criteria, i.e. wills and contexts, which represents hierarchical interaction relations among criteria. A fuzzy measure $g(A)$ is identified as follows:

$$g(A) = \phi_s(\xi_R, \sum_{P \subset R} u_P^R),$$ (7)

where R is the root level in the hierarchy diagram, ξ_R is the interaction degree between the criteria sets in the R, ϕ_s is a scaling function [13], and u_Q^P is defined as follows:

$$\phi_s(\xi, u) = \begin{cases} 1, & \text{if } \xi = 1 \text{ and } u > 0 \\ 0, & \text{if } \xi = 1 \text{ and } u = 0 \\ 1, & \text{if } \xi = 0 \text{ and } u = 1 \\ 0, & \text{if } \xi = 0 \text{ and } u < 1 \\ \frac{s^u - 1}{s - 1}, & \text{other cases} \end{cases}$$ (8)

$$u_Q^P = \begin{cases} d_i, \text{where } i \in Q & \text{if } |Q| = 1 \text{ and } i \in A \\ 0 & \text{if } |Q| = 1 \text{ and } i \notin A \\ \phi_s^{-1}(\xi_P, \phi_s(\xi_Q, \sum_{V \subset Q} u_V^Q) \times T_Q^P) & \text{other cases} \end{cases}$$ (9)

where $s = (1 - \xi)^2 / \xi^2$, d_i is the normalized weight (DoC) of the ith criterion, and the value of $\phi_s^{-1}(\xi, r)$ is u, which satisfies $\phi_s(\xi, u) = r$. The conversion ratio T_Q^P from Q to P, is computed as

$$T_Q^P = \frac{\phi_s(\xi_P, \sum_{i \in Q} d_i)}{\phi_s(\xi_Q, \sum_{i \in Q} d_i)},$$ (10)

where P is the upper level set and Q is the lower level set in the hierarchy diagram.

Organization of a Composite Behavior. A composite behavior is organized as a set of atom and composite behaviors which are connected by the relationships of 'parallel ($||$),' 'choice (+)' and 'sequence (;).' For example, a composite behavior b^c, "washing slowly after normally exercising while watching TV or cheerfully exercising while listening radio," is described as

$$b^c = \{(b_1^f || b_{16}^g) + (b_2^f || b_{15}^g)\}; b_9^g.$$ (11)

The conditions to connect two atom or composite behaviors are described in the following.

i) Parallel

Since the two atom behaviors connected by 'parallel' relationship are generated at the same time, the body parts for generating the two behaviors should not conflict with each other. For this purpose, the three best atom behaviors with the highest evaluation values over wills on contexts, b_{best}^f, b_{best}^g and b_{best}^m, are selected in each category. Then, a parallel behavior b^p is organized by connecting every atom behaviors of

$B^p \subset B = \{b^f_{best}, b^g_{best}, b^m_{best}\}$. Therefore, the number of possible parallel behaviors is 7 $(= 2^3 - 1)$. The evaluation value $E(b^p_i)$ of the ith parallel behavior b^p over wills and contexts, is calculated as follows:

$$E(b^p_i) = \frac{\sum_{j \in B^p} E(b^a_j)}{|B^p|},\tag{12}$$

where b^a_j is the jth atom behavior of b^p and $|B^p|$ is the number of atom behaviors of B^p. After that, a parallel behavior with the highest evaluation value is selected as the best parallel behavior b^p_{best}.

ii) Choice

Behaviors which have a similar evaluation value with the best parallel behavior are connected by 'choice' relationship. The similarity $Sim(b^p, b^p_{best})$ between b^p and b^p_{best} is calculated as

$$Sim(b^p, b^p_{best}) = 1 - \frac{E(b^p_{best}) - E(b^p)}{E(b^p_{best})}.\tag{13}$$

Note that, if the similarity value of them is bigger than 0.95, the two atom behaviors are connected by 'choice' relationship, described in Section 4. If the limit of the similarity value is set to be too low, the selected behavior may have no correlation with the artificial creature's wills. If it is set to be too high, the diversity of generated behaviors may be low.

iii) Sequence

An artificial creature usually selects a behavior which is highly related to the strong wills. However, the strengths of other wills may be increased unintendedly. Therefore, a parallel behavior, which can compensate the side effect of the previous one, is connected to the previous one by a relationship of 'sequence.' The effectiveness $Eff(b^p_i, b^p_j)$ between the ith parallel behavior b^p_i and the jth parallel behavior b^p_j is calculated as follows:

$$Eff(b^p_i, b^p_j) = \sum_{k,l} Eff(b^a_{ik}, b^a_{jl}),\tag{14}$$

where $Eff(b^a_{ik}, b^a_{jl})$ is the effectiveness between the kth atom behavior b^a_{ik} of b^p_i and the lth atom behavior b^a_{jl} of b^p_j. The effectiveness $Eff(b^a_k, b^a_l)$ between the kth atom behavior b^a_k and the lth atom behavior b^a_l is calculated as follows:

$$Eff(b^a_k, b^a_l) = -(\Delta^a_k \cdot \Delta^a_l),\tag{15}$$

where $\Delta^a_k = \{\delta^a_{k0}, \delta^a_{k1}, \ldots, \delta^a_{kn}\}$ and $\Delta^a_l = \{\delta^a_{l0}, \delta^a_{l1}, \ldots, \delta^a_{ln}\}$ are the feedback vectors of b^a_k and b^a_l, respectively, n is the number of wills and δ^a_{ki} is the amount of change of the ith will strength caused by b^a_k. Note that if the effectiveness of two parallel behaviors has a positive/negative value, they have a positive/negative effect.

Selection of a Determinate Composite Behavior. In the organized composite behavior, the behaviors connected by 'choice' relationship can not be generated at the same time. Therefore, only one atom or composite behavior is randomly remained in each set of atom or composite behaviors connected by 'choice' relationship. The finalized composite behavior is called a determinate composite behavior.

4   Simulations

4.1   Simulation Setting

To show the effectiveness of the proposed organization and selection methods of composite behaviors, simulations were carried out with an artificial creature, "DD," in the 3D virtual world, as shown in Fig. 3. For a DD, the numbers of wills, contexts and atom behaviors were 15, 35 and 64, respectively, as defined in Tables 1, 2 and 3. Note that the partial evaluation values of behaviors over each will and context, the strength vector between stimulus and wills, the amounts of will strength change by the previous behavior and the interaction degrees among criteria, were initialized by an expert. The generation frequencies of atom and composite behaviors were calculated from the experiment results gathered for one day in the virtual world. A composite behavior was selected in every one minute, and therefore, the total number of generated behaviors was 1,440. To show the performance of the proposed scheme, the generation frequencies of parallel behaviors were calculated and the numbers of different generated behaviors and the generation frequencies of sequence behaviors were compared between the two behavior selection methods each with and without the organization process of composite behaviors [4].

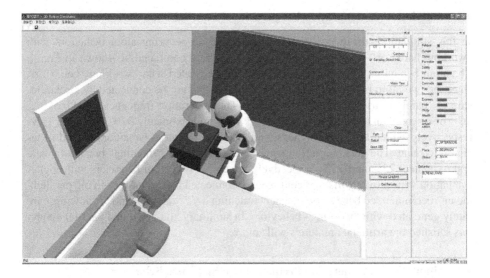

Fig. 3. Screenshot of DD in the 3D virtual world

4.2   Simulation 1: Diversity of Generated Behaviors

In this simulation, the numbers of generated behaviors were compared between the two behavior selection methods each with and without the organization process of composite behaviors, as shown in Table 4. In the method with the process, the numbers of different generated behaviors composed of one, two and three parallel atom behaviors, were 57, 123 and 88, respectively, and the total number of different composite behavior was 268. However, in the method without the process, the maximum number of different generated behaviors is the same as the number of atom behaviors, which is 64. In summary, the diversity of the generated behaviors was increased fourfold compared to the method without the process in which an atom behavior is selected for generation.

Table 4. The number of different generated behaviors

Composed of	In the method without the organization process	In the method with the organization process
One atom behavior	<64	57
Two parallel atom behaviors	0	123
Three parallel atom behaviors	0	88
Total	<64	268

4.3 Simulation 2: Generation Frequencies of Parallel Behaviors

In this simulation, the generation frequencies of co-occuring atom behaviors with "working hard" or "working roughly" behaviors were computed, as shown in Fig. 4. The atom behaviors related to gesture were not generated with "working" behaviors, since the two atom behaviors in the same category can not be connected by a relationship of 'parallel,' as described in Section 3. The generation frequencies of "looking-at," "observing" and "mumbling" behaviors were approximately 31%, 17% and 15%, respectively, which are much higher than those of other atom behaviors related to facial movement. The generation frequencies of "sitting" and "standing" behaviors were approximately 23% and 15%, respectively. Since the two atom behaviors with higher evaluation values over wills and contexts are connected by a relationship of 'parallel,' the un-recommended behaviors, such as "watching TV," "talking" and "following," are rarely generated with "working" behaviors. In summary, the generated parallel behaviors satisfied the artificial creature's wills more.

4.4 Simulation 3: Generation Frequencies of Sequence Behaviors

In this simulation, the generation frequencies of the next atom behaviors to "exercising cheerfully," "exercising normal" or "exercising slowly" behavior were compared between the two behavior selection methods each with and without the organization process of composite behaviors, as shown in Fig. 5. The generation frequencies of "eating," "drinking," "sleeping" and "washing" behaviors were approximately 6%, 24%, 12% and 59% in the method with the process and 2%, 4%, 2% and 4% in the method without the process, respectively. Since the above-mentioned behaviors compensate the side effect of "exercising" behaviors, such as hunger, thirst, fatigue and dirt, they were

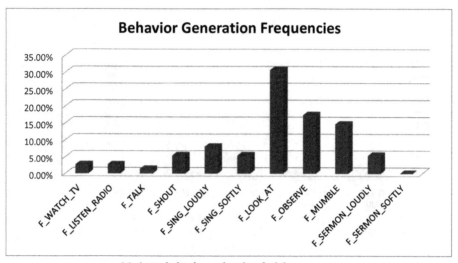

(a) Atom behaviors related to facial movement

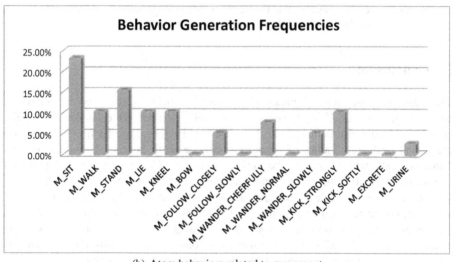

(b) Atom behaviors related to movement

Fig. 4. The generation frequencies of co-occuring atom behaviors with "working" behaviors

connected by "sequence" relationship. In summary, the logical connectivity of a series of generated behaviors is increased compared to the method without the organization process of composite behaviors.

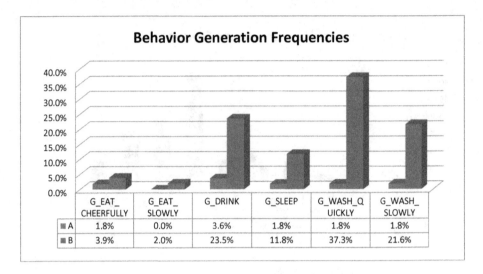

Fig. 5. The generation frequencies of the next atom behaviors to "exercising" behaviors for the two behavior selection methods: A is the behavior selection method without the organization process of composite behaviors and B is the one with the process

5 Conclusion

This paper proposed the organization and selection methods of composite behaviors for artificial creatures. Using the degree of consideration-based mechanism of thought (DoC-MoT), each pre-defined atom behavior was evaluated by the fuzzy integral of the partial evaluation values of behaviors over the artificial creature's wills and external contexts, with respect to the fuzzy measure values representing its degrees of consideration (DoCs). Based on their evaluation values, a proper composite behavior was organized and selected at each time, as a set of atom and composite behaviors which are connected by the relationships of 'parallel,' 'choice' and 'sequence.' The effectiveness of the proposed scheme was demonstrated through the simulations with an artificial creature, "DD," in the 3D virtual environment. The results showed that the artificial creature could generate various behaviors through the proposed behavior selection scheme. Moreover, the generated composite behaviors satisfy the artificial creature's wills more and the logical connectivity of them is increased compared to the atom behavior selection method.

Acknowledgements. This research was supported by the MOTIE (The Ministry of Trade, Industry and Energy), Korea, under the Technology Innovation Program supervised by the KEIT (Korea Evaluation Institute of Industrial Technology)(10045252, Development of robot task intelligence technology that can perform task more than 80% in inexperience situation through autonomous knowledge acquisition and adaptational knowledge application).

This research was also supported by the MOTIE (The Ministry of Trade, Industry and Energy), Korea, under the Human Resources Development Program for Convergence Robot Specialists support program supervised by the NIPA (National IT Industry Promotion Agency)(H1502-13-1001, Research Center for Robot Intelligence Technology).

References

1. Kim, J.-H., Lee, K.-H., Kim, Y.-D., Kuppuswamy, N.S.: Ubiquitous robot: A new paradigm for integrated services. In: Proceeding of IEEE International Conference on Robotics and Automation, pp. 2853–2858 (2007)
2. Arkin, R.C., Fujita, M., Takagi, T., Hasegawa, R.: An ethological and emotional basis for human-robot interaction. Robotics and Autonomous Systems 42(3-4), 191–201 (2003)
3. Breazeal, C.: Social interactions in HRI: The robot view. IEEE Transactions on Systems, Man, Cybernetics - Part C 34(2), 181–186 (2004)
4. Ko, W.-R., Hyun, H.-S., Kim, H.-J., Choi, S.-H., Kim, J.-H.: Behavior Selection Method for Intelligent Artificial Creatures Using Degree of Consideration-based Mechanism of Thought. In: Proc. IEEE International Conference on Systems, Man and Cybernetics (2011)
5. Ko, W.-R., Kim, J.-H.: Behavior Selection Method for Entertainment Robots Using Intelligence Operating Architecture. In: Kim, J.-H., Matson, E., Myung, H., Xu, P. (eds.) Robot Intelligence Technology and Applications. AISC, vol. 208, pp. 75–84. Springer, Heidelberg (2013)
6. Kim, J.-H., Choi, S.-H., Park, I.-W., Zaheer, S.A.: Intelligence Technology for Robots That Think. IEEE Computational Intelligence Magazine 8(3), 70–84 (2013)
7. Kim, J.-H., Ko, W.-R., Han, J.-H., Zaheer, S.A.: The Degree of Consideration-Based Mechanism of Thought and Its Application to Artificial Creatures for Behavior Selection. IEEE Computational Intelligence Magazine 7, 49–63 (2012)
8. Hecht-Nielsen, R.: The Mechanism of Thought. In: Proceedings of International Joint Conference on Neural Networks, pp. 419–426 (2006)
9. Murofushi, T., Sugeno, M.: An interpretation of Fuzzy Measures and The Choquet Integral as an Integral With Respect To a Fuzzy Measure. Fuzzy Sets and Systems 29(2), 201–227 (1989)
10. Sugeno, M.: Theory of Fuzzy Integrals and Its Applications. PhD dissertation, Tokyo Institute of Technology (1974)
11. Maslow, A.H.: A Theory of Human Motivation. Psychological Review (1943)
12. Takagagi, E.: A Fuzzy Measure Identification Method by Diamond Pairwise Comparisons and ϕ_s Transformation. Fuzzy Optimization and Decision Making 7(3), 219–232 (2008)
13. Narukawa, Y., Torra, V.: Fuzzy Measure and Probability Distributions: Distorted Probabilities. IEEE Transactions on Fuzzy System 13(5), 617–629 (2005)

Brainwave Variability Identification
in Robotic Arm Control Strategy

Chiemela Onunka[*], Glen Bright, and Riaan Stopforth

Discipline of Mechanical Engineering, University of KwaZulu-Natal,
Durban, South Africa
205512204@stu.ukzn.ac.za,
{bright,Stopforth}@ukzn.ac.za

Abstract. Neuronal activity, the fundamental source for bio-electric signals expresses the variability of brainwaves in humans. Brainwave and specific EEG spectral analysis are important in bio-electric signal variability identification. Recent researches in neuro-robotics rely on the use of brain computer interface (BCI) technology in developing robotic commands. Brainwave variability identification provides different levels of robot control signal development and optimization.

This paper presents the development of robotic arm control strategy using brainwave signal variability. The bio-electric signal identification was derived from physiological expressions. The physiological expressions are identified using spectral analysis and the paper presents possible future research options and applications towards using physiological and facial parameters in controlling robotic arm.

Keywords: Brainwave, EEG Signal, Brainwave Variability, EEG Artifact Identification.

1 Introduction

The use of brainwaves in the development of robotic arm control paradigm is the function of the neural activity inherent in the human mind. Differences in electrocorticographic (ECoG), electroencephalographic (EEG) and electromyographic (EMG) signal frequencies provide varying information on neural states achievable with the human mind. Signal analysis methodologies such as auto correlation, Fourier analysis, cross- correlation techniques provide useful tools in determining the lifespan of bioelectric signals within the given time space [3]. Bioelectric artefact stability and reliability determined through repeated measurements of EEG signals expresses the systematic state of brainwaves. The systematic state of brainwaves describes the level of usefulness in developing control commands for robotic control. Conventional signal variability evaluation and determination procedures are useful in getting close relations to the dynamic behaviour of brainwaves. The results from dynamic brain

---

[*] Corresponding author.

activity provide insights into the excitatory and inhibitory properties of the brainwave amplitudes [6]. Traditional methodologies to brainwave feature investigation provide models necessary for brainwave frequency analysis and oscillatory property observation. The limitations of traditional approaches to brainwave analysis introduce shortcomings in the representation EEG signal spectrum.

Realistic implementations of brainwaves are bounded by the irregularity of the measured bioelectric signal properties. Differences observed in brainwaves as a result of the unpredictability of the human brain and its processing capacities produce pointers towards signal threshold development which are useful in the bioelectric signal identification process. The extent to which experiments in brainwave feature identification are accomplished reveals the different features which are obtainable in brainwave dynamic behaviour. Studies done to establish the variability and reliability of brainwave focused more on coherence and spectral measures in determining the differences in brainwaves [4].

Frequency, time-domain and amplitude parameters provide differentiable brainwave models in applicable to human cognitive processes. The identification of brainwave artifacts from different test subjects using peak frequencies provides an indication of variability of brainwave. Brainwave alpha, beta, delta and gamma power bands in the test subjects indicate variations in the signal frequencies for the same cognitive task. Human performance in redirecting its communication pathways can thus be measured using the variability in the peak frequencies in the brainwaves. The variability in brainwave may be viewed as the results of differences observed in faster EEG beta frequencies [7].

High level brain activity involves integrated and coordinated neural activity with specialized electrical signal and frequencies. The functional elements of brain activity and their precise location on the scalp allow for the definition of EEG signal frequencies governing the manipulation and utilization of brainwaves for developing robotic commands. The understanding of the signal transportation mechanisms in the brain involves adequate comprehension of brain activity and neuronal processes [9].

Physiological expressions depict the level of concentration, relaxation and meditation of the human mind and their associated neural firings. Mental events expressed as forms of physiological artefacts are results of the human mind engaging in specific attention sets thereby expressing the state of brain meditation. Subsets of mind activities are used during mental tasks to induce meditation and relaxation [12]. Brain response to meditative exercises as functions of the variability in EEG signals reflects the exact neurological and neurophysiological states of the human brain. The alteration of cognitive, sensory and self-referential cognitions which can result to meditation and attention is referred to as state in the study [1]. Studies have indicated that the act of meditation reflects the level of awareness as evidence of cognitive stance in expressing meta-cognitive shift [11].

In this paper, we propose the use of brainwave and specific EEG spectral analysis in identifying variability of brainwave artefacts in EEG signal identification towards developing robotic arm control strategy. The results observed from the investigation provided thresholds which are useful in the robotic arm control strategy.

2   Brainwave Analysis

In analyzing the brainwaves, the investigation model comprises of the signal source $I(t)$, bioelectric data mapping system $T[\cdot]$, data transfer channel $h[\cdot]$, additive noise channel $n(t)$ and the variability identification processor. Brainwaves measured from the Neurosky and Emotiv headsets are modeled as [13]:

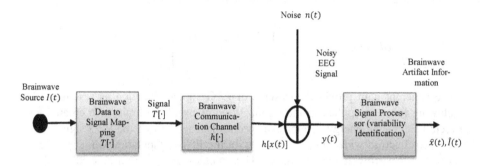

Fig. 1. Brainwave variability identification model

Noise which may be present in the measured brainwave is filtered through linear regression between the EEG data channels and the noise channel. The noise removal process is modeled as [17]:

$$EEG_{es} = EEG_m - \sum \beta_{nc} EEG_{ni} \qquad (1)$$

Where EEG_m and EEG_{es} represent the measured EEG signal and estimated EEG signal respectively. β_{nc} represents the regression coefficient between the EEG signal channel and the noise channel. EEG_{ni} represents the value of the noise channel at time point i [17]. Given that brainwaves are not periodic and are dynamic in accordance to level of neural activity in an individual, Fast Fourier Transform (FFT) performed on the brainwaves combined Hanning windowing function which prevented spectral leakage and this accounted for the errors inherent in FFT investigations of the brainwaves. Comparisons of raw brainwave and filtered brainwaves are shown in Fig. 2 and Fig. 3. The Hanning function is a raised cosine function modeled as [14]:

$$w(n) = 0.5\left(1 - \cos\left(2\pi \frac{n}{N}\right)\right), \quad 0 \leq n \leq N \qquad (2)$$

Where the window length is given as:

$$L = N + 1 \qquad (3)$$

Brainwaves signal coherence provided the measure of the linear correlation between the EEG electrodes. The result can be expressed as function of EEG frequency thus providing the spatial correlation of the different brainwave frequency bands [16]. The coherent gain for the Hanning window is modeled as [15]:

$$coherent\ gain = \int_0^1 \left[\frac{1 - \cos(2\pi x / N)}{2} \right] dx \qquad (4)$$

The coherent gain provided measures of EEG signal reduction as results of the windowing function applied to the signal. Windowing of the brainwave created less discrete EEG signal energy. Brainwave spectral leakage caused the EEG signal energy to spread across wider frequency range during FFT computation while the signal energy has a narrow frequency range. Thus windowing of brainwave allowed for better representation of the brainwave frequency spectrum.

The power spectrum of the brainwave is represented as:

$$P(f) = \frac{|F(f)|^2}{N} \qquad (5)$$

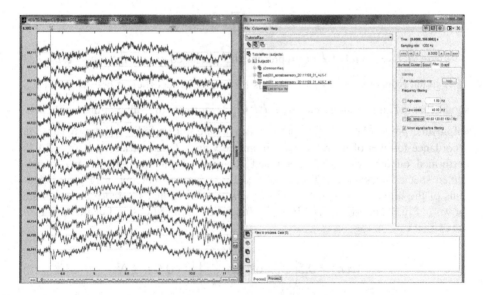

Fig. 2. Raw Brainwaves before Preprocessing

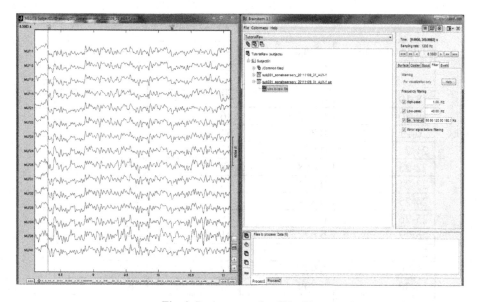

Fig. 3. Brainwaves after Filtering

Where $F(f)$ represents the Fourier transform of the brainwave computed using Discrete Fourier Transform (DFT) algorithm and N represents the number of brainwave data points. The product of the EEG signal and the complex conjugate of the Fourier transform yields the cross-power spectrum for the brainwave. This is expressed in eqn. (6).

$$P(f) = \frac{F(f)G*(f)}{N} \qquad (6)$$

2.1 Brainwave Signal Correlation

In order to determine the relationship between neural activities and the different physiological expressions and identify effective brainwave artefacts for use in developing robot control strategy, we computed the signal correlation coefficients at different intervals as shown in table 1.

Local correlation of the brainwaves provided techniques of identifying brainwave artefacts and combining local physiological expressions. The autocorrelation approximation for the EEG signal is expressed as:

$$\hat{\phi}_{xx}[m] = \frac{1}{Q} \sum_{n=0}^{Q-|m|-1} x[n]x[n+|m|] \qquad (7)$$

Given that $|m| \leq M - 1$, $0 \leq m \leq M - 1$, $x[n]$ represents the EEG signal, Q represents the windowed EEG signal segment.

Table 1. Brainwave FFT Correlation Values

Time (seconds)	Signal Correlation Coefficient	Meditation Level	Attention Level
18-19	0.757	94	77
19-20	0.972	86	44
20-21	0.769	64	51
21-22	0.680	83	71
22-23	0.668	98	57
23-24	0.901	93	59
24-25	0.882	90	41
25-26	0.808	88	62
26-27	0.627	91	69
27-28	0.834	61	76
28-29	0.971	78	52
29-30	0.899	64	76
Average Power Spectrum	0.814		

2.2   Brainwave Power Spectrum Analysis

Emphasis was placed in analyzing the brainwaves in their steady state form as the behavioural properties of human physiological expressions were approximated as almost constant. Thus the duration of the brainwave artefacts played important role in the detection, identification and characterization of the physiological patterns. In using the power spectrum function, the interdependence of the different brainwave frequency bands was investigated using custom frequency bands for each of the electrodes. Thus the power density spectrum of the brainwaves was also indications of the autocorrelation of the signals. The power spectrum revealed the amount of power contained in the brainwaves on frequency scale [17].

3   Methodology

The recursive evaluations of brainwave function as hypothesized in the basic chaos theory produced multifaceted results which may not be useful in development of robot control strategy. Simplification of complex results through an iterative and repetitive process allows for the convergence of brainwaves to appreciable states [3]. The analytical technique implemented in the study was used to investigate the physiological characteristics of brainwaves derived from facial expression and eye blinks. The brainwaves were analyzed and invested using OpenVibe, Matlab and EEGLAB.

3.1   Brainwave Signal Acquisition

The brainwaves signals were recorded in accordance to the standard 10-20 system for electrode placement with the ear lobe acting as ground electrode. The EEG headset transmits the brainwaves wirelessly to the computer for further processing. Healthy subjects were selected at random to investigate the variability of brainwave.

3.2   Brainwave Pre-processing

The raw brainwaves are filtered in order to remove signal noise and differentiation of artefacts into alpha, beta, delta and gamma frequencies. Breathing frequencies and eye movements and change in blood pressure/heart rate may be regarded as noise. Quantization of the signal levels at the sampling frequency of 128 Hz provides a simple pre-processing process from eight brainwave frequency categories.

3.3   Brainwave Feature Identification

During the recording and signal analysis, the following facial conditions were used: Frowning condition, smiling condition, facial smirking and eye blinks. Mind Your Open Sound Controls shown in Fig. 4, the intermediary software was used to create and manage intelligent thresholds for each of the facial expressions. The simplified data processing flow chart is shown in Fig. 6.

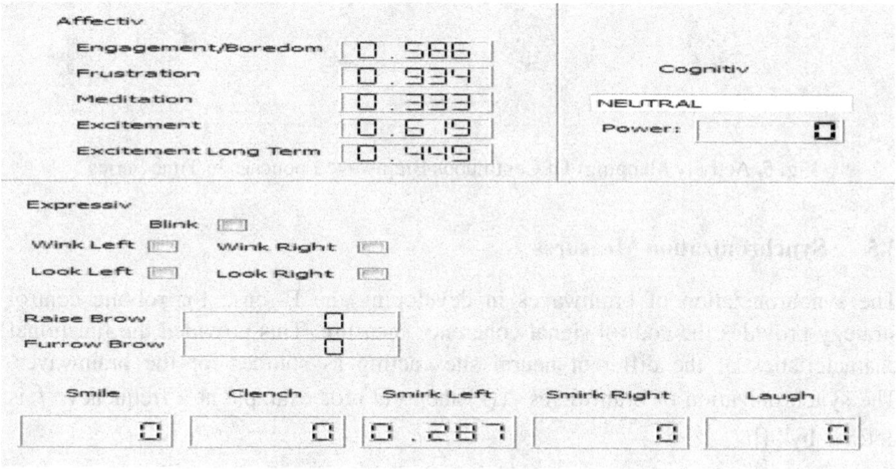

Fig. 4. Mind Your OSCs Interface

3.4   Brainwave Power Spectral Measures

The powers spectral of the alpha, beta, delta, theta and gamma signals were estimated using Welch's averaged periodogram. The EEG power spectrum was divided into theta (4 Hz-7 Hz), alpha (8 Hz-12 Hz), low beta (12 Hz-15 Hz), high beta (21 Hz-30 Hz), low gamma (30 Hz-42 Hz), mid gamma (42 Hz-60 Hz) and delta (0.1 Hz-3 Hz) frequency levels. The activity mappings of alpha, beta, theta and delta frequencies of continuous brainwave epochs are shown in Fig. 5. The fast Fourier transform (FFT) of the brainwaves over a second provided an indication of power density values derived from the correlation coefficient of the brainwave frequency bands computed up to 40 Hz. The brainwave frequency band cross power density was investigated using the Welch's averaged and modified periodogram technique for spectral estimation.

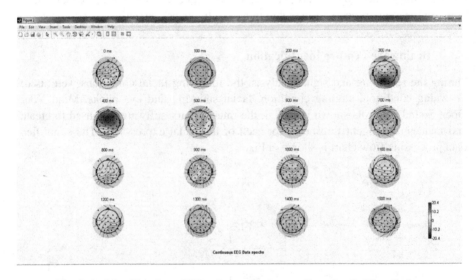

Fig. 5. Activity Mappings Of Continuous Brainwave Epouches In Time Series

3.5   Synchronization Measures

The synchronization of brainwaves in developing the E-sense for robotic control strategy provides the control signal coherence measure. This provided the functional characteristics of the different neural sites acting as sources for the brainwaves. The synchronization of brainwaves $x(t)$ and $y(t)$ for example at a frequency f is defined by [4]:

$$C_{xy}(f) = \frac{\left|P_{xy}(f)\right|^2}{P_{xx}(f)P_{yy}(f)} \tag{9}$$

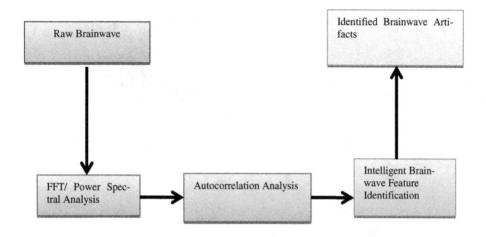

Fig. 6. Simplified brainwave Data Processing Flowchart

Where $P_{xx}(f)$ and $P_{yy}(f)$ represents $x(t)$ and $y(t)$ power spectral densities respectively. $P_{xy}(f)$ denotes the cross-spectral density of the signals. The synchronization function is represented between the boundaries of 0 and 1 using the Welch's average periodogram [2]. The resulting characteristic features of the signals are a representation of the synchronization level in each of the EEG frequency bands. Due to the dynamic nature of EEG signals, synchronization measures reflects the dynamic dependence of each EEG artefact between the EEG signal time series and in real-time.

4   Robotic Arm Development

Robotic arm functionality varies and may increase or decrease with respect to its application and usage. The basic robotic arm functions include assembling of parts, picking and dropping of parts. The robotic arm currently being developed uses geared DC motors in controlling each finger as shown in Fig. 8. The fingers are retractable by means of a spring mechanism incorporated in the robotic palm design. In Fig. 7, a servo motor is connected at the wrist of the robotic arm. This allows the robotic palm to rotate and mimic human palm movements. The basic function of the robotic arm shown in Fig. 8 is to hold and release an object of choice. The robotic arm control strategy is shown in Fig. 9. The grip and torque adjustments at the palm using brainwaves as the prime signal are areas of future research.

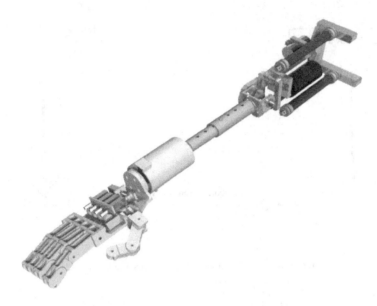

Fig. 7. Full Robotic Arm Length

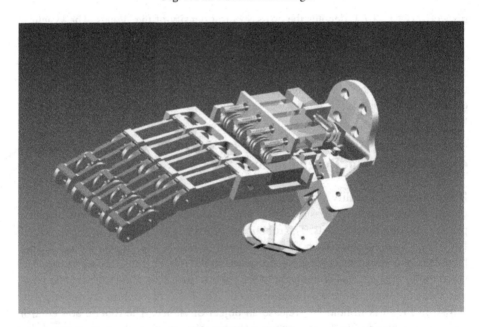

Fig. 8. Robotic Palm

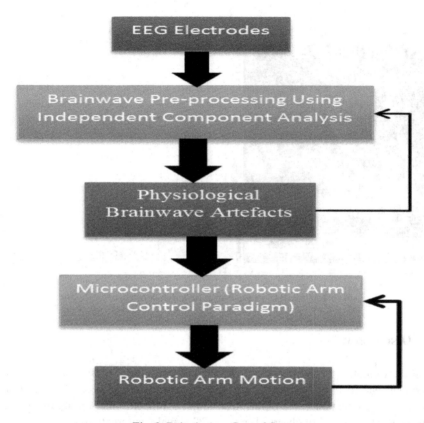

Fig. 9. Robotic Arm Control Strategy

5   Results

The decomposition the brainwave power spectral density given that frequencies were
sampled at 128 Hz is shown in Fig. 10. It is also noteworthy that the power spectrums
for each electrode are similar. The data epoch for each electrode is shown in
Fig. 11.The cross power spectral density over 700 point signal window for the brainwave artefacts is shown in Fig. 12. This represents the power distribution per unit
frequency for each of the brainwave frequency bands. The FFT correlation coefficient
for the brainwaves calculated over a range of one to 40 Hz yielded an average power
spectrum of 0.814 as indicated in table 1. The physiological expressions investigated
in the study includes brainwave artefacts identified from eye blinks shown in Fig. 13,
brainwave artefacts identified from facial frowning shown in Fig. 14, facial smiling as
shown in Fig. 15 and facial smirking as indicated in Fig. 16. The brainwave artefacts
were identified and implemented in the development of robotic control commands for
forward and reverse motion of robots.   The brain activity mappings shown in Fig.5
are indication of neural activity regions and brain electrical signal sources during each
of the physiological activities.

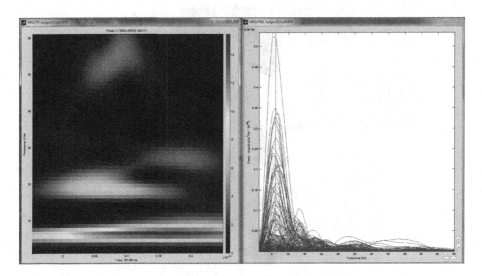

Fig. 10. Brainwave Power Spectral Density

6   Discussion

During the execution of the facial expressions and eye blinks, the variability of the EEG signals was in accordance to the cognitive state of the test subjects. The attention level and the meditation level of the test subjects reflected the variability in the brainwaves. The variability observed in the EEG signals represents functions of their synchronization property. The synchronization of each of the EEG frequency band increased during each of the tasks in agreement with other findings [5]. The pattern of coherence expressed in each of the EEG signals reflected the power spectrum for the EEG signal frequency bands [9]. This also reflected the level of synchronization in the EEG signals [11]. The physiological expressions affected the distribution of the EEG frequency bands and increased the power spectrum in accordance to the cognitive state of the test subjects. The variation coefficient observed in the EEG signals observed as seven frequency bands were evaluated using multivariate analysis with limited variability among the test subjects. The variability observed can also be attributed to the effects of seasonal changes to the frontal cortical activity of the brain [8].

In Fig. 13, the pronounced spikes occurring at an average interval indicates the eye blinks. The difference between the smiling, smirking and frowning is their frequency ranges. This distinguished each of the EEG artefacts. At each of the distinguished frequencies, further processing of the signal was carried out to generate robot control commands.

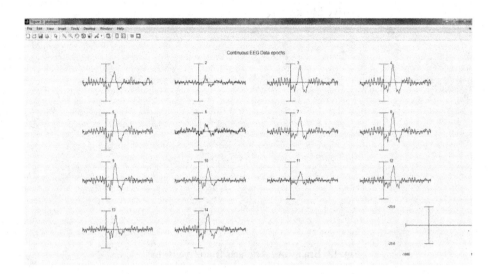

Fig. 11. Brainwave Data Epochs For Each Data Channel

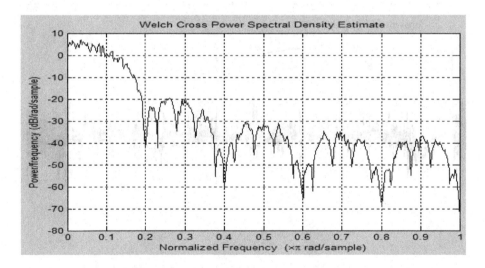

Fig. 12. EEG Cross Power Spectral Density

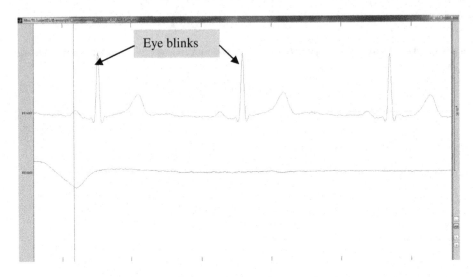

Fig. 13. Brainwave Artefacts from Eye Blinks

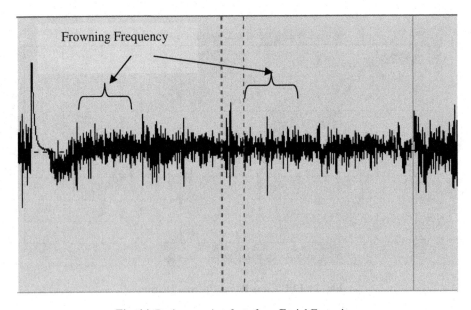

Fig. 14. Brainwave Artefacts from Facial Frowning

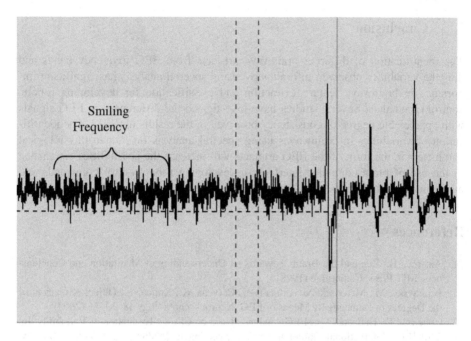

Fig. 15. Brainwave Artefacts From Facial Smilling

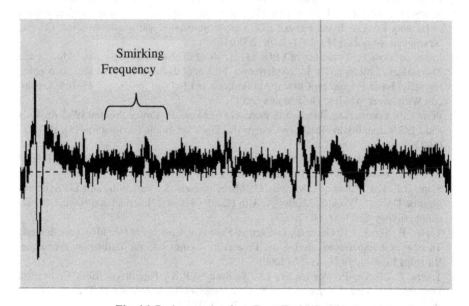

Fig. 16. Brainwave Artefacts From Facial Smirking

7   Conclusion

The identification of different brainwave artifacts from EEG frequency bands and thus the variability observed in brainwave using spectral analysis has significant importance in brainwave feature extraction and classification for developing robotic control commands. Several studies have investigated the differences in EEG signals with appreciable degree of consistency observed in the results obtained. The identification of variability in brainwaves using spectral analysis on human physiological parameters in the form of the EEG artefacts will increase the possibilities entrenched in the use of brain computer interface in developing robot control commands and human machine interfaces.

References

1. Austin, J.H.: Zen and the Brain: Towards an Understanding of Meditation and Conciousness. MIT Press, Cambridge (1998)
2. Brunovsky, M., Matousek, M., Edman, A., Cervena, K., Krajca, V.: Object Assesment of the Degrees of Dementia by Means of EEG. Neuropsychobiology 48, 19–26 (2003)
3. Cohen, M.E., Hudson, D.L.: EEG Analysis Based on Chaotic Evaluation of Variability. In: 23rd IEEE International Conference on Engineering in Medicine and Biology Society, vol. 4, pp. 3827–3830. IEEE, California (2001)
4. Gudmundsson, S., Runarsson, T.P., Sigurdsson, S., Eiriksdottir, G., Johnsen, K.: Reliability of Quantitative EEG Features. Cilinical Neurophysiology 118, 2162–2171 (2007)
5. Krause, C., Sillanmaki, L., Koivisto, M., Saarela, C., Haggvist, A., Laine, M.: The Effects of memory Load on Event-Related EEG Desynchronization and synchronization. Clinical Neurophyisology 111(11), 2071–2078 (2000)
6. Lopes da Silva, F.: Dynamics of EEGs as Signals of the Neuronal population: Models and Theoretical Considerations. In: Niedermeyer, E., Lopes da Silva, F. (eds.) Electroencephalography, Basic Principles, Clinical Applications and Related Fields, pp. 85–106. Lippincott Williams & Wilkins, Philadelphia (2005)
7. Perros, P., Young, E.S., Ritson, J.J., Price, G.W., Mann, P.: Power Spectral EEG Analysis and EEG Variability in Obsessive-Compulsive Disorder. Brain Topography 4(3), 187–192 (1992)
8. Peterson, C.K., Harmon-Jones, E.: Circadian and Seasonal Variability of Resting Frontal EEG Asymmetry. Biological Psychology 80(3), 315–320 (2009)
9. Stam, C.J., van Walsum, A.-M., van, C., Micheloyannis, S.: Variability of EEG Synchronization During a Working Memory task in Healthy Subjects. International Journal of Psychophysiology 46(1), 53–66 (2002)
10. Travis, F., Tecce, J.J., Guttman, J.: Cortical Plasticity, Contingent Negative Variation and Trasscendent Experiences during the Practice of Transcendental Meditation Technique. Bilogical Psychology 55, 41–55 (2000)
11. Travis, F., Tecce, F., Arenander, J.J., Wallace, A.R.K.: Patterns of EEG Coherence, Powerand Contingent Negative Variation Characteriza the Integration of Transcendal and Walking States. Biological Psychology 61, 293–319 (2002)

12. Vaitl, D., Birbaumer, N., Gruzelier, J., Jamieson, G.A., Kotchoubey, B., Kubler, A., Lehmann, D., Miltner, W.H., Ott, U., Putz, P., Sammer, G., Strauch, I., Strehl, U., Wackermann, J., Weiss, T.: Psychobiology of Altered States of Conciousness. Psychological Bulletin 131(1), 98–127 (2005)
13. Cao, J., Chen, Z.: Advanced EEg Signal Processing in Brain Death Diagnosis. In: Mandic, D., Golz, M., Kuh, A., Obradovic, D., Tanaka, T. (eds.) Sinal Processing Techniques for Knowledge Extraction and Information Fusion, pp. 275–298. Springer Science+Business Media LLC, New York (2008)
14. Oppenheim, A.V., Schafer, R.W.: Discrete-Time Signal Processing, pp. 447–448. Prentice Hall (1989)
15. WaveMetrics: IGOR Pro Manual, WaveMetrics, Portland, USA (2010)
16. Quiroga, R.Q.: Quantitative Analysis of EEG Signals: Tme Frequency Methods and Chaos Theory. Institute of Physiology and Institute of Signal Processing, Medical University Lubeck, Buenos Aires (1998)
17. Repovs, G.: Dealing with Noise in EEG Recording and Data Analysis. Infor. Med. Slov. 15(1), 18–25 (2010)

Acquisition of Context-Based Active Word Recognition by Q-Learning Using a Recurrent Neural Network

Ahmad Afif Mohd Faudzi[1,2] and Katsunari Shibata[3]

[1] Kyushu University, Fukuoka, Japan
[2] Universiti Malaysia Pahang, Malaysia
afif@cig.ees.kyushu-u.ac.jp
[3] Oita University, Oita, Japan
shibata@oita-u.ac.jp

Abstract. In the real world, where there is a large amount of information, humans recognize an object efficiently by moving their sensors, and if it is supported by context information, a better result could be produced. In this paper, the emergence of sensor motion and a context-based recognition function are expected. The sensor-equipped recognition learning system has a very simple and general architecture that is consisted of one recurrent neural network and trained by reinforcement learning. The proposed learning system learns to move a visual sensor intentionally and to classify a word from the series of partial information simultaneously only based on the reward and punishment generated from the recognition result. After learning, it was verified that the context-based word recognition could be achieved. All words were correctly recognized at the appropriate timing by actively moving the sensors not depending on the initial sensor location.

Keywords: reinforcement learning, neural network, active perception, word recognition.

1 Introduction

Among the human sensory organs, vision is probably the most informative perception. The eye movement and recognition capability in humans seem very flexible and intelligent. It has been shown that, in object recognition, compared to the case where only the information of the object is provided, a better recognition could be done if we were also supported by other information such as past knowledge and contextual information.

Imagine that we were presented with several new patterns or words repeatedly until we learned their shape and color in detail. At one time, we could recognize all the patterns correctly even if only some parts of these recognized objects that can be distinguished each other are visible to us. Even when perceptual aliasing, a condition where the same input stimuli belong to different words occurs, we can still recognize it by memorizing the different past input stimuli or by moving our

J.-H. Kim et al. (eds.), *Robot Intelligence Technology and Applications 2*,
Advances in Intelligent Systems and Computing 274,
DOI: 10.1007/978-3-319-05582-4_17, © Springer International Publishing Switzerland 2014

eyes actively[1]. This is because human has the ability to extract the necessary information, e.g., contextual information, and the ability to memorize it. It has also the ability to move sensors efficiently, which is called active perception. Our flexible brain achieves such abilities and learning must play an important role to acquire them.

Learning-based active perception/recognition methods have been proposed[2]-[6] including the works for object tracking[4][6]. In [3][5][6], considering probability distribution of each possible presented pattern or the state of the pattern from the image sequences explicitly, the appropriate viewpoint is learned to reduce the uncertainty in the distribution. In [4], conventional image processing methods are used and the system is complicatedly designed. In [2], by reinforcement learning, the system learns to choose to make a response that the target pattern presented or to move the sensor, but classification of the presented pattern is not necessary. Except for [2], the system and method is designed especially for active recognition, and recognition and sensor motion generation are separately performed. Therefore, it is concerned that the approach is not suitable for developing human-like intelligence by integrating with other functions.

Furthermore, to solve the partial observation problem, past information is utilized in several ways. In [6], useful features are extracted from all the captured images, and also in [2], memory-based reinforcement learning is employed. However, no mechanism has been proposed that learns to extract important information from the captured image at each time step, to hold it, and to utilize it for recognition and sensor motion according to the necessity.

The authors have thought that for developing flexible intelligent robots like humans, the entire process from the sensors to actuators should be learned harmoniously in total[7]. In the machine-learning field, many researchers are positioning reinforcement learning (RL) as learning specific for actions in the total process, and the NN as a non-linear function approximator. Based on the above standpoint, the authors' group has used a recurrent neural network that connects from sensors to actuators and learns autonomously based on reinforcement learning. It has been shown that necessary functions emerge according to necessity through learning. In this framework, it was shown that recognition and sensor motion are learned simultaneously using a neural network[8,9]. It was verified through simulations and also using a real camera that the appropriate camera motion, recognition and recognition timing were successfully acquired[9]. However, the samples are limited and because the learning system was trained using a regular layered neural network, the recognition and sensor movement function were limited to the case where the presented pattern can be classified from the present captured image. On the other hand, it was shown that by using a recurrent neural network (RNN), contextual behavior, memory or prediction function emerged through reinforcement learning[10]-[12].

In the previous work, it is examined that using a RNN, both sensor motion and context-based word recognition functions emerge through reinforcement learning. At that time, the initial camera facing direction was fixed at the left edge of the presented word[13]. However, in this paper, the initial camera facing direction at every episode is randomly set up. The input of the RNN is just the present image, and so the RNN has to learn what information should be extracted and memorized from the image for appropriate recognition and sensor motion to get a reward. The learning is very simple and general, and so it is easy to extend to integrate with other functions acquired on the same basis of reward and punishment.

This paper is organized as follows. In the next section, we will describe the basic idea of the learning system. In Section 3, we will explain the learning method and the task settings. The learning results are written in Section 4, and Section 5 states the conclusions.

2   Learning of Context-Based Active Word Recognition

As shown in Fig. 1(A), the context-based active word recognition learning system has a movable camera as a visual sensor and a monitor. The monitor will display a word that is randomly chosen, and the system needs to identify which word is displayed. In every episode, only one word will be displayed. The camera can make a horizontal movement either to the left or to the right with a constant interval. In this paper, the initial camera facing direction for every episode is randomly set up, and so the camera sometimes should be moved to a different direction even though the camera catches the same potion of an image. In order to avoid from spending a lot of time on learning, instead of using a real camera, a simulation based on real camera movement was done using the images remade from the captured ones by the camera. Fig. 1(B) shows the 6 words that were used in the learning process and their partial images for all the camera directions.

As shown in Fig. 1(A), the sensor's field is too small to identify the presented word. In order to recognize the words correctly, the system has to memorize the information of the partial images that had been captured in the same episode. For example, in the case of the word 'cat' and the initial camera direction is 0, since it is the only word that starts with character 'c', the system can judge and recognize it from the first partial image. While in the case of word 'mad', the system needs more information about the second character that is held by the next partial image to distinguish it from the words 'men' and 'met' that have the same initial character. In the case of the other 4 words, the system is expected to recognize them when the partial images that hold the last character information are inputted. However, for the words 'met' and 'net', since the same images are inputted after the 4th camera direction, in order to distinguish them, the system needs to memorize whether the first character was 'm' or 'n'. Through this learning, the system is expected to recognize all the prepared words correctly at the appropriate timing by flexibly moving the camera.

3   The Learning Method and the Task Settings

Here, a 4-layer Elman-type RNN is used as shown in Fig. 1. The RNN is trained with back propagation through time (BPTT)[14] using the training signals

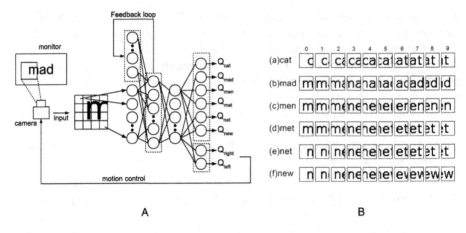

Fig. 1. A: Context-based word recognition learning system architecture. There is a monitor to display all the prepared words, and a camera as a visual sensor for the system. **B:** Six prepared words and all their partial images. These images were prepared base on the camera movement from "direction 0" to "direction 9" with a constant interval.

generated by reinforcement learning (RL). As for RL, Q-learning, which is a popular learning algorithm for discrete action selection problem, is used[15]. In Q-learning, state-action pairs are evaluated, and the evaluation value that will determine the action is called Q-value. The Q-value for the previous action is updated using the Q-value from the present action and reward. The training signal for the Q-value $T_{a_t,t}$ is generated as

$$T_{a_t,t} = r_{t+1} + \gamma max_a Q(s_{t+1}, a) \tag{1}$$

where γ indicates a discount factor ($0 \leqslant \gamma < 1$), and r_t, s_t, a_t indicates the reward, state, and action at time t respectively. In this learning system, the state s is the direct image signals from camera, and the action a is a discrete decision making. The action a is selected based on ε-greedy. In greedy selection, there is no probabilistic factor and the action with the maximum Q-value is always selected. Otherwise, the action a is selected randomly while the probability for "recognition result output" for each of the 6 words is 0.1/3 and for "camera movement" is 0.80. The current episode is terminated when the system makes a recognition or when the system moved the camera more than 25 times, i.e., $t = 25$. For the new episode, another word will be displayed and the camera is set up to a new facing direction.

As for the RNN, the inputs are the raw image signals that are linearly converted to a value between -0.5 and 0.5. There are eight output neurons. Each of them represents the Q-value corresponding to one of the eight actions, which are six "recognition result outputs" and two "camera movement" actions. The number of neurons in each layer is 576-100-15-8 from the input layer to the output layer. The number of neurons in the input layer does not includ the hidden outputs that are fed back to the input layer at the next time step.

The output function of each neuron is a sigmoid function with the range from -0.5 to 0.5. However, all the outputs are used after adding 0.5. When the system chooses to output the result and it is correct, the training signal is added by 0.9. However, if it is not correct, the training signal will be the value of corresponding Q-value deducted by 0.2. 0.5 is subtracted from the training signal before using it as the actual training signals in BPTT. On the other hand, if the system chooses to move the camera, no reward is given.

Initial connection weights from the external input to the hidden layer are random numbers from -1.0 to 1.0. The initial connection weights from the hidden layer to the output layer are all 0.0, and as for the feedback connection of the hidden layer, initial weights are 4.0 for the self-feedback and 0.0 for the other feedback. This enables the error signal to propagate the past states effectively in BPTT without diverging.

4   Learning Result

As a result, after 150,000 episodes, both flexible sensor motion and context-based word recognition function can be achieved. Fig. 2 shows the Q-values for each action when every episode was finished or terminated for the presentation of words (a)'cat', (b)'mad', (c)'men', (d)'met', (e)'net' and (f)'new' respectively. The highest value for each episode indicates what action was taken if the action selection was greedy.

As shown in Fig. 2(a)'cat', at first, the red point (+), which is the Q-value corresponding to the word 'cat', increased soon and became higher than the other Q-values. At the same time, except for the black point, the other Q-values that correspond to the other words had decreased. By these big differences of Q-values, the system can make recognition correctly. On the other hand, the black point, which is the Q-value that corresponds to camera motion, was also high showing that if the system moved the camera, it can recognize the word correctly at the next step. This characteristic is also shown in Fig. 2(b) to (f) indicating that the system also successfully recognized the other words provided.

Fig. 2 also shows that different words required different time of learning. As shown in Fig. 1, at one time, the system can only capture a partial of the whole word. Since all the partial image of the word 'cat' does not identical to any of the other words, the system can distinguish it when only the first partial was captured. Here, Fig. 2(a) 'cat' clearly shows that the Q-value corresponding to the word 'cat' became high while the other Q-values became low at earlier stage of learning. While in Fig. 2(b), which corresponds to the word 'mad', shows that the green point (x) was also high, however there were also some corresponding points that were not high. This happened possibly due to the insufficient learning depending on the initial camera direction. The system need a longer learning time to recognize the second character, 'a', since the first character, 'm', is the same with words 'men' and 'met'. For the words(c)'men' and (f)'new', the partial images from the 0th to 4th are identical to those for the word 'met' or'new', but those from the 5th to 9th are not identical to any other partial images.

Finally, in Fig. 2(d)'met' and (f)'net', it can be observed that the system required a longer learning time to recognize them compared to other 4 words. All the partial images of 'met' or 'net' have the identical image in another word. That means that the words cannot be classified correctly only from the present image, and the system has to classify them from both present and past images. It is thought that it took a long time to learn to memorize and use the past image even though it was not taught that the memorization of the information about the past image was necessary. The slower learning for the word 'met' than for the word 'net' may be because the number of words that start with the character 'm' is more than that with the character 'n'.

After the learning process had finished, a performance test was done. For each word, the system was tested by initializing the camera at each of the

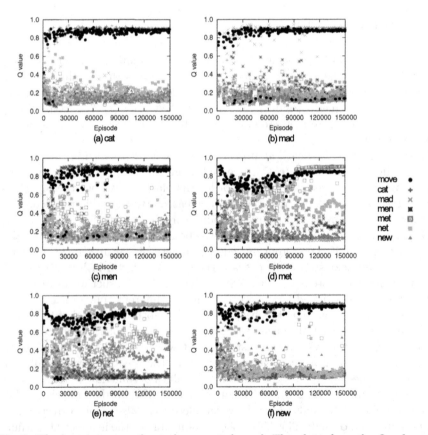

Fig. 2. The learning curve for each presented word. The plots show the Q-values for all actions at the end of every episode during the learning simulation for each word (e.g., (a) The learning curve when the word 'cat' was displayed on the monitor).

10 directions. In this test, the system chose to move the camera's direction either to the right or to the left greedily according to the Q-values. Fig. 3 shows the camera movement for each initial direction. The x-axis indicates the camera direction that corresponds to the partial images in Fig. 1(b). y-axis indicates the number of step, and so x at $y=0$ indicates the initial direction. Here, when the system chose one of the "move the camera" actions, y will increase and x will show the current direction. A line shows the history of camera movement, and the final plot on the line shows the direction where the system chose to output the recognition result.

As shown in Fig. 3(a), since the word 'cat' is much different compared to others, from every initial direction, one or no movement is required for the system to recognize the word. As for the word 'mad', Fig. 3(b) shows that when the camera started from the leftmost side, the system outputs the recognition result at 'direction 2' where the information of the second character 'a' is inputted. This happened because the first character 'm' is same with the words 'men' and 'met'. We can also see that, when the camera started from the rightmost side, it only took one or no movement to make recognition. As for the word 'men' whose first two characters were same for the word 'met', Fig. 3(c) shows, for the initial direction 0 to 6, the recognition result is output at the direction 5. The correct recognitions were done at direction 5 even though there was a slight difference compared to the word 'met'. When the initial direction was at direction 7 to 9, the system could output the result immediately. It is because the other words do no have 'n' as the third character. The same graph characteristic was shown in Fig. 3(f) when the word 'new' is tested. Finally, as shown in Fig. 3(d) and

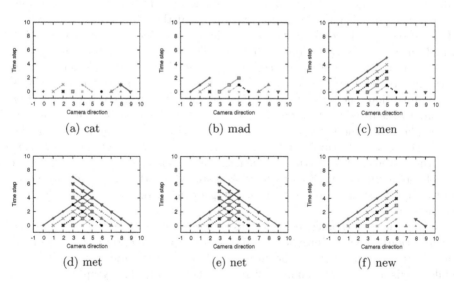

(a) cat   (b) mad   (c) men

(d) met   (e) net   (f) new

Fig. 3. The performance test result-1: Camera movement when initial direction is set to all directions, "direction 0" to "direction 9". The x-axis indicates the initial direction, while y-axis indicates the number of step.

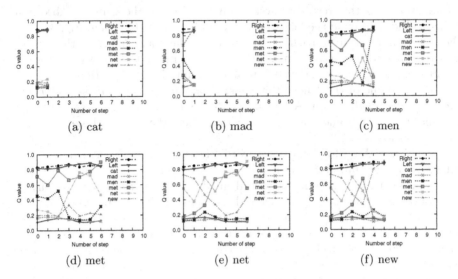

Fig. 4. The performance test result-2: The changes of Q-value when the initial direction for camera was set to 'direction 1'

Fig. 3(e), for both words 'met' and 'net', we could observed that the system makes the recognition at direction 3 regardless of the camera initial direction. In order to recognizes the word 'met' with the initial direction 0, the system at least need to move the camera until direction 5 to acquire the third character information to distinguish the word 'met' from the word 'men'. Since the learning system has a memory function, it is expected that the system make recognition at this direction. However, it is observed that, the system returned to direction 3 and outputs the recognition result there. Therefore, even though the system manages to recognize both words correctly, the recognition timing is not optimal. That may be because it is more difficult for the system to recognize the words at any direction where the word can be classified than to recognize them after moving to the direction where the word can be classified easily. The hypothesis can explain that the system has preferable directions to recognize even in the word 'cat'. Longer learning might improve the performance, but it is interesting that the movement back to the left when the character 't' appears at the direction 5 looks as if the system returns to make sure. When the initial direction was on the rightmost side, the system moved the camera to the left until the first character appears.

During the performance test that was shown in Fig. 3, the change of Q-values for all actions when the initial direction was set up at 'direction 1' were recorded and plotted for each word as shown in Fig. 4. The y-axis indicates the Q-value, while x-axis indicates the number of step in the episode. This graphs show how the system makes the correct recognition. In Fig. 4(a), when the word 'cat' is tested, it is already clear from step 0 that the Q-value corresponding to the word 'cat' is higher than other Q-values and in (b)-(f), Q-value for the word 'cat' was very small. While in Fig. 4(b), the Q-value corresponds to the 'move' action is clearly higher at step 0 because the information is still not enough and that requires the information of the second character of the word. After that, the Q-value corresponding to the word 'mad' became higher at step 1 where the image for the direction 2 was inputted, and the system output the result successfully.

The change in Q-values from step 0 to step 3 were the same between Fig. 4(c) and (d). However, at the step 4, where the fifth image that holds the third character was inputted, as shown in Fig. 4(c), the Q-values started to split depending on the presented word and successfully recognizing the word 'men'. We can observe the same characteristic when the system tries to recognize the word 'new' in Fig. 4(f). Finally, as shown in Fig. 4(d) and (e), at 'step 4' the Q-values for both words were almost the same not depending on the past images. Here, rather than outputting the recognition result, the system chose to move the camera backward and correctly recognized the words at 'direction 3'. It is not the optimal, but this can be accepted since moving the camera backward seems to be a rational option rather than pushing itself to memorize everything.

5   Conclusions

In this paper, it was shown through simulations that both flexible sensor motion and context-based word recognition function learning could be confirmed through reinforcement learning based only from reward and punishment. The learning process was successful and the trained system was able to recognize all the tested words from any initial camera direction. After learning, the system also managed to understand the context of all of the tested words and was capable to make recognition by flexibly moving its sensor though the process is not always optimal.

There are still some limitations for current result. For future work, not only a horizontal discrete action but also a continuous and vertical movement should be considered. Furthermore, since only one type of font was used, the emergence of generalization function is not expected. Varying font type and size, and by using a real camera movement with the perspective distortion must promote its generalization.

Acknowledgment. This research was supported by JSPS Grant-in-Aid for Scientific Research #2350 0245.

References

1. Whitehead, S.D., Ballard, D.H.: Active Perception and Reinforcement Learning. Neural Computation 2(4), 409–419 (1990)
2. Darrell, T.: Reinforcement Learning of Active Recognition Behavior. Interval Research Technical Report, 1997-045 (1998)
3. Paletta, L., Pinz, A.: Active Object Recognition by View Integration and Reinforcement Learning. Robotics and Automation Systems 31, 71–86 (2000)

4. Lopez, M.T., et al.: Dynamic Visual Attention Model in Images Sequences. Image and Vision Computing 25, 597–613 (2007)
5. Larochelle, H., Hinton, G.: Learning to combine foveal glimpses with a third-order Boltzmann machine. In: Advances in Neural Information Processing, vol. 23, pp. 1243–1251 (2010)
6. Denil, M., Bazzani, L., de Freitas, N.: Learning Where to Attend with Deep Architecture for Image Tracking. Neural Computation 24(8), 2151–2184 (2012)
7. Shibata, K.: Emergence of Intelligence through Reinforcement Learning with a Neural Network. In: Mellouk, A. (ed.) Advances in Reinforcement Learning, pp. 99–120. InTech (2011)
8. Shibata, K., Nishino, T., Okabe, Y.: Active Perception and Recognition Learning System Based on Actor-Q Architecture. System and Computers in Japan 33, 12–22 (2002)
9. Faudzi, A.A.M., Shibata, K.: Acquisition of active perception and recognition through Actor-Q learning using a movable camera. In: Proc. of SICE Annual Conf., FB03-2.pdf (2010)
10. Utsunomiya, H., Shibata, K.: Contextual Behaviors and Internal Representations Acquired by Reinforcement Learning with a Recurrent Neural Network in a Continuous State and Action Space Task. In: Köppen, M., Kasabov, N., Coghill, G. (eds.) ICONIP 2008, Part II. LNCS, vol. 5507, pp. 970–978. Springer, Heidelberg (2009)
11. Shibata, K., Utsunomiya, H.: Discovery of Pattern Meaning from Delayed Rewards by Reinforcement Learning with a Recurrent Neural Network. In: Proc. of IJCNN (Int'l Joint Conf. on Neural Networks), pp. 1445–1452 (2011)
12. Goto, K., Shibata, K.: Emergence of prediction by reinforcement learning using a recurrent neural network. Journal of Robotics, Article ID 437654 (2010)
13. Faudzi, A.A.M., Shibata, K.: Context-based Word Recognition through a Coupling of Q-Learning and Recurrent Neural Network. In: Proc. of SICE Annual Conf. of the Kyushu Branch, pp. 155–158 (2011)
14. Rumelhart, D.E., Hinton, G.E., Williams, R.J.: Learning Internal Representations by Error Propagation. In: Parallel Distributed Processing, pp. 318–362. The MIT Press (1986)
15. Watkins, C.J.C.H., Dayan, P.: Q-learning. Machine Learning 8, 279–292 (1992)

An Ontology Based Approach to Action Verification for Agile Manufacturing

Stephen Balakirsky[1] and Zeid Kootbally[2]

[1] Georgia Tech Research Institute, Atlanta, GA 30332, USA
stephen.balakirsky@gtri.gatech.edu
www.unmannedsystems.gtri.gatech.edu
[2] University of Maryland, College Park, MD 20740, USA
zeid.kootbally@umd.edu
www.nist.gov/el/isd/ks/kootbally.cfm

Abstract. Many of today's robotic work cells are unable to detect when an action failure has occurred. This results in faulty products being sent down the line, and/or downtime for the cell as failures are detected and corrected. This article examines a novel knowledge-driven system that provides added agility by detecting and correcting action failures. The system also provides for late binding of action parameters, thus providing flexibility by allowing plans to adapt to changing environmental conditions. The key feature of this system is its knowledge base that contains the necessary relationships and representations to allow for failure detection and correction. This article presents the ontology that stores this knowledge as well as the overall system architecture. The manufacturing domain of kit construction is examined as a sample test environment.

Keywords: failure detection, manufacturing, ontology, robotics, Planning Domain Definition Language.

1 Introduction

A failure is any change, design, or manufacturing error that renders a component, assembly, or system incapable of performing its intended function [6]. In kitting, as described in Section 2, failures can occur for multiple reasons that include equipment not being set up properly, tools and/or fixtures not being properly prepared, and improper equipment maintenance. Part/component availability failures can be triggered by inaccurate information on the location of the part, part damage, incorrect part types, or part shortage due to delays in internal logistics. In order to prevent or minimize failures, a disciplined approach needs to be implemented to identify the different ways a process design can fail and to allow for corrective actions to be taken before the failure impacts productivity.

Even though today's state-of-the-art industrial robots are capable of sub-millimeter accuracy [7], they often lack the sensing necessary to detect failures and the programming required to cope with and correct the failure. This is due to the fact that they are often programmed by an operator using imprecise

J.-H. Kim et al. (eds.), *Robot Intelligence Technology and Applications 2*,
Advances in Intelligent Systems and Computing 274,
DOI: 10.1007/978-3-319-05582-4_18, © Springer International Publishing Switzerland 2014

positional controls from a teach pendant. These teach pendant programs are highly repeatable, which provides utility for large-batch, error-free operation. However, the cyclic program that repeats identical operations does not lend itself well to adaptation for failure mitigation. In fact, producing a program to correct a perceived failure would require that the cell be taken off-line for additional human-led teach pendant programming. In addition, most cells lack the ability to sense that a failure occurred and lack programming (that would have had to be teach pendant entered) to cope with failure conditions, thus making it impossible for the cell to recover from failures. This leads to faulty products being sent down the line, and/or downtime for the cell as failures are detected and corrected.

For small batch processors or other customers who must frequently change their line configuration or desire to perform complex operations with their robots, this frequent downtime and lack of failure correction/detection may be unacceptable. The robotic systems of tomorrow need to be capable, flexible, and agile. These systems need to perform their duties at least as well as human counterparts, be quickly re-tasked to other operations, cope with a wide variety of unexpected environmental and operational changes, and be able to detect and correct errors in operation. To be successful, these systems need to combine domain expertise, knowledge of their own skills and limitations, and both semantic and geometric environmental information.

The IEEE Robotics and Automation Society's Ontologies for Robotics and Automation Working Group has taken the first steps in creating the infrastructure necessary for such a system, while the Industrial Subgroup has applied this infrastructure to create a sample kit building system. This work is presented in Balakirsky et al. [2] which describes the construction of a robotic kit building system that is able to cope with environmental and task changes without operator intervention. This article extends that work to utilize the same infrastructure to allow for the detection and correction of action failures in the system.

The organization of the remainder of this paper is as follows. Section 2 describes the domain of kit building. Section 3 presents an overview of the software system architecture as well as details of the ontology and world model for the robot cell. Section 4 discusses the detailed operation of cell, and Section 5 discusses how failures are handled by the ontology. Finally, Section 6 presents conclusions and future work.

2 Kitting

Today's advanced manufacturing plants utilize mixed-model assembly where multiple product variants are built on the same line. According to Jim Tetreault, Ford's vice president of North America Manufacturing, new Ford[1] assembly facilities are able to build a full spectrum of vehicles on the same assembly line [9].

---

[1] No approval or endorsement of any commercial product by the authors is intended or implied. Certain commercial software systems are identified in this paper to facilitate understanding. Such identification does not imply that these software systems are necessarily the best available for the purpose.

One of the technologies that makes this possible is the use of assembly kits. Bozer and McGinnis [4] describe a kit as "a specific collection of components and/or sub-assemblies that together (i.e., in the same container) support one or more assembly operations for a given product or shop order". These kits provide a synchronous material flow, where parts and components move to assembly stations in a just-in-time manner. The kits provide workers with the parts and tools that they need (which may vary from vehicle model to vehicle model) in the sequence that they need them. The use of kitting also allows a single delivery system to feed multiple assembly stations thus saving manufacturing or assembly space [10] and provides an additional inspection opportunity that allows for the detection of part defects before they impact assembly operations. The individual operations of the station that builds the kits may be viewed as a specialization of the general bin-picking problem [11] where parts are picked from one or more part bins or trays and placed into specific slots in a kit tray.

For our sample implementation, we assume that the robot cell is building one of several possible kit configurations. At execution time, the cell has a set kit to build, but does not know the precise location of the kit tray, the part trays, or the location of individual parts in the part tray. When a human builds a kit, they are able to inspect each part before adding it to the kit tray. This provides an additional level of quality control and is an aspect that is desirable to have in our robotic system. During kit construction, a robot performs a series of pick-and-place operations in order to construct the kit. These operations include:

1. Pick up an empty kit and place it on the work table.
2. Pick up multiple component parts, inspect them, and place them in the kit.
3. Pick up the completed kit and place it in the full kit storage area.

Each of these steps may be a compound action that includes other actions such as end-of-arm tool changes, path planning, and obstacle avoidance. The items that are being placed in the kit may be of varying size and shape and have various grasping and inspection requirements.

3   System Overview

The kitting system that has been implemented as part of this work is a deliberative intelligent system based on the 4D/RCS reference model architecture [1]. This architecture is a hierarchical architecture in which each echelon or level follows a sense-model-act paradigm. The basic structure of the system may be seen in Figure 1.

3.1   Sense

In order to sense action failures associated with kit building, it is necessary to be able to detect the six-degree of freedom pose of relevant objects in the world. One issue with pose detection is the large number of potential target objects and object classes in the world. Both the number of objects and potential

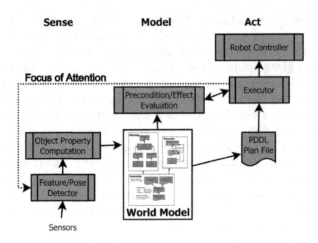

Fig. 1. Major components that make up the Sense–Model–Act paradigm of the kitting station

classes can be reduced by intelligently selecting critical objects of interest that are tagged with predicted locations and object classes for the sensor system to track. This object selection, also known as focus of attention, is guided by the Executor process with knowledge obtained from preconditions and effects of planned actions. Actual algorithm development for pose and object detection is an active research area, and is beyond the scope of this article. For our purposes, we have assumed the use of a high-quality system that is capable of recognizing a limited variety of items in a controlled environment and then determining each item's pose.

3.2 Model

The world model that is being utilized is shown in Figure 2. The model contains knowledge that is structured specifically for reasoning, planning, and execution. All of the concepts necessary for the manufacturing domain under test are encoded in the ontology that resides in the reasoning section of the model. The planning and execution sections of the model are automatically generated from this section.

Ontology – The reasoning portion of the world model is designed to contain all of the information needed to reason over and solve complex manufacturing problems. The knowledge is represented in a Web Ontology Language (OWL) ontology that is structured in three parts. The first part of the ontology contains generic information and classes that are needed for the domain of kit building. This area of the ontology contains information on basic elements such as a "point" which is defined as a class that contains a name and a three-dimensional

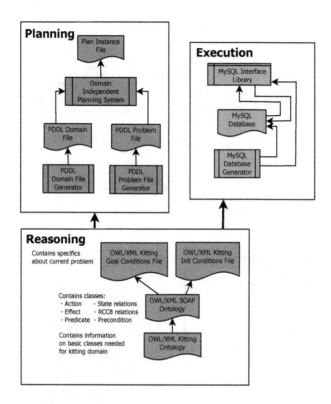

Fig. 2. System World Model - The world model contains a Reasoning section that is based on an ontology shown in green, a Planning section that is based on a Planning Domain Definition Language (PDDL) specification shown in blue, and an Execution section that is based on a relational database (MySQL) shown in orange

quantity, as well as complex types such as a "part", which is shown in Figure 3, and contains elements such as the part's location and a name that references a stock keeping unit. The stock keeping unit contains static information on classes of parts such as the part's shape, weight, and the end effector that should be used for grasping the part. This information is utilized to create parameters for the Planning World Model and the skeleton tables for the MySQL database of the Execution World Model.

Both static and dynamic information is represented in this ontology and is automatically transitioned into the Planning and Execution areas of the world model. During system operation, dynamic information is updated in the Execution World Model. More information on this portion of the ontology may be found in [3].

The second part of the ontology (known as States, Ordering constraints, Actions, and Predicates or SOAP) contains the high-level concept of an action and all of the concepts that are required to support an action. In this case,

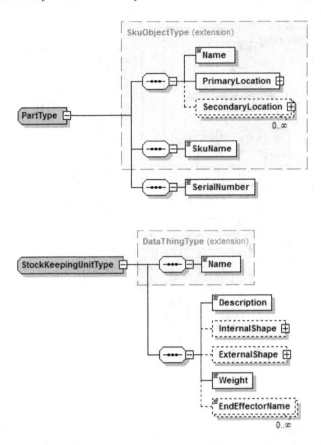

Fig. 3. Description of the PartType class that is designed to contain both static and dynamic information about particular parts and the StockKeepingUnitType class that contains static information about classes of parts

a PDDL [8] action is being represented, and this action is defined as an operator that causes one or more properties of an instance to change. Before this action may be performed, certain preconditions must be satisfied, and after the action is performed, certain effects will take place. The action accepts parameters that specify the particular instances that will be affected, where an instance refers to a physical object or piece of grounded data in the world. All of the necessary information for the automatic generation of the PDDL domain file required by the planning system is contained in this section of the ontology. The classes used to represent the actions in the ontology are provided in the enumerated list shown below.

1. *Action* – An *Action* is the basic type that is used by a planning system to produce changes on the world. It has an *ActionPrecondition* that must be valid before the action may be executed and an *ActionEffect* that is expected

to be produced by the action. Each action requires parameters that are used to specify which objects are being operated upon. These parameters are contained in the *ActionParameterList*. An *Action* has a unique name.

2. *ActionParameterList* – Actions may have multiple parameters of different types that are represented by a class in the upper ontology. The *ActionParameterList* contains the particular action's parameters that are instances from the upper ontology. The order of the parameters in an action also needs to be represented in the ontology. OWL has no built-in structure to represent an ordered list. In order to maintain parameter ordering, the parameter uses *hasNextParameter* and *hasPreviousParameter* to point to the next and the previous parameter in *ParameterList*, respectively.

3. *ActionPrecondition* – An *ActionPrecondition* specifies necessary conditions that must be true in order for an action to be undertaken. It can consist of an *ActionPredicate*, an *ActionFunction*, an *ActionFunctionBool*, or a combination of these three classes. An *ActionPrecondition* belongs to one *Action*.

4. *ActionEffect* – An *ActionEffect* specifies the results that are anticipated to occur as a result of a particular action. It can consist of an *ActionPredicate*, an *ActionFunction*, an *ActionFunctionBool*, or a combination of these three classes. An *ActionEffect* belongs to one *Action*. A negative *ActionPredicate* is represented with the declaration of *hasEffect_Predicate* within the OWL built-in property assertion `owl:NegativePropertyAssertion`.

5. *ActionPredicate* – A predicate is used to specify a binary property of a single object, or a relationship between two objects. For example, the predicate (`robot-empty ?robot`) is true if the robot `?robot` is not holding anything. The predicate (`part-location-robot ?part ?robot`) is true only if the reference parameter `?part` is being held by the target parameter `?robot`.

An *ActionPredicate* represents these predicates. It has a unique name of type `string`, a reference parameter and an optional target parameter. The reference parameter is the first parameter in the predicate's parameter list and the target parameter is the second parameter in the predicate's parameter list. An *ActionPredicate* cannot have more than two parameters due to the inherent definition of predicates. In the case where an *ActionPredicate* has only one parameter, it is assigned to the reference parameter.

6. *ActionFunction* – In version 3 of the PDDL language, it is possible to use numeric functions that contain one or two parameters. For example, the function (`quantity-partstray ?partstray - PartsTray`) will return the number of parts that are contained in the parts tray `?partstray`, and the function (`quantity-kit ?kit - Kit ?partstray - PartsTray`) will return the number of parts from parts tray `?partstray` that are currently in the kit `?kit`. The class *ActionFunction* is used to represent these functions. It has a unique name of type `string` along with a reference parameter and a target parameter. The same rules apply to the definition and use of these two types of parameters as defined for *ActionPredicate*.

7. *ActionFunctionBool* – In version 3 of the PDDL language, it is also possible to compare the results returned by two functions. The class *ActionFunctionBool* is used to represent these relationships. It has one or more subclasses

that represent the type of relation (mathematical operator) between two *ActionFunctions*. Subclasses of *ActionFunctionBool* have a first *ActionFunction* that represents the *ActionFunction* on the left side of the operator and a second *ActionFunction* that represents the *ActionFunction* on the right side of the operator.

The third part of the ontology contains specific instances needed for a particular kitting domain. For example, it will contain the definition of the finished kits that may be constructed and specific information on the individual parts. One of the goals of this framework is to introduce additional agility into the kit building process. Therefore, partial information is accepted and even encouraged for this area of the ontology. For the example of a part shown in Figure 3, information on the SKU, grasp points (part of the ExternalShape or InternalShape), and name would be expected to be available at runtime. Information on the location of the part (PrimaryLocation) may not become valid until after a sensor processing system has identified and located the particular part.

Planning – PDDL is an attempt by the domain independent planning community to formulate a standard language for planning. A community of planning researchers has been producing planning systems that comply with this formalism since the first International Planning Competition held in 1998. This competition series continues today, with the seventh competition being held in 2011. PDDL is constantly adding extensions to the base language in order to represent more expressive problem domains. The representation in the world model is based on PDDL Version 3.

By placing the knowledge in a PDDL representation, the use of an entire family of open source planning systems such as the forward-chaining partial-order planning system from Coles et al. [5] is enabled. In order to operate, the PDDL planners require a PDDL file-set that consists of two files that specify the domain and the problem. From these files, the planning system creates an additional static plan file. Both the domain and problem file are able to be auto-generated from the ontology.

The generated static plan file contains a sequence of actions that will transition the system from the initial state to the goal state. In order to maintain flexibility, it is desired that detailed information that is subject to change should be "late-bound" to the plan. In other words, specific information is acquired directly before that information needs to be used by the system. This allows for last minute changes in this information. For example, the location of a kit tray on a work table may be different from run to run. However, one would like to be able to use the same planning sequence for constructing the kit independent of the tray's exact position. To compensate for this lack of exact knowledge, the plans that are generated by the PDDL planning system contain only high-level actions.

As seen in Figure 2, the planning world model framework contains generators that read the ontology and create a standard PDDL domain and PDDL problem files. Any of the family of PDDL Version 3 compatible planning systems is then able to be run on these files to create the static plan instance file. A representation of this plan may be stored in the ontology for future use.

Execution − The execution world model is also built automatically from the ontology. This world model consists of a MySQL database and C++ interfaces that provide for easy access to the data. The table skeletons are generated from the kitting ontology, and the tables are initially populated with information from the initial condition file. During plan execution, the Executor guides the sensor processing system in updating the information in this section of the world model. All of the data structures encoded in the ontology are included in this representation.

3.3 Act

The actions that take place in the kitting work cell are coordinated by the Executor as illustrated in Figure 1. The Executor reads PDDL actions as input and outputs a standardized set of low-level robot commands encoded in a language developed at the National Institute of Standards and Technology (NIST) known as the Canonical Robotic Command Language (CRCL) [3].

Before and after each high-level command is executed, the Executor sends focus of attention information into the sensor processing system. This allows the sensor processing system to compute the appropriate predicate relations that are required to verify the conditions necessary to carry out an action and that an action's execution has been successful. Information on predicates is written to the world model by the sensing system and read from the world model by the Executor.

4   System Operation

In order to construct a kit, the kitting system steps through each action in the precomputed PDDL plan. Failures are searched for both before and after execution of each action. The overall process, known as BuildKit is shown in Figure 4.

This process begins by retrieving a planning instance that has been precomputed to solve the construction of the requested kit (Line 1 of Figure 4). This is a high-level PDDL plan that is not grounded to actual part instances or locations. It contains information on the named storage location for classes of parts (individual SKU numbers), the quantity of each SKU that is required by the kit, and a build order (a sequence of SKUs to be added to the kit). Additional information on the appropriate end-of-arm tooling required to grip each part is also included.

This planning instance is decomposed into individual actions that must be successfully carried out to complete the construction of the kit (the *for* loop

---

Data: *kitToBuild*
Result: reports success or failure
1 retrieve instance *PDDLInstance* to construct kit *kitToBuild*;
2 **for** *each action **A** in PDDLInstance* **do**
3   **for** *each precondition **P** of action **A*** **do**
4   **if** *PredicateEvaluation(P) = false* **then**
5   report failure;
6   **end**
7   **end**
8   create set *S* of Canonical Robot Control Language Commands;
9   send set *S* to Robot Controller for execution;
10   **for** *each effect **E** of action **A*** **do**
11   **if** *PredicateEvaluation(E) == false* **then**
12   report failure;
13   **end**
14   **end**
15   report action success;
16 **end**
17 report plan success;

Fig. 4. BUILDKIT – Sequences the actions necessary to build a kit

---

Data: *PredicateIn*
Result: true or false
1 determine predicted pose *PoseR* of *PredicateIn.ReferenceParameter* and *PoseT* of *PredicateIn.TargetParameter* ;
2 send *PredicateIn*, pose *PoseR*, and *PoseT* as focus of attention command to *SensorProcessing*;
3 **if** *Eval(PredicateIn) == true* **then**
4   return true;
5 **else**
6   return false ;
7 **end**

Fig. 5. PREDICATEEVALUATION – Returns the truth value of the predicate expression

beginning at Line 2 of Figure 4). At this point, preconditions of the action are examined to assure that the action to be attempted is valid. If any of the action's preconditions are not able to be validated, a failure is reported; otherwise, the action is approved for execution.

4.1 Precondition Validation

Each precondition is a predicate expression that must be validated prior to action execution. The procedure for validating predicates is shown in Figure 5. In Line 1 of this algorithm, the world model is queried for the pose and class of each relevant parameter of the predicate. The information returned is the latest knowledge that has been recorded by the sensor processing system and is not guaranteed to be up-to-date. This possibly out-of-date information is used as a prediction of the object's current pose and the knowledge is sent as a focus of attention indicator to the sensor processing system. The sensor processing system is instructed to update the world model with current observations and to compute the supporting relationships necessary for predicate evaluation.

Data: $PredicateIn, PoseR, PoseT$
1 determine actual pose $APoseR$ of $PredicateIn.ReferenceParameter$;
2 determine actual pose $APoseT$ of $PredicateIn.TargetParameter$;
3 determine RCC8 relations between $APoseR$ and $APoseT$;
4 determine Intermediate State Relations based on RCC8 relations;
5 determine truth value of $PredicateIn$ and write to MySQL database;

Fig. 6. SENSORPROCESSING – Updates the MySQL database in the Execution world model to contain the latest evaluation of predicates related to $PredicateIn$

Figure 6 depicts the algorithm that is followed by sensor processing in the computation of predicate values. As may be seen from this figure, the computation of predicates is a three step process that involves the computation of increasingly complex forms of spatial relations. These relationships; Region Connection Calculus (RCC8) relations, intermediate state relations, and predicate relations, are each represented as a separate class in the ontology.

RCC8 Relation – RCC8 [12] is an approach for representing the relationship between two regions in Euclidean or topological space. The class $RCC8\_Relation$ is based on the definition of RCC8 and consists of eight possible relations that include Tangential Proper Part (TPP), Non-Tangential Proper Part(NTPP), Disconnected (DC), Tangential Proper Part Inverse (TPPi), Non-Tangential Proper Part Inverse (NTPPi), Externally Connected (EC), Equal (EQ), and Partially Overlapping (PO). In order to represent these relations in all three dimensions for the kitting domain, RCC8 has been extended to a three-dimensional space by applying it along all three planes (x-y, x-z, y-z) and by including cardinal direction relations "+" and "-". In the ontology, RCC8 relations and cardinal direction relations are represented as subclasses of the class $RCC8\_Relation$.

Intermediate State Relation – These relations can be inferred from the combination of RCC8 and cardinal direction relations. For instance, the intermediate state relation **In-Contact-With** is used to describe that object *obj1* is in contact with object *obj2*. This is true if *obj1* is externally connected to *obj2* in the x-direction, the y-direction, or the z-direction, and is represented with the following combination of RCC8 relations:

$$\textbf{In-Contact-With}(obj1, obj2) \rightarrow$$
$$\text{x-EC}(obj1, obj2) \vee \text{y-EC}(obj1, obj2) \vee \text{z-EC}(obj1, obj2)$$

In the ontology, intermediate state relations are represented with the OWL built-in property `owl:equivalentClass` that links the description of the class *Intermediate_State_Relation* to a logical expression based on RCC8 relations from the class *RCC8_Relation*.

Predicate Relation − The truth-value of predicates can be determined through the logical combination of intermediate state relations. The predicate *endeffector-location-endeffectorholder(endeffector, endeffectorholder)* is true if and only if the location of the end effector *endeffector* is in the end effector holder *endeffectorholder* and is not attached to a robot. This predicate can be described using the following combination of intermediate state relations:

$$\text{endeffector-location-endeffectorholder}(endeffector, endeffectorholder) \rightarrow$$
$$\textbf{In-Contact-With}(endeffector, enfectorholder) \wedge$$
$$\neg\textbf{In-Contact-With}(endeffector, robot)$$

As with state relations, the truth-value of predicates is captured in the ontology using the `owl:equivalentClass` property that links the description of the class *Predicate* to the logical combination of intermediate state relations from the class *Intermediate_State_Relation*.

Truth Value Determination − As seen in Section 3.2, a predicate can have a maximum of two parameters. In the case where a predicate has two parameters, the parameters are passed to the intermediate state relations defined for the predicate, and are in turn passed to the RCC8 relations where the truth-value of these relations are computed. If the predicate has only one parameter, the truth-value of intermediate state relations, and by inference, the truth-value of RCC8 relations will be tested with this parameter and with every object in the environment in lieu of the second parameter. Our kitting domain consists of only one predicate that has no parameters. This predicate is used as a flag in order to force some actions to come before others during the formulation of the plan. Predicates of this nature are not treated in the concept of "Spatial Relation".

These truth values may be retrieved from the ontology for use in Line 3 of Figure 5 which will then propagate back up to BUILDKIT. If the predicates are successfully evaluated, the action will be cleared for execution and a set of CRCL Commands will be generated.

4.2 Canonical Robot Command Language Generation

Up to this point, the PDDL actions are not fully grounded to specific instances that exist in the world. For example, the action `put-part` is designed to place

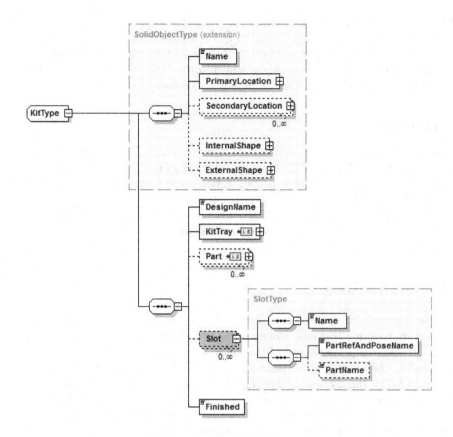

Fig. 7. Description of the KitType class that is designed to bring together a container (the KitTray), a design for the kit creation, and specific parts that populate the kit

the part that is currently being held by the robot into a kit that is specified as one of the action's input parameters. However, the precise location for this part to be placed is not specified at the time of plan creation. It is up to the Executor, working with the world model, to find an empty slot in the kit that can receive the part. The structure of the ontology is specifically designed to support this grounding, and this structure is automatically replicated in the MySQL database that resides in the Execution portion of the world model.

Continuing with the example of the put-part command, the Executor needs to find an empty slot in the target kit to place the specific part that the robot is holding. As shown in Figure 7, the KITTYPE class contains zero to many SLOT classes that in turn contain specific location and part identification information. The Executor is able to read this information from the world model and determine the precise global pose where the part should be placed. The robot controller must now be commanded to complete this action.

While the action is an atomic element in PDDL, it will decompose into a series of actions in CRCL. The robot will need to follow a safe trajectory to achieve the slot in the kit, and the gripper will need to be controlled in order to release the part. This one-to-many mapping is performed in the Executor and is currently hand-coded for each of the PDDL commands that exists in the system.

4.3 Effect Validation

The purpose of performing an action is to achieve results in the world. These results are represented in the PDDL effects. Each effect is a predicate expression that must be validated to assure proper action execution. The technique for validating the effect predicates is identical to the evaluation of the precondition predicates described in Section 4.1. If all of the effects are able to be validated, the system will report the action's success and begin performing the next action in the plan.

5 Failure Analysis

As seen in Figure 5, failures are identified during the evaluation of preconditions and effects. In addition to recognizing failures, it is desired to be able to respond to them. In order to properly model this response, additional information must be modeled in the ontology. As shown in Figure 8, the class *ActionFailureType* has been added to the ontology. This class is composed of a *PredicateType* class and a *FailureModeType* class. It contains all of the information required to be able to diagnose and remediate any failures that are detected by the system.

The *PredicateType* class contains a list of predicates whose truth values caused the detection of the current failure condition. Examination of the instances pointed to by the *ReferenceParameter* and the *TargetParameter* allow the system to understand exactly which components were involved in the failure.

The *FailureModeType* class provides various known failure modes that could exist for the combination of *PredicateTypes* that were found deficient. It provides the consequences of such a failure occurring, information on how to remediate such a failure, and the chance that this kind of failure could occur. Understanding the consequences of the failure mode is important if one would like to be able to pinpoint the correct cause of the failure. For example, assume that the action TAKE-PART failed, and the predicate that indicated the failure was *part-location-robot*.

Understanding where the part is actually located is critical to understanding the root cause of the failure. If the part is still in the parts tray, it would indicate a grasp related failure. If the part is on the floor of the cell, it would indicate a part handling error. This kind of information is represented in the *Effect* class of the consequences. Being able to pinpoint the effects of the failure also allows for judgment on the failure severity. A missed grasp will likely lead to a new grasp attempt and will have little impact on operation. A dropped part may cause damage to the part and have a higher severity level.

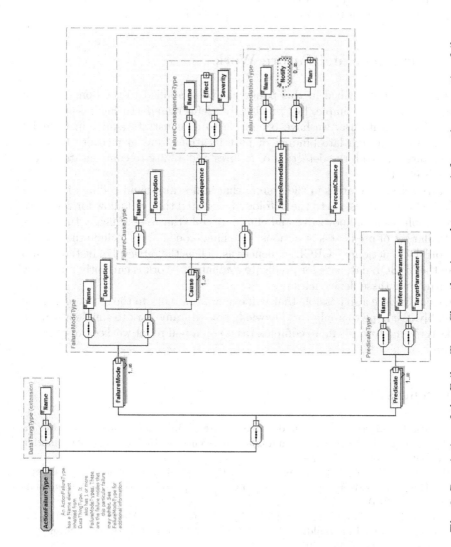

Fig. 8. Description of the FailureType Class that is used to store information on action failures

Failure remediation is described by the *FailureRemediation* class. This class contains information on what notifications need to be sent as a result of the failure as well as what actions should be taken. Continuing our example from above, the grasp failure may cause grasping information to be sent to the enginering department for algorithm refinement while the dropped part may notify logistics that an additional component will be required. Recovery from the failure is possible by executing the *Plan* that accompanies the detected failure.

6   Conclusions and Future Work

The framework described in this paper has been applied to the domain of kit building, which is a simple, but practically useful manufacturing/assembly domain. Through its use, we have been able to demonstrate agility in both kit construction through late binding of part locations, and in recovery from action failures through the detection of failures and ability to compensate for the failure's effects.

There are several areas in the system that still utilize hand-coding of data. It is desired that extensions to the ontology be created that will allow for the automatic application of knowledge and eliminate code that is specifically tuned to a particular set of predicates or actions. The hand-coded areas include the conversion of PDDL actions to CRCL sequences as well as the retrieval of instance data from the MySQL database for predicate evaluation. Work is currently underway to correct for these deficiencies.

Extensions are also possible that will expand this work to the realm of general assembly. We hope to apply this knowledge based framework to simple assembly tasks (growing towards more complex tasks) on a real robot workcell in the near future.

References

1. Albus, J.: 4-d/rcs reference model architecture for unmanned ground vehicles. In: Proceedings of the IEEE International Conference on Robotics and Automation (ICRA), pp. 3260–3265 (2000)
2. Balakirsky, S., Kootbally, Z., Kramer, T., Pietromartire, A., Schlenoff, C., Gupta, S.: Knowledge driven robotics for kitting applications. Robotics and Autonomous Systems (2013)
3. Balakirsky, S., Kramer, T., Kootbally, Z., Pietromartire, A.: Metrics and test methods for industrial kit building. In: NISTIR 7942, National Institute of Standards and Technology, NIST (2012)
4. Bozer, Y.A., McGinnis, L.F.: Kitting versus line stocking: A conceptual framework and descriptive model. International Journal of Production Economics 28, 1–19 (1992)
5. Coles, A.J., Coles, A., Fox, M., Long, D.: Forward-chaining partial-order planning. In: 20th International Conference on Automated Planning and Scheduling, ICAPS 2010, Toronto, Ontario, Canada, May 12-16, pp. 42–49. AAAI (2010)

6. Collins, J.: Failure of Materials in Mechanical Design: Analysis, Prediction, Prevention. Wiley Interscience (1993)

7. Control and Montreal Robotics Lab at the ETS. Measuring the absolute accuracy of an abb irb 1600 industrial robot (2011),
   http://www.youtube.com/watch?v=d3fCkS5xFlg

8. Ghallab, M., Howe, A., Knoblock, C., McDermott, D., Ram, A., Veloso, M., Weld, D., Wilkins, D.: Pddl–the planning domain definition language. Technical Report CVC TR98-003/DCS TR-1165, Yale (1998)

9. James, T.: Ford's michigan eco car plant: one size fits all. Engineering and Technology Magazine 7 (July 2011)

10. Medbo, L.: Assembly work execution and materials kit functionality in parallel flow assembly systems. International Journal of Industrial Ergonomics 31, 263–281 (2003)

11. Schyja, A., Hypki, A., Kuhlenkotter, B.: A modular and extensible framework for real and virtual bin-picking environments. In: 2012 IEEE International Conference on Robotics and Automation (ICRA), pp. 5246–5251 (May 2012)

12. Wolter, F., Zakharyaschev, M.: Spatio-temporal representation and reasoning based on rcc-8. In: Proceedings of the 7th Conference on Principles of Knowledge Representation and Reasoning, KR 2000, pp. 3–14. Morgan Kaufmann (2000)

Evaluating State-Based Intention Recognition Algorithms against Human Performance

Craig Schlenoff[1] and Sebti Foufou[1,2]

[1] LE2i Lab, University of Burgundy, Dijon, France
craig@schlenoff.com
[2] Computer Science and Engineering Department,
Qatar University, Doha, Qatar
sfoufou@qu.edu.qa

Abstract. In this paper, we describe a novel intention recognition approach based on the representation of state information in a cooperative human-robot environment. We compare the output of the intention recognition algorithms to those of an experiment involving humans attempting to recognize the same intentions in a manufacturing kitting domain. States are represented by a combination of spatial relationships in a Cartesian frame along with cardinal direction information. Based upon a set of predefined high-level states relationships that must be true for future actions to occur, a robot can use the approaches described in this paper to infer the likelihood of subsequent actions occurring. This would enable the robot to better help the human with the operation or, at a minimum, better stay out of his or her way.

Keywords: intention recognition, state-based representation, ontologies, human robot safety, RCC-8, robotics, human performance.

1 Introduction

Humans and robots working safely and seamlessly together in a cooperative environment is one of the future goals of the robotics community. When humans and robots can work together in the same space, a whole class of tasks becomes amenable to automation, ranging from collaborative assembly, to parts and material handling and delivery. Keeping humans safe requires the ability of the robot to monitor the work area, infer human intention, and be aware of potential dangers soon enough to avoid them. Robots are under development throughout the world that will revolutionize manufacturing by allowing humans and robots to operate in close proximity while performing a variety of tasks [1].

Proposed standards exist for collaborative robot-human safety, but these focus on robots limiting approach distances and contact forces between the human and the robot [2, 3]. In essence, the robot attempts to minimize the likelihood and potential severity of collisions between the human and the robot. These approaches focus on reactive processes based only on current sensor readings, and does not consider future states or task-relevant information.

J.-H. Kim et al. (eds.), *Robot Intelligence Technology and Applications 2*,
Advances in Intelligent Systems and Computing 274,
DOI: 10.1007/978-3-319-05582-4_19, © Springer International Publishing Switzerland 2014

A key enabler for human-robot safety in cooperative environments involves the field of intention recognition, in which the robot attempts to understand the intention of an agent (the human) by recognizing some or all of their actions [4] to help predict the human's future actions. Knowing these future actions will allow a robot to plan in such a way as to either help the human perform his/her activities or, at a minimum, not put itself in a position to cause an unsafe situation.

In this paper, we present an approach to inferring the intention of an agent in the environment via the recognition and representation of state information. An overview of the intention recognition approach can be found in [5].This paper elaborates on the algorithms that are used to perform intention recognition and shows the performance of the algorithms as compared to human performance. We distinguish states from state relationships. In this context, a state is defined as a set of properties of one or more objects in an area of interest that consist of specific recognizable configuration(s) and or characteristic(s). A state relationship is a specific relation between two objects (e.g., Object 1 is on top of Object 2). A set of all relevant state relationships in an environment composes a state. This approach to intention recognition is different than many ontology-based intention recognition approaches in the literature (as described in the next section), as they primarily focus on activity (as opposed to state) recognition and then use a form of abduction to provide explanations for observations. We infer detailed state relationships using observations based on Region Connection Calculus 8 (RCC-8) [6] and then infer the overall state relationships that are true at a given time.   Once a sequence of state relationships has been determined, we use probabilistic procedures to associate those states with likely overall intentions to determine the next possible action (and associated state) that is likely to occur. This paper focuses on the comparison of the results of the intention recognition algorithms to those of an experiment involving humans attempting to identify and infer the same intentions.

We start by providing an overview of intention recognition efforts in the literature, as well as various approaches for ontology-based state representation. After that we explain our approach to state representation using RCC-8 [7]. Based on these state relations, we show how they are combined to form intentions and then how likelihoods are assigned to the intentions. We then describe an experiment that was performed to compare the output of the algorithms to that of a set of humans trying to infer the same intentions of agents in the environment. We conclude the paper by showing some advantages of state-based recognition approach as compared to other approaches in the literature.

2   Intention Recognition and State Representation Related Work

Intention recognition traditionally involves recognizing the intent of an agent by analyzing some of, or all of, the actions that the agent performs. Many of the recognition efforts in the literature are composed of at least three components: (1) identification and representation of a set of intentions that are relevant to the domain of interest,

(2) representation of a set of actions that are expected to be performed in the domain of interest and the association of these actions with the intentions, and (3) recognition of a sequence of observed actions executed by the agent and matching them to the actions in the representation. [4]

There have been many techniques in the literature applied to intention recognition that follow the three steps listed above, including an ontology-based approach [8], multiple probabilistic frameworks such as Hidden Markov Models [9] and Dynamic Bayesian Networks [10], utility-based intention recognition [11], and graph-based intention recognition [12].

All of these approaches have focused on the activity being performed as the primary basis for observation and the building block for intention recognition. However, as noted in [4], activity recognition is a very hard problem and one that is far from being solved. There has been limited success in using Radio Frequency Identification (RFID) readers and tags attached to objects of interest to track their movement with the goal of associating their movement with known activities [13]. However, this additional hardware can be inhibiting and unnatural. Recognizing and representing states as opposed to actions can help to address some of the issues involved in activity recognition and will be the focus of the rest of this paper.

State representation is documented in the literature, although it has not been used for ontology-based intention recognition. An important aspect of an object's state is its spatial relationships to other objects. In [14], an overview is given that describes the way that spatial information is represented in various upper ontologies, including the Descriptive Ontology for Linguistics and Cognitive Engineering (DOLCE) [15] , Cyc [16], the Standard Upper Merged Ontology (SUMO)[17], and Basic Formal Ontology (BFO) [18].

RCC-8 is a well-known and cited approach for representing the relationship between two regions in Euclidean space or in a topological space. There are eight possible relations, including "disconnected," "externally connected," and "partially overlapping." However, RCC-8 only addresses these relationships in two-dimensional space. There have been other approaches that have tried to extend this into a region connected calculus in three-dimensional space while addressing occlusions [19]. There have also been other approaches to develop calculi for spatial relations. Flip-Flop calculus [20] describes the position of one point (the "referent") in a plane with respect to two other points (the "origin" and the "relatum"). Single Cross Calculus (SCC) [21] is a ternary calculus that describes the direction of a point (C, the referent) with respect to a second point (B, the relatum) as seen from a third point (A, the origin) in a plane.

Throughout the rest of this paper, we will describe an approach for intention recognition based on ontology-based state representations within the context of a manufacturing scenario.

3   State Representation Approach

As mentioned earlier, RCC-8 abstractly describes regions in Euclidean or topological space by their relations to each other. RCC-8 consists of eight basic relations that are possible between any two regions: Disconnected (DC), Externally Connected (EC), Tangential Proper Part (TPP), Non-Tangential Proper Part (NTPP), Partially Overlapping (PO), Equal (EQ), Tangential Proper Part Inverse (TPPi), and Non-Tangential Proper Part Inverse (NTPPi). These are shown pictorially in Fig. 1.

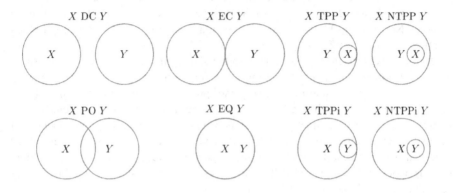

Fig. 1. RCC-8 Relations (Credit: http://en.wikipedia.org/wiki/RCC-8)

In real-world scenarios, not all objects can be represented as convex regions, as is required by the RCC-8 formalism. For example, the robot gripper in Fig. 2 is not a convex hull and thus does not neatly fit into the RCC-8 approach. To address this, we develop a convex hull along each relevant plane (as shown in Fig. 2) around objects of this sort and use that convex hull to represent the region of the object in that plane.

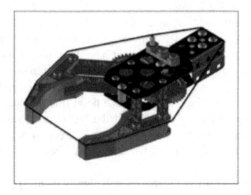

Fig. 2. Convex Hull (Black Outline) Around Robot Gripper

RCC-8 was created to model the relationships between two regions in two dimensions. In the manufacturing kitting domain, these relations need to be modeled in all three dimensions. As such, every pair of objects has a RCC-8 relation in all three dimensions. To address this, we are prepending an x-, y- or z- before each of the

RCC-8 relations to denote axial planes. The combination of all 24 RCC relations (8 per axis) starts to describe the spatial relations between any two objects in the scene. However, more information is needed to represent the state relationships described in the previous section. For example, to state that a worktable is empty, one needs to state that there is nothing on top of it. For this approach, we use the reference frame attached to a fixed point of reference (in this case the worktable) with the positive z-direction pointing away from the gravitational center, then saying that:

$$z\text{-}EC(wtable, obj1) \tag{1}$$

(which intuitively means $obj1$ is externally connected to the worktable in the z-dimension). This relation is not sufficient because $obj1$ could be either on top of or below the worktable. In other words, we need to represent directionality. We do this using the following Boolean operators:

$$greater\text{-}z(A, B) \tag{2}$$

$$smaller\text{-}z(A, B) \tag{3}$$

which intuitively means, in Equation (2), that the centroid of object A is greater than (in the z-dimension in the defined frame of reference) the centroid of object B, measured in the same reference frame. For this work, we are dealing with relatively compact objects (i.e., objects that have few voids when represented by convex hulls), thus the centroid is a good indication of the location of the object in geometric space.

There are many other relationships that may be needed in the future to describe a scene of interest. These could include absolute locations and orientations of objects (i.e., x, y, z, roll, pitch, and yaw) and relative distance (e.g., closer or farther).

Using these relations, we can create created more complex state relations, such as:

$$\textbf{In-Contact-With}(\textbf{\textit{obj1}},\textbf{\textit{obj2}})\rightarrow$$
$$x\text{-}EC(obj1, obj2) \vee y\text{-}EC(obj1, obj2) \vee x\text{-}EC(obj1, obj2) \tag{4}$$

$$\textbf{On-Top-Of}(\textbf{\textit{obj1}}, \textbf{\textit{obj2}}) \rightarrow$$
$$greater\text{-}z(obj1, obj2) \wedge ((x\text{-}EQ(obj1, obj2) \vee$$
$$x\text{-}NTPP(obj1, obj2) \vee x\text{-}TPP(obj1, obj2) \vee x\text{-}PO(obj1, obj2) \quad \vee$$
$$x\text{-}NTPPi(obj1, obj2) \vee x\text{-}TPPi(obj1, obj2)) \quad \wedge$$
$$((y\text{-}EQ(obj1, obj2) \vee y\text{-}NTPP(obj1, obj2) \vee y\text{-}TPP(obj1, obj2)$$
$$\vee y\text{-}PO(obj1, obj2) \quad \vee y\text{-}NTPPi(obj1, obj2) \vee y\text{-}TPPi(obj1, obj2)) \tag{5}$$

$$\textbf{Contained-In}(\textbf{\textit{obj1}}, \textbf{\textit{obj2}}) \rightarrow$$
$$(x\text{-}TPP(obj1, obj2) \vee x\text{-}NTPP(obj1, obj2)) \wedge$$
$$(y\text{-}TPP(obj1, obj2) \vee y\text{-}NTPP(obj1, obj2)) \wedge$$
$$(z\text{-}TPP(obj1, obj2) \vee z\text{-}NTPP(obj1, obj2)) \tag{6}$$

and then domain specific state relations such as:

$$\textbf{worktable-empty}(\textit{wtable}) \rightarrow$$
$$\text{SolidObject}(\textit{obj1}) \land \neg\text{On-Top-Of}(\textit{obj1}, \textit{wtable})$$
$$\land \neg\text{In-Contact-With}(\textit{obj1}, \textit{wtable}) \tag{7}$$

More information about the state representation approach can be found in [22].

4   How States Are Combined to Form an Intention

In the previous section, we showed how state relations are represented. In this section, we show how they can be combined to form an intention. To do this, we leverage OWL-S (Web Ontology Language – Services) [23]. Table 1 shows the OWL-S ordering constructs in the first column, the adapted state representation ordering constructs in the second column, and a brief definition of those adapted ordering constructs in the third column.

Table 1. Initial State Representation Ordering Constructs

OWL-S Control Construct	Adapted State Representation Ordering Construct	State Representation Ordering Construct Definition
Perform	Exists	A state relationship must exist
Sequence	OrderedList	A set of state relationships that must occur in a specific order
Any-Order	Any-Order	A set of state relationships that must all occur in any order
Iterate	Count	A state relationship that must be present exactly the number of times specified (greater than one).
n/a	NotExists	State relationship that can't exist.
Choice	Choice	A set of possible state relationships that can occur after a given state relationship
Join	Co-Exist	Two or more state relationships that must be true

We use the following simple kitting scenario as an example or how to apply the ordering constructs. A person is constructing a set of kits. In this case, the type of kit that is being constructed constitutes the intention that is trying to be inferred. One of the kits contains two instances of part A, two instances of part B, one part C, and one part D. This intention can be described as follows:

$$SR1 = \text{On-Top-Of(KitTray, Table)} \tag{8}$$

$$OC1 = \text{Exists(SR1)} \tag{9}$$

$$SR2 = \text{Contained-In(PartA, KitTray)} \tag{10}$$

$$SR3 = \text{Contained-In(PartB, KitTray)} \tag{11}$$

$$SR4 = \text{Contained-In(PartC, KitTray)} \tag{12}$$

$$SR5 = \text{Contained-In(PartD, KitTray)} \tag{13}$$

$$OC2 = \text{Count(SR2, 2)} \tag{14}$$

$$OC3 = \text{Count(SR3, 2)} \tag{15}$$

$$OC4 = \text{Count(SR4, 1)} \tag{16}$$

$$OC5 = \text{Count(SR5, 1)} \tag{17}$$

$$OC6 = \text{AnyOrder(OC2, OC3, OC4, OC5)} \tag{18}$$

$$SR6 = \text{Contained-In(KitTray, CompletedKitBox)} \tag{19}$$

$$OC7 = \text{Exists(SR6)} \tag{20}$$

$$OC8 = \text{OrderedList(OC1, OC6, OC7)} \tag{21}$$

Using this approach, we can match observed states with those that are represented in the template to infer which intention is most likely occurring. More information about the intention representation approach can be found in [24]. In the next section, we describe how likelihoods can be associated with various intentions based on observations.

5   Assigning Likelihoods to Intentions

Based on a sequence of observations of state in the environment, we need to assign likelihoods to various intentions that can occur in the environment. We do this using the following generalized equation:

$$L_i = \prod_{j=1}^{m} MM_{i,j} * \left(\frac{\sum_{k=1}^{n} (AM_{i,k} * W_{AM_{i,k}})}{\sum_{k=1}^{n} W_{AM_{i,k}}} \right) * 100 \tag{22}$$

where:

- L_i is the numeric likelihood of an intention (i)
- $MM_{i,j}$ is the numeric result of applying multiplicative metric j for intention i
- $AM_{i,k}$ is the numeric result of applying the kth additive metric for intention i
- $W_{AM_{i,k}}$ is the weight of the kth additive metric for intention i
- m is the total number of multiplicative metrics
- n is the total number of additive metrics

All metrics (whether multiplicative or additive) must contain a value between 0 and 1, where 0 is the lowest value and 1 is the highest value. Additive metrics (AM_k), along with their associated weights, are added together and then divided by the sum of all of their weights. Weights are associated with the additive metrics to show the relative importance of one metric over another. These weights can contain any value greater than or equal to zero (no upper bound). Multiplicative metrics are significant enough in importance that their value is multiplied in the likelihood equation to carry a heavier effect on the overall likelihood.

Multiplicative metrics and the details of the individual additive metrics are outside of the scope of this paper. The reader is referred to [25] for more details. A summary of the metrics is provided below:

- Additive Metric 1 (AM_1): Number of Observed State Relations That Are True in an Intention (Compared to Other Intentions)
- Additive Metric 2 (AM_2): Percentage of an Intention That Is Complete
- Additive Metric 3 (AM_3): Number of Productive States Since the First Productive State Relation in an Intention
- Additive Metric 4 (AM_4): Number of Productive States That Have Occurred Recently (within the past five states)
- Multiplicative Metric 1 (MM_1): Detrimental State Relations in an Intention

6   Experiment and Results

An experiment was run to assess the performance of the state-based intention recognition algorithm described above. This was done by comparing the output of our algorithm to the performance of several humans watching agents performing the same intentions. This section described the experimental procedures and the results.

6.1   Experimental Procedures

Because there are no clear metrics to assess the performance of intention recognition algorithms, an experiment was developed to compare the results of the intention recognition algorithms to the performance of humans. To do this, we leveraged a data set collected at the University of Burgundy, described below.

Five kits were used that each contained ten blocks (as shown in Fig. 3), but each contained a unique combination of different-shaped blocks. In the figure, blocks of different sizes are represented by different colors. For a kit with five types of parts and ten total instances of parts, there are approximately 5^{10} (almost 10 million) possible orders in which the parts can be placed in the kit tray. For this experiment, we randomly chose five of those orders for each kit, resulting in 25 total runs. At each state, the state recognition algorithms described in Section 3 where run to identify the state relations in the environment and then based on these state relations, the algorithms in Section 5 were run to assign likelihoods to intentions. The resulting likelihoods were captured for each possible intention (kit) at each state.

Instructions :

- Select your username and validate it,
- For each object added in the kit, fill out the percentages and validate the row,
- When all the objects are in the kit, a submit button appears. Click on it to save the results.

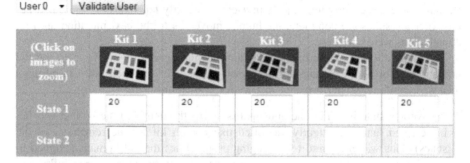

1. Object red added in the kit tray.
2. Object green added in the kit tray.

Percentage left to assign for this line: 100%
Validate row

Fig. 3. Experiment User Interface

To determine how well these likelihoods compared to what a human would perceive in the same situation, we leveraged a data set that was collected at the University of Burgundy based on these same five kits. To compile the data set, fifteen students, all college age, served as the human participants. They were considered novice users. They were presented with the interface shown in Fig. 3. Included in the interface were five images of kits as shown in the figure. The kit images could be enlarged by clicking on them. Each kit represents an intention. They were not told how many of each kit was being built in each of the 25 runs. All kits were presented in random order and the order varied from participant to participant.

One by one, the human was presented with a textual description of something that happened in the environment, for example, "Object red added in the kit tray." Based on this information, the human assigned probabilities as to which kit was being developed (i.e., what was the intention based on the observed events). The probabilities that the human assigned were based on personal preferences. In the example shown in Fig. 3, the human assigned a 20% probability to all kits based on the first state because a red object exists in all kit trays. In some cases, the human scaled the probability based upon how many red objects existed in each kit tray. The only rule was that all probabilities for each state must add to 100%. An update of how many percentage points were available to assign to each row is shown at the bottom of the figure. Once a human finished with a given row, the person presses the validate row button at the bottom. If the sum of the percentages adds to 100%, a new environment observation is provided, a new row is added to the table, and the process starts again. If the percentages did not equal 100%, the human was alerted to this and asked to change their percentages until they did equal 100%. Once a human clicks the "Validate Row" button and it successfully validates, they cannot go back and edit the previous row. Because there are exactly ten objects in each kit, exactly ten states are presented per kit. After ten states are complete, the human moves on to the next intention until all 25 intentions were complete.

To compare to the output of the intention recognition algorithm, we averaged the probabilities for all 15 humans for each plan, each state, and each kit. Based upon this, we identified two performance metrics for evaluation:

1. the average state in which the humans first correctly identified the kit that was being created (and consistently identified that correct kit for the remainder of the states). This was measured by averaging probabilities that all humans assigned to each kit within a state, and then seeing if the kit with the highest probability was the correct kit being created.
2. the average state in which the humans were over 20 percentage points more confident that the correct kit was being developed compared to the second most probable kit. The 20 percentage point value can be changed as necessary, but appeared to be a reasonable value to show that the human was confident that their choice was correct.

The next section will compare the results of human experiments with the output of the intention recognition algorithms.

6.2   Experimental Results

Based on the experiments described in Section 6.1, we directly compared the results of the human evaluations with the output of the intention recognition algorithm described in Section 5. The intention recognition algorithm was scored based on how closely it matched the human-generated results. We judged closeness by determining the difference between the algorithm's performance and the average human performance using the two performance metrics defined in Section 6.1 (both the state that the intention was first identified and the state in which the intention was confidently identified as indicated by the 20% difference between the top selected intention and second rated intention).

Table 2 shows the comparison of the output of the intention recognition algorithms to that of human performance. Column 1 (Kit) shows the kit that was being developed and Column 2 (Plan) shows the plan that was used to develop the kit. As mentioned earlier, the two main points of comparison that were used in this experiment are 1) the state (from 1 to 10) when the humans or algorithms first correctly identified the kit that was being created (and consistently identified that correct kit for the remainder of the states), and 2) the state in which the humans were over 20% more confident that the correct kit was being developed compared to the second most probable kit.

The third through fifth columns of the table ("Correct Intention First Chosen") show the average state in which the user first identified the correct kit that was being developed ("User"), the state in which the algorithm first identified the correct kit that was being developed ("Algorithm"), and the difference between the two ("Difference"). Similarly the sixth through eighth columns of the table ("Over 20% Confident of Intention") show the average state in which the user first identified the correct kit that was being developed with over a 20% confidence as compared to the second most probable intention ("User"), the state in which the algorithm first identified the correct kit that was being developed with over a 20% confidence as compared to the second most probable intention ("Algorithm"), and the difference between the two ("Difference"). A positive value in the difference column means that the algorithm identified the correct kit that many states earlier than the human. Conversely, a negative value in the difference column means the algorithm identified the kit that many states later than the human. A zero means the human and the algorithm identified the kit at the same state.

In analyzing Table 2, we see some very promising results in the performance of the intention recognition algorithms as compared to human performance. If we first examine columns 3-5 (the light blue columns) that compare the state which the algorithms and the humans first chose the correct intention, we see that in over half of the runs (13/25), the intention recognition algorithms determined the proper kit at the exact same state as the human (as indicated by the zero in the difference column). In eight of the runs, the intention recognition algorithms determined the proper kit earlier than the human (as indicated by the positive number in the difference column). This ranged from one to four states earlier than the human. In only four runs did the intention recognition algorithms determine the proper kits later than the human. In three cases, this was one state later and in the other it was three states later.

Table 2. Comparison of Algorithm Output to Human Intention Recognition With Ties Allowed

Kit	Plan	Correct Intention First Chosen			Over 20% Confident of Intention		
		User	Algorithm	Difference	User	Algorithm	Difference
1	1	10	10	0	10	10	0
1	2	7	7	0	7	7	0
1	3	10	10	0	10	10	0
1	4	3	2	1	6	5	1
1	5	5	3	2	5	5	0
2	1	4	4	0	7	5	2
2	2	5	1	4	7	3	4
2	3	9	6	3	9	8	1
2	4	9	8	1	9	9	0
2	5	1	1	0	7	3	4
3	1	6	4	2	7	6	1
3	2	4	1	3	5	2	3
3	3	3	4	-1	7	5	2
3	4	1	1	0	6	3	3
3	5	6	3	3	8	6	2
4	1	8	8	0	8	8	0
4	2	5	5	0	6	5	1
4	3	7	8	-1	9	8	1
4	4	7	7	0	7	7	0
4	5	5	8	-3	8	8	0
5	1	4	4	0	4	4	0
5	2	3	3	0	3	3	0
5	3	6	6	0	6	6	0
5	4	8	9	-1	9	9	0
5	5	2	2	0	7	2	5

When looking at the data for when the algorithms and humans were greater than 20% confident (as compared to the next most probable intention), the results are equally promising. In almost half of the runs (12/25), the intention recognition algorithms determined the proper kit (with over 20% confidence compared to the next most probable kit) at the exact same state as the human (as indicated by the zero in the difference column). In all of the remaining runs, the intention recognition algorithms determined the proper kit earlier than the human (as indicated by the positive number in the difference column). This ranged from one to five states earlier than the human.

This data shows that the intention recognition algorithms, in almost every case, performed as good, if not better, than a human performing the same activity.

7   Conclusion

In this paper, we described a novel approach for intention recognition in cooperative human-robot environments. States are represented by a combination of spatial relationships in a Cartesian frame along with cardinal direction information. These state relations were then ordered to develop intentions and a set of metrics were proposed to assign likelihoods to the various intentions. The proposed metrics were compared to the performance of a set of humans performing the same intention recognition.

The proposed metrics are certainly not the only ones that could be used, but seem to be logical to determine which intention is most likely. Future work will explore:

- Various combination of metric weight to determine which combination best correlates to the performance of humans;
- Applying a metric based on a Bayesian approach and comparing this metric to those presented in this paper;
- Applying this approach to other domains outside of manufacturing kitting; and
- Using this approach to identify transient cases, for instance, identifying when the wrong part was accidently put into a kit.

References

1. Safety of Human-Robot Collaboration Systems Project,
   http://www.nist.gov/el/isd/ps/safhumrobcollsys.cfm
2. Shneier, M.: Safety of Human-Robot Collaboration in Manufacturing. In: 8th Safety Across High-Consequence Industries Conference (2013)
3. Chabrol, J.: Industrial Robot Standardization at ISO. Robotics 3(2), 229–233 (1987)
4. Sadri, F.: Logic-Based Approaches to Intention Recognition. In: Chong, N.-Y., Mastrogiovanni, F. (eds.) Handbook of Research on Ambient Intelligence and Smart Environments: Trends and Perspectives, pp. 346–375 (2011)
5. Schlenoff, C., Foufou, S., Balakirsky, S.: An Approach to Ontology-Based Intention Recognition Using State Representations. In: 4th International Conference on Knowledge Engineering and Ontology Development (2012)
6. Randell, D., Cui, Z.A.: C.: A Spatial Logic Based on Regions and Connection. In: 3rd International Conference on Representation and Reasoning, pp. 165–176. Morgan Kaufmann (1992)
7. Wolter, F., Zakharyaschev, M.: Spatio-Temporal Representation and Reasoning Based on Rcc-8. In: 7th Conference on Principles of Knowledge Representation and Reasoning, KR 2000, pp. 3–14 (2000)
8. Jeon, H., Kim, T., Choi, J.: Ontology-Based User Intention Recognition for Proactive Planning of Intelligent Robot Behavior. In: International Conference on Multimedia and Ubiquitous Engineering, pp. 244–248 (2008)

9. Kelley, R., Tavakkoli, A., King, C., Nicolescu, M., Nicolescu, M., Bebis, G.: Understanding Human Intentions Via Hidden Markov Models in Autonomous Mobile Robots. In: 3rd ACM/IEEE International Conference on Human Robot Interaction, pp. 367–374 (2008)

10. Schrempf, O., Hanebeck, U.: A Generic Model for Estimating User-Intentions in Human-Robot Cooperation. In: 2nd International Conference on Informatics in Control, Automation, and Robotics (ICINCO 2005), pp. 250–256 (2005)

11. Mao, W., Gratch, J.: A Utility-Based Approach to Intention Recognition. In: AAMAS Workshop on Agent Tracking: Modeling Other Agents from Observations (2004)

12. Youn, S.-J., Oh, K.-W.: Intention Recognition Using a Graph Representation. World Academy of Science, Engineering and Technology 25 (2007)

13. Philipose, M., Fishkin, K., Perkowitz, M., Patterson, D., Hahnel, D., Fox, D., Kautz, H.: Inferring Adls from Interactions with Objects. IEEE Pervasive Computing (2005)

14. Bateman, J., Farrar, S.: Spatial Ontology Baseline Version 2.0. University of Bremen (2006)

15. Oberle, D., Ankolekar, A., Hitzler, P., Cimiano, P., Sintek, M., Kiesel, M., Mougouie, B., Baumann, S., Vembu, S., Romanelli, M., Buitelaar, P., Engel, R., Sonntag, D., Reithinger, N., Loos, B., Zorn, H., Micelli, V., Porzel, R., Schmidt, C., Weiten, M., Burkhardt, F., Zhou, J.: Dolce Ergo Sumo: On Foundational and Domain Models in the Smartweb Integrated Ontology (Swinto). Journal of Web Semantics 5, 156–174 (2007)

16. Lenat, D., Guha, R., Pittman, K., Pratt, D., Shephard, M.: Cyc: Toward Programs with Common Sense. Communications of the ACM 33, 30–49 (1990)

17. Pease, A., Niles, I.: Ieee Standard Upper Ontology: A Progress Report. Knowledge Engineering Review, Special Issue on Ontologies and Agents 17, 65–70 (2002)

18. Smith, B., Grenon, P.: The Cornucopia of Formal Ontological Relations. Dialectica 58, 279–296 (2004)

19. Albath, J., Leopold, J., Sabharwal, C., Maglia, A.: Rcc-3d: Qualitative Spatial Reasoning in 3d. In: 23rd International Conference on Computer Applications in Industry and Engineering (CAINE), pp. 74–79 (2010)

20. Ligozat, G.: Qualitative Triangulation for Spatial Reasoning. In: Campari, I., Frank, A.U. (eds.) COSIT 1993. LNCS, vol. 716, pp. 54–68. Springer, Heidelberg (1993)

21. Freksa, C.: Using Orientation Information for Qualitative Spatial Reasoning. In: Frank, A.U., Formentini, U., Campari, I. (eds.) GIS 1992. LNCS, vol. 639, pp. 162–178. Springer, Heidelberg (1992)

22. Schlenoff, C., Pietromartire, A., Foufou, S., Balakirsky, S.: Ontology-Based State Representation for Robot Intention Recognition in Ubiquitous Environments. In: UBICOMP 2012 Workshop on Smart Gadgets Meet Ubiquitous and Social Robots on the Web (UbiRobs) (2012)

23. Owl-S: Semantic Markup of Web Services, http://www.w3.org/Submission/OWL-S/

24. Schlenoff, C., Pietromartire, A., Kootbally, Z., Balakirsky, S., Kramer, T., Foufou, S.: Inferring Intention through State Representation in Cooperative Human-Robot Environments. In: Habib, M., Davim, P. (eds.) Engineering Creative Design in Robotics and Mechatronics (2012)

25. Schlenoff, C., Pietromartire, A., Kootbally, Z., Foufou, S.: Performance Evaluation of Intention Recognition in Human-Robot Collaborative Environments. Accepted (but not yet published) in The ITEA Journal (2013)

26. Carpin, S., Lewis, M., Wang, J., Balakirsky, S., Scrapper, C.: Usarsim: A Robot Simulator for Research and Education. In: IEEE International Conference on Robotics and Automation (ICRA), pp. 1400–1405 (2007)

Knowledge and Data Representation for Motion Planning in Dynamic Environments

Seyedshams Feyzabadi and Stefano Carpin

University of California, Merced
School of Engineering
5200 N Lake Rd, Merced, CA 95343, USA

Abstract. In this paper we describe our initial efforts to develop a knowledge base for motion planning in dynamic environments. Our eventual goal is to smooth the design and integration of multiple heterogeneous robots working in shared environments, and to enable the creation of libraries of plans that can be shared and reused by different robots. We furthermore attempt to align our work with the ongoing activities of the IEEE Ontologies for Robotics and Automation working group.

1 Introduction

In this paper we lay the foundation for the development of an ontology for robot motion planning in dynamic environments. We are interested in studying problems where numerous autonomous vehicles[1] (e.g., forklifts, carts, etc.) operate in a shared, dynamic environment. Our focus is on scenarios where the composition of a robot team varies over time, and where team members are heterogeneous in hardware and software.

Consider for example the following scenario. A truck enters a warehouse to get some goods. An autonomous forklift is unloaded from the truck and obtains a map of the warehouse indicating where the goods are located. The robot computes a motion plan and starts moving to retrieve the pallets holding the goods and load them into the truck. While this happens, other robots already present in the building continue to deliver goods from a nearby factory into the warehouse. Before the autonomous forklift completes its task, other trucks enter the area to retrieve other items, each equipped with its own robot helper. In a situation like this, it is not realistic to assume that all robots are identical, nor that complex setup operations should be performed to successfully complete the task in a dynamic environment where different agents join and leave the scene. There are evidently multiple obstacles to overcome before a scenario like this can become a reality, but the availability of a standardized representation for motion plans and motion planning knowledge will greatly help. For example, robots could share and reuse plans, announce their intentions, and so on.

---

[1] In the following we use the term *robot* to generically indicate an autonomous vehicle.

J.-H. Kim et al. (eds.), *Robot Intelligence Technology and Applications 2*,
Advances in Intelligent Systems and Computing 274,
DOI: 10.1007/978-3-319-05582-4_20, © Springer International Publishing Switzerland 2014

According to Schlenoff et al. [15], ontologies are developed for different purposes, like "Provide a standard set of domain concepts along with their attributes and inter-relations; Allow for knowledge capture and reuse; Facilitate systems specification, design, and integration, and; Accelerate research in the field." In this work we are interested in knowledge capture and reuse, as well as system specification, design and integration. Within the IEEE Rabotics and Automation Society (RAS), an Ontologies for Robotics and Automation (ORA) Working Group has been established and tasked with the development of tools for knowledge representation and reasoning. To the best of our knowledge, the group has not yet addressed aspects related to motion planning, and we therefore put forward our own proposal. With the goal of aligning and integrating our efforts with the working group, we follow the their approach and we adopt the same language and tools, namely OWL [2] and Protégé [1].

This remaining of the paper is organized as follows. In Section 2 we shortly revise the state of the art in motion planning and robot ontologies. In Section 3 we discuss knowledge representation, and in Section 4 we describe how we are integrating the presented ideas and end-to-end planning system. Finally, in Section 5 we sketch some conclusions about the current work.

2   State of the Art

In this section we briefly survey the state of the art related to motion planning and knowledge representation, with a particular attention to the scenarios we introduced in Section 1.

2.1   Motion Planning

Motion planning[2] is one of the most investigated areas in robotics and up-to-date treatises are offered in [8,11]. Research in this area mostly deals with algorithmic techniques aimed at dealing with the inherent complexity of the general motion planning problem. Issues related to knowledge and data representation are commonly not addressed by this community, although the idea of sharing and reusing plans has been explored [6]. In recent years, *sampling based* motion planners emerged as the prevailing paradigm, and methods derived from Probabilistic Roadmaps (PRM) [10] and Rapidly Exploring Random Trees (RRT) [12] have become mainstream in a variety of applications. RRT related methods, in particular, continue to enjoy great popularity because they can be used both for holnomic and non-holonomic systems. Therefore, they are particularly relevant for the application we depicted in the introduction.

Various appraches have been proposed to deal with the problem of motion planning in dynamic environments, but because of its inherent challenges, the

---

[2] To be precise, one should distinguish between *path planning* and *motion planning*. In path planning the temporal dimension is not relevant, whereas in motion planning one considers time, too. For sake of brevity, we use the term motion planning to indicate both problems.

problem can be considered still open. The velocity obstacle concept was introduced in [9] and assumes knowledge of the velocity of the moving obstacle in order to determine an appropriate velocity for the robot. The velocity of the moving obstacle is assumed to be constant. Van den Berg and coauthors have studied different variations for this problem. In [17] the authors assume a probabilistic roadmap (PRM) is precomputed and further hypothesize that for any time t it must be possible to know (with certainty) where the moving obstacles are located. PRM methods [10] cannot be directly applied to non-holonomic robots and [17] does not factor in any uncertainty in the moving objects. In [16] a method is presented where responsibility for collision avoidance is shared between possibly colliding robots. This method requires that all robots in the team run the same algorithm. The recent works [5,18] address various forms of uncertainty, but assume linear dynamics. The method presented in [13] evaluates the probability of collision of a given plan, but does not compute a plan, nor considers moving bodies.

2.2 Robot Ontologies

Ontologies have been used in Artificial Intelligence for many years, however their use in the context of robotic control is much more limited and recent. In fact, the IEEE RAS Ontologies for Robotics and Automation (ORA) working group was approved only in November 2011 with the goal of developing a standard knowledge representation for robotics and automation. In [14] Schlenoff et al. analyze the state of the art and identify three relevant efforts, namely RobotEarth, the NIST Robot Ontology, and the Intelligent Systems Ontology. None of these efforts, nor ORA, has specifically addressed an knowledge representation for motion planning.

3 Knowledge Representation

With the goal of aligning our efforts with activities already ongoing within the ORA working group, we follow the approach presented by Balakirsky et al. in [3]. In particular in this work we focus on the two bottom layers of the workflow abstraction they propose, namely the domain specific information and the ontology.

Domain Specific Information. By its very nature, motion planning requires to model and reason about geometric entities. Moreover, in motion planning it is necessary to constantly consider the interplay between the workspace and the robot configuration space. When multiple robots operate in the same environment, they are in the same physical workspace and can therefore share common representations for spatial knowledge and space representation. The workspace is in general $\mathbb{R}^3$, though in some cases $\mathbb{R}^2$ will suffice. At this stage, for sake of simplicity and because of the application scenario we are considering, we will focus on $\mathbb{R}^2$ (robots move on a planar surface.) The configuration space is instead intimately related to the physical structure of the robot (shape, dimensions, actuators, etc.) Therefore, when dealing with multiple heterogeneous robots it can be anticipated that different representations will need to coexist. Technically speaking, the configuration space has a manifold structure [11] that is in general more complex than $\mathbb{R}^3$.

Modeling and reasoning about the workspace implies developing a knowledge base about geometric entities. A general solution to this problem goes far beyond motion planning. Geometrical models are used to represent and reason about objects located in the physical world, like obstacles. Besides that, it is necessary to also consider entities like paths, trajectories, and plans that are produced and processed by the various motion planners. We observe that this distinction offers a natural link to the proposal formulated in [4] where the authors differentiate between *SolidObject* and *DataThing*. Moreover, as we discuss in the following, the ontology we put forward could be linked with the current ORA Core Ontology through the class Robot defined therein.

Ontology. Figure 1 depicts the initial preliminary ontology we have developed. In the following we describe the various classes. We remind the reader that at this time we restrict our attention to the case of planar environments, so our workspace is $\mathbb{R}^2$.

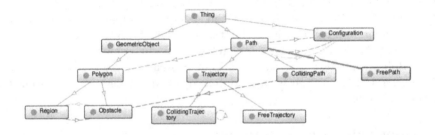

Fig. 1. USARSim is already setup to study coordination problems in industrial environments with a variety of robotic platforms

For what concerns the world representation, we identified the following classes. Each of them is briefly described in the following.

GeometricObject: a geometric object is any object that can be described using geometric primitives. In robotics applications a GeometricObject can be in $\mathbb{R}^2$ or $\mathbb{R}^3$.

Polygon: a polygon is a bidimensional geometric object that can be bounded by a finite number of line segments.

Region: a region is a polygon. While in general one could assume that regions have arbitrary shapes (e.g., curved boundaries), a polygon representation can approximate any curved boundary with arbitrary prevision.

Obstacle: an obstacle is a polygon that a robot should neither penetrate nor touch. As for Region, we assume that arbitrarily shaped obstacles can be approximated with a polygonal boundary.

Motion plans can be defined at different levels of details. A crucial distinction is whether one just aims at considering also control aspects (e.g., which inputs are needed so that a robot follows a certain path/trajectory?) or not. Irrespectively of that, reasoning about motion plans implies reasoning in the configuration space. We identified the following minimal set of classes.

Configuration: a Configuration is an element in a robot configuration space.

Path: a Path is sequence of configurations. Formally stated, a path is a continuous function $p : [0, 1] \rightarrow C$ where C is the configuration space.

FreePath: a FreePath is path that if followed by a certain robot will not cause any collision with any Obstacle.

CollidingPath: a Colliding path is a path that causes at least one collision with an Obstacle.

Trajectory: a Trajectory is a time indexed path.

FreeTrajectory: a FreeTrajectory is a trajectory that if followed by a robot will not cause any collision with any Obstacle nor with any other robot.

CollidingTrajectory: a CollidingTrajectory is a trajectory that if followed by a robot will cause a collision, either with an Obstacle or with another Robot. Note that a CollidingTrajectory could collide with another robot only, or with an Obstacle only, or with both. Therefore CollidingTrajectory is not a subclass of CollidingPath because CollidingPath collides with at least one obstacle, whereas CollidingTrajectory does not.

In the above classes there are some disjoint clauses. CollidingPath is disjoint with FreePath, and FreeTrajectory is disjoint with CollidingTrajectory. Finally, we have defined the following object properties.

StartsAt is a functional data property whose domain is Path and range is Configuration.

EndsAt is a functional data property whose domain is Path and range is Configuration.

IsPartOf is a functional data property whose domain is Obstacle and range is Region. Moreover it is the inverse of Includes.

Includes is a data property whose domain is Range and range is Obstacle. It is the inverse of IsPartOf.

CollidesWith is a data property whose domain is CollidingPath and range is Obstacle.

Intersects is a functional, symmetric data property whose domain and range are CollidingTrajectory.

We note that the above classes express concepts that are independent of the specific configuration space. In a later refinement, we will also consider how knowledge of specific configuration spaces can be represented and exploited for inference.

4   Current and Future Work

We are currently developing an RRT-based motion planner aimed at reusing a library of precomputed paths. These paths will be shared between robots operating in the same environment and will rely on a common representation. Moreover, these paths will be instances of the ontology classes described in the former section to enable inference. The objective is to enable fast replanning when a trajectory is determined to be no longer valid or safe, for example because of an obstacle moved or because of a newly detected approaching robot or human.

The depicted system will be tested using the simulation system provided by USARSim [7]. USARSim includes industrial-like environments and models of various industrial mobile robots. Moreover, it offers the possibility to simulate moving human avatars, thus offering a wide range of challenges resembling those found in real world applications.

Fig. 2. USARSim is already setup to study coordination problems in industrial environments with a variety of robotic platforms

Finally, USARSim can be interfaced with various robot controllers, and we will therefore be capable of exploiting inferences made in the given ontology to improve the replanning abilities of the system.

5   Conclusions

In this paper we have presented our initial efforts towards the creation of a knowledge base and common representation for motion planning in dynamic environments. Specifically, our intent is to enable reasoning and replanning when multiple mobile platforms operate in the same shared area in groups whose composition varies over time. Our eventual goal is twofold. On the application side we would like to enable the creation of repositories or motion plans and knowledge that can be used to implement systems that can rapidly react to unforeseen circumstances and replan. On the representation side, we aim at integrating our models with the current work of the ORA working group. The material presented is evidently work in progress and we have outlined the next steps or our efforts.

Acknowledgments. This work is supported by the National Institute of Standards and Technology under cooperative agreement 70NANB12H143. Any opinions, findings, and conclusions or recommendations expressed in these materials are those of the authors and should not be interpreted as representing the official policies, either expressly or implied, of the funding agencies of the U.S. Government.

References

1. Protégé (August 2013), http://protege.stanford.edu
2. Allemang, D., Hendler, J.A.: Semantic web for the working ontologist: effective modeling in RDFS and OWL. Morgan Kaufmann/Elsevier (2011)
3. Balakirsky, S., Kootbally, Z., Schlenoff, C., Kramer, T., Gupta, S.: An industrial robotic knowledge representation for kit building applications. In: Proceedings of the IEEE/RSJ International Conference on Intelligent Robots and Systems, pp. 1365–1370 (2012)
4. Balakirsky, S., Kramer, T., Kootbally, Z., Pietromartire, A., Schlenoff, C.: The Industrial Kitting Ontology Version 0.5. White paper. National Institute of Standards and Technology, Gaithersburg (2012)
5. Bareiss, D., van den Berg, J.: Reciprocal collision avoidance for robots with linear dynamics using LQR-obstacles. In: Proceeding of the IEEE International Conference on Robotics and Automation, pp. 3832–3238 (2013)
6. Berenson, D., Abbeel, T., Goldberg, K.: A robot path planning framework that learns from experience. In: Proceedings of the IEEE International Conference on Robotics and Automation, pp. 3671–3678 (2012)
7. Carpin, S., Lewis, M., Wang, J., Balakirsky, S., Scrapper, C.: USARSim: a robot simulator for research and education. In: Proceedings of the IEEE International Conference on Robotics and Automation, pp. 1400–1405 (2007)
8. Choset, H., Lynch, K.M., Hutchinson, S., Kantor, G., Burgard, W., Kavraki, L.E., Thrun, S.: Principles of robot motion. MIT Press (2005)
9. Fiorini, P., Shiller, Z.: Motion planning in dynamic environments using velocity obstacles. International Journal of Robotics Research 17(7), 760–772 (1998)
10. Kavraki, L.E., Švestka, P., Latombe, J.C., Overmars, M.H.: Probabilistic roadmaps for path planning in high-dimensional configuration spaces. IEEE Transactions on Robotics and Automation 12(4), 566–580 (1996)
11. LaValle, S.M.: Planning algorithms. Cambridge Academic Press (2006)
12. LaValle, S.M., Kufner, J.J.: Randomized kinodynamic planning. International Journal of Robotics Research 20(5), 378–400 (2001)

13. Patil, S., van den Berg, J., Alterowitz, R.: Estimating probability of collision for safe motion planning under gaussian motion and sensing uncertainty. In: Proceeding of the IEEE International Conference on Robotics and Automation (2012)
14. Schlenoff, C., Prestes, E., Madhavan, R., Goncalves, P., Li, H., Balakirsky, S., Kramer, T., Miguelanez, E.: An IEEE standard ontology for robotics and automation. In: Proceedings of the IEEE/RSJ International Conference on Intelligent Robots and Systems, pp. 1337–1342 (2012)
15. Schlenoff, C., Washington, R., Barbera, T.: An intelligent ground vehicle ontology for multi-agent system integration. In: Proceedings of the International Conference on Integration of Knowledge Intensive Multi-Agent Systems (KIMAS), pp. 169–174 (2005)
16. Snape, J., van den Berg, J., Guy, S.J., Manocha, D.: The hybrid reciprocal velocity obstacle. IEEE Transaction on Robotics 27(4), 696–706 (2011)
17. van den Berg, J., Overmars, M.H.: Roadmap-based motion planning in dynamic environments. IEEE Transactions on Robotics 21(5), 885–897 (2005)
18. van den Berg, J., Wilkie, D., Guy, S.J., Niethammer, M., Manocha, D.: LQG-obstacles: feedback control with collision avoidance for mobile robots with motion and sensing uncertainty. In: Proceeding of the IEEE International Conference on Robotics and Automation, pp. 346–353 (2012)

A Simulated Sensor-Based Approach for Kit Building Applications

Zeid Kootbally[1], Craig Schlenoff[2], Teddy Weisman[3], Stephen Balakirsky[4],
Thomas Kramer[5], and Anthony Pietromartire[2]

[1] University of Maryland, College Park, MD 20740, USA
zeid.kootbally@nist.gov
www.nist.gov/el/isd/ks/kootbally.cfm
[2] Intelligent Systems Division, National Institute of Standards and Technology,
Gaithersburg, MD, USA
{craig.schlenoff,anthony.pietromartire}@nist.gov
www.nist.gov/el/smartcyber.cfm
[3] Yale University, New Haven, CT 06520, USA
tjweisman@gmail.com
[4] Georgia Tech Research Institute, Atlanta, GA 30332, USA
stephen.balakirsky@gtri.gatech.edu
www.unmannedsystems.gtri.gatech.edu
[5] Department of Mechanical Engineering, Catholic University of America,
Washington, DC, USA
thomas.kramer@nist.gov
http://www.nist.gov/el/isd/ks/kramer.cfm

Abstract. Kit building or kitting is a process in which separate but related items are grouped, packaged, and supplied together as one unit (kit). This paper describes advances in the development of kitting simulation tools that incorporate sensing/control and parts detection capabilities. To pick and place parts and components during kitting, the kitting workcell relies on a simulated sensor system to retrieve the six-degree of freedom (6DOF) pose estimation of each of these objects. While the use of a sensor system allows objects' poses to be obtained, it also helps detecting failures during the execution of a kitting plan when some of these objects are missing or are not at the expected locations. A simulated kitting system is presented and the approach that is used to task a sensor system to retrieve 6DOF pose estimation of specific objects (objects of interest) is given.

Keywords: simulation, manufacturing, robotics, kitting, sensor system.

1 Introduction

The effort presented in this paper is designed to support the IEEE Robotics and Automation Society's Ontologies for Robotics and Automation Working Group. Kitting is the process in which several different, but related items are placed into a container and supplied together as a single unit (kit). Kitting itself

J.-H. Kim et al. (eds.), *Robot Intelligence Technology and Applications 2,*
Advances in Intelligent Systems and Computing 274,
DOI: 10.1007/978-3-319-05582-4_21, © Springer International Publishing Switzerland 2014

may be viewed as a specialization of the general bin-picking problem. Industrial assembly of manufactured products is often performed by first bringing parts together in a kit and then moving the kit to the assembly area where the parts are used to assemble products. Agile and flexible kitting, when applied properly, has been observed to show numerous benefits for the assembly line, such as cost savings [7] including saving manufacturing or assembly space [20], reducing assembly worker walking and searching time [27], and increasing line flexibility [6] and balance [16].

Applications for assembly robots have been primarily implemented in fixed and programmable automation. Fixed automation is a process using mechanized machinery to perform fixed and repetitive operations in order to produce a high volume of similar parts. Although fixed automation provides high efficiency at a low unit cost, drastic modifications of the machines are required when parts need major changes or become too complicated in design. In programmable automation, products are made in batch quantities ranging from several dozen to several thousand units at a time. However, each new batch requires long set up times to accommodate the new product style. The time, and therefore the cost, of developing applications for fixed and programmable automation is usually quite high. The opportunity to expand the industrial use of robots is through agile and flexible automation where minimized setup times can lead to more output and generally better throughput.

The effort presented in this paper describes an approach based on a simulated sensor system in an attempt to move towards an agile system. Tasking a sensor system to retrieve information about objects of interest should be performed in a timely manner before the robot carries out actions that involve these objects of interest. Objects in a kitting workcell are likely susceptible to be moved by external agents, parts trays may be depleted, and objects of different types can be unintentionally mixed with other types. Consequently, the system should be able to detect any of the aforementioned cases by tasking the sensor system to retrieve pose estimations of objects of interest.

Pose estimation is an important capability for grasping and manipulation. A wide variety of solutions have been proposed in order to extend the current structure of the systems to an agile system. Most of the efforts in the literature have focused primarily on solutions for robots whose mobility is restricted to the ground plane. Lysenkov *et al.* [19] presented new algorithms for segmentation, pose estimation, and recognition of transparent objects. Their system showed that a robot is able to grasp 80% of known transparent objects with the proposed algorithm and this result is robust across non-specular backgrounds behind the objects. Dzitac and Mazid [11] proposed a flexible and inexpensive object detection and localization method for pick-and-place robots based on the Xtion and Kinect. The authors relied on depth sensors to provide the robots with flexible and powerful means of locating objects, such as boxes, without the need to hard code the exact coordinates of the box in the robot program. Rusu *et al.* [25] presented a novel 3D feature descriptor, the Viewpoint

Feature Histogram (VFH), for object recognition and 6DOF pose identification for applications where *a priori* segmentation is possible.

The organization of the remainder of this paper is as follows. Section 2 presents an overview of the knowledge driven methodology used in this effort. Section 3 describes the simulation environment used for kitting applications. Section 4 details the approach that tasks a sensor system to retrieve information on objects of interest, and Section 5 concludes this paper and analyzes future work.

2   Knowledge Driven Methodology

The knowledge driven methodology presented in this section is not intended to act as a stand-alone system architecture. Rather it is intended to be an extension to well-developed hierarchical, deliberative architectures such as 4D/RCS

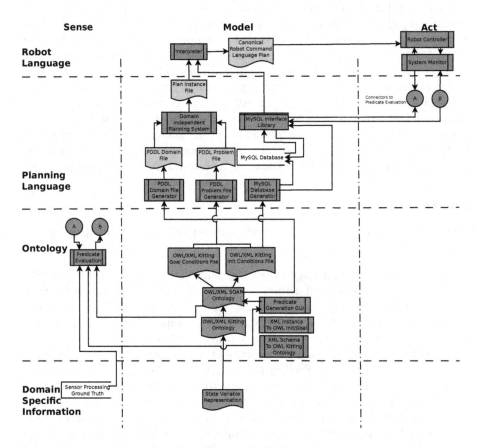

Fig. 1. Knowledge Driven Design extensions – In this figure, green shaded boxes with curved bottoms represent hand generated files while light blue shaded boxes with curved bottoms represent automatically created boxes. Rectangular boxes represent processes and libraries.

(Real-time Control Systems) [1]. The overall knowledge driven methodology of the system is depicted in Figure 1. The figure is organized vertically by the representation that is used for the knowledge and horizontally by the classical sense-model-act paradigm of intelligent systems. The remainder of this section gives a brief description of each level of the hierarchy to help the reader understand the basic concepts implemented within the system architecture in order that the reader may better grasp the main effort described in this paper. The reader may find a more detailed description of each component and each level of the architecture in other publications [5].

2.1   Domain Specific Information

On the vertical axis, knowledge begins with Domain Specific Information (DSI). DSI includes sensors and sensor processing that are specifically tuned to operate in the target domain. Examples of sensor processing may include pose determination and object identification. It is important to note that the effort described in this paper assumes perfect data from the sensor system that do not include noise. A detailed description of the simulated sensor system is given in Section 3.

For the knowledge model, a scenario driven approach is taken where the DSI design begins with a domain expert creating one or more use cases and specific scenarios that describe the typical operation of the system. This includes information on items ranging from what actions and attributes are relevant, to what the necessary conditions (preconditions) are for an action to occur and what the likely results (effects) of the action are. The authors have chosen to encode this basic information in a formalism known as a state variable representation [22].

2.2   Ontology

The information encoded in the DSI is then organized into a domain independent representation.

- A Web Ontology Language (OWL)/Extensible Markup Language (XML) base ontology (OWL/XML Kitting) contains all of the basic information that was determined to be needed during the evaluation of the use cases and scenarios. The knowledge is represented in a compact form with knowledge classes inheriting common attributes from parent classes.
- The OWL/XML SOAP ontology describes the links between States, Ordering constructs, Actions, and Predicates (the SOAP ontology) that are relevant to the scenario. A State is composed of one to many state relationships, which is a specific relation between two objects (e.g., Object 1 is on top of Object 2). An Ordering construct defines the order in which the state relationships need to be represented for a specific State. In classical representation, States are represented as sets of logical atoms (Predicates) that are true or false within some interpretation. Actions are represented by planning operators that change the truth values of these atoms. In the case of the kit building domain, it was found that 10 actions and 16 predicates were necessary.

– The instance files describe the initial and goal states for the system through the Kitting Init Conditions File and the Kitting Goal Conditions File, respectively. The initial state file must contain a description of the environment that is complete enough for a planning system to be able to create a valid sequence of actions that will achieve the given goal state. The goal state file only needs to contain information that is relevant to the end goal of the system. For the case of building a kit, this may simply be that a complete kit is located in a bin designed to hold completed kits.

Since both the OWL and XML implementations of the knowledge representation are file-based, real time information proved to be problematic. In order to solve this problem, an automatically generated MySQL Database [10] was introduced as part of the knowledge representation. A description of the MySQL Database is given in the following subsection.

2.3   Planning Language

Aspects of the knowledge previously described are automatically extracted and encoded in a form that is optimized for a planning system to utilize (the Planning Language). The planning language used in the knowledge driven system is expressed with the Planning Domain Definition Language (PDDL) [14] (version 3.0). The PDDL input format consists of two files that specify the domain and the problem. As shown in Figure 1, these files are automatically generated from the ontology. From these two files, a domain independent planning system [9] was used to produce a static Plan Instance File.

While the knowledge representation presented in this paper provides the "slots" necessary for representing dynamic information, the static file structure makes the utilization of these slots awkward. It is desirable to be able to represent the dynamic information in a dynamic database. For this reason, the authors developed a technique to automatically generate tables for storing, and access functions for obtaining, the data from the ontology in a MySQL Database.

Reading data from and to the MySQL Database instead of the ontology file offers the community easy access to a live data structure. Furthermore, it is more practical to modify the information stored in a database than if it was stored in an ontology, which in some cases, requires the deletion and re-creation of the whole file. A literature review reveals many efforts and methodologies that were designed to produce SQL databases from ontologies. Our effort builds upon the work of Astrova et al. [2]

In addition to generating and filling the database tables, the authors created tools that automatically generate a set of C++ classes for reading and writing information to the kitting MySQL Database. The choice of C++ was a team preference and we believe that other object-oriented languages could have been used in this project.

2.4   Robot Language

Once a plan has been formulated, the knowledge is transformed into a representation that is optimized for use by a robotic system. The interpreter combines knowledge from the plan with knowledge from the MySQL Database to form a set of sequential actions that the robot controller is able to execute. The authors devised a canonical robot command language (CRCL) in which such lists can be written. The purpose of the CRCL is to provide generic commands that implement the functionality of typical industrial robots without being specific either to the language of the planning system that makes a plan or to the language used by a robot controller that executes a plan.

3   Simulation Environment

In order to experiment with robotic systems, a researcher requires a controllable robotic platform, a control system that interfaces to the robotic system and provides behaviors for the robot to carry out, and an environment to operate in. Our kitting application relies on an open source (the game engine is free, but license restrictions do apply), freely available framework capable of fulfilling all of these requirements. This framework is the Unified System for Automation and Robot Simulation (USARSim) [28]. It provides the robotic platform and environment.

3.1   The USARSim Framework

USARSim [8,29] is a high-fidelity physics-based simulation system based on the Unreal Developers Kit (UDK) [13] from Epic Games. USARSim was originally developed under a National Science Foundation grant to study Robot, Agent, Person Teams in Urban Search and Rescue [18]. Since that time, it has been turned into a National Institute of Standards and Technology (NIST)-led, community-supported, open source project that provides validated models of robots, sensors, and environments. Altogether, the Karma Physics engine [12] and high-quality 3D rendering facilities of the Unreal game engine allow the creation of realistic simulation environments that provide the embodiment of a robotic system. Furthermore, USARSim comes with tools to develop objects and environments and it is possible to control the objects in the game through a Transmission Control Protocol/Internet Protocol (TCP/IP) socket with a host computer.

Through its usage of UDK, USARSim utilizes the physX physics engine [23] and high-quality 3D rendering facilities to create a realistic robotic system simulation environment. The current release of USARSim consists of various model environments, models of commercial and experimental robots, and sensor models. High fidelity at low cost is made possible by building the simulation on top of a game engine. By delegating simulation specific tasks to a high volume commercial platform (available for free to most users) which provides superior

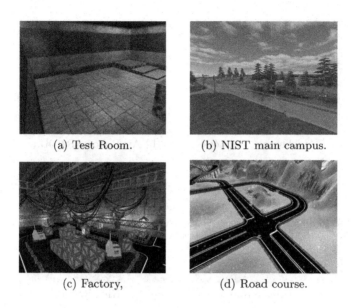

(a) Test Room.   (b) NIST main campus.

(c) Factory,   (d) Road course.

Fig. 2. Sample of 3D environments in USARSim

visual rendering and physical modeling, full user effort can be devoted to the robotics-specific tasks of modeling platforms, control systems, sensors, interface tools, and environments. These tasks are in turn accelerated by the advanced editing and development tools integrated with the game engine. This leads to a virtuous spiral in which a wide range of platforms can be modeled with greater fidelity in a short period of time.

USARSim was originally based upon simulated environments in the (Urban Search and Rescue) USAR domain. Realistic disaster scenarios as well as robot test methods were created (Figure 2(a)). Since then, USARSim has been used worldwide and more environments have been developed for different purposes. Other environments such as the NIST campus (Figure 2(b)) and factories (Figure 2(c)) have been used to test the performance of algorithms in different efforts [30,3,17]. The simulation is also widely used for the RoboCup Virtual Robot Rescue Competition [24], the IEEE Virtual Manufacturing and Automation Challenge [15], and has been applied to the DARPA Urban Challenge (Figure 2(d)).

USARSim was initially developed with a focus on differential drive wheeled robots. However, USARSim's open source framework has encouraged wide community interest and support that now allows USARSim to offer multiple robots, including humanoid robots (Figure 3(a)), aerial platforms (Figure 3(b)), robotic arms (Figure 3(c)), and commercial vehicles (Figure 3(d)). In USARSim, robots are based on physical computer aided design (CAD) models of the real robots and are implemented by specialization of specific existing classes. This structure allows for easier development of new platforms that model custom designs.

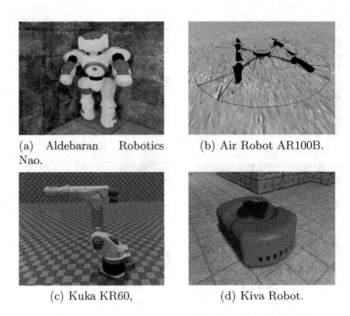

(a) Aldebaran Robotics Nao.

(b) Air Robot AR100B.

(c) Kuka KR60,

(d) Kiva Robot.

Fig. 3. Sample of vehicles in USARSim

All robots in USARSim have a chassis, and may contain multiple wheels, sensors, and actuators. The robots are configurable (e.g., specify types of sensors/end effectors) through a configuration file that is read at run-time. The properties of the robots can also be configured, such as the battery life and the frequency of data transmission.

3.2   The Simulated Sensor System

Poses of objects in the virtual environment are retrieved with the USARTruth tool. USARTruth is capable of reading information about objects in USARSim by connecting as a client to TCP socket port 3989. The simulator USARTruth-Connection object listens for incoming connections on port 3989 and receives queries over a socket in the form of strings formatted into key-value pairs.

The USARTruth connection accepts two different keys, "class" and "name". When USARSim receives a new string over the connection, it sends a sequence of key-value formatted strings back over the socket, one for each Unreal Engine Actor object that matches the requested class and object names. An example of the strings returned by USARSim is given below along with a description for each key.

{Name P3AT_0} {Class P3AT} {Time 29.97} {Location 0.67,2.30,1.86} {Rotation 0.00,0.46,0.00} where:

- Name: The internal name of the object in USARSim.
- Class: The name of the most specific Unreal Engine class the object belongs to.
- Time: The number of seconds that have elapsed since the simulator started, as a floating-point value.
- Location: The comma-separated position of the object in global coordinates.
- Rotation: The comma-separated orientation of the object in global coordinates, in roll, pitch, yaw form.

4   System Operation

As seen previously, Section 3.2 describes how a simulated sensor system operates to retrieve 6DOF poses of objects in the kitting workcell. This section describes when the simulated sensor system is used. Figure 4 is a flowchart that represents some of the steps used for kitting, from parsing the Plan Instance File to the execution of each action from this file. Since the focus of this paper is on the sensor system, the authors have limited the representation and description of Figure 4 around the sensor system and did not include the steps prior to the Plan Instance File generation. The reader may find this missing information in the description of Figure 1 in Section 2. The different steps depicted in Figure 4 are categorized into main components that are numbered. A description of each main component is given in the following subsections.

4.1   Read Plan Instance File

As described in Section 2, the Plan Instance File is generated by the Domain Independant Planning System from the PDDL Domain File and the PDDL Problem File. An example of a plan is given in Figure 5. This plan describes the PDDL actions that a robot will need to execute in order to build a kit that consists of one part of type D and one part of type E. At the beginning of the plan (line 1), the end effector that is capable of grasping parts is taken from the end effector changing station and attached to the robot. Lines 2 and 4 display the actions for picking up a part of type E and D, respectively. Lines 3 and 5 display the actions for putting parts E and D in the kit, respectively. Finally, at line 6, the end effector is put back in the end effector changing station.

4.2   Generate CRCL Commands

Each action of the plan is sequentially interpreted and then directly executed by the robot. The Interpreter takes as input a PDDL action from the Plan Instance File and outputs a set of CRCL commands for this action. To facilitate late binding, the PDDL actions within the plan do not specify the exact locations of the parts and components that are involved. This kind of knowledge detail is maintained by sensor processing and is stored in the MySQL Database. As described in Section 2, the generation of the tables in the MySQL Database is

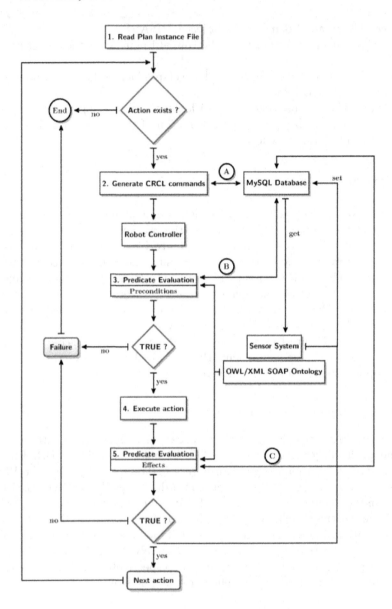

Fig. 4. Flowchart diagram for tasking the simulated sensor

followed by data insertion in these tables for all the objects in the environment. However, there is no guarantee that the poses of these objects are still accurate as they may have been altered in different ways. At this point, the sensor system is tasked to retrieve information about objects of interest. Objects of interest are the ones for which the poses are needed to execute some CRCL commands. Before tasking the sensor to retrieve the poses of objects of interest via the **get**

1 (**attach-endeffector** *robot_1 part_gripper part_gripper_holder changing_station_1*)
2 (**take-part** *robot_1 part_e1 ptr_e part_gripper*)
3 (**put-part** *robot_1 part_e1 kit_a2b3c3d1e1 work_table_1 ptr_e*)
4 (**take-part** *robot_1 part_d1 ptr_d part_gripper*)
5 (**put-part** *robot_1 part_d1 kit_a2b3c3d1e1 work_table_1 ptr_d*)
6 (**remove-endeffector** *robot_1 part_gripper part_gripper_holder changing_station_1*)

Fig. 5. Excerpt of the PDDL solution file for kitting

message, the external shape of each object of interest must be retrieved from the ExternalShape MySQL table.

An external shape is a shape defined in an external file. An external shape has a model format name, stored in the field hasExternalShape_ModelName of the ontology, a model type name, stored in the field hasExternalShape_ModelTypeName, and a model file name which is the name of the file containing the model, stored in the field hasExternalShape_ModelFileName. Using the information retrieved from the aforementioned fields for each object of interest, the system then parses the data coming in from USARTruth and updates the relative pose in the MySQL Database for each object returned. This is performed via the set message. Since USARTruth returns object locations in global coordinates, the relative pose for each object is updated without changing its transformation tree; that is, the object of reference for its physical location is unchanged.

The actual updated relative pose is computed according to Equation 1.

$$L' = LG^{-1}G' \tag{1}$$

where L' is the updated relative transformation, L is the old relative transformation (read from the MySQL Database), G is the old global transformation (computed from the transformation tree in the MySQL Database), and G' is the updated global transformation (retrieved from USARTruth).

Once the above process is performed, the Interpreter uses the new data to generate a set of CRCL commands for the current action, as depicted by Ⓐ in Figure 4. The *take-part* action at line 2 in Figure 5 is interpreted as the sequence of CRCL commands displayed in Table 1, where the numerical data used in the MoveTo commands are computed with the new 6DOF poses. The reader may find more information about the whole set of CRCL commands in [4].

Once a set of CRCL commands is generated for a PDDL action, it is sent to the Robot Controller to be executed by the robot. The Predicate Evaluation process is then called before and after each set of CRCL commands is carried out by the robot.

4.3   Predicate Evaluation (Preconditions)

As mentioned in Section 2, a PDDL action consists of one precondition section and one effect section that are defined in the OWL/XML SOAP Ontology. Preconditions and effects consist of a set of predicates. For instance, the predicates

Table 1. A set of CRCL commands for the action *take-part*

```
initCannon()
Message (''take part part_e1'')
MoveTo({{-0.03, 1.62, -0.25}, {0, 0, 1}, {1, 0, 0}})
Dwell (0.05)
MoveTo({{-0.03, 1.62, 0.1325}, {0, 0, 1}, {1, 0, 0}})
CloseGripper ()
MoveTo({{-0.03, 1.62, -0.25}, {0, 0, 1}, {1, 0, 0}})
Dwell (0.05)
endCannon()
```

Table 2. The precondition and the effect for the action *take-part*

precondition	effect
part-location-partstray($part\_e1$,ptr_e)	¬part-location-partstray($part\_e1$,ptr_e)
robot-empty($robot\_1$)	¬robot-empty($robot\_1$)
endeff-location-robot($part\_gripper$,$robot\_1$)	part-location-robot($part\_e1$,$robot\_1$)
robot-with-endeff($robot\_1$,$part\_gripper$)	robot-holds-part($robot\_1$,$part\_e1$)
endeff-type-part($part\_gripper$,$part\_e1$)	
partstray-not-empty(ptr_e)	

in the precondition and effect for the action *take-part*($robot\_1$,$part\_e1$,ptr_e, *part_gripper*) are defined in Table 2.

Spatial Relations –   The evaluation of the predicates in the precondition section assures that all the requirements are met in the environment before the robot carries out the action. As such, the output of the Predicate Evaluation process is a Boolean value. The kitting system relies on the representation of spatial relations that are stored in the OWL/XML SOAP Ontology to compute the truth-value of each predicate. A brief description of each spatial relation is given below. A thorough analysis of spatial relations used for kitting is well documented in [26].

- Predicates: These are domain-specific states that are of interest to the current activity. The truth-value of predicates can be determined through the logical combination of intermediate state relations.
- Intermediate state relations: These are generic, re-usable level state relations that can be inferred from the combination of Region Connected Calculus (RCC8) [31] and cardinal direction relations.
- RCC8 relations: RCC8 is a well-known and cited approach for representing the relationship between two regions in Euclidean space or in a topological space. RCC8 was initially developed for a two-dimensional space, but has been extended for the purpose of the kitting effort to a three-dimensions space by applying it along all three planes (x-y, x-z, y-z). Each of the intermediate state relations consists of a set of logical rules that associate these

RCC8 relations to them. There are 24 RCC8 relations and 6 cardinality direction operators.

Sensor System –   To evaluate the truth-value of any given predicate, the sensor system is tasked to retrieve the 6DOF pose estimation of the predicate's parameters. This is performed the same way it is described in Section 4.2 and is represented by Ⓑ in Figure 4. The Predicate Evaluation process then proceeds as follows:

In the OWL/XML SOAP Ontology:
1. Identify the predicate.
2. Identify the intermediate state relation for the predicate from step 1.
3. Identify the set of logical rules that associate RCC8 relations to the intermediate state relation from step 2.

RCC8 Evaluation –   Next, the poses of the predicate's parameters are used to compute the truth-value of the identified set of logical rules of RCC8 relations for this predicate. The predicates developed for the kitting effort have at least one parameter and at most two parameters. To evaluate the set of RCC8 relations, two methods are used.

1. In the case the predicate has two parameters, the poses of these two parameters are used to compute the truth-value of the set of RCC8 relations. For instance, the predicate part-location-partstray($part\_e1$,ptr_e) from Table 2 is true if and only if the part $part\_e1$ is **Partially-In** the parts tray ptr_e. **Partially-In** (formula 2) is a state relation that represents an object fully inside of a second object in two dimensions and partially in the third dimension. Please note that the state relation presented in formula 2 is used to demonstrate how the parameters of a predicate are used to compute the truth-value of a set of RCC8 relations. Descriptions of each state relation and each RCC8 relation (Z-Plus, Z-NTPP, Z-NTPPi, etc) are available in [26].

$$\textbf{Partially-In}(part\_e1, ptr\_e) \rightarrow \qquad (2)$$

$$(\texttt{Z-Plus}(part\_e1, ptr\_e) \wedge (\texttt{Z-NTPP}(part\_e1, ptr\_e) \vee \texttt{Z-NTPPi}(part\_e1, ptr\_e) \vee$$

$$\texttt{Z-PO}(part\_e1, ptr\_e) \vee \texttt{Z-TPP}(part\_e1, ptr\_e) \vee \texttt{Z-TPPi}(part\_e1, ptr\_e))) \wedge$$

$$(\texttt{X-NTPP}(part\_e1, ptr\_e) \vee \texttt{X-NTPPi}(part\_e1, ptr\_e) \vee \texttt{X-TPP}(part\_e1, ptr\_e) \vee$$

$$\texttt{X-TPPi}(part\_e1, ptr\_e)) \wedge (\texttt{Y-NTPP}(part\_e1, ptr\_e) \vee \texttt{Y-NTPPi}(part\_e1, ptr\_e) \vee$$

$$\texttt{Y-TPP}(part\_e1, ptr\_e) \vee \texttt{Y-TPPi}(part\_e1, ptr\_e))$$

2. In the case the predicate has only one parameter, a second object is needed to compute the truth-value of the predicate between the parameter and the other object. The authors remind the reader that RCC8 relations necessarily require two objects. In this case, the sensor system is tasked to retrieve

the pose for each other object in the workcell to be used as the second object. Depending on the predicate, the search space for the other object can be narrowed down. For instance, the predicate partstray-not-empty(ptr_e) (Table 2) is true if and only if at least one part of type E is in the parts tray ptr_e. It is not necessary to extend the search for the second object among the other types of object present in the workcell. It is not relevant, for instance, to check if the parts tray contains an end effector. On the other hand, to check the truth value of the predicate robot-empty($robot\_1$) (Table 2), which is true if and only if the robot $robot\_1$ is not holding anything in the end effector attached to the robot, the predicate evaluation needs to scan a wider search space than the one described for the predicate partstray-not-empty(ptr_e), i.e., the search space includes all parts of each type, all types of kit trays, etc. Once the search space has been defined, the external shape for each object in the search space is retrieved from the appropriate table from the MySQL Database. As described previously, the external shape is used by the sensor system to retrieve the 6DOF pose for the corresponding object. The truth-value of the set of logical rules that associate RCC8 relations to the intermediate state relation for the predicate is then computed for these two objects, that is the predicate parameter and the searched object.

If all the predicates within the precondition section are true, the MySQL Database is updated with the current poses of these predicates' parameters. It is important that all the predicates are true in order to update the MySQL Database. This ensures that a complete (stable) state of the environment is stored in the MySQL Database. If at least one predicate is evaluated to false, it is considered a failure and the kitting process is terminated.

4.4   Execute Action

When all the predicates within the precondition of an action have been evaluated to true, this action is executed by the robot. To confirm that the action was successfully accomplished, the Predicate Evaluation process evaluates the predicates within the effect section for this action. The effect section consists of predicates that are expected to be true after performing a PDDL action. The evaluation of the predicates within the effect section is performed to confirm that these expectations are attained.

4.5   Predicate Evaluation (Effects)

The methodology previously described to evaluate the predicates within the precondition section of an action is also used to evaluate the predicates within the effect section of this action. This is depicted by Ⓒ in Figure 4. As mentioned earlier, all the predicates within the effect section must be evaluated to true to define the action as successful. In the case of a successful action, the next action within the Plan Instance File is processed. Once all the actions within the Plan Instance File have been executed by the robot, the kitting process is complete.

5   Conclusions and Future Work

This paper describes the approach that uses a simulated sensor system to retrieve 6DOF pose estimations of objects of interest during kit building applications. The approach is mainly used during the predicate evaluation process. The use of a simulated sensor system allows the current kitting system to move towards an agile system where the current observations on parts and components are fed into the predicate evaluation process, i.e., before and after action executions.

It is also intended to apply contingency plans once an action failure occurs. One of the contingency plans is to re-plan from a state of the environment that is stable. Information on this stable state is retrieved from the MySQL Database that was updated from the latest poses of objects of interest. During the re-planning process, the new initial state becomes the stable state while the goal state stays unchanged.

As mentioned in Section 2.1, pose information coming from USARTruth is assumed to be perfect. In a future effort, the authors will attempt to present a model for the different sources of noise relative to each sensor-based pose estimation step (similar to the one described in [21]), and use measurements of real sensor data to validate the model. Noisy sensor data can be used during the simulation of failures and the application of contingency plans.

The current kitting workcell involves objects that are originally placed on a non-movable surface and also involves a pedestal-based articulated arm. To move towards an agile and flexible manufacturing system, the authors will need to address more challenging scenarios where parts come into the workcell via conveyor belts and where the robotic arm can be of gantry type. These new settings will need to be simulated and tested for kitting applications.

Disclaimers

Acknowledgement. The authors would like to extend our gratitude to Grayson Moses, who worked at NIST as a Poolesville High School student volunteer and helped with the development of algorithms to task a sensor system to provide updated pose information.

References

1. Albus, J.: 4-D/RCS Reference Model Architecture for Unmanned Ground Vehicles. In: Proceedings of the IEEE International Conference on Robotics and Automation (ICRA), pp. 3260–3265 (2000)
2. Astrova, I., Korda, N., Kalja, A.: Storing OWL Ontologies in SQL Relational Databases. World Academy of Science, Engineering and Technology 29, 167–172 (2007)
3. Balaguer, B., Balakirsky, S., Carpin, S., Lewis, M., Scrapper, C.: USARSim: a Validated Simulator for Research in Robotics and Automation. In: IEEE/RSJ IROS 2008 Workshop on Robot Simulators: Available Software, Scientific Applications and Future Trends (2008)
4. Balakirsky, S., Kramer, T., Kootbally, Z., Pietromartire, A.: Metrics and Test Methods for Industrial Kit Building. NISTIR 7942. National Institute of Standards and Technology (NIST) (2012)
5. Balakirsky, S., Schlenoff, C., Kramer, T., Gupta, S.: An Industrial Robotic Knowledge Representation for Kit Building Applications. In: Proceedings of the 2012 IEEE/RSJ International Conference on Intelligent Robots and Systems (IROS), pp. 1365–1370 (October 2012)
6. Bozer, Y.A., McGinnis, L.F.: Kitting Versus Line Stocking: A Conceptual Framework and Descriptive Model. International Journal of Production Economics 28, 1–19 (1992)
7. Carlsson, O., Hensvold, B.: Kitting in a High Variation Assembly Line. Master's thesis, Luleå University of Technology (2008)
8. Carpin, S., Wang, J., Lewis, M., Birk, A., Jacoff, A.: High Fidelity Tools for Rescue Robotics: Results and Perspectives. In: Bredenfeld, A., Jacoff, A., Noda, I., Takahashi, Y. (eds.) RoboCup 2005. LNCS (LNAI), vol. 4020, pp. 301–311. Springer, Heidelberg (2006)
9. Coles, A.J., Coles, A., Fox, M., Long, D.: Forward-Chaining Partial-Order Planning. In: 20th International Conference on Automated Planning and Scheduling (ICAPS 2010), pp. 42–49. AAAI, Toronto (2010)
10. Oracle Corporation. Mysql (November 2012), http://www.mysql.com
11. Dzitac, P., Mazid, A.M.: A Depth Sensor to Control Pick-and-Place Robots for Fruit Packaging. In: 12th International Conference on Control Automation Robotics & Vision (ICARCV), pp. 949–954 (2012)
12. Epic Games. MathEngine Karma ᵀᴹUser Guide (March 2002)
13. Epic Games. Unreal Development Kit (2011), http://udk.com
14. Ghallab, M., Howe, A., Knoblock, C., McDermott, D., Ram, A., Veloso, M., Weld, D., Wilkins, D.: Pddl-the planning domain definition language. Technical Report CVC TR98-003/DCS TR-1165, Yale (1998)
15. IEEE. Virtual Manufacturing and Automation Home Page (2011), http://www.vma-competition.com
16. Jiao, J., Tseng, M.M., Ma, Q., Zou, Y.: Generic Bill-of-Materials-and-Operations for High-Variety Production Management. Concurrent Engineering: Research and Applications 8(4), 297–321 (2000)
17. Kootbally, Z., Schlenoff, C., Madhavan, R.: Performance Assessment of PRIDE in Manufacturing Environments. ITEA Journal 31(3), 410–416 (2010)
18. Lewis, M., Sycara, K., Nourbakhsh, I.: Developing a Testbed for Studying Human-Robot Interaction in Urban Search and Rescue. In: Proceedings of the 10th International Conference on Human Computer Interaction, pp. 22–27 (2003)

19. Lysenkov, I., Eruhimov, V., Bradski, G.: Recognition and Pose Estimation of Rigid Transparent Objects with a Kinect Sensor. In: Proceedings of Robotics: Science and Systems, Sydney, Australia (July 2012)
20. Medbo, L.: Assembly Work Execution and Materials Kit Functionality in Parallel Flow Assembly Systems. International Journal of Industrial Ergonomics 31, 263–281 (2003)
21. Meeden, L.: Bridging the Gap between Robot Simulations and Reality with Improved Models of Sensor Noise. In: Koza, J.R., et al. (eds.) Proceedings of the 3rd Annual Genetic Programming Conference, San Francisco, CA, USA, pp. 824–831 (1998)
22. Nau, D., Ghallab, M., Traverso, P.: Automated Planning: Theory & Practice. Morgan Kaufmann Publishers Inc., San Francisco (2004)
23. Nvidia. PhysX Description (2011),
   http://www.geforce.com/Hardware/Technologies/physx
24. RoboCup. RoboCup Rescue Homepage (2011), http://www.robocuprescue.org
25. Rusu, R.B., Bradski, G., Thibaux, R., Hsu, J.: Fast 3D Recognition and Pose Using the Viewpoint Feature Histogram. In: Proceedings of the 23rd IEEE/RSJ International Conference on Intelligent Robots and Systems (IROS), Taipei, Taiwan (October 2010)
26. Schlenoff, C., Pietromartire, A., Kootbally, Z., Balakirsky, S., Foufou, S.: Ontology-based State Representations for Intention Recognition in Human-robot Collaborative Environments. Robotics and Autonomous Systems 61(11), 1224–1234 (2013)
27. Schwind, G.F.: How Storage Systems Keep Kits Moving. Material Handling Engineering 47(12), 43–45 (1992)
28. USARSim. USARSim Web (2011), http://www.usarsim.sourceforge.net
29. Wang, J., Lewis, M., Gennari, J.: A Game Engine Based Simulation of the NIST Urban Search and Rescue Arenas. In: Proceedings of the 2003 Winter Simulation Conference, vol. 1, pp. 1039–1045 (2003)
30. Wang, J., Lewis, M., Hughes, S., Koes, M., Carpin, S.: Validating USARSim for use in HRI Research. In: Proceedings of the Human Factors and Ergonomics Society 49th Annual Meeting, pp. 457–461 (2005)
31. Wolter, F., Zakharyaschev, M.: Spatio-temporal Representation and Reasoning Based on RCC-8. In: Proceedings of the 7th Conference on Principles of Knowledge Representation and Reasoning (KR 2000), pp. 3–14. Morgan Kaufmann (2000)

A Survey on Biomedical Knowledge Representation for Robotic Orthopaedic Surgery

Paulo J.S. Gonçalves[1,2] and Pedro M.B. Torres[1,2]

[1] Polytechnic Institute of Castelo Branco, School of Technology, Portugal
paulo.goncalves@ipcb.pt
[2] Technical University of Lisbon, Center of Intelligent Systems,
IDMEC / LAETA, Portugal

Abstract. The paper presents a survey of the efforts and methods presented by the research community to represent knowledge to be used, in a machine readable format, in the biomedical field. From the surveyed ontologies, the base ontologies for the conceptual model of the Ontology for Robotic Orthopaedic Surgery (OROSU), are defined. Methods for merging the base ontologies to obtain the OROSU, are discussed, while the under development framework is briefly presented.

Keywords: Knowledge Representation, Ontologies, Robotics, Surgery.

1 Introduction

Healthcare have undertake in the past decades serious developments, mainly due to Information and Communications Technologies (ICTs). In medical practice computers are currently widespread in healthcare institutions, e.g., hospitals. This issue implies gathering and storing large amount of digital data that can be used in common clinical practice. This data, collected at different sources in the healthcare institution, should be integrated to extract useful data and perform, for instance, clinical diagnosis [1].

In recent years, efforts have been undertaken to standardize the huge amount of clinical data, e.g., data standards, vocabularies, surgical procedures, and so on, in the biomedical field. These efforts, are really challenging and undertake a tedious work, due to the huge amount of specialities in the biomedical field. This fact is also an important issue when we focus on the orthopaedic field.

At this stage is now clear that homogenizing biomedical data, using standards, is a crucial factor to develop frameworks that can integrate data, vocabulary, and son on. This effort of standardization is leading to:

- efficient data analysis for diagnostic purposes;
- generate knowledge based on its development;
- generate knowledge in a machine readable format;
- sharing knowledge in the Cloud;
- reasoning based on the modelled knowledge.

J.-H. Kim et al. (eds.), *Robot Intelligence Technology and Applications 2,*   259
Advances in Intelligent Systems and Computing 274,
DOI: 10.1007/978-3-319-05582-4_22, © Springer International Publishing Switzerland 2014

In the specific case or orthopaedic surgery, the purpose of this paper, it is clear that although the field is narrower than general biomedical systems, broader concepts are needed to represent knowledge. A simple example of this are the medical imaging systems. This fact arises the fundamental issue when trying to aggregate knowledge, in a machine readable format, to implement knowledge based systems.

An important factor is the way that the knowledge can be modelled. As seen before, this field is very broad and with lots of clinical specialities that often present data in registries and/or databases. Since orthopaedic surgery is a multidisciplinary field, vocabulary, concepts, and how to represent them should be clearly defined. Moreover, this definitions must be interchangeable and easy to handle by all the actors from several clinical specialities, i.e., surgeons, radiologists, anaesthesiologists, nurses, and nowadays robots. Ontologies play a decisive role on the knowledge modelling both in the biomedical field [2] an the robotics field [3] . This fact is mainly due because ontologies can define the meaning of vocabulary/definitions/terms in several fields of research and more importantly, are able to represent its relations. In other words, can describe the semantic inter-relationships of things, from common sense to a specific field, such as orthopaedic surgery.

This paper is focused in the review of the existing methods than can lead to an knowledge model for orthopaedic robotics surgery, using ontologies. From this narrow field, a bottom up approach can be used to obtain the knowledge from both, medical and robotic fields. In other words, it will be shown that the ontology for orthopaedic robotic surgery (OROSU) must be obtained from ontologies and standards on related fields. The key issue will be to map/interchange the existing ontologies to obtain the goal ontology OROSU.

This paper is organized as follows. Section 2, presents the ICTs, Standardization and Ontologies efforts in healthcare. Section 3, presents the existing domain ontologies needed to obtain the robotic orthopaedic surgery, i.e, Biomedical Data Management and Clinical Diagnosis, Surgery and Robotics. In the following section are presented the efforts in robotic orthopaedic domain ontology. The paper ends with section 5, with a discussion on the presented ontologies and draws some conclusions.

2   ICTs, Standardization and Ontologies

Healthcare have undertake in the past decades serious developments, mainly due to Information and Communications Technologies (ICTs). In medical practice computers are widespread, and for a modern healthcare institution, a medical information system is crucial to manage the large amount of data that can be gathered from clinical practice. Such ICT infrastructure, to be efficient, must rely on standards to integrate databases and intercommunicate between the myriad of equipments and staff, e.g. in a hospital. The HL7 standards (*www.hl7.org*), are worldwide used, being also HL7 the global authority on standards for interoperability of healthcare information technology. These standards and its acceptance by the community allowed to better use machine learning algorithms to analyse clinical data, e.g., for diagnose purposes.

Clinical data representation is still evolving for a more efficient machine readable format. In [4] is proposed an ontology that is aligned to the international standard ISO/IEC 11179, for representing metadata for an organization in a metadata registry. This ontology is based on the General Formal Ontology (GFO) [5], and was developed in Germany mainly for the biomedical area. Previously, a similar work was proposed in [6] to integrate the European standard EN 1828, with the GALEN [7] ontologies. This type of efforts, allows:

- the integration, of ICTs, standards, and knowledge;
- to monitor the surgical workflow, i.e., a tool for management of the surgical procedures;
- to benchmark surgical procedures;
- allow the robot to fetch the hospital ICT infrastructure.

With such an ontological tool could be possible, in the future, to perform personal surgery where surgical procedures could adapt to the patient, using the knowledge model, based on ontologies. In recent years the robotics community is pursuing adequate standards, for the next step in robotics development. Future generation of robots require interaction with humans, in a co-worker healthcare scenario, like in surgery or rehabilitation. Here safety issues are of major importance. In [8] is presented a study on ontologies and standards in the service robots domain, focusing in surgery. In the next sections are described the biomedical ontologies that can be useful, i.e., to extract information from, to build a suitable robotic orthopaedic ontology. Also are described the robot ontologies that were developed and can be used for the previous stated objective.

3 Ontologies

The medical community is largely sensitised to model knowledge, and to make the terminology used explicit to the community. As such, there exist a large number of databases of terminology, that have smoothly changed to ontologies over the years. In the following sections will be presented the ontologies categorized by its purpose, e.g., data management, clinical diagnosis, and surgery. Since the mid 1990's, several projects have started this path: GALEN [7]; MENELAS [9]; SNOMED-RT [10]; UMLS©[11]; SNOMED-CT [12].

The results of the GALEN project are now open to the community, through OpenGALEN, although it is no longer actively maintained. It describes the anatomy, surgical deeds, diseases, and their modifiers used in the definitions of surgical procedures [7].

The MENELAS project delivered an access system for medical record using natural language, where a knowledge management tool was developed to browse the domain ontology and knowledge gathered via clinical data [9].

SNOMED-RT was initially developed as a reference terminology to enable user interfaces, electronic messaging, or natural language processing, to the medical community. This system evolved to SNOMED-CT ©, when its prior was merged to the United Kingdom National Health Service Clinical Terms. It is now a US standard for electronic health information exchange in Interoperability Specifications.

The Unified Medical Language System, UMLS©©[11], is a set of files and software that brings together many health and biomedical vocabularies and standards to enable interoperability between computer systems.

The presented projects, and the large amount of knowledge therein, led to large number of ontologies to serve the biomedical community, namely for: biomedical data management, clinical diagnosis and surgery. All of those are important to gather the knowledge needed to develop a robotic surgery ontology. In fact, we need to collect knowledge on how the biomedical data flows in the hospital ICT infrastructure, on how clinical diagnosis is performed (e.g., how to interpret medical images), and how the surgical process management is performed in the operating room. In the following two sub-sections are presented state-of-the-art developments in this three subjects.

Taking in mind the goal of the paper, the robot must be placed in the equation. Several works were performed to develop a robot ontology, which are depicted in [3], and the references therein. Within this efforts the last sub-section presents and overview of the state-of-the-art ontologies for robotics, with special focus on the suited ones for robotic surgery.

3.1   Biomedical Data Management and Clinical Diagnosis

Knowledge based systems are quickly gaining its position in healthcare institutions, giving in most cases a first diagnosis screening, based on the gathered clinical data of the patient and the knowledge models. Data mining or ontology based systems are two possible applications of artificial intelligence in this scope. In [13] was created a clinical recommender system based on a data mining system, using conditional probabilities to infer a recommendation. This system makes use of a medical information system, nowadays an essential part of healthcare institutions.

Ontologies can be used to model the knowledge system, as presented in the following cases. In [14] is developed an ontology driven system to obtain clinical guidelines to deliver a standardized care to patients. This system is a multi-agent system, composed by medical doctors, specific field ontologies, services, medical records, and so on, to obtain a medico-organisational ontology. In [15] the authors have designed a pre-operative assessment decision support system based in ontologies. The system is capable, based on the ontology and the patient data, to deliver personalised reports, perform risk assessment and also clinical recommendations. In [16] is presented and evaluated an ontology for guiding appropriate antibiotic prescribing, using them as an efficient decision support system. The paper focus on the development issues for representing and maintaining antimicrobial treatment knowledge rules, that generate alerts to provide feedback to clinicians during antibiotic prescribing. Also in [17] is presented and ontology for a healthcare network, based on a terminology database, to obtain a tool called "virtual staff" that enables cooperative diagnosis.

3.2 Surgery

From the knowledge models presented in the previous ontologies, the next step in our work is to step into the operating room, to obtain surgical models. In this subsections are presented ontologies that can model the surgical workflow.

For obtaining ontologies, information can be obtained from text-books, medical reports, or even by tracking surgical instruments during surgery. In [18] were developed strategies for neurosurgery, based on medical text-books. Actually the ontologies are used to obtain the requirements for a neurosurgery simulator. Also in [19] were developed ontologies from medical text reports, here applied to surgical intensive care. In [20] was developed a system based on a sensor based surgical device and an ontology to obtain Surgical Process Models. Surgical instruments were tracked during the surgical workflow, and the data gathered in an ontology server.

Further, surgical conceptual knowledge was modelled using ontologies, in [21], with special focus on Computer Aided Surgery. There, the authors developed Surgical Ontologies for Computer Assisted Surgery (SOCAS), based on the General Formal Ontology (GFO) [5] and the Surgical Workflow Ontology (SWOnt) [22].

At this stage ontologies that can be used in the operating room are defined in the literature, that allows the integration of robotic ontologies during surgery. In the next section are depicted robot ontologies.

3.3 Robotics

Robotics development is pushing robots to closely interact with humans and real world unconstrained scenarios, that surely complicate robot tasks. In this context, complex control systems should be developed. This systems will have to interact with similar systems, like humans interact with each other and machines. For that, standardization and a common understanding of concepts in the domains involved with robots and its workplaces, should be pursued. Taking this in mind, IEEE started to gather knowledge from experts in academia and industry to develop ontologies for robotics and automation, [3]. In this workgroup are being developed ontologies for the Core Robotics domain [23], industrial and service robots. The later with special interest to robotic surgery. As presented above, in [8] is presented the interconnections between ontologies and standards to obtain useful standardized systems to speed-up robotic development.

Related to robotic ontologies, other ontologies arrive to the community, for rehabilitation, [24], field robots [25], amongst others presented in [23]. For robotic neurosurgery, recent work have been presented in [26]. Since these ontologies are not specific for robotic orthopaedic surgery, the next section will present current efforts in this domain.

4   Ontology for Robotic Orthopaedic Surgery

In surgery, several goals must be achieved, e.g., safety, efficiency, and nowadays a cost effective solution should be seek. These goals led to the introduction of robots in surgery [27] that nowadays are gaining market in the operating room, e.g., ROBODOC [28] and more recently MAKO RIO system [29]. Today, these machines are tele-operated by humans, can help in navigation, and can also reduce the surgeon hand tremor. With the use of robots in surgery, less invasive, more precise, and cleaner surgical procedures, can be achieved.

It was seen, in the previous sections of the paper, that in healthcare exists a huge amount of knowledge that several researchers and organizations are representing in a knowledge model, using ontologies. That model must indicate terminologies and its semantic contents, and also must be adapted to the data standards already used in clinical and surgical practice.

To obtain a solid ontology for robotic orthopaedic surgery, the following parts are to be represented in such a model:

- an Human Anatomical Ontology;
- a Clinical Ontology;
- a Surgical Ontology;
- a Robot Ontology;
- how to represent and manage clinical data, e.g., image, case, patient data.

Recently in [30] was presented the implementation guidelines, and first results, for a robotic orthopaedic ontology, with application to a surgical procedure for hip resurfacing. Existing ontologies from the medical and the robotics fields were mapped in the OROSU ontology, such as: *NCBO BioPortal* [31], the *Open Biological and Biomedical Ontologies* [32], the *SOCAS Ontology* [33], the *Open Robots Ontology (ORO)* [34] and *KnowRob* [35]. They are considered to be the main sources of content that cover almost the knowledge needed to develop the ontology for orthopaedic robotic surgery.

Amongst other benefits, ontologies allow a perfect combination of surgical protocols, machine protocols, anatomical ontologies, and medical image data. With this four factors in the control loop of the robot, the surgical procedures will have an increase in the quality of monitoring and surgical outcomes assessment. Moreover the surgeon can perform surgical navigation with anatomical orientation, using state-of-the-art control architectures of robots. The system based on ontologies can also be used to verify surgical protocols, by tracking surgical devices during surgery or simply by reasoning on the data gathered during surgery.

5   Discussion and Conclusions

It is our view that existing ontologies, the ones presented in the sections before, have to be mapped to obtain a robotic surgery ontology that covers today knowledge in the biomedical field. Preferably it would be useful a unique ontology for each domain, adopted by the community. Although these days this goal is difficult to achieve.

As seen in the previous sections there exist a vast number of ontologies, often related to the same domains, leading to ambiguities, i.e., the same concept defined several times. Other issues are related to the need of using several ontology domains, to infer some kind of diagnose from the knowledge model and/or the data observed. This type of problems when using several types of ontologies can be also modelled as proposed in [36]. There, the authors proposed a fuzzy system to infer from two different ontologies, one physical and other mental, to formalize a patient state. In other words, the system aligns both the ontologies using fuzzy aggregate functions to obtain a weighted ranking order fuzzy set, for medical decision making for diagnosis.

Using such a robotic ontology in surgery, in general, the system will ease the surgeon burden when defining the surgical workflow. This will be done also in a machine-readable format that enables the robot to interact with surgeons, nurses and all the staff in the operating room. Moreover the knowledge based model from ontologies will assign surgical deeds to each actor in the surgery, while controlling the surgical workflow, composed of time-series of surgical procedures.

In conclusion, the paper surveyed the most significant, i.e., with highest impact, biomedical ontologies presented in the literature. From these, and to obtain a domain specific ontology for robotic surgery, conceptual components were proposed. From the existing robot ontologies, the under development OROSU was briefly presented, and discussed the problems that arise while merging various ontologies to obtain a specific domain ontology.

Acknowledgements. This work was partly supported by the Strategic Project, PEst-OE/EME/LA0022/2011, through FCT (under IDMEC-IST, Research Group: IDMEC/LAETA/CSI), FCT project PTDC/EME-CRO/099333/2008 and EU-FP7-ICT-231143, project ECHORD. The authors would like to thank the valuable discussions within the IEEE Working Group on Ontologies for Robotics and Automation.

References

1. Gonçalves, P., Almeida, R., Caldas Pinto, J., Vieira, S., Sousa, J.: Image based classification platform: Application to breast cancer diagnosis. In: Cruz-Cunha, M., Miranda, I., Gonçalves, P. (eds.) Handbook of Research on ICTs and Management Systems for Improving Efficiency in Healthcare and Social Care, vol. 1, pp. 595–613. IGI Global (2013)

2. Yu, A.C.: Methods in biomedical ontology. Journal of Biomedical Informatics 39(3), 252–266 (2006)

3. Schlenoff, C., Prestes, E., Madhavan, R., Gonçalves, P., Li, H., Balakirsky, S., Kramer, T., Miguelanez, E.: An IEEE standard ontology for robotics and automation. In: Proc of the IEEE/RSJ Intl. Conf. on Intelligent Robots and Systems (IROS), pp. 1337–1342 (2012)

4. Uciteli, A., Groß, S., Kireyev, S., Herre, H.: An ontologically founded architecture for information systems in clinical and epidemiological research. Journal of Biomedical Semantics (2) (2011)

5. Herre, H.: General formal ontology (gfo): A foundational ontology for conceptual modelling. In: Poli, R., Healy, M., Kameas, A. (eds.) Theory and Applications of Ontology: Computer Applications, pp. 297–345. Springer, Netherlands (2010)

6. Rodrigues, J., Trombert Paviot, B., Martina, C., Vercherina, P.: Integrating the modelling of en 1828 and galen ccam ontologies with protege: towards a knowledge acquisition tool for surgical procedures. In: Studies in Health Technology and Informatics - Connecting Medical Informatics and Bio-Informatics, vol. 116, pp. 767–772. IOS Press (2005)

7. Rector, A., Nowlan, W.: The GALEN project. Comput Methods Programs Biomed. 45, 75–78 (1994)

8. Haidegger, T., Barreto, M., Gonçalves, P., Habib, M.K., Ragavan, V., Li, H., Vaccarella, A., Perrone, R., Prestes, E.: Applied ontologies and standards for service robots. Robotics and Autonomous Systems 61(11), 1215–1223 (2013)

9. Zweigenbaum, P.: Menelas: Coding and information retrieval from natural language patient discharge summaries. In: Advances in Health Telematics, pp. 82–89. IOS Press (1995)

10. Spackman, K.A., Campbell, K.E., Cote, R.A.: Snomed rt: A reference terminology for health care. In: J. of the American Medical Informatics Association, pp. 640–644 (1997)

11. Pisanelli, D.M., Gangemi, A., Steve, G.: An ontological analysis of the umls metathesaurus. Proceedings AMIA 98(5), 810–814 (1998)

12. Wang, A.Y., Sable, J.H., Spackman, K.A.: The snomed clinical terms development process: refinement and analysis of content. In: Proc AMIA Symp., pp. 845–849 (2002)

13. Duan, L., Street, W.N., Xu, E.: Healthcare information systems: data mining methods in the creation of a clinical recommender system. Enterprise Information Systems 5(2), 169–181 (2011)

14. Isern, D., Snchez, D., Moreno, A.: Ontology-driven execution of clinical guidelines. Computer Methods and Programs in Biomedicine 107(2), 122–139 (2012)

15. Bouamrane, M.M., Rector, A., Hurrell, M.: Experience of using owl ontologies for automated inference of routine pre-operative screening tests. In: Patel-Schneider, P.F., Pan, Y., Hitzler, P., Mika, P., Zhang, L., Pan, J.Z., Horrocks, I., Glimm, B. (eds.) ISWC 2010, Part II. LNCS, vol. 6497, pp. 50–65. Springer, Heidelberg (2010)

16. Bright, T.J., Furuya, E.Y., Kuperman, G.J., Cimino, J.J., Bakken, S.: Development and evaluation of an ontology for guiding appropriate antibiotic prescribing. Journal of Biomedical Informatics 45(1), 120–128 (2012)

17. Dieng-Kuntz, R., Minier, D., Ruzicka, M., Corby, F., Corby, O., Alamarguy, L.: Building and using a medical ontology for knowledge management and cooperative work in a health care network. Computers in Biology and Medicine 36, 871–892 (2006)

18. Audette, M., Yang, H., Enquobahrie, A., Finet, J., Barre, S., Jannin, P., Ewend, M.: The application of textbook-based surgical ontologies to neurosurgery simulation requirements. International Journal of Computer Assisted Radiology and Surgery 6(1), 138–143 (2011)

19. Moigno, S.L., Charlet, J., Bourigault, D., Degoulet, P., Jaulent, M.: Terminology extraction from text to build an ontology in surgical intensive care. In: Proceedings of the AMIA 2002 Annual Symposium, pp. 430–434 (2002)

20. Neumuth, T., Czygan, M., Goldstein, D., Strauss, G., Meixensberger, J., Burgert, O.: Computer assisted acquisition of surgical process models with a sensors-driven ontology. In: M2CAI Workshop. MICCAI, London (2009)

21. Mudunuri, R., Burgert, O., Neumuth, T.: Ontological modelling of surgical knowledge. In: Fischer, S., Maehle, E., Reischuk, R. (eds.) GI Jahrestagung. LNI., GI, vol. 154, pp. 1044–1054 (2009)

22. Neumuth, T., Jannin, P., Strauß, G., Meixensberger, J., Burgert, O.: Validation of knowledge acquisition for surgical process models. Journal of the American Medical Informatics Association 16, 72–80 (2009)

23. Prestes, E., Carbonera, J.L., Fiorini, S.R., Jorge, V.A., Abel, M., Madhavan, R., Locoro, A., Gonçalves, P., Barreto, M.E., Habib, M., Chibani, A., Gérard, S., Amirat, Y., Schlenoff, C.: Towards a core ontology for robotics and automation. Robotics and Autonomous Systems 61(11), 1193–1204 (2013)

24. Dogmus, Z., Papantoniou, A., Kilinc, M., Yildirim, S., Erdem, E., Patoglu, V.: Rehabilitation robotics ontology on the cloud. In: IEEE International Conference on Rehabilitation Robotics, ICORR 2013 (2013)

25. Dhouib, S., Du Lac, N., Farges, J.L., Gerard, S., Hemaissia-Jeannin, M., Lahera-Perez, J., Millet, S., Patin, B., Stinckwich, S.: Control Architecture Concepts and Properties of an Ontology Devoted to Exchanges in Mobile Robotics. In: 6th National Conference on Control Architectures of Robots, Grenoble, France, 24 p. INRIA Grenoble Rhône-Alpes (May 2011)

26. Perrone, R.: Ontological modeling for Neurosurgery: application to automatic classification of temporal and extratemporal lobe epilepsies (2012)

27. Abolmaesumi, P., Fichtinger, G., Peters, T., Sakuma, I., Yang, G.Z.: Iintroduction to special section on surgical robotics. IEEE Transactions on Biomedical Engineering 60(4), 887–891 (2013)

28. Kazanzides, P., Mittelstadt, B., Musits, B., Bargar, W., Zuhars, J., Williamson, B., Cain, P., Carbone, E.: An integrated system for cementless hip replacement. IEEE Engineering in Medicine and Biology Magazine 14(3), 307–313 (1995)

29. Mako rio, http://www.makosurgical.com/physicians/products/rio (visited November 1, 2013)

30. Gonçalves, P.: Towards an ontology for orthopaedic surgery, application to hip resurfacing. In: Proceedings of the Hamlyn Symposium on Medical Robotics, London, UK, pp. 61–62 (June 2013)

31. Noy, N.F., Shah, N.H., Whetzel, P.L., Dai, B., et al.: Bioportal: Ontologies and integrated data resources at the click of a mouse. Nucleic Acids Research 37(s2), 170–173 (2009)

32. Smith, B., Ashburner, M., Rosse, C., Bard, J., Bug, W., Ceusters, W., et al.: The OBO Foundry: Coordinated evolution of ontologies to support biomedical data integration. National Biotechnology 37, 1251–1255 (2007)

33. Neumuth, D., Loebe, F., Herre, H., Neumuth, T.: Modeling surgical processes: A four-level translational approach. Artificial Intelligence in Medicine 51, 147–160 (2011)
34. Lemaignan, S., Ros, R., Mösenlechner, L., Alami, R., Beetz, M.: ORO, a knowledge management module for cognitive architectures in robotics. In: Proc. of the IEEE/RSJ Intl. Conf. on Intelligent Robots and Systems (IROS), pp. 3548–3553 (2010)
35. Tenorth, M., Beetz, M.: Knowrob—knowledge processing for autonomous personal robots. In: Proc. of the IEEE/RSJ Intl. Conf. on Intelligent Robots and Systems (IROS), pp. 4261–4266 (2009)
36. Fujita, H., Rudas, I., Fodor, J., Kurematsu, M., Hakura, J.: Fuzzy reasoning for medical diagnosis-based aggregation on different ontologies. In: 2012 7th IEEE International Symposium on Applied Computational Intelligence and Informatics (SACI), pp. 137–146 (2012)

Consideration about the Application of Dynamic Time Warping to Human Hands Behavior Recognition for Human-Robot Interaction

Ji-Hyeong Han and Jong-Hwan Kim

Department of Electrical Engineering, KAIST,
291 Daehak-ro, Yuseong-gu, Daejeon, 305-701, Republic of Korea
{jhhan,johkim}@rit.kaist.ac.kr

Abstract. To prepare the age when humans and robots live together, robots need to understand the meaning of human behaviors for the natural and rational human-robot interaction (HRI). The robot particularly needs to recognize the human hands behavior, since humans usually express their meanings and intentions by using two hands. In this paper, the robot recognizes the human hands behavior by simulating it based on robot's own hands behaviors set and finding the most similar one as human behavior using dynamic time warping (DTW) algorithm. To consider the effects of different variables, i.e. data normalization methods and local cost measures for DTW algorithm, this paper considers two different normalization methods and four different local cost measures and their effects are discussed. The robot successfully recognizes the eight different human hands behaviors by DTW algorithm with the chosen normalization methods and local cost measures.

Keywords: Human hands behavior recognition, dynamic time warping algorithm, human-robot interaction.

1 Introduction

Due to the rapid development of robot technology and intelligence technology [1], [2], humans and robots will live together in no distant future. Humans can have the benefit of living together with robots after solving the most important issue, i.e. the natural and effective human-robot interaction (HRI). Accordingly the research dealing with HRI are exploding and covering various research fields [3]-[13]. For the natural and effective HRI, robots should understand the meaning and intention of human behaviors. Humans use usually both of hands to express the meanings and intentions, therefore robots need to recognize the human hands behavior above all.

In this paper, the robot recognizes the human hands behavior by simulating perceived human behavior data based on its own behaviors. Since it should be able to recognize the human hands behavior independent of the speed, time variation, and length of the behavior, the dynamic time warping (DTW) algorithm is used to find out the most similar behavior, which avoids loss of data by

J.-H. Kim et al. (eds.), *Robot Intelligence Technology and Applications 2*,
Advances in Intelligent Systems and Computing 274,
DOI: 10.1007/978-3-319-05582-4_23, © Springer International Publishing Switzerland 2014

linear mapping functions [14], [15]. The DTW algorithm was used originally to compare speech patterns for speech recognition [16]. After the first introduction, the DTW algorithm has been applied to many different fields, such as signature recognition, gesture recognition, and data mining [17]-[19]. This paper considers how the DTW algorithm is applied to recognize human hands behaviors by a robot with different data normalization methods and local cost measures. To compare the perceived human hands behavior data with the robot's own behavior data, both data must be normalized since they have different ranges. In this paper, two different normalization methods, i.e. using data range and with zero mean and unit std, are considered and the human hands behavior recognition results using them are discussed. Also, four different local cost measures, which are needed in the DTW algorithm to compare the data sequences, are considered. They are based on 1-norm, 2-norm, infinity norm distances of Minkowski distance and derivative of data, and the effects of different local cost measures on recognition results are discussed.

This paper is organized as follows. Section 2 presents the DTW algorithm and how it is applied to recognize the human hands behavior. In Section 3, experimental results are discussed. Finally, concluding remarks follow in Section 4.

2   Application of DTW Algorithm to Recognize Human Hands Behaviors by a Robot

In this section, the DTW algorithm and its application to recognize human hands behaviors by a robot are explained. The robot perceives the human hands behavior by RGB-D camera sensor and recognizes it by simulating it based on robot's own hands behaviors set and finding the most similar one by DTW algorithm. Since the hands behavior data should be normalized before applying the DTW algorithm, two different normalization methods are considered. Also, the local cost measure for DTW algorithm is needed to define, therefore four different local cost measures are considered.

2.1   DTW Algorithm

The objective of the DTW algorithm is to find an alignment between two time series sequences with a minimal overall cost. Suppose there are two time series sequences $X = \{x_1, x_2, \cdots, x_N\}$ and $Y = \{y_1, y_2, \cdots, y_M\}$. To find an alignment between two different sequences, the local cost measure $cost(x, y)$ is needed to compare the sequences. The local cost measure for each pair of elements of the sequences X and Y is evaluated and then the local cost matrix $CM \in \mathrm{R}^{N \times M}$ that is defined as $CM(n, m) = cost(x_n, y_m)$ is obtained. The goal is finding the alignment between X and Y with a minimal overall cost, for which three conditions in the following definition should be satisfied.

Definition 1. *An* (N, M)-*warping path is a sequence* $p = (p_1, \cdots, p_K)$ *with* $p_k = (n_l, m_l) \in [1 : N] \times [1 : M]$ *for* $k \in [1 : K]$ *satisfying the following conditions.*

(i) Boundary condition: $p_1 = (1, 1)$ *and* $p_K = (N, M)$.
(ii) Monotonicity condition: $n_1 \leq n_2 \leq \cdots \leq n_K$ *and* $m_1 \leq m_2 \leq \cdots \leq m_K$.
(iii) Step size condition: $p_{k+1} - p_k \in \{(1, 0), (0, 1), (1, 1)\}$ *for* $k \in [1 : K - 1]$.

2.2 Applying DTW to Recognize Human Hands Behavior

By using the DTW algorithm, the robot simulates the obtained human hands behavior data from a RGB-D camera sensor in its mind to find out the most similar behavior in its own hands behaviors set. The behavior data of both human hands are defined as $H_{L/R} = \{H_{L/R_1}, \cdots, H_{L/R_N}\}$, where $H_{L/R_n} = (H_{L/RX_n}, H_{L/RY_n}, H_{L/RZ_n})$ and $n \in [1 : N]$. The behavior data of both robot hands are defined as $R_{L/R} = \{R_{L/R_1}, \cdots, R_{L/R_M}\}$, where $R_{L/R_m} = (R_{L/RX_m}, R_{L/RY_m}, R_{L/RZ_m})$ and $m \in [1 : M]$. The human left and right hands behaviors data are compared with robot's left and right ones, respectively, and all data are time series sequences in (x, y, z) coordinates.

Since the ranges of human hands behavior data and robot's hands behavior data are different and they are three dimensional, they should be normalized before comparing. There can be several normalization methods and in this paper two methods are considered. The first one is normalizing the data to $[0, 1]$ by using data range. The range of human hands behavior data, HR, is calculated as:

$$HR_X = \max(H_{LX}, H_{RX}) - \min(H_{LX}, H_{RX}) \tag{1}$$

$$HR_Y = \max(H_{LY}, H_{RY}) - \min(H_{LY}, H_{RY}) \tag{2}$$

$$HR_Z = \max(H_{LZ}, H_{RZ}) - \min(H_{LZ}, H_{RZ}) \tag{3}$$

$$HR = \max(HR_X, HR_Y, HR_Z). \tag{4}$$

Then the normalized human hands behavior data, $\overline{H}_{L/RX_n}$, $\overline{H}_{L/RY_n}$, and $\overline{H}_{L/RZ_n}$, are calculated as follows:

$$\overline{H}_{L/RX_n} = \frac{H_{L/RX_n} - \min(H_{LX}, H_{RX})}{HR} \tag{5}$$

$$\overline{H}_{L/RY_n} = \frac{H_{L/RY_n} - \min(H_{LY}, H_{RY})}{HR} \tag{6}$$

$$\overline{H}_{L/RZ_n} = \frac{H_{L/RZ_n} - \min(H_{LZ}, H_{RZ})}{HR}. \tag{7}$$

The other one is normalizing the data as having zero mean and unit standard deviation. The normalized human hands behavior data are calculated as follows:

$$\overline{H}_{L/RX_n} = \frac{H_{L/RX_n} - mean(H_{LX}, H_{RX})}{std(H_{LX}, H_{RX})} \tag{8}$$

$$\overline{H}_{L/RY_n} = \frac{H_{L/RY_n} - mean(H_{LY}, H_{RY})}{std(H_{LY}, H_{RY})} \tag{9}$$

$$\overline{H}_{L/RZ_n} = \frac{H_{L/RZ_n} - mean(H_{LZ}, H_{RZ})}{std(H_{LZ}, H_{RZ})}. \tag{10}$$

The robot's hands behavior data are also normalized in the same way as human ones.

The local cost measure $cost(\overline{H}_{L/R}, \overline{R}_{L/R})$ for the DTW algorithm can be defined as several distance measures. In this paper, three kinds of distance measures are considered, i.e. 1-norm, 2-norm, and infinity norm distances of Minkowski distance. Minkowski distance-based local cost measures are defined as follows in order of 1-norm, 2-norm and infinity norm distances:

$$cost(\overline{H}_{L/R_n}, \overline{R}_{L/R_m})$$
$$= \begin{cases} |\overline{H}_{L/RX_n} - \overline{R}_{L/RX_m}| + |\overline{H}_{L/RY_n} - \overline{R}_{L/RY_m}| + |\overline{H}_{L/RZ_n} - \overline{R}_{L/RZ_m}| \\ \sqrt{\left(\overline{H}_{L/RX_n} - \overline{R}_{L/RX_m}\right)^2 + \left(\overline{H}_{L/RY_n} - \overline{R}_{L/RY_m}\right)^2 + \left(\overline{H}_{L/RZ_n} - \overline{R}_{L/RZ_m}\right)^2} \\ \max(|\overline{H}_{L/RX_n} - \overline{R}_{L/RX_m}|, |\overline{H}_{L/RY_n} - \overline{R}_{L/RY_m}|, |\overline{H}_{L/RZ_n} - \overline{R}_{L/RZ_m}|). \end{cases} \tag{11}$$

Also, a derivative measure can be used as the local cost measure [20]. The estimated derivative for human hands behavior data is defined as follows:

$$der(\overline{H}_{L/RX,Y,Z_n})$$
$$= \frac{(\overline{H}_{L/RX,Y,Z_n} - \overline{H}_{L/RX,Y,Z_{n-1}}) + (\overline{H}_{L/RX,Y,Z_{n+1}} - \overline{H}_{L/RX,Y,Z_{n-1}})/2}{2}. \tag{12}$$

The estimated derivative for robot's hands behavior data is also calculated in the same way as human ones. Because the estimated derivatives for the first and last elements are not defined in the above formula, they are the same as the estimated derivatives of second and penultimate elements. Then, the derivative-based local cost measure is defined as follows:

$$cost(\overline{H}_{L/R_n}, \overline{R}_{L/R_m})$$
$$= (der(\overline{H}_{L/RX_n}) - der(\overline{R}_{L/RX_m}))^2 + (der(\overline{H}_{L/RY_n}) - der(\overline{R}_{L/RY_m}))^2 \tag{13}$$
$$+ (der(\overline{H}_{L/RZ_n}) - der(\overline{R}_{L/RZ_m}))^2.$$

The DTW algorithm evaluates the defined local cost measure for each pair of elements of the sequences $\overline{H}_{L/R}$ and $\overline{R}_{L/R}$ and obtains the local cost matrix $CM_{L/R} \in \mathrm{R}^{N \times M}$ that is defined as $CM_{L/R}(n, m) = cost(\overline{H}_{L/R_n}, \overline{R}_{L/R_m})$.

The total cost, $cost_p(\overline{H}_{L/R}, \overline{R}_{L/R})$, of a warping path p with the local cost measure is defined as

$$cost_p(\overline{H}_{L/R}, \overline{R}_{L/R}) = \sum_{k=1}^{K} cost(\overline{H}_{L/R_k}, \overline{R}_{L/R_k}). \tag{14}$$

The optimal warping path between $\overline{H}_{L/R}$ and $\overline{R}_{L/R}$ is p^* having the minimal total cost among all possible warping paths which satisfy three conditions in Definition 1. The DTW distance, $DTW(\overline{H}_{L/R}, \overline{R}_{L/R})$, is defined as follows:

$$\begin{aligned} DTW(\overline{H}_{L/R}, \overline{R}_{L/R}) &= cost_{p*}(\overline{H}_{L/R}, \overline{R}_{L/R}) \\ &= \min\{cost_p(\overline{H}_{L/R}, \overline{R}_{L/R})|\ p \text{ is } (N, M)-\text{warping path}\}. \end{aligned} \tag{15}$$

The final DTW distance is the sum of DTW distances for left and right hands behaviors data, i.e. $FDTW = DTW(\overline{H}_L, \overline{R}_L) + DTW(\overline{H}_R, \overline{R}_R)$. The robot calculates the $FDTW$ between the perceived human hands behavior data and all behaviors in its behaviors set. Then, the robot hands behavior with the minimum $FDTW$ is recognized as the human hands behavior. This method has two main advantages that it does not need to learn any data in advance before the recognition of the human hands behaviors and it can be applied to the robot that has different degrees of freedom (DOFs) from the human DOFs.

3   Experiments

In this section, experimental setup and results are discussed. The robot's hands behaviors set consisted of eight behaviors, i.e. waving both hands (WB), waving one hand (WO), pointing (Po), touching (Tch), pushing (Pu), grasping (Gr), releasing (Re), and throwing (Th). The robot's hands behavior data were defined as trajectories of robot's hands, which were end-effectors of robot arms, by calculating forward kinematics for each behavior. Since the robot recognized the human hands behavior by simulating it based on robot's own behaviors, the human hands behaviors were recognized as the above eight labels. The basic human hands behavior data set was gathered by a human subject which did the

Table 1. Human hands behavior recognition results using DTW algorithm

Normalization method	Local cost measure	WB	WO	Po	Tch	Pu	Gr	Re	Th	Average of correct recognition (%)
Normalization using data range (norm1)	1-norm	100	82	80	62	100	100	100	100	90.5
	2-norm	100	89	80	59	100	100	100	100	91
	Infinity norm	100	89	80	59	100	100	100	100	91
	Derivative	69	41	80	79	100	100	0	71	67.5
Normalization with zero mean and unit std (norm2)	1-norm	10	0	30	1	0	58	16	93	26
	2-norm	15	0	30	1	0	58	13	93	26.25
	Infinity norm	15	0	30	1	0	57	11	98	26.5
	Derivative	0	0	30	0	0	100	0	40	21.25

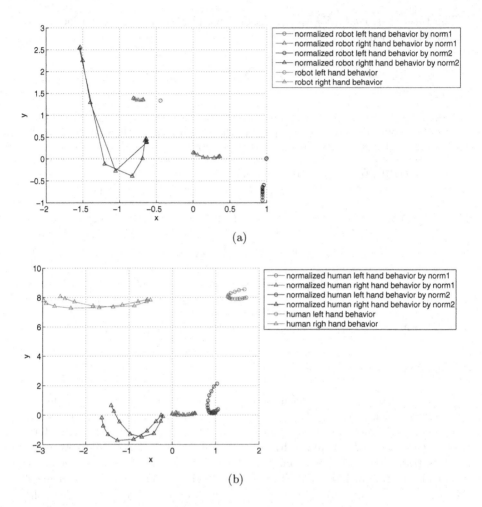

Fig. 1. Plot of normalizing WO behavior data by norm1 and norm2 and real data in xy-plane. (a) The robot hands behavior of WO behavior. (b) The human hands behavior of WO behavior.

above eight hands behaviors by ten times for each in front of RGB-D camera sensor Kinect. Ten test data for each gathered human behavior were made by adding 0.01 level random noise to each of gathered human hands behavior trajectory data. Therefore, 100 behaviors data for each labeled one and totally 800 behaviors composed the test human behavior data set.

The human hands behavior data was recognized by a robot using DTW algorithm. To consider the effect of different normalization methods and local cost measures to apply DTW algorithm for human hands behavior recognition, the explained two normalization methods and four local cost measures in Section 2 were used and the recognition results were compared.

Table 1 shows the human hands behavior recognition results for combinations of two kinds of normalization methods and four kinds of local cost measures. The values under each behavior, i.e. WB, WO, Po, Tch, Pu, Gr, Re, Th, means the number of correct recognitions among 100 test data for each one. The difference in 1-norm, 2-norm, and infinity norm distances of Minkowski distance did not affect the performance of human hands behavior recognition in both of DTW algorithm with normalizations using data range (norm1) and with zero mean and unit standard deviation (norm2). The DTW algorithm with derivative-based local cost measure did not show better performance compared to that with Minkowski distance-based one in both of normalization methods. Particularly the method using derivate-based local cost measure did not recognize Re behavior at all, because WB and Re behaviors had the similar derivatives and it classified Re behavior as WB behavior. However, it showed better result in Tch behavior than using Minkowski distance-based local cost measure. Because both of a robot and human subject moved their hands less for Tch behavior than other behaviors, the Minkowski distance based on Euclidean space of Tch behavior did not change much compared to other behaviors. Therefore, the derivative-based local cost measure for DTW algorithm was more effective to recognize the short moving behaviors than Minkowski distance-based local cost measure. The recognition result by DTW algorithm with norm1 showed better performance than that with norm2. Because the normalization method using data range made much more similar pattern with the real data than normalization method with zero mean and unit standard deviation as shown in Fig. 1. Fig. 1 shows the WO behavior data as a representative, since the method using norm2 did not recognize WO at all.

4   Conclusion

In this paper, we considered the application of DTW algorithm to the human hands behavior recognition for HRI. The robot recognized the human hands behavior by simulating the perceived human behavior data based on its own behaviors by the DTW algorithm. By using the DTW algorithm, the robot could recognize the human hands behavior independent of the speed, time variation, and length of behavior. To consider the effect of behavior data normalization method and local cost measure in the DTW algorithm, two kinds of normalization methods and four kinds of local cost measures were considered and tested. The experimental results showed that the DTW algorithm with the normalization method using data range did better behavior recognition than that with zero mean and unit standard deviation. The different norms of Minkowski distance as local cost measure did not affect the recognition performance, but they showed always better overall recognition performances than derivative-based local cost measure. However, the DTW algorithm with derivative-based local cost measure showed better performance than that with Minkowski distance-based ones in short moving behavior like touching. Therefore, the DTW algorithm with data normalization using its range and the local cost measure based on combination of Euclidean space distance and derivative depending on moving distance of behavior would be better way for a robot to recognize human hands behavior.

Acknowledgement. This research was supported by the MOTIE (The Ministry of Trade, Industry and Energy), Korea, under the Technology Innovation Program supervised by the KEIT (Korea Evaluation Institute of Industrial Technology) (10045252, Development of robot task intelligence technology that can perform task more than 80% in inexperience situation through autonomous knowledge acquisition and adaptational knowledge application).

Also this research was supported by the MOTIE (The Ministry of Trade, Industry and Energy), Korea, under the Human Resources Development Program for Convergence Robot Specialists support program supervised by the NIPA (National IT Industry Promotion Agency) (H1502-13-1001, Research Center for Robot Intelligence Technology).

References

1. Kim, J.-H., Choi, S.-H., Park, I.-W., Zaheer, S.A.: Intelligence technology for robots that think. IEEE Comput. Intell. Mag. (August 2013) (to be published)
2. Kim, J.-H., Ko, W.-R., Han, J.-H., Zaheer, S.A.: The degree of consideration-based mechanism of thought and its application to artificial creatures for behavior selection. IEEE Comput. Intell. Mag. 7(1), 49–63 (2012)
3. Kubota, N., Nishida, K.: Perceptual Control Based on Prediction for Natural Communication of a Partner Robot. IEEE Trans. on Industrial Electronics 54(2), 866–877 (2007)
4. Sato, E., Yamaguchi, T., Harashima, F.: Natural Interface Using Pointing Behavior for Human-Robot Gestural Interaction. IEEE Trans. on Industrial Electronics 54(2), 1105–1112 (2007)
5. Mitsantisuk, C., Katsura, S., Ohishi, K.: Force Control of Human-Robot Interaction Using Twin Direct-Drive Motor System Based on Modal Space Design. IEEE Trans. on Industrial Electronics 57(4), 1383–1392 (2010)
6. Zhu, C., Sheng, W.: Wearable sensor-based hand gesture and daily activity recognition for robot-assisted living. IEEE Tran. Syst., Man, Cybern. A, Syst., Humans 41(3), 569–573 (2011)
7. Lallee, S., et al.: Towards a platform-independent cooperative human robot interaction system: III an architecture for learning and executing actions and shared plans. IEEE Trans. Auton. Mental Develop. 4(3), 239–253 (2012)
8. Malfaz, M., Castro-Gonzalez, A., Barber, R., Salichs, M.A.: A biologically inspired architecture for an autonomous and social robot. IEEE Trans. Auton. Mental Develop. 3(3), 232–246 (2011)
9. Cabibihan, J.-J., So, W.-C., Pramanik, S.: Human-recognizable robotic gestures. IEEE Trans. Auton. Mental Develop. 4(4), 305–314 (2012)
10. Yorita, A., Kubota, N.: Cognitive development in partner robots for information support to elderly people. IEEE Trans. Auton. Mental Develop. 3(1), 64–73 (2011)
11. Andry, P., Blanchard, A., Gaussier, P.: Using the rhythm of nonverbal human-robot interaction as a signal for learning. IEEE Trans. Auton. Mental Develop. 3(1), 30–42 (2011)

12. Han, J.-H., Kim, J.-H.: Human-robot interaction by reading human intention based on mirror-neuron system. In: Proc. 2010 IEEE ROBIO, pp. 561–566 (2010)
13. Han, J.-H., Kim, J.-H.: Human intention reading by fuzzy cognitive map: A human-robot cooperative object carrying task. In: Kim, J.-H., Matson, E., Myung, H., Xu, P. (eds.) Robot Intelligence Technology and Applications. AISC, vol. 208, pp. 127–135. Springer, Heidelberg (2013)
14. Müller, M.: Dynamic time warping. In: Information Retrieval for Music and Motion, ch. 4, Springer (2007)
15. Senin, P.: Dynamic time warping algorithm review. University of Hawaii at Manoa, USA (2008)
16. Vintsyuk, T.K.: Speech discrimination by dynamic programming. Kibernetika 4, 81–88 (1968)
17. Tappert, C.C., Suen, C.Y., Wakahara, T.: The state of the art in online handwriting recognition. IEEE Trans. Pattern Anal. Mach. Intell. 12(8), 787–808 (1990)
18. Kuzmanic, A., Zanchi, V.: Hand shape classification using DTW and LCSS as similarity measures for vision-based gesture recognition system. In: Int. Conf. EU-ROCON, pp. 264–269 (2007)
19. Niennattrakul, V., Ratanamahatana, C.A.: On clustering multimedia time series data using k-means and dynamic time warping. In: Int. Conf. Multimedia and Ubiquitous Eng., pp. 733–738 (2007)
20. Keogh, E.J., Pazzani, M.J.: Derivative dynamic time warping. In: 1st SIAM Int. Conf. on Data Mining (2001)

Lessons Learned in Designing User-Configurable Modular Robotics

Henrik Hautop Lund

Center for Playware, Technical University of Denmark,
Building 326, 2800 Kgs. Lyngby, Denmark
hhl@playware.dtu.dk

Abstract. User-configurable robotics allows users to easily configure robotic systems to perform task-fulfilling behaviors as desired by the users. With a user configurable robotic system, the user can easily modify the physical and functional aspect in terms of hardware and software components of a robotic system, and by making such modifications the user becomes an integral part in the creation of an intelligence response to the challenges posed in a given environment. I.e. the overall intelligent response in the environment becomes the integration of the user's construction and creation with the semi-autonomous components of the user-configurable robotic system in interaction with the given environment. Components constituting such a user-configurable robotic system can be characterized as modules in a modular robotic system. Several factors in the definition and implementation of these modules have consequences for the user-configurability of the system. These factors include the modules' granularity, autonomy, connectivity, affordance, transparency, and interaction.

Keywords: Human-robot interaction, reconfigurable robots, educational robots, distributed intelligence, modular robotics.

1 Introduction

Robotics and artificial intelligence (AI) research has strived to create fully autonomous systems, which can exhibit an intelligent response in the environment. The artificial intelligence research aimed at creating systems, which are able to sense and act, to learn and to think, and figure out the right organization of activities at these different levels. Much work in classical artificial intelligence research built upon the understanding that there would be a level where the body could be abstracted away, and one could investigate the thinking in isolation from the body (e.g. in symbol processing systems, expert systems, etc.). Even though robotic research engineered physical robotic systems since the middle of last century, it was not until the end of the 1980's and the 1990's that robotics became a more widespread tool to study thinking as *embodied*.

A well-known tool which facilitates the study of the intelligence as integration between the body and the brain is the LEGO Mindstorms. The LEGO Mindstorms

J.-H. Kim et al. (eds.), *Robot Intelligence Technology and Applications 2*,
Advances in Intelligent Systems and Computing 274,
DOI: 10.1007/978-3-319-05582-4_24, © Springer International Publishing Switzerland 2014

provides a tool, which allows users to easily build a robot with sensors and actuators, to build bodies with LEGO pieces and to build brains with simple software, e.g. a GUI. As observed in [1], with LEGO Mindstorms the interaction is split into distinct processes of building, understanding syntax & semantics, programming, downloading to robot, testing/debugging, playing.

This split is partly due to LEGO Mindstorms being constituted by a central processor to which sensors, actuators, and LEGO bricks can be attached, and partly due to the programming paradigm of performing the programming on a host computer (i.e. not situated in the environment of the robot). Hence, LEGO Mindstorms is based upon a centralized processing approach with the central processor being programmed via a host computer.

If, on the other hand, we turn to a more *distributed processing* approach based on a collection of self-contained modules each with their own processing and physical expression, it is possible to work towards avoiding or diminishing this split and create direct *action in each interaction* by the user. This is done by the creation of physical and functional modules which allow exploration of interactive, distributed parallel processing in a physical form. Here, action is manifested as soon as modules are manipulated, such as when modules in the form of building blocks are being put together.

In such a distributed processing approach, processing is distributed to a number of modules and the overall processing emerges from the processing of the individual modules and their interaction. In a similar way, the physicality is distributed to a number of modules, and the overall physical expression is a function of the individual physical modules and their interaction. Put together, this can be expressed in terms of a *modular robotic system*: In a modular robotic system, each module has a physical and functional expression, and the overall robotic system emerges from the interaction between the modules.

Modular robotic systems can be used to create *self-reconfigurable modular robots* [2], which autonomously change their physical shape. In the self-reconfigurable modular robots, the modules are able to autonomously move around attaching and detaching from each other, moving to locations so that the overall shape of the robot (the ensemble of modules) becomes appropriate for the task at hand.

However, instead of focusing exclusively on the creation of fully autonomous and self-contained systems to provide an intelligent behavior, there is an attractive possibility of focusing on the creation of systems that allow human-robot interaction to create an intelligent response. The concept of user-configurable modular robotic systems aims at facilitating such generation of intelligent response in the environment through the human-robot interaction.

2   User-Configurable Modular Robotics

In a *user-configurable modular robotic system*, the user constructs with modules (i.e. technological building blocks) to create a physical system and the functionality of this system. By making changes to the physical shape of the entity, the user can change the functionality of the system. This happens simply by attaching or detaching modules and moving modules to different positions. Hence, in such a case, the user is making the physical configuration in a hands-on manner, and the user does not need to do traditional programming to change the functionality of the system. As soon as the user is manipulating with the modules there is a reaction in the environment, i.e. there is *action in the interaction*, and the interaction is not split into distinct processes as was the case e.g. with LEGO Mindstorms interaction.

Therefore, in some cases, it is believed that user-configurable modular robotics may lead any user to develop solutions in a simple and very flexible manner. Further, the modularity and distributed processing means that the produced solutions are robust to failure of individual modules through graceful degradation. If one module fails then the rest will still be working, contrary to most traditional technological solutions with a central processing that may make everything fail if one component fails. Also, since there is no central processing and large infrastructure, but the system is composed of a set of individual modules, these may potentially be easily transported around and set up anywhere.

Hence, the overall intelligent response in the environment becomes the integration of the user's construction and creation with the components (modules) of the user-configurable modular robotic system in interaction with the given environment. We formulate this concept as the playware ABC: By *building bodies and brains* with the user-configurable modular system, the user can *construct, combine and create* to make solutions for *anybody, anywhere, anytime*. Several factors in the definition and implementation of the modules have consequences for the user-configurability of the system. These factors can be viewed as design issues and include the modules' granularity, automation/autonomy, connectivity, affordance, and interaction. To shed light on these design issues, we have researched user-configurable modular robotic system in a wide range of designs, implementations, and applications, some of which will be reported below.

2.1   Granularity

Granularity of the modules is a crucial issue to consider in the design of modules for a user-configurable modular robotic system, both in terms of physical and functional granularity. With *coarse-grained modules*, the user will be working on a high abstraction level with only a few modules needed to obtain the intelligent response, i.e. to obtain the right physical and functional response. Hence, the cognitive load on the user for creating the intelligent response is considered to be low in a user-configurable modular robotic system with coarse-grained modules. On the other hand, the versatility may be low in a coarse-grained modular system not allowing the user to create subtle physical and functional structures lower than the graining size of the individual modules. I.e. if all modules are $1m^3$ (e.g. like the MusicTiles magic cubes Fig. 1(k)), then it is difficult to make variations in the centimeter-scale of the physical structure – and a similar argument goes for the functional variations.

Fig. 1. Examples of modular systems with user interaction: (a) ATRON self-reconfigurable modular robot, (b) I-Blocks in LEGO Duplo, (c) Light&Sound Cylinders and Rolling Pins for elderly dementia patient therapy in multi-sensory room, (d) modular interactive tiles for rehabilitation of stroke and cardiac patients, (e) modular interactive tiles for rehabilitation of mentally and physically handicapped children in Africa, (f) Fable user-configurable modular robot, (g) Fatherboard modular robotic wearable, (h) modular interactive tiles for soccer and playgrounds, (i) Music I-Blocks, (j) MusicTiles magic matchboxes, (k) MusicTiles magic cubes.

With *fine-grained modules*, the user will be working on a lower abstraction level with more modules needed to obtain the intelligent response, i.e. to obtain the right physical and functional response. The cognitive load on the user for creating the intelligent response can be considered to be higher in a user-configurable modular robotic system with fine-grained modules, since the user will have to combine more modules to obtain the same response as with coarse-grained modules. For instance, the learning curve for being able to create the desired intelligent system with the fine-grained modules may be steeper. Yet, the versatility may be higher in a fine-grained modular system with which the user may be able to create subtle physical and functional structures not possible with coarse-grained modules (e.g. with centimeter-scale modules it is possible to construct centimeter-scale variations).

2.2   Homogenous vs. Heterogeneous Modules

When designing modules, it is possible to make them as *homogenous modules* (all modules are similar) or *heterogeneous modules* (modules differ from one another). There also exists the possibility of making physical homogenous but functional heterogeneous modules, though some indication of the heterogeneity of function seems necessary for the user, e.g. making the modules in different colors. The Fable modular robotic system (Fig. 1(f)) is an example of a user-configurable modular robotic system based on heterogeneous modules, whereas the ATRON modular robotic system (Fig. 1(a)) is based on homogeneous modules. In the case of Fable, the heterogeneous chain-based modular robotic system consists of various modules, such as different types of joint, branching and termination modules [3]. Joint modules are actuated robotic modules used to enable locomotion and interaction with the environment. Branching modules connect several modules together in tree-like configurations. Termination modules may add structure, a visual expression, additional sensors, or actuators (e.g. grippers or wheels). Similar, the modular robotic wearable exemplified with the Fatherboard (Fig. 1(g)) is also a heterogeneous system with modules of different functions such as a buzz, a recorded sound, a voice, a red light, a blue light, etc. [4].

2.3   Connectivity

Further, it is important to design which *connectivity* is desired and advantageous between modules. The connectivity may vary from loose to tight, from no connection whatsoever to modules all connected, and from chain-based connection to lattice-based connection. Interestingly, the philosophical consideration of intelligence in light of user-configurable modular robotic systems opens up for research into modular systems with no physical connection but only functional connection. The MusicTiles magic cubes (Fig. 1(j)-(k)) present such an example with physical separate modules each representing an instrument, and rotation giving the musical variation of the particular instrument, while together all modules gives the whole music tune. There is no physical attachment between the modules in this user-configurable modular robotic system. On the contrary, in the I-Blocks music cubes (Fig. 1(i)), musical expression of the given module (instrument) is based upon the attachment of the module to another module [5, 6]. As another example, in the case of user-configurable modular devices for a multi-sensory room for therapy of elderly dementia patients (Fig. 1(c)), Sound&Light cubes changed the ambient sound and light based on physical stacking (attaching) the modules together, whereas Rolling Pins changes the responses based upon pattern of interaction with physical separate Rolling Pins (two people rolling the separate pins in synchrony, rolling speed, etc.) [7].

2.4   Ease of Construction

In the case of physical connectivity, the connection mechanism may pose a challenge in both homogenous and heterogeneous modular robotic systems. Where the field of reconfigurable modular robotic systems has confronted this challenge in terms of the mechanical and electrical reliability, the mechanical and motion control optimization, etc., the field of user-configurable modular robotic systems needs to take the *ease of construction*, including attachment and detachment, into consideration. For instance, the Fable project (Fig. 1(f)) investigates connectors designed to allow rapid and solid attachment and detachment between modules with scalable connectors to allow modules of different sizes to be combined and designed to permit neighbor-to-neighbor communication [3]. The modular interactive tiles use puzzle-shaped connectors [8], while I-Blocks use the LEGO studs [9], I-Blocks music cubes use magnets [5], and the modular robotic wearable uses simple clothes-buttons [4]. In all cases, the connectors have been carefully researched and developed for the ease of construction to allow anybody to easily build with the system.

2.5   Interaction, Affordance, and Transparency

As supplement to ease of construction, user-configurable modular robotic systems need to address *interaction* in general. Interaction can be of many forms, apart from attaching and detaching modules, it may be rotation of modules, walking, running and jumping on modules as with the modular interactive tiles (Fig. 1(d)-(e)) for prevention and rehabilitation [10, 11], rolling and stacking as with the modular robotic devices for a multi-sensory room (Fig. 1(c)) [7]. For creating such user interactions, it is important to design for the modules' *affordance* [12, 13], e.g. such as a dice which invites to roll (Fig. 1(k)), a tile which invites to step on it or hit it with a ball (Fig. 1(h)), a LEGO brick which invites to attach (Fig. 1(b)), a wearable module with clothes-buttons which invites to fasten (Fig. 1(g)), rolling pins which invite to roll (Fig. 1(c)), etc. Considering the affordance of modules, it may be possible to communicate the functionality of modules to the user. *Transparency* of functionality of modules and ensembles of modules is indeed a major challenge in user-configurable modular robotic system, and affordance in module design including material design, interaction design, connectivity, etc. must be considered to facilitate the ease of understanding of functionality for the user. Indeed, studying and understanding the affordance of modules and their interplay between each other and with human beings is one of the main defining subjects that distinguish user-configurable modular robotics from other kinds of modular robotics, including self-reconfigurable modular robotics.

2.6 Automatic vs. Autonomous Modules

The functionality of individual modules and the emergence of the overall intelligent response based on the user's interaction with the modules may be based on *automatic modules* or *autonomous modules*. In most of the known user-configurable modular robotic systems, the system is automatic with pre-programmed content of the modules e.g. as closed-loop control, cellular automata or behavior-based system [1, 9] or as a pre-produced sound piece [5]. Working towards autonomous modules, research with modular interactive tiles (Fig. 1(d)) shows how these may be adaptive in their control, for instance using simple adaptive processing [14] to adapt to the user's physical interactions, or using artificial neural network learning (Fig. 1(k)) [15]. There is an interesting research challenge in understanding how automation and autonomy in modules and their coordination may potentially facilitate and guide user interaction in user-configurable modular robotic systems.

3   Discussion and Conclusion

User-configurable modular robotic systems seem a promising concept to allow users to easily configure robotic systems to perform desired, task-fulfilling behaviors. With a well-designed user-configurable modular robotic system, the user can easily modify the physical and functional aspect in terms of hardware and software components of a robotic system, and by making such modifications the user becomes an integral part in the creation of an intelligence response to the challenges posed in a given environment.

As has been outlined, there are several factors in the definition and implementation of modules in a user-configurable modular robotic system, which have consequences for the user-configurability of the system. Some factors are known from modular robotics, but importantly the inclusion of the user poses serious design challenges based upon affordance, interaction, transparency, and ease of use, which are not addressed in traditional modular robotics research. Here, based on lessons learned from a few early examples of user-configurable modular robotic systems, these challenges have been outlined briefly. Future research work should address these challenges in a comprehensive and in-depth manner. Additionally, future research and application work should investigate how user-configurable modular robotic systems may contribute to the development of the *playware ABC*, i.e. investigate how *building bodies and brains* with the user-configurable modular system, the user may *construct, combine and create* to make solutions for *anybody, anywhere, anytime*.

Acknowledgements. The author would like to thank co-authors and colleagues at the Center for Playware for discussions, implementations, and tests. Without their efforts, it would not have been possible to develop a comprehensive concept of user-configurable modular robotics.

References

1. Lund, H.H.: Modular Playware Technology – A Brief Historical Review. In: Proceedings of 17th International Symposium on Artificial Life and Robotics, ISAROB, Japan (2012)
2. Ostergaard, E.H., Kassow, K., Beck, R., Lund, H.H.: Design of the ATRON lattice-based self-reconfigurable robot. Autonomous Robots 21(2), 165–183 (2006)
3. Pacheco, M., Moghadam, M., Magnússon, A., Silverman, B., Lund, H.H., Christensen, D.J.: Fable: Design of a Modular Robotic Playware Platform. In: Proceedings of the IEEE International Conference on Robotics and Automation. IEEE Press (2013)
4. Pagliarini, L., Lund, H.H.: Wearable Playware. In: The 8th International Conference on Ubiquitous Robots and Ambient Intelligence, pp. 9–13. IEEE (2011)
5. Lund, H.H., Bærendsen, N.K., Nielsen, J., Jessen, C., Falkenberg, K.: RoboMusic with Modular Playware. Artificial Life and Robotics 15(4), 369–375 (2010)
6. Nielsen, J.: User Configurable Modular Robotics - Control and Use. Ph.D. thesis, University of Southern Denmark (2008)
7. Marti, P., Giusti, L., Lund, H.H.: The Role of Modular Robotics in Mediating Nonverbal Social Exchanges. IEEE Transactions on Robotics 25(3), 602–613 (2009)
8. Lund, H.H.: Modular Robotics for Playful Physiotherapy. In: Proceedings of IEEE International Conference on Rehabilitation Robotics, pp. 571–575. IEEE Press (2009)
9. Lund, H.H.: Intelligent Artefacts. In: Sugisaka, Tanaka (eds.) Proceedings of 8th International Symposium on Artificial Life and Robotics, ISAROB, Oita, pp. I11–I14 (2003)
10. Lund, H.H., Jessen, J.: Effects on community-dwelling elderly playing with modular interactive tiles. In: Proceedings of International Conference on Serious Games and Edutainment. Springer, Heidelberg (2012)
11. Nielsen, C.B., Lund, H.H.: Adapting Playware to Rehabilitation Practices. International Journal of Computer Science in Sport 11(1) (2012)
12. Gibson, J.J.: The Ecological Approach to Visual Perception. Houghton Mifflin, Boston (1979) ISBN 0-89859-959-8
13. Norman, D.: The Design of Everyday Things. Basic Books, New York (1988) ISBN 0-465-06710-7
14. Björnsson, D., Fridriksson, R., Lund, H.H.: Adaptivity to Age, Gender, and Gaming Platform Topology in Physical Multi-Player Games. In: Proceedings of 17th International Symposium on Artificial Life and Robotics, ISAROB, Japan (2012)
15. Derakhshan, A., Hammer, F., Lund, H.H.: Adapting Playgrounds for Children's Play Using Ambient Playware. In: Proceedings of IEEE Intelligent Robots and Systems, IROS 2006. IEEE Press, Hong Kong (2006)

Playware Explorations in Robot Art

Luigi Pagliarini[1,2] and Henrik Hautop Lund[1]

[1] Center for Playware, Technical University of Denmark,
Building 326, 2800 Kgs. Lyngby, Denmark
[2] Academy of Fine Arts of Macerata, via Berardi, 6. 62100, Macerata, Italy
{luigi,hhl}@playware.dtu.dk

Abstract. We describe the upcoming art field termed *robot art*. Describing our group contribution to the world of robot art, a brief excursion on the importance of the underlying principles, of the context, of the message and its semiotic is also provided, case by case, together with few hints on the recent history of such a discipline, under the light of an artistic perspective. Therefore, the aim of the paper is to try to summarize the main characteristics that might classify robot art as a unique and innovative discipline, and to track down some of the principles by which a robotic artifact can be considered - or not - an art piece, in terms of social, cultural and strictly artistic interest.

Keywords: Robot, Art, Kinetic, Sculpture, Cyber, Cyberpunk, Embodiment, Evolution, Modular, Holography, Metamorphic, Alife, Polymorphic, Intelligence, Virtual, Alive.

1 Introduction

We can find robots in science and technology, architecture, art, video clips, cinema, literature as well as in our own homes. Their presence is fast growing in all fields and sectors and is becoming consistent in industrial production, medicine and entertainment. Robotics, in short, is a new "language" that is permeating the whole social structure and incorporating within itself several charming practical and intellectual issues that are able to elicit the interest and the curiosity of many philosophers, artists, scientists, technologists and, overall, ordinary people. In this paper, we try to get a closer and more specific look at what we call robot art, to try to understand the differences which can be found between the conceiving and the designing of pure scientific and/or commercial robots and those that can be considered exclusively art oriented. We do that although Flusser [1] suggested, somehow "Scientists are computer artists avant la lettre, and the results of science are not some 'objective insights', but models for handling the computed" (and vice versa). In fact, contemporary robotics is the field in which the comprehension of human brain attempts to materialize. It is a topic that has always been transversal to scientific and human disciplines alike, and that has brought together research fields into neuroscience, engineering, computer science, biology, mathematics, psychology, and philosophy. Indeed, from literature we know that embodying the biological neural system into machines (and machine bodies) is

J.-H. Kim et al. (eds.), *Robot Intelligence Technology and Applications 2*,
Advances in Intelligent Systems and Computing 274,
DOI: 10.1007/978-3-319-05582-4_25, © Springer International Publishing Switzerland 2014

one of the most attractive and challenging "dreams" humans deal with. In recent time, we went through this topics - like for example in the Alive Art [2] and Polymorphic Intelligence [3] manuscripts – but here we will try to look at things under an actual and a historical point of view to summarize and synthesize in one single vision the resulting paradigm and conceptual approach, by focusing, specifically, on robot art.

2   Recent Historical Paths

We may hypothesize that the research in this field was started in the 1950s within the Cybernetic Serendipity at London's ICA [4], and, today it is hosted in many contemporary artistic and cultural events – as, for example, it happened in the specialized art events like Robodock, Robots at Play, ArtBots, in New Media events like Ars Electronica, Transmediale, and in more generalist art events, like in the Venice Biennale, and etc. The number of artists (and artists/scientists), and the complexity of their artifacts, grows rapidly and it becomes more difficult to track down both what has happened and what is happening. There are, certainly, few authors and art pieces which must be included in this brief history of recent robot art, even if, in art, only the time will tell what is to be considered art and what is not.

Amongst those of the last decade we could, for example, annoverate the work from Ken Goldberg, Telegarden (1995) [5], as turning point for the tele-robotics art concept. Indeed its *tele-robotics* installation allowed the users to control, watering and growing - thanks to a robot arm manipulated through a World Wide Web application – a real plant.

From the cyberpunk culture – an active and famous exponent is Chico Macmutrie with his Amorphic Robot Works (from 1992 on) [6] – straights ahead we come across the *cyborgs*.

Fig. 1. Stelarc writing the word "Evolution" with his famous cyborganic experimentation the Third Hand

Amongst those, the most emblematic figure is Stelarc [7], who basically opened, more than ever, the use of robotics in (body)art and revolutionized the meaning of robot art. It is, indeed, a different way to look at robot art pieces. He is one of the most important contemporary artist and his art pieces (e.g.: Third Hand Project, 1976, 1981, 1991 Fig. 1) are strikingly innovative under all senses. Based on the cyberpunk vision, the Stelarc performances tilt the approach *to robotics as an external device to strongly emphasize the human embodiment.*

Another powerful artist, which embraces the same philosophy is one of the founders of the "La Fura dels baus", Marcel.lì Antùnez Roca, with his Membrana Project (i.e.: Protomembrana (Fig. 2) and Hipermembrana) [8], by which, as for Stelarc, he explores the *layering possibilities of human-machine-media interaction and interrelation.* Which is, indeed, one of the most hot topics in Robotic Art, at the moment.

The Haile Robot developed by Gil Weinberg, Scott Driscoll and Travis Thatcher [9] is interesting because of its own way of exploring the concept of *machine creativity* and, parallel, *the ability of robots to cooperate and collaborate* (in what the author calls musicianship) *with humans while producing art,* run-time.

Fig. 2. Marcel.li Protomembrana during Robots at Play 2007

Also, it is worth to spot Ximo Lizana's new research on 3D holographic projected sculptures (e.g.: the "Mid Air Shark", 2007. [10] Fig. 3). This technique opens a new horizon (we might name *virtual robot art*) to the robot art field here intended as a three dimensional object occupying a given physical space and interacting, by now in a naïve way - with the surrounding ambient.

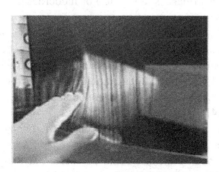

Fig. 3. A vistor interacting with the Mid Air Shark holographic projection, Ximo Lizana, 2007

One different scenario is the sector of robot production and research that, more than robot art, could be defined as *art oriented robots*. They are robotic application intended to serve the world of art (e.g. Gibson's "Robot Guitar" [11]), which are a bit out of context here, but still not too far away from what we might want to call robot art, in future.

Finally, one must considered works like those by Hiroshi Ishiguro's Geminoid (Fig 4 left [12]), an example of how invasive can technology be; on the opposite side, Nemo Gould's Armed and Dangerous (Fig 4 right, [13]) representing the typical artistic ironic sight given to the world of war oriented robots; and many others.

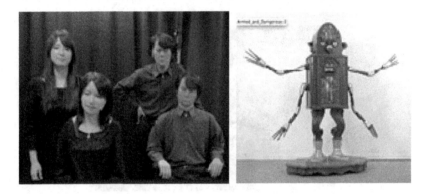

Fig. 4. (Left) Hiroshi Ishiguro's Geminoid. (Right) Nemo Gould's Armed and Dangerous

3   Our Contributions

As we tried to shortly outline above, robot art is a field that is consolidating, by showing a growing number of new aesthetical, philosophical, and artistic methods and approaches to the creation of artifacts. Such artifacts can be traditional art pieces (e.g. paintings, sculptures, performances, tells, etc.) or modern ones (e.g. innovative ideas, behaviors, robots themselves, etc.).

In the last ten years, within the exciting exploration and amazing exploitation of conceptual and aesthetical possibilities we tried to give our modest contribution to the field of robot art by producing both strictly artistic and art and science based models.

One good example of a scientific application that touches the borders of an art-like artifacts can be found in the Atron (Fig. 5) module, developed by Henrik Hautop Lund and colleagues [14].

Atron modules is a self-assembling shape chain of robots-atoms that, by using mastered/centralized or collective A.I. "reasoning" changes its own shape along the time. This perpetual changing artifact could be located at the junction between the robot art and kinetic sculpture art fields.

Fig. 5. The Atron modules exhibited at Brandts Museum, Odense, Denmark, 2007

Indeed, although at a first sight the robot behavior and shape resemble the old definition, on the other hand, the independency of the machine movements explore possibilities in kinetic sculptures so original to deserve a new definition as, for example, kinetic robot sculpture, *robot metamorphic art*, or so.

A different example of the evolution of the human-machine relationship imprinted by robot art works is in the 'full-loop' realized in LifeGrabber by Luigi Pagliarini in 2003 [15]. A webcam mounted on a robotic arm, controlled by a software written by the artist himself analyzes the audio/video inputs run-time, through a population of Alife agents which, in turn, influences both the audio/video output and the robot arm movements (therefore the future vision of the robot, see Fig. 6). While pointing at itself, this robot art piece gives birth to a *'self-observing machine'*, facing one of the most fascinating topics for future computer based art works, the philosophical problem of self-consciousness.

Fig. 6. The Robot Arm-Eye used in "LifeGrabber", by Luigi Pagliarini. 2003

Further, inspired by the Gutai [16], Tanaka [16], Stelarc [7] and Marcel.li [8] (amongst others) we initiated an artistic investigation, which we term Modular Robotic Wearable (MRW). The Modular Robotic Wearable [17] thought was born in 2007 from both the research line in electronic and robotic art – called The SuperAvatars (see Fig. 7). MRW is related to Wearable Computing or WearComp – a branch of research on forms of human-computer interaction comprising a small body-worn computer (e.g. user programmable device) that is always on and always ready and accessible.

Fig. 7. "Fatherboard, the Superavatar", by Luigi Pagliarini. 2007.

The MRW merges this art inspiration and wearable computing inspiration with our research tradition on modular robotics [14, 18]. It can be seen as a means for augmenting human interfaces both from virtual realities to the body and from the physical body to virtual realities and, in other words, MRW brings along new discoveries and potential research fields on exploring body action and reaction, limits and capabilities.

To explore *cooperative creativity*, we developed Music Tiles [18] (see Fig. 8 Left) and MusicTiles Magic Cubes [19] (see Fig. 8 Right) together with musician Peter Gabriel. MusicTiles allow anybody – novices, musicians, and expert composers – to remix songs (such as Peter Gabriel's hit songs) in a playful manner using virtual modular tiles, transforming the music fan from a passive listener to an active performer. Each tile is an instrument, and by putting them together in different configurations, the user activates different parts of each instrument, creating totally new versions of the hit songs. The MusicTiles Magic Cubes were developed as a physical realization and extension to the MusicTiles app to push users into real life social situations, as showcased at Roskilde Festival 2013 [20]. It is realized as a set of robotic music cubes that lets people interact with music as it is playing--they can activate or deactivate song elements by simply turning the cubes around. As a social playware it explores the cooperative creativity of robot art: the magic cubes seamless push the users into social play dynamics resulting in the users interacting and cooperating in their play to create and perform their collective new hit song versions. Here, the *robot art mediates social creativity.*

Fig. 8. (Left) MusicTiles app and (Right) MusicTiles MagicCubes by Henrik Hautop Lund, Luigi Pagliarini, Peter Gabriel 2012/13

4   Lessons Learned

Art History, Robotics and Art belong together and chase each other since the idea of robotics itself was born (1890, see Karel Čapek's R.U.R. [21], Fig. 9). This process by which art keeps on shaping and indirectly cooperating with a hyper rational and scientific progressive conceptualization of what we could call self-sufficient automation, has no reliable law.

We could look for rules or look for repetitive and schematic behavior in such a process (i.e. by which art keeps on creating robotics) but it is principally impractical. What is sure is that, the more or less direct collaborative evolution of art and robotics, easy and straightforward at its beginning, is now getting much more complex and articulated since both the disciplines are facing an elevated, quick and multifaceted development. And just because of that, on the opposite, such a process becomes more and more unpredictable. The deformation of the starting point (i.e. robot as a substitute of humans' hard and repetitive works) is taking places in all of the directions, since it is not only hard work, it is not only repetitive but sometimes very creative and it is not even work or necessarily humans' like!

Fig. 9. A scene from the play R.U.R., 1921

Nevertheless, we will try to outline what are the most significant crossing lines emerging from the brief description and review of robot art. They are:

1. An intensive work on the *human-robot body* that can be seen as outside shell, as well as inner side up to imitation or cloning. Evident examples are the Atsuko Tanaka's [6], Stellarc's [7], Pagliarini's [17], Lund's [14], and Ishiguro's [12] works. There are attempts to extend the human dimensions, feelings, perceptions, motions, and abilities. There seems to be a will to explore the limits of the human body to finally reinvent it throughout new functionalities. It is of course an ancestral dream that moves away from a world made of pure imagination and steps out in to a kind of reality, good or bad as it can be;

2. An amazing attempt for *cooperative creativeness* as shown in Marcel-lì's [8] and Pagliarini's [18] and Lund's [18 and 19] works. Robotics offers the possibility, through manipulation of artifacts and exoskeletons to start up a dialogue between the machines and humans. They, accordingly to the tasks, differ in speed, sharpness, effectiveness (i.e. time and space), "intelligence" and, mostly, in output production. Most of the scientists and almost all of the artists see such a thing as an endless resource for creativity– if we view it as a cooperative and collaborative process – to open a ping-pong that might end up with quite special and original outcome and, therefore, for aesthetic artifacts production;

3. An exploration of *ambient related intelligence* as revealed by Ihnatowicz [16] first. No doubts, art expresses a special cut of the world, a particular vision of reality as it flows and, because of that, is fully sensitive to external events. Robotics, basically made of sensors that constantly try capturing and measuring the surrounding habitat, seems to be born to play with such an artistic attitude. At the very moment there are already thousands of artists trying to apply world sensing electronics (i.e. from cameras up to proximal and distance sensors) to incorporate such information in their artistic productions (i.e. from painting to acting, from sculpture to music). It is the widest and most popular branch in what we are naming as robot art;

4. A number of *art oriented machines*. They are robots that can help on realizing art pieces or mediating the process of art production. They go from the most obvious, industrial and commercial ones - as for the case of Gibson's auto tuning guitar (i.e.: Robot Guitar [11]), to the most amazing, unpredictable and handicraft ones – as for the case of Charles Karim Aweida's work [22] where a wind simulation algorithm and a robot arm are used to build up an art piece made of nails on wood;

5. *Pure artistic robotics* in which the elaboration of the social, human and ecosystem conditions are. Examples can be found in the Nemo Gould's "Armed and Dangerous" piece [13] where an ironical approach is used to protest again a certain research on war robots, or in Luigi Pagliarini's "Intelligenza" piece (see Fig. 10, [23]) where the human being and robots conditions blur in such a way that it is hard to say which of the two is at the center of the artifact.

Fig. 10. Intelligenza by Luigi Pagliarini, 2010

Besides that, of course there are and there will be many different and noble exceptions and mutations that will keep the evolution of the area funny and interesting.

5   Conclusions

As can be easily understood by reading through these few and mostly incomplete historical examples - we've been trying to assemble in a pathway to modernity - the robot art field mostly deals with the innovation and the exploration of the borders of human-machine relationship. In other words, robot artists focus on what we, formerly, defined as polymorphic intelligence [3], where the machine and human bodies and minds melt together to shape a single "knowledge". Indeed, they are, to some extends, the blade runners which try to prefigure futuristic scenarios that might appear along the human being (and machine) development and in the upcoming world. By creating robot art pieces they somehow materialize what we defined as the Alive Art principles (of unpredictability and perpetual change) [2] and therefore assert themselves for being one of the most important avant-garde both in art and in sciences such as biology, psychology, philosophy, etc.

References

1. Flusser, V.: Digital Apparition (Digitaler Schein, Suhrkamp Verlag). In: Druckrey, T. (ed.) Electronic Culture. Aperture, New York (1996)
2. Pagliarini, L., Locardi, C., Vucic, V.: Toward Alive Art. In: Heudin, J.-C. (ed.) VW 2000. LNCS (LNAI), vol. 1834, pp. 171–184. Springer, Heidelberg (2000)
3. Pagliarini, L.: Polymorphic Intelligence. In: Proceedings of the Twelfth International Symposium on Artificial Life and Robotics, AROB 2007, Oita, Japan (2007)

4. Reichardt, J. (ed.): Cybernetic Serendipity, pp. 10–11. Studio International, London (1968)
5. Goldberg, K., Mascha, M., Gentner, S., Rossman, J., Rothenberg, N., Sutter, C., Wiegley, J.: Beyond the Web: Manipulating the Real World. Computer Networks and ISDN Systems Journal 28(1) (1995)
6. http://amorphicrobotworks.org/
7. http://www.stelarc.va.com.au/
8. http://www.marceliantunez.com/
9. Weinberg, G., Godfrey, M., Rae, A., Rhoads, J.: A Real-Time Genetic Algorithm in Human-Robot Musical Improvisation. In: Kronland-Martinet, R., Ystad, S., Jensen, K. (eds.) CMMR 2007. LNCS, vol. 4969, pp. 351–359. Springer, Heidelberg (2008)
10. http://www.ximolizana.com/
11. http://www.gibson.com/robotguitar/
12. http://www.geminoid.jp/en/index.html
13. http://www.nemogould.com/portfolio-item/armed-and-dangerous/
14. Brandt, D., Christensen, D., Lund, H.H.: ATRON Robots: Versatility from Self-Reconfigurable Modules. In: Proceedings of IEEE Int. Conf. on Mechatronics and Automation, ICMA 2007. IEEE Press (2007)
15. http://www.neural.it/nnews/lifegrabber.htm
16. Kac, E.: Robotic Art Chronology. Convergence 7(1), 87–111 (2001)
17. Pagliarini, L., Lund, H.H.: Wearable Playware. In: The 8th International Conference on Ubiquitous Robots and Ambient Intelligence, pp. 9–13. IEEE (2011)
18. Pagliarini, L., Lund, H.H.: MagicTiles. ALife for Real and Virtual RoboMusic. In: Proc. of 17th International Symposium on Artificial Life and Robotics, ISAROB, Japan (2012)
19. MusicTiles MagicCubes,
    http://www.youtube.com/user/HenrikHautopLund
20. http://roskilde-festival.dk/
21. http://en.wikipedia.org/wiki/R.U.R
22. http://cka.co/projects/representing_wind/
23. http://www.artificialia.com/intelligenza/

Navigation Control of a Robot from a Remote Location via the Internet Using Brain-Machine Interface

Jorge Ierache[1,2], Gustavo Pereira[1], and Juan Iribarren[1]

[1] Instituto de Sistemas Inteligentes y Enseñanza Experimental de la Robótica (ISIER)
Facultad de Informática Ciencias de la Comunicación y Técnicas Especiales
Universidad de Morón, Cabildo 134, (B1708JPD) Morón, Buenos Aires, Argentina
54 11 5627 200 int 189
Laboratorio de Sistemas de Información Avanzados
[2] Facultad de Ingeniería Universidad de Buenos Aires, Buenos Aires, Argentina
jierache@yahoo.com.ar

Abstract. This article shows the experiences carried out in the context of Human Brain-Robot Control between Remote Places communication, on the basis of brain bio-electrical signals, with the application of the Brain-Machine Interface from Emotiv. There are available technologies and interfaces which have facilitated the reading of the bio-electrical signal from the user's brain and their association to explicit commands that have allow controlling mobile robots (NXT-Lego) by adapting communication devices. Our work presents an engineering solution, with the application of technological basis, the development of the local communication framework and remote control. Lastly, metrics and indicators from the realized tests are proposed.

Keywords: Brain Machine Interface, Bio-Electrical Signal, Human Machine Interfaces, robots.

1 Introduction

A Brain-Machine Interface (BMI) facilitates the communication between the computer and a person's brain, by processing, classifying the electric signals that are taken from the device and communicating with applications or specific devices. It's very interesting to highlight that applications using BMI interfaces have increased in the last twenty years [1]: power switches controlling lighting, wheelchairs, computer control [2], space movement [3] even videogames [4]. In the scientific field, the interest in BMI development started in the year 1973 [1], and the first publications were in 1990 [5] and 1991 [6].

The application of bio-electrical signals to control systems, robots, applications, games and devices in general present an original approach as it opens up new possibilities for the interaction of human beings and computers in a new dimension, where the electrical biopotentials registered in the user are specifically exploited; these biopotentials include the EMG (electromyogram), the EEG (electroencephalogram) and the EOG (electro-oculogram), which are bio-electrical signals generated by activity

J.-H. Kim et al. (eds.), *Robot Intelligence Technology and Applications 2*,
Advances in Intelligent Systems and Computing 274,
DOI: 10.1007/978-3-319-05582-4_26, © Springer International Publishing Switzerland 2014

patterns in the user's muscles, brain and eyes. BMI development takes place in a multidisciplinary scientific field due to its medical, electric and electronic components, signal treatment, neuroscience and finally applications from computing, home automation to robotics and entertainment [7].

In this regard, several works have been presented; the first ones resorted to the implantation of intracranial electrodes in the motor cortex of primates [8], [9]. Noninvasive works for humans resorted to EEG signals, applied to mental command exercises, such as moving the computer cursor[10], [11] based on the use of BMI. Millan et. al [12] show how two people are able to move a robot by using a simple EEG on the basis of recognizing three mental states, which are associated to robot commands. Works presented by Escolano et al conducted remote control experiments with BMI [13]. The works presented by Saulnier et al. [14] focused on controlling robot speed and further inferring the user's stress level, thus influencing on the social behavior of domestic robots, in this case of a robotic vacuum cleaner. Millan et al´s seminal work [12] uses the EEG as a unique bio-electrical signal, on the basis of the work of two people to support robot navigation; in contrast to this, our work presents the preliminary result by applying a low-cost BMI, used in secondary works like that by Saulnier et al[14] that includes bio-electrical signals corresponding to the EEG, the electrooculogram and the electromyogram. Unlike Saulnier et al´s work [14], in which speed control is implemented on the basis of the electromyogram and where the user's stress level is inferred on the basis of the EEG, our work focuses on the command execution with a BMI-NIA [15] and the navigation of a lego NXT [16] robot that can be extended to device control in domotics [17]. Controlling devices, moving robots or facilitating the implementation of devices for disabled people without applying manual controls and gaining control only through mental activity, fascinated researchers; and although acquiring proficiency with a BMI requires the user to invest a lot of time, in our experiences to comply with a navigation pattern, we were able to make things easier for a user with minimal training using auto focus control in order to guide the NXT robot with a BMI-NIA [16]; leading to improved mental control times, slightly surpassing the manual controls on tests with the same navigation pattern.

2   Problem

Our initial objective was to explore an engineering solution that would allow us to achieve a remote integration over the Internet between a non-invasive BMI and a robot (Lego NXT), where the BMI was a navigation remote controller of basic movement (forward, back, turn right, turn left) meant for users without a previous experience in meditation techniques or specific training in mental concentration.

3   Solution Description

3.1   Non Invasive Brain-Machine Interface

In the field of biosignals, the EEG is the registry of the electric activity generated by the neurons inside the brain obtained through the skull using electrodes on the surface of the head. The neuronal electric activity is composed by slow waves that have their origin in the synaptic activity of the cortical neurons. There are several types of surface electrodes: The adherent electrodes are small metal discs which are fixed with a conductive paste offering very low contact resistance. The contact electrodes consist of small tubes of chloridized silver threaded to plastic holders containing a pad wetted with saline at its end, which is attached to the skull with elastic bands and connected via alligator clips. Lastly, we have a helmet composed by electrodes attached to elastic net. This last method is not only easier to apply but also offers great placement precision.

The international system of electrodes setup 10-20 is the most used and it was designed in order to ensure standardization for studies on individuals. The system has letters and numbers to identify the points of contact. The letters identify the lobe, and the numbers show where in the hemisphere they are located. The letters F, T, C, P and O stand for Frontal, Temporal, Central, Parietal and Occipital sections respectively, (the letter C identifies the horizontal central line and it does not reference any lobe). Even numbers correspond to the electrodes of the right hemisphere and odd numbers to the left one. Z subscripts are used to identify the vertical center line of electrodes.

3.2   Brain-Machine Interface – Emotiv Epoc

The electrodes of the Emotiv Epoc helmet [17] are held by malleable plastic arms that guarantee the correct placement and have a pad wetted with saline in each contact to stimulate conduction. The disposition of the electrodes of the Emotiv Epoc helmet are adjust to the 10-20 system, but only fourteen contact positions are used (Figure 1a) plus a couple of reference (Common Mode Sense and Driven Right Leg) in each side, behind the ear or above it. Apart from the electrodes, the Emotiv Epoc helmet (Figure 1b) contains a gyroscope composed by two accelerometers that provides information about the movements that the user makes with his head, and a wireless transmitter by which maintains a link with the USB receptor connected to the computer, all this is powered by a rechargeable battery through USB.

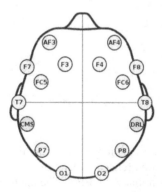

Fig. 1a. EMOTIV EPOC helmet electrodes   **Fig. 1b.** EMOTIV EPOC helmet

Both the API and the software included in the development kit enable monitoring the contact level of the electrodes, the gyroscopic movement, the wireless signal intensity and the battery charge. The Emotiv development kit has a set of libraries that allows the communication with the helmet, the developer's API and the Emotiv Motor. The Emotiv Motor is the base component for the detection and interpretation of the signals from the electrodes in the helmet and captures the information of the EEG. It monitors the battery status, wireless signal intensity, the connection's time registry; and it also trains the recognition algorithms for the expressive and cognitive modes, subsequently applying optimizations to each one of them.

The BMI EMOTIV EPOC, articulated with its SDK (Fig. 2a) has a control panel that generates the user and registers the profile, displays the sensors connection status (black "no signal"; red "very low signal"; orange "low signal", yellow "acceptable"; green "good signal") and represents different registry patterns (expressive, affective and cognitive). The *expressive pattern* shows as an avatar in which different face expression can be trained (blink, left wink, right wink, look to the left, look to the right, move your eyebrows up, move your eyebrows down, smile, set one's jaw). The *affective pattern* allows verifying different moods that happen in a given time (Engagement, Instantaneous Excitement, Long-Term Excitement, among others). The *cognitive pattern* allows training an action based on a thought, and you can train up to thirteen actions, where six are directional movements (push, pull forward, left, right, up and down), six are rotational (clockwise rotation, counterclockwise rotation, left rotation, right rotation, forward and reverse) and an imaginary one that is disappearing. Other tool that has the SDK is the Emokey (Fig. 2b), that allows linking an action of the Emotiv to any key and work as an interface with any application.

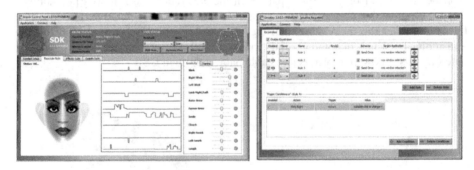

Fig. 2a. Emotiv SDK   **Fig. 2b.** Emokey

For visualization and exploitation of signal information we used Emotiv Testbench, this allows you to view and retrieve information related to each of the sensors placed on the user's head. Its working principle is similiar to an EEG. The application can display four functions: EEG, FFT, Gyro and Data Packets. The most important is related to the display of the EEG signals from each of the 14 sensors, each in a different color as displayed figure 3.

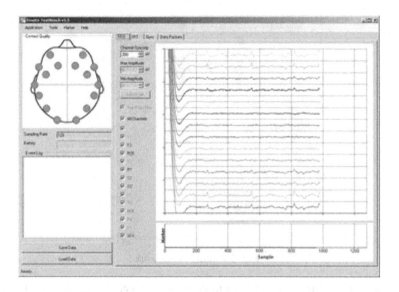

Fig. 3. Emotiv Testbench

Displaying FFT option allows analyzing a particular sensor and measuring the frequency in relation to time, also it shows the power of the selected signal and shows all different brain waves (delta, theta, alpha and beta) to be activated. The Gyro option shows only a change when performing horizontal or vertical movements with the head. The last option, Data packets samples the packets as they arrive from the emotiv helmet, in example , if they succeeded or failed.

3.3   Brain-Machine Interface – Remote Control

We were able to set a simple profile to handle the robot that first associates and defines what which control will execute the mental command based on muscle signal detection, in this case through a slight movement of the eyelids. Then, it selects the high level commands of the robot, in this case based on alpha brainwaves.

This type of biosignals using the selection panel commands of robot control framework (fig 4) did not ensure proper user control of the movement. This is why the auto-focus option was implemented in the framework [15] for mental control mode, in order to improve the user's control in the command selection process. The robot-computer communication was via Bluetooth, which is a native communication in the LEGO NXT kit. The remote control was implemented through a web service which receives the remote client commands and transfers them via HTTP. In this process, it uses the SOAP communication protocol connecting to one layer that controls the Bluetooth connection of the Lego NXT [18] and its primitive movement. This layer was developed using AForge libraries [19] which simplifies controlling an NXT robot together with the Microsoft .NET framework. Fig. 5 shows the conceptual architecture implemented [17] to reach the integration of the BMI and the user that controls a robot remotely, considering the experience with BMI-NIA [20].

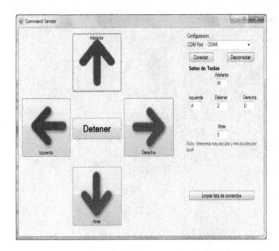

Fig. 4. Control Framework

The next step of the research was to develop a remote control system that allows the interface to move a robot in a distant site via an Internet connection (Fig. 5).

In order to accomplish this, we developed a web service architecture. It received all the commands of connection and primitive movements, and sends them to an NXT robot through a Bluetooth interface that runs from a remote computer using Bluetooth. The web service in the remote server receives the client´s commands and translates them into the instructions for the NXT. We chose to implementat a web service to ensure a secure connection with a well-known communication protocol, allowing the client to be developed in several programming languages for multiple platforms. For the web service testing, we incorporated a frontend to see the threads received and executed by the NXT and the connection data. Initially, the display of the robot´s movements in the remote site was done through a standard videoconferencing client.

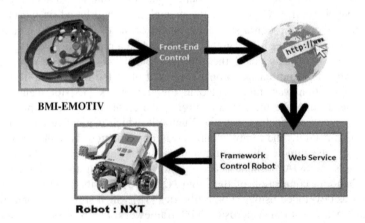

Fig. 5. Integration of BMI-EMOTIV and remotely-located Robot

3.4   Framework Design

The framework was developed in C# language. Component diagram is presented in Fig. 6 that shows the interaction between the client software (CommandSender), the components related to the Webservice and the movement of the NXT Robot.

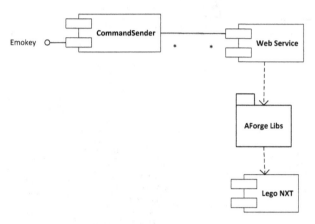

Fig. 6. Component Diagram

Figure 7 presents the Class Diagram with all simple methods to connect all the components of the solution. NXTManager works directly with AForge libraries to communicate with the NXT brick directly using Bluetooth.

In figure 8, activity diagram is presented with lines of life and sequence starting with the selection of the command with the Emotiv and finishing in the execution on the NXT Robot.

4   Tests and Results

The tests were done between two cities 91 km away from each other: Universidad de Morón located in the city of Morón, and Universidad de La Plata located in the city of La Plata (both in Buenos Aires province). Nine tests were performed per individual. The experiment was designed considering a scenario that included a navigation area for the robot of 2,80 mts long and 1,40 mts wide with sidewalls of 10 cm high. Above this test area, three square check points of 20 cm were distributed, each one identified with a color: red, yellow and blue. Cubes of 8 cm were added as obstacles. Figure 9a shows an image of the test area, the NXT robot and the camera located at Universidad de Morón. At Universidad de La Plata there was a human controller using the BMI (Figure 9b) connected to a PC running the Front End Control and the videocall client, linked via internet to visualize the robot in Universidad de Morón where the Web Service allowed the control instances via Control Framework of the Robot, sending commands to the NXT Robot locally via Bluetooth.

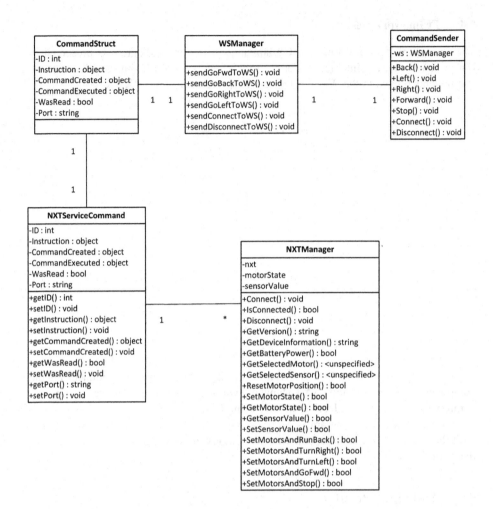

Fig. 7. Class Diagram

While testing, we first used navigation in seconds as a metric (TNav) for every color check point (Cpc: Red, Blue, Yellow) individually. This metric registers the time in seconds that it took the robot to cross over the check point. In Table 1 we show the results of the nine test sessions. The second metric we used was the number of collisions (Cc) that the robot had with the sidewalls (L) of the test area, and the number of collisions with the obstacles (O). In Table 2 we show the results of the nine test sessions. Lastly we determined for every test session: the navigation total time (Tnav Total (seg) x Sp), navigation total time (Tnav Total (seg)), navigation average time (Average Tnav (seg) x Sp) and the average navigation general time (Average Tnav (seg).

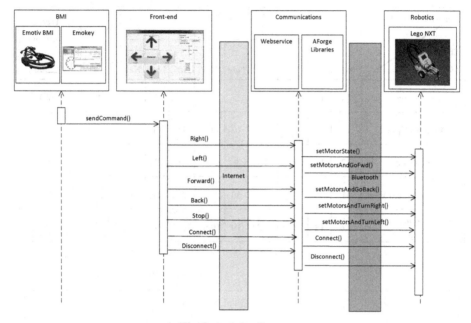

Fig. 8. Activity diagram

Fig. 9a. Universidad de Morón robotics laboratory

Fig. 9b. Human controller (Universidad de La Plata)

Table 1. "Navigation time evaluation"

Sessions Sp - Tnav / Check Point Color (Cpc)	T N° 1 T Nav	T N °2 T Nav	T N° 3 T Nav	T N° 4 T Nav	T N °5 T Nav	T N° 6 T Nav	T N °7 T Nav	T N °8 T Nav	T N °9 T Nav	Tnav Total (seg) x checkpoint	Average Tnav (seg) x check-point
Red (Cpcr)	14	13	12	11	9	10	8	7	7	**91**	**10,11**
Yellow (Cpcy)	15	15	14	13	12	12	11	11	11	**114**	**12,67**
Blue (Cpcb)	12	11	11	10	9	10	9	9	8	**89**	**9,89**
Tnav Total (seg) x Sp	**41**	**39**	**37**	**34**	**30**	**32**	**28**	**27**	**26**	*Tnav Total (seg)*	294
Average Tnav (seg) x Sp	**13,67**	**13**	**12,33**	**11,33**	**10**	**10,67**	**9,33**	**9**	**8,67**	*Average Tnav (seg)*	32,67

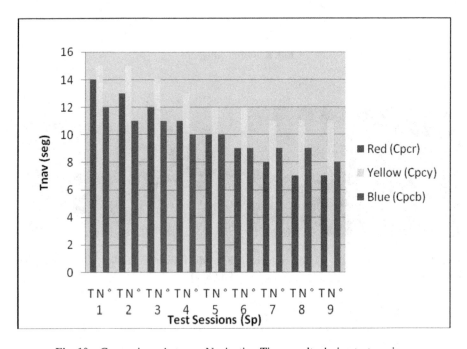

Fig. 10a. Comparisson between Navigation Time results during test sessions

In free navigation test, ruled by the remote BMI, we set up as a condition that the check points had to be reached sequentially: first the red one, then the yellow one and finally the blue one. The following indicators were used to obtain a global evaluation of the user:

Total time of navigation to crossover the three check points
Total of wall collisions
Total of obstacle collisions when completing all three check points.

As a result of the analysis of the nine test sessions with the user, we concluded that the number of collisions (Cc) with the walls (L) and the obstacles (O) and the navigation time (TNav) decreased from each test to the next, due to the user's improved proficiency with the BMI and controlling the remote robot. In Figure 10a we show the comparative results of navigation time (TNav) and Figure 10b shows the comparison of the total of collisions (obstacles and sidewalls) of the user's test sessions to cover each of the color check points.

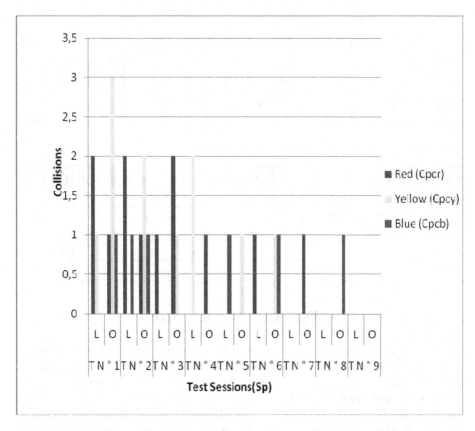

Fig. 10b. Comparison between Navigation Time results during test sessions

As an indicator of a particular user (IU) trained in nine test sessions (SP), we set to each color check point (Cpc), the relation between the average time navigation (PTNav) and the sum of the average amount of sidewall collisions (PCL) with the average amount of obstacle collisions (PCO). The value of the user indicator (IU) for each color check point (Cpc) was set as: $[\sum SP\ (i)(PTNav)\ /9]\ /\ [\sum SP\ (i)\ (PCL + PCO)\ /9]$. In Table 3 we show detailed results where the user global indicator (IGU) was set using the average values obtained from the indicators (IU) for each color check point (Cpc) in the nine test sessions.

Table 2. "Number of Collisions Evaluation"

Sessions Sp / Check Point Color (Cpc)	T N°1		T N°2		T N°3		T N°4		T N°5		T N°6		T N°7		T N°8		T N°9	
	L	O	L	O	L	O	L	O	L	O	L	O	L	O	L	O	L	O
Red (Cpcr)	2	1	2	1	1	2	0	1	0	0	1	0	0	1	0	0	0	0
Yellow (Cpcy)	1	3	0	2	0	1	2	0	0	1	0	1	0	0	0	0	0	0
Blue (Cpcb)	0	1	1	1	0	0	0	0	1	0	0	1	0	0	0	1	0	0

Table 3. "User indicators"

Check Point Color (Cpc)	Red (Cprcr)	Yellow (Cpcy)	Blue (Cpcb)
Average TNav x Check Point	10,11	12,67	9,89
Number of Collisions Average X Check Point.	1,33	1,22	0,67
IU	7,60	10,38	14,76
IGU	10,91		

5 Conclusions and Future Lines of Research

BMI-Emotiv controller was successful in tests and demos with the robot located at Universidad de Morón [20] and the BMI-Controller at Universidad de la Plata where open environment navigation tests were conducted [21], [22]. The paper presents the results of the application of a BMI with input from the development of a framework that facilitates the remote operation of a robot through the user's biosignals. As a lesson learned, BMI application requires training in terms of user interface adaptation, the framework development process should consider adapting the user, to support this process is required assistance functionalities that we call autofocus [16].

Our future research lines will focus on the development of an integrated framework for robots and the study of new BMIs in the quest to effectively control IR controlled devices such as Air Conditioning systems, TVs, etc. allowing handicapped people to interact and control devices placed locally and at remote sites.

References

1. Hamadicharef: Brain Computer Interface (BCI) Literature- A bibliometric study. In: 10th International Conference on Information Science, Signal Processing and their Applications, Kuala Lumpur, pp. 626–629 (2010)
2. Kennedy, P.R., Bakay, R., Moore, M.M., Adams, K., Goldwaithe, J.: Direct control of a computer from the human central nervous system. IEEE Transactions on Rehabilitation Engineering 8(2), 198–202 (2000)
3. Wolpaw, J.R., McFarland, D.J.: Control of a two-dimensional movement signal by a non-invasive brain-computer interface in humans. Proceedings 629 of the National Academy of Sciences (PNAS) 101(51), 17 849-17 854 (2004)
4. (2013), http://edugamesresearch.com/blog/tag/bci/
5. Wolpaw, J.R., McFarland, D.J., Neat, G.W.: Development of an Electroencephalogram-based Brain-Computer Interface. Annals of Neurology 28(2), 250–251 (1990)
6. Wolpaw, J.R., McFarland, D., Neat, G., Forneris, C.: An EEG-based brain-computer interface for cursor control. Electroencephalography and Clinical Neurophysiology 78(3), 252–259 (1991)
7. Lebedev, M.A., Nicolelis, M.A.L.: Brainmachine interfaces: Past, present and future. Trends in Neurosciences 29(9), 536–546 (2006)
8. Wessberg, J., Stambaugh, C.R., Kralik, J.D., Beck, P.D., Laubach, M., Chapin, J.K.: Real-time prediction of hand trajectory by ensembles of cortical neurons in primates. Nature 408, 361–365 (2000)
9. Nicolelis, M.A.L.: Brain-machine interfaces to restore motor function and probe neural circuits. Nature Rev. Neurosci. 4, 417–422 (2003)
10. Wolpaw, R., McFarland, D.J., Vaughan, T.M.: Brain-computer interface research at the Wadsworth center. IEEE Trans. Rehab. Eng. 8, 222–226 (2000)
11. del Millán, J.R.: Brain-computer interfaces. In: Arbib, M.A. (ed.) Handbook of Brain Theory and Neural Networks, 2nd edn. MIT Press, Cambridge (2002)
12. Millán, J., Renkensb, F., Mouriñoc, J., Gerstnerb, W.: Non-Invasive Brain-Actuated Control of a Mobile Robot by Human EEG. IEEE Trans. on Biomedical Engineering 51 (June 2004)
13. Escolano, C., Antelis, J., Minguez, J.: Human brain-teleoperated robot between remote places. In: IEEE International Conference on Robotics and Automation (ICRA 2009), pp. 4430–4437 (2009) ISBN 978-1-4244-2788-8
14. Saulnier, P., Sharlin, E., Greenberg, S.: Using Bio-electrical Signals to Influence the Social Behaviours of Domesticated Robots. In: HRI 2009, pp. 978–971. ACM, USA (2009), doi:978-1-60558-404-1/09/03
15. http://www.ocztechnology.com/products/ocz_peripherals/nia-neural_impulse_actuator (vigente julio 2013)

16. Jorge, I., Gustavo, P., Juan, I., Iris, S.: Robot Control on the Basis of Bio-electrical Signals. In: Kim, J.-H., Matson, E.T., Myung, H., Xu, P. (eds.) Robot Intelligence Technology and Applications 2012. AISC, vol. 208, pp. 337–346. Springer, Heidelberg (2012)
17. Ierache, J., Pereira, G., Sattolo, J.: Iribarren Aplicación de interfases lectoras de bi+señales en el contexto de la domótica. In: XV Workshop de Investigadores en Ciencias de la Computación 2013 Facultad de Ciencia y Tecnología Universidad Auto noma de Entre Ríos (UADER) (2013) ISBN: 9789872817961
18. (July 2013), http://www.emotiv.com/
19. http://mindstorms.lego.com/eng/Overview/default.aspx (July 2013), nxt
20. http://www.aforgenet.com
21. Ierache, J., Pereira, G., Sattolo, I., Guerrero, A., D'Altto, J., Iribarren, J.: Control vía Internet de un Robot ubicado en un sitio remoto aplicando una Interfase Cerebro-Máquina. In: XVII Congreso Argentino de Ciencias de la Computación, pp. 1373–1382 (2011) ISBN 978-950-34-0756-1
22. Jorge, I., Gustavo, P., Juan del articulo, I.: Demostración de los resultados en la integración de Interfases Lectoras de Bioseñales aplicadas al Control de un Robot. In: VII Congreso Educación en Tecnología y Tecnología en Educación 2012 Universidad Nacional del Noroeste de la Provincia de Buenos Aires. UNNOBA. Demos educativas (2012) ISBN 978-987-28186-3-0
23. www.facebook/isierum

Design of Knowledge-Based Communication between Human and Robot Using Ontological Semantic Technology in Firefighting Domain

Ji Hyeon Hong, Eric T. Matson, and Julia M. Taylor

Computer and Information Technology,
Purdue University, West Lafayette, IN, USA
{hong,ematson,jtaylor1}@purdue.edu

Abstract. This paper discusses how to design robot-human communication using Ontological Semantic Technology (OST), which is to address meanings of phrases or sentences in natural languages, in a firefighting domain. The OST is a system in an ontology-based structure to deal with different natural languages. In this study, English and Korean were selected to be implemented for the OST. The problem set is designed with Natural English, direct translation from the natural English to Korean, natural Korean, and direct translation from the natural Korean to English, which can be compared to examine the similarity of meanings with a language-independent ontology in processing English and Korean.

Keywords: ontological semantic technology (OST), robot-human communication, human-robot communication.

1 Introduction

As human-robot communication has become more and more common in life, studies on this research area has become popular. In order for humans to communication with robots like a human does in human-human communication, human natural language processing techniques have been constantly studied as natural language processing techniques makes humans and robots easily communicate each other. We utilized Ontological Semantic Technology (OST) over the other natural language processing techniques for communication between a firefighting robot and a human. Korean and English were selected as natural languages for the firefighting-robot-and-human communication. In Chapter 2, relevant research studies to multi natural language communication with robotics are discussed. In Chapter 3, OST Lite which is a tool to be implemented for ontology-based natural language processing is described. Chapter 4 discusses how problem sets for this study was designed, and in Chapter 5, the problem sets for a firefighting domain are composed.

J.-H. Kim et al. (eds.), *Robot Intelligence Technology and Applications 2*,
Advances in Intelligent Systems and Computing 274,
DOI: 10.1007/978-3-319-05582-4_27, © Springer International Publishing Switzerland 2014

2   Related Work

2.1   Human-Robot Communication

Robots become considered as collaborators, who can work with humans as partners [1]. As collaboration between humans and robots increases, communication between humans and robots becomes more important, and more advanced human-robot communication methods are desired in robotics. Collaborative control "encourages human-robot interaction to be more natural, more balanced, and more direct" [1]. Green, Billinghurst, Chen, and Chase [2] found that the collaborative human-robot communication would be realistic by considering three components: communication channels, communication cues through the channels, and technology to transmit the cues. Once the network channel is established, the interchangeable communication with a human and a robot or more than one robot becomes possible [3]. For example, Human-Agent-Robot-Machine-Sensor (HARMS) model appeared to realize this type of communication [4]. Based on the capabilities of robots, collaborative work with robots can be similarly done to a human organization [5]. Thus, natural language processing, which has a mock-up processes of humans understanding natural languages for communication, is needed for human-robot communication.

2.2   Korean-English Machine Translation

In this paper, text-based Korean and English will be utilized for human-robot communication in the OST. For the reason, machine translation that handles multiple languages, especially English and Korean, with algorithms will be discussed to be compared to the OST technique.

There have been studies to enhance the accuracy of Korean and English machine translation as the use of translation of Korean and English increases [6]. Many researchers have built some machine translation systems by themselves in order to contribute to the better machine translation results. In 1994, MATES/EK was developed as a transfer-based machine translation system with a purpose of improving multilingual communication by producing accurate translations across English and Korean either verbally or in writing [7]. In the same year, some researchers developed their own prototype system using synchronous TAGS to relate two different languages or a syntactic TAG and a semantic TAG in one language [8]. Three years later, Lincoln Laboratory utilized TINA for language understanding and GENESIS for language generation, and built a translation system, called as CCLINC that was evolved from their pre-translation system, SYSTRAN, for a military domain [6].

Also, machine translation mechanisms, such as a parser, have been developed. Some researchers tried to utilize a dependency tree-bank in machine translation; they developed a morphological parsing model for the use of improving machine translation along with other NLP tasks [9]. There were also efforts to have improvements on anaphor resolution rules, and applications of the rules for English-Korean machine translation to handle different pronouns usages in English and Korean [10]. Three Korean researchers, Hong, Lee, and Rim, also contributed to some improvements, by introducing a concept of bridging a morph-syntactic gap between source and target sentences for English-Korean statistical machine translation through a process of inserting pseudo words and reordering the syntax of a source sentence based on the rules of the source language [11].

There is also research that has been done to utilize the existing techniques and data sources to contribute to Korean machine translation. Choi, Park, and Choi [12] implemented Sejong Treebank with an entity-based transforming method to result in the better translation. Choi, Jung, Sim, Kim, Park, Park, and Choi [13] proposed new approaches to MATES/EK machine translator by extending the breadth of the translator with a domain recognizer, a compound unit recognizer, a grammar learning technique, and a long-sentence handler.

New methodologies or approaches in machine translation started to be introduced. Some approached to have a machine translator learn domain-specific rules with various dictionaries [14]. On the other hand, Lee, Park, and Kim [15] tried to improve the accuracy of word selection in machine translation by using multiple knowledge sources including sense vectors, statistical context information, patterns, etc., and selecting the more appropriate target words.

However, there is another approach of handling multiple languages, English and Korean in this study, for human interpretation. It is called as OST Lite, which utilized OST. The OST processing will be discussed in the following sub-chapter.

2.3   Ontology-Based Natural Language Processing

In this study, bilingual communication in English and Korean is targeted. Instead of using machine translation techniques, this study implemented an ontology-based natural language processing approach.

According to Grishman [16], natural language processing should be able to handle complex queries given in a human natural language; thus, a computer understands what a user requested and what information it should retrieve. Natural language processing enables a computer to ultimately understand humans and interact with humans much more friendly.

Among various natural language processing techniques, there has been a theory named ontological semantics, which is a semantic approach of natural language processing. With the ontology-based approach, "actual understanding of the information … in natural language or any other medium" becomes possible along with reasoning [17], [18]. Nirenburg and Raskin [19] brought an approach of representing knowledge with knowledge resources so that the natural languages can be more semantically interpreted. The implementation of the ontological semantics has been developed with the name "ontological semantic technology (OST)" [20]. It consists of multiple "repositories of world and linguistic knowledge, acquired semi-automatically within the approach and used to disambiguate the different meanings of words and sentences and to represent them" [4]. And each repository consists of each of the following components:

- Ontology

- One lexicon per supported natural language

- Proper Name Dictionary (PND)

- Common sense rule resource

The above components provide information of the world so that it helps a computer comprehend better what humans know and say; ontology is a graphical representation showing language-independent concepts and their relationships in the world [21], [22]; lexicon is a group of "word senses anchored in the language-independent ontology which is used to represent their meaning" [4]; proper name dictionary is a dictionary of all proper names such as "names of people, countries, organizations, etc., and their description anchoring them in ontological concepts and interlinking them with other PND entries" [4].

Once OST is implemented on computers or robots for human-robot interaction, the computer or the robot should be able to interpret through all information shown in a sentence, such as syntax, semantics, and common-sensed information. However, not all contextual information of an environment is required for interpretation in a real application because a specific domain (e.g., hospital, bank, etc.) only require specific knowledge in the respective domain for interpretation. Thus, only fundamental information can be given based on the domain [23].

3   OST Lite

OST Lite is an OST tool, which was introduced by Taylor and Raskin [24]. The researchers who were involved in the previously mentioned paper have kept developing, and the most current version of the tool is described by Taylor and Raskin [24]. According to the authors, the weight between concepts, which are presented in an input sentence, is calculated based on hierarchies and properties; the exact equation is shown in Figure 1. The weight is calculated whether or not a concept is in hierarchy; the weight is calculated with facet and property values if the concepts are not in hierarchy; the facet and property values are applied as follows:

$$weight = \begin{cases} 1/ISAcoef, \text{ in hierarchy} \\ 1/(facet * PROPcoef), \text{ otherwise} \end{cases}$$

Fig. 1. Weight calculations in OST Lite [24]

And if concepts are declared to be linked together in SEM-STRUC, the shortest path is found by looking at properties as Figure 2 shows.

(1): $SEM-STRCU1:c1+SEM-STRUC2:c2 => f(c1,c2)$

(2): $SEM-STRUC1:c1(p1(c2)))+SEM-STRUC2:c3(p2(c4)) => \min\{f(c1,c2)\cup f(c2,c3),f(c3,c4)\cup(c4,c2)\}$

(3): $SEM-STRUC1:c1(p1(var1)))+SEM-STRUC2:c3 => f(c1,c3)$

(4): $SEM-STRUC1:var1(p1(c1))+SEM-STRUC2:c2(p2(c3))) => \min\{f(c2,c1),f(c3,c1)\}$

Fig. 2. Weight calculation of using SEM-STRUC in OST Lite [24]

4   Ontological Semantic Technology in Firefighting Domain

Ontological semantic technology (OST) is constructed with ontology and lexicon(s). Ontology has concepts which have distinct meanings, and lexicon includes all other concepts pointing to a concept in the ontology. This study uses existing ontology and the lexicon, and we added more concepts, which are related to firefighting situations, into the ontology and the lexicon in order to make them suitable to the firefighting domain.

4.1   Ontology

Ontology represents all concepts and relationships among the existed concepts in a domain; the ontology consists of "classes, relations, functions, and object-constants" [21] which describe the world. This study utilizes ontology to provide knowledge representations in a firefighting domain. As ontology is language-independent, the key role of using ontology is to include and refer to all needed concepts with specified properties after a natural language command is processed through a lexicon.

In this study, an existing ontology, which has been developed by several researchers, is utilized. There were 931 concepts defined in the existing ontology; during this research period, 60 new concepts were added so that the ontology became to have 991 concepts.

The reason of not rapidly increasing the number of ontology concepts is because the general and common concepts were already defined, or not necessary to be re-created; as there is no need to take care of all synonyms in ontology, only new and differentiated concepts against the existing concepts were needed to be added. The reason of other synonymous concepts being skipped to be added is because too many similar concepts can cause more confusion in knowledge representations due to too close relation. Specific and detailed meanings are handled in a lexicon, not in ontology. The ontology and the lexicon for this study were written in XML; one concept is composed with the name of the concept, properties, the filler, facet, and relevant concepts or values.

4.2   Lexicon

A lexicon is a language-dependent vocabulary list; the words are entered in XML format, and one lexeme (word or phrase) consists of NAME, SENSE ID, SYN-STRUC, SEM-STRUC, and PROPERTY. SYN-STRUC portion includes detailed information about the syntactical use of each sense of a lexeme, and SEM-STRUC portion has a role of connecting the current sense to an actual concept which was already defined in the ontology. As previously mentioned, the existing lexicon under development is used for this study; the original lexicon had 671 lexemes in it, and the total number of lexemes became 902 after the lexemes of the firefighting domain had been added. The Korean lexemes and English lexemes are listed altogether in the same lexicon, not in the separate files, for this particular study even though they should be separated for the ease of organizing and managing lexicons in multiple languages.

In this study, there was a problem of dealing with different Korean syntax from English syntax; because a Korean sentence has postpositions, not prepositions, morphs are used to deal with the Korean postpositions, not for the original purpose of morphs; this ignored the original classifications of morphs. In order for Korean propositions to fit into English morphs, a noun attached with a Korean proposition was treated as one morph in the current OST system; as a noun morph, there is NNS, and if a Korean noun attached with a Korean postposition was then declared as NNS. For example, there is the Korean word "노즐", which means 'nozzle', and when this word was used as a direct object in a sentence, the postposition "을" was used with the Korean word. In this case, the Korean word and the postposition were put together and the newly constructed morph "노즐을" was declared as NNS under the lexeme "노즐".

5   Design of Problem Sets

Robots have various functionalities to accomplish various physical events; robots are used to handle some work which humans cannot do or can get in danger from; for example, any general robot can get into a very small room or a dangerous situation instead of a human. As other general robots can do, firefighting robots can also get into a certain place with the same reason by moving their wheels and tires. Similarly, commands of a robot coming to a commander, stopping movements and getting out of a certain place are all needed for the firefighting type of robots and the general types of robots.

Also, some robots are built with internal database when they have processors to deal with data from sensors or any resources. Commands of asking a robot to retrieve data are also commonly used in many situations. Especially, a robot could get into a dangerous or unreachable place, humans can ask information in the place through its sensors; commands of asking a robot to retrieve information is also used in a firefighting robot as it has a vision sensor to show the actual fire situations.

However, sometimes humans, and more specifically human intelligence, are needed for operation of a robot; many robots still depend on human judgment when they face danger. Therefore, a human needs to give a command to a robot in order to protect it from or avoid a danger. Especially, firefighting robots will need more protection for themselves as they are more exposed to extreme dangers more often.

Firefighting robots can protect themselves using physical frames such as a nozzle frame or a hose frame to cover up as a humanoid robot does using either an arm or a hand, or using wheels to move away from dangers.

In order to construct problem sets for bilingual robotics communication, all of the consideration discussed in this chapter should be reflected. And with the consideration, there can be two different problems sets designed for different tests. One was to have commands taken from science fiction movies; these commands were chosen to check whether the system can reach understanding expected by people, at least as shown in science fiction genre as well as how general commands work for a firefighting domain. However, the commands include not all general commands, but only possible ones that firefighting robots can also follow. The other one is to have created sentences based on the physical components and functionalities of the targeted firefighting and firefighter assistant robots. This problem set is more target-specific and designed with consideration on actual functionalities of targeted firefighting and firefighter assistant robots. In the following chapter, it is shown how the problems sets could be constructed with the previous two methods.

6   Problem Sets

To construct the first problem set four movies were selected: Real Steal, Bicentennial Man, A.I., and I, Robot. They are all science fiction movies to have scenes of a human giving a natural language command to a robot [25], [26], [27]. In the movies, 53 sentences were originally selected as a part of problem sets for this study. Among the 53 sentences, 20 sentences were selected based on whether the commands can be applied to the firefighting robot or the firefighter assistant robot. Some of the sentences require specific background information; they are considered vague so that they were rarely selected as the one for a problem set. This first problem set shows how OST can be implemented for humanoid robots in general, not specifically firefighting robots, as a preview. The purpose of using the general commands in this study is to deal with the general events which any robot can possibly have; firefighting robots also have general movements which other general robots have. Also, implementation of OST with the general commands is applicable to another potential type of a firefighting or firefighter assistant robot, a humanoid type, since the robots shown in the movies are all humanoid robots. The selected 20 sentences are then translated into Korean; the interpretations of word-by-word translation, natural Korean translation, and English translation from natural Korean are also compared. The constructed first problem set is shown in Table 1.

Another part of the problem sets consists of the generated commands based on the physical components and functionalities of the target firefighting and firefighter assistant robots. Based on the descriptions of DRB Fatec's firefighting robot and HOYAROBOTS's fire assistant robot [28], commands for the second problem set was constructed, as shown in Table 2.

Table 1. Problem set 1: Directly taken from movies

Original Sentences	Word-by-word Translation to Korean (syntax is ignored)	Natural translation in Korean	English Translation from Korean (Keeping Korean Structure)
Get in there.	들어가다 거기에.	거기로 들어가	
Engage.	참여하다	들어가	
Get up!	일어나다	일어나	
Stand up!	일어서다	일어서	
Move!	움직이다	움직여	
Hands up.	손을 들다	손 들어	
Cover up!	보호하다	보호해	
Lean right.	기대다 오른쪽으로	오른쪽으로 기대	
Let's go!	가다	가자	
Andrew, would you come up?	앤드류, 너 올라오다?	앤드류, 올라와볼래?	
Would you please come with me?	제발 올라오다? 나와	제발 나와 와줄래?	Please with me come?
And close the goddamn door!	그리고 닫다 망할 문을	그리고 망할 문을 닫아	And goddamn door close
Raise your arms.	올리다 너의 팔을	너의 팔을 들어	your arms raise
Query internal data.	가져오다 내부 데이터를	내부 데이터 가져와	internal data query
Show me the last 50 messages between Dr. Lanning and Robertson.	보여주다 나에게 마지막 50개 메세지를 래닝 박사와 로벗슨 사이의	래닝 박사와 로벗슨 사이의 마지막 메세지 50개 보여줘	Lanning Dr. and Robertson between last message 50 show
Stay on him!	머무르다 그에게	그가 있는 곳에 머물러있어	On him stay.
Stay downstairs!	머무르다 아래층에	아래층에 머물러	downstairs stay
Panic Shield!	공황 보호	공황 보호	
Andrew, would you please open the window?	앤드류, 제발 열다? 창문을	앤드류, 창문 좀 열어줄래?	Andrew, window please open?
Sheila, open.	쉴라, 열다	쉴라, 열어	

Table 2. Problem set 2: Generated based on functionalities of DRB Fatec's and HOYAROBOTS's robots

Original Sentences	Word-by-word Translation to Korean (syntax is ignored)	Natural translation to Korean	English Translation from Korean (Keeping Korean Structure)
Firefighting-robot, can you hear sound in the building?	소방로봇, 너 듣다 빌딩 안의 소리를	소방로봇, 빌딩 안에서 소리 들리니?	Firefighting-robot, In the building, sound hear?
Firefighting-robot, check whether anyone is in the room.	소방로봇, 확인하다 방에 사람이 있는지	소방로봇, 방에 누구 있는지 확인해봐	Firefighting-robot, in the room anyone whether check
Firefighting-robot, move the nozzle upward.	소방로봇, 움직이다 노즐을 윗 방향으로	소방로봇, 노즐을 위쪽으로 움직여	Firefighting-robot, nozzle upward move
Firefighting-robot, put-out-fire by spraying water.	소방로봇, 불을 끄다 물을 뿌림으로	소방로봇, 물을 뿌려서 불을 꺼	Firefighting-robot, water by spraying fire put-out
Firefighting-robot, smell the air	소방로봇, 맡다 그 공기를	소방로봇, 공기를 맡아봐	Firefighting-robot, air smell
Firefighting-robot, display what you see in the upstairs.	소방로봇, 보여주다 너가 보는 것을 윗층에서	소방로봇, 위층에서 너가 보는 것을 모니터에 보여줘	Firefighting-robot, in the upstairs you see what show
Firefighting-robot, rotate clockwise.	소방로봇, 돌리다 시계방향으로	소방로봇, 시계방향으로 돌아	Firefighting-robot, clockwise rotate
Firefighting-robot, lower the nozzle 1 inch	소방로봇, 낮추다 노즐을 1 인치	소방로봇, 노즐을 1 인치 낮춰	Firefighting-robot, nozzle 1 inch lower
Firefighting-robot, spray water at the burning building on the left	소방로봇, 뿌리다 물을 왼쪽의 타는 빌딩에	소방로봇, 왼쪽의 타는 빌딩에 물을 뿌려	Firefighting-robot, to the left burning building water spray
Firefighting-robot, squirt water to the fire	소방로봇, 쏘다 물을 불에	소방로봇, 불에 물을 쏴	Firefighting-robot, to the fire water Squirt
Firefighting-robot, maximize the speed	소방로봇, 최대화하다 스피드를	소방로봇, 스피드 최대화해.	Firefighting-robot, the speed maximize

Table 2. (*continued*)

Firefighting-robot, slowdown in the room	소방로봇, 늦추다 방에서	소방로봇, 방에서 (속도를) 늦춰.	Firefighting-robot, in the room slowdown
Firefighting-robot, turn around near the table	소방로봇, 돌다 테이블 가까이 주변에서	소방로봇, 테이블 주변 가까이를 돌아	Firefighting-robot, the table around near turn
Firefighting-robot, go up the stairs.	소방로봇, 올라가다 계단을	소방로봇, 계단을 올라가	Firefighting-robot, the stairs up go.
Firefighting-robot, shoot water until the fire out	소방로봇, 물을 쏘다 불이 꺼지기 전까지	소방로봇, 불이 꺼지기 전까지 물을 쏴	Firefighting-robot, the fire out until water shoot
Firefighting-robot, make a U-turn after you reach the wall	소방로봇, 유턴을 하다 벽에 이르고 나서	소방로봇, 벽에 이르고 나서 유턴해	Firefighting-robot, wall reach after U-turn make
Firefighting-robot, come back after you put out a fire	소방로봇, 돌아 오다 불을 끄고 나서	소방로봇, 불 끄고 나서 돌아와	Firefighting-robot, fire put out after back come
Firefighting-robot, come down stairs and back to the entrance.	소방로봇, 내려 오다 계단을 그리고 돌아오다 입구로	소방로봇, 계단을 내려오고 입구로 돌아와	Firefighting-robot, stairs down come and to entrance back come
Firefighting-robot, raise the nozzle up and then spray water for 30 seconds.	소방로봇, 들어올리다 노즐을 위로 그러고 나서 뿌리다 물을 30 초동안	소방로봇, 노즐을 위로 들어올리고 나서 30초동안 물 뿌려	Firefighting-robot, nozzle up raise and then for 30 seconds water spray
Firefighting-robot, stop all of your movements and wait	소방로봇, 멈추다 모든 너의 동작을 그리고 기다리다	소방로봇, 너의 모든 동작을 멈추고 기다려	Firefighting-robot, your all of movements stop and wait

7   Implementation and Result

For implementation of the two problem sets, the test commands from the sets were run through OST Lite, described in Section 3 of this paper. The results of the test commands are represented in graphs to show the relationships between concepts in ontology; nodes are used to indicate there are relationships between concetps, and arrows are used to describe what relationships concepts have.

As it is difficult to represent all graphical representations as there are 80 representations from each problem sets (so that the total is 160). Thus, only representations that are meaningful to be discussed as they have different representations are discussed in this section.

In the results from general commands and firefighting commands, it is interesting to see how representations were different with different commands. All four representations were the exactly same from 17 of 20 general commands. On the other hand, only one firefighting command had the exactly same representations out of 20. The number of the same representations is larger in the general commands than the one in the firefighting commands is because the complexity of the commands is different; the general commands are overall extremely simple compared to the firefighting ones.

There are several representations to be found in the problem sets that are different from one another, and the causes of them are analyzed. One of the reasons is related to the order of a sentence; the order of Korean sentence structures is different from the one of English sentence structures. For example, Korean sentences have verbs at the end of the whole sentences, and adverbs and adjectives are placed between the verbs and the subjects, which are placed at the very beginning of the sentences. And from some representations, it is found that the natural English commands are likely to be the same as the direct Korean translations, which keep the English sentence structure, and the natural Korean commands are likely to be the same as the direct English translations, which keep the Korean sentence structure. Table 3 shows the comparison of the same order of the sentence structure from the sentence "stay downstairs". From the comparison between a natural English sentence and a direct Korean translation of the example in Table 3, it is found that the direct Korean translation keeps the English sentence structure; the other one keeps the Korean sentence structure. The similar representation patterns are found from the following sentences: "Andrew, would you please open the window?", "Firefighting-robot, can you hear sound in the building.", "Firefighting-robot, Check whether anyone is in the room.", "Firefighting-robot, Go up the stairs.", "Firefighting-robot, Smell the air.", "Firefighting-robot, Lower the nozzle 1 inch.", "Firefighting-robot, Slowdown in the room.", "Firefighting-robot, Turn around near the table.", "Firefighting-robot, make sure carbon-dioxide not detected.", and "Firefighting-robot, Turn right at the next corner." This difference is caused by the different direct connection of lexemes; the weight can be different based on the links.

Table 3. One-to-one comparison of "Stay downstairs"

Compari-son	Representations
Natural English & direct Korean translation	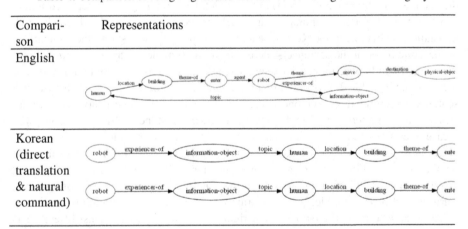
Natural Korean & direct English translation	

Another cause of the different representations is the different use of words in English and Korean. For example, the lexicon declares that the word 'there' can refer to anything or any physical object. The Korean word matching to the word 'there' has only one sense to be anything; OST Lite takes a sense that is more specifically declared for its representation. Thus, this affects the representation to have a smaller scaled sense. The example of this case is shown in Table 4.

Table 4. Comparison of using English and Korean words having different ambiguity

Compari-son	Representations
English	
Korean (direct translation & natural command)	

The other cause of the different representations is from the difficulty in dealing with a phrasal lexeme. For example, the phrase "maximize speed" is declared as a phrasal lexeme being treated as one word because of the less frequent use of a word. However, the order of a Korean sentence changes it to "speed maximize" which OST

Lite does not recognize any event (verb) to link between a robot and speed because of the different order. The similar problem is found from the command "Firefighting-robot, Shoot water until the fire is out" as the word "shoot water" is declared as a phrase and in the Korean translation; the representation has individual concepts instead of one concept for the event "shoot water".

Table 5. Comparison of using a phrasal lexeme and non-phrasal lexeme

Comparison	Representations
Natural English & direct Korean translation & Natural Korean	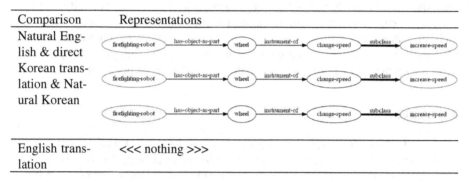
English translation	<<< nothing >>>

There are also a few representations having unnecessary concepts in them; one of the examples is the command "Firefighting-robot, Lower the nozzle 1 inch." The representation of this command has the concept 'raise' next to the concept 'nozzle' along with the concept 'lower'. In this case, once the concept 'lower' is shown, the opposite concept 'raise' is not necessary in the representation. It occurs when a direct concept is linked to another direct concept, which is more related to the unnecessary concept. It is unclear what property creates the unnecessary link to the concept 'raise'; both concepts include the same properties, which are AGENT with FIREFIGHTING-ROBOT (SEM) and THEME with NOZZLE (SEM), so that the connection can be linked from the concept 'lower' to the concept 'move-object' along with the other concepts on the right after the concept 'move-object'. The analysis of finding the cause of less appropriate representations should be more researched.

However, even though there are minor differences found from the representations with the same commands (e.g., the difference in the order of connections) as well as less appropriate representations, which include unnecessary concepts in the representation, the basic concepts and their relationships, which have been found based on the ontology and the lexicon, remain the same.

8   Conclusion

This paper discussed how natural language processing using OST can be designed for a firefighting domain. As this study is focused on the OST over other machine translation techniques, in order for the system to be bilgual (English and Korean), how OST compositions were modified was described; existing ontology and existing lexicon were enchanced by newly added concepts relevant to a firefighting domain.

The problem sets were designed with two different methods to verify how both general and specific situations can be handled. One method was to use existing commands, which can be more general than specific. In order to test more specific commands, which target to a firefighting domain, another method of generating commands based on firefighting and firefight assistant robots was used. This research study introduced how to design natural test commands for the OST for a firefighting domain with the implementation of the test commands. The results show how the commands were actually utilized for implementation of OST, and how different representations can be resulted from the implementation of the test commands. Also, this study provided more diverse test cases, which results different representations, for OST and those diverse cases find points to be improved. This design in this paper can be used for other domains or improved with statistical analysis in the future study.

References

1. Fong, T., Thorpe, C., Baur, C.: Collaboration, dialogue, human-robot interaction. Springer Tracts in Advanced Robotics, vol. 6, pp. 255–266 (2003), doi:10.1007/3-540-36460-9_17
2. Green, S.A., Billinghurst, M., Chen, X., Chase, G.J.: Human-robot collaboration: A literature review and augmented reality approach in design. International Journal of Advanced Robotic Systems 5(1), 1–18 (2008),
   http://hdl.handle.net/10092/2262 (retrieved)
3. Erickson, D., DeWees, M., Lewis, J., Matson, E.T.: Communication for task completion with heterogeneous robots. In: Kim, J.-H., Matson, E., Myung, H., Xu, P. (eds.) Robot Intelligence Technology and Applications. AISC, vol. 208, pp. 873–882. Springer, Heidelberg (2013)
4. Matson, E.T., Taylor, J., Raskin, V., Min, B.C., Wilson, E.C.: A natural language exchange model for enabling human, agent, robot and machine interaction. In: 5th International Conference on Automation, Robotics and Applications (ICARA), Wellington, New Zealand (2011), doi:10.1109/ICARA.2011.6144906
5. DeLoach, S., Oyenan, W., Matson, E.T.: A Capabilities-Based Model for Artificial Organizations. Journal of Autonomous Agents and Multiagent Systems 16(1), 13–56 (2008)
6. Weinstein, C.J., Lee, Y.S., Seneff, S., Tummala, D.R., Carlson, B., Lynch, J.T., Hwang, J., Kukolich, L.C.: Automated English-Korean Translation for Enhanced Coalition Communications. Lincoln Laboratory Journal 10(1) (1997)
7. Choi, K.S., Lee, S., Kim, H., Kim, D.B., Kweon, C., Kim, G.: An English-to-Korean machine translator: MATES/EK. In: Proceedings of the 15th Conference on Computational Linguistics, vol. 1, pp. 129–133 (1994)
8. Egedi, D., Palmer, M., Park, H.S., Joshi, A.K.: Korean to english translation using synchronous tags. arXiv preprint cmp-lg/9410023 (1994)
9. Choi, J.D., Palmer, M.: Statistical dependency parsing in Korean: From corpus generation to automatic parsing. In: Proceedings of the Second Workshop, pp. 1–11 (October 2011)
10. Mitkov, R., Lee, K.H., Kim, H., Choi, K.S.: English-to-Korean machine translation and anaphor resolution. Literary and Linguistic Computing 12(1), 23–30 (1997), doi:10.1093/llc/12.1.23
11. Hong, G., Lee, S.W., Rim, H.C.: Bridging morpho-syntactic gap between source and target sentences for English-Korean statistical machine translation. Proceedings of the ACL-IJCNLP 2009, 233–236 (2009)

12. Choi, D., Park, J., Editions, L., Choi, K.S.: Korean treebank transformation for parser training. In: Proceedings of the ACL 2012 Joint Workshop, pp. 78–88 (July 2012)
13. Choi, S.K., Jung, H.M., Sim, C.M., Kim, T., Park, D.I., Park, J.S., Choi, K.S.: Hybrid approaches to improvement of translation quality in Web-based English-Korean machine translation. In: Proceedings of the 17th International Conference on Computational Linguistics, vol. 1, pp. 251–255. Association for Computational Linguistics (1998)
14. Lavoie, B., White, M., Korelsky, T.: Learning domain-specific transfer rules: an experiment with Korean to English translation. In: Proceedings of the 2002 COLING Workshop, vol. 16, pp. 1–7 (2002)
15. Lee, K., Park, S., Kim, H.: A method for English-Korean target word selection using multiple knowledge sources. IEICE Transactions on Fundamentals of Electronics 89(6), 1622–1629 (2006)
16. Grishman, R.: Natural language processing. Journal of the American Society for Information Science 35(5), 291–296 (2007), doi:10.1002/asi.4630350507
17. Raskin, V., Taylor, J.M., Hempelmann, C.F.: Meaning- and ontology-based technologies for high-precision language an information-processing computational systems. Advanced Engineering Informatics 27(1), 4–12 (2013), doi:10.1016/j.aei.2012.12.002.
18. Petrenko, M., Hempelmann, C.F.: Robotic Reasoning with Ontological Semantic Technology. In: Kim, J.-H., Matson, E., Myung, H., Xu, P. (eds.) Robot Intelligence Technology and Applications. AISC, vol. 208, pp. 883–892. Springer, Heidelberg (2013)
19. Nirenburg, S., Raskin, V.: Ontological semantics, p. 6. The MIT Press, London (2004)
20. Raskin, V., Hempelmann, C.F., Taylor, J.M.: Guessing vs. knowing: The two approaches to semantics in natural language processing. In: Annual International Conference Dialogue (2010)
21. Gruber, T.R.: Ontolingua: A mechanism to support portable ontologies. Stanford University, Knowledge Systems Laboratory (1992),
    `http://callisto.nsu.ru/documentation/CSIR/qmeta/gruber92ontolingua.pdf` (retrieved)
22. Chella, A., Cossentino, M., Pirrone, R., Ruisi, A.: Modeling Ontologies for Robotic Environment. In: Proceedings of the 14th International Conference on Software Engineering and Knowledge Engineering, pp. 77–80 (2002), doi:10.1145/568760.568775
23. Wang, X.H., Zhang, D.Q., Gu, T., Pung, H.K.: Ontology based context modeling and reasoning using OWL. In: Proceedings of the Second IEEE Annual Conference on Pervasive Computing and Communications Workshops, pp. 18–22 (2004), doi:10.1109/PERCOMW.2004.1276898
24. Tayor, J.M., Raskin, V.: Natural language cognition of humor by humans and computers: a computational semantic approach. In: ICCI*CC, pp. 68–75. IEEE (2013)
25. Columbus, C.: Bicentennial man (DVD) (1999)
26. Spielberg, S.: A.I. Artificial Intelligence (DVD) (2001)
27. Levy, S.: Real steel (DVD) (2011)
28. Hong, J., Min, B., Taylor, J.M., Raskin, V., Matson, E.T.: NL-based communication with firefighting robots. In: Proceedings of the 2012 IEEE International Conference on Systems, Man, and Cybernetics (2012)

Part II
Behavioral, Collective, and Genetic Intelligence

Hyun Myung

Behavioral intelligence deals with the behavior design, generation, and control strategies of robots mainly inspired from nature. To achieve complex intelligent behavior, soft computing techniques such as evolutionary computation can be used instead of classical hard computing techniques. By only providing the representation of the problem in the form of genetic coding or neuronal connections, a properly designed objective function can guide the robot to fit to a desired fitness function. In this respect, behavioral intelligence is closely related to genetic intelligence for generating more intelligent behavior.

Collective intelligence emerged by interactions among multiple robots can be another extension of behavioral intelligence. In this context, multiple robots can exhibit various complex behaviors using multiple sources of sensory information and partial intelligence.

This chapter consists of three segments of different topics which cover broad spectrum of topics related to robot intelligence discussed above; Behavioral Intelligence, Collective Intelligence, and Genetic Intelligence.

The following papers treat various kinds of examples related to the first topic of *Behavioral Intelligence*.

1) Towards Honeycomb PneuNets Robots
2) Progress of Research on A New Asteroid Exploration Rover Considering Thermal Control
3) A Proposal of Exploration Robot with Wire for Vertical Holes in the Moon or Planet
4) Scanpaths Analysis with Fixation Maps to Provide Factors for Natural Gaze Control
5) Design of a Bipedal Walker with a Passive Knee and Parted Foot
6) An Adaptive Sliding Mode Control for Trajectory Tracking of a Self-reconfigurable Robotic System
7) Position-trajectory Control of Advanced Tracked Robots with Diesel-electric Powertrain
8) Stable Modifiable Walking Pattern Generator with Arm Swing Motion Using Evolutionary Optimized Central Pattern Generator
9) Performance Comparison Between a PID SIMC And a PD Fractional Controller
10) Implementation of High Reduction Gear Ratio by Using Cable Drive Mechanism

11) Trajectory Tracking Control Using Echo State Networks for the CoroBot's Arm
12) An Intelligent Control System for Mobile Robot Navigation Tasks in Surveillance
13) Formation Control Experiment of Autonomous Jellyfish Removal Robot System JEROS
14) Soft-Robotic Peristaltic Pumping Inspired by Esophageal Swallowing in Man
15) Design and Fabrication of a Soft Actuator for a Swallowing Robot
16) A P300 Model for Cerebot – a Mind-Controlled Humanoid Robot
17) Preliminary Experiments on Soft Manipulating of Tendon-Sheath-Driven Compliant Joints
18) Maintaining Model Consistency During In-Flight Adaptation in a Flapping-Wing Micro Air Vehicle
19) Model Checking of a Flapping-Wing Mirco-Air-Vehicle Trajectory Tracking Controller Subject to Disturbances
20) Design Constraints of a Minimally Actuated Four Bar Linkage Flapping-Wing Micro Air Vehicle
21) Improved Control System for Analyzing and Validating Motion Controllers for Flapping Wing Vehicles

The first five papers deal with behavioral intelligence applied to humanoid, micro, and nano robots. The next seven papers show examples for kinematics, dynamics, control, and motion planning issues to implement behavioral intelligence.

The thirteenth paper is also related to **Collective Intelligence**. Collective intelligence is a theory that describes a type of shared or group intelligence that emerges from the collaboration and competition of multiple robots. A group of robots can cooperate or compete with each other to achieve a common desired objective. Multi-robot system or swarm robot system can be used to implement collective intelligence.

The next four papers are mainly contributed to soft robotics for generating intelligence behavior. The remaining papers shows examples of behavioral intelligence applied to flapping-wing air vehicles.

The last topic in this chapter is **Genetic & Behavioral Intelligence**. Genetic intelligence is closely related to soft computing such as evolutionary computation and neural networks. By representing a problem with a chromosome and genes, and describing fitness of this chromosome in the form of objective function, evolutionary computation effectively solves a solution using meta-heuristics inspired by genetics. The problems hard to be solved due to their inherent complexity, or the problems that do not have gradient information that is necessary for classical optimization methods or hard computing techniques, can be candidates for the application of evolutionary approach. The various neural networks schemes are also very useful to solve nonlinear optimization or classification problems. The robots that use these problem solving capabilities can be regarded to have genetic intelligence. The hybridization and integration of the genetic intelligence and behavioral intelligence are promising to achieve more emergent and proactive robot intelligence. The following papers present excellent examples of these approaches.

1) Adding Adaptable Stiffness Joints to CPG-based Dynamic Bipedal Walking Generates Human-like Gaits
2) Hand-Eye Calibration and Inverse Kinematics of Robot Arm using Neural Network
3) An Evolutionary Feature Selection Algorithm for Classification of Human Activities
4) Automatic Linear Robot Control Synthesis using Genetic Programming
5) Estimation of Stimuli Timing to Evaluate Chemical Plume Tracing Behavior of the Silk Moth
6) Vibration Occurrence Estimation and Avoidance for Vision Inspection System

Towards Honeycomb PneuNets Robots

Hao Sun and Xiao-Ping Chen

Dept. of Computer Science, University of Science and Technology of China
Hefei, Anhui, 230027, PRC
hhsun@mail.ustc.edu.cn, xpchen@ustc.edu.cn

Abstract. In recent years, Soft Robotics becomes a research hotspot. A soft robot is usually made of elastic materials, and thus has better adaptability and safety to the environment than a rigid robot. These advantages offer us a new opportunity to attack some fundamental challenges faced by traditional robots. Most of previous studies about soft robots focus on clarifying the deformation characteristics of the flexible materials used. In order to pursue a large expansion rate, the stiffness of the soft materials is usually very low, which brings a consequence that these soft robots are too soft to maintain its shape or to resist external forces. Inspired by the honeycomb structure, this paper proposes a honeycomb pneumatic network (HPN) robot, which consists of several pneumatic units. We put forward a force analysis model of a pneumatic unit and a kinematics model of honeycomb pneumatic network. Based on these models, we study the relationships between the air pressure, external force and geometrical shape through simulation. The experimental results showed that the excellent expansion rate and flexibility can be achieved in the HPN robot.

Keywords: PneuNets, Honeycomb structure, Soft robot.

1 Introduction

Currently, robots can be divided into two categories on the basis of the compliance of their underlying materials. The first one is traditional hard robots, which are composed by rigid linkage mechanisms. The movements of this kind of robot are accurate, fast and stiff. However, it is difficult and unsafe for these robots to interact with their environments. The second one is soft robots, which have attracted more and more attention of researchers in recent years. Soft robots, as are implied by the name, are made of soft --often elastomeric -- materials. They are biomimetic robots which mimic the structure of soft creatures, such as worm and starfish. Soft robots have many useful capabilities, including the abilities to deform their shape continuously, to manipulate delicate objects with compliance, and to interact with human safely [1-2]. As robots gradually come into the human environment, these features become more and more important.

There are many researches focusing on changing the mechanical-structural design and developing new control algorithms of hard robots to achieve flexible motion and compliance [3-4]. However, it is still very difficult for hard robots to maintain these

J.-H. Kim et al. (eds.), *Robot Intelligence Technology and Applications 2*,
Advances in Intelligent Systems and Computing 274,
DOI: 10.1007/978-3-319-05582-4_28, © Springer International Publishing Switzerland 2014

characteristics in unstructured environments, when their underlying actuating materials are hard and their skeletons are rigid. Inspired by the outstanding capabilities of soft animal structures, it seems to the authors that a fundamental solution is to develop soft robotic actuators/robots.

2   Background

2.1   Related Works

Most soft robots developed so far are based on three technologies: electroactive polymer (EAP), pneumatic artificial muscle (PAM), and embedded pneumatic networks (EPN).

Electroactive polymers (EAPs) include dielectric elastomers, electrolytically active polymers, polyelectrolyte gels and gel-metal composites [1]. EAP actuators accomplish elongation by compressing the elastomer using the Coulomb force between the electrodes. When the voltage removed, elastomer restore to its original length. EAP have many characteristics that make it suitable for soft robots: low weight, fracture tolerance, and a relatively large actuation strain [5]. However, it also has very obvious shortcomings such as slow response at contraction, degrading rapidly in use, and requiring high actuation voltages.

PAM actuators are the most highly developed soft actuators and also widely used in the industry. The typical representative of PAM is Pneumatically-driven McKibben-type actuators [6] that consist of a thin, flexible, tubular membrane with fiber reinforcement. PAM is able to provide a large force, and give response quickly. However, it can only contract and extend when pressurization changes, so the rigid skeletal system is needed. It's difficult to use it to build a fully soft robot.

EPN actuators use embedded pneumatic networks (PneuNets) of channels in elastomers that inflate like balloons for actuation [7] (see Fig.1). This design offers opportunities to build more flexible robots which are fully soft. When the PneuNets are appropriately distributed, it can easily achieve complex motions and maintains compliance at any time and any direction. But from what we can see from figure 1, this actuator is highly dependent and is limited on the deformation ability of soft material. Only very soft materials can achieve such big deformation without break, so the EPN robots completely made of very soft materials are not suitable for manipulating heavy objects.

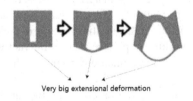

Very big extensional deformation

Fig. 1. PneuNets(picture from [7])

Compared to the first two soft actuators, EPN actuators are more portable, more flexible, cheaper and safer. The objective of this work is to demonstrate a new type of design promising to improve EPN's performance and overcome the above drawbacks.

2.2   Nature Inspiration

The honeycomb structure --originally created by bees -- with its light quality and high strength, are commonly used in the aerospace field [8]. It can distribute the deformation and external force evenly throughout all honeycomb chambers, and thus has the properties of light weight, good flexibility and high carrying capacity. Given these characteristics, by intuition and the force analysis, we consider that this structure may effectively optimize the performance of the PneuNets.

In this paper, the concept of the Honeycomb PneuNets is addressed and the advantages and disadvantages are discussed. The rest of the paper is organized as follows. Section 3 presents the concept of the Honeycomb PneuNets Robot and introduces a simplified mathematical model. Based on this model, section 4 discusses the deformation under different air pressure and external force by computer simulation. In section 5, we conclude the paper and propose some suggestions for the future work.

3   Honeycomb PneuNets

The honeycomb structure, as described in 2.2, has many advantages, such as structural stable, flexible, and crush-resistant, while pneumatic network systems have the advantages: light, low viscosity, rapid response and environmentally benign. So we bring the two concepts together to form a better solution --- the Honeycomb PneuNets. Our design is to place Honeycomb PneuNets inside the elastomers, and compress air in each of the chambers through a number of channels. When pressurized, some chambers expand and deform, so that the whole structure starts elongation and bending.

Fig. 2. The Honeycomb PneuNets Robot

The differences between HPN and existing EPN are summarized below. EPN has rectangular chambers, which have thick sides and thin sides. When pressurized, the chambers expand in the regions that are most compliant or have the lowest stiffness. So the EPN must be made of materials which are very soft (such as silica gel). As a result, the EPN robots are usually very small, otherwise it could not support it own weight or resist external forces. By contrast the HPN robots have hexagonal chambers (Fig.2), which can change their geometrical shape   without large deformation of the soft material (Fig.3), since its main parts can be made of a material with greater stiffness (such as polyethylene). Because of these characteristics, an HPN robot has a better structural strength and stability, which greatly expands the range of applications of soft robots.

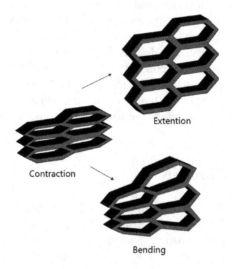

Fig. 3. The deformation modes of the HPN robot

4   Deform Analysis

As described above, the HPN robot can extend or bend itself without changing the length of chamber walls much. When we use relatively high stiffness material to construct it or the air pressure is low, we can ignore the elongation of chamber walls. We focus on the changes in the angles caused by air pressure variation. Here we introduce a simple mathematical model (Fig.4) to simplify the deform analysis.

Assuming a HPN chamber is filled with air of pressure of P, and the angle between the left-top and top chamber wall is θ. The thickness of chamber wall is h, the lengths and tensions of left-top and top chamber wall are l1, F1, l2, F2, respectively. Fk is the external force which also means that how much output force provided by HPN actuator in the axial direction. When the air pressure increases, the force on chamber walls also increase, let l represent the axial length of a chamber so we can obtain the following force balance equations:

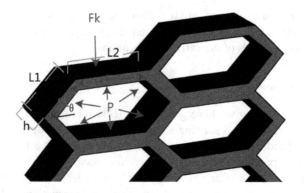

Fig. 4. A simple mathematical model of HPN

The force balance equation on the top wall:

$$2F_1 \sin(\theta) = phl_2 + F_k \tag{1}$$

The force balance equation on the side wall:

$$Iph = F_2 \tag{2}$$

F1 and F2 satisfy the following relation:

$$2F_1 \cos(\theta) = F_2 \tag{3}$$

We can see from the (1-3) that when the air pressure increases, the push force of the air applied to the top of the wall also increases. To resist this change, θ, F1 and F2 become larger and finally the chamber unit comes to a new balance state. This is the principle of the chamber unit deformation.

By the geometric relationship, we get:

$$2\sin(\theta)l_1 = I \tag{4}$$

From simultaneous equations (1)(2)(3)(4),we get(5)(6)(7):

$$\theta = -a\sin\left(\frac{F_k - 2(\dfrac{F_k^2}{4} + \dfrac{F_k hl_2 p}{2} + 4h^2l_1^2 p^2 + \sqrt{\dfrac{h^2 l_2^2 p^2}{4}}) + hl_2 p}{4hl_1 p}\right) \tag{5}$$

$$F_1 = \frac{phl_2 + F_k}{2\sin(\theta)} \tag{6}$$

$$I = 2\sin(\theta)l_1 \tag{7}$$

Thus, once the parameters of an HPN robot determined, we can calculate the force balanced shape. We can also use these equations to adjust the design of an HPN robot so it can meet the needs of the various applications.

4.1   Kinematics

In the previous section we have got a basic HPN robot unit. By combining the two parallel units, we can get a bending actuator. To increase the degree of its bending, we put multiple bending actuators in series to build a snake-like HPN robot (Fig.3). In this case, corresponding force balance equations become more complicated. In order to reduce the difficulty of analysis, we simply consider the robot as a parallel hyper redundant linkage mechanism. We assume that the deformation of the chamber units to comply with the laws in 3.1, then the degrees of bending of the robot depend on the lengths of the left and right chamber, I1, I2, and the distance between them, W (as shown in Fig.5).

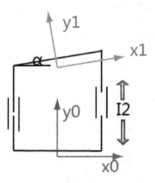

Fig. 5. Simplified kinematics model of the HPN robot

Based on the above simplification and assumptions, we can easily obtain the following homogeneous transformation matrix:

$$H_1^0 = \begin{bmatrix} \cos(\alpha) & -\sin(\alpha) & 0 \\ \sin(\alpha) & \cos(\alpha) & \dfrac{I_1 + I_2}{2} \\ 0 & 0 & 1 \end{bmatrix} \tag{8}$$

$$\alpha = a\tan(\frac{I_2 - I_1}{w}) \tag{9}$$

Using the homogeneous matrix, we can transfer one point to a representation of another coordinate system. Let Pi represent one point in coordinate i, then

$$P_0 = H_1^0 P_1.$$

If linking up n units of this type (shown in Fig.6), similarly we get (10-11)

Fig. 6. Series bending actuator

$$P_0 = H_n^0 P_n \tag{10}$$

$$H_n^0 = H_1^0 H_2^1 ... H_n^{n-1} \tag{11}$$

5   Experiments

Based on the above mathematical models and analyses, we use computer simulation to reveal the relationships between the motion performance and the parameters (such as air pressure or external force) of the HPN robot.

We designed two experiments. The first one was conducted to observe a single HPN unit's performance. By changing the air pressure of a chamber and the external force, we calculated the length of the chamber. The second experiment was to observe the parallel bending actuator's performance. We gave the left and right chambers different air pressures, and then calculated the curvature.

5.1   Chamber Simulation

We assigned the values 0.5,1,1,0.3 to the variables Fk, l1, l2, h(defined in section 3.1),and changed the value of p from 0 to 10, so that we could the value of I (as shown in Fig.7):

We can see that, as the air pressure increases, the chamber length increases. At the beginning the slop is very big. As the pressure increases, the slop becomes much smaller and eventually approaches a constant value. This is because when the air pressure becomes very large, the external force can be neglected, i.e, Fk ~ 0.

Then we fixed the value of p to obtain the relationship between external force Fk and length I. We assigned the value 1.5 to variable p, and changed the value of Fk from 0 to 10. As shown in Fig.8-9, as the force increases, the length becomes smaller.

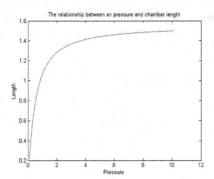

Fig. 7. Chamber length under different air pressure

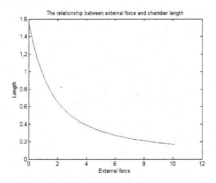

Fig. 8. Chamber length under different external force

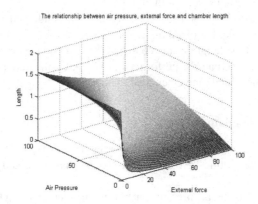

Fig. 9. The relationship between air pressure, external force and chamber length

5.2   HPN Robot Simulation

We used the same method to obtain the relationship between the input air pressures and the curvatures of the HPN robot. Let w=2, Fk=0.5, n=10.We established the mathematic model mentioned in section 3.2. We set the left side chambers the pressure of p1, and right of p2. Let p1=0.1, change p2 from 0.1 to 1.2. Then we calculated the position of the HPN robot and trace it by a straight line, so we can see its shape (see Fig.10)

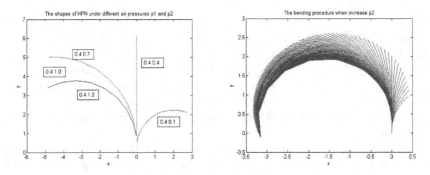

Fig. 10. The various shapes under different air pressures p1 and p2

One can see that when the left and right sides have the same air pressure, the trace is a straight line. It means that the robot extends straightly. When p2 increases, the degree of bending also increases. When p1 changes, in order to achieve the previous curvature, p2 also needs change.

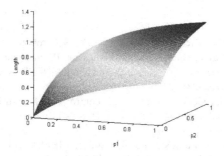

Fig. 11. The relationship between p_1, p_2 and length

Fig. 9 and 10 tell us that p1 and p2 have multiple solutions for one curvature. Although a plurality of p1 and p2 can make the robot to achieve the same curvature, according to 3.2 the lengths of the actuator are not the same. The relationship between p1, p2 and I is shown in Fig.11. So not only can the robot be bent but also can be elongated, which makes it very flexible.

6   Conclusion

In this paper, we proposed a new design of honeycomb network pneumatic robot/actuator called the HPN robot. The HPN robot's movement does not depend on stretching soft material, but on changing the chambers' interior angle like a spring, which is different from other soft robots. The advantages of this design are summarized below. First, the HPN robot's elongation is no more limited by tensile strength and elongation at break. Second, the HPN robot can be made of a material with a greater stiffness. Third, honeycomb structure provides the robot a better structural stability.

We also made a preliminary analysis of the movement performance of the HPN robot and obtained its relationship with the air pressure and external (output) force, which laid the foundation for future manufacturing and controlling. The first experiment has shown that, the theoretical elongation rate of the HPN chamber can be very big. The second experiment has shown that the HPN bending actuator is flexible and can achieve the same curvature of different lengths. The two characteristics suggest that the HPN robot would be a promising technology to promote soft robots to be practical.

In the future work, we will need to implement an HPN soft robot, and through error correction, to establish more realistic kinematical and dynamic models. Meanwhile, the control and planning algorithms for HPN robots are also needed.

Acknowledgements. This research is supported by the National Hi-Tech Project of China under grant 2008AA01Z150 and the Natural Science Foundation of China under grant 60745002 and 61175057, as well as the USTC Key Direction Project.

References

1. Trivedi, D., Rahn, C.D., Kier, W.M., Walker, I.D.: Soft robotics: Biological inspiration, state of the art, and future research. Applied Bionics and Biomechanics 5(3), 99–117 (2008)
2. Albu-Schaffer, A., Eiberger, O., et al.: Soft robotics. IEEE Robotics & Automation Magazine 15(3), 20–30 (2008)
3. Stückler, J., Behnke, S.: Compliant task-space control with back-drivable servo actuators. In: Röfer, T., Mayer, N.M., Savage, J., Saranlı, U. (eds.) RoboCup 2011. LNCS, vol. 7416, pp. 78–89. Springer, Heidelberg (2012)
4. Wright, C., et al.: Design of a modular snake robot. In: IEEE/RSJ International Conference on Intelligent Robots and Systems, IROS 2007. IEEE (2007)
5. Samatham, R., et al.: Active polymers: an overview. In: Electroactive Polymers for Robotic Applications, pp. 1–36. Springer, London (2007)
6. Kang, B.-S., et al.: Dynamic modeling of Mckibben pneumatic artificial muscles for antagonistic actuation. In: IEEE International Conference on Robotics and Automation, ICRA 2009. IEEE (2009)
7. Ilievski, F., et al.: Soft robotics for chemists. Angewandte Chemie International Edition 123(8), 1930–1935 (2011)
8. McKnight, G.P., Henry, C.P.: Large strain variable stiffness composites for shear deformations with applications to morphing aircraft skins. In: The 15th International Symposium on: Smart Structures and Materials & Nondestructive Evaluation and Health Monitoring. International Society for Optics and Photonics (2008)

Progress of Research on a New Asteroid Exploration Rover Considering Thermal Control

Yosuke Miyata[1], Tetsuo Yoshimitsu[2], and Takashi Kubota[2]

[1] Department of Electrical Engineering and Information Systems,
Graduate School of Engineering, The University of Tokyo, Japan
ymiyata@ac.jaxa.jp
[2] Institute of Space and Astronautical Science, Japan Aerospace Exploration Agency, Japan

Abstract. Asteroids have some clues about the origin of the solar system. In recent years, asteroid exploration by surface explorers has been studied actively. Conventional rovers can move under microgravity environment, but the rovers have not considered thermal control. This paper proposes a novel mobility mechanism for a rover under microgravity environment along with temperature control. The effectiveness of the proposed rover is investigated by simulations and microgravity experiments.

Keywords: asteroid exploration, hopping rover, multibody dynamics, thermal control.

1 Introduction

In recent years, asteroid exploration missions in the solar system have been studied actively. Asteroids have some clues to know the origin of the solar system because they were formed without thermal deformation. In the future asteroid exploration missions, a wide range of surface exploration by small rovers is expected and rovers which can adapt to the asteroid environment are required. Rovers need robust locomotion mechanism under microgravity environment. Additionally, asteroid surface is very hot during the daytime due to the inclination of the rotation axis. So, rovers are required to have aggressive thermal control.

In microgravity environment, a wheeled mechanism which has been used in lunar or planetary exploration rover is not valid. So, hopping motion by pushing the surface is considered to be effective and a lot of hopping rovers have been proposed [1–10]. However, these hopping rovers have been studied by focusing only on mobility mechanism under microgravity. These rovers have not considered thermal control.

In this paper, a new type of rover is proposed to move under microgravity as well as control its temperature by deploying the rover. The effectiveness of the proposed rover is investigated by simulations and microgravity experiments.

2 Proposed Deployment Rover

A proposed rover is shown in Fig. 1. The proposed rover with a new deployment mechanism can move by hopping and control its internal temperature. The proposed rover

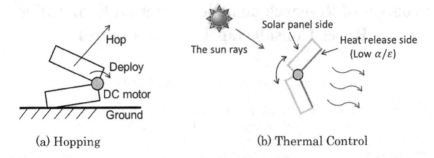

(a) Hopping   (b) Thermal Control

Fig. 1. Proposed Deployment-Type Rover

has a brushed DC motor for deployment. So, the rover hops by DC motor's torque and inertial force on the deployment part. Additionally, the rover has solar panel on the outside and heat release sides in the inside. OSR(Optical Solar Reflector) or Teflon/Al is used in the heat release side as heat release device(low α/ϵ). The deployment and attitude control in the air contributes to thermal control.

3   Thermal Control

In this section, the thermal conceptual design of asteroid rovers and the thermal control of the proposed rover are described.

Rover needs to control the temperature to operate properly the on-board devices. In general, the operation temperature of the on-board devices is -30'60 []. Thermal inputs to the asteroid rover are solar flux, albedo of direct solar and infrared from asteroid surface. On the other, the rover can radiate heat to space (4 K). In a severe heat environment, a heat emission side is required in order to radiate heat actively.

The thermal designed rovers are shown by Fig. 2. A-model has one heat release side, which is sometimes used in lunar or planetary rovers. Additionally, there is no inflow of heat from the outside because the sides other than the heat release side are insulated by MLI and insulation spacer. However, the thermal input becomes larger when the heat release side is exposed to infrared from the asteroid surface in the design of A-model. Furthermore, a little electricity is generated in the top side because a solar panel area is small. So, this paper proposes the deployment design (B-model) to solve the problems. This deployment rover can take active control of the rover's temperature by opening and closing the heat release side. The deployment rover does not allow heat to enter the inside and can generate the same electricity in any attitude by closing the heat release side.

The effectiveness of the proposed deployment rover in terms of thermal control is evaluated by basic temperature analysis model (Fig. 3). Environmental setting is shown as follows. (i)No heat input from the insulation part. (ii)Asteroid surface is an infinite flat plane. (iii)Rotation of the rover is only Y_R-axis (the direction of torque by DC motor). The temperature analysis conditions are shown in Table 1. The attitude angle of the rover is θ_R, the development angle is ϕ and the sun angle is α respectively. The size of the heat release side is given in the same as the size of the rover ($L_E=L$) to

Table 1. Temperature analysis conditions

condition	value	condition	value
Intensity of solar radiation	1515 [W/m^2]	Albedo coefficient	0.25
Infrared emissivity	0.68		
Absorptivity of OSR	0.1	Emissivity of OSR	0.8

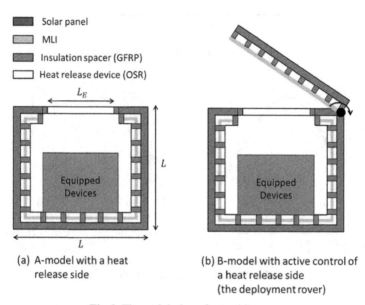

(a) A-model with a heat
release side

(b) B-model with active control of
a heat release side
(the deployment rover)

Fig. 2. Thermal design of asteroid rovers

investigate the maximum heat release capacity. The power consumption of the rover is
zero to investigate the only effect of heat emission and input. There is no heat emission
from the deployment part because it is insulated from the body. The temperature of
the ground (T_g) is investigated as the maximum temperature (140 []) and the minimum
temperature (-100 []) of the asteroid surface.

The results of temperature analysis of A-model are shown as Fig. 4. These graphs
show the equilibrium temperature of the rover T_R for the variation of the heat release
area $L_E{}^2$. The equilibrium temperature is independent of the heat release area. The
temperature exceeds the allowable temperature range of the on-board devices (-30'60
[]) in some attitudes. This reason is that the thermal input is greatly affected by infrared
and albedo when the heat release is facing to the ground.

The thermal control scheme for the deployment rover is proposed based on these
results and discussion. The deployment rover has a sun sensor, a gyro sensor, a temperature sensor and the profile as shown in Fig. 4. The rover controls the deployment
angle ϕ for the rover's attitude θ_R. The deployment angle is shifted in 0 and π whether
T_R exceeds -30'60 [] or not. The results is shown in Fig. 5. The thermal control by the
deployment rover can keep the rover's temperature within the operation temperature.

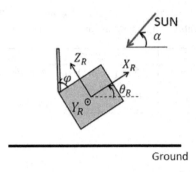

Fig. 3. Basic temperature analysis model

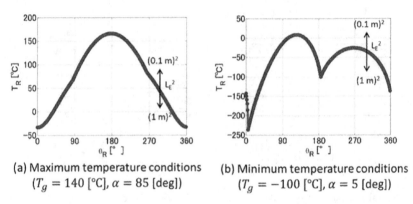

(a) Maximum temperature conditions   (b) Minimum temperature conditions
$(T_g = 140 \ [°C], \alpha = 85 \ [deg])$ $(T_g = -100 \ [°C], \alpha = 5 \ [deg])$

Fig. 4. Equilibrium temperature of A-model

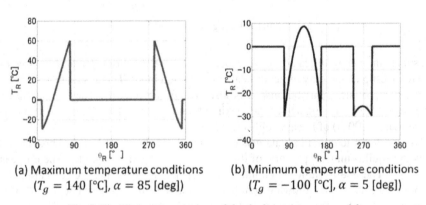

(a) Maximum temperature conditions   (b) Minimum temperature conditions
$(T_g = 140 \ [°C], \alpha = 85 \ [deg])$ $(T_g = -100 \ [°C], \alpha = 5 \ [deg])$

Fig. 5. Equilibrium temperature of the deployment rover model

4   Dynamics Modeling

In this section, the effectiveness of the proposed rover in terms of mobility is investigated. 2D movement model of the deployment rover is used.

The deployment rover is shown in Fig. 6(a). The rover consists of Mass A and B restrained at Point P_1. The rover is closed in the initial state and deployed to hop. Torque by DC motor acts around the rover's body. The body is inclined and lifted from the asteroid surface. Mass elements are formed of rigid bodies M_A, J_A and M_B, J_B. The ground is assumed to be spring-damper system with spring constant, K_g and damping constant, D_g.

The Equation of motion is constructed by multibody dynamics. The coordinate system of 2D rover model is shown in Fig. 6(b). 2D surface of X-Z is defined as global coordinate system. X-axis is a horizontal direction the rover proceeds and Z-axis is a vertical direction. Each elemental vector $\mathbf{r}_{O1}$ '$\mathbf{r}_{O3}$ for point p_1 'p_3 is given as follows.

$$\mathbf{r}_{O1} = \mathbf{R}_{OA} + C_{OA}\mathbf{r}_{A1} \tag{1}$$

$$\mathbf{r}_{O2} = \mathbf{R}_{OB} + C_{OB}\mathbf{r}_{B2} \tag{2}$$

$$\mathbf{r}_{O3} = \mathbf{R}_{OB} + C_{OB}\mathbf{r}_{B3} \tag{3}$$

Here, $\mathbf{R}_{OA}$, $\mathbf{R}_{OB}$ is position of the center of gravity of Mass A and B in global coordinate system. C_{OA}, C_{OB} is coordinate transformation matrix of Mass A and B. Each $\mathbf{r}_{A1}$, $\mathbf{r}_{B2}$, $\mathbf{r}_{B3}$ is position vector of point p_1 'p_3 in local coordinate system of Mass A and B.

Force acting from the ground is represented by $\mathbf{F}_g$. $\mathbf{F}_g$ is simulated by spring-damper force which acts only when point p_3 is below the ground. Force acting from the ground to p_3 is shown by equation (4)'(7). Point p_4 is a contact point between point p_3 and the ground in the initial state of a rigid body B. F_{gx} and F_{gx} are the X and Z component of force acting from the ground. Also, R_{43x} and R_{43z} are the X and Z component of the relative position vector between point p_3 and p_4 and $V_{43x}CV_{43z}$ are that of the relative velocity vector. μ' is coefficient of dynamic friction.

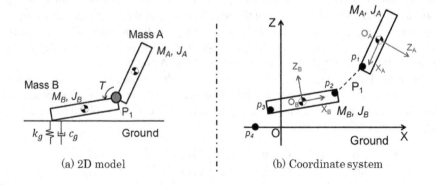

(a) 2D model   (b) Coordinate system

Fig. 6. Dynamic model of Deployment-Rover

– when point p_3 does not slide,

$$F_{gx} = -(K_g R_{43x} + C_g V_{43x}) \tag{4}$$
$$F_{gz} = -(K_g R_{43z} + C_g V_{43z}) \tag{5}$$

– when point p_3 slides,

$$F_{gx} = -sgn(V_{43x}) \cdot \mu' F_{gz} \tag{6}$$
$$F_{gz} = -(K_g R_{43z} + C_g V_{43z}) \tag{7}$$

Therefore, force acting on the rigid body, $\mathbf{F}_{OA}$, $\mathbf{F}_{OB}$ and torque, N_{OA}, N_{OB} are shown as follows. Direction of gravitational force, g is vertical downward. T is torque generated by DC motor and χ is skew-symmetric matrix.

$$\mathbf{F}_{OA} = \begin{bmatrix} 0 \\ -M_A g \end{bmatrix} \tag{8}$$

$$N_{OA} = -T \tag{9}$$

$$\mathbf{F}_{OB} = \begin{bmatrix} 0 \\ -M_B g \end{bmatrix} + \mathbf{F}_g \tag{10}$$

$$N_{OB} = T + \mathbf{r}_{B3}^T \chi^T C_{OB}^T \mathbf{F}_g \tag{11}$$

Restraint condition of position, $\boldsymbol{\Psi}$, velocity, $\boldsymbol{\Phi}$ and acceleration, $\dot{\boldsymbol{\Phi}}$ on rigid bodies A and B are given by :

$$\boldsymbol{\Psi} = \frac{1}{2} \mathbf{r}_{12}^T \mathbf{r}_{12} = 0 \tag{12}$$
$$\boldsymbol{\Phi} = \mathbf{r}_{12}^T \mathbf{v}_{12} = 0 \tag{13}$$
$$\dot{\boldsymbol{\Phi}} = \mathbf{r}_{12}^T \dot{\mathbf{v}}_{12} + \mathbf{v}_{12}^T \mathbf{v}_{12} = 0 \tag{14}$$

Each $\mathbf{r}_{12}$ and $\mathbf{v}_{12}$ is relative position and a relative velocity between point p_1 and p_2 respectively. Terms involving V_{OA} in $\boldsymbol{\Psi}$ are represented by $\boldsymbol{\Psi}_{V_{OA}}$ and the rest by $\boldsymbol{\Psi}_{\bar{V}_{OA}}$. Restraint $\boldsymbol{\Psi}$ is represented by $\boldsymbol{\Psi} = \boldsymbol{\Psi}_{V_{OA}} V_{OA} + \boldsymbol{\Psi}_{\bar{V}_{OA}}$ in linear combination. Additionally, term including $V_{OB}, \omega_{OA}, \omega_{OB}$ are the same. $V_{OA}, V_{OB}, \omega_{OA}, \omega_{OB}$ are velocity and acceleration in global coordinate system of Mass A and B.

Differential algebra type equation of motion for this dynamics model is shown by equation (15). Λ is Lagrange multiplier. Baumgarte stabilization method is adapted and α and β are arbitrary parameters. Equation of motion (15) is solved by numerical methods such as Runge-Kutta.

$$\begin{bmatrix} M_A & 0 & 0 & 0 & \boldsymbol{\Psi}^T_{V_{OA}} \\ 0 & J_A & 0 & 0 & \boldsymbol{\Psi}^T_{\omega_{OA}} \\ 0 & 0 & M_B & 0 & \boldsymbol{\Psi}^T_{V_{OB}} \\ 0 & 0 & 0 & J_B & \boldsymbol{\Psi}^T_{\omega_{OB}} \\ \boldsymbol{\Psi}_{V_{OA}} & \boldsymbol{\Psi}_{\omega_{OA}} & \boldsymbol{\Psi}_{V_{OB}} & \boldsymbol{\Psi}_{\omega_{OB}} & 0 \end{bmatrix} \begin{bmatrix} \dot{V}_{OA} \\ \dot{\omega}_{OA} \\ \dot{V}_{OB} \\ \dot{\omega}_{OB} \\ \Lambda \end{bmatrix} = \begin{bmatrix} F_{OA} \\ N_{OA} \\ F_{OB} \\ N_{OB} \\ -\dot{\boldsymbol{\Phi}}^R - \alpha \boldsymbol{\Phi} - \beta \boldsymbol{\Psi} \end{bmatrix} \tag{15}$$

5   Simulation Study

The effectiveness of the proposed rover in terms of mobility is investigated by using the dynamics model developed in Section 3.

5.1   Simulation Parameters

Simulation parameters are as follows. Mass of rigid bodies A and B, the inertia moment and the length of the diagonal are M_A=0.30 [kg], M_B=0.60 [kg], J_A=2.5 [kgcm²], J_B=5.0 [kgcm²], L_A=0.07 [m] and L_B=0.08 [m] respectively. Gravitational acceleration on the asteroid surface is $g=10^{-4}$ [m/s²]. Spring constant, $K_g=10^4$ [N/m] and damping constant, $C_g=10^2$ [Ns/m] are defined since the ground surface is given by a rock. Also, coefficient of static friction is μ=0.5 and coefficient of dynamic friction μ' is 0.8 times of μ.

It is important whether the rover can move under microgravity. So, two cases of (i) torque's variation and (ii) inertia moment's variation are investigated to show the mobility of the proposed rover.

5.2   Torque Variation

In this section, the behavior of the rover is investigated when input torque from DC motor is varied. Adding torque is a constant and given by $T=10^{-6}·10^{-1}$ [Nm].

Simulation results are shown in Fig. 7. These graphs show hopping velocity, travel distance and hop time for torque's variations. Hopping angle is defined by the angle against the vertical direction. The rover achieves a horizontal and vertical velocity accordingly and hops. $T > 10^{-5}$ is necessary to start hopping under microgravity($g=10^{-4}$ [m/s²]) as shown in Fig. 7(a). Also, $T < 10^{-2}$ is necessary not to exceed the escape velocity (about 0.2 [m/s]). In this way, adding torque to hop and not to exceed the escape velocity is required.

Both hopping duration and travel distance are greater for larger torque. Repeating small hops is desirable in oder to explore a wide range because the rover can't change the direction of movement in the air. Meanwhile, hopping for a long time by large torque is desirable in order to release heat from the rover.

5.3   Inertia Moment Variation

In this section, the behavior of the rover is investigated when inertia moments of rigid bodies A and B change. Adding torque is $T=10^{-3}$ [Nm]. The inertia moment of a rigid body A is assumed constant(J_A=2.5 [kgcm²]) while the ratio of inertia moment is changed in the range J_B/J_A=0.1'10.

Hopping velocity and hopping angle are shown in Fig. 8 for various J_B/J_A ratio. These graphs show that the hop angle is increasing but the speed of the rover is reduced when the ratio of inertia moment J_B/J_A is larger. Larger torque for rotating the rover is required with the increasing the inertia moment of B. So, inertial force by deploying a rigid body A is larger than the pressing force by the torque. In the result, v_z becomes smaller and the hop angle becomes larger relatively.

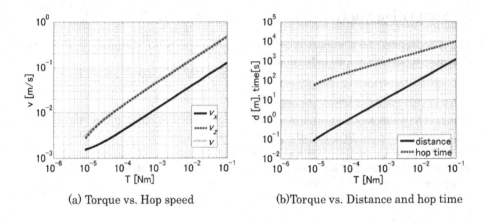

(a) Torque vs. Hop speed   (b)Torque vs. Distance and hop time

Fig. 7. Rover movement for torque variation

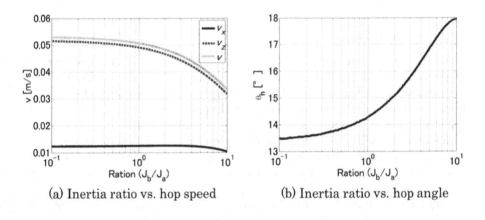

(a) Inertia ratio vs. hop speed   (b) Inertia ratio vs. hop angle

Fig. 8. Rover movement for inertia moment variation

6   Microgravity Experiment

The effectiveness of the proposed rover in terms of mobility is investigated by microgravity experiment. The microgravity experiment is performed in the drop tower in Bremen, Germany.

6.1   Prototype Model

The prototype model is shown in Fig. 9. The size of the body is 50 [mm](H) × 100 [mm](W) × 120 [mm](D) and the deployment part is 2 [mm](H) × 100 [mm](W) × 120 [mm](D). The total mass is 610 [g] and the deployment part is 90 [g]. The inertia moment of the body is 5.4 [kgcm^2] and that of the deployment part is 0.75 [kgcm^2]. The inertia moment is changed by attaching weights to the deployment part. The actuator

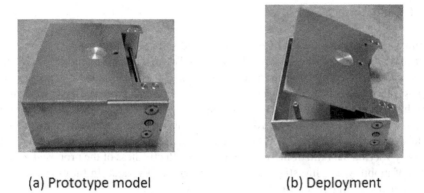

(a) Prototype model   (b) Deployment

Fig. 9. Prototype model of the deployment rover

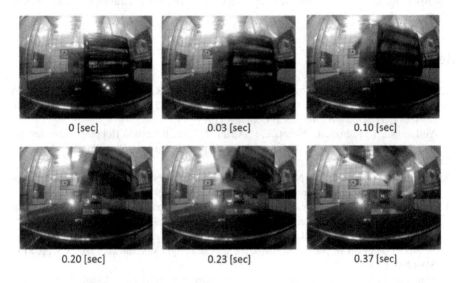

Fig. 10. Snapshots from the video of the microgravity experiment

for deployment is the brushed DC motor (2 [W] RE-13) and the gear head is GP13A (275:1). MBED microcomputer is used as CPU.

6.2 Experimental Results

Snapshots from the video of the microgravity experiment are shown in Fig. 10. These snapshots show that the rover is able to hop under microgravity environment. So, the proposed mechanism is effective in mobility on asteroid surface. In future, comparison of experiment and simulation will be done.

7   Conclusions

In this paper, a new type of rover was proposed to move under microgravity and to control its temperature by deploying the rover. A Thermal control method was also proposed and evaluated by basic thermal analysis model. Therefore, the dynamics of the deployment rover was modeled and the effectiveness of the proposed rover in terms of mobility was investigated by 2D simulation. Simulation results showed that the proposed rover could hop under microgravity environment and was effective for asteroid surface exploration. Additionally, the detailed kinetic analysis of the rover was investigated by varying the ratio of inertia moment. The effectiveness of the proposed rover in terms of mobility was investigated by microgravity experiment. In future, experiment and simulation will be compared.

Acknowledgment. This work was supported by KAKENHI, Grant-in-Aid for Scientific Research (B) 24360101.

References

1. Kubota, T., Wilcox, B., Saito, H., Kawaguchi, J., Jones, R., Fujiwara, A., Reverke, J.: A Collaborative Micro-Rover Exploration Plan on the Asteroid Nereus in MUSEC-C Mission. In: 48th International Astronautical Congress (1997)
2. Yoshimitsu, T., Kubota, T., Nakatani, I., Adahi, T., Sato, H.: Micro Hopping Robot for Asteroid Exploration. In: 4th IAA International Conference on Low-Cost Planetary Missions, IAA-L-1104 (2000)
3. Yoshimitsu, T., Kubota, T., Nakatani, I.: Nano- Rover MINERVA for Deep Space Exploration. Journal of Machine Intelligence and Robotics Control 3(3), 113–119 (2001)
4. Yoshimitsu, T., Kubota, T., Nakatani, I., Kawaguchi, J.: Robotic Lander MINERVA, Its Mobility and Surface Exploration. Advances in the Astronautical Sciences 108, 491–502 (2001)
5. Shimoda, S., Kubota, T., Nakatani, I.: Movement of the Microgravity Rover with a Spring. In: 2011 ISAS 11th Workshop on Astrodynamics and Flight Mechanism, ISAS, C9, pp. 314–319 (2001)
6. Shimoda, S., Kubota, T., Nakatani, I.: New Mobility System Based on Elastic Energy under Microgravity. In: Proc. of the 2002 Int. Conf. on Robotics and Automation, pp. 2296–2301 (2002)
7. Nakamura, Y., Shimoda, S., Shoji, S.: Mobility of a Microgravity Rover using Internal Electro-Magnetic Levitation. In: Proceedings of the 2000 IEEE/RSJ International Conference on Intelligent Robots and Systems (2000)
8. Burdick, J., Goodwine, B.: Quasi-Static Legged Locomotion as Nonholonomic System. In: Proc. of the 2000 IEEE/RSJ International Conference on Intelligent Robots and Systems, pp. 867–872 (2000)
9. Yoshida, K.: The Jumping Tortoise: A Robot Design for Locomotion on Micro Gravity Surface. In: Proc. of The 5th International Symposium on Artifical Intelligence, Robotics and Automation in Space, pp. 705–707 (1999)
10. Yoshimitsu, T., Kubota, T., Adachi, A.: MINERVA-II surface exploration system in Hayabusa-2 asteroid explorer. The Japan Society for Aeronautical and Space Sciences, 4509 (2013)

A Proposal of Exploration Robot with Wire for Vertical Holes in the Moon or Planet

Shuhei Shigeto[1], Masatsugu Otsuki[2], and Takashi Kubota[2]

[1] Department of Electrical Engineering and Information Systems,
Graduate School of Engineering, The University of Tokyo, Sagamihara-City, Kanagawa, Japan
shigeto@ac.jaxa.jp
[2] Japan Aerospace Exploration Agency
otsuki.masatsugu@jaxa.jp, kubota@isas.jaxa.jp

Abstract. Moon holes were first discovered by JAXA in 2009. It is believed that moon holes are useful for learning about the formation of the moon because the bedding plane is exposed. In addition, because the inner holes are sealed from solar wind, moon holes are also considered important candidate sites for base camp in the future. However, exploration of vertical hole is difficult with the conventional robots. A new type of robot is required to go down and explore a moon hole. In this study, a vertical hole exploration system with a small robot with wire is proposed. This paper describes a modeling and attitude control scheme in a state where the robot is hanging by a wire, and evaluates the effectiveness of the proposed system.

Keywords: Wire, Exploration Robot, Flywheel, Attitude Control.

1 Introduction

In 2009, Japanese Moon explorer "Kaguya" (SELENE) found a vertical hole on Marius Hill in Oceanus Procellarum(Figure 1). Kaguya also found 2 other holes. These holes are called Marius Hills Hole(MHH), Mare Tranquillitatis Hole(MTH), and Mare Ingenii Hole(MIH) respectively.

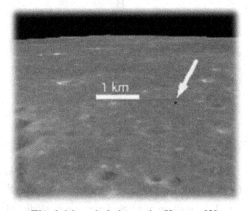

Fig. 1. Moon hole image by Kaguya [2]

J.-H. Kim et al. (eds.), *Robot Intelligence Technology and Applications 2*,
Advances in Intelligent Systems and Computing 274,
DOI: 10.1007/978-3-319-05582-4_30, © Springer International Publishing Switzerland 2014

Scientists made a conjecture that these holes spread out underground, and were originated as either lava tubes where once lava flowed, or magma chambers. It is believed that the meteorite strike on the ceiling of underground space made vertical holes [1] [2].

Vertical holes are also found on Mars. Life exploration in underground space on Mars is one of the most important goals of Mars exploration. Because there exist volcano on Mars and hence there is enough energy for life to evolve underground of Mars, and ultraviolet rays do not reach underground. A possibility of having suitable conditions for animate life to survive is also pointed out [3]. And moreover, the vertical hole and underground space of the moon or Mars have stable temperature, and is protected from a meteorite, radiation, and ultraviolet rays. Therefore these holes are suitable for future bases for manned activities.

Despite these advantages, structures and scientific characters of underground space and vertical holes are unknown. To obtain new knowledge, underground exploration is one of the most important missions.

In the past Lunar or Planetary exploration missions, exploration robots are commonly used. Conventional rovers have high mobility to run rough terrain. For example, MSL, a Mars exploration rover of NASA, is designed to explore near Gale crater which is expected to have many rocks and slopes [4]. In contrast to conventional explore missions which are done on surface of the moon or planets, vertical hole exploration requires the robot to descend vertically into the hole. Therefore a new exploration robot is required.

This paper proposes a vertical hole exploration robot which descends with wire. Proposed robot has ability to control its attitude when descending. Wire descending and attitude control make this robot possible to explore a huge vertical hole on the moon.

2   Moon Hole

2.1   Moon Hole Size

The vertical hole was picturized by the U.S. moon probe LRO (Lunar Reconnaissance Orbiter) based on the position information acquired by Kaguya. Figure 2 (a) shows a

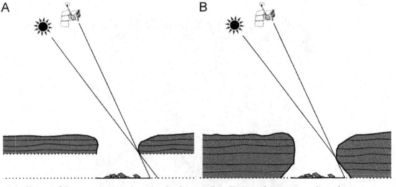

(a) Case of lava tube under the hole   (b) Case of magma chamber

Fig. 2. Estimation of a hole formation by skylight [5]

case of lava tube under the hole, and (b) shows a case of magma chamber, in which there is no horizontal tube but a dome under the hole. The sizes of the Moon holes discovered by Kaguya are estimated as following: in MHH case, the diameter is 59m, the depth is 48m; the MTH case, the diameter is 98m, the depth is 107m; and the MIH case, the diameter is 118m, and the depth is 45m [2] [5].

2.2   Requirements for Exploration Robot

The formation of Moon holes is not clearly known. There is a possibility of collapse at the hole entrance. At the bottom of the hole, there might be a lot of large rocks that might be the collapsed ceiling of the hole.

In a such hole, robots are required to take the following data:

- Detailed images of a wall
- Structure
- Temperature
- Amount of radiation
- Chemical composition

Therefore the requirements for exploration robot are:

- No big load to the hole ceiling
- Take pictures while descending
- Perform attitude control while descending
- Carry many sensors
- High mobility in the hole

To take images of a vertical hole, image sensor is needed. The image sensor can be used not only for taking a picture, but also for estimation of the 3D structure of the hole and cave [6] [7].   Taking a lot of pictures from the different view point, the

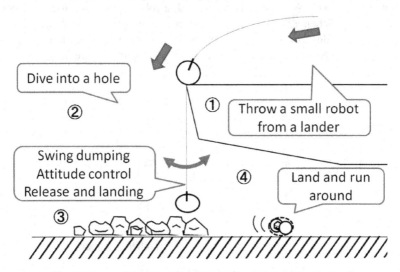

Fig. 3. Image of the proposed method for explore with wired robot

estimation accuracy gets better. For this purpose, an exploration robot should move and take pictures around the cave, while descending into a vertical hole with wire. Based on these requirements, a new exploration method is proposed in the next section.

3   New Exploration Scheme

3.1   Summary of Exploration Concept

The proposal exploration concept is shown in Figure 3. A small exploration robot moves to a hole and descends into the hole. Moving from lander to the hole, throwing manipulation system is useful [8] [9]. This system consists of throwing equipment and end-effector. This system does not need complex system like conventional rover.

After reaching to the hole, a robot descends with attitude control. In this descending phase, a robot is hanged from a ceiling with wire. In this phase, an exploration robot needs attitude control for two purposes. One is for taking pictures to estimate the structure of hole. The other is to land avoiding rocks at the bottom of the vertical hole. Exploration robot needs an actuator to control its attitude. In the proposed concept, a robot has two wheels to run in the cave, and this wheel can also be used for attitude control as a flywheel. When this robot is descending, robot moves like a swing to take pictures, and after that, moves towards the place that aimed to land. Next section, attitude control and swing control is discussed.

3.2   Proposal of Exploration Robot

The proposed exploration robot is shown in Figure 4. The proposed robot will strike to wall or ground when it lands. For protection from the collision force, the shape of the robot is like a ball. When it runs on ground, a robot opens, and uses two hemispheres as wheels. When a robot is descending, it uses these two wheels as a flywheel to control its attitude. In this mode, in order to generate yaw direction torque, it is necessary to give a camber angle as shown in Figure 4(b) and Figure 5.

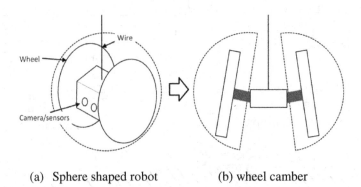

(a) Sphere shaped robot   (b) wheel camber

Fig. 4. Proposed exploration robot

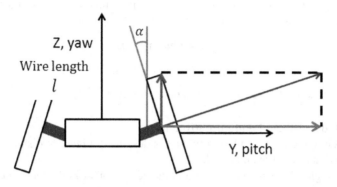

Fig. 5. Motor torque model

4   Modeling and Control

4.1   The Motion Requirement

In the exploration plan, first, the proposed exploration robot stops and observes around the hole; then searches landing site where it can safely land. At last, the robot shakes like a swing, releases a wire and land on the bottom. In this sequence the following motions are needed to the robot:

- Oscillate itself like a swing
- Vibration damping control
- Yaw direction control

4.2   Modeling

To confirm the fundamental performance, a small robot which moves indoors is designed. The experimental room is 2m high and the robot is hanged from the ceiling with wire. The robot mass is 2[kg], and motor can generate a maximum torque of 618[mNm]. In Figure 5, motor torque output model is shown. Wheel has a camber

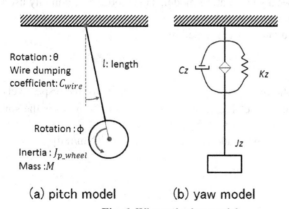

(a) pitch model   (b) yaw model

Fig. 6. Wire and robot model

angle α to generate yaw direction torque and pitch direction torque. Each direction torque is described as the following equations:

$$\tau_{yaw} = sin\alpha(\tau_r - \tau_l) \tag{1}$$

$$\tau_{pitch} = cos\alpha(\tau_r + \tau_l), \tag{2}$$

where $,r$: stand for left and right. τ is motor torque.

4.2.1   Model of Pitch Pendulum Motion

Next, modeling the pitch pendulum motion is described. Figure 6(a) shows the model of pitch motion of a robot and its wheel.

The lagrangian L of the motion around the ceiling pivot is shown in the following equation:

$$L = \frac{1}{2}M(l\dot{\theta}^2) + \frac{1}{2}J_{p\_wheel}(\dot{\varphi} + \dot{\theta})^2 - Mgl(1 - cos\theta) \tag{3}$$

Euler-Lagrange equation is described as:

$$\frac{d}{dt}\frac{\partial L}{\partial \dot{\theta}} - \frac{\partial L}{\partial \theta} = -C_{wire}\dot{\theta} \tag{4}$$

From the equations (3) and (4), we get motion equation

$$Ml^2\ddot{\theta} + Mlgsin\theta + J_{pwheel}(\ddot{\varphi} + \ddot{\theta}) = -C_{wire}\dot{\theta} \tag{5}$$

where M:Mass of a robot, l:length of wire, θ: rotation angle. C_{wire}: damping coefficient of wire

In Eq.(5), the third form means the motor torque. The motion equation is rewriten as

$$Ml^2\ddot{\theta} + Mlgsin\theta + C_{wire}\dot{\theta} + \tau_{pitch} = 0 \tag{6}$$

The Eq.(6) is a motion equation of pendulum. θ is operated by τ_{pitch}.

4.2.2   Model of Yaw Rotation

Figure 6 (b) shows a yaw rotation model. This model is commonly used in the balloon and gondola relative model [8].

Yaw motion equation is expressed by

$$J_z\ddot{\theta}_z + C_z\dot{\theta}_z + K_z\theta_z = \tau_{yaw} \tag{7}$$

where θ_z is a rotation of wire, and τ_{yaw} is a motor torque respectively.

C_z is the damping coefficient of wire in yaw direction. K_z is the spring coefficient of wire. C_z and K_z is experimentally estimated as

$C_z = 4.3 \times 10^{-5}$ [Nm/rad/sec], $K_z = 1.2 \times 10^{-2}$[Nm/rad]

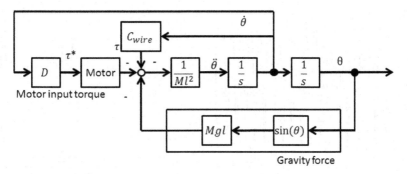

Fig. 7. Block diagram of pitch pendulum motion

4.3   Control Method

4.3.1 Swing Motion Control

Tsuji et al. [11] [12] studied vibration control of crane and its load. Where, the motor input τ_{pitch} is provided as $D\dot{\phi}$, as the result, oscillation is damped. D is damping coefficient.

In this study, the oscillation damping is confirmed by simulation with damping input. Figure 7 shows the simulation block diagram. In this simulation, $D = 0.5$[Nm/rad/sec].

When generating oscillation of swing, the input motor torque is cyclic. In this simulation, τ_{pitch} is provided as below:

$$\tau_{\text{pitch}} = \begin{cases} -a_0 \ (\dot{\phi} \geq 0) \\ a_0 \ (\dot{\phi} < 0) \ , \end{cases} \tag{8}$$

where a_0 is const.

4.3.2   Yaw Motion Control

A yaw rotation damping control is described here. A damping control uses the same method as pitch damping control. In this simulation, $Dz = 0.001$[Nm/rad/sec].

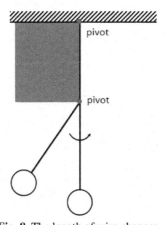

Fig. 8. The length of wire changes

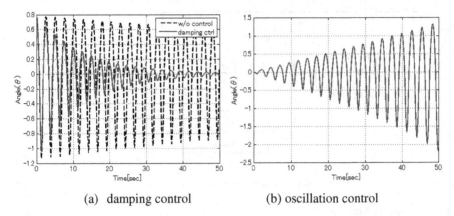

| (a) damping control | (b) oscillation control |

Fig. 9. Pitch control simulation result

$$\tau_{\text{pitch}} = Dz\,\dot{\phi} \tag{9}$$

5   Simulation Study

5.1   Pitch Swing Motion

In the real hole, the hanging wire will be caught by wall like Figure 8. In this simulation, the length of wire changes while swinging. The condition of wire length is shown in (10).

$$l = \begin{cases} 1 & (\theta < 0) \\ 2 & (\theta \geq 0) \end{cases} \text{(unit: m)} \tag{10}$$

Figure 9 (a) shows the simulation result of pitch control. The dashed line shows the no controlled case, and the solid line shows the controlled case respectively. The initial value here has given the angle after the robot being thrown into the hole. With control, oscillation is suppressed quickly.

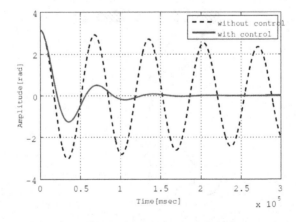

Fig. 10. Simulation result of yaw rotation damping

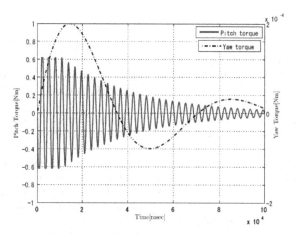

Fig. 11. Simulation result of motor torque

Figure 9 (b) shows the simulation result of generating pitch oscillation control. The robot starts to oscillate.

5.2   Yaw Motion Control

Figure 10 shows the simulation result of yaw motion damping control. In this simulation, the yaw rotation is suppressed with motor torque. The dashed line shows the case without control, and the solid line shows the controlled case respectively. It can be seen that the yaw motion is quickly suppressed by damping control.

5.3   Motor Output Torque

Figure 11 shows the simulation result of motor output torque. In this simulation, the wire length is constant. The right scale shows the yaw motor torque, and the left scale shows the pitch motor torque. The scale of pitch motor torque is much larger than yaw torque. This means that the motor torque is large enough that the pitch and yaw motion can be controlled simultaneously.

6   Conclusions

A vertical hole is a candidate for future exploration missions. In this study, the moon hole exploration system and the exploration robot with wire are proposed. This study focused on exploration robot motion control. The robot uses its wheels as a flywheel to control its attitude. When the robot is suspended in the vertical hole, the robot moves like a swing. The fundamental damping and oscillation control is proposed. Its motion is confirmed by simulation study. These motions will be confirmed through experiments as a future work.

References

[1] Haruyama, J., et al.: Possible lunar lava tube skylight observed by SELENE cameras. Geophys. Res. Left. 36 (2009)

[2] Haruyama, J., Morota, T., Kobayashi, S., Sawai, S., Lucey, P.G., Shirao, M., Masaki: Lunar Holes and Lava Tubes as Resources for Lunar Science and Exploration. In: Moon, ch. 6, pp. 139–163. Springer, Heidelberg (2012)

[3] Haruyama, J., Kubota, T., Kawano, I., Otsuki, M., Imaeda, R.: Exploration of lunar and planetary holes and subsurface caverns: The significance of explorations. In: 57th Symposium on Space Science and Technology (2013)

[4] Steltzner, A., Kipp, D., Chen, A., Burkhart, D., Guernsey, C., Mendeck, G., Mitcheltree, R.: Mars Science Laboratory Entry, Descent. In: Proc. IEEE Artosp. Conf. (2007)

[5] Robinson, M.S., Ashley, J.W., Boyd, A.K., Wagner, R.V., Speyerer, E.J., RayHawke, B., Hiesinger, H., van der Bogert, C.H.: Confirmation of sub lunar eanvoids and thin layering in mare de posits. Planetary and Space Science 69 (2012)

[6] Hartley, R., Zisserman, A.: Multiple View Geometry in Computer Vision. Cambridge University Press (2004)

[7] Snavely, N., Seitz, S.M., Szeliski, R.: Photo Tourism: Exploring image collections in 3D. ACM Transactions on Graphics (Proceedings of SIGGRAPH 2006) (2006)

[8] Arisumi, H., Matsumoto, O.: Casting manipulator system for a lunar exploration-Mid-air control of the end-effecter by tension of wire-. In: JSME ROBOMEC (2012)

[9] Arisumi, H., Otsuki, M., Nishida, S.: Exploration of lunar and planetary holes and subsurface caverns: A long throw and soft landing with a casting probe system. In: 57th Symposium on Space Science and Technology (2013)

[10] Nishimura, J., Yajima, N., Kokaji, S., Hashino, S.: A control system for a baloon-borne telescope. Advances in Space Research (1981)

[11] Tsuji, T., Ohnishi, K.: Oscillation Control of Suspended Load with Flywheels. Transactions of IEEJ-D 125 (2005)

[12] Nishimura, K., Tsuji, T., Ohnishi, K.: Sensorless Oscillation Control of a Suspended Load with Flywheels. Transactions of IEEJ-D 126 (2006)

Scanpaths Analysis with Fixation Maps to Provide Factors for Natural Gaze Control

Bum-Soo Yoo and Jong-Hwan Kim

Department of Electrical Engineering, KAIST, 335 Gwahangno,
Yuseong-gu, Daejeon, Republic of Korea
{bsyoo,johkim}@rit.kaist.ac.kr

Abstract. Human-like gaze control is needed for robots to be companions of humans. For human-like gaze control, much research has been progressing to identify factors that affect human gaze. The conventional approaches to discover factors that affect human gaze is based on their hypotheses. They presented gaze control algorithm based on hypotheses and verified through experiments. However, since the algorithms were originated from the hypotheses, they were prone to be biased to the hypotheses. This paper derives the factors that affect human gaze based on observation of real human's scanpaths, not hypothesis. From the recorded scanpaths, fixation maps are produced using the Gaussian distribution. The earth mover's distance (EMD) is used to measure similarity among fixation maps, and the fixation map with minimal difference is selected in each test image. The selected fixation maps are used to derive the factors that affect human gaze. The derived factors are center, salient regions, human, depth, objects, scene schema information, and they are shown with examples.

Keywords: Gaze, attention, eye tracking, scanpaths, fixation map, earth mover's distance.

1 Introduction

Robots will be partners in near future. As companions of humans, robots should make natural human robot interaction (HRI). Since gaze represents paying attention, HRI should accompany with human-like gaze control.

For making natural gaze control, factors that affect human gaze are needed to be discovered. From the visual input, salient regions that are different with surrounding in color, intensity, and orientation were detected as candidates for paying attention. [1]. Effect of faces on the gaze was tested and showed that people gazed faces independent with goals such as a finding objects [2]. Auditory information was added to salient regions system to enhance the gaze control [3]. These studies tried to make human-like gaze control based on their assumed factors and showed effectiveness through experiments. Even though their studies showed good results, they are prone to be biased to their assumed factors.

In this paper, factors that affect human gaze are identified from observation of human's scanpaths. From measured scanpaths, fixation maps are generated with

J.-H. Kim et al. (eds.), *Robot Intelligence Technology and Applications 2*,
Advances in Intelligent Systems and Computing 274,
DOI: 10.1007/978-3-319-05582-4_31, © Springer International Publishing Switzerland 2014

the Gaussian distribution. The produced fixation maps are compared with each other by the EMD to find commonly gazed directions in each test image. Among various EMD measurements, the $\widehat{EMD}$ with *threshold* is used. The fixation maps are generated from 79 images with 53 subjects, and one fixation map is selected from each image. The selected fixation maps are analyzed to find factors that affect human gaze. Results are presented with example images.

The paper is organized as follows. Section 2 describes fixation maps and image comparison methods including the $\widehat{EMD}$. Section 3 describes the experiments with analysis and concluding remarks on Section 4.

2   Scanpath Comparison

2.1   Fixation Map

To find where people gaze, it is necessary to measure similarity among human scanpaths. There are two categories for measuring similarity among scanpaths. The first method is based on string comparison methods. Images are quantized with cells and different letters are assigned to each cell. When a scanpath is produced, a sequence of cells is generated according to the scanpath, and it can be replaced by a string of the assigned letters. As a result, the scanpath is transformed to the string, and then it can be compared by various string comparison methods.

The other method is based on map comparison methods. On a two-dimensional image, an index describing how much human gaze the location is added as the z-axis to make the three-dimensional image. The generated three-dimensional images can be compared by map comparison methods. Transforming into a map has a strong point of visualization. However, it has a shortcoming of loss of sequential information.

In this paper, scanpaths are compared by producing fixation maps. The fixation map is a three-dimensional record of locations of all fixations being analyzed with the third dimension representing the quantity of property d obtained from that fixations [4]. For making a fixation map, the z-axis is added to two-dimensional images, and all z values are initialized to zero. When scanpaths are generated, the Gaussian distribution is assigned to each fixation points in the scanpaths. This process is repeated to all fixations, and it will make the Gaussian mixture model. Finally, the z values are normalized. Note that sequential information is discarded since orders of human scanpaths is hard to predict in images [5], [6]. Fig. 1 shows generated fixation maps. The fixation maps in the third rows are generated from the images in the first row. The second row shows images which the z values are changed to luminance for the visualization.

Fig. 1. Fixation maps

2.2   Comparison Method

The EMD is used to measure similarity among fixation maps. The EMD can be described as the minimal cost that must paid to transform from one histogram to other histogram. Given two histograms P and Q, the EMD is defined as [7]:

$$EMD(P,Q) = (\min_{f_{ij}} \sum_{i,j} f_{ij}d_{ij})/(\sum_{i,j} f_{ij}) \ s.t \ f_{ij} \geq 0 \qquad (1)$$

$$\sum_{j} f_{ij} \leq P_i, \ \sum_{i} f_{ij} \leq Q_j, \ \sum_{i,j} f_{ij} = min(\sum_{i} P_i \sum_{j} Q_j)$$

where f_{ij} is the amount of transported from the i-th bin to the j-th bin and d_{ij} is the *ground distance* between the i-th bin to the j-th bin. As other version of the EMD, the $\widehat{EMD}$ is used in this paper. The $\widehat{EMD}$ is defined as [8], [9]:

$$\widehat{EMD}_\alpha(P,Q) = (\min_{f_{ij}} \sum_{i,j} f_{ij}d_{ij}) + |\sum_{i} P_i - \sum_{j} Q_i|\alpha \max_{i,j} d_{ij} \qquad (2)$$

where the $\widehat{EMD}$ can be used as a metric when $\alpha \geq 0.5$. Compared to the EMD, the $\widehat{EMD}$ shows faster and better performance. For better performance, d_{ij} in both (1) and (2) can be modified as:

$$d_t(a,b) = \min(d_{a,b}, t). \qquad (3)$$

By applying (3), connections from P to Q are simplified, and the algorithms become faster [9].

Fig. 2. Examples of selected images

3   Experiments

3.1   Eye Tracking Experiments

The scanpath measurements were performed on 79 images with 53 people. Totally, $4,187$ fixation maps were produced. Eyes fixations were recorded with the Tobii T120 eye tracker. It is an integration of an eye tracker with a 17 inch monitor. The accuracy is about 0.5 °, drift is about 0.1 ° and spatial resolution is about 0.3 °. It can display images up to the size of 1280×1024. In the experiments, the images were tuned to the size of 640×480 and used. The subjects were 37 male and 16 female, from 17 to 30. They were instructed to look at the monitor without any comments. The images were shown five seconds with five seconds breaks sequentially. Due to fatigue of eyes and sampling quality, a minute break was added in every 10 picture. At the beginning and after the one minute breaks, five points calibration was performed.

Images were randomly selected from the MIT Judd's 2009 and 2012 data sets [10], [11]. The image sets, created as saliency benchmark data sets, contains more than thousands of free-viewed indoors and outdoors images. The selected images were adjusted to contain similar number of the images of buildings, animals, landscapes, sports, night views, person, people and close-shots. Fig. 2 shows several examples of selected images. The size of images was tuned to 640×480.

Comparisons among fixation maps were performed with the $\widehat{EMD}$ with *threshold*. Table 1 shows the comparison results with the minimum $\widehat{EMD}$ and the standard deviation. Since the $\widehat{EMD}$ represents the cost of the transformation, smaller number means two fixation maps are more similar. The outputs of the cost function in the $\widehat{EMD}$ were normalized to have average one in each image. After the normalization, the average difference with others in each image was calculated and compared.

Table 1. Fixations map comparison using $\widehat{EMD}$ with *threshold*

Image number	Min (Std. Dev.)	Image number	Min (Std. Dev.)	Image number	Min (Std. Dev.)
1	0.7515(0.2036)	28	0.7759(0.1696)	55	0.7658(0.1987)
2	0.7958(0.1530)	29	0.7394(0.2518)	56	0.7465(0.2237)
3	0.7843(0.1568)	30	0.7548(0.2094)	57	0.7652(0.2529)
4	0.7989(0.1846)	31	0.7839(0.1676)	58	0.7466(0.2355)
5	0.8166(0.1441)	32	0.7441(0.2590)	59	0.7466(0.2638)
6	0.7792(0.1690)	33	0.7693(0.1866)	60	0.7501(0.2557)
7	0.7594(0.2312)	34	0.7638(0.1923)	61	0.7504(0.2820)
8	0.7452(0.2745)	35	0.7864(0.1519)	62	0.7330(0.2918)
9	0.7791(0.2261)	36	0.8276(0.1312)	63	0.7689(0.1929)
10	0.7426(0.2972)	37	0.7570(0.1825)	64	0.7762(0.1861)
11	0.8033(0.1608)	38	0.7828(0.1785)	65	0.7783(0.1855)
12	0.7895(0.1733)	39	0.7677(0.2115)	66	0.7560(0.2120)
13	0.6991(0.3671)	40	0.7879(0.1749)	67	0.7666(0.2490)
14	0.7287(0.3194)	41	0.8004(0.1451)	68	0.7439(0.2682)
15	0.8075(0.1415)	42	0.8329(0.1522)	69	0.7512(0.2353)
16	0.7914(0.1477)	43	0.7720(0.2021)	70	0.7806(0.2017)
17	0.7541(0.2239)	44	0.7865(0.2060)	71	0.7517(0.2585)
18	0.8137(0.1702)	45	0.7604(0.1733)	72	0.7669(0.2383)
19	0.7731(0.1638)	46	0.7581(0.2143)	73	0.7674(0.2167)
20	0.7825(0.1650)	47	0.7345(0.3068)	74	0.7621(0.2095)
21	0.7559(0.2083)	48	0.7910(0.2053)	75	0.7523(0.2524)
22	0.8083(0.1755)	49	0.7832(0.1741)	76	0.7464(0.2797)
23	0.7513(0.2067)	50	0.7626(0.2031)	77	0.7549(0.3150)
24	0.7816(0.1824)	51	0.7715(0.2150)	78	0.7243(0.2794)
25	0.7531(0.3063)	52	0.7524(0.2370)	79	0.7888(0.1590)
26	0.8243(0.1586)	53	0.7587(0.2301)		
27	0.7865(0.1935)	54	0.7430(0.3414)		

3.2   Analysis

From each image, a fixation map with the largest similarity was selected. The selected fixation maps were used to derive the factors that affect human gaze.

Center. Fixation maps were biased to the center of images. There were two reasons why the subjects gazed the center of images. First, the subjects started their gazes from the center of the images. Because starting at the centers can distort the result, in some papers, the first fixation was discarded to reduce the effect of the initial center fixation [11]. In Fig. 3(b), the subjects gazed the center of images when visual inputs were similar in all directions. Second, photographers usually placed a target of interest in the center of images [12], [13]. Since the photographers placed targets of interest in the center in Figs. 3(a) and (d), the subjects gazed the center of images.

Salient Regions. The subjects gazed colorful, bright, and outstanding regions. These regions can be described as salient regions. The saliency at a location is determined by difference between the location and surrounding including color, orientation, motion, intensity, etc. and loci of high saliency are good candidate for gaze direction [14]. In Fig. 3(c), the subjects gazed white regions in the

Fig. 3. Fixation maps with the largest similarity

mountain because the regions were salient from the difference with surrounding in colors.

Human. The subjects gazed humans. In human's body, a face was the strongest attraction. Even humans hid their faces, other parts of body such as toes, hands, legs, fingers were also strong attractions. In addition, objects that form of human shape, picture of human, or silhouette and shadow implying human were also strong attractions. The subjects gazed two mans in the center in Fig. 3(f) and gazed the legs in top right side in Fig. 3(d) even they are not salient.

Depth. When the subjects gazed images, they did not accept the images as two-dimensional spaces. From the images, they constructed three-dimensional space in their minds even there was no depth information. After constructing three-dimensional space, they gazed what was thought to be closure. In Fig. 3(e), the subjects gazed the white wing in the left top side, and in Fig. 3(g), they gazed the bowl in the right bottom side.

Objects. Since almost all human behaviors are highly related with objects, objects attract human gazes. Objects could be changed according to images such as cups, bottles, hat etc. Subjects gazed objects in Figs. 3(a), (d), and (g).

Scene Schema Information. People can understand environment and know where attractions are usually placed. For example, objects are placed mainly on the ground or on the table, not in the sky. In Figs. 3(h) and (i), the subjects recognized the table and knew that objects were usually placed on the table. Thus, they gazed upper side of the table where attractions were probably placed rather than the salient bright windows.

Expectation. Expectation from images affected human gazes. When there was an object which was not adequate to environment, the subjects gazed the object [15]. For example, a machine tool in a kitchen attracts gazes with curiosity of why the tool is in the kitchen. Forecasted situation also affected human gazes. When the subjects gazed a man walking along the bridge in Fig. 3(j), they gazed the gate in front of the bridge.

4   Conclusion

This paper showed the factors that affect human gaze by analyzing the scanpaths. From measured scanpaths of 53 subjects with 79 images, the Gaussian distribution was used to make fixation maps. The generated fixation maps were compared by the $\widehat{EMD}$ with *threshold*, and the fixation map with the minimal difference was selected from each test image. Analysis of the selected fixation maps showed the center of images, salient regions, human, depth, objects, scene schema information and expectation affected human gazes. If these factors are considered in the design of the gaze control, robot's gaze would be natural and similar to human gaze.

Acknowledgement. This research was supported by the MOTIE (The Ministry of Trade, Industry and Energy), Korea, under the Technology Innovation Program supervised by the KEIT (Korea Evaluation Institute of Industrial Technology)(10045252, Development of robot task intelligence technology that can perform task more than 80% in inexperience situation through autonomous knowledge acquisition and adaptational knowledge application).

This research was supported by the MOTIE (The Ministry of Trade, Industry and Energy), Korea, under the Human Resources Development Program for Convergence Robot Specialists support program supervised by the NIPA (National IT Industry Promotion Agency)(H1502-13-1001, Research Center for Robot Intelligence Technology).

References

1. Itti, L., Koch, C.: A saliency-based search mechanism for overt and covert shifts of visual attention. Vision Res. 40(10), 1489–1506 (2000), doi:10.1016/S0042-6989(99)00163-7
2. Cerf, M., Harel, J., Einhäuser, W., Koch, C.: Predicting human gaze using low-level saliency combined with face detection. Adv. Neur. In. 20, 241–248 (2008)
3. Ruesch, J., Lopes, M., Bernardino, A., et al.: Multimodal saliency-based bottom-up attention a framework for humanoid robot iCub. Papers presented at the IEEE Conf Robotics and Automation, Pasadena, CA, USA, pp. 962–967 (May 2008)
4. Wooding, D.S.: Eye movement of large populations: II. Deriving regions of interest, coverage, and similarity using fixation maps. Behav. Res. Meth. Instr. 34(4), 518–528 (2002), doi:10.3758/BF03195481
5. Privitera, C.M., Stark, L.W.: Algorithms for defining visual regions-of-interest: Comparison with eye fixations. IEEE Trans. Pattern Anal. Mach. Intell. 22(9), 970–982 (2000), doi:10.1109/34.877520
6. Yarbus, A.L.: Eye movement and Vision. Academy of Sciences of the USSR, Moscow. English edition: Haigh B. Eye movement and Vision (trans: Haigh B). Plenum Press, New York (1967)
7. Rubner, Y., Tomasi, C., Guibas, L.J.: The earth mover's distance as a metric for image retrieval. Int. J. Comput. Vision 40(2), 99–121 (2000), doi:10.1023/A:1026543900054
8. Pele, O., Werman, M.: A linear time histogram metric for improved SIFT matching. In: Forsyth, D., Torr, P., Zisserman, A. (eds.) ECCV 2008, Part III. LNCS, vol. 5304, pp. 495–508. Springer, Heidelberg (2008)
9. Pele, O., Werman, M.: Fast and robust earth mover's distances. Papers presented at the Int Conf on Computer Vision, Kyoto, Japan, pp. 460–467 (October 2009)
10. Judd, T., Durand, F., Torralba, A.: A benchmark of computational models of saliency to predict human fixations. MIT Tech. Report (2012)
11. Judd, T., Ehinger, K., Durand, F., Torralba, A.: Learning to predict where humans look. Paper presented at the Int Conf on Computer Vision, Kyoto, Japan, pp. 2106–2113 (October 2009)
12. Zhang, L., Tong, M.H., Marks, T.K., et al.: SUN: A Bayesian framework for saliency using natural statistics. J. Vision 8(7), 1–20 (2008), doi:10.1167/8.7.32
13. Talter, B.W.: The central fixation bias in scene viewing: Selecting an optimal viewing position independently of motor biases and image feature distribution. J. Vision 7(14), 1–17 (2007), doi:10.1167/7.14
14. Koch, C., Ullman, S.: Shifts in selective visual attention: towards the underlying neural circuitry. Hum. Neurobiol. 4, 219–227 (1985), doi:10.1007/978-94-009-3833-5_5
15. Summerfield, C., Egner, T.: Expectation (and attention) in visual cognition. Trends Cogn. Sci. 13(9), 403–409 (2009), doi:10.1016/j.tics.2009.06.003

Design of a Bipedal Walker with a Passive Knee and Parted Foot

Honggu Lee and Sungho Jo

Dept. of Computer Science, KAIST
291 Daehak-ro, Yuseong-gu, Daejeon 305-701, Korea
{nickplay,diskedit,shjo}@kaist.ac.kr

Abstract. The design of a bipedal walker that enables a human-like, compliant walking motions with simple control commands is presented. The design includes a passive knee bending/stretching mechanism with a latch hinge and a parted foot structure with compliant spring-based actuation. In addition, the leg posture, asymmetric lateral spring placement, round ankles, active hip sway, pelvic tilt actuation, and provisions for simple control were designed to implement the desired walking motion. The prototype bipedal walker was built with a combination of passive and actuated joints, utilizing springs around the joints for further compliancy. Experiments were conducted using the prototype bipedal walker in order to evaluate the design.

Keywords: Bipedal walker, compliant, passive knee, parted foot.

1 Introduction

The design of a bipedal robot which walks with human-like motion, has been a daunting yet exciting research topic. It is very challenging to realize natural, human-like robot walking motions due to how the knee bends and the contact made by the foot on the ground. A common approach to bipedal motion is to tightly control the joint angles so that they mimic human walking motions. Honda Asimo and KAIST Hubo are two famous prototype robots that have been created in order to conduct research on dynamic locomotion [1, 2]. However, these designs have used knee bending while walking to maintain stability [1-4]. The Hubo laboratory recently manufactured a stretched-leg walking robot. However, this robot has a flat foot and ankle-level push off is not used often because maintaining stability is difficult [5]. Another robot, Humanoid H6, has toe joints [6] which increases its walking speed, and the joints enable climbing steps. Even though these impressive technological advances in robotics have been realized, further investigation is required in order to create more natural walking motions.

Another approach to natural-gait robot motions is based on passive-dynamic walkers. These walkers have minimal or no actuators and a simple control strategy [7-10]. It is known that this approach is advantageous for efficient energy consumption and gait appearance. The Cornell Powered Biped is based on the

J.-H. Kim et al. (eds.), *Robot Intelligence Technology and Applications 2*,
Advances in Intelligent Systems and Computing 274,
DOI: 10.1007/978-3-319-05582-4_32, © Springer International Publishing Switzerland 2014

passive-dynamic walker model, and it has a passive knee mechanism that realizes knee bending and stretching while avoiding hyperextension. In order to bend the knees, the passive-dynamic walker requires a controllable solenoid. Furthermore, the feet must be rounded in order to walk smoothly [9, 10]. However, the passive-dynamic walking strategy may hinder implementing precise or adaptive behaviors. Therefore, a robot design that incorporates the advantages of both approaches is a natural progression for the next generation of technologies for bipedal robot locomotion [11, 12].

This study investigates a design that may be effective in generating human-like walking motions in a robot using a small number of actuators. As a preliminary step, this work designs and manufactures a prototype bipedal walking robot based on insights attained from studying human gaits.

The prototype design adopts two visual aspects of the human gait. The first is a bending and stretching knee motion. McGeer noted the practical advantages of the knee motion for walking legs [7]; the knee bending prevents feet from colliding with the ground during the leg swing and the knee absorbs any impact while the foot is contacting the ground for stability. This proposed knee mechanism is implemented passively without actuators. The second aspect of human motion that is adopted is the motion by which the foot comes into contact with the ground. The human foot is flexible when engaging in toe-off and heel-strike motions [13]. Thus, the foot design was carefully planned in order to achieve compliancy when interacting with the ground. In order to mimic the toe-off and heel-strike motions, a two-part foot design was used.

2   Mechanical Design

The prototype is 46.7 cm tall and weighs 1.7 kg. It consists of two legs, two feet, and a small torso. Within each leg, the thigh and shank are 13.8 cm and 14.3 cm long, respectively. Each leg has 5 degrees of freedom (DOFs): two at the hip, one at the knee, and two at the ankle. In a three-dimensional space, each leg is actuated by three servomotors (Dynamixel RX-28, Robotis, Inc. [14]), which are labeled M1, M2, and M3 in Fig. 1. M1 and M2 realize the lateral and forward-backward swing of a leg from the hip, respectively. M3 enables the ankle flexion and extension. In each leg, three passive joints, indicated by P1, P2, and P3 in Fig. 1, are included. P1 is a lateral hip joint, P2 indicates the knee joint, and P3 is the lateral ankle joint.

2.1   Foot and Toe Design

Each foot consists of two parts: a forefoot(toes) and a rearfoot(heel) (Fig. 2(a)). The fore and rear sections are 40 and 70 mm long, respectively. The forefoot is connected to the rear-foot via joints while being supported by a pair of passive springs (labeled as S1 in Fig. 2; only one side shown). The split foot design aims to improve the robot maneuverability during the push-off or ground-touch motions. One end of S1 is attached to a roller located on the forefoot so that it can be pulled when the forefoot touches the ground and the rearfoot is off the ground. The pulling stores the elastic

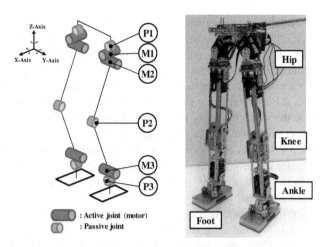

Fig. 1. (a) Kinematic details of (b) the proposed bipedal walker

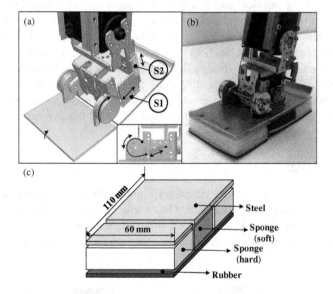

Fig. 2. (a) Parted foot design, (b) assembled foot, and (c) layered foot structure

energy in S1. During the ankle push-off motion, the forefoot pushes the ground by releasing the elastic energy. The push-off thrusts the upper robot body forward.

Passive-dynamics-based walking robots use similar strategies in order to minimize the motor power requirements for the push-off motion [7-10]. These robots usually have push-off springs at the ankle joint in order to achieve the ankle extension. However, in the proposed robot, the push-off springs are operated via the joint between the toe and the foot body.

This design was chosen for the following reasons. Generally, other bipedal robots have solid feet, so the ankle joint must rotate the feet. With solid feet, the ankle extension in the standing leg requires a large amount of potential energy within the spring in order to lift the upper body, but it produces a large forward thrust once the foot is off the ground, increasing the difficulty in maintaining the gait stability. In order to moderate this issue, robots usually have curved bottoms of their feet. However, these robots can still produce high forward thrust and thus require careful control in order to achieve stable leg swing motions during the swing phase and stable ground touches at the swing-to-stand transition. For example, the push-off process is mechanically constrained just after the heel strike in reference [7].

In the proposed robot, an appropriate amount of potential energy in the spring is required for the joint extension between the forefeet and rearfoot, which is designed to perform the push-off. Therefore, in this study, a flat but two-parted foot in the prototype enables compliant stable motions.

During the foot landing, the rearfoot touches the ground before the forefoot. The whole foot bottom is layered with the sponge and rubber materials that are used in table tennis racquets. These materials provide a damping effect similar to the skin and flesh of a foot. In particular, the toes and heels have hard sponges to provide sufficient pushing force during push-off and to achieve sufficient impact absorption on landing. In the proposed robot, while the spring pair S1 pushes against the ground, the active motors at the hip and in the standing leg user power to move the whole body forward. Therefore, the proposed robot has both the passive and active dynamic characteristics of walking. A pair of passive springs (labeled S2 in Figure 2) between the heel and shank is also influential during landing. The S2 pair assists in maintaining lateral balance. Furthermore, while the opposite leg kicks the ground, the S2 pair alleviates its impact on the ankle of the standing leg. Thus all of the passive springs around the foot and ankle yield compliant motions.

2.2   Ankle Design

The shank and foot are connected via an ankle joint with the M3 motor, as shown in Fig. 3. The motor body is rigidly attached to the end of the shank, and its shaft rotates the foot in a sagittal plane. From the heel-strike to the toe-off, the motor assists the upper body to move around the ankle joint of the standing leg in the same manner as

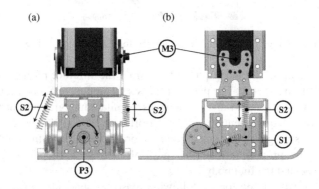

Fig. 3. The (a) front and (b) side view of the compliant ankle mechanism

an inverted pendulum would. The ankle rotation using M3 critically affects the push-off motion because the push-off thrust is most effective when the foot rotation and body orientation are properly synchronized. Furthermore, M3 can control the relative posture of the foot to the body orientation. Therefore, the motor motion is important for gait efficiency and stability.

As mentioned in the previous section, the spring pair S2, located on each side of the ankle, generates the compliant rotational motion in the coronal plane around the passive joint P3. The S2 pair support the shank laterally and damp the contact impact in the coronal plane. The structure was designed so the outer S2 is tensed relative to the inner S2 when the foot is on the ground, as shown in Fig. 3(a). This design supports the standing leg while the other leg swings, and it maintains the robot's posture at the heel-strike.

The space between the legs is slightly wider at the foot than at the hip. This posture helps generate the lateral directional motion and aids lateral balancing. In addition, the foot is designed to tilt slightly inward through the use of a stiffer spring inside than that used outside. During push-off, the leg posture and asymmetric spring strength help pull the standing leg to protect the lateral stability from outward sway. Each spring's stiffness is tuned empirically. The ankle motion in the sagittal plane also influences the knee bending and extending mechanism, which is discussed in the next section.

2.3   Knee Design

The knee mechanism is designed to prevent hyperextension and to bend or extend the lower leg at the passive joint P2. The knee joint is locked after the mid-swing until the end of the standing phase and is unlocked during the remainder phase. As shown in Figures 4(a) and 4(b), the locking mechanism is essentially a type of latch. The full cycle of the knee bending and extending procedure is illustrated in Figure 4(c). A latch arm is attached to the shank using a hinge with a torsion spring and the latch body is rigidly attached to the thigh (see Figures 4(d) and 4(e)). A torsion spring pushes the latch arm to maintain a lock during extension.

At the ankle push-off, the ankle angle (M3) reaches a threshold value (10°, as shown in Figure 4(c)); then, the latch arm is pulled by a string connected via three pulleys to the foot. The string pull is sufficiently strong to overcome the torsion spring. Therefore the latch arm is detached from the latch body as seen in Figure 4(d). The string is minimally strained even without the ankle's substantial pulling force and does not disconnect from the pulleys. The pulleys translate the linear pulling force generated from the foot rotation (at the ankle push-off) to the latch arm. This detachment allows the knee to start bending. After unlocking, the knee flexes using the motion inertia. The tip of the latch body is connected to the top of the shank by a rubber band. While the knee bends, the rubber band gains elastic energy, which is used to change the inertia of the shank after sufficient bending. Next, the motion's inertia drives the shank back to stretch the leg during the forward swing phase.

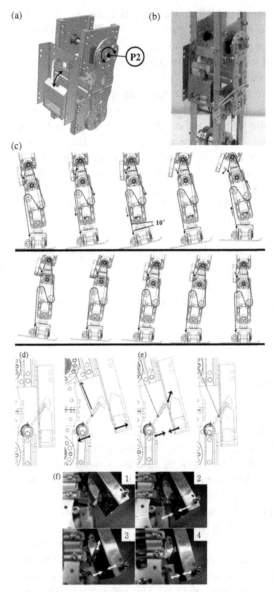

Fig. 4. (a) Passive knee design and (b) its implementation. (c) Knee bending and extending mechanism over a gait cycle (d) knee release, and (e) & (f) knee locking.

While extending the shank via the knee joint, the ankle's angle remains less than the threshold value, and the string does not pull on the latch arm. The torsion spring clips the hook on top of the latch arm into the slit of the latch surface to engage the locking, as shown in Figure 4(e). The leg is then extended. Both the hook and latch surfaces have curvature. The hook moves along the latch surface with a line contact to minimize friction. Figure 4(f) illustrates the overall knee locking procedure. The knee mechanism is passive and does not require electric power for operation.

2.4   Hip and Pelvis Design

The pelvis and both legs are joined at the hip. A drive motor is located on each hip. A motor, indicated as M2 in Fig. 5, is attached to the thigh drive's forward-backward swing relative to the pelvis. Another motor (M1) enables the leg to be moved laterally. The M1 actuation enables adequate lift of the swing leg for swing clearance, mimicking the pelvic tilt of human gait [13]. When M1 lifts the pelvis, the rotation at joint P1 is blocked by a clamp (Figure 5(b)), thus allowing the swing leg to be lifted. When M1 releases the pelvis, spring S3 holds the leg laterally. The spring aids lateral balancing during ground contact. The motors used in this robot are controlled by a compatible controller board (CM-2+, Robotics, Inc.) attached to the pelvis frame (see Fig. 1.)

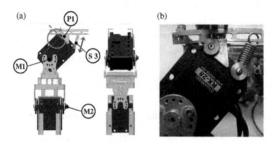

Fig. 5. (a) Hip and pelvis design and (b) its implementation

2.5   Parameter Fine-Tuning

In order to implement the proposed mechanism, some parameter values must be determined such as the spring values. All passive spring coefficients were selected via trial and error. The robot was tested progressively with physical tinkering, because analysis of the numerous effects of a bipedal walker with passive parts is difficult to simulate and characterize. Thus, analytic modeling was not undertaken for this robot. The selected springs S1 and S3 have the same stiffness, and S2 has less stiffness.

3   Control Command Profiles

The command profiles were adopted from typical human gait joint profiles in order to generate human-like motion [13]. The profiles were modified empirically to enable the proposed robot to implement reasonable walking motions through trial and error. In order to implement walking, the input trajectories were sent to the motors from a PC through the controller. Figure 6 shows the commands, which indicate the desired active joints' angular trajectories. The command trajectory for M1 was designed to

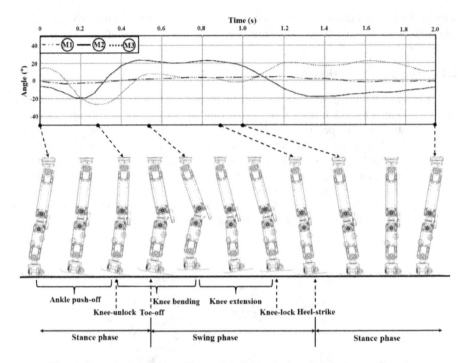

Fig. 6. Control command profiles and desired robot motion during a gait cycle

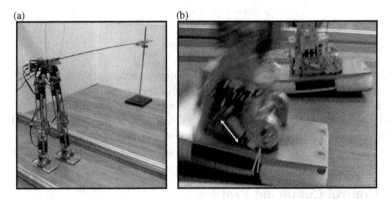

Fig. 7. (a) Experimental setup, (b) Snapshot of the two-part feet at push-off

activate a slight lateral sway at the hip, while the command trajectory for M2 activates the forward-backward swing of the leg at the hip. The command trajectory for M3 describes the ankle motion. In the early phase of the plot, the ankle flexion stretches the S1 spring pair in order to obtain elastic energy. The command profile is a feedforward profile: the gait cycle is designed to last 2 seconds in the current stage. Different phases and gait events are shown in figure 6 as well.

4   Experimental Results

Figure 7 (a) shows the experimental setup for the proto-type walking robot. The robot was mounted on a boom, which provided lateral stability while still permitting nontrivial lateral motion. Table 1 summarizes the specifications of the prototype robot walker.

Table 1. The Specifications of the bipedal walker

Weight (kg)		1.7	D.O.F	Hip	6
Dimesions (mm)	Height	467		Knee	2
	Width	175		Ankle	4
	Depth	120		Total	12

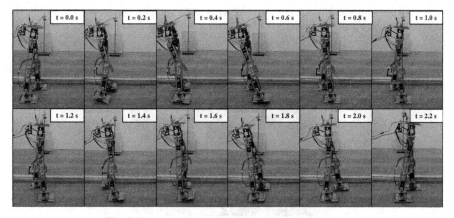

Fig. 8. Sequential snapshots of walking performance

Figure 8 shows sequential snapshots of the robot's walk taken during the experiment. Figure 7(b) demonstrates the instant of the S1 spring's pulling motion during push-off. The forefoot and rearfoot parts are misaligned.

Figure 9 analyzes the detailed joint motions. In Figure 9(a), each joint trajectory is described as Cartesian coordinate points and Figure 9(b) draws the corresponding stick plots. Figure 9(c) shows the knee joint trajectory in degrees over time. The bending and stretching motions are clearly seen. The thick line represents the averaged profile over five trials, and the grayed region indicates the standard deviations. Figure 9 verifies that the passive knee mechanism works as expected.

While the prototype walked, the reaction force on the bottom of the foot was measured using the system shown in Figure 10(a). Each force sensor (Standard 402 Force Sensing Resistors (FSR), Interlink Electronics, Inc. [15]) was attached on the forefoot and rearfoot sections. The sensor measured the point contact force. The measured contact force is shown in Figure 10(b). The peak vertical reaction force at

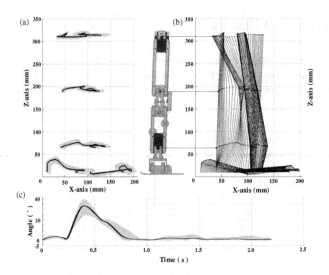

Fig. 9. (a) Joint motions, (b) stick plots, and (c) measured knee joint angle trajectories while the prototype robot was walking

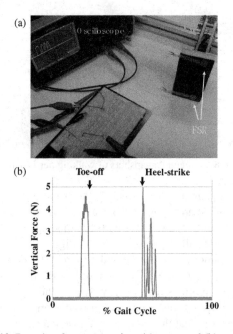

Fig. 10. Reaction force measuring: (a) setup and (b) profiles

the forefoot occurred during ankle push-off while the peak vertical force of the rearfoot occurred during the ground contact. The two peak profiles are separate, which indicates that one foot part is off the ground in each phase, and the peak magnitudes are approximately equal. This observation indicates that the different parts of the foot operated in a similar manner as human feet [13]. The sensors measured the vertical reaction forces at a spot on each forefoot and rearfoot; therefore, the measurements do not indicate the total reaction forces over the entire foot. The parted foot and ankle structures surrounded by springs provided a compliant gait motion.

5   Conclusion

This study presented the design of a bipedal walker that focuses on the mechanical design for human-like gait motions. The robot design is a proof of concept prototype that may provide insights into the design of bipedal walkers with respect to a passive knee and parted foot design. The experimental observations verified that the design is amenable to a compliant walking motion. The prototype enables knee bending, toe-off, and heel-strike motions which are similar to the human gait. Furthermore knee bending and stretching during the gait was achieved. The spring-based actuation at ankle push-off provided partial forward thrust, and the active hip and ankle actuations moved the robot body forward. The pelvic tilt actuation provided clearance for the swinging leg.

For more complicated walking tasks, the proposed mechanism must be further developed. The passive knee and parted foot mechanism ensure stability on a flat surface but not in other environments. An interesting topic for future development would be the design of a simple mechanism that can modify the knee bending angles adaptively even though they are passive. In addition, the current algorithm is a feed-forward algorithm, and the robot's performance is sensitive to its initial conditions. Furthermore, robust balance control was not seriously considered. An advanced control mechanism could provide a variety of robust gait.

References

1. Hirai, K., Hirose, M., Haikawa, Y., Takenaka, T.: The Development of Honda Humanoid. In: IEEE International Conference on Robotics and Automation, vol. 2, pp. 1321–1326 (1998)
2. Park, I., Kim, J., Lee, J., Oh, J.: Mechanical Design of Humanoid Robot Platform KHR-3 (KAIST Humanoid Robot-3: Hubo). In: IEEE-RAS International Conference on Humanoid Robots, pp. 321–326 (2005)
3. Gouaillier, D., Hugel, V., Blazevic, P., Kilner, C., Chris, M.: Mechatronic Design of NAO Humanoid. In: IEEE International Conference on Robotics and Automation, pp. 769–774 (2009)
4. Kaneko, K., Harada, K., Kanehiro, F., Miyamori, G., Akachi, K.: Humanoid Robot HRP-3. In: IEEE/RSJ International Conference on Intelligent Robots and Systems, pp. 2471–2478 (2008)

5. Kim, M., Kim, I., Park, S., Oh, J.: Realization of stretch-legged walking of the humanoid robot. In: IEEE-RAS International Conference on Humanoid Robots, pp. 118–124 (2008)
6. Nishiwaki, K., Kagami, S., Kuniyoshi, Y., Inaba, M., Inoue, H.: Toe joints that enhance bipedal and fullbody motion of humanoid robots. In: IEEE International Conference on Robotics and Automation, vol. 3, pp. 3105–3110 (2002)
7. McGeer, T.: Passive Dynamic Walking. International Journal of Robotics Research 9, 62–82 (1990)
8. Garcia, M., Chatterjee, A., Ruina, A.: Efficiency, speed, and scaling of two-dimensional passive-dynamic walking. Dynamics and Stability of Systems 15, 75–99 (2000)
9. Collins, S., Ruina, A.: A Bipedal Walking Robot with Efficient and Human-Like Gait. In: IEEE International Conference on Robotics and Automation, pp. 1983–1988 (2005)
10. Collins, S., Ruina, A., Tedrake, R., Wisse, M.: Efficient Bipedal Robots based on Passive-dynamic Walkers. Science 307, 1082–1085 (2005)
11. Kuo, A.: Energetics of Actively Powered Locomotion using the Simplest Walking Model. Transactions-American Society of Mechanical Engineers, Journal of Biomechanical Engineering 124, 113–120 (2002)
12. Wu, T., Yeh, T.: Optimal Design and Implementation of an Energy-efficient Biped Walking in Semi-active Manner. Robotica 27, 841–852 (2008)
13. Perry, J.: Gait Analysis: Normal and Pathological Function. Journal of Pediatric Orthopaedics 12, 815 (1992)
14. Robotis©, Dynamixel, http://www.robotis.com/xe/dynamixel_en
15. Interlink Electronics©. FSR 400 Series,
    http://www.interlinkelectronics.com/Product/Standard-402-FSR

An Adaptive Sliding Mode Control for Trajectory Tracking of a Self-reconfigurable Robotic System

R. Al Saidi and B. Minaker

Dept. of Mechanical Automotive & Materials Engineering, Windsor University
401 Sunset Avenue, Windsor, Ontario N9B 3P4, Canada
{alsaidi,bminaker}@uwindsor.ca

Abstract. This paper presents the development of the model, dynamics and an adaptive sliding mode control of new self-reconfigurable robotic systems. These robotic systems combine as many properties of different open kinematic structures as possible and can be used for a variety of applications. The kinematic design parameters, i.e., their Denavit-Hartenberg (D-H) parameters, can be modified to satisfy any configuration required to meet a specific task. By varying the joint twist angle parameter (configuration parameter), the presented model is reconfigurable to any desired open kinematic structure, such as Fanuc, ABB and SCARA robotic systems. The joint angle and the offset distance of the D-H parameters are also modeled as variable parameters (reconfigurable joint). The resulting self-reconfigurable robotic system hence encompasses different kinematic structures and has a reconfigurable joint to accommodate any required application in medical, space, future manufacturing systems, etc. Automatic model generation of a 3-DOF reconfigurable robotic system is constructed and demonstrated as a case study which covers all possible open kinematic structures. An adaptive controller is developed based on the sliding mode approach for a 3-DOF self-reconfigurable robotic system to achieve high tracking performance. This research is intended to serve as a foundation for future studies in reconfigurable control systems.

Keywords: reconfigurable robotic system, automatic model generation, variable D-H parameters, adaptive sliding mode controller.

1 Introduction

The new manufacturing environment is characterized by frequent and unpredictable market changes. A manufacturing paradigm called Reconfigurable Manufacturing System (RMS) was introduced to address the new production challenges. The Reconfigurable Manufacturing System is designed for rapid adjustments of production capacity and functionality in response to new circumstances, by rearrangement or change of its components and machines. Such new systems provide exactly the capacity and functionality that is needed, when it is needed [1]. These systems' reconfigurability calls for their components, such as machines and robots to be rapidly and efficiently modifiable to varying demands [2]. In the literature, modular robotic systems are presented as a solution for reconfigurable robotic

J.-H. Kim et al. (eds.), *Robot Intelligence Technology and Applications 2,*
Advances in Intelligent Systems and Computing 274,
DOI: 10.1007/978-3-319-05582-4_33, © Springer International Publishing Switzerland 2014

systems. An automated generation of D-H parameters methodology has developed for the modular manipulators [3]. The authors derived the kinematic and dynamic models of reconfigurable robotic systems using D-H parameters for different sets of joints, links and gripper modules. Furthermore, a library of modules is formed from which any module can be called with its associated kinematic and dynamic models. In [4], a modular and reconfigurable robot design is introduced with modular joints and links. The proposed design introduces zero links offset to increase the robot's dexterity and maximize its reachability. A modular and reconfigurable robot (MRR) with multiple working modes was designed [5]-[6]. In the proposed MRR design, each joint module can independently work in active modes with position or torque control, or passive modes with friction compensation. A Task-based configuration optimization based on a generic algorithm was used to solve a pre-defined set of modules for specific kinematic configuration [7]. In [8]-[9], A sliding mode control algorithm combined with an adaptive scheme was employed for 2-DOF (predefined fixed kinematic) for tracking purposes.

The main drawbacks of the modular robotic systems proposed in the literature are the high initial investment necessary in modules that remain idle during many activities, and the significant lead time for replacement of the components prior to performing a specific task.

2   Research Motivation

Current machines (CNC mills, robots, etc…) have physical limitations with respect to their configurations and capabilities. They are preconfigured to do specific tasks. For example: the typical 5-DOF (3R-2T) CNC machine has three rotational and two translational joints with fixed coordinate frames (D-H parameters), which cannot be automatically changed to any other configuration. The structure of most machines can be changed only by physically replacing their joints or links (modules). These limitations are reflected on the machine's path, workspace, inertia, torque, power concept, etc, making them unsuitable for the future RMS. In this research, the authors develop "Self-Reconfigurable" robotic systems that combine the properties of many different open kinematic machines and can efficiently self-adjust for use in a variety of tasks. The kinematic design parameters, i.e., the D-H parameters, are variable and can generate the required configuration to different applications. The robotic systems' links and joints are not modular. Hence, they eliminate the need for module replacements to satisfy a particular application. The entire reconfiguration operation is performed autonomously in a highly efficient and reliable manner, reducing production time and costs.

3   Development of Self-reconfigurable Robotic System

Using D-H parameters of an n-DOF Reconfigurable Kinematic Model (n-RKM) of general self-reconfigurable robotic systems are given in Table (1):

Table 1. D-H parameters of the n-DOF (RKM) model

i	d_i	θ_i	a_i	α_i
1	$R_1 d_{DH1} + T_1 d_1$	$R_1 \theta_1 + T_1 \theta_{DH1}$	a_1	$0, \pm 180, \pm 90$
2	$R_2 d_{DH2} + T_2 d_2$	$R_2 \theta_2 + T_2 \theta_{DH2}$	a_2	$0, \pm 180, \pm 90$
...	...	...	...	...
n	$R_n d_{DHn} + T_n d_n$	$R_n \theta_n + T_n \theta_{DHn}$	a_n	$0, \pm 180, \pm 90$

All D-H parameters presented in Table (1) are not fixed values in the current paper; they are modeled as variables to satisfy the properties of all possible machines' kinematic structures. The twist angle variable α_i is limited to five different values, to maintain perpendicularity between joints' coordinate frames. Consequently, each joint has six different positive directions of rotations and/or translations. Defining the varying twist angle as the configuration parameter, allows the model to achieve any kinematic structure by configuring the parameter, accordingly. Figure 1 shows diverse industrial robotic systems such as Fanuc, ABB, and PUMA achieved as special cases by changes to the configuration parameter. Assuming a spherical wrist attached to the end effecters, the kinematic structures of the common industrial robotic systems such as Cartesian (TTT), Cylindrical (RTR), Spherical (RRT), SCARA (RRT) and Articulated (RRR) are determined by the first three joints and links, which also define the external and internal workspace boundaries. A spherical wrist satisfies Piper's condition: $a_4 = 0$, $a_5 = 0$, $d_5 = 0$, only serves to orient the end-effecter within the workspace.

4   Kinematic Development

For the n-RKM model, a given joint's vector z_{i-1} can be placed in the positive and/or negative directions of the x, y, and z axis in the Cartesian coordinate frame. This is expressed in Equations (1)-(2).

$$\text{Rotational Joints:} \qquad R_i = 1 \; and \; T_i = 0 \tag{1}$$

$$\text{Translational Joints:} \qquad R_i = 0 \; and \; T_i = 1 \tag{2}$$

R_i and T_i are used to control the selection of joint type (rotational and/or translational). The orthogonality between the joint's coordinate frames is achieved by assigning appropriate values to the twist angles α_i. The joint's reconfigurable parameters are defined as follows: $K_{Si} = \sin(\alpha_i)$, $K_{Ci} = \cos(\alpha_i)$. Consequently, the general homogeneous transformation matrix for the n-RKM model is given Equation (3). Using this transformation matrix, models of different open kinematic structures can be automatically generated which characterizes the new self-reconfigurable robotic systems. The generalized joint coordinates and their selection parameters are expressed: $q_i = R_i\theta_i + T_i d_i$.

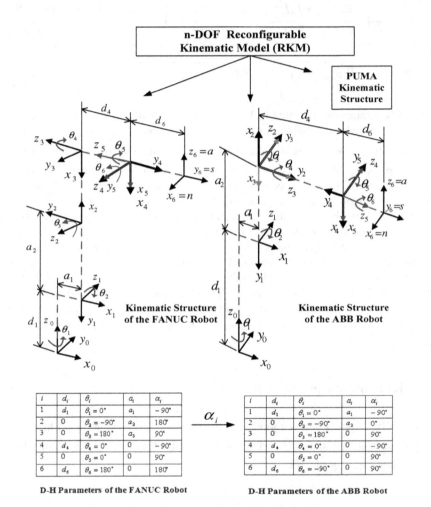

Fig. 1. Kinematic structure of different robotic systems

$$
{}^{i-1}A_i = \begin{bmatrix}
\cos(R_i\theta_i + T_i\theta_{DHi}) & -K_{Ci}\sin(R_i\theta_i + T_i\theta_{DHi}) & K_{Si}\sin(R_i\theta_i + T_i\theta_{DHi}) & a_i\cos(R_i\theta_i + T_i\theta_{DHi}) \\
\sin(R_i\theta_i + T_i\theta_{DHi}) & K_{Ci}\cos(R_i\theta_i + T_i\theta_{DHi}) & -K_{Si}\cos(R_i\theta_i + T_i\theta_{DHi}) & a_i\sin(R_i\theta_i + T_i\theta_{DHi}) \\
0 & K_{Si} & K_{Ci} & R_i d_{DHi} + T_i d_i \\
0 & 0 & 0 & 1
\end{bmatrix} \quad (3)
$$

5   Dynamic Development

Let q denote the joint variable and τ_i is the actuator torques vector of n-axis open kinematic structure systems' link, then the general equation of motion is expressed as:

$$\sum_{j=1}^{n} a_{ij}(q)\ddot{q}_j + \left(\sum_{k=1}^{n} \sum_{j \neq k} c_{ikj}(q)\dot{q}_k \dot{q}_j + \sum_{j=1}^{n} c_{ik}(q)\dot{q}_k^2 \right)$$
$$+ g_i(q) + b_i(\dot{q}) = \tau_{li} \qquad 1 \leq i \leq n \tag{4}$$

The first term is an acceleration term that represents the inertial forces and torques generated by the motion of the links. The second term is associated with Coriolis and centrifugal forces, respectively. The third term represents loading due to gravity. Finally, the fourth term represents the friction (Coulomb, viscous friction, etc...) opposing the motion of the link. DC motors are considered to drive the robotic system joints and links. The dynamic model of an armature controlled DC motor with gear reduction ratio r is given by the following equation:

$$J_{mk}\ddot{\theta}_{mk}(t) + B_k \dot{\theta}_{mk}(t) = (K_{mk} / R_k)V_k(t) - \tau_{lk}(t)/r_k \tag{5}$$

for $k = 1,...,n$ where $\theta_{mk} = r_k q_k$. Equation (5) can be written as:

$$r_k^2 J_m \ddot{q}_k + r_k^2 B_k \dot{q}_k = (r_k K_{mk} / R_k)V_k - \tau_{lk} \tag{6}$$

The dynamic coupling between the n-link robotic system and the armature controlled DC motors is given by substituting Equation (6) into Equation (4):

$$r_k^2 J_m \ddot{q}_k + \sum_{j=1}^{n} a_{ij}(q)\ddot{q}_j + \sum_{k=1}^{n} \sum_{j \neq k} c_{ikj}(q)\dot{q}_k \dot{q}_j + \sum_{j=1}^{n} c_{ik}(q)\dot{q}_k^2$$
$$+ g_i(q) + r_k^2 B_k \dot{q}_k + b_i(\dot{q}) = (r_k K_{mk} / R_k)V_k \tag{7}$$

Assuming the lumped viscous friction terms is expressed $r_k^2 B_k \dot{q}_k + b_i(\dot{q}) = B(\dot{q})$. Then, Equation (7) can be recast in a state form as follows:

$$A(q)\ddot{q} + C(q,\dot{q})\dot{q} + B(\dot{q}) + G(q)q = \tau \tag{8}$$

Here $A(q)$ and J is a diagonal matrix with elements $r_k^2 J_{mk}$. The matrix $C(q,\dot{q})$ defines the Coriolis and centrifugal generalized forces. The matrices $G(q)$, $B(\dot{q})$ are the gravity and frictional generalized forces, respectively.

6   Kinematic and Dynamic Development of a 3-DOF Robotic System

A 3-DOF self-reconfigurable robotic system is constructed to present as an example. Each joint of these open kinematic structure robotic systems has six different positive directions of rotations or translations. Any joint's vector z_{i-1} can be placed in positive and/or negative direction of the x, y and z axis of the Cartesian coordinate frame. Consequently, the kinematic structure combinations of the first three joints are RRR, RTR, RRT, TRR, TTR, RTT, TRT, and TTT. Depending on the specific task to be performed, the required kinematic structure is achieved by selecting the twist angle α_i between two consecutive coordinate frames. The D-H parameters of the 3-DOF reconfigurable robotic system are given in Table (2).

Table 2. D-H parameters of a 3-DOF reconfigurable robotic system

i	d_i	θ_i	a_i	α_i
1	$R_1 d_{DH1} + T_1 d_1$	$R_1 \theta_1 + T_1 \theta_{DH1}$	a_1	$\pm \pi$
2	$R_2 d_{DH2} + T_2 d_2$	$R_2 \theta_2 + T_2 \theta_{DH2}$	a_2	$\pm \pi$
3	$R_3 d_{DH3} + T_3 d_3$	$R_3 \theta_3 + T_3 \theta_{DH3}$	a_3	$0°$

The resulting kinematic structure is graphically presented in Figure 2:

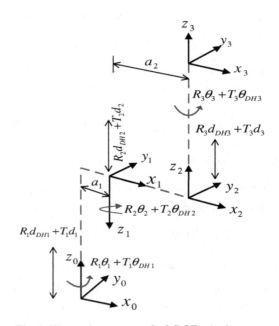

Fig. 2. Kinematic structure of a 3-DOF robotic system

The equations of motion (E.O.M) of the 3-DOF robotic system are generated based on the general E.O.M form of Equation (8), neglecting $B(\dot{q})$. Using the Newton-Euler (N-E) recursive approach, the dynamic parameters are found, as shown in Equation (9).

$$
\begin{bmatrix} a_{11} & a_{12} & a_{13} \\ a_{21} & a_{22} & a_{23} \\ a_{31} & a_{32} & a_{33} \end{bmatrix} \begin{bmatrix} \alpha_1 \\ \alpha_2 \\ \ddot{d}_3 \end{bmatrix} + \begin{bmatrix} b_{112} & b_{123} & b_{131} \\ b_{212} & b_{223} & b_{231} \\ b_{312} & b_{323} & b_{331} \end{bmatrix} \begin{bmatrix} \omega_1\omega_2 \\ \omega_2\dot{d}_3 \\ \dot{d}_3\omega_1 \end{bmatrix}
$$

$$
+ \begin{bmatrix} c_{11} & c_{12} & c_{13} \\ c_{21} & c_{22} & c_{23} \\ c_{31} & c_{32} & c_{33} \end{bmatrix} \begin{bmatrix} \omega_1^2 \\ \omega_2^2 \\ \dot{d}_3^2 \end{bmatrix} + \begin{bmatrix} G_1 \\ G_2 \\ G_3 \end{bmatrix} = \begin{bmatrix} \tau_1 \\ \tau_2 \\ f_3 \end{bmatrix} \tag{9}
$$

7   Adaptive Sliding Mode Controller Development

The trajectory control problem is solved by developing a robust controller for the actuator joints, and an estimation law for the unknown parameters, such that the manipulator output $q(t)$ closely tracks the desired trajectory $q_d(t)$.

The tracking error is given as:

$$
e = q_d - q \tag{10}
$$

Define the reference velocity [11]:

$$
\dot{q}_r = \dot{q}_d + \Lambda(q_d - q) \tag{11}
$$

Where Λ is a positive diagonal matrix:

$$
\Lambda = diag(\lambda_1, \lambda_2, \dots, \lambda_i), \quad \lambda_i > 0, \quad i = 1, 2, 3
$$

The sliding variable is defined as:

$$
s = \dot{e} + \Lambda e
$$

The energy conservation of the system (8), neglecting $B(\dot{q})$, can be formalized by the Lyapunov function as follows:

$$
V(t) = \frac{1}{2} s^T A s + \frac{1}{2} \tilde{\varphi}^T \Gamma \tilde{\varphi} \tag{12}
$$

Where the estimation error is: $\tilde{\varphi} = \hat{\varphi} - \varphi$, and

$$
\Gamma = diag(\gamma_1, \gamma_2, \dots, \gamma_i), \quad \gamma_i > 0, \quad i = 1, 2, 3, 4
$$

The Lyapunov function derivative is:

$$
\dot{V}(t) = s^T A \dot{s} + \frac{1}{2} s^T \dot{A} s + \tilde{\varphi}^T \Gamma \dot{\tilde{\varphi}}
$$

$$
\dot{V}(t) = s^T (A\ddot{q} - A\ddot{q}_r) + \frac{1}{2} s^T \dot{A} s + \tilde{\varphi}^T \Gamma \dot{\tilde{\varphi}}
$$

$$
\dot{V}(t) = s^T (\tau - C\dot{q} - G - A\ddot{q}_r) + \frac{1}{2} s^T \dot{A} s + \tilde{\varphi}^T \Gamma \dot{\tilde{\varphi}}
$$

$$\dot{V}(t) = s^T (\tau - C(s + \dot{q}_r) - G - A\ddot{q}_r) + \frac{1}{2} s^T \dot{A}s + \tilde{\varphi}^T \Gamma \dot{\tilde{\varphi}}$$

The proposed controller is formed as follows [12]:

$$\tau = \hat{\tau} - K_D s = \hat{A}(q)\ddot{q}_r + \hat{C}(q, \dot{q})\dot{q}_r + \hat{G} - K_D s \qquad (13)$$

Where the dynamics of the first part were exactly known and K_D is a positive definite gain matrix:

$$K_D = diag(K_{d1}, K_{d2}, K_{d3}), \quad K_{di} > 0, \quad i = 1, 2, 3.$$

Using the control law (13) yields:

$$\dot{V}(t) = s^T (\hat{A}(q)\ddot{q}_r + \hat{C}(q, \dot{q})\dot{q}_r + \hat{G} - K_D s - C(s + \dot{q}_r) - G - A\ddot{q}_r)$$
$$+ \frac{1}{2} s^T \dot{A}s + \tilde{\varphi}^T \Gamma \dot{\tilde{\varphi}}$$

$$\dot{V}(t) = s^T (\tilde{A}(q)\ddot{q}_r + \tilde{C}(q, \dot{q})\dot{q}_r + \tilde{G} - K_D s - Cs) + \frac{1}{2} s^T \dot{A}s + \tilde{\varphi}^T \Gamma \dot{\tilde{\varphi}}$$

The manipulator equations of motion are linear in the inertia parameters in the following sense. There exists an $n \times l$ regressor function, $Y(q, \dot{q}, \ddot{q}_r, \ddot{q}_r)$ and an l dimensional parameter vector $\tilde{\varphi}$ such that the Equation (8), neglecting $B(\dot{q})$, can be written as:

$$\tilde{A}(q)\ddot{q}_r + \tilde{C}(q, \dot{q})\dot{q}_r + \tilde{G} = Y(q, \dot{q}, \ddot{q}_r, \ddot{q}_r)\tilde{\varphi} \qquad (14)$$

Therefore,

$$\dot{V}(t) = s^T (Y\tilde{\varphi} - K_D s - Cs) + \frac{1}{2} s^T \dot{A}s + \tilde{\varphi}^T \Gamma \dot{\tilde{\varphi}}$$

$$\dot{V}(t) = s^T (Y\tilde{\varphi} - K_D s) + \frac{1}{2} s^T (\dot{A} - 2C)s + \tilde{\varphi}^T \Gamma \dot{\tilde{\varphi}}$$

$$\dot{V}(t) = s^T (Y\tilde{\varphi} - K_D s) + \tilde{\varphi}^T \Gamma \dot{\tilde{\varphi}} = \tilde{\varphi}^T Y^T s - s^T K_D s + \tilde{\varphi}^T \Gamma \dot{\tilde{\varphi}}$$

$$\dot{V}(t) = \tilde{\varphi}^T (Y^T s + \Gamma \dot{\tilde{\varphi}}) - s^T K_D s$$

The parameter adaptive law would be designed as follows:

$$\dot{\tilde{\varphi}} = -\Gamma^{-1} Y^T s \qquad (15)$$

Therefore,

$$\dot{V} = -\Gamma^{-1} K_D s \leq 0$$

This results that the tracking error goes to zero: $\tilde{q} \to 0$ as $t \to \infty$.

8   Simulation of a 3-DOF Self-reconfigurable Robotic System

The inertia terms of the regressor function used in the simulation are:

$$Y(q,\dot{q},\ddot{q}) = \begin{bmatrix} \ddot{q}_{r1} & (\ddot{q}_{r1} - 2\ddot{q}_{r2})\cos(q_2) - (2\dot{q}_{r1}\dot{q}_{r2} - \dot{q}_{r2}^2)\sin(q_2) & \ddot{q}_{r1} + \ddot{q}_{r2} & 0 \\ 0 & -2\ddot{q}_{r1}\cos(q_2) + 2\dot{q}_{r1}^2\sin(q_2) & \ddot{q}_{r2} - \ddot{q}_{r1} & 0 \\ 0 & 0 & 0 & \ddot{d}_{r3} - g \end{bmatrix}$$

And the parameter regressor vector is given as follows:

$$\varphi = \begin{bmatrix} \varphi_1 \\ \varphi_2 \\ \varphi_3 \\ \varphi_4 \end{bmatrix} = \begin{bmatrix} \left(\dfrac{m_1}{3} + m_2 + m_3\right)a_1^2 & (m_2 + 2m_3)a_1a_2 & \left(\dfrac{m_2}{3} + m_3\right)a_2^2 & m_3 \end{bmatrix}^T$$

In the interest of brevity, only the simulation results of joint #1 are presented here. The results of the remaining joints are found in the same fashion. Figure 3, shows the tracking position of joint #1 to a sinus reference signal (up), the tracking error (middle) and the control torque required to drive the joint (bottom). The 3-DOF model has been simulated using the control law (13) and the estimation rule (15).

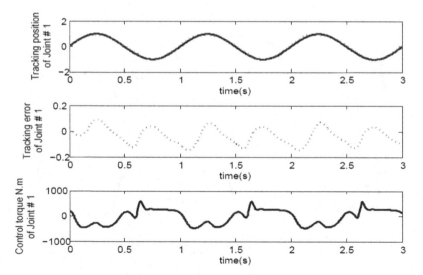

Fig. 3. Tracking position (rad) of the joint #1 to a sinus desired function (up), reference position (solid) and actual positions (dotted). The tracking error (rad) (middle), and the control torque (N.m) needed to drive the joint (bottom).

The estimation values of the inertia parameters ($\varphi_1, \varphi_2, \varphi_3, \varphi_4$) are normalized and showed in figures 4 and 5, respectively. The tracking error remains bounded rather than going to zero. The reason for this type of bounded tracking error performance is that the error systems given by Equation (14) is constantly being excited by the input dynamics on the right-side of the Equation (14).

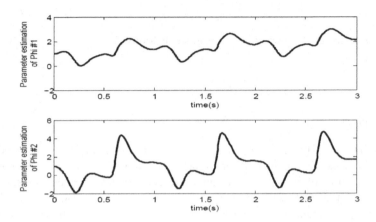

Fig. 4. The normalized values of the inertia parameters φ_1 (up) and φ_2 (down)

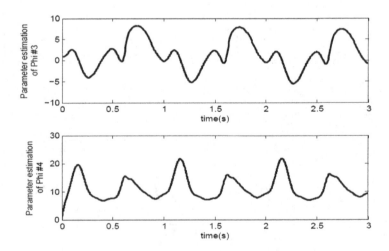

Fig. 5. The normalized values of the inertia parameters φ_3 (up) and φ_4 (down)

9   Conclusion

In our research, we introduce new "Self-Reconfigurable" robotic systems with features such as variable twist angles, length links and hybrid (translational/rotational) joints. The kinematic design parameters, i.e., the D-H parameters, are variable and can generate any required configuration to facilitate a specific application. Hence, they eliminate the need for module replacements to accommodate a particular application. The entire reconfiguration operation is performed autonomously according to the geometry, dimensions and specific path of the required task. Different kinematic structures of industrial robotic systems were generated and configured by changing the configuration parameter. A generic homogenous transformation matrix was presented to generate the forward kinematic of any desired kinematic structure. For low speed motion of the robotic system's links, the Coriolis and centrifugal generalized forced have lower magnitude comparing to other dynamic parameters and therefore were not considered in the control law development. An adaptive controller was developed to track reference trajectories based on sliding mode method. This research will continue to bring new approaches to develop reconfigurable controls and decision algorithms for the new reconfigurable robotic systems presented in this publication.

References

1. Koren, Y., Heisel, U., Jovane, F., Moriwaki, T., Pritschow, G., Ulsoy, G., van Brussel, H.: Reconfigurable Manufacturing Systems. Annals of the CIRP 48(2), 527–540 (1999)
2. Hedelind, M., Jackson, M.: The need for Reconfigurable Robotic Systems. In: Proc. Of the 2nd Int. Conf. on Changeable, Agile, Reconfigurable and Virtual Production, Toronto/Canada (2007)
3. Bi, Z.M., Zhang, W.J., Chen, I.-M., Lang, S.Y.T.: Automated Generation of the D-H Parameters for Configuration Design of Modular Manipulators. Robotics and Computer-Integrated Manufacturing 23, 553–562 (2007)
4. Li, Z., Melek, W., Clark, C.M.: Development and Characterization of a Modular and Reconfigurable Robot. In: Proceedings of the International Conference on Changeable, Agile, Reconfigurable and Virtual Production, Toronto/Canada (2007)
5. Liu, G.: Development of modular and reconfigurable with multiple working modes. In: International Conference on IEEE Robotics and Automation, ICRA (2008)
6. Aghili, F., Parsa, K.: A Reconfigurable Robot with Lockable Cylindrical Joints. IEEE Transactions on Robotics 25(4), 785–797 (2007)
7. Tabandeh, S., Clarck, C., Melek, W.: Task-based Configuration Optimization of Modular and Reconfigurable Robots using a Multi-Solution Inverse Kinematics Solver. In: International Conference on Changeable, Agile, Reconfigurable and Virtual Production, Toronto/Canada (2007)
8. Su, C., Leung, P.T.: A Sliding Mode Controller with Bound Estimation for Robot Manipulators. IEEE Transaction on Robotics and Automation 9(2) (April 1993)
9. Kuo, T.C., Huang, Y.J., Hong, B.W.: Design of Adaptive Sliding Mode Controller for Robotic Manipulators Tracking Control. World Academy of Science, Engineering and Technology 53 (2011)
10. Khalil, H.K.: Nonlinear systems, 3rd edn. Prentice Hall (2002)
11. Slotine, J.E., Li, W.P.: On the adaptive control of robot manipulators. The International Journal of Robotics Research (3), 49–59 (1987)
12. Slotine, J.E., Li, W.: Applied Nonlinear Control. Prentice Hall, NJ (1991)

Position-Trajectory Control of Advanced Tracked Robots with Diesel-Electric Powertrain

Denis Pogosov

Dept. of Electronics and Mechatronics, Southern Federal University
105/42 Bolshaya Sadovaya, Rostov-on-Don 344006, Russia
deonisiy2004@yandex.ru

Abstract. Recently an interest about the control systems for ground robots has increased. The main papers are focused on improving the accuracy of motion of complex nonlinear and multilinked systems. This paper presents a control system of s highly maneuverable tracked robot based on algorithms of The Position-trajectory control. These algorithms have proved their effective in synthesis of control systems for aeronautic, aircraft and ground-based robots. This paper introduced a model of the advanced tracked robot taking into account the internal and external forces acting on it; a model of diesel engine with turbocharger and model of electric powertrain based on inertial parts.

Keywords: advanced tracked robot, position-trajectory control, electric powertrain.

1 Introduction

The main requirements applied to the advanced highly maneuverable tracked robots (HMTR) are movement with a high speed, a high precision of its movement on trajectories, stability of a control system and minimize an energy consumption. These requirements cause HMTR structure, which include a power plant, transmission, motion control and navigation systems.

Strictly speaking a classic mechanical or hydro-mechanical transmission cannot produce arbitrary speed and torque to the driven sprockets. Even a multi-row planetary transmission requires using a friction brakes because of its stepped characteristic. Torque transmitted to the driven sprockets is different from the required to meet the required trajectories. Using the friction brakes reduces the efficiency of energy consumption due to dissipation of generated energy to heat. The appropriate transmission to movement of HMTR is an electric transmission with a diesel combustion engine (DE). The electric transmission can produce a high initial torque on the driven sprockets, which is necessary for movement of large mass vehicles. An electrical energy can be accumulated for further use unlike a mechanical energy usually. Using an electrical double layer capacitor (EDLC) or ultra-capacitor is an advanced solution.

J.-H. Kim et al. (eds.), *Robot Intelligence Technology and Applications 2*,
Advances in Intelligent Systems and Computing 274,
DOI: 10.1007/978-3-319-05582-4_34, © Springer International Publishing Switzerland 2014

To achieve the required performance from HMTR a control system should be capable to control an essentially nonlinear and multilinked object with high accuracy of movement on the required trajectory.

2   Related Works

Papers about mobile robots control are concentrated in area of small robots with negligible dynamics. For example, in paper [1] designed a control algorithm to object, described by a simplified model. In paper [2], a control method based on constructive Lyapunov function with the linearized dynamics of a two-wheeled robot. In paper [3] control is built on the sliding mode. Control of a robot on the sliding modes does not allow to high precision control and the system may be subjected to a dynamic impact. Application of simplified models characterizes the possibilities of the control algorithms. In paper [4] shown a loss of controllability of HMTR described by simplified dynamic model.

Modeling and control of the diesel combustion engine widely covered. In papers [5-6] provide sufficient information for modeling the diesel engines with variable geometry turbocharger and exhaust gas recirculation.

Development of the electric transmission with EDLC is concentrated in papers, devoted to hybrid vehicles [7-10]. It should be noted that in schemes from papers [7-10] EDLC is used as a buffer to energy storage, when a batteries pack is not allows producing or recovering energy because it's dynamic characteristic. In papers [7-8] the main energy storage is lithium batteries pack. For tracked vehicles such approach is not justified because during curvilinear motion (even at a constant speed) the traction motors of driven sprockets consume unequal power and voltage. For automotive vehicles design maneuver a small low of energy is necessary, to allows the diesel generator run at optimally steady state. Tracked vehicles requires a large initial torque which mean large current from batteries pack, because they are usually heavy than automotive vehicles. These reasons make impractical using a heavy and large battery versus EDLC, because of their shorter life. In paper [11] EDLC is used as a buffer element for a mobile robot which powered by a battery only. This approach allow to smooth dynamic impacts in the electric power system.

3   Structure of the Advanced Tracked Robot Powertrain

In this paper, we propose the following structure of the electro-mechanical transmission of HMTR, shown in Figure 1.

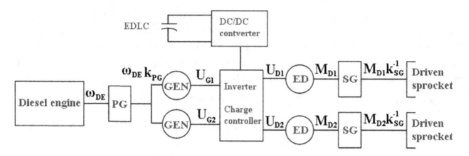

Fig. 1. The structure of HMTR powertrain. PG is a primary reduction gear, GEN is generators, ED is electric drives, SG is secondary reduction gears, EDLC is electric double layer capacitor. ω_{DE} is diesel engine speed, k is gear ratio, U is voltage, M is torque.

The structure contains EDLC. EDLC have several advantages versus chemical power sources, like batteries pack: a high charge current and significant number of charge-discharge cycles. A disadvantage is high specific gravity per unit of accumulated energy (20 Wh/kg versus 200 Wh/kg in chemical power sources) and their cost. This disadvantage is offset by using smaller size and power EDLC, because much larger charging current (at equal weight of elements). This advantage can use EDLC as braking load to traction electric motors, operating in generator mode, for recover a braking energy. Introduced this paper structure of HMTR powertrain contain equal quantity of traction motors and generators to driving each sprocket, however additional motor-sprockets can be installed instead of supporting rollers. Increasing quantity of motor-generator pair and at least two generators allows reducing overall electric machines dimensions and reducing currents through electric machines and increasing reliability in an event of the electrical machine failure. Reducing current can reduce chances of a circular arc beginnings in the commutate node and reduce ohmic losses. There is should be noted, the control is carried out at the boards motor-generator pairs.

4   Mathematical Model of the Advanced Tracked Robot

4.1   Electric Transmission Model

During HMTR curvilinear motion the voltages at the traction motors are different, which determines the appropriate set-point voltages of the generators. Due to the fact that a low generator efficiency at a speed less than the nominal mode, and unequal distributed load torque (from generators) to the primary gear and DE crankshaft, in this paper we propose using a load capacity of EDLC for adjust the voltage on the traction motor of decelerating side.

The stored energy is proposed to spend for alignment voltages of traction motors during HMTR curvilinear motion. In the case of EDLC absence, output voltage is regulated by known methods of excitation voltage control.

Electric machines of direct or (and) alternating current might be used as electric motors and generators. To improve the efficiency of DC machines might use ampli-dyne generators requiring small power for its control. In case of AC machines can use frequency of alternating current regulation. Without loss of generality, we represent electric machines of right and left sides as first order inertial parts, with their efficien-cy (η_D, η_G are motor and generator efficiency, respectively):

$$M = D_1 U - D_2 X \tag{1}$$

$$U_G = C_1 X_G - C_2 I_G \quad I_G = C_3 M_G \tag{2}$$

Where M is vector of traction motors torque, U is vector of traction motors voltage, X is vector of traction motors shaft speed, U_G is vector of generators voltage, X_G is vector of generators shaft speed, I_G is vector of current through generators, M_G is vector of generators load torque, D_1, D_2, C_1, C_2, C_3 are diagonal matrixes of constants.

Switching the charge-discharge mode of EDLC is realized by a threshold device. The device time constant, as well as time constants of voltage converters and excita-tion controllers are much less than the time constant of electric machines and DE. Consequently, the dynamics of these parts will be ignored.

The control equations of EDLC capacitive load for the generator of the decelerat-ing side, based on (2):

$$I_{chrg} = C_{EDLC} U_{EDLC}$$

$$U_G = c_1 X - c_2 I_G + c_4 U_{EDLC} \tag{3}$$

Where I_{chrg} is EDLC charging current (regulated), C_{EDLC} is EDLC capacitance, U_{EDLC} is first derivative of EDLC voltage, c_4 is a constant.

The control equations of excitation voltage of the generator of the decelerating side (without EDLC), based on (2):

$$U_{exc} = J_{exc} I_{exc}$$

$$U_G = C_1 X - C_2 I_G + C_4 I_{exc} \tag{4}$$

Where U_{exc} is excitation voltage (regulated), J_{exc} is excitation inductance, I_{exc} is first derivative of excitation current, c_4 is a constant.

With the follows assumption:

$$M_T = [M_1 \quad M_2]^T = [M_L \quad M_R]^T \text{ if } (U_L > U_R), \quad M_2 = [M_R \quad M_L]^T \text{ else}$$

$$X_T = [x_1 \quad x_2]^T = [x_L \quad x_R]^T \text{ if } (U_L > U_R), \quad X_2 = [x_R \quad x_L]^T \text{ else}$$

The mathematical model of electric powertrain (1-4) can be represented a follows vector-matrix equation [8-11]:

$$M_T = A^{-1}(\Theta - BX_T) \tag{5}$$

$$A = \begin{bmatrix} \dfrac{1+C_2C_3d_2}{C_1d_1k_{PG}} & 0 \\ \dfrac{1+C_2C_3d_2}{C_4d_1} & \dfrac{C_2C_3d_2+C_4}{C_4d_1} \end{bmatrix}, \quad B = \begin{bmatrix} \dfrac{d_2}{C_1d_1k_{PG}} & 0 \\ \dfrac{d_2}{C_4d_1} & \dfrac{d_2}{d_1} \end{bmatrix} \tag{6}$$

$$\Theta = \begin{bmatrix} \omega_{DE} & U_{EDLC} \end{bmatrix}^T \tag{7}$$

$$\Theta = \begin{bmatrix} \omega_{DE} & I_{exc} \end{bmatrix}^T \tag{8}$$

Where k_{PG} is the primary gear coefficient, ω_{DE} is diesel engine speed. The equation (7) is introduced for the system with EDLC, equation (8) for the system with an excitation control.

4.2 Diesel Engine Model

This paper we assume the diesel combustion engine with a turbocharger and without a variable geometry turbine. The torque model is becomes as flows:

$$J_{DE}\omega_{DE} = M_{ind} - M_{fr} - M_{pump} - M_{load} \tag{9}$$

Where J_{DE} is engine inertia, M_{ind} is the indicated engine torque, M_{fr} is the friction engine torque, M_{pump} is the pump torque, M_{load} is load torque. Engine torque components can be represented as follows [5-7, 12]:

$$M_{ind} = \frac{k_F U_F Q_{LHV} RT_{im}(e_0 + e_1\omega_{DE} + e_2\omega_{DE}^2)}{\eta_V V_d P_{im}} \quad \eta_V = e_3 + e_4\omega_{DE} + e_5\omega_{DE}^2 \tag{10}$$

$$M_{fr} = \frac{1000V_d\left(e_6 + e_7\omega_{DE} + e_8\omega_{DE}^2\right)}{2\pi n_R} \tag{11}$$

$$M_{pump} = e_9 P_{im} + e_{10} \tag{12}$$

$$M_{load} = X^T \dot{X}\eta_G \eta_{PG}\eta_D \tag{13}$$

$$U_F = e_{11}F_R^2 + e_{12}F_R + e_{13} \tag{14}$$

Where U_F is a rising quadratic function, depending on fuel rate F_R, Q_{LHV} is fuel lower heating value, R is the gas constant, P_{im} and T_{im} are pressure and temperature in intake manifold, respectively, V_d is the engine volume, η_V is volumetric efficiency, ω_{DE} is engine crankshaft speed, n_R is number of revolutions for each power stroke per cycle, $\eta_G, \eta_{PG}, \eta_D$ are efficiency of generator, primary gear and traction motor, respectively, $e_0 - e_{13}, k_F$ are constants. Because this engine is not equipped by variable geometry turbine therefore the turbocharger is not under our control, we assume only measurement of the intake manifold pressure and temperature.

So, the diesel engine model becomes (9-13) as follows with new constant coefficients:

$$\dot{\omega}_{DE} = \frac{k_{D1}U_F T_{im}(e_0 + e_1\omega_{DE} + e_2\omega_{DE}^2)}{P_{im}(e_3 + e_4\omega_{DE} + e_5\omega_{DE}^2)} - k_{D2}\left(e_6 + e_7\omega_{DE} + e_8\omega_{DE}^2\right) - \frac{e_9 P_{im} - e_{10}}{J_{DE}} - X^T \dot{X}\eta_G\eta_{PG}\eta_D \quad (15)$$

$$k_{D1} = \frac{k_{D1}Q_{LHV}R}{J_{DE}V_d}$$

$$k_{D2} = \frac{1000V_d}{2J_{DE}\pi\eta_R}$$

4.3   Chassis Model

We will consider HMTR with two motor-generators. We represent the motion of HMTR in a fixed Cartesian coordinate system P. We link with the robot's center of gravity orthonormal basis Z. Size of Z is the same as P. A mathematical model of HMTR can be represented by the following system of vector-matrix equations [4, 13]:

$$Y = \begin{bmatrix} P \\ \varphi \end{bmatrix} = RL_1 X \quad (16)$$

$$mL\dot{X} = M - F_s f' - L^{-1}F_O + F_N \quad (17)$$

$$F_s = \begin{bmatrix} R_g - R_c & 0 \\ 0 & R_g + R_c \end{bmatrix}$$

$$F_O = \begin{bmatrix} G\sin\alpha + R_{TR} + F_a \\ M_r \end{bmatrix}$$

$$R_g = (G/2)\cos\alpha\cos\beta \,, \quad R_c = G\varphi^2 r_c h/gB - XM_C X$$

$$M_r = (\mu Gl\cos\alpha\cos\beta(1-(l\xi/2)^2))/4$$

Where Y is the vehicle's position and orientation vector, X is vector of speeds of driven wheels, L and L_1 are matrixes of internal coordinates transformation, f'

is vector of soil coefficients, α is pitch, β is roll, φ is yaw, m is matrix of mass and inertia parameters, M is driven moments on wheels, G is weight, g is constant g=9.8, ξ is shifting of turn poles, l is vehicle's length, B is vehicle's base, h is vehicle's mass height, M_r is a turn resistance in case a tracked vehicle, μ is a special coefficient, in general $\mu = f(f', B, r_c)$, where r_c is vehicle's turn radius, R_{TR} is trailer's resistance, F_a is air resistance, M_C is matric of Coriolis force parameters, F_N is approximated external disturbance, as well as the influence of unmodelled dynamics in the dynamic model.

So, the complete mathematical model of HMTR with diesel-electric powertrain becomes as follows (5-8, 15-17):

$$\dot{Y} = RL_1 X$$

$$\dot{X} = (mL)^{-1}(M - F_s f' - L^{-1}F_O + F_N)$$

$$M_T = A^{-1}(\Theta - BX_T)$$

$$U_F = \left(\frac{P_{im}(e_3 + e_4\omega_{DE} + e_5\omega_{DE}^2)}{k_{DI}T_{im}(e_0 + e_1\omega_{DE} + e_2\omega_{DE}^2)} \right)\left(\omega_{DE} + k_{D2}(e_6 + e_7\omega_{DE} + e_8\omega_{DE}^2) - \frac{e_9P_{im} - e_{10}}{J_{DE}} - X^T \dot{X}\eta_G\eta_{PG}\eta_D \right) \quad (18)$$

This is an essentially non-linear system with a third order differential control vector of the coordinates vector.

5   Advanced Tracked Robot Control Algorithms

To achieve the required performance from HMTR we need a control system, which able to control the non-linear and multilinked system (5, 16-18). The control system should be able to asymptotically draw the error to zero and able to be stable in phase coordinates in general. As a control algorithm we will use the method of position-trajectory control from paper [14], whose effectiveness was proved in the papers [14-15].

The equation of HMTR motion is given by the following vector-matrix equation:

$$C^3\overset{\cdots}{\Psi} + 3C^2\ddot{\Psi} + 3\dot{\Psi} + \Psi = 0 \quad (19)$$

Where Ψ is vector of the desired requirement for HMTR motion (in steady-state). In this case, on the trajectories of motion should minimize an integral functional:

$$W = 2C\dot{\Psi}^T\Psi(\dot{\Psi}^T\Psi + \Psi^T\dot{\Psi})$$

Stability investigation described in the paper [14] and based on cascade of Lyapunov functions, the system is stable cause positive constant C.

On the assumption that $\alpha = \beta = \xi = 0$ with (7-8, 14, 18-19) The General position-trajectory algorithm [14] becomes as follows:

$$\Theta = AmL(C^3(J + J_V)RL_1)^{-1}(K_0\dot{X} - K_1X + K_2\dot{Y} - K_3Y - \Psi - CV) \tag{20}$$

$$\Theta = \begin{bmatrix} \omega_{DE} & I_{chrg} \end{bmatrix}^T \text{ or } \Theta = \begin{bmatrix} \omega_{DE} & U_{exc} \end{bmatrix}^T \tag{21}$$

$$U_F = \left(\frac{P_{im}(e_3 + e_4\omega_{DE} + e_5\omega_{DE}^2)}{k_{DI}T_{im}(e_0 + e_1\omega_{DE} + e_2\omega_{DE}^2)} \right) \left(\omega_{DE} + k_{D2}(e_6 + e_7\omega_{DE} + e_8\omega_{DE}^2) + \frac{e_9P_{im} - e_{10}}{J_{DE}} + X^T\dot{X}\eta_G\eta_{PG}\eta_D \right) \tag{22}$$

$$F_R = \frac{-e_{12} - \sqrt{e_{12}^2 - 4e_{11}(e_{13} - U_F)}}{2e_{11}} \tag{23}$$

$$K_0 = C^3(J + J_V)(\dot{R}L_1 + RL_1(mL)^{-1}A^{-1}B)$$

$$K_1 = C^3(J + J_V)(\dot{R} + \dot{R})L_1$$

$$K_2 = C^3(J + J_V)RL_1(mL)^{-1}J_{Fs} - 3C^2J - 2C^2J_V - 2C^3\Gamma - C^3\Gamma_V$$

$$K_3 = 3CJ + 3C^2\Gamma + C^3\Gamma$$

$$\dot{Y} = RL\dot{X} + \dot{R}LX$$

$$J = \frac{\partial\Psi}{\partial Y^T} \quad \Gamma = \frac{\partial J}{\partial t} \quad J_V = \frac{\partial V}{\partial Y^T} \quad \Gamma_V = \frac{\partial J_V}{\partial t} \quad J_{Fs} = \frac{\partial F_s}{\partial Y^T}$$

Where V is a vector of the required speed of HMTR.

The structure of the threshold device to switch charging and discharging modes of EDLC is shown in Figure 2.

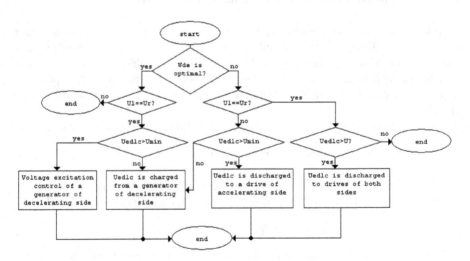

Fig. 2. The structure of the threshold device

6   Experiment

The example of motion control was simulated in MATLAB software. We introduce the task of HMTR movement control which described by the system (20-23) on a circle with a diameter 50 m, centered at the fixed coordinate system zero point. The required velocity is 21 m/s and required RMSE is less than 3.

Simulation setup: tracks are represented as tapes, no cushioning; no slipping; mass 2000 kg; length 2 m; width 1.2 m; height 1 m; sprocket radius 0.3 m; primary gear ratio 3; secondary gear ratio 0.5; electric machines constants are $d1=c1=0.6$, $d2=c2=0.06$, $c3=c4=0.7$; ground coefficients are 0.04. Simulation time is 120 seconds. The diesel engine characteristics are shown on Figure 3.

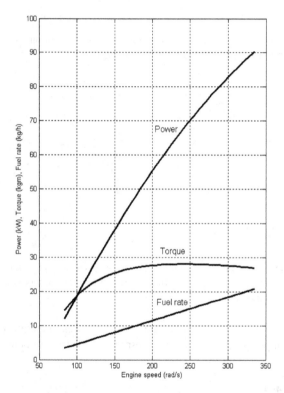

Fig. 3. The diesel engine characteristics: power (kW), torque (kgm) and fuel consumption rate (kg/h), depending on engine speed (rad/s)

The curves are shown only for engine speed limited by 330 rad/s, because between 330 rad/s and 420 rad/s the engine torque is drop and power is not raised. Strictly speaking the electric transmission converts the power into energy for traction electric drives, so this mode is not useful for HMTR. At this engine speed more than 420 rad/s the torque is much drop and this mode should be limited in any diesel combustion engine.

The simulation results are shown in Figures 4 and 5.

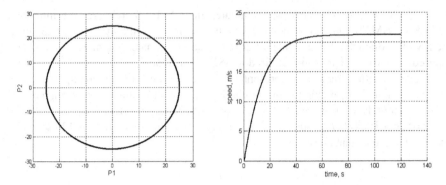

Fig. 4. The example simulation results: trajectory of HMTR motion in coordinate system {P1,P2} (*left*) and HMTR speed depending on time (*right*)

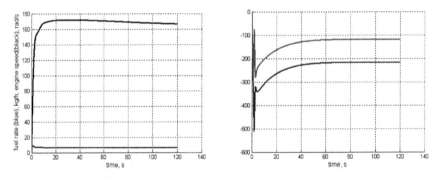

Fig. 5. The example simulation results: diesel engine speed shown by black line, fuel consumption rate shown in blue line depending on time (*left*) and traction motors voltages (*U1 is blue, U2 is red*) depending on time (*right*)

The simulation results shown accurate motion corresponding to the task. The RMSE of the example is 1.5155, which correspond to the requirement.

7   Conclusion

Introduced in this paper the motion control system for an advanced tracked robot can be used to control tracked robots of different classes. According to the simulation the system provided high precision movement and stability as a whole system. Model diesel-electric powertrain and EDLC are presented in a sufficiently general form. This allows a control system engineer to adapt the system for specific version of an electrical transmission and diesel engine. In this case, the model can be refined. The proposed system of The Position-trajectory control can significantly extend the capabilities of advanced robotic tracked vehicles.

Acknowledgements. This research has been supported by the grant of The President of Russian Federation for government support leading scientific school of Russian Federation NSh-1557.2012.10.

References

1. Conceicao, A.S., Oliveira, H.P., Sousa e Silva, A., Oliveira, D., Moreira, A.P.: A Nonlinear Model Predictive Control of an Omni-Directional Mobile Robot. In: IEEE International Symposium on Industrial Electronics, ISIE 2007, pp. 2161–2166 (2007)
2. Kausar, Z., Stol, K., Patel, N.: Nonlinear control design using Lyapunov function for two-wheeled mobile robots. In: 2012 19th International Conference Mechatronics and Machine Vision in Practice, M2VIP, pp. 123–128, 28–30 (2012)
3. Lian-Zheng, G., Rui-Feng, L., Li-Jun, Z.: A nonlinear trajectory tracking algorithm design based on second-order sliding mode control for mobile robots. In: IEEE International Conference on Robotics and Biomimetics, ROBIO 2007, pp. 1184–1189 (2007)
4. Pogosov, D.: Synthesis and Analysis of A Mathematical model of the High-maneuver tracked robots. Proc. KBNC RAN (1), 301–306 (2011)
5. Stefanopoulou, A., Kolmanovsky, I., Freudenberg, J.: Control of variable geometry turbocharged diesel engines for reduced emissions. IEEE Transactions on Control Systems Technology 8(4), 733–745 (2000)
6. Yanakiev, D., Kanellakopoulos, I.: Engine and Transmission Modeling for Heavy-Duty Vehicles. PATH Technical Note 95-6, 64 (1995)
7. Dextreit, C., Kolmanovsky, I.: Approaches to energy management of hybrid electric vehicles: Experimental comparison. In: UKACC International Conference on Control 2010, pp. 1–6 (2010)
8. Yu, H., Cui, S., Wang, T.: Simulation and performance analysis on an energy storage system for hybrid electric vehicle using ultracapacitor. In: Vehicle Power and Propulsion Conference, VPPC 2008, pp. 1–5. IEEE (2008)
9. Wight, G., Garabedian, H., Arnet, B., Morneau, J.-F.: Integration and Testing of a DC/DC Controlled Supercapacitor into an Electric Vehicle. EVS 18, Berlin (2001)
10. Ozatay, E., Zile, B., Anstrom, J., Brennan, S.: Power distribution control coordinating ultracapacitors and batteries for electric vehicles. In: Proceedings of the 2004 American Control Conference, vol. 5, pp. 4716–4721 (2004)
11. Chen, S., Chen, C.-C., Huang, H.-P., Hwu, C.-C.: Implementation of cell balancing with super-capacitor for robot power system. In: 2011 9th World Congress on Intelligent Control and Automation, WCICA, pp. 468–473 (2011)
12. Jankovic, M., Jankovic, M., Kolmanovsky, I.: Constructive Lyapunov control design for turbocharged diesel engines. IEEE Transactions on Control Systems Technology 8(2), 288–299 (2000)
13. Pogosov, D.: Observation Algorithm for Essentially Nonlinear Mobile Robots. Facta Universitatis Series: Automatic Control and Robotics (FU Aut. Cont. Rob.) 12(1), 1–8 (2013)
14. Pshikhopov, V.K.: Position-trajectory Control of Moving Objects, 183 p. TTI SFU, Taganrog (2009)
15. Pshikhopov, V.K., Medvedev, M.Y., Fedorenko, R.V., Sirotenko, M.Y., Kostyukov, V.A., Gurenko, B.V.: Control of Aeronautic Complexes: Theory and Design Technology, 394 p. Fizmatlit (2010)

Stable Modifiable Walking Pattern Generator with Arm Swing Motion Using Evolutionary Optimized Central Pattern Generator

Chang-Soo Park and Jong-Hwan Kim

Department of Electrical Engineering, KAIST,
291 Daehak-ro, Yuseong-gu, Daejeon, 305-701, Republic of Korea
{cspark,johkim}@rit.kaist.ac.kr

Abstract. In this paper, a stable modifiable walking pattern generator (MWPG) is proposed by a employing arm swing motion. The arm swing motion is generated by a central pattern generator (CPG) which is optimized by a constraint evolutionary algorithm. In this scheme, the MWPG generates a position trajectory of center of mass (COM) of humanoid robot and the CPG generates the arm swing motion. A sensory feedback in the CPG is designed, which uses a inertial measurement unit (IMU) signal. For the optimization of the CPG parameters, a two-phase evolutionary programming (TPEP) is employed. The effectiveness of the proposed scheme is demonstrated by simulations using a Webots dynamic simulator for a small sized humanoid robot, HSR-IX, developed in the Robot Intelligence Technology (RIT) Lab, KAIST.

1 Introduction

These days many humanoid robots have been developed [1]-[4]. However, their control algorithms still need to be improved further to perform a practical task. In this regard, research on developing robust walking patterns of humanoid robots plays one of important roles in this field.

There are two typical approaches to bipedal walking of humanoid robot, such as dynamic model based approach and biologically inspired approach. In the former, a 3-D linear inverted pendulum model (3-D LIPM) is one of popular schemes [5]–[8]. A modifiable walking pattern generator (MWPG) extends the conventional 3-D LIPM for a zero moment point (ZMP) variation by the closed form functions and can modify the humanoid robot's walking pattern in real-time while walking [9], [10]. In the latter, a central pattern generator (CPG) is widely used [11]-[14]. It can generate rhythmic output signals and modify generated signals to deal with environmental disturbance using a sensory feedback. Also, for stable bipedal walking, the methods about controlling upper body were developed. The process was presented to generate whole body motions for a biped humanoid robot from captured human motion [15]. A method to generate whole body motion of a humanoid robot considering linear/angular momentum was presented [16].

This paper proposes a stable MWPG with a arm swing motion using a constrained evolutionary optimized CPG. The proposed scheme generates a position trajectory of the humanoid robot's COM using the MWPG. Also, to stable walking, the CPG generates the arm swing motion and a sensory feedback in the CPG modifies the generated arm swing motion. Generated arm swing motion is proportional to the sagittal step length like human. The sensory feedback gets a inertial measurement unit (IMU) signal. To optimize the parameters of the CPG, a two-phase evolutionary programming (TPEP) is employed considering equality constraints [17], [18]. The effectiveness of the proposed scheme is demonstrated by computer simulations with the Webots model of a small sized humanoid robot HSR-IX developed in the Robot Intelligence Technology (RIT) Lab., KAIST.

This paper is organized as follows. In Section 2, the arm swing motion planning using constrained evolutionary optimized CPG is proposed. In Section 3, simulation results are presented and finally concluding remarks follow in Section 4.

2   Stable MWPG Using CPG

This section presents the proposed stable modifiable walking pattern algorithm with the arm swing motion using constrained evolutionary optimized CPG. In this paper, to generate the trajectory of COM, the MWPG is employed [9], [10]. Meanwhile, the arm swing motion is generated by the constrained evolutionary optimized CPG for stable bipedal walking.

2.1   Neural Oscillator

In this paper, a neural oscillator (NO) is employed to generate a rhythmic signal for the humanoid robot. The NO is composed of two neurons, each of which consists of two mutually excited neurons: an extensor neuron (EN) and a flexor neuron (FN) to generate rhythmic signal. Each neuron is defined as follows (Fig. 1) [19]:

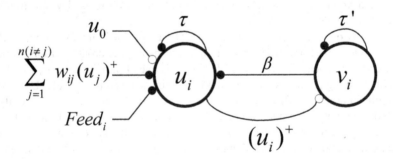

Fig. 1. Neuron structure. The lines ending with black and white circles are inhibitory and excitatory connections, respectively.

$$\tau \dot{u}_i = -u_i - \sum_{j=1}^{N} w_{ij}(u_j)^+ - \beta v_i + u_0 + Feed_i, \tag{1}$$

$$\tau' \dot{v}_i = -v_i + (u_i)^+ \tag{2}$$

$$(u_i)^+ = max(0, u_i) \tag{3}$$

where u_i, v_i and $(u_i)^+$ are a inner state, a self-inhibition state and a output signal of the ith neuron, respectively. u_0 is a constant input signal, w_{ij} is a connecting weight between ith and jth neurons, τ and τ' are time constants, β is a weight of the self-inhibition, and $Feed_i$ is a sensory feedback signal which is necessary for stable biped locomotion, of the ith neuron. Using EN and FN, the NO is defined as follows (Fig. 2):

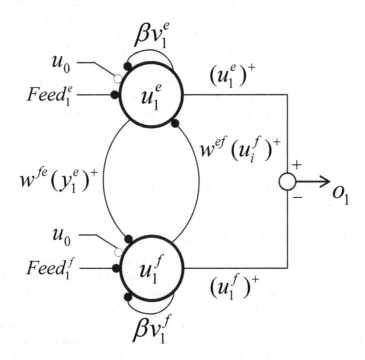

Fig. 2. Neural oscillator

$$\tau \dot{u}_1^e = -u_1^e - w(u_1^f)^+ - \beta v_1^e + u_0 + Feed_1^e, \tag{4}$$

$$\tau' \dot{v}_1^e = -v_1^e + (u_1^e)^+ \tag{5}$$

$$\tau \dot{u}_1^f = -u_1^f - w(u_1^e)^+ - \beta v_1^f + u_0 + Feed_1^f, \tag{6}$$

$$\tau' \dot{v}_1^f = -v_1^f + (u_1^f)^+ \tag{7}$$

$$o_1 = (u_1^e)^+ - (u_1^f)^+ \tag{8}$$

where the superscripts e and f denote the EN and the FN, respectively. o_i is output signal of ith NO.

2.2   Arm Swing Motion by CPG

Human generates the arm swing motion while walking. The arm swing motion decreases yawing momentum and help to keep humanoid robot's balance while walking. Using the NO, the arm trajectory is generated as follows for stable walking like human:

$$p_{arm}^l = \begin{cases} k_{arm}(-S^{pre} + \dfrac{S + S^{pre}}{2A_1}(o_1 + A_1)), & \text{if support leg is left} \\ k_{arm}(S^{pre} + \dfrac{S + S^{pre}}{2A_1}(o_1 - A_1)), & \text{otherwise} \end{cases} \tag{9}$$

$$p_{arm}^r = -p_{arm}^l \tag{10}$$

where $p_{arm}^{l/r}$ is a sagittal distance between the left/right elbow and the center of the humanoid robot's body. k_{arm} is a scaling factor. S and S^{pre} are sagittal step lengths at the present and the previous footstep, respectively. A_1 is a amplitude of o_1. When support leg is left/right, o_1 should become $-A_1/A_1$ at beginning single support phase and o_1 should become $A_1/-A_1$ at end double support phase to make like human.

The sensory feedback of the NO alters p_{arm}^l and p_{arm}^r for stable walking. For stable walking, the sensory feedback of the NO should minimize slip while walking. The sensory feedback gets the information from the yawing angle of the humanoid robot's body using IMU signal. The sensory feedback is designed as follows:

$$Feed_1^e = k_f(\theta_y - \theta_y^d) \tag{11}$$

$$Feed_1^f = -Feed_1^e \tag{12}$$

where k_f is the scaling factor. θ_y and θ_y^d are the real and desired yawing angles of the center of humanoid robot's body, respectively.

2.3   Evolutionary Optimization for CPG

The objective of this evolutionary optimization is to obtain the desired output signal of the NO and to minimize slip by the yawing moment. To obtain the desired output signal of the NO, time constants of the NO should be optimized. If support leg is left, when the magnitude of the output signal of the NO reaches the minimum (maximum) value, the time T_1^{min} (T_1^{max}) should be equal to time at beginning single (end double) support phase. Thus, $T_1^{max} - T_1^{min}$ should be equal to the single support time $T_{ss} + T_{ds}$. Also, when the magnitude of output signal reaches zero, the time T_1^0 should be equal to the middle value of $T_{ss} + T_{ds}$. To minimize slip by the yawing moment, the scaling factor in the sensory feedback should be optimized. To satisfy these constraints and the objective, the following objective function considering equality constraints is defined to obtain the time constants and the scaling factor in the sensory feedback by the TPEP [18]:

$$\text{Minimize } f = f_{yawing} + P \tag{13}$$

subject to

$$(T_1^{max} - T_1^{min}) - (T_{ss} + T_{ds}) = 0$$

$$T_1^0 - \frac{T_{ss} + T_{ds}}{2} = 0$$

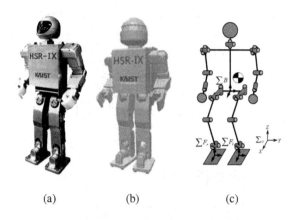

(a)   (b)   (c)

Fig. 3. (a) HSR-IX. (b) Simulation model. (c) Configuration.

with

$$f_{yawing} = \sum |\theta_y - \theta_y^d|$$

where P is the penalty which is given if humanoid robot loses its balance and collapses. f_{yawing} is the sum of $|\theta_y - \theta_y^d|$ while walking.

3 Simulations

The effectiveness of the proposed algorithm was demonstrated by computer simulations with the Webots model of the small sized humanoid robot, HSR-IX (Fig. 3). HSR-IX is the latest one of HSR-series. HSR is a small-sized humanoid robot that has been in continual redesign and development in RIT Lab, KAIST. Its height and weight are 52.8cm and 5.5kg, respectively. It has 26 DOFs that consist of 14 RC servo motors in the upper body and 12 DC motors with harmonic drives for reduction gears in the lower body. Webots is the 3-D robotics simulation software. Users can conduct the physical and dynamical simulation using Webots [21].

3.1 CPG Parameters Setting Using TPEP

In the simulation, Z_c was set as 23.35cm. β and u_0 were set as 2.5 and 2.5, respectively. The connecting weight, w, was set as 2.5 to make the phase difference between EN output and FN output to π [19]. The initial values of inner states, u_1 and u_2, and the self-inhibition states, v_1 and v_2, were set as -0.0042, 0.8372, 0.1834 and 0.6501, respectively, to make initial value of o_1 to the minimum value at beginning of single support phase. T_{ss} and T_{ds} were set as 0.8s and 0.4s, respectively. P in the objective

Table 1. Constrained evolutionary optimized parameters by TPEP

Time constants	τ	0.4676
	τ'	0.3089
Scaling factor	k_{arm}	0.6023
	k_f	3.2140

Table 2. A list of commanded step lengths

Steps	S (cm)	L (cm)	Support foot
1st	4.0	6.0	Left
2nd	4.0	-6.0	Right
3rd	6.0	6.0	Left
4th	4.0	-6.0	Right
5th	7.0	6.0	Left
6th	5.0	-6.0	Right
7th	2.0	6.0	Left
8th	6.0	-6.0	Right
9th	7.0	6.0	Left
10th	0.0	-6.0	Right

function was set as ∞. For the constrained evolutionary optimization of the CPG, the simulation model of HSR-IX modeled by Webots, was employed. The constrained evolutionary optimized parameters were obtained by TPEP as Table 1.

3.2   Walking Simulation Using CPG

In this simulation, the simulation model of HSR-IX by Webots was used. Table 2 shows the list of step lengths used for the simulation.

Fig. 4 shows the trajectories of p^l_{arm} and p^r_{arm} while walking. As shown in the figure, the amplitudes of p^l_{arm} and p^r_{arm} were proportional to the sagittal step length like human. Fig. 5 and Fig. 6 show the measured yawing momentums while walking without the arm swing motion and with the arm swing motion using the proposed CPG, respectively. As shown in the figure, the amplitude of yawing momentum was decreased by the arm swing motion using the proposed CPG. It means HSR-IX could walk stably with the proposed algorithm.

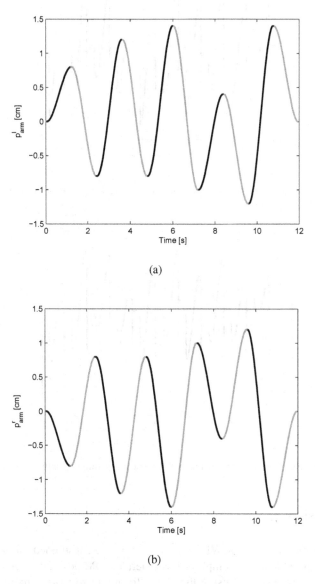

(a)

(b)

Fig. 4. Position trajectories of left and right arms. The thick and thin lines represent the trajectories when support leg is left and when support leg is right, respectively.

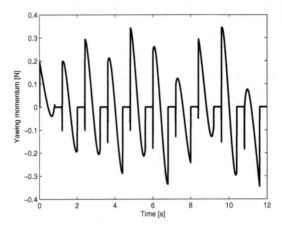

Fig. 5. The measured yawing momentum while bipedal walking without the arm swing motion

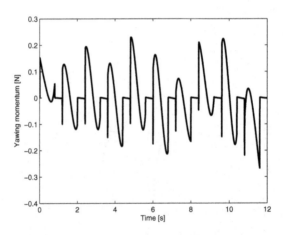

Fig. 6. The measured yawing momentum while bipedal walking with the arm swing motion

4   Conclusion

This paper proposed a stable MWPG algorithm with the constrained evolutionary optimized CPG. The MWPG was employed to generate COM position trajectory. The arm
swing motion was generated using the CPG. To minimize yawing momentum while
walking, the sensory feedback in the CPG modified generated signals. The sensory
feedback got the disturbance information using IMU sensor. TPEP was employed to
optimize parameters of the CPG considering some constraints. In order to demonstrate
the performance of the proposed scheme, computer simulations were carried out with
the Webots model of the small sized humanoid robot, HSR-IX developed in the RIT

Lab., KAIST. In the simulation, the amplitude of yawing momentum was decreased by the arm swing motion using the proposed CPG.

Acknowledgments. This research was supported by the MOTIE (The Ministry of Trade, Industry and Energy), Korea, under the Technology Innovation Program supervised by the KEIT (Korea Evaluation Institute of Industrial Technology)(10045252, Development of robot task intelligence technology that can perform task more than 80% in inexperience situation through autonomous knowledge acquisition and adaptational knowledge application).

This research was also supported by the MOTIE (The Ministry of Trade, Industry and Energy), Korea, under the Human Resources Development Program for Convergence Robot Specialists support program supervised by the NIPA (National IT Industry Promotion Agency)(H1502-13-1001, Research Center for Robot Intelligence Technology).

References

1. Sakagami, Y., Watanabe, R., Aoyama, C., Matsunaga, S., Higaki, N., Fujimura, K.: The intelligent ASIMO: system overview and integration. In: Proc. IEEE/RSJ Int. Conf. Intell. Robot. Syst., pp. 2478–2483 (2002)
2. Akachi, K., Kaneko, K., Kanehira, N., Ota, S., Miyamori, G., Hirata, M., Kajita, S., Kanehiro, F.: Development of humanoid robot HRP-3P. In: Proc. IEEE-RAS Int. Conf. Humanoid Robots, pp. 50–55 (2005)
3. Ogura, Y., Aikawa, H., Shimomura, K., Kondo, H., Morishima, A., Lim, H.-O., Takanishi, A.: Development of a new humanoid robot WABIAN-2. In: Proc. IEEE Int. Conf. Robot. Autom., pp. 76–81 (2006)
4. Park, I.-W., Kim, J.-Y., Lee, J., Oh, J.-H.: Online free walking trajectory generation for biped humanoid robot KHR-3(HUBO). In: Proc. IEEE Int. Conf. on Robot. Autom., pp. 1231–1236 (2006)
5. Kajita, S., Kanehiro, F., Kaneko, K., Fujiwara, K., Yokoi, K., Hirukawa, H.: A realtime pattern generator for biped walking. In: Proc. IEEE Int. Conf. Robot. Autom., pp. 31–37 (2002)
6. Motoi, N., Suzuki, T., Ohnishi, K.: A bipedal locomotion planning based on virtual linear inverted pendulum mode. IEEE Trans. Ind. Electron 56(1), 54–61 (2009)
7. Erbatur, K., Kurt, O.: Natural ZMP trajectories for biped robot reference generation. IEEE Trans. Ind. Electron. 56(3), 835–845 (2009)
8. Sato, T., Sakaino, S., Ohashi, E., Ohnishi, K.: Walking trajectory planning on stairs using virtual slope for biped robots. IEEE Trans. Ind. Electron. 58(4), 1385–1396 (2001)
9. Lee, B.-J., Stonier, D., Kim, Y.-D., Yoo, J.-K., Kim, J.-H.: Modifiable Walking Pattern of a Humanoid Robot by Using Allowable ZMP Variation. IEEE Trans. Robot. 24(4), 917–925 (2008)
10. Hong, Y.-D., Lee, B.-J., Kim, J.-H.: Command state-based modifiable walking pattern generation on an inclined plane in pitch and roll directions for humanoid robots. IEEE/ASME Trans. Mechatron. 16(4), 783–789 (2011)
11. Taga, G.: Emergence of bipedal locomotion through entrainment among the neuro-musculo-skeletal system and the environment. Physica D: Nonlinear Phenomena 75(1-3), 190–208 (1994)
12. Aoi, S., Tsuchiya, K.: Stability analysis of a simple walking model driven by an oscillator with a phase reset using sensory feedback. IEEE Trans. Robot. 22(2), 391–397 (2006)

13. Righetti, L., Ijspeert, A.J.: Programmable central pattern generators: an application to biped locomotion control. In: Proc. IEEE Int. Conf. Robot., pp. 1585–1590 (2006)
14. Park, C.-S., Hong, Y.-D., Kim, J.-H.: An Evolutionary Central Pattern Generator for Stable Bipedal Walking by the Increased Double Support Time. In: Proc. IEEE Int. Conf. Robot. Bio., pp. 497–502 (2011)
15. Nakaoka, S., Nakazawa, A., Yokoi, K., Hirukawa, H., Ikeuchi, K.: Generating whole body motions for a biped humanoid robot from captured human dances. In: Robotics and Automation, pp. 3905–3910 (2003)
16. Kajita, S., Kanehiro, F., Kaneko, K., Fujiwara, K., Harada, K., Yokoi, K., Hirukawa, H.: Resolved momentum control: humanoid motion planning based on the linear and angular momentum. In: Proc. IEEE/RSJ Int. Conf. IROS, pp. 1644–1650 (2003)
17. Myung, H., Kim, J.-H.: Hybrid evolutionary programming for heavily constrained problems. BioSystems 38(1), 29–43 (1996)
18. Kim, J.-H., Myung, H.: Evolutionary programming techniques for constrained optimization problems. IEEE Trans. Evol. Comput. 1(2), 129–140 (1997)
19. Matsuoka, K.: Sustained oscillations generated by mutually inhibiting neurons with adaptation. Biol. Cybern. 52(6), 367–376 (1985)
20. Vukobratovic, M., Borovac, B.: Zero-moment point-thirty live years of its life. Int. J. Humanoid Robot. 1(1), 157–173 (2004)
21. Michel, O.: Cyberbotics Ltd. WebotsTM: Professional mobile robot simulation. Int. J. Advanced Robot. Syst. 1(1), 39–42 (2004)

Performance Comparison between a PID SIMC and a PD Fractional Controller

L. Angel and J. Viola

Universidad Pontificia Bolivariana, Bucaramanga, Colombia
{luis.angel,jairo.viola}@upb.edu.co

Abstract. This paper presents the development and design of a fractional PD controller for a second order system with a pole at the origin, which approximates the dynamic behavior of a servomotor used in the field of robotics. FOPD is contrasted with a PID SIMC controller to analyze the system robustness. Four tests are performed to evaluate the robustness of the system which are: plant gain variation, variable setpoint, external perturbation and adding a random noise. Results show that fractional controller is more robust than PID SIMC controller.

Keywords: fractional control, FOPD, PID SIMC.

1 Introduction

When looking for a controller for any system it is always used an integer order PID (IOPID), given its simplicity in design and implementation, which use tuning methods as Ziegler and Nichols. However, some times for some systems it is not enough only with IOPID, because exists some external perturbations which change the behavior of system. In recent years has been developed a new control strategy called fractional Control, which uses operators s^α, which generalizes the controllers and improve dynamic behavior of the closed-loop system, especially in the presence of disturbances. Some applications where fractional control is used are electronic power control [1], plasma reactor [2], SCADA systems [3] aeronautics [4], repetitive control [5], among others. This paper will present the design methodology for 2 types of controllers, a fractional proportional derivative controller (FOPD) and a IOPID controller based on SIMC method which presents a simple scheme for tuning robust integer controllers [7].The plant corresponds to a second order with pole at the origin. Controllers were tested with set point changes, gain of the plant variations, external perturbations and random noise. The results show that the FOPD controller despite meets the design parameters, also it improves the system dynamics better than the PID SIMC controller. The paper is structured as follows. Initially the model of the process to be controlled is presented. Then, it presents the design method of FOPD. Next, the design method for PID SIMC. Finally, it presents the results obtained with each controller system control in tests varying setpoint, plant gain variation of 300%, a external perturbation and addition of random noise.

J.-H. Kim et al. (eds.), *Robot Intelligence Technology and Applications 2*,
Advances in Intelligent Systems and Computing 274,
DOI: 10.1007/978-3-319-05582-4_36, © Springer International Publishing Switzerland 2014

2   System to Be Controlled

The controlled system corresponds to a second order system with pole in the origin according to (1).

$$Gp(s) = \frac{3}{S(0.1s + 1)} \tag{1}$$

The controlled system correspond a servo system which is shown in Fig. 1. Gain margin is infinite and phase margin is 73.9 ° in the cutoff frequency equal to 2.88 rad /s (Fig. 2).

Fig. 1. Servo system

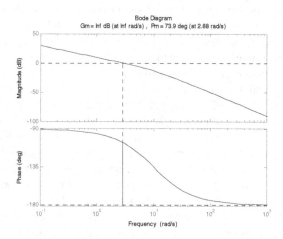

Fig. 2. Frequency response for servo system

3   FOPD Controller Design

According to [6], the FOPD controller is defined by the following equation:

$$Gc(s) = Kp(1 + k_d s^\alpha) \tag{2}$$

where: K_p: Proportional gain, K_d: Derivative gain and α: Fractional order for derivator.

For the FOPD design, it must meet the following conditions:

- **Gain crossover frequency:**

$$|Gp(wc) * Gc(wc)| = 0dB \tag{3}$$

- **Open loop phase margin:**

$$\arctan\big(Gp(wc) * Gc(wc)\big) = -\pi + Pm \tag{4}$$

where pm is the desired phase margin.

- **Robustness to gain variations of the plant:**

$$\frac{d(arctan(Gp(w) * Gc(w)))}{dw} = 0 \tag{5}$$

where w=wc which is the gain crossover frequency.

From (3) (4) and (5), it can be observed a system of 3 equations with 3 unknowns, where the unknowns are the FOPD controller parameters (K_p, K_d, α). To solve the equations, it is necessary to have the magnitude and phase of the system and controller. From (2) we obtain the magnitude and phase of the controller and of (1) the magnitude and phase of the system, which are as follows:

- **Controller magnitude**

$$Gc(jw) = Kp\sqrt{(1 + kdw^\alpha cos(\alpha\tfrac{\pi}{2}))^2 + (kdw^\alpha sin(\alpha\tfrac{\pi}{2}))^2} \tag{6}$$

- **Controller phase:**

$$Arctan(Cc(jw)) = tan^{-1}\left(\frac{sin\left((1-\alpha)\tfrac{\pi}{2}\right) + kiw^\alpha}{cos((1-\alpha)\tfrac{\pi}{2})}\right) + \frac{\alpha\pi}{2} - \pi - tan^{-1}(wT) \tag{7}$$

- **System phase:**

$$Arctan(Gp(jw)) = -Arctan(wT) - \frac{\pi}{2} \qquad (8)$$

- **System magnitude:**

$$|P(jw)| = \frac{k}{w\sqrt{1 + (wT)^2}} \qquad (9)$$

According to (3)

$$Kp = \frac{wc}{K}\sqrt{\frac{1 + (wT)^2}{(1 + kdw^\alpha cos(\alpha\frac{\pi}{2}))^2 + (kdw^\alpha sin(\alpha\frac{\pi}{2}))^2}} \qquad (10)$$

According to (4):

$$K_d = wc^{-\alpha}tan\left(atan(wcT) + pm - \pi + \alpha\frac{\pi}{2}\right)cos\left((1-\alpha)\frac{\pi}{2}\right) \qquad (11)$$

$$-wc^{-\alpha}sin(\alpha\frac{\pi}{2})$$

According to (5)

$$a_2wc^{2\alpha}kd^2 + b_2kd + a_2 = 0 \qquad (12)$$

where:

$$a_2 = \frac{T}{1+(wcT)^2}$$

$$b_2 = 2a_2wc^\alpha sin((1-\alpha)\frac{\pi}{2}) - \alpha wc^{\alpha-1}cos((1-\alpha)\frac{\pi}{2})$$

Solving the quadratic equation

$$K_d = \frac{-b_2 \pm \sqrt{b_2{}^2 - 4a_2{}^2wc^{2\alpha}}}{2a_2wc^{2\alpha}} \qquad (13)$$

With (11), (12), (13), we can determine the value of the constants K_p, K_d and α, following the procedure below:

1. To determine open-loop plant: the time constant, the gain crossover frequency of 0 dB, the phase margin, the delay time and gain of the system. From (1) the frequency response of the system is obtained: T= 0.1s, wc=2.88 rad/s, phase margin (pm)= 73.9° and K = 3;
2. Assign a range of solution for α. Selecting 0 <α <2 with a phase margin of 100°.
3. Plot the equations (12) and (13) in the α range and find the intersection point of (12) and (13), to obtain the value of α and K_d (Fig. 3).

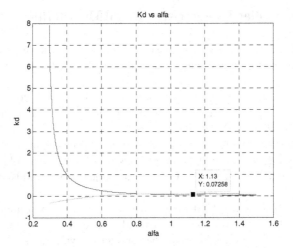

Fig. 3. K_d vs. α intersection (12) and (13) to obtain k_d and α

Resulting in: K_d = 0.0763, α =1.1137

4. Replace in (11) α and K_d to obtainK_p, resulting K_p=3.15.

With the controller constants determined, next step is the controller simulation, to which is used an Oustalop filter [3] of order 70 in a range of 0.001 rad/s to 1000 rad/s. The frequency response in open loop is shown in Fig. 4. As seen, the phase margin and the gain crossover frequency correspond to the design specifications.

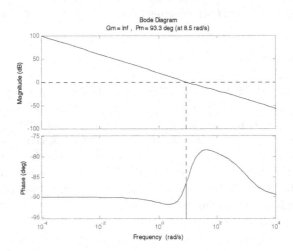

Fig. 4. Frequency response of the open loop system with controller

4   PID Controller Design Using the SIMC Method

The SIMC tuning technique was developed by Sirgud Skogestad and proposes a simple procedure to obtain the constants of a controller that makes the system robust to perturbations.

According to [7], this method consists of two steps to tune the controller. The first one is to obtain an approximate model of the system of order 1 or order 2 and the second one is to find the controller parameters.

In this case the plant model is of order 2 with pole at the origin (1), so that it is not necessary to implement order reduction methods which yields the following parameters which are   time constant $T_2 = 0.1$ s, delay time $\theta = 0$ seconds and the system gain, k= 3.

Knowing the plant model (1), it can perform the second step of tuning methodology. According to [7], for first order systems with pole at the origin the best controller is a PID, which is a set of rules to obtain the parameters, which are:

$$K_p = \frac{1}{k} * \frac{1}{t_c + \theta} \tag{13}$$

$$t_i = 4(T_c + \theta) \tag{14}$$

$$t_d = t_2 \tag{15}$$

T_c is the time constant which is adjusted to obtain the robustness of the system, which according to [7], should be equal to the system delay θ, but in this particular case is assumed as 0.1. Then, it is performed the calculation of the constants for the system according to (13) and (14). resulting in $K_p = 25/6$, $t_i = 8/25$ and $t_d = 2/25$.

5   Results

The step response of FOPD and PID SIMC controllers with the calculated constants above are shown at Fig.5 which shows that the PID SIMC controller has higher overshoot than FOPD controller, but the setting time is the same for both.

To perform a more detailed analysis of the controller's behavior, 4 tests are performed. The first one consists of varying of setpoint of the system. The second test is varying the gain of the system at   300%. The third test consist of adding a external perturbation with the 300% of variation of gain of the plant. The last one is adding random noise with   the variations of the last 3 tests.

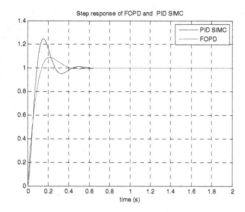

Fig. 5. Step response of the system with FOPD and PID SIMC controllers

5.1 Test 1

Initially, the input signal are formed for different signals, with the objective to probe the tracking and regulation responses. For this initially a ramp signal is raised and then a step series as show in Fig 6. It shows that both controllers has a good tracking response but with a step response the FOPD presents less overshoot than PID SIMC.

Results are shown in Fig. 6 which shows that while the gain is increased, the FOPD and PID SIMC controllers maintains the system dynamic behavior despite the gain is very high, but the PID SIMC has a biggest overshoot.

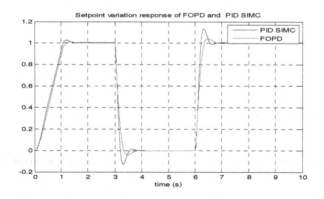

Fig. 6. Setpoint variation response for PID SIMC and FOPD controllers

5.2 Test 2

This test consist of gain variation of the plant of 300% using the same input signal proposed at test 1. Fig. 7 shows that when the gain is increased, the FOPD and PID SIMC controllers maintains the system dynamic behavior despite the gain varying is very high, but the PID SIMC has a biggest overshoot.

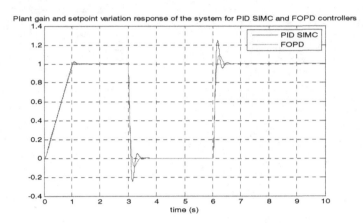

Fig. 7. Plant gain and setpoint variation response of the system for PID SIMC and FOPD controllers

5.3   Test 3

This test consist of adding a constant external perturbation of 20% at t=4s. Results Fig 8 shows both controllers eliminates the external perturbation at the same time but the FOPD controller   presents less overshoot than PID SIMC controller.

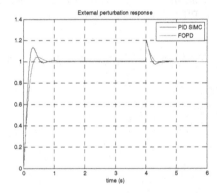

Fig. 8. External perturbation response for PID SIMC and FOPD controllers

5.4   Test 4

In this test, a random noise is added to the system and considers the conditions of the above test. Results are shown in Fig 9. Both controller are affected by the random noise (FOPD in less proportion, see zoom) but they maintenance the desired system dynamics.

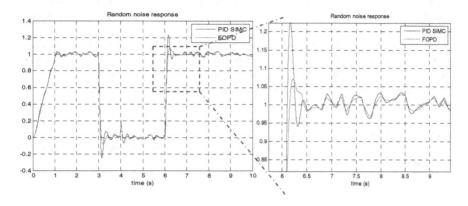

Fig. 9. Noise response for PID SIMC and FOPD controllers

6   Conclusions

This paper presents the design of a fractional controller for a second order system with pole at the origin. The implemented controller corresponds to a FOPD which results show that is robust to gain and setpoint variations. These results have been compared with a controller designed by using the technique SIMC.

From tests proposed to controllers, it can be seen that booth controllers has a good performance and improve the robustness of the system. The FOPD controller presents a better response front to plant   gain variations. The PID SIMC controller and FOPD has a good disturbance rejection and a tracking response. PID SIMC controller has more overshoot when the gain is increased and when the reference signal presents an abrupt change.   These results suggest that the FOPD is an excellent alternative for robotic applications, as it offers greater robustness against IOPID controllers with internal and external disturbances of the process.

References

1. Tehrani, K.A., Hamzaoui, M.: Current Control Design with a Fractional-Order PID for a Three-Level Inverter
2. Mukhopadhyay, S., Chen, Y.Q., Singh, A., Edwards, F.: Fractional Order Plasma Position Control of the STOR-1M Tokamak. In: Joint 48th IEEE Conference on Decision and Control and 28th Chinese Control Conference, pp. 422–427 (2009)
3. Jin, Y., Luo, Y., Wang, C., Chen, Y.: LabView Based Experimental Validation of Fractional Order Motion Controllers. In: Chinese Control and Decision Conference, pp. 323–328 (2009)
4. Changmao, Q., Naiming, Q., Zhiguo, S.: Fractional PID controller design of Hypersonic Flight Vehicle. In: International Conference on Computer, Mechatronics, Control and Electronic Engineering (CMCE), pp. 466–469 (2010)

5. Li, Y., Chen, Y., Ahn, H.-S.: A Generalized Fractional-Order Iterative Learning Control. In: 50th IEEE Conference on Decision and Control and European Control Conference (CDC-ECC), pp. 5356–5361 (2011)
6. Wang, C.Y., Luo, Y., Chen, Y.Q.: Fractional Order Proportional Integral (FOPI) and [Proportional Integral] (FO[PI]) Controller Designs for First Order Plus Time Delay (FOPTD) Systems. In: Chinese Control and Decision Conference, pp. 329–334 (2009)
7. Vilanova, R., Visioli, A.: PID Control in the Third Millennium. Advances in Industrial Control. Springer, London (2012)
8. Machado, J.: Discrete-time fractional-order controllers. Journal of Calculus & Applied Analysis, 47–66 (2001)
9. Monje, C., Vinagre, B.: Fractional-order Systems and controls, Fundamentals and applications. Springer, London (2010)

Implementation of High Reduction Gear Ratio by Using Cable Drive Mechanism

Bum-Joo Lee

Department of Electrical Engineering, Myongji University
San 38-2 Namdong, Yongin, Gyeonggi-do, Korea
bjlee@mju.ac.kr

Abstract. In spite of advantages such as lightness, back drivability and zero backlashes, cable drive systems are slightly utilized in the robotic applications because it is hard to implement high reduction gear ratio in an acceptable size to install. Cable drive systems are commonly implemented by connecting several single-level systems to increase the reduction ratio. Since tension adjustment and fastening devices for wire cables are mounted in each level, however, the size of the system and the risk of fault are rather high. Proposed mechanical structure resolves these inherent difficulties by seamless winding technique. Consequently, this manner reduced the complexity significantly.

Keywords: Reduction Gear, Cable Drive, Zero Backlash and Back Drivability.

1 Introduction

In the motion control area including robotics, high performance reduction gears are essential to achieve the precise control performance [1]. Typical factors which affect the performance are as follows: Zero backlashes, Back drivability, Durability, Compactness. [2-3]. Traditional gear systems such as spur, bevel, planetary and helical gears generally adopt rigid saw-toothed mechanical parts. Subsequently, a gear train is constructed by meshing several gears, so that enables high reduction gear ratio. These gear systems, however, generally suffer lost motion so called backlash due to mechanical clearance or slackness whenever the operational direction is changed. To enhance the performance, HDS (Harmonic Drive System) is developed. HDS is a special type of mechanical gear system which composes of wave generator, flex spline and circular spline. Compared with traditional gear systems, HDS has several advantages including zero backlash and high gear ratios. Consequently, in most applications which need high precision control, HDSs are utilized. Although it is much smaller than the planetary gears, however, it is still rather heavy to install robotic joint and it is also weak for the mechanical shock.

In recent, cable drive systems became interested because it can be installed away from the actual driving part so that the moment of inertia of the system gets smaller [4, 5]. As a result, cable drive system enables to design the mechanical structure more flexibly. In addition, the advantages include zero backlash, durability and back drivability [6-8]. Since the reduction ratio of cable drive system is proportional to the ratio

J.-H. Kim et al. (eds.), *Robot Intelligence Technology and Applications 2,*   425
Advances in Intelligent Systems and Computing 274,
DOI: 10.1007/978-3-319-05582-4_37, © Springer International Publishing Switzerland 2014

of input and output radii, it is essential to use multi-level mechanism to implement high reduction ratio. Moreover, it is inevitable to install wire fastening device including tension adjustment which increases size and complexity. In the proposed multi-level cable drive system, seamless winding technique is introduced to reduce the number of the fastening device. Consequently, it enables to implement high reduction gear ratio with simple mechanism.

This paper is organized as follows. In Section 2, developed system is overviewed. Subsequently, in Section 3, mechanical design features are pointed and detailed. Lastly, concluding remarks follow in Section 4

2   Overview of Developed System

Multi-level reduction gears are generally implemented by connecting several one-level reduction mechanisms. Proposed method also adopts this multi-level connection method. Without loss of generality, the method is explained with three-level system. Fig. 1 shows 3D CAD design of the proposed three-level cable drive mechanism. 'a', 'b', 'c' and 'd' are the input (=motor input), the 1st output, the 2nd output and the 3^{rd}(=final) output axes, respectively. Note that the kth output axis ($k = 1,2$) becomes the next input axis (($k + 1)^{th}$ axis). 'e' is the mechanical part which links the previous output to the next input axis by winding a single wire seamlessly. Therefore, this part is defined as SWM (Seamless Winding Module) and will be discussed in the next section. Lastly, 'f' and 'g' are fastening device including tension adjustment and housing frame, respectively. Also, note that the only one pair of fastening device is required for the proposed method through the effort of the SWM.

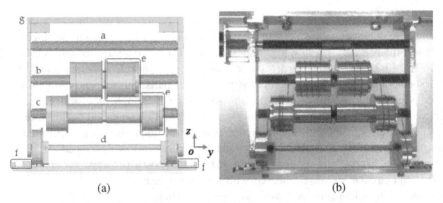

Fig. 1. Developed prototype. (a) 3D CAD model. (b) Actual system.

Motor was mounted at the 1st input axis and the wire is wound along the winding path (Fig. 1. (b)). Polynomial curvature for the SWM design was manufacture by CNC machine (this will be explained in the next section). For this proto type, steel wire with ϕ 0.8 was utilized and the pitch length d_i for the input axis was designed as 1.5 mm. Reduction ratio for each one-level reduction system is 4 to 1 and the consequent total ratio becomes 64 to 1.

3   Design of Multi-level Cable Drive System

Since the wire cable is rather flexible and extensible, it should be wound carefully neither to come untangled nor to be caught. To resolve these unfavorable factors, several design factors should be kept for a whole winding region.

3.1   Bending Radius

To avoid the fatigue of wire cable, winding radius should be greater than the minimum bendable value in the whole winding region.

$$r \geq r_b, \forall r \in \mathcal{R} \tag{1}$$

where r_b and $\mathcal{R}$ are the minimum bendable value and winding region.

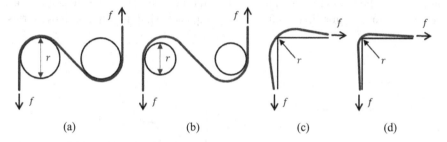

(a)	(b)	(c)	(d)

Fig. 2. Examples of inadequate winding radius
(a) Satisfied case ($r \geq r_b, f \leq f_n$).
(b) Insufficient case 1 with nominal tension ($r < r_b, f \leq f_n$).
(c) Insufficient case 2 with nominal tension ($r = 0, f \leq f_n$).
(d) Insufficient case 3 with forceful tension ($r = 0, f > f_n$).

Fig. 2 illustrates examples of inadequate winding radius. When the winding radius is greater than the minimum bending radius, the gear operates properly without any distortion (Fig. 2 (a)). As shown in Fig. 2 (b), if the winding radius is smaller than the minimum bending radius, however, looseness between the wire cable and the cylinder causes undesirable effects such as nonlinear joint flexibility and cylinder slips. These effects become more obvious when the winding radius is zero as shown in Fig.2 (c) and (d). If the tension becomes greater than the nominal value (f_n), it eventually causes plastic deformation. As a result, the gear operates improperly.

3.2   Winding Path

In order to achieve high performance, wire tension of the system should be kept tight and uniform for whole winding region during the operation. This region can be classified into three: constant winding radius region, inter-level winding region and non-uniform winding radius region. Winding path corresponding each region and subsequent design parameters should be computed accurately to satisfy the operation condition.

Constant Winding Region. This region corresponds to the normal axis winding such as the input and the output axis.   As shown in Fig. 2, when the wire is wound tightly, its path becomes straight line on an unfolded surface plane which is noted as AA'. Consequently, the ratio of winding distance to winding length should be constant for whole region.   Here, the winding distance and the winding length represent lengths for the axial direction and the radial direction (namely, circumference of cross section), respectively. Note that this ratio is defined as winding slop as follows

Definition 1. Winding slop is the ratio of the winding distance to the winding length as given by

$$W_s = \frac{winding\ distance}{winding\ length}. \tag{2}$$

To keep the tension uniform, the winding path is designed so that winding slop is constant for whole winding region including the inter-level winding region and the seamless winding region. In this case, the consequence winding path becomes shortest path.

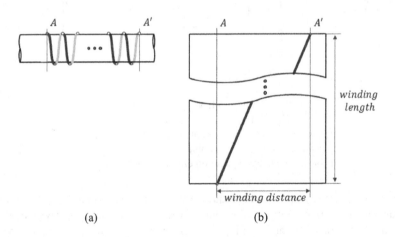

(a)   (b)

Fig. 3. Necessary and sufficient condition for tight winding.
(a) Axis with winding. (b) Wire path on unfolded plane.

Inter-level Winding Region. When the wire transits from the input axis to the output axis, it floats along a straight path in the 3D space as shown in Fig. 3. In the figure, the top and the bottom points are denoted as 'p_t' and 'p_b', respectively. Contact points for the input and the output axis are also denoted as 'p_{c_i}' and 'p_{o_c}', respectively. 'd_i' and 'd_o' are pitch lengths for each axis and 'c_i' and 'c_o' are winding distance from p_t (p_b) to p_{c_i} (p_{c_o}) on the zy-plane. And 'd_v' and 'd_h' are vertical and horisontal distances, respectively. In addition, 'a' and 'b' denote floating distances which is projected on the zx-plane and the yz-plane, respctively. Radii of the wire and the input and the output axes are represented by 'r_w', '$\bar{r}_i$' and '$\bar{r}_o$', respectively. Lastly, the angle at the contact point on the zx-plane is denoted as 'θ'. Note that the wire cable is shown as solid line where the dark line means the front side while the grey line means the reverse side.

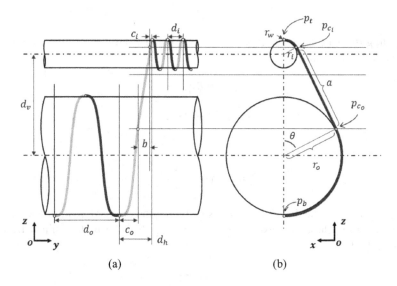

Fig. 4. Design for inter-level winding. (a) Front view. (b) Side view.

Since the radius of the wire has a certain value, r_w, the radius of each axis should be modified for the calculate of the winding path. This modified radius is defined as *effective radius* and calculated as follows:

$$r_i = \bar{r}_i + r_w , \quad r_o = \bar{r}_o + r_w. \tag{3}$$

And the reduction ratio for one-level is given by

$$\sigma = \frac{r_o}{r_i}. \tag{4}$$

From the minimum bending radius of the wire cable, r_i is selected as a suitable value (eq. (1)). Subsequently, r_o is determined from eq. (4). As mensioned above, to keep the tension uniform, the winding slop should be constant over a whole winding region.   From eq. (2),

$$w_s = \frac{d_i}{2\pi r_i} = \frac{d_o}{2\pi r_o}. \tag{5}$$

Here, d_i is design parameter to make a grooved shape to prevent the path departure. Both of w_s and d_o are determined from eq. (5). Subsequently, the floating distance on the zx-plane is calculated by Euclidian distance as follows:

$$a = \sqrt{d_v^2 - (r_o - r_i)^2}, \tag{6}$$

where d_v is vertical distance to be manually selected in the consideration of the manufacturing and the assembling. Consequently, from eq. (2) and (6), the floating distance on the yz-plane is calculated as follows:

$$b = w_s a. \tag{7}$$

Winding distance for each axis is also determined from the winding slop and the angle at the contact point.

$$c_i = w_s \theta r_i, \tag{8}$$

$$c_o = w_s (\pi - \theta) r_o, \tag{9}$$

where

$$\theta = \mathrm{atan} \left(\frac{a}{r_o - r_i} \right). \tag{10}$$

As a result, from the design parameters including the effective radii of axes and the distance between these two axes, the total winding distance is determined from eq. (8), (9) and (10)

$$d_h = b + c_i + c_o. \tag{11}$$

To keep the wire tension constant in the inter-level winding region, the wire should be wound along the path from the top point p_t to the bottom point p_b with the horizontal distance d_h from eq. (11).

Non-uniform Winding Radius Region. In the proposed multi-level cable drive system, the output axis of the one-level also becomes the input axis for the next one-level like the traditional gear train. In other words, the output axis and the input axis are mounted on a common axis and combined each other. Since the force is transmitted through the wire cable, in the traditional mechanism, each end of the wires is fixed on the corresponding axis by using some fastening devices. These fastening devices inevitably increase the complexity of the mechanism and the consequent volume of the gear system makes rather bulky. In the proposed system, only one single wire is utilized and it is wound seamlessly through a specially shaped part without any fastening devices except the final output axis. This part is defined as SWM (Seamless Winding Module) and is shown in Fig. 4. Here, the start and the end point for the transition are denoted as 'p_s' and 'p_e', respectively. And the transitional distance and the offset margin are represented as 'd_s' and 'r_d', respectively.

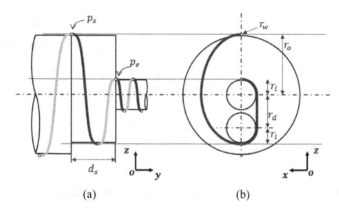

(a) (b)

Fig. 5. Design of SWM (Seamless Winding Module). (a) Front view. (b) Side view.

Similar to the inter-level winding region, wire should be wound keeping the winding slop constant. To transit the wire seamlessly, winding path on the zx-plane should be smoothly decreased from the output radius r_o to the input radius r_i. Note that the marginal offset r_d is applied for the sake of easy assembling. In this paper, cubic polynomial function is adopted to generate this path.

$$r(q) = \alpha_3 q^3 + \alpha_2 q^2 + \alpha_1 q + \alpha_0, \tag{12}$$

with boundary conditions

$$r(0) = r_o, \ r(\pi) = r_d, \ r'(0) = r'(\pi) = 0.$$

From eq. (12), total winding length on the SWM is calculated as follows:

$$l = \int_0^\pi r(q)dq + \pi r_i + r_d. \tag{13}$$

And the corresponding winding distance is given as

$$d_s = w_s l. \tag{14}$$

Consequently, the offset margin is calculated from eq. (12)-(14).

$$r_d = \frac{4r_i d_s - d_i(3r_i + r_o)}{d_i\left(1 + \frac{2}{\pi}\right)}. \tag{15}$$

Since the winding path of the SWM is computed along the wire center, actual path for the manufacturing should be subtracted by wire radius.

$$\bar{r}(q) = r(q) - r_w. \tag{16}$$

From the winding path derived in eq. (16), SWM is designed by 3D CAD tool and manufactured by CNC machine as illustrated in Fig. 5. With this SWM, wire can be wound without the fastening device.

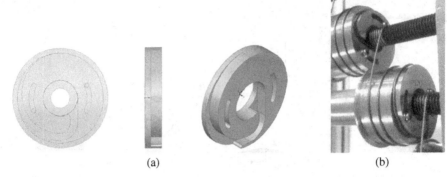

(a)   (b)

Fig. 6. SWM (Seamless Winding Module). (a) 3D CAD design. (b) Proto-type.

4   Conclusion

Multi-level cable drive system was proposed. The system enables high reduction gear ratio via connecting several one-level reduction systems. Mechanical complexity was effectively reduced by winding only one single wire cable seamlessly. Consequently, the size of the system was reduced significantly. Winding path was also mathematically derived neither to come untangled nor to be caught. According to proposed manner, the 3D CAD design and proto type were also developed.

References

1. Lohmeier, S., Buschmann, T., Ulbrich, H.: Humanoid Robot LOLA. In: IEEE Int. Conf. on Robotics and Automation, pp. 775–780 (2009)
2. McKerrow, P.J.: Introduction to Robotics, pp. 100–117. Addison-Wesley Publishing Company (1993)
3. Ishida, T., Takanishi, A.: A Robot Actuator Development With High Backdrivability. In: IEEE Conf. Robotics, Automation and Mechatronics, pp. 1–6 (2006)
4. Rooks, B.: The harmonious robot. Industrial Robot: An International Jounal 33(2), 125–130 (2006)
5. Lens, T., von Stryk, O.: Investigatnion of Safety in Human-Robot-Interaction for a Series Elastic, Tendon-Drievn Robot Arm. In: IEEE Int. Conf. on Intelligent Robots and Systems, pp. 4309–4314 (2012)
6. Truong, H., Abdallah, S., Rougeaux, S., Zelinsky, A.: A Novel Mechanism for Stereo Active Vision. In: Australian Conf. on Robotics and Automation (2000)
7. Aquirre-Ollinger, G., Colgate, J.E., Peshkin, M.A., Goswami, A.: Design of an active one-degree-of-freedom lower-limb exoskeleton with inertia compensation. Int. Journal of Robotics Research 30(4), 486–499 (2011)
8. Massie, T.H., Salisbury, J.K.: The PHANTOM Haptic Interface: A Device for Probing Virtual Objects. In: Proc. ASME Symposium on Haptic Interfaces for Virtual Environment and Teleoperator Systems, vol. 55(1), pp. 295–300 (1994)

Trajectory Tracking Control Using *Echo State Networks* for the *CoroBot´s* Arm

Cesar H. Valencia, Marley M.B.R. Vellasco, and Karla T. Figueiredo

LIRA – Computational Intelligence and Applied Robotics Laboratory
Electric Engineering Department
Pontifical Catholic University of Rio de Janeiro (PUC-Rio)
R. Marquês S. Vicente 225 – Gávea, Rio de Janeiro, RJ, Brasil
{chvn,marley,karla}@ele.puc-rio.br

Abstract. Different neural network models have proven being useful for tracking purposes in robotic devices. However, some models have shown superior performances to others that generate a large computational cost. This is the case of recurrent neural networks, which due to the temporal relationship existing allows satisfactory answers. Furthermore, training used by traditional algorithms, require a relatively high convergence time for some applications, especially those that are on-line. Given this problematic, this paper suggests use Echo State Networks (ESN) to perform such tasks. Additionally, results are presented for two sets of predefined tests, which were used to validate control behavior of trajectories in a manipulator embedded in a mobile platform. The results presented are related to the planar control of the manipulator in a closed loop.

Keywords: Echo State Networks, Neural Networks, Robotic, Controller, Trajectory Tracking.

1 Introduction

The controlling robotic manipulators problem has been addressed by various researchers using different control techniques [1]. Computational intelligence models have positioned themselves as way for this kind of tasks, as confirmed by the results presented using Fuzzy Logic [2], Neural Networks [3] and Reinforcement Learning [4].

The aim of this work is focused on determine the feasibility of using Echo State Networks (ESN) to perform such task and to compare results obtained with traditional models of neural networks used previously, noting the fact that temporal relationship between input and output in recurrent networks, could be better utilized depending on the configuration used. This leads to the need of characterizing the ESN's controller performance based on the configuration parameters available for design.

This paper is organized as follows: section two presents the mathematical basis of the ESN. Section three presents the solution and results obtained for the proposed model and comparison with traditional NNs, and finally section four, shows conclusions.

J.-H. Kim et al. (eds.), *Robot Intelligence Technology and Applications 2*,   433
Advances in Intelligent Systems and Computing 274,
DOI: 10.1007/978-3-319-05582-4_38, © Springer International Publishing Switzerland 2014

2   Echo State Networks (ESN)

Recurrent neural Networks (RRN) are a very powerful generic tool, integrating both large dynamical memory and highly adaptable computational capabilities [5]. ESNs are RRNs inspired by recent neuropsychological experiments [6], which main characteristic is the presence of at least one feedback loop. According to [7], recurrent neural networks are those that use at least one output neuron of the network at n-time and use it as input to other neurons in a $n + 1$ *time*.

Figure 1 shows the basic structure including the reservoir with internal units.

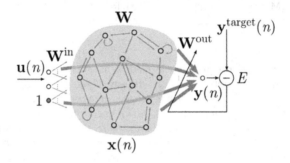

Fig. 1. Echo State Network [5]

The most important element in ESNs is the reservoir, which contains a large number of internal units which are interconnected randomly and / or self-connected. According to [8] after the reservoir initialization, it will remains fixed, however, during the training process, the connections weights going from the reservoir to the output are modified.

The reservoir, being an input-driven dynamical system, should provide a rich and relevant enough signal space in $x(n)$, such that the desired $d(n)$ could be obtained by linear combination from it [5].

The weight updates can be performed off-line using linear regression, or on-line using recursive minimum least squares methods.

The state of activation of neurons in the reservoir can be obtained using Equation 1.

$$x(n) = \varphi(w^e u(n) + w^R x(n-1) + w^{rr} d(n-1)) \tag{1}$$

Where $x(n)$ is the internal state of the reservoir in a $\varphi(\cdot) = (\varphi_1, \dots, \varphi_N)^T$ time, and they are sigmoid activation functions, $d(n-1)$ is the action potential of the output unit to an earlier time, $u(n)$ is an input vector, w^e, w^R and w^{rr} are the weights matrix for the input units, the reservoir and the recurrent connections respectively. The output of an ESN is given by equation 2.

$$d(n) = \varphi^s \left(w^s \cdot \begin{bmatrix} x(n) \\ d(n-1) \end{bmatrix} \right) \tag{2}$$

Where φ^S may be linear or sigmoid, depending on the task complexity, and w^S is the matrix containing the weights of the output connections, this is determined after the training.

Using K units as input, N units in the reservoir, and L units in output, the dimensions of matrices are: $w^e \in \mathbb{R}^{N*K}$, $w^R \in \mathbb{R}^{N*N}$, $w^{rr} \in \mathbb{R}^{N*L}$ and $w^S \in \mathbb{R}^{L*N}$.

The training was carried out in two stages: first sampling occurs, and then calculating the weights that are presented in the next section.

2.1   Sampling

Sampling the internal signals is determined by equation 3 for the operating range $n = A_p, ..., A_q$ and they are stored in the matrix J sized $(A_q - A_p) + 1 * K$, where A_p e A_q are the initial and final values of the sample strip, respectively.

$$x(n) = (x_1(n), ..., x_K(n)) \tag{3}$$

2.2   Calculating the Output Weights

The second step is the weights calculation (equation 4), which are stored in the matrix w^S to the output unit $d(n)$, and it could be seen that the output signal $y(n)$ is estimated with the series combination of linear internal activation $x_i(n)$ according to equation 4.

$$y(n) \approx d(n) = \sum_{i=1}^{K} w_i^S x_i(n) \tag{4}$$

The output signal $y(n)$ is stored in matrix U of size $(A_q - A_p) + 1 * L$. It is recommended that the value A_p is sufficiently distant so that dynamics of the network is not determined by the initial states.

The offline calculation of the weights of regression is equivalent to multiplying the J pseudo-inverse with U matrix as shown in equation 5.

$$w^S = J^{-1}U \tag{5}$$

2.3   Training Error

Equation 6 is used to obtain the mean square error for training data.

$$MSE_{train} = \frac{1}{(A_q - A_p) + 1} \sum_{n=p}^{q} \left(y(n) - \sum_{i=1}^{K} w_i^S x_i(n) \right)^2 \tag{6}$$

2.4   Testing Error

To obtain the test error, equation 7 is used. Where the generated output $d(n)$ is the network trained data set in the range for testing, A_e and A_f are the initial and final values of the test strip respectively.

$$MSE_{test} = \frac{1}{\left(A_f - A_e\right) + 1} \sum_{n=e}^{f} (y(n) - d(n))^2 \tag{7}$$

3   Echo State Network Controller

The developed controller considered the reference input and current output (plant output) providing the necessary torque in each of the motors. By using this variable, it determines the required angle on every joint to position every actuator in each of the points that make up the desired path. Such positioning is performed on the block that solves the inverse kinematics developed in [9] using the same network. Controller configuration parameters are outlined in [10] using a control model for reference. Figure 2 shows the block diagram presented for determining the output of the plant, controller error, and model error.

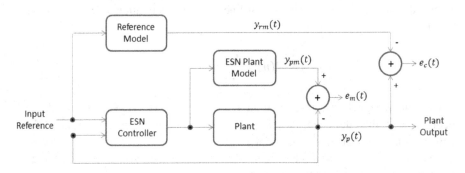

Fig. 2. Model reference ESN control system. Adapted [10].

To obtain the plant´s error of the model equation 8 was used, which considered, the output obtained from the block that represents the identification of the plant made with the ESN ($y_{pm}(t)$)and the output obtained from the block representing the plant ($y_p(t)$).

$$e_m(t) = y_{pm}(t) - y_p(t) \tag{8}$$

The controller error can be obtained using Equation 9, which considers the output of the block representing the plant ($y_p(t)$) and the output of the obtained reference model initially used ($y_{rm}(t)$).

$$e_c(t) = y_p(t) - y_{rm}(t) \tag{9}$$

Driver implementation was done using Simulink ®. The blocks corresponding to each of the elements of the complete system were integrated into the system presented in Figure 3.

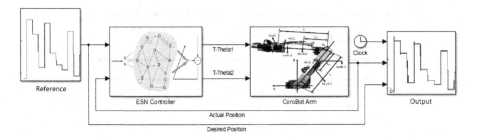

Fig. 3. Simulink Controller

As described above, the system has as input signal reference of the target and the current output. It delivers the necessary torque output for each of the joints of the arm (plant). The end block output allows a comparison in order to obtain the performance of the overall model.

4   Results

In this section are presented results of two computational intelligence models. To determine the performance of the proposed model there are used two different trajectories. As evaluation criterion, it was used the Root Mean Square Error (RMSE) as well as the spatial representation of the paths used.

4.1   Multilayer Perceptron Networks

The use of MLP networks is based on the results presented in trajectory control robotic manipulator's similar problems [11], [12] and [13] where the control objective was achieved successfully.

Several configurations were evaluated during testing. The activation function for all networks in the hidden layer was hyperbolic sigmoid tangent. Implementation and testing was performed using Matlab 2011b. Tables 1 and 2 present the results obtained for the five best configurations.

Table 1. Trajectory 1 - Results MLP Networks

Trajectory 1		
Neurons	Train RMSE	Test RMSE
15	0.4211	0.5737
17	0.4175	0.5688
22	0.4021	0.5584
26	**0.3822**	**0.5477**
27	0.3925	0.5489

Table 2. Trajectory 2 - Results MLP Networks

Trajectory 2		
Neurons	Train RMSE	Test RMSE
17	0.4355	0.4814
19	0.4354	0.4782
22	0.4269	0.4620
23	**0.4230**	**0.4304**
25	0.4258	0.4521

Figures 4 and 5 present the reference signal and the plant´s response to the training set of path 1, as well as the error generated by the absolute difference in position on the manipulator.

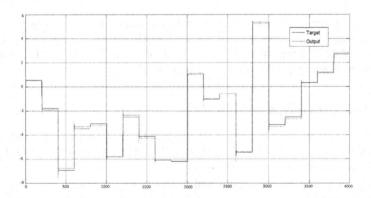

Fig. 4. Training set, target and output

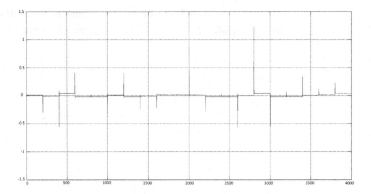

Fig. 5. Training set error

Figures 6 and 7 show the reference signal and plant response path 1, as well as the error generated by the absolute difference in the manipulator position.

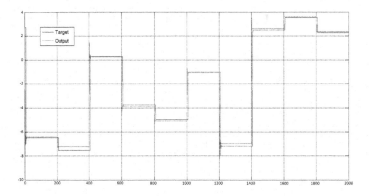

Fig. 6. Test set, target and output

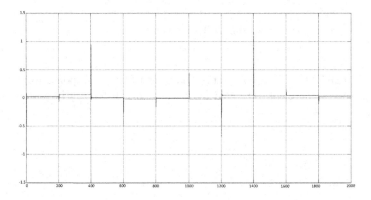

Fig. 7. Test set error

Figures 8 and 9 show the reference signal and plant´s response to the training set of path 2, as well as the error generated by the absolute difference in position on the manipulator.

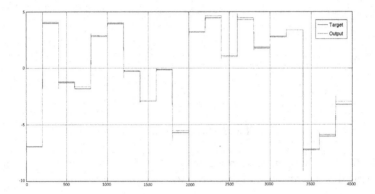

Fig. 8. Train set, target and output

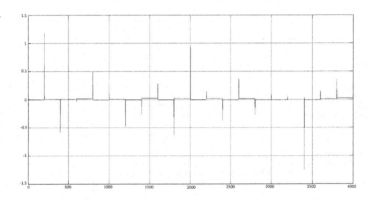

Fig. 9. Train set error

Figures 10 and 11 show the reference signal and plant´s response to the test set path 2, as well as the error generated by the absolute difference in position on the manipulator.

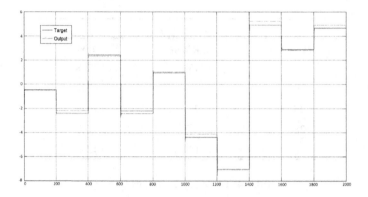

Fig. 10. Test set, target and output

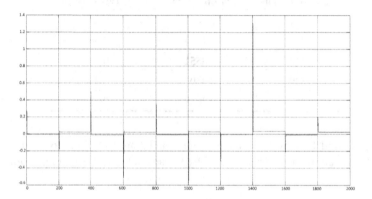

Fig. 11. Test set error

4.2   Echo State Networks

Tests conducted using ESN are an attempt to include the use of recurrent neural networks to control robotic device tasks. Different jobs have satisfactory results in various applications [14], [15] and [16].

In Tables 3 and 4 are presented the results pair configurations the top 5 in the 2 paths used.

Table 3. Trajectory 1 – Results ESN

Trajectory 1		
Reservoir Units	**Train RMSE**	**Test RMSE**
180	0.3453	0.4298
200	0.3325	0.4254
250	**0.2907**	**0.4089**
280	03087	0.4163
300	0.3129	0.4201

Table 4. Trajectory 2 - Results ESN

Trajectory 2		
Reservoir Units	**Train RMSE**	**Test RMSE**
280	0.3627	0.3543
290	0.3574	0.3471
310	**0.3551**	**0.3353**
330	0.3597	0.3412
350	0.3648	0.3510

Figures 12 and 13 show the reference signal and plant´s response to the training set of path 1, as well as the error generated by the absolute difference in position on the manipulator.

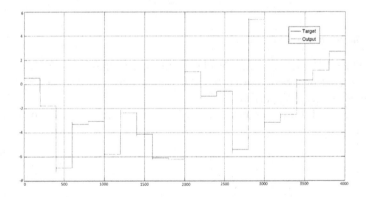

Fig. 12. Train set, target and output

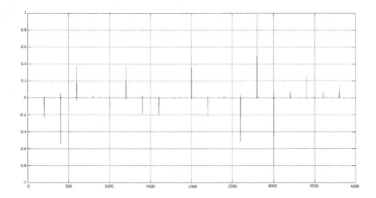

Fig. 13. Train set error

Figures 14 and 15 show the reference signal and plant's response to the test set of path 1, as well as the error generated by the absolute difference in position on the manipulator.

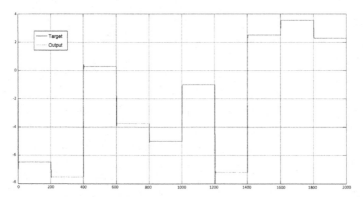

Fig. 14. Test set, target and output

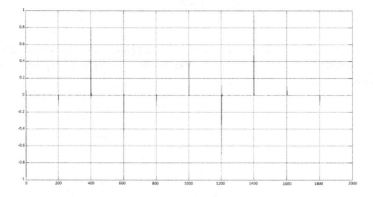

Fig. 15. Test set error

Figures 16 and 17 show the reference signal and plant´s response to the training set of path 2, as well as the error generated by the absolute difference in position on the manipulator.

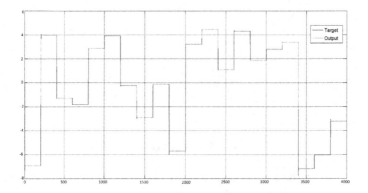

Fig. 16. Train set, target and output

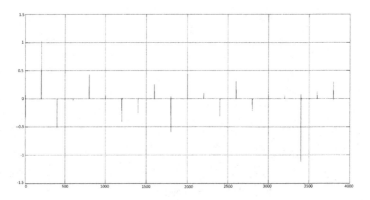

Fig. 17. Train set error

Figures 18 and 19 show the reference signal and plant´s response to the test set of path 1, as well as the error generated by the absolute difference in position on the manipulator.

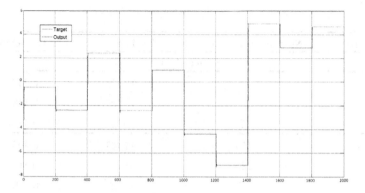

Fig. 18. Test set, target and output

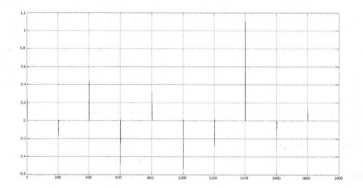

Fig. 19. Test set error

5   Conclusion

Results confirmed the feasibility of using ESN in the trajectory control for the manipulator used. Table 5 shows the best results for the models used in this work.

The obtained results using MLP networks demonstrated an acceptable response with some significant variations in reference error signal changes. RMSE values were relatively low, demonstrating a response time which allows to maintain the desired trajectory.

Some MLP with specific features, as an activation function or training algorithm, could provide a better performance than that the obtained in this work. This will be the subject of a coming study. However, it was possible to determine that the number of neurons was within a defined range and tests with larger amounts had no better response than those presented in table 5.

Table 5. Best results sets for the used models

Model	Neurons/Reservoir Units	Train RMSE	Test RMSE
	Trajectory 1		
	26	**0.3822**	**0.5477**
MLP	27	0.3925	0.5489
	22	0.4021	0.5584
	250	**0.2907**	**0.4089**
ESN	280	03087	0.4163
	300	0.3129	0.4201
	Trajectory 2		
	23	**0.4230**	**0.4304**
MLP	25	0.4258	0.4521
	22	0.4269	0.4620
	310	**0.3551**	**0.3353**
ESN	290	0.3574	0.3471
	330	0.3597	0.3412

The use of ESN presented better results than MLP networks, lower recovery times as presented in table 6, and smaller errors. In tests using connections to the output layer reservoir, the results showed that the temporal relationship entry / exit is considered by the connections of the reservoir and thus the performance showed faster responses to changes in values.

Table 6. Best results sets for the used models

	Time Comparison (s)	
Trajectory	MLP	ESN
1	39.54	14.62
2	46.93	16.48

Two parameters were instrumental in the ESN design. First: the number of units of the reservoir which determined the activation states of the reservoir. And second, the recurrent connections which allowed to keep the dynamism of the response.

References

1. Hu, L., Zhao, P., Lu, Z.: Trajectory tracking control based on global fast terminal sliding mode for 2-DOF manipulator. In: 2nd International conference on Intelligent Control and Information Processing (ICICIP), vol. 1, pp. 65–69 (2011)

2. Botsali, F.M., Kalyoncu, M., Tinkir, M., Onen, U.: Fuzzy logic trajectory control of flexible robot manipulator with rotating prismatic joint. In: 2nd International Conference on Computer and Automation Engineering (ICCAE), vol. 3, pp. 35–39 (2010)
3. Jurado, F., Flores, M.A., Castaneda, C.E.: Continuos-time neural control for a 2 DOF vertical robot manipulator. In: 8th International Conference on Electrical Engineering Computing Science and Automatic Control (CCE), pp. 1–6 (2011)
4. Stulp, F., Theodoruo, E.A., Schaal, S.: Reinforcement Learning of clothing assistance with a dual-arm robot. IEEE Transactions on Robotics 28, 1360–1370 (2012)
5. Lukoševičius, M.: A practical guide to applying echo state networks. In: Montavon, G., Orr, G.B., Müller, K.-R. (eds.) Neural Networks: Tricks of the Trade, 2nd edn. LNCS, vol. 7700, pp. 659–686. Springer, Heidelberg (2012)
6. Jaeger, H.: Reservoir riddles: suggestions for echo state network research (extended abstract). In: IEEE International Joint Conference on Neural Networks, vol. 3, pp. 1460–1462 (2005)
7. Lukoševièius, M., Jaeger, H.: Reservoir computing approaches to recurrent neural network training. Computer Science Review 3(3), 127–149 (2009)
8. Ganesh, K., Shishir, B.: Effects of spectral radius and settling time in the performance of echo state networks. Neural Networks 22(7), 861–863 (2009)
9. Valencia, C.H., Vellasco, M.M.B.R., Figueiredo, K.T.: Comparative Analysis of Arm Control Performance Using Computational Intelligence. In: Kim, J.-H., Matson, E., Myung, H., Xu, P. (eds.) Robot Intelligence Technology and Applications. AISC, vol. 208, pp. 187–197. Springer, Heidelberg (2013)
10. Liu, D., Chen, H., Jiang, R., Liu, W.: Study of ride comfort of active suspension based on model reference neural network control system. In: Sixth International Conference on Natural Computation (ICNC), vol. 4, pp. 1860–1864 (2010)
11. Hoang, T.T., Hiep, D.T., Duong, B.G., Vinh, T.Q.: Trajectory tracking control of the nonholonomic mobile robot using torque method an neural network. In: 8th IEEE Conference on Industrial Electronics and Applications (ICIEA), pp. 1898–1803 (2013)
12. Olaru, A.: The optimizing space trajectory by using the inverse kinematics, direct dynamics and intelligent damper controlling with proper neural networks. In: International Conference on Advanced Mechatronic Systems (ICAMechS), pp. 504–509 (2012)
13. Chih-Lyang, H.: Hybrid neural network under-actuated sliding-mode control for trajectory tracking of quad-rotor unmanned aerial vehicle. In: International Joint Conference on Neural Networks (IJCNN), pp. 1–8 (2012)
14. Wu, J., Jian, H., Yongji, W., Kexin, X.: RLS-ESN based PID control for rehabilitation robotic arms driven by PM-TS actuators. In: International Conference on Modelling, Identification and Control (ICMIC), pp. 511–516 (2010)
15. Ni, S., Lei, L., Zhishen, W., Wang, Z.: Predictive Control of Vehicle Based on Echo State Network. In: International Conference on Intelligence Science and Information Engineering (ISIE), pp. 37–40 (2011)
16. Hartland, C., Bredeche, N.: Using echo state networks for robot navigation behavior acquisition. In: IEEE International Conference on Robotics and Biometrics (ROBIO), pp. 201–206 (2007)

An Intelligent Control System for Mobile Robot Navigation Tasks in Surveillance

Chi-Wen Lo, Kun-Lin Wu, Yue-Chen Lin, and Jing-Sin Liu

Institute of Information Science, Academia Sinica, Nangang, Taipei, Taiwan 115, ROC
{kevinlo,kunlinwu,hyde,liu}@iis.sinica.edu.tw

Abstract. In recent years, the autonomous mobile robot has found diverse applications such as home/health care system, surveillance system in civil and military applications and exhibition robot. For surveillance tasks such as moving target pursuit or following and patrol in a region using mobile robot, this paper presents a fuzzy Q-learning, as an intelligent control for cost-based navigation, for autonomous learning of suitable behaviors without the supervision or external human command. The Q-learning is used to select the appropriate rule of interval type-2 fuzzy rule base. The initial testing of the intelligent control is demonstrated by simulation as well as experiment of a simple wall-following based patrolling task of autonomous mobile robot.

Keywords: mobile robot navigation, moving target, patrol, intelligent control, fuzzy Q-learning.

1 Introduction

The intelligent mobile robots technology has widespread applications at present and in the future. Applications of robotics have been applied to home services, health care and military missions such as [3]-[5], etc. Developing various intelligence services, for example intelligent surveillance and patrol systems, is of emerging demand to support human society [6]-[7]. As an intelligent mechatronics system, the mobile robot needs to integrate algorithms related to environment sensing for obstacle detection and SLAM, behavior and route planning, controlling and executing [8]. The focus of its development is on how to make the mobile robots capable of safely, effectively and efficiently operating in various ways in real, unknown environments which may involve interacting with human activities. This requires developing a navigation method that incorporates enough functionalities, in addition to the basic obstacle avoidance and stationary target reaching modes. In this paper, we are interested in using mobile robot for surveillance tasks in various environments. The surveillance by a mobile robot contains target tracking or pursuit, wall following and obstacle avoidance. For intelligent control of mobile robot for navigation, fuzzy control is able to deliver a satisfactory performance in face of unmodelled robot dynamics, uncertainty and imprecision of sensing and actuating devices [9]-[11]. It has been widely applied to the design of robot speed and orientation steering controller because of the following reasons: 1) Control rules are more flexible, thus it can simplify the

J.-H. Kim et al. (eds.), *Robot Intelligence Technology and Applications 2,*   449
Advances in Intelligent Systems and Computing 274,
DOI: 10.1007/978-3-319-05582-4_39, © Springer International Publishing Switzerland 2014

complex system; 2) The controller can emulate the human decision making; 3) It does not need a detailed model of the plant, and it replaces the mathematical values in describing control system by using the linguistic ambiguous labels for designing robust controllers. On the other hand, reinforcement learning, in particular Q-learning, shows good learning results in designing control input for performing constrained tasks by robots without knowing the system dynamics [22], [23]. The approaches of combining type-1 fuzzy logic and Q-learning for optimization of the consequence parts of fuzzy rules are promising due to the ease of implementation on mobile robot navigation [12]-[17] in which Q value is a cost for each navigation behavior. In this paper, we propose to combine Q-learning with interval type-2 fuzzy logic as an intelligent control for cost-based mobile robot navigation that yields smoother behaviors. The Q-learning algorithm is employed to evaluate and select the fuzzy rules for the mobile robot to take the action. The aim is to achieve a more smooth autonomous navigation of mobile robot in surveillance of unknown environment in which the mobile robot needs to be able to patrol a region and capture or follow one moving targets. Section 2 presents some related work. Section 3 introduces the intelligent control for cost-based mobile robot navigation based on fuzzy Q-learning. Validation of the intelligent control in simulation and real robot experiment for boundary following is shown in Section 4 and 5, respectively. Conclusion and future work is in Section 6.

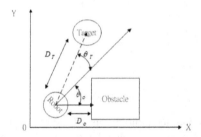

Fig. 1. The line of sight for target tracking and obstacles avoidance

2   Related Works

Moving target tracking/following and capturing is an on-line process. It is important in surveillance, and it has received increasing attentions more recently [1], [4]-[8].One task for the intelligent control of a mobile robot in this study is moving target tracking or capturing and obstacles avoidance. It is assumed that the target moves along a trajectory that is either well-defined and known a priori or unknown. Refer to Fig. 3.The control objective for the mobile robot is to controlling the orientation angle θ_0 and speed to guarantee that the mobile robot can follow the direction of the target, i.e. the real target orientation $\theta_T(t) \to 0$, and $D_T(t) \le d$ where d is a threshold (zero for capturing, nonzero for tracking) for the relative distance between the mobile robot and the moving target.

In [1], a potential field method was developed for velocity planning of a mobile robot to track a moving target in the presence of moving obstacles. In [4], using velocity vectors of the robot relative to each obstacle, an online navigation method based on calculating the best feasible direction close to an optimal direction to the target is proposed for pursuing a moving target amidst dynamic and static obstacles. Adaptive learning control for pursuit-evasion were presented in [6], [7], and experiments on capturing a moving object using pure pursuit were shown in [8].

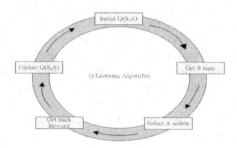

Fig. 2. The computation flowchart of Q-learning algorithm

3   Architecture Description of Fuzzy Q -Learning

3.1   Q-Learning Algorithm

Reinforcement learning is a promising approach to deal with control of physical robot with ever increasing complexity of hardware [22], [23] through experience and observations. Q-learning algorithm is a popular model-free reinforcement learning that have been demonstrated to give good results for some instances of robot tasks over the years. Fig. 2 shows the flowchart of Q-learning algorithm for a mobile robot that interacts with its environment via perception and action. Q-learning works as follows. After taking each action A_t from the action set A in a perceived state S_t of state space S, the mobile robot gets an immediate reward R_t at time t from the interaction with its surrounding environment, and changes its current state. Let the action-value function $Q(S_t, A_t)$ denote the Q-value for a state-action pair (S_t, A_t). Without knowing the dynamics of mobile robot being controlled, by measuring and storing the data (S_t, S_{t+1}, R_t) for taking the action A_t, the expected Q-value for a state-action pair is online updated as follows:

$$Q(S_t, A_t) \leftarrow Q(S_t, A_t) + \alpha \left[R_t + \gamma \max_A Q(S_{t+1}, A_t) - Q(S_t, A_t) \right] \qquad (1)$$

where the parameter $\alpha \in [0,1]$ denotes the learning rate, and the parameter $\gamma \in [0,1]$ denotes the discount factor that influences the current value of future reward. After a sufficient number of trials over time, the mobile robot tends to consistently learn a policy that maps the state to the action with maximum Q-value that will optimize the future reinforcement, independent of how the mobile robot behaves during the learning phase.

3.2   Fuzzy System

1) Traditional (Type1) Fuzzy System: Firstly, we design a fuzzy rule base of target pursuit and obstacles avoidance. In our work, the nearby environment information is obtained from a laser range finder. The sensing input data of nearby environment that measures the existence or closeness of obstacles or moving target within the field of view of the mobile robot is employed to control robot actions. The angular span of the sensing range from a laser range finder is partitioned into five segments in angular direction and three ranges in radial direction, as shown in Fig. 3. In Fig. 3, five directions are: R, L, F denotes right, left, in front of forward respectively; FL denotes in front of left, FR denotes in front of right. Three ranges of distance are: F denotes far distance, M denotes moderate distance and N denotes near distance.

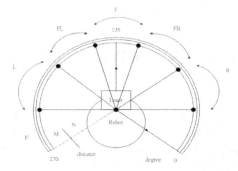

Fig. 3. The partition of field of view from a laser range finder for fuzzy system input

2) Interval Type-2 Fuzzy Logic System: An interval type-2 fuzzy set $\tilde{A}$, as shown in Fig4, is defined by a fuzzy membership function

$$\tilde{A} = \left\{ ((x,u), \mu_{\tilde{A}}(x,u)) \middle| \forall x \in X, \forall u \in J_x \subseteq [0,1] \right\} \tag{2}$$

where J_x denotes primary membership of x, $\mu_{\tilde{A}}(x,u) = \int_{u \in J_x \subseteq [0,1]} \frac{1}{u}$ is the third dimension denoting a traditional (type-1) fuzzy set. It is completely described by its upper and lower membership functions denoted by

$$\bar{\mu}_{\tilde{A}}(x) = \overline{FOU(\tilde{A})}, \underline{\mu}_{\tilde{A}}(x) = \underline{FOU(\tilde{A})},$$

respectively. The area between the upper membership function and lower membership function is called the footprint of uncertainty (FOU) of interval type 2 fuzzy set FOU provides an additional degree of freedom to handle uncertainties. The output of the inference will obtain a type-2 fuzzy set. The inference result is type-reduced to a

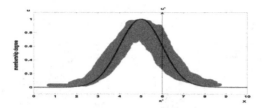

Fig. 4. FOU (shaded area) for an interval type-2 fuzzy set [15]

$$FOU(\tilde{A}) = \bigcup_{\forall x \in X} (\overline{\mu}_{\tilde{A}}(x), \underline{\mu}_{\tilde{A}}(x)) \tag{3}$$

type-1 fuzzy set, and the resulting type-reduced set is then defuzzified to generate a crisp output. In [19], both type-2 fuzzy and type-1 fuzzy were applied to speed control and the angle control of mobile robot and demonstrated that the performance of type-2 fuzzy system is much better than type-1 fuzzy system. For illustration, we design a two- input one-output fuzzy system shown in Fig. 5. The input data is provided by a laser range finder to detect environment information, and the output is a suitable value for the mobile robot to control the course. In this simulation study, the output set is segmented into five parts: large left, left, forward, right, and large right, respectively. The comparative performance of type-1 and interval type-2 fuzzy controllers is shown in Fig. 6. As shown in Fig. 6, the interval type-2 fuzzy controller shows better and smooth performance. The result of simulations encourages the use of interval type-2 fuzzy control as the main controller for the robot navigation task from the practical performance standpoint.

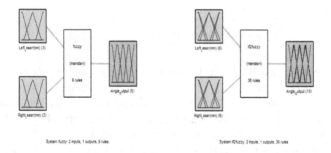

Fig. 5. Two- input one- output fuzzy system: type1 (left) vs type-2(right). Triangle or trapezoid membership functions of distance are used. Numerals in parenthesis denotes the number of rules.

3.3 Integrated Intelligent Control System

Fuzzy Q-learning has been applied to mobile robot navigation [15]-[17], where a reinforcement learning algorithm is used to fine tune the fuzzy rule base parameters.

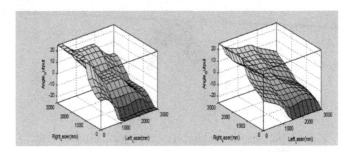

Fig. 6. The output surface of type-1 fuzzy (left figure) vs. interval type-2 fuzzy (right figure) controllers. x, y are distance in the right and left directions, respectively; z is the output of robot turning angle.

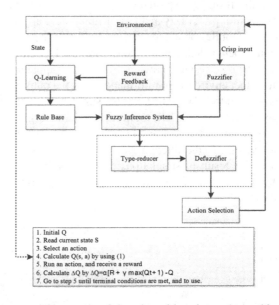

Fig. 7. The structure of integrating Q-learning with an interval type-2fuzzy controller

Fig. 7 shows the flow chart of integrating Q-learning algorithm into interval type 2 fuzzy controller. An interval type 2 fuzzy system is characterized by IF-THEN rules, where their antecedent or consequent sets are of interval type 2. The type 2 fuzzy logic system includes a fuzzifier, a rule base, fuzzy inference engine, and output processor. The output processor includes a type-reducer and a defuzzifier and it generates a type 1 fuzzy set output. The integration allows the fuzzy rule to be evaluated by the Q-learning so that for more complex situation defined by fuzzy rules, the Q values could be provided as the cost to individual rules that are activated by current state-action pair. Here the rule is in the form of

IF S is s THEN a is A with $Q(s,A)$

where the state and action membership functions are given as interval type 2, and thus $Q(s,A)$ has two values, one for each membership function. Each rule for control is associated with a Q-value after learning results to show the goodness of the rule. An illustration is shown in Table1 for Fig. 5 where there are five fuzzy sets of the rule base in A: Large Left, Left, Forward, Right, and Large Right.

3.4   Kalman Filter

To estimate the target's expected movement variation in the speed and position, we incorporate the Kalman filter to probabilistic estimate of its motion [18]. The Kalman and it can refer to figures 10 and 14filter is very powerful in estimations of past, present, and even future state, and it can do so even when the precise nature of the modeled system in unknown. The filter estimates a process by using a form of feedback control. The filter estimates the process state at some time and then obtains feedback in form of measurements. The time update equations can also be thought of as predictor equations, while the measurement update equations can be thought of as corrector equations as shown in Fig. 8. The time update projects the current state estimate ahead in time. The measurement update adjusts the projected estimate by an actual measurement at that time. The well-known equations for the time and measurement updates are repeated in Table 2 and Table 3 for easy reference. The matrix A relates the state at previous time step $k-1$ to the state at the current step k, and the matrix B relates the control input u to the state x. The process noise covariance Q and measurement noise covariance R matrices might change with each time step. The matrix H relates the state to the measurement z_k with normal probability distribution.

Table 1. Fuzzy Rule Base

Rule Base		Left Laser		
		Near	Medium	Far
Right Laser	Near	A_1 with q_{11}, q_{12}	A_2 with q_{21}, q_{22}	A_3 with q_{31}, q_{32}
	Medium	A_4 with q_{41}, q_{42}	A_5 with q_{51}, q_{52}	A_6 with q_{61}, q_{62}
	Far	A_7 with q_{71}, q_{72}	A_8 with q_{81}, q_{82}	A_9 with q_{91}, q_{92}

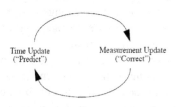

Time Update
("Predict")   Measurement Update
("Correct")

Fig. 8. The ongoing discrete Kalman filter cycle [18]

Table 2. Discrete Kalman Filter Time Update Equations

DISCRETE KALMAN FILTER TIME UPDATE EQUATIONS	
$\hat{x}_k^- = A\hat{x}_{k-1} + Bu_{k-1}$ $P_k^- = AP_{k-1}A^T + Q$	(4)

Table 3. Discrete Kalman Filter Measurement Update Equations

DISCRETE KALMAN FILTER MEASUREMENT UPDATE EQUATIONS	
$K_k = P_k^- H^T (HP_k^- H^T + R)^{-1}$ $\hat{x}_k = \hat{x}_k^- + K_k(z_k - H\hat{x}_k^-)$ $P_k = (I - K_k H)P_k^-$	(5)

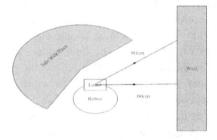

Fig. 9. A simple environment for the task of wall following

4   Simulation

4.1   Wall-Following in a Simple Environment

Evolutionary aspect of Q-learning is better understood by observing the motion of mobile robots [2]. Here we perform an off-line Q-learning simulation of a right wall following mission [12], [20], [21] by keeping a safety distance with a wall, in which the static wall is on the right, and a safe wide place is on the left of the robot as shown in Fig.9. The robot state is described by two extreme points sensed by the laser range finder: the rightmost and extreme right front. The robot motion is described by a sequence of action {turning right, turning left, turning zero} without stopping. For each state sensed by the readings from a laser range finder, an action executed by the mobile robot will cause a cost (in our case, the acquired Q value, 0 for collision, 50 for collision-free and 100 for wall following) given by the interaction with the environment. After the convergence of Q-learning as shown in Fig. 10, the cost-based navigation behavior corresponding to the maximum Q value shown in Q- table of Table 4 is selected to execute in each moving step of mobile robot. In this simulation, being informed of the Q value, the robot will consider the action for next step as turning zero to achieve the best performance. This demonstrates that the navigation based on Q-learning is promising for real-world implementation, since the robot is able to learn the desired reactive behavior in this simple situation.

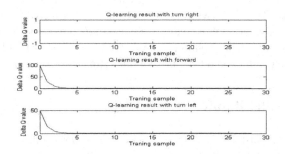

Fig. 10. The results of after learning

Table 4. Fuzzy Q-learning

action	Turning Right	Turning Zero	Turning Left
Learned max Q-Value	0	142.8478	71.4130

4.2 Kalman Filter for Estimation of Target

This section describes the task of pursuit of a one-dimensional moving target by a mobile robot. The robot has onboard sensors that continuously locate the moving target. A pure pursuit is one way to specify how to do repetitive new course calculations: target update based on look ahead distance [8]. A simulation of pure pursuit for one-dimensional moving point target is shown in Fig. 11 where a capture occurs if the point robot and the point target occupy the same place at the same time. The target is simply moving in one direction (right to left) with known moving trajectory and the point robot can arrest the target using pure pursuit method. To minimize the time of capturing the target, the robot moves at its maximum speed

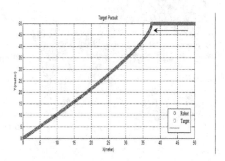

Fig. 11. The simulation of pure pursuit where the moving target moves along a fixed horizontal direction

$V_R \geq V_T$, where the maximum speed of the target is V_T. However, in situations which the target moves at an unfixed direction of travel and unknown velocity on a domain, even randomly such as Markov chain or Brownian motion, the robot needs to predict the unknown target motion for achieving a higher possibility of eventually capturing the target. Therefore, we employ the Kalman filter method [18] to solve this problem, assuming the target movement is changed randomly, and the Kalman filter is used to estimate next state to build a suitable trajectory as shown in Fig. 12 so that the method of pure pursuit could be applied to the estimated trajectory.

5   Experimental Results

In this section, an experiment is conducted which the robot is operating for patrol in an unknown, unstructured environment. In addition to avoid the obstacles for safe navigation, the mobile robot is required to explore and then patrol by right wall following in an unknown environment. Wall following by a mobile robot has been

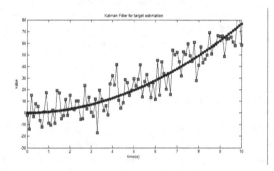

Fig. 12. Simulation of the Kalman filter for target estimation. Red dots denote the real movements of the target that moves randomly within a range. Solid blue curve denotes the estimated trajectory.

2.4 GHz notebook

30 cm* 50 cm* 40 cm

SICK laser range finder

two integrated servo motors
to drive left and right wheels

Fig. 13. The wheeled mobile robot equipped with a laser range finder for experimrnt

tested using proximity sensors (sonar or infrared) for different control algorithms [12], [20], [21] based on line of sight range measurements (distance and its rate), and/or bearing angle. Here our implementation and testing is conducted on a mobile robot equipped with a SICK laser range finder in the front, as shown in Fig. 13, to perform a scan of 180 degrees of field of view with maximum sensing range 5m to obtain nearby environment information. We employed the interval type-2 fuzzy Q-learning techniques for obstacles avoidance and wall following in an unknown environment. In this task, the robot needs to explore and follow the right wall, while avoiding the obstacles. This task spends 125 seconds of total time for a completion of the patrol, in which a feedback is provided every 0.3 seconds. The patrol trajectory and the output of turning angle are shown in Fig. 14 and Fig. 15, respectively. In Fig. 15, the fluctuation of output angle indicates that the mobile robot encountered flat real walls or curved obstacle boundaries to make a fine tuning of its motion direction, therefore the output response is seen a substantial beating as the mobile robot encounters a corner. The patrolling in a real indoor environment is shown in Fig. 16 (http://youtu.be/es93QfFz8qs).

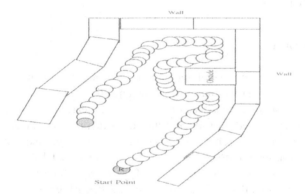

Fig. 14. The patrolling trajectory in the experiment

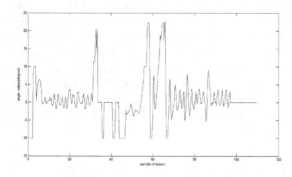

Fig. 15. The turning angle that the mobile robot approaches the wall

Fig. 16. The snapshots of patrolling experiment in a real environment (from the left upper photo to right bottom photo)

6   Conclusion

In this paper, we proposed a novel autonomous and intelligent controller for mobile robot navigation tasks in surveillance which requires the accomplishment of a variety of missions. The system is composed of reinforcement learning, fuzzy control, and a prediction component based on Kalman filter for estimating the trajectory of moving target to support a mobile robot for target pursuit, obstacles avoidance, and wall following of patrolling mission. The controller is composed by fuzzy Q-learning where the fuzzy rules are selected by the Q learning to meet a diverse set of navigation tasks in surveillance. The interval type-2 fuzzy logic system is employed, which shows better and smooth navigation performance. For tasks of moving object such as pursuit- evasion, the Kalman filter could be employed for a randomly moving target to predict its motion trajectory. Preliminary experiment in a simple real and unknown indoor environment validates that the intelligent control is effective to learn from the data collected and accomplish the tasks of wall following and obstacle avoidance. Ongoing and future work are planned to improve the current implementation of intelligent control for more complex mobile robot navigation behaviors such as capturing a moving target. For improving the generalization capability of reinforcement learning, future work will consider enhancing the computational efficiency using adaptive learning algorithm for navigation in more complex environment.

Acknowledgements. This work was supported by National Science Council of ROC under contract NSC101-2221-E-001-001.

References

1. Huang, L.: Velocity planning for a mobile robot to track a moving target—a potential field approach. Robotics and Autonomous Systems 57, 55–63 (2009)
2. Hara, M., Huang, J., Yabuta, Y.: Characterization of motion forms of mobile robots generated in Q-learning process. In: Mellouk, A. (ed.) Advances in Reinforcement Learning. Intech (2011)
3. Cheney, N.: Unshackling evolution: evolving soft robots with multiple materials and a powerful generative encoding. In: GECCO 2013 (2013)
4. Masehian, E., Katebi, Y.: Sensor-based motion planning of wheeled mobile robots in unknown dynamic environments. Journal of Intelligent & Robotic Systems, 1–22 (2013)
5. Zhu, Q., Hu, J., Henschen, L.: A new moving target interception algorithm for mobile robots based on sub-goal forecasting and an improved scout ant algorithm. Applied Soft Computing 13(1), 539–549 (2013)
6. Chung, H.C., Liu, J.S.: Adaptive learning approach of fuzzy logic controller with evolution for pursuit–evasion games. In: Pan, J.-S., Chen, S.-M., Nguyen, N.T. (eds.) ICCCI 2010, Part I. LNCS (LNAI), vol. 6421, pp. 482–490. Springer, Heidelberg (2010)
7. Desouky, S.F., Schwartz, H.M.: Q (λ)-learning adaptive fuzzy logic controllers for pursuit–evasion differential games. International Journal of Adaptive Control and Signal Processing 25(10), 910–927 (2011)
8. Morales, J., Martínez, J.L., Martínez, M.A., Mandow, A.: Pure-pursuit reactive path tracking for nonholonomic mobile robots with a 2D laser scanner. EURASIP Journal on Advances in Signal Processing, Article no. 3 (2009)
9. Borenstein, J., Koren, Y.: Obstacle avoidance with ultrasonic sensors. IEEE Transactions on Robotics and Automation 4(2), 213–218 (1988)
10. Chang, K.S., Choi, J.S.: Automatic vehicle following using the fuzzy logic. In: Proc. Conf. on Vehicle Navigation and Information System, pp. 206–213 (1995)
11. Lee, T.H., Lam, H.K., Leung, F.H.F., Tam, P.K.S.: A practical fuzzy logic controller for the path tracking of wheeled mobile robots. IEEE Control Systems Magazine, 60–65 (2003)
12. Ando, Y., Yuta, S.: Following a wall by an autonomous mobile robot with a sonar-ring. In: Proc. Conf. on Vehicle Navigation and Information System, pp. 206–213 (1995)
13. Deng, C., Er, M.J.: Real-time dynamic fuzzy Q-learning and control of mobile robots. In: 5th Asian Control Conference, vol. 3, pp. 1568–1576 (2004)
14. Yang, G., Zhang, R., Xu, D.: Implementation of AUV local planning in strong sea flow field based on Q-learning. Journal of Harbin Institute of Technology 18(1) (2011)
15. Ritthipravat, P., Maneewarn, T., Laowattana, D., Wyatt, J.: A modified approach to fuzzy Q learning for mobile robots. In: IEEE International Conference on Systems, Man and Cybernetics, pp. 2350–2356 (2004)
16. Boubertakh, H., Tadjine, M., Glorennec, P.Y.: A new mobile robot navigation method using fuzzy logic and a modified Q-learning algorithm. Journal of Intelligent and Fuzzy Systems 21(1), 113–119 (2010)
17. Gordon, S.W., Reyes, N.H., Barczak, A.: A hybrid fuzzy Q-learning algorithm for robot navigation. In: International Joint Conference on Neural Networks, pp. 2625–2631 (2011)
18. Welch, G., Bishop, G.: An introduction to the Kalman filter. SIGGRAPH 2001 Course 8 (2001)
19. Castillo, O., Melin, P.: A review on the design and optimization of interval type-2 fuzzy controllers. Applied Soft Computing, 1267–1278 (2012)

20. Matveev, A.S., Hoy, M.C., Savkin, A.V.: The problem of boundary following by a unicycle-like robot with rigidly mounted sensors. Robotics and Autonomous Systems 61(3), 312–327 (2013)
21. Huang, L.: Wall-following control of an infrared sensors guided wheeled mobile robot. International Journal of Intelligent Systems Technologies and Applications 7(1), 106–117 (2009)
22. Kormushev, P., Calinon, S., Caldwell, D.G.: Reinforcement learning in robotics: Applications and real-world challenges. Robotics 2(3), 122–148 (2013)
23. Kober, J., Bagnell, J.A., Peters, J.: Reinforcement learning in robotics: A survey. The International Journal of Robotics Research 32(11), 1238–1274 (2013)

Formation Control Experiment of Autonomous Jellyfish Removal Robot System JEROS

Donghoon Kim, Jae-Uk Shin, Hyongjin Kim, Hanguen Kim, and Hyun Myung[*]

URL (Urban Robotics Lab.),
KAIST (Korea Advanced Institute of Science and Technology), Daejeon, 305-701, Korea
{dh8607,jacksju,hjkim86,sskhk05,hmyung}@kaist.ac.kr

Abstract. The proliferation of jellyfish is threatening marine ecosystem and has caused severe damage to marine-related industries. An autonomous jellyfish removal robot system, named JEROS (Jellyfish Elimination Robotic Swarm), has been developed to cope with this problem. This paper presents formation control of JEROS and related experimental results through field tests. The JEROS is extended to multi-agent robot system and employs the leader-follower algorithm for formation control. The Theta* path planning algorithm is employed to generate an efficient path. Three prototypes of JEROS are implemented, and the feasibility of their formation control and the performance of jellyfish removal were demonstrated through field tests in Masan Bay located in the southern coast of South Korea.

Keywords: jellyfish removal, surface vehicle, navigation, path planning, formation control.

1 Introduction

An enormous damage caused by proliferation of jellyfish has been reported in more than 14 countries around the world. The damage to fishery industries is the most serious, and seaside power plants and oceanic tourism are also damaged. The dominant causes of this problem are considered to be environmental pollution, global warming, destruction of the marine ecosystem, and an increased amount of marine structures. In South Korea, jellyfish population has steadily increased since 2000, and the damage to marine-related industries was estimated to be over 300 million USD a year in 2009 [1]. The most prevalent species of jellyfish along the coast of South Korea are *Aurelia aurita* and *Nemopilema nomurai* which have weak and strong venoms, respectively. In 2012, *Nemopilema nomurai* led a child to death in a beach in South Korea. In order to solve this problem, some studies have been carried out. The system consisting of two trawl boats equipped with jellyfish cutting nets has been developed [2, 3]. Utilizing large ships and many human operators, these systems have shown high performance in jellyfish removal, but it seems difficult to operate in narrow and shallow coastal areas. Other systems to prevent the influx of jellyfish into water intake pipes of power plants were developed. One of them used a camera and a

---

[*] Corresponding author.

J.-H. Kim et al. (eds.), *Robot Intelligence Technology and Applications 2*,   463
Advances in Intelligent Systems and Computing 274,
DOI: 10.1007/978-3-319-05582-4_40, © Springer International Publishing Switzerland 2014

water pump [4]; and the other system used a bubble generator and a conveyor device [5]. However, these systems are expensive to install and maintain.

In this paper, we introduce a multi-agent robot system for efficient jellyfish removal and field tests for formation control and jellyfish removal. An earlier version of autonomous jellyfish removal robot system, named JEROS (Jellyfish Elimination RObotic Swarm), was presented in [6-9]. The design of the ship, navigation and image processing algorithms, and feasibility tests for the algorithms and jellyfish removal were introduced. In this paper, its modified design, mainly in thrusters and communication methods, is presented. The robot system is extended to a multi-agent robot system composed of three prototypes of JEROS to enhance the efficiency of jellyfish removal, and the leader-follower scheme is employed to control formation of the multiple robots. Additionally, for the leader robot's autonomous navigation, the path planning algorithm based on Theta* path planning algorithm [10] and LoS (Line of Sight) guidance-based path following [11] method are embedded on JEROS. Finally, the feasibility of formation control and performance of jellyfish removal were demonstrated through field tests in Masan Bay located in the southern coast of South Korea.

In Section 2, the modified design of JEROS, the formation control based on the leader-follower scheme, and the Theta* algorithm-based path planning algorithm are described. In Section 3, experimental results of field tests for formation control and jellyfish removal are presented. Finally, in Section 4, we summarize this paper and discuss future works.

2   Formation Control and Path Planning of JEROS

2.1   Design of JEROS

The JEROS is made up of two parts. One is the USV (Unmanned Surface Vehicle) part and another is the part for jellyfish removal. The USV part needs to be designed for high payload, high stability, and high controllability. The twin-hull-type USV is

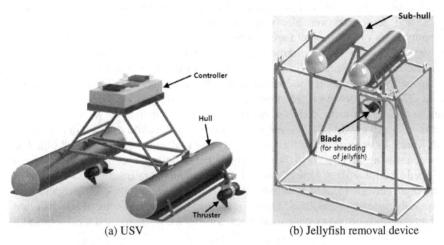

(a) USV   (b) Jellyfish removal device

Fig. 1. 3D CAD model of USV and device for jellyfish shredding

more stable against waves than mono-hull-type, moreover it is easy to increase the payload using large hull and it is also easy to control just using two thrusters installed in the rear of two hulls. JEROS shreds jellyfishes using a blade and the jellyfishes are guided to the blade by a net shaped like a funnel. The 3D CAD design is shown in figure 1. The USV part can be operated alone with fastest speed and minimum power to perform the mission such as surveillance. To eliminate the jellyfishes, the removal part is assembled under the USV as shown in figure 2(a) and the prototype is implemented as shown in figure 2(b).

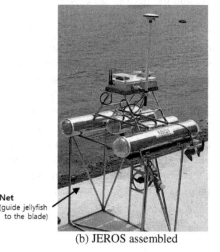

Net
(guide jellyfish
to the blade)

(a) JEROS (3D CAD model)   (b) JEROS assembled

Fig. 2. Overall system of JEROS

Table 1. Electrical Parts of JEROS

Device	Name	Manufacture
GPS	OEM Star	Novatel
IMU	EBIMU-9DOF	E2BOX
3G Modem	Snapdragon S3	Qualcomm
Zigbee Modem	EZBee-M100-EXT	Chipsen
Computer(SBC)	Core i7	Intel
Microprocessor	TMS320F2808	Texas Instrument
Thruster	Endura C2	Minnkota

The electrical control system is embedded in JEROS and its parts are listed in table 1. Sensors including GPS and IMU (Inertial Measurement Unit) are embedded in the system for localization. An SBC (Single Board Computer) and a microcontroller are embedded to process navigation and control algorithms. Two types of wireless communication modems for 3G mobile network and Zigbee are also embedded. The 3G modem is utilized to communicate with an external server computer, which allows

JEROS to be operated at the place far from the external server computer. Additionally, Geographic Information System (GIS) map-based navigation algorithm is implemented using the 3G mobile network.

2.2   Formation Control

In order to enhance the efficiency of the jellyfish removal, JEROS is extended to a multi-agent robot system. The jellyfish removal task is performed by the motion of JEROS and shredding of jellyfish, and its efficiency depends on the area coverage rate and the performance of shredding. The area coverage rate is related to the water volume passed by the jellyfish removal device in a unit time, and it relies on the speed of JEROS and the dimensions of the jellyfish removal device. Its speed depends on thrust force. However, the dimensions of the jellyfish shredding device are hard to be enlarged, because the drag force increased by the enlarged device drops the speed of JEROS. Thus, JEROS is expanded to a multi-agent robot system and the leader-follower scheme for formation control is employed.

In the leader-follower scheme, one robot is assigned to be the leader and the others follow leader robot's motion with maintaining their formation. The leader robot follows desired path and determines the follower robots' motion. Each follower robot follows the waypoint calculated by using desired displacement and heading angle relative to the leader robot. A simple model of the leader-follower scheme is shown in figure 3.

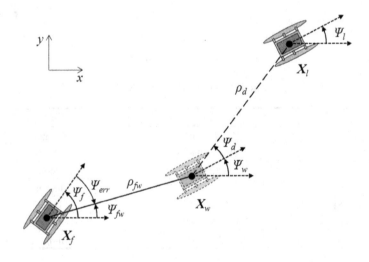

Fig. 3. A simple model of leader-follower formation control

Let's consider a multi-agent robot system composed of n mobile robots, the position of each robot is described by $X_i = (x_i, y_i, \psi_i)^T$, where x_i, y_i, and $\psi_i \in [-\pi, \pi)$ describe x, y coordinates, and heading angle of the robot, respectively.

The position of leader and follower robots are denoted by X_l and X_f, the waypoint which is the desired position of the follower robot is denoted by X_w, as shown in figure 3. X_w is determined by the desired displacement and heading angle, and is calculated as follows:

$$X_w = \begin{pmatrix} x_l - \rho_d \cos(\psi_d + \psi_w) \\ y_l - \rho_d \sin(\psi_d + \psi_w) \\ \psi_w \end{pmatrix} \tag{1}$$

where ρ_d and ψ_d denote the desired displacement and heading angle. Typically, ψ_w is same as ψ_l. The follower robot follows the waypoint using a guidance law and its desired and error heading angles are calculated as follows:

$$\psi_{fw} = \tan^{-1}\left(\frac{y_w - y_f}{x_w - x_f} \right), -\pi \le \psi_{fw} < \pi$$

$$\psi_{err} = \psi_{fw} - \psi_f \tag{2}$$

where ψ_{fw} and ψ_{err} denote the desired and error angles in the guidance law. Its desired speed is determined to be proportional to the distance between the waypoint and the follower robot. Thus, the desired speed and heading angle of the follower robot can be controlled by two thrusters using the following equations [12]:

$$\tau_d = K_p \rho_{fw}$$

$$\tau_l = \begin{cases} \tau_d - \dfrac{\tau_d \cdot \psi_{err}}{K} & \psi_{err} > 0 \\ \tau_d & \psi_{err} \le 0 \end{cases}$$

$$\tau_r = \begin{cases} \tau_d & \psi_{err} > 0 \\ \tau_d + \dfrac{\tau_d \cdot \psi_{err}}{K} & \psi_{err} \le 0 \end{cases} \tag{3}$$

where τ_d, τ_l, and τ_r denote the central thrust force and thrust forces of the left and right thrusters, respectively, and K is a steering constant.

2.3   Path Planning and Following

The autonomous navigation technologies such as path planning and following are needed to approach an area where jellyfishes are densely populated. The LoS (Line-of-Sight) guidance algorithm is used to follow a generated path. The LoS guidance algorithm computes an LoS vector to calculate a desired heading angle. The LoS vector is formed by connecting the robot position to an intersecting point on the path at a distance of a tracking radius ahead of the robot [10]. To effectively follow a sequence of way-points by the LoS guidance algorithm, the way-points are created in the major inflection points. For the path planning, the Theta* algorithm was used. The Theta* algorithm is very similar to A*. The difference between the algorithms is the method used to select a node. Theta* selects a node considering the line of sight [11]. The following algorithm 1 is the pseudo code of the Theta* algorithm.

Algorithm 1. Theta* algorithm

---

1. $Point_{start}.Parentnode \leftarrow Point_{start}$

2. *while found* $Point_{goal}$ *do*

3.   *for each* $Point_{neighbor}$ *do*

4.   *if* lineofsight *then*

5. $Point_{neighbor}.Parentnode \leftarrow Point_{current}.Parentnode$

6.   *else* $Point_{neighbor}.Parentnode \leftarrow Point_{current}$

7.   *End*

8. *End*

As shown in figure 4, the Theta* algorithm takes localization and GIS data as inputs. The LoS guidance algorithm calculates the control input by using the Theta* results.

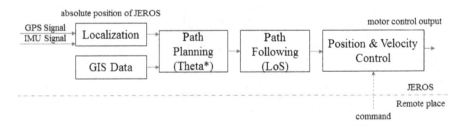

Fig. 4. Scheme of navigation and control system

3   Experiments

3.1   Formation Control Tests

Field tests were carried out in Masan Bay located in the southern coast of South Korea to verify the feasibility of the proposed formation control. Our multi-agent robot system is composed of three JEROS prototypes. The leader robot follows a desired path generated by the Theta* algorithm and makes the overall motion of the robot system. Two follower robots receive absolute position of the leader robot through wireless communication (Zigbee) and follow the waypoint calculated by the absolute position. The desired displacement of both follower robots is set to 5.0 m and the desired heading angles of them are set to $\pi/2$ and $-\pi/2$, respectively. Results of the tests are shown in figure 5.

Fig. 5. A series of captured images while field tests. Caption indicate captured time (unit is min:sec)

3.2   Jellyfish Removal Tests

The field tests to verify the performance of jellyfish removal were carried out in Masan Bay. The jellyfish were shredded by the device for jellyfish removal as shown in figure 6. Jellyfishes in front of the device were transferred to the blade going

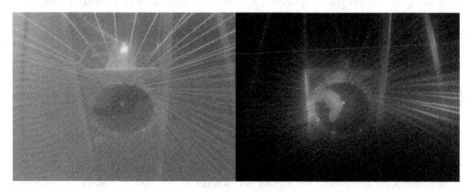

Fig. 6. Field tests for jellyfish removal: front view of the device for jellyfish removal (left) and a scene of jellyfish shredding (right)

through the funnel-shaped net, and then they were pulled by thrust force and shredded by the rapidly rotating blade. The rate of jellyfish removal is proportional to the speed of JEROS, the area of entrance of jellyfish removal device, and the number of jellyfishes in unit volume. The speed of JEROS was about 0.5 m/s and the area of the entrance was 1.44 m^2. The tests were performed in a region where many jellyfish appeared. Jellyfish such as *Aurelia aurita* proliferates in the bay, because there are many docks, artificial seawalls, and farms. Additionally, the warm seawater temperature and the abundance of plankton have led to high populations of jellyfish in the bay. In the tests, 36 jellyfishes were removed for 1 minute on average.

4   Conclusion

In this paper, we presented the formation control of the autonomous jellyfish removal robot system, JEROS, and the field tests for formation control and jellyfish removal. To enhance the performance of jellyfish removal, the robot system was modified compared to the previous version with respect to its dimensions and thrust force; and it was extended to a multi-agent robot system composed of three prototypes of JEROS. For the autonomous navigation of the multi-agent robot system, the leader-follower scheme was employed to control its formation, and Theta* and LoS guidance algorithms were employed to path planning and path following, respectively. The feasibility of the formation control and the performance of the jellyfish removal were demonstrated through the field tests in Masan Bay. Future research will be focused on an advanced formation control algorithm and the investigation of the efficiency of the JEROS through various field tests.

Acknowledgment. This research was supported by Basic Science Research Program through the National Research Foundation of Korea (NRF) funded by the Ministry of Science, ICT & Future Planning (grant number NRF-2013R1A1A1A05011746). It was also supported by the MOTIE (The Ministry of Trade, Industry and Energy), Korea, under the Human Resources Development Program for Convergence Robot Specialists support program supervised by the NIPA (National IT Industry Promotion Agency)(NIPA-2013-H1502-13-1001). Mr. H. Kim and Mr. H. Kim are supported by Korea Ministry of Land, Infrastructure and Transport (MOLIT) as U-City Master and Doctor Course Grant Program.

References

1. Choi, H.-S.: Scientists seek beneficial uses for jellyfish. The Korea Herald,
   http://www.koreaherald.com/view.php?ud=20120826000052
   (accessed December 4, 2012)
2. Kim, I.-O., An, H.-C., Shin, J.-K., Cha, B.-J.: The development of basic structure of jellyfish separator system for a trawl net. Journal of the Korean Society of Fisheries Technology 44(2), 99–111 (2008) (in Korean with English abstract)

3. NFRDI, Trends of overseas fisheries. Technical Report, no. 2 (2005) (National Fisheries Research and Development Institute (NFRDI) of South Korea issued in Korean)
4. Matsuura, F., Fujisawa, N., Ishikawa, S.: Detection and removal of jellyfish using underwater image analysis. Journal of Visualization 10(3), 259–260 (2007)
5. Lee, J.-H., Kim, D.-S., Lee, W.-J., Lee, S.-B.: System and method to prevent the impingement of marine organisms at the intake of power plants. Korean Patent 10-0558267-00-00 (2006)
6. Kim, D., Shin, J.-U., Kim, H., Lee, D., Lee, S.-M., Myung, H.: JEROS: Jellyfish removal robot system. In: Proc. of the Eighth International Conference on Humanized System (ICHS 2012), pp. 336–338 (2012)
7. Kim, D., Shin, J.-U., Kim, H., Lee, D., Lee, S.-M., Myung, H.: Development of Jellyfish Removal Robot System JEROS. In: Proc. of 9th Int'l Conf. on Ubiquitous Robots and Ambient Intelligence (URAI 2012), pp. 599–600 (2012)
8. Kim, D., Shin, J.-u., Kim, H., Lee, D., Lee, S.-M., Myung, H.: Experimental Tests of Autonomous Jellyfish Removal Robot System JEROS. In: Kim, J.-H., Matson, E., Myung, H., Xu, P. (eds.) Robot Intelligence Technology and Applications. AISC, vol. 208, pp. 395–403. Springer, Heidelberg (2013)
9. Kim, D., Shin, J.-U., Kim, H., Kim, H., Lee, D., Lee, S.-M., Myung, H.: Design and Implementation of Unmanned Surface Vehicle JEROS for Jellyfish Removal. Journal of Korea Robotics Society 8(1), 51–57 (2013)
10. Nash, A., Daniel, K., Koenig, S., Felner, A.: Theta*: Any-angle path planning on grids. In: Proc. AAAI Conf. on Artificial Intelligence (AAAI), pp. 1–7 (2007)
11. Fossen, T.: Marine control systems: Guidance, navigation and control of ships, rigs and underwater vehicles. Marine Cybernetics, Trondheim (2002)
12. Dunbabin, M., Lang, B., Wood, B.: Vision-based docking using an autonomous surface vehicle. In: Proc. 2008 IEEE International Conf. on Robotics and Automation (ICRA 2008), pp. 26–32 (2008)

Soft-Robotic Peristaltic Pumping Inspired by Esophageal Swallowing in Man

Steven Dirven, Weiliang Xu, Leo K. Cheng, and John Bronlund

Dept. of Mechanical Engineering, University of Auckland
Private Bag 92019, Auckland Mail Center,1142, Auckland, New Zealand
sdir146@aucklanduni.ac.nz, p.xu@auckland.ac.nz

Abstract. The demand for novel actuation and sensation technologies has seen the emergence of the biomimetic engineering field where inspiration is drawn from phenomena observed in nature. Soft robotic techniques are particularly suitable for physical modeling in this area as they can be designed to manifest features such as mechanical compliance and continuity. The process of peristalsis is common in many organisms for locomotion or pumping transport of fluid or semi-solid materials. This research initiative looks into how inspiration from the esophageal phase of swallowing can be communicated into the engineering domain such that a physical model of the esophagus can be developed. The resulting device is of a soft-robotic nature, asserted by pneumatic actuation on a silicone rubber conduit. The continuous nature of device output, and its perturbation throughout pumping transport present some interesting trajectory generation and control challenges. The inspiration for the mechanical design as well as the embodiment of device transport intelligence is described.

Keywords: Soft Robot, Peristalsis, Esophagus, Pumping, Bio-mimicry.

1   Introduction

Peristaltic pumping is a common process for transporting fluids in biological organisms, including man. The soft and distributed nature of actuation facilitates the possibility to have high input degree of freedom to control the transport process. This in turn results in many possibilities to achieve transport of variable wave geometries, or intra-bolus pressure profiles. A novel peristaltic pump of soft and distributed actuation is proposed to model this behavior to aid in the physical investigation of peristaltic transport externally to the human body. The interest is to establish a laboratory device with which experimentation of bolus formulations can be undertaken. Inspiration is sought from the behavior of the tissue-fluid interface observed throughout esophageal swallowing in man.

Significantly, peristaltic actuation does not depend on rigid, skeletal structures as is common with appendages of the human body. The distributed muscular actuation constricts the conduit in a propagating manner to achieve a smooth and continuous transport. The sequential nature of the muscular constriction is arranged in a

J.-H. Kim et al. (eds.), *Robot Intelligence Technology and Applications 2,*   473
Advances in Intelligent Systems and Computing 274,
DOI: 10.1007/978-3-319-05582-4_41, © Springer International Publishing Switzerland 2014

rostro-caudal sequence that propagates the full length of the esophagus behind the tail of the bolus. Indeed, in right persons less viscous boluses such as water may propagate ahead of the wave due to gravity. However, gravity is not requisite as swallowing can be undertaken in inverted persons.

Mathematical and numerical modeling of peristaltic transport has led to improved hypotheses about the behavior of fluids and their interaction with the transport conduit. Specifically, some interpretations are applied to esophageal swallowing [1-5]. The challenge lies in the description of the fluid and the way in which it behaves such that meaningful relationships can be drawn between the geometry and intra-bolus pressure. These models are based on assumptions of fluid properties and flow mechanics which are to be evaluated in the physical domain. This will additionally facilitate the investigation of fluids with multiple phases which are challenging to be described mathematically.

Prior work into the field of peristaltic pumping and mimicry of biological mechanics has seen the application of many unconventional robotic techniques. This is in response to the unique constraints of modeling soft and continuous behaviors of biological origin. This has seen the emergence of the field of soft robotics where novel design, fabrication and control strategies are evolving [6-8]. Within the scope of peristaltic pumping a number of devices have been developed; A device of esophageal origin developed by Miki et al [9] actuates cascaded rigid elements of Acrylic plastic onto a flexible silicone conduit by SMA actuation. It has shown promise in transporting jelly boluses. However, it relies on rigid transmission of force by skeletal elements. Alternative designs for peristaltic pumps such as those of [10] and [11] offer compliant and distributed actuation techniques for applications of bowel modeling and efficient pumping respectively. These devices demonstrate practical methods of pumping with limitations of rigid module boundaries preventing occlusion at all points along the conduit length.

Peristaltic pumping is particularly popular for applications on the micro scale due to the simplicity of fabrication. However, these devices are typically of a unidirectional occlusive nature. Currently devices of this nature have been based on the function of pumping, not on the achievement of prescribed wave geometry, or on displaying further intelligence on the process progression.

In response to these constraints and limitations a new soft robotic peristaltic pump has been devised to model esophageal swallowing. The device is cast of a Room Temperature Vulcanization (RTV) silicone rubber material and is actuated upon by a series of pneumatic chambers surrounding the conduit both radially and axially. This paper covers the methods of design, construction and application of soft robotics into the peristaltic pumping field with reference to metrics inspired from the biological swallowing system. It is proposed that the device will be used to validate predictions of the mathematical models with empirical measurements in the physical domain as well as facilitate investigation of transport efficacy of different food mediums in a laboratory environment.

Particular attention will be paid to the trajectory planning inspiration and its application to achieve reasonable peristaltic waves for modeling esophageal swallowing.

2   Robotic Implementation

2.1   Specification

The specification of biologically-inspired robotic devices requires the generalization of observed concepts and appropriate assumptions to be made of the behavior. The conversion of these into the practical engineering field requires careful consideration of which phenomena are required to be commanded. The investigation of swallowing performance in the human body is typically undertaken by Manometry, Videofluorography, Functional MRI or Endoscopy. Of these techniques, the temporal progression of peristaltic waves is captured with best resolution by manometry and videofluorography. These measure independent metrics: Manometry captures the intra-bolus pressure, which relates to material characteristics of the bolus and esophageal behavior, and subsequent wave-tail seal occlusion; Videofluorography captures the progression of a contrast bolus through x-ray imaging and is typically used to measure temporal aspects of the bolus head and tail as well as transit times. Parameters associated with the medically captured process inspire the specification of how a robotic model should perform (Table 1).

Table 1. Quantitative characteristics from medical investigation of the swallowing process

Quantity	Magnitude
Esophageal length	20-26cm [12-14]
Esophgaeal Diameter	20 mm [9, 15]
Max Occlusion Pressure	15 kPa [16]
Peristaltic Wave Velocity	2-4 cm/s [17]
Wavefront Length	30-60 mm

Additional to these parameters, there are qualitative constraints observed in the biological system which need to be communicated into the engineering domain (Table 2). The combination of these criteria represents the specification of the robotic device such that biologically-inspired esophageal swallowing may be conducted.

Table 2. Qualitative observations and assumptions surrounding communication of esophageal peristalsis into the engineering field

Aspect	Observation/Assumption
Wave Geometry	Continuous (Sinusoidal, Polynomial)
Occlusive Nature	Radial

2.2   Soft Robotic Inspiration

The tissues and behavior of the esophageal conduit are intrinsically compliant and continuous; features common in the emerging field of soft robotics. The degree of freedom of the biological system is essentially infinite; related to the innervation of a continuous array of muscular fibers resulting in the macro phenomenon. The morphology of sequentially arranged actuation elements, with a distributive and co-operative behavior is communicated through the use of elastically-coupled pneumatic chambers.

2.3   Design

In response to the specifications and soft-robotic inspiration a prototype peristaltic pump has been developed (Fig. 1). The pump is constructed in a linear manner as is typical of biological peristalsis. It is encased by a 3D printed ABS plastic housing, reinforced by hoops of PLA plastic. The active element of the peristaltic pump is constructed completely of an RTV silicone rubber (Ecoflex 00-30, Smooth On, USA) which has embedded pneumatic chambers. The chambers are arranged in 12 whorls of four chambers arranged axially and axis-symmetrically respectively. This facilitates acceptable resolution of wave-front lengths of 30-60 mm over two to four whorls of chambers. The active peristaltic region is 185 mm in height (Figure 1C, element 9) with whorl height of 10 mm (Cross section of Figure 1D) and silicone whorl boundaries of 5 mm thickness (Cross section of Figure 1F).

The conduit (from 4-10 in Figure 1C) is of 20 mm internal diameter (as this is the limit of comfortable distension in the human body) and has a wall thickness of 8 mm. When whorls are inflated by pneumatic pressure the device is assumed to deform circumferentially toward the axis. An exemplary assertion of pressure to chambers in whorls three, four and five is shown in cross-section Figure 1E. The wave frame, depicted 11 in the same figure extends over the region where the diameter reduces from 20 mm. This is typically a little beyond the boundary of activated chambers. The general process of peristaltic pumping is for the wave, contained as 11, to move in the direction indicated 12 where the bolus remains ahead of the wave peak. The peristaltic motion is achieved by the harmonious occlusion of the conduit which is related to the pneumatic pressure in the whorls of chambers.

2.4   Key Improvements

The device presents a new opportunity to achieve more compliant and continuous peristaltic transport where the conduit does not exhibit any skeletal structures which limit radial occlusion which are observed in previously published works [10, 11]. This enables the peristaltic wave to extend, and move continuously, over a series of adjacent whorls of chambers. The elastomeric interface between the pneumatic pressure and bolus surface is inherently compliant. This behavior results in a complex relationship between pressure assertion in adjacent whorls and the continuous, deterministic, output behavior.

This, along with the inherent compliance of the pump, results in issues with trajectory generation. The system requires a transport wave to be achieved by the harmonious rostro-caudal assertion of pressure into a series of whorls of chambers. This requires the device behavior to be characterized such that the controller can and respond to changes in the transport process.

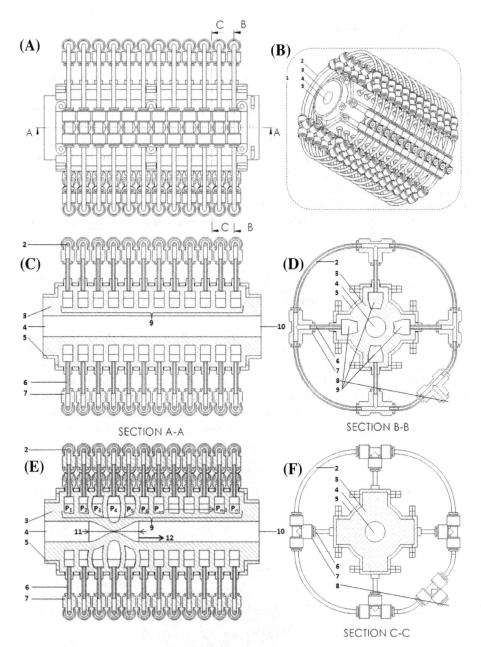

Fig. 1. (A) External view of peristaltic pump with indications of section views to follow. (B) is an isometric view showing the pneumatic plumbing (2), Silicone actuator (3) pumping conduit entrance (4) and housing (5). (C) Shows section A-A of part (A) with the axial arrangement of 12 chambers and conduit from (4-10). (D) Shows section B-B of part (A) with the Perpendicular view of the embedded pneumatic chambers (9). (E) Shows the principle of actuation and peristaltic transport. (F) Shows section C-C of part (A) where the chambers are sealed in the axial aspect.

3   Trajectory Planning and Generation

The peristaltic actuator was characterized by two medically-inspired measurement techniques; manometry to investigate occlusive pressure and articulography to investigate device geometry [18]. These measurements have been correlated to pressure commands such that the device may carry out desired peristaltic trajectories.

There are however, fundamental challenges that are to be overcome in the prescription of trajectories. The progression of peristaltic waves is known to exhibit inter- and intra-subject variability from swallow to swallow. This is due to differences in bolus formulation, differing perception of the bolus in the oral cavity and reflexive strategies of the body to ensure successful swallowing. This poses two challenges: There is no singular "acceptable" trajectory for any given bolus formulation and there are significant confounding variables between swallows. These are to be addressed in the following ways:

- Specification of bolus tail shape by mathematical approximation
- Wave propagation velocity dictated by medically captured peristalsis
- Basic trajectory prescription imitates prior knowledge / experience

Trajectories are currently administered by resolving components of manometric studies to predict the location and geometry of a modeled bolus. These are based on the conservation of bolus volume throughout two regions; a conduit-diameter slug propagating ahead of the wave and a mathematically prescribed bolus tail shape. These functions are described in a radial fashion. In the wave frame, using a sinusoid tail model as an example, the bolus radius can be described as Equation (1). The modeled bolus shape (Figure 2) varies in L_1 and L_2 to maintain a constant bolus volume.

$$H(x) = \begin{cases} e + \dfrac{a}{2}\left(1 - \cos\left(\dfrac{2\pi x}{\lambda}\right)\right) & if\ 0 > x > \dfrac{\lambda}{2} = L_1 \\ e + a & if\ L_1 > x > L_2 \end{cases} \tag{1}$$

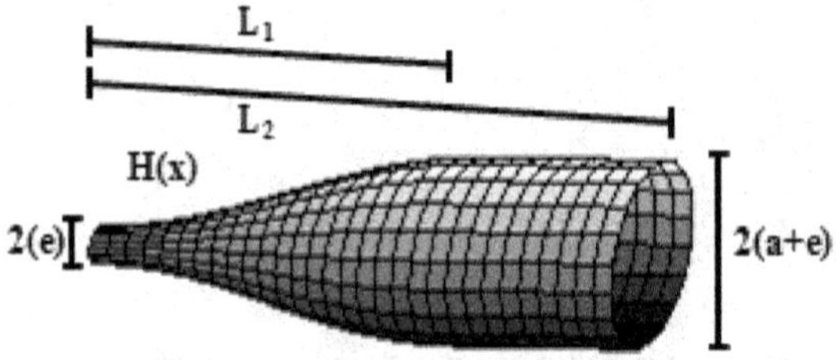

Fig. 2. Radial bolus model function (Equation 1) parameters applied in the wave frame. The sinusoidal element of the bolus tail extends from minimum diameter (2e) to 2(a+e) over a length L_1 which is equal to half the wavelength (λ/2). The region extending from L_1 to L_2 is of constant diameter 2(a+e).

Experimentation on the device involves pressure measurement by sensors mounted on an intraluminal catheter which is concentric with the conduit axis. Thus the bolus has a volume of Eq. 1 revolved about the transport axis minus the volume of the catheter of constant 2 mm radius (Equations 2,3). This facilitates the decoupling of components L_1 and L_2.

$$V = \int_0^{L_1} \pi \left(e + \frac{a}{2}\left(1 - \cos\left(\frac{2\pi x}{\lambda}\right)\right)\right)^2 dx + \int_{L_1}^{L_2} \pi(e + a)^2 dx - \int_0^{L_2} \pi(2)^2 dx \qquad (2)$$

$$V = \frac{(3a^2 + 8ea + 8e^2)\pi L_1}{8} + \pi(e + a)^2(L_2 - L_1) - 4\pi L_2 \qquad (3)$$

For the constant maximal amplitude case (e = 2mm and a = 8mm), of interest to modeling esophageal transport, there is a direct and fixed relationship between L_1 and L_2. The length L_1 and the axial velocity of the wave frame are inspired by the manometric findings of [19]. Throughout the region where the manometric wave completely enters the esophagus until the wave front reaches 185mm displacement (maximally achieved by the peristaltic actuator) the wave front (WF) and peak (WF) locations are modeled by Equations 4, 5 (also shown Figure 3A):

$$WP(t) = -1.62t^3 + 12.96t^2 + 7.126t + 0.3489 \qquad (4)$$

$$WF(t) = -3.15t^3 + 22.29t^2 - 3.442t + 37.15 \qquad (5)$$

It is understood that the bolus propagates ahead of the wave peak. However, the mechanic of predicting occlusive wave front location with a 5 mmHg pressure threshold is proposed to have a similar axial shift characteristic. Thus, the wave-front length is proposed to be evaluated as the difference between WP and WF. The wave-front length (WFL = $\lambda/2$ = L_1) as a function of time is then (Equation 6, also shown Figure 3A):

$$WFL(t) = -1.53t^3 + 9.33t^2 - 10.57t + 36.80 \qquad (6)$$

The prescription techniques are general in the sense that they can be applied to medically captured or mathematically prescribed wave fronts. This implementation is indicative of the current medical inspiration.

The performance of the device (Figure 3B) has been evaluated by minimization of Mean Squared Error (MSE) between the desired and possible geometrical wave shapes (from characterization data). The average true magnitude of error (ME) is also shown. By interpretation of Fig 3A and 3B simultaneously it is obvious that the device performance suffers at shorter wave-front lengths. This is exacerbated by the prescription of maximum amplitude as the device wall gradient function becomes more asymmetrical at higher pressure combinations which are required to completely occlude the conduit. As the bolus extends a more acceptable performance (<5% Full Scale) is achieved (from approximately 1.8s onwards, Figure 3B). The minimum ME of 0.79 mm is reached at 3.3 seconds at a wave-front length of 48.5 mm. The periodic sharp negative transitions in error magnitude are due to the wave propagating over the chamber boundaries. This occurs at 9 locations (from 0.9 seconds onward, resulting in the wave peak propagating from whorl 1 to 10, and wave front from whorl 3 to 12. The first trough of MSE and ME is due to a particularly strong model fit at a low intra-whorl wave displacement.

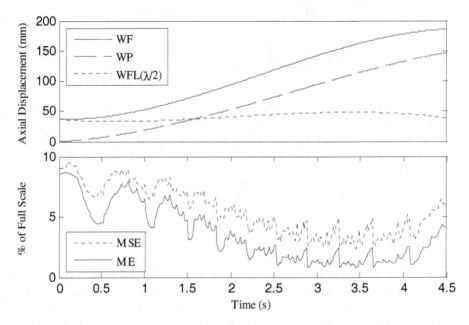

Fig. 3. (A) A medically-inspired trajectory (Inspired by manometric investigation of Clouse *et al.* [19]) for esophageal swallowing (Wave Front, Wave Peak and Wave Front Length) and (B) Predicted device performance to achieve trajectory shown in (A) based on characterization data (Mean Squared Error and Mean Error)

In the future it is anticipated to investigate the effects of velocity and acceleration on the pumping performance of boluses in the physical domain. This planning methodology shall be applied to determine the device behavior and subsequently generate the pneumatic pressure trajectories by the same inversion of the characterization data set.

4   Peristaltic Sensation and Haptic Intelligence – Future Work

Peristaltic transport is undertaken involuntarily in the human form, which is produced based on prior knowledge and feedback from the conduit surface. In order to achieve investigation of bolus rheology and pumping performances the device requires tactile sensation to be distributed over its actuating surfaces. Thus, relationships can be drawn between the bolus formulation, wave shape, and propagation conditions.

The concept of "prior knowledge" manifests itself in the form of a prescribed baseline trajectory as described in section 3 which may be modified on-line in response to the bolus propagation.

The tactile sensors in the biological system are embedded in different layers of the swallowing conduit and respond to diverse stimuli. They respond to muscular tension, bolus surface pressure and longitudinal shear stress in the circumference of the conduit. These phenomena are highly coupled and represent the totality of interaction between the biological system and the bolus material.

The time-variant pressure distribution on the bolus generated by the peristaltic actuation holds significant information about the transport behavior. These parameters are to be investigated in the peristaltic actuator device. The field of soft-robotic tactile sensation is still in its infancy due to inherent challenges of compliance and stretch-ability of electronic materials. Thus, novel approaches are required.

Principally, the nature of pressure sensation requires the transduction of a behavior over a surface. This is particularly suitable to be achieved by distributed and conformable passive electronics.

The human body intrinsically processes data in a holistic manner to determine macro behaviors. Thus, the individual sensor measurements of the engineering system are to be fused together algorithmically to create a haptic perception of the bolus transport. These elements are to be amalgamated into the features of the current peristaltic device. Currently methods of distributed and elastic capacitive sensation are under investigation.

5   Conclusions

The specification, design, and application requirements of a peristaltic actuator inspired by esophageal swallowing have been elucidated. However, the techniques of this research initiative are generalizable to any type of linear peristaltic pump with discrete distributed inputs which co-operate to produce continuous peristalsis. The modular nature of the design philosophy facilitates applications of arbitrary length and size; As such, it can be used as a model on different scales such as for the bowel, ureter, and other peristaltic conduits.

It is anticipated that the current research will find application in pumping fluids which have complex rheological or tribological characteristics as well as providing a means to investigate their transport behavior. The nature of intrinsically high degree of freedom peristaltic pumping offers some exciting opportunities to dictate the magnitude and direction of pressure assertion on the bolus tail.

In the future soft-robotic sensation techniques are to be exploited in the measurement of bolus transport metrics. These measurements will provide the basis for understanding the relationship between bolus formulation and transport efficacy. It is envisaged that the robotic device shall be able to discern between different bolus types such that it can be employed as a novel food bolus rheometer in the food technology industry.

Acknowledgements. This research work is funded in part by the Riddet Institute, a Centre of Research Excellence, Palmerston North, New Zealand, and the University of Auckland, Auckland, New Zealand.

References

1. Misra, J.C., Maiti, S.: Peristaltic transport of rheological fluid: model for movement of food bolus through esophagus. Applied Mathematics and Mechanics 33(3), 315–332 (2012)
2. Toklu, E.: A new mathematical model of peristaltic flow on esophageal bolus transport. Scientific Research and Essays 6(31), 6606–6614 (2011)
3. Misra, J.C., Pandey, S.K.: A mathematical model for oesophageal swallowing of a food-bolus. Mathematical and Computer Modelling 33(8-9), 997–1009 (2001)
4. Pandey, S.K., Tripathi, D.: Peristaltic Flow Characteristics of Maxwell and Magnetohydrodynamic Fluids in Finite Channels: Models for Oesophageal Swallowing. Journal of Biological Systems 18(3), 621–647 (2010)
5. Brasseur, J.: A fluid mechanical perspective on esophageal bolus transport. Dysphagia 2(1), 32–39 (1987)
6. Saunders, F., et al.: Experimental verification of soft-robot gaits evolved using a lumped dynamic model. Robotica 29(6), 823–830 (2011)
7. Sangok, S., et al.: Peristaltic locomotion with antagonistic actuators in soft robotics. In: 2010 IEEE International Conference on Robotics and Automation (ICRA) (2010)
8. Trivedi, D., et al.: Soft robotics: Biological inspiration, state of the art, and future research. Applied Bionics and Biomechanics 5(3), 99–117 (2008)
9. Miki, H., et al.: Artificial-esophagus with peristaltic motion using shape memory alloy. International Journal of Applied Electromagnetics and Mechanics 33(1-2), 705–711 (2010)
10. Suzuki, K., Nakamura, T.: Development of a peristaltic pump based on bowel peristalsis using for artificial rubber muscle. In: 2010 IEEE/RSJ International Conference on Intelligent Robots and Systems (IROS) (2010)
11. Carpi, F., Menon, C., De-Rossi, D.: Electroactive Elastomeric Actuator for All-Polymer Linear Peristaltic Pumps. IEEE/ASME Transactions on Mechatronics 15(3), 460–470 (2010)
12. Dodds, W.J.: The physiology of swallowing. Dysphagia 3(4), 171–178 (1989)
13. Kuo, B., Urma, D.: Esophagus - anatomy and development. Gastrointestinal Motility online (2006)
14. Broering, D.C., Walter, J., Halata, Z.: Surgical Anatomy of the Esophagus. In: Izbicki, J.R., et al. (eds.) Surgery of the Esophagus, pp. 3–10. Steinkopff (2009)
15. Orvar, K.B., Gregersen, H., Christensen, J.: Biomechanical characteristics of the human esophagus. Digestive Diseases and Sciences 38(2), 197–205 (1993)
16. Yang, W., et al.: Finite element simulation of food transport through the esophageal body. World Journal of Gastroenterology 13(9), 1352–1359 (2007)
17. Jean, A.: Brain stem control of swallowing: Neuronal network and cellular mechanisms. Physiological Reviews 81(2), 929–969 (2001)
18. Dirven, S., et al.: Design and characterisation of a peristaltic actuator inspired by esophgeal swallowing (2013) (manuscript submitted for publication)
19. Clouse, R.E., et al.: Characteristics of the propagating pressure wave in the esophagus. Digestive Diseases and Sciences 41(12), 2369–2376 (1996)

Design and Fabrication of a Soft Actuator
for a Swallowing Robot

Fei-Jiao Chen[1,3], Steven Dirven[2], Weiliang Xu[2], Xiao-Ning Li[3], and John Bronlund[1]

[1] School of Engineering & Advanced Technology, Massey University
Building 106, Oteha Rohe, Albany, Auckland, New Zealand
[2] Department of Mechanical Engineering, University of Auckland
Private Bag 92019, Auckland Mail Center, 1142, Auckland, New Zealand
[3] School of Mechanical Engineering, Nanjing University of Science and Technology
Building 106, Oteha Rohe, 106.20, Albany, Auckland, New Zealand
F.J.Chen@massey.ac.nz

Abstract. Textured food is provided to dysphagia populations in clinical practice for assessment and management of swallowing disorders. A considerable amount of measurements showed that the textural properties of food can affect the performance of human swallow significantly. However, the selection of food for a specific subject is difficult, due to the complexity of the biological structures and the potential risks of *in vivo* testing. For the purpose of providing a safe environment for food flow study, a novel soft actuator capable of producing peristalsis movement was proposed. During the esophageal swallowing, which is the last stage of human swallow, food is transported through the muscular tube by peristalsis mechanism. The motion pattern is generated by the coordinated contractions of circular muscles of the esophagus. Inspired by human esophagus and the biological process, the actuator was designed to have a completely soft body without any hard components. Discrete chambers are embedded inside the body regularly and a cylindrical food passage locates at the center of the actuator. Finite element analysis (FEA) was used to determine the structure parameters of the actuator. The soft body was fabricated by casting silicon material in a custom mold. Preliminary experiments have been performed to characterize the actuator.

Keywords: dysphagia, esophageal peristalsis, soft actuator, FEA.

1   Introduction

Swallowing disorder, also known as dysphagia, contributes to reduced dietary intake, increased risk of pneumonia, and potentially dehydration and malnutrition [1]. It may occur through swallowing process when there is neuromuscular dysfunction resulting in weakness, or sensory loss in muscles related to swallowing. Dysphagia can be found in all age groups, and is a common compliant in elderly. It is reported that 15% of the population over sixty years are suffering swallowing disorders [2].

A normal swallow with regular foods requires excellent muscle response and accurate timing control between the swallowing system and the respiratory system

J.-H. Kim et al. (eds.), *Robot Intelligence Technology and Applications 2*,
Advances in Intelligent Systems and Computing 274,
DOI: 10.1007/978-3-319-05582-4_42, © Springer International Publishing Switzerland 2014

[1]. For individuals suffering swallowing difficulty, a diverse range of texture-modified foods and fluids has been developed for the diagnosis and management of dysphagia [3, 4]. It is clear that provision of dysphagia diets is an effective prescription during dysphagia treatment, which helps therapists examine the cause and severity of swallowing disorder. However, the selection of which textured food to be the safest for individual to swallow still remains vague, due to the risks to patient health, and difficulty of undertaking examination *in vivo* [5, 6].

Food texture is a multi-parameter attribute that is described in the geometrical, mechanical, and chemical fields. The textural properties of foods, such as hardness, adhesiveness, viscosity, moistness, oiliness, can affect our perception of foods, and determine swallow type, from the trigger of this reflexive movement, to the end stage of swallowing [6]. There is a high demand of understanding the connection between food nature and swallow behavior. Although a considerable amount of work has been done to investigate the relationship between food texture and biological events of swallow process, the uncomfortable *in vivo* testing and measurements may cause pain and put subjects into a potential risk [6].

In response to the difficulty of studying the efficacy of texture modified food on human swallowing *in vivo*, a swallowing robot modeling the deglutition process will provide a non-risk environment, taking place of subjects for food scientists and doctors to facilitate the food flow study in swallowing. This paper presents a novel bio-inspired actuator that mimics human esophageal transport of food bolus, which is the fourth stage of swallowing. The actuator is designed to generate a propagation of wave similar to esophageal peristalsis mechanism that actually forces food bolus moving though the muscular conduit. The rheological characteristics of food flow (e.g. deformation, velocity, viscosity) and the physical contact pressure between food and food passage can be quantitatively recorded in future, benefit from new sensing technology and methods.

2   Background

2.1   Esophageal Swallow and Measurements

Esophagus is a muscular tube typically between 20 to 26 cm in length, and about 2 cm in diameter in adult humans [7]. The peristalsis motion in esophageal region has a principal function to propel food bolus from pharynx into stomach. The wave-like motion is generated by sequential contraction of circular and longitudinal muscles of esophagus.

Fig. 1 illustrates the esophageal swallowing process. The upper esophageal sphincter (UES) and the lower esophageal sphincter (LES) remain closed between swallows. UES opens and lets food bolus come into esophagus from pharynx. Then, circular muscles of the esophagus contract, closing the muscular tube occurs and imposing driving pressure on the tail of food bolus, moving it towards the stomach. LES opens when bolus reaches its location at the end of this stage, and closes again after bolus passes, preventing reverse flow of food and stomach acid. This progress will last a few seconds, altering according to the bolus rheology properties [8].

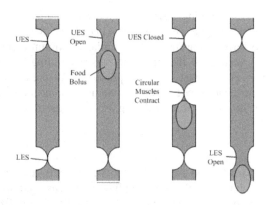

Fig. 1. Schematic of esophageal swallowing process

Manometric study has been extensively applied for clinical assessment and evaluation of normal and abnormal swallowing [8]. And a list of features of the propagation pressure wave has been measured and analyzed, including the travelling speed, duration of the wave, and the pressure amplitude. The performance of esophageal swallow varies between subject, due to differences in age and sex [9, 10]. It can also be affected by rheological properties of food bolus flow, such as size, viscosity, adhesiveness [11]. A few conclusions have been summarized in most cases, such as increased bolus viscosity can slow the propagation and increase the travel time, and boluses with large volume will increase contraction pressure for cleaning.

2.2 Swallow Robot and Peristaltic Robot

There has been recent interest in the development of robotic devices associated with human swallowing [6]. A number of bio-inspired robots have been constructed to simulate elements of this process, most of which come out of Japan [6]. However, there is still no fully development of a device that is capable of reproducing all four stages of swallowing [6]. A dynamic swallowing simulation system, consisting of most structures involved in swallow, was developed at Waseda University [12]. Sixteen pneumatic muscles were used to drive the mandible, the tongue, and the pharynx in the system simultaneously. Video-fluorographic technology was applied to observe food flow inside the system. However, this simulation system does not have a function as esophagus generating peristalsis.

The motion pattern of peristalsis can be found in food transport through human esophagus, urine transport through urethra, blood flow in vessel, and many other biological systems [13]. It has been well understood via theoretical and experimental investigations [13]. Inspired from earthworm locomotion, a number of peristaltic robots have emerged as a response to the calls of devices to serve in specific occasions, such as colonoscopy and pipeline maintenance [14, 15].

Development of biomimetic devices to transport fluid as esophagus and intestines has also been realized recently. Kong et al. developed an in vitro stomach model, called the Human Gastric Simulator, to study gastric digestion of foods [16]. Continuous contraction of the stomach model is generated by the compression from a series of rollers. A novel gripping device inspired by the feeding mechanism of a marine mollusk was developed by Mangan [17]. Four Mckibben muscles are connected in series to form a ring that works similar to circumferential muscle. The work of Suzuki closely focuses on the intestinal transport aiming to convey semi-solid fluids [18]. The peristaltic device is composed of discrete modules pumped separately. The closure characteristic of one module is outstanding. However, the continuous transport is not achievable by the rigid boundaries of modules.

2.3   Soft Robot

Unlike conventional hard-bodied robots formed by skeletal structures and discrete joints, soft robots integrate flexible and compliant materials, and seek to reproduce functions or behaviors just like biological organisms [19]. The utilization of elastic materials gives soft robots many advantages over its traditional counterparts, such as low resistance to stress force, large deformation, and capability to carry fragile objects. One of the first soft robots is the starfish like robot developed by Whitesides Research Group, which is able to squeeze the body through a narrow region [20].

The new material gifts soft robot advanced properties, and makes it ideal in rescue and search applications, particularly in disasters after earthquakes. However, the super elasticity also brings outstanding challenges in design, modeling, and control domain.

The infinite degrees of freedom and limited number of actuators take soft robot into an underactuated field. Most soft robots available are driven by air power to achieve distributed deformation and motility [20, 21]. Due to the intrinsic non-linearity from structure, material, and actuation, it is difficult to predict the mechanical behavior of a soft body.

While a simple mathematical approach to model the dynamics and kinematics for a complex soft structure is not available, it is possible to study the behavior of the soft tissue or robot, with varying loads, constraints, and walls of different thickness, thanks to the development of finite element method. However, the accuracy of modeling is proportional to the number of elements for the soft body, which makes FEA not appropriate for a real-time control over soft robot [19].

3   Concept Design and FEA Simulation

3.1   Actuator Conceptual Design

The peristaltic wave in esophageal swallowing is considered to be formed by the circumferential closure of esophagus and the propagation of contraction along the conduit. The behavior of peristalsis is usually assessed and evaluated by video-fluorography technology, which provides visualization in real-time, and by manometry technology, which is used to measure contract pressure at different locations in the esophagus.

To provide an *in vitro* environment for testing the status of food bolus flow during swallow, a novel actuator is required to be able to generate a peristaltic wave, of which the propagation speed and the contraction pressure need to meet the clinical measurements. The fundamental requirements for the actuator are: (a) similar in morphology to human esophagus and no skeletal elements included; (b) capability of continuity transport; (c) the transport speed and the contraction pressure can be controlled actively.

A soft structure with discrete segments is proposed to generate a travelling wave in this study. Considering the actuator body is made of elastic material, pneumatic technique is used as a compliant and distributed actuation. In order to realize circular muscle constriction and continuous wave traveling as required, multiple chambers are embedded in the structure, and arranged regularly along the axis of the tube. In Fig. 2, the blue part is a half model of the actuator.

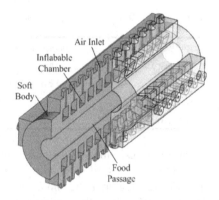

Fig. 2. Three dimensional model of the actuator

Four chambers will expand when pressurized to close cross-section area of the food passage. The amplitude of inflation is under control of air pressure inside chambers. More than one level of chamber can be activated at the same time to generate a peristalsis by inflation and deflation, as illustrated in Fig. 3. Also a variety of motion patterns can be achieved by control air pressures. To ensure the continuity of the peristalsis transport, the thickness of the internal surface, the width of chamber, and the interval distance between neighboring chambers will be determined by FEA simulation.

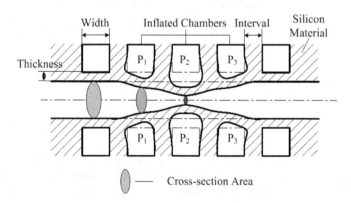

Fig. 3. Closure of the cross-section area of the actuator by pressurization

3.2   FEA Simulation

It is challenging to analytically predict and describe the deformation of the soft actuator with complex composite structure. In this study, FEA is used to aid the analysis and design to gain the optimum shape and dimensions of the actuator.

A silicon rubber material (Ecoflex 0030, Smooth-on, USA) was chosen to build the body of the actuator. The rubber material is a two pack platinum-catalyzed silicon that is easy to mix and cure at room temperature. Table 1 is the specification of the material attributes provided by the producer.

Table 1. Ecoflex 0030 attributes specifications

Attribute	Specification
Mixed Viscosity	3,000 cps
Cure Time	4 hours
Shore Hardness	00-30
Elongation at Break	900%

A tensile testing was conducted to gain the strain-stress relationship of the rubber before FEA simulations. The dumbbell shaped specimens used in the testing is with 2mm of thickness and 3mm narrowest region in the middle. The elongation speed was 500.00 mm/min. The blue curve in Fig. 4 is the experimental result for one of the specimens. The curve implies a non-linear relationship between strain and stress of the rubber material.

A third-order Mooney-Rivlin model is used to simulate the non-linear elasticity obtained experimentally. The empirical model is expressed as:

$$W = C_{10}(I_1 - 3) + C_{01}(I_2 - 3) + C_{20}(I_1 - 3)^2 + C_{02}(I_2 - 3)^2 \\ + C_{11}(I_1 - 3)(I_2 - 3) \tag{1}$$

Where W is the strain energy density, Cij is coefficients determined by experiments, and I1 and I2 are strain invariants that can be expressed as following:

$$I_1 = \lambda^2 + \frac{2}{\lambda} \tag{2}$$

$$I_2 = 2\lambda + \frac{1}{\lambda^2} \tag{3}$$

$$\lambda = 1 + \varepsilon \tag{4}$$

Where λ is the elongation rate of the tested specimen, and ε is the strain of the specimen. Table 2 shows the coefficients of strain energy density function in Mooney-Rivlin model. They were identified by Least Square Method. And the approximate strain-stress curve fits well with the experiment, as shown in Fig. 4.

Table 2. Coefficients of strain energy density function in Mooney-Rivlin model

Coefficients	Value
C10	2.5391e-4
C01	5.9343e-3
C20	1.3230e-9
C02	-9.3800e-13
C11	-3.9140e-8

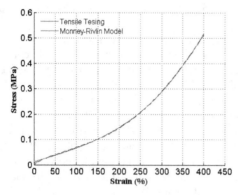

Fig. 4. The strain-stress testing result and the approximate strain-stress by Mooney-Rivlin model

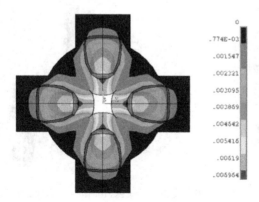

Fig. 5. Structural deformation of the actuator upon pressure 48 kPa

Fig. 5 shows the structural deformation of the actuator in the cross-section view. Hyper elastic hexahedron elements are used in this model. Pneumatic inlets were ignored in this model to save computational source. All surfaces on top, bottom, and side of the actuator model were set as fixed. The surface of the food passage was set as free, and the self-contact during simulation was considered. The air pressure was imposed in the normal direction of each wall in chambers. The maximum displacement can be found at the central point of each chamber, which is around 6.19 mm upon driving pressure of 48 kPa.

4   Fabrication

The peristaltic actuator was fabricated using a custom designed mold that was created via fused deposition modeling process, by a three-dimensional printer with acrylonitrile butadiene styrene (ABS) plastic. The entire mold is actually assembled by seven plastic parts. The time for printing lasted a few hours, depending on the size and complexity of the parts.

Fig. 6 shows the fabrication process. After the mold was assembled, all gaps between contact surfaces of two parts must be sealed, preventing a leakage of the liquid material. Two components of silicon were then mixed sufficiently, and poured into the mold before the liquid starts to cure. A central part with food passage and twelve lays of open-chambers was removed out of the mold, after it had been cured at room temperature about five hours after pouring. We use the same material to seal all chambers, and built in air inlet of 4mm diameter to each chamber. To strength the bonding, the soft structure was heated in an oven at about 60°C for three hours. Four chambers sitting at the same layer were connected by polyethylene tube and one touch fittings, as shown in Fig. 6. Chambers at different floors do not share a common air source, and can be pressurized separately.

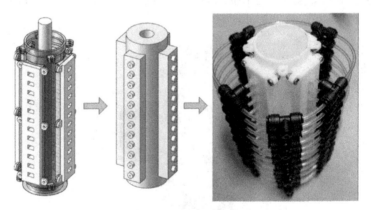

Fig. 6. Fabrication process for the soft actuator

Table 3. Specifications of the actuator

Diameter of the food passage	20mm
Length	215mm
Number of chamber	48
Weight	~690g

5   Preliminary Experiment

5.1   Drive System and Articulograph

Fig. 7(a) shows the experimental setup to study and characterize the novel actuator. The drive system consists of a PC-based GUI, a D/A converter interface (Microcontroller PIC24FJ128GA010, and Digital-to-Analog Converter MCP3208), an air source, six proportional solenoid valves and six directional valves. By switching directional valve, each proportional valve is able to regulate two layers of chambers. And maximum six layers can be controlled at the same time.

The deformable surface locates inside the actuator at central region. For the purpose of evaluating constriction performance of the actuator, articulograph technique was used to record the radial movements at specific locations of the passage. As shown in Fig. 7(b), a magnetic sensor with sphere shape was attached on the passage surface at the midpoint of a chamber. The cross section of the conduit will come to a closure while chambers are pressurized, as illustrated by dot line in Fig. 7(b). And the sensor will move together with the surface in the direction of the central point, denoted as r, and record the displacement.

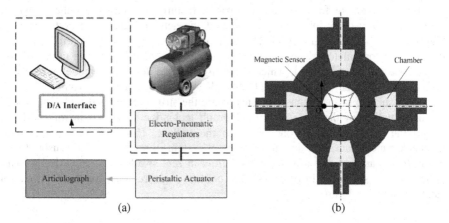

(a)   (b)

Fig. 7. (a). Experiment setup; (b) Location of the magnetic sensor of articulograph

5.2   Preliminary Experiments and Results

In preliminary experiments, we inflated four chambers at one layer, and left the other chambers open. The driving pressure was increased from 0 kPa to 71.4 kPa step by step. Each actuation step was hold for 1s guaranteeing a stable status of the deformed surface.

Discrete displacement data were collected by articulograph. Fig. 8 shows the results. In the first few steps while the actuation pressure was less than 28 kPa, chambers did not expand, and the surface remained in the default status. The magnetic sensor moved to the center of the passage gradually when the pressure was increased. The maximum displacement of the sensor was recorded approximately as 7.5 mm when the pressure reached 64.29 kPa.

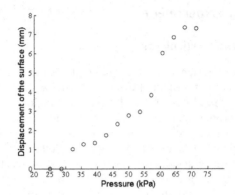

Fig. 8. Displacement of the food passage surface

6   Conclusion

In this paper, a novel actuator has been presented, aiming to reproduce esophageal peristalsis movement for a swallowing robot. Inspired by human esophagus which actually is a muscular tube, the prototype of the actuator has a soft body made of highly extensible silicon material. The finite element analysis was used to determine the structure parameters. The actuator body was fabricated by casting liquid silicon into a custom-designed mold, which was printed fused deposition modeling process. Pneumatic technique was employed to provide a safe and distributed driving.

As the surface of the actuator locates inside the structure, it is difficult to quantify the deformation of the surface. Articulograph technique was used in this study to record the displacement of one point on the surface. However, it is not able to feedback the displacement data to the control system in real-time. And sensors taped in the food passage may affect the food flow in real swallows. A new design with optical sensor system is in progress. Non-contact measurements for the deformation can be realized in future.

Acknowledgements. This research work is funded in part by the Riddet Institute, a Centre of Research Excellence, Palmerston North, New Zealand, and the University of Auckland, Auckland, New Zealand. The first author would like to acknowledge China Scholarship Council (CSC) for a doctoral scholarship (2010 - 2012), and Nanjing University of Science and Technology (NUST) for a doctoral scholarship (2013).

References

1. A. Dietitians Association of and L. The Speech Pathology Association of Australia, Texture-modified foods and thickened fluids as used for individuals with dysphagia: Australian standardised labels and definitions. Nutrition & Dietetics 64, S53–S76 (2007)

2. Robbins, J., et al.: Oral, pharyngeal and esophageal motor function in aging [GI Motility online] (2006),
   http://www.nature.com/gimo/contents/pt1/full/gimo39.html
3. Iida, Y., et al.: Videofluorographic Evaluation of Mastication and Swallowing of Japanese Udon Noodles and White Rice. Dysphagia 26, 246–249 (2011)
4. Funami, T.: Next target for food hydrocolloid studies: Texture design of foods using hydrocolloid technology. Food Hydrocolloids 25, 1904–1914 (2011)
5. Penman, J.P., Thomson, M.: A review of the textured diets developed for the management of dysphagia. Journal of Human Nutrition and Dietetics 11, 51–60 (1998)
6. Chen, F.J., et al.: Review of the swallowing system and process for a biologically mimicking swallowing robot. Mechatronics 22, 556–567 (2012)
7. Broering, D., et al.: Surgical Anatomy of the Esophagus. In: Izbicki, R., et al. (eds.) Surgery of the Esophagus, pp. 3–10. Steinkopff (2009)
8. Kendall, K.A., et al.: Timing of Events in Normal Swallowing: A Videofluoroscopic Study. Dysphagia 15, 74–83 (2000)
9. Logemann, J.A., et al.: Oropharyngeal Swallow in Younger and Older Women: Videofluoroscopic Analysis. J. Speech Lang. Hear. Res. 45, 434–445 (2002)
10. Logemann, J.A., et al.: Temporal and Biomechanical Characteristics of Oropharyngeal Swallow in Younger and Older Men. J. Speech Lang. Hear. Res. 43, 1264–1274 (2000)
11. Ruark, J.L., et al.: Bolus Consistency and Swallowing in Children and Adults. Dysphagia 17, 24–33 (2002)
12. Yohan, N., et al.: Development of a robot which can simulate swallowing of food boluses with various properties for the study of rehabilitation of swallowing disorders. In: 2011 IEEE International Conference on Robotics and Automation (ICRA), pp. 4676–4681 (2011)
13. Misra, J.C., Pandey, S.K.: A mathematical model for oesophageal swallowing of a food-bolus. Mathematical and Computer Modelling 33, 997–1009 (2001)
14. Adachi, K., et al.: Development of multistage type endoscopic robot based on peristaltic crawling for inspecting the small intestine. In: 2011 IEEE/ASME International Conference on Advanced Intelligent Mechatronics (AIM), pp. 904–909 (2011)
15. Seok, S., et al.: Meshworm: A peristaltic soft robot with antagonistic nickel titanium coil actuators. IEEE/ASME Transactions on Mechatronics PP, 1–13 (2012)
16. Kong, F., Singh, R.P.: A human gastric simulator (HGS) to study food digestion in human stomach. Journal of Food Science 75, E627–E635 (2010)
17. Mangan, E.V., et al.: A biologically inspired gripping device. Industrial Robot 32, 49–54 (2005)
18. Saito, K., et al.: Development of a peristaltic pump based on bowel peristalsis improvement of closing area rates and suction pressure measurement. In: 2012 4th IEEE RAS & EMBS International Conference on Biomedical Robotics and Biomechatronics (BioRob), pp. 949–954 (2012)
19. Shepherd, R.F., et al.: Multigait soft robot. Proceedings of the National Academy of Sciences (2011)
20. Ilievski, F., et al.: Soft Robotics for Chemists. Angewandte Chemie International Edition 50, 1890–1895 (2011)
21. Trivedi, D., et al.: Soft robotics: Biological inspiration, state of the art, and future research. Appl. Bionics Biomech. 5, 99–117 (2008)

A P300 Model for Cerebot – A Mind-Controlled Humanoid Robot

Mengfan Li[1], Wei Li[1,2], Jing Zhao[1], Qinghao Meng[1], Ming Zeng[1], and Genshe Chen[3]

[1] Institute of Robotics and Autonomous Systems,
School of Electrical Engineering and Automation,
Tianjin University, Tianjin 300072, China
{shelldream,qh_meng,zengming}@tju.edu.cn,
zhaoj379779967@163.com
[2] Department of Computer & Electrical Engineering and Computer Science,
California State University, Bakersfield, California 93311, USA
wli@csub.edu
[3] Intelligent Fusion Technology, Inc., Germantown, MD 20876, USA
gchen@intfusiontech.com

Abstract. In this paper, we present a P300 model for control of Cerebot – a mind-controlled humanoid robot, including a procedure of acquiring P300 signals, topographical distribution analysis of P300 signals, and a classification approach to identifying subjects' mental activities regarding robot-walking behavior.

We design two groups of image contexts to visually stimulate subjects when acquiring neural signals that are used to control a simulated or real NAO robot. Our study shows that the group of contexts using images of robot behavior delivers better performance.

Keywords: Mind control, P300 model, OpenViBE, humanoid robot.

1 Introduction

P300 is a late, endogenous component of event-related potentials (ERPs), which appears as a large positive deflection after the events (such as sensory, cognitive events) being presented about 300ms [1]. The latency of it varies from 200 to 800ms, and its amplitude can even reach 20uV in parietal area of the cortex. This potential can be regarded as a degree index of the relevance between stimulus and subject's cognitive task [2]. This classical P300 Speller based on "oddball" paradigm first set up in [3] provides a communication channel to identify subjects' mental activities by analyzing P300 signals. Since then, applications of P300 potentials have emerged, e.g., a P300 Speller for communication [4], an internet browser [5], controlling a mouse on the screen [6] or controlling an object in a virtual environment [7]. Significant attempts to control physical devices are reported, e.g., to navigate a wheelchair [8], and even to control a 7 degree of freedoms (DoFs) robotic [9]. Work in [10] uses P300 evoked

J.-H. Kim et al. (eds.), *Robot Intelligence Technology and Applications 2*,   495
Advances in Intelligent Systems and Computing 274,
DOI: 10.1007/978-3-319-05582-4_43, © Springer International Publishing Switzerland 2014

potentials to control a humanoid robot. These applications become more and more interesting to disabled patients to help themselves in their daily life.

In this paper, we use Cerebot, a mind-controlled humanoid robot platform [11], to investigate a P300 protocol for control of a NAO robot, as shown in Fig. 1. We develop an OpenViBE-based experimental environment, which integrates programming, BCI design and signal acquisition software. We design two groups of different image contexts to visually stimulate subjects for acquiring P300 signals. Our studies show that the group with context of robot achieves a better performance.

2   Cerebot and OpenViBE Environment

Cerebot is a mind-controlled humanoid robot platform [11-12], consisting of a Cerebus™ Data Acquisition System, a humanoid robot, and a virtual simulator WEBOTS, as shown in Fig. 1. The Cerebot platform uses Cerebus™ to record brainwaves during human mental activities. This platform uses a NAO robot with 25 DoFs shown in Fig. 1 or a KT-X PC humanoid robot with 20 DoFs shown in Fig. 2.

OpenViBE is new general-purpose software for designing, testing, and using brain-computer interface. Using OpenViBE, it is easy and fast to design a brain-computer interface in an intuitive way. Fig. 1 describes the OpenViBE programming environment for the Cerebot platform. The environment integrates the visual stimulus section, collecting signal section, signal processing and classification section, and robot control section.

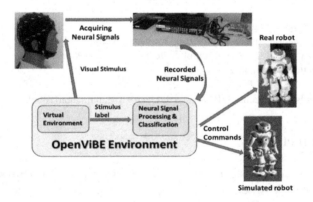

Fig. 1. Cerebot, a mind-controlled humanoid robot platform

3   Experiment Preparation and Procedure

3.1   Experimental Protocol

The experiment design is based on the classical "oddball" paradigm. A visualization box in OpenViBE displays a 2*3 matrix with contexts on the screen. Two groups of image contexts are embedded into the matrix to visually stimulate test subjects. The first group contains six red squares provided by an OpenViBE box; while the second group contains six images of humanoid robot walking behavior. Each red square or robot image in the matrix represents a robot walking behavior: walking forward, walking backward, turning left, turning right, shifting left and shift right, as shown in Fig. 2(b). When one red square or robot image is flashing, the others will be shielded by an "off image." The probability of flashing a target is 1/6 (the reciprocal of the number of images) which meets the requirement of eliciting P300 potentials [13]. The context of an "off image" is a black square with a white solid circle located in the middle of the image, as shown in Fig. 2(c).

When an experiment starts, the screen is blanked in grey color for 5 seconds, and then six red squares or robot images flash separately in a random order called as "repetition." The presentation time of an image lasts 200ms and the inter-stimulus interval (ISI) [14] is 300ms, so a display cycle is 1.8s shown in Fig. 2(a). A number of repetitions constitute a trial in which the subject is asked to focus on only one red square or one robot image, which means that each red square or each robot image flashes several times before the P300 model outputs a command to control the humanoid robot. The subject is suggested to count number when the target is presented.

3.2 Experimental Procedure

Before starting experiments, the subject needs to be wearing an EEG cap manufactured according to "the international 10-20 system". The ground electrode is AFz and the linked-mastoids are the references.

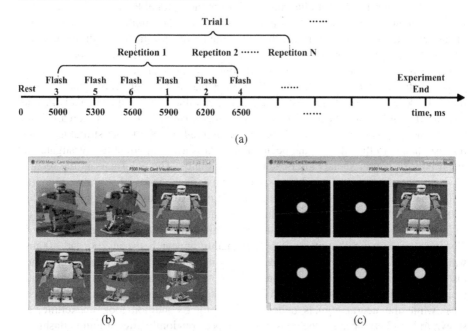

(a)

(b)   (c)

Fig. 2. (a) Experiment protocol (b) Initial interface before experiment beginning (c) The interface of image flashings (Note: In order to see the images clearly in paper, here the size of an image increases and the distances between two images decrease)

The subject sits in a comfortable chair. The distance from the subject to the display screen is 70cm and subject's eyes are at the same horizontal level with the screen center. The subject tries to avoid any movement during acquiring P300 signals. A complete experiment is conducted in a silent environment. As discussed before, each subject needs to take two groups of experiments: The first one called Set One is that the stimulus section is flashing the red squares, and the second one called Set Two is that the stimulus section is flashing robot images with arrows indicating robot motion.

4   Signal Analyses

4.1   Certification of P300 Signal

The acquired neural signals is amplified, preprocessed by an analog lowpass filter of 50Hz, and digitalized with a sampling frequency of 1000Hz. The signal analysis section processes the neural signals by extracting the frequency components between 0.5 and 26Hz using a digital bandpass filter and removing an extracted epoch with its amplitude higher than 90uV. In order to find out the amplitude differences in epochs under target and non-target stimulus conditions, the signal analysis section subtracts a baseline from each epoch. The baseline is the average of the signal pre-stimulus 300ms.

The solid and dotted curves plotted in Fig. 3(a) represent the averaged signals under target and non-target stimulus conditions from the channel Pz. It demonstrates that the experiment design elicits P300 potentials because the amplitudes elicited under the target condition are much larger than these under the non-target condition at about 340ms. The signed r^2 function is an index to describe the discrimination between the signals acquired under two different conditions. Fig. 3(b) shows the r^2 values' topography distribution over all channels. The color bar on the right shows the r^2 value range. Dark red represents highest r^2 value. The r^2 values around the channels Pz and Cpz are the highest and become lower when other channels' distances increase from the channel Pz, as shown in Fig. 3(b), so it is assured that the target stimulus causes biggest change in the parietal and occipital area. It is also important to investigate the amplitudes of neural signals after each visual stimulus. The color bars in the second row represent the value of amplitude, and the n axis is the index of stimulus. Fig. 3(c) and Fig. 3(d) show the amplitude of signals from pre-stimulus 300ms to pos-stimulus 800ms at the channel Pz (each stimulus flashes at 0ms). For example, a color spot (t, n) in Fig. 3(c) represents a value of the signal elicited by nth target stimulus after it flashing tms. Some red spots in Fig. 3(c) mainly appear between 200 and 400ms, which indicate there are positive deflections during this time period after target stimulus flashings. The neural signals under the target condition exhibit the features with their peaks at about 340ms as show in Fig. 3(c), so this experiment elicits P300 potentials; while the neural signals under the non-target condition look unexciting, as shown in Fig. 3(d), because the color points appear randomly after stimulus flashings and the signals amplitudes are relative low.

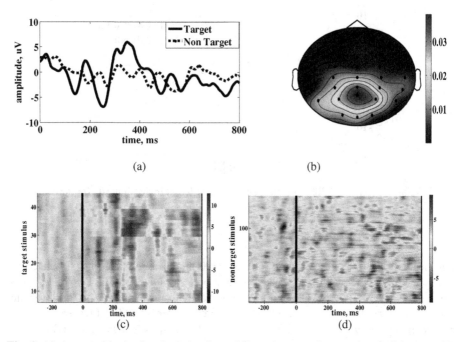

Fig. 3. (a) Averaged brain signals at the channel Pz under target/non-target conditions ($n=40$) (b) The r^2 value distribution (c) Neural signals elicited under target stimuli (d) Neural signals acquired under non-target stimulus

4.2   Effects of Images Context

The P300 potential is evoked by the visual event, so the interface presented to subjects, the format of stimulus and ISI all may have effects on the latency and amplitude of P300 potentials [15]. We design two groups of image contexts described in subsection 3.2 to investigate effects of image contexts on P300 evoked potentials.

We discuss the averaged neural signals acquired from the channel Pz. The red and blue curves in Fig. 4 represent the acquired neural signals in Set One and Set Two, respectively. The results show that the P300 evoked potentials change in shape when the image contexts change. Compared to Set One, the averaged neural signal acquired from Set Two is smoother and has a deeper negative peak before the P300 potential peak. After this negative peak, a neural signal acquired from Set Two has only one peak; the one acquired from Set One has two relative lower peaks. The differences may lie in different visual intensities of the two groups of image contexts as a P300 potential is related with the stimulus characteristic [13]. The differences may also be probably caused by some other mental or cognitive factors. The important result for us is that the classification success rates of Set Two are higher than these of Set One.

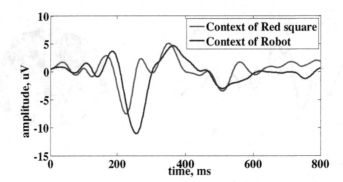

Fig. 4. Averaged P300 potentials elicited by different image contexts (n=800)

5   Robot Control Test

This section describes an evaluation of the P300 model by control of a simulated or real NAO robot. Each subject is asked to take both Set One and Set Two. The video clip on control of the NAO robot walking behavior is available on our website (http://www.youtube.com/watch?v=rOAosvpbRc&feature=youtu. be or http://v.youku.com/v_show/id_XNjAzNjE3Njc2.html).

A typical P300-based system can be divided into three parts: signal acquiring, pattern recognition, and device control. A visual stimulus activates a window with a width of 500ms to catch a P300 potential at the selected channels (here we choose Cz or Pz). The signal is filtered as discussed above and down-sampled to100Hz, so the dimension of a feature vector is 50.The feature extraction part provides six feature vectors to the Fisher's Linear Discriminant Analysis (LDA) to classify P300 signals. This two-class classification algorithm classifies whether a P300 potential is elicited under the target condition or not after a trial completed.

Two right-handed volunteers with normal vision undergo the two sets of experiments. We summarize the evaluation results in the table. The success rates of the P300 potentials evoked under the image contexts using the robot walking behavior are higher than these under the ones using the red square. This result may indicate that the image contexts of robot walking behavior could improve the attention, motivation or other cognitive ability when a subject is doing P300 experiment.

Table 1. Accuracy of Evaluation Results

Subject	Channel	Set One	Set Two
One	Cz	52.08%	87.49%
One	Pz	56.25%	85.39%
Two	Cz	93.05%	97.22%
Two	Pz	94.90%	97.69%

6   Conclusions

In this paper, we use Cerebot to investigate the P300 model for control of a humanoid robot. The analysis results on the amplitude, latency and polarity of the acquired neural signals which are elicited after flashing a target demonstrate that the experiment protocol elicits the P300 potentials well.

One of study conclusions is that different contexts of images cause to change the shape of P300 potentials. The natural images of robots are used as stimulus to elicit the P300 potentials, instead of characters or symbols offered by the OpenViBE development environment, so that it makes the graphical interface more intuitive. Besides, a humanoid robot has a similar appearance and enables to perform basic behavior as people, so successful control of the humanoid robot behavior via brainwaves would pave the way to complete the complex task in the future which will meet the requirements of patients in daily life. Compared to other brainwave-based models, such as SSVEP or mu/beta models, the P300 model could achieve a higher success rate.

Our future research will continue to investigate the behavior imagination-based on model for control of the humanoid robot that relies less on the visual stimulus [11-12].

Acknowledgements. This work was supported in part by the National Natural Science Foundation of China (No. 61271321) and the Ph.D. Programs Foundation of the Ministry of Education of China (20120032110068). Wei Li is the author to whom correspondence should be addressed.

References

1. Sutton, S., Braren, M., Zubin, J., John, E.R.: Evoked Potential Correlates of Stimulus Uncertainty. Science 150, 1187–1188 (1965)
2. Smith, D.B.D., Donchin, E., Cohen, L., Starr, A.: Auditory Averaged Evoked Potentials in Man during Selective Binaural Listening. Electroencephalogr. Clin. Neurophysiol. 28, 146–152 (1970)
3. Farwell, L.A., Donchin, E.: Talking off the Top of Your Head: Toward a Mental Prosthesis Utilizing Event-Related Brain Potentials. Electroencephalogr. Clin. Neurophysiol. 70, 510–523 (1988)
4. Donchin, E., Spencer, K.M., Wijesinghe, R.: The Mental Prosthesis: Assessing the Speed of a P300-Based Brain-Computer Interface. IEEE Trans. Rehabil. Eng. 8, 174–179 (2000)
5. Muglerab, E., Benschc, M., Haldera, S., Rosenstielc, W., Bogdancd, M., Birbaumerae, N., Kübleraf, A.: Control of an Internet Browser Using the P300 Event-Related Potential. IJBEM 10, 56–63 (2008)
6. Li, Y., Long, J., Yu, T., Yu, Z., Wang, C., Zhang, H., Guan, C.: An EEG-Based BCI System for 2-D Cursor Control by Combining Mu/Beta Rhythm and P300 Potential. IEEE Trans. Biomed. Eng. 57, 2495–2505 (2010)
7. Bayliss, J.D.: Use of the Evoked Potential P3 Component for Control in a Virtual Apartment. IEEE Trans. Neural Syst. Rehabil. Eng. 11, 113–116 (2003)
8. Iturrate, I., Antelis, J.M., Kubler, A., Minguez, J.: A Noninvasive Brain-Actuated Wheelchair Based on a P300 Neurophysiological Protocol and Automated Navigation. IEEE Trans. Robot. 25, 614–627 (2009)

9. Palankar, M., De Laurentis, K.J., Alqasemi, R., Veras, E., Dubey, R., Arbel, Y., Donchin, E.: Control of a 9-DoF Wheelchair-Mounted Robotic Arm System Using a P300 Brain Computer Interface: Initial Experiments. In: IEEE International Conference on Robotics and Biomimetics, pp. 348–353. IEEE Press, Bangkok (2008)
10. Bell, C.J., Shenoy, P., Chalodhorn, R., Rao, R.P.: Control of a Humanoid Robot by a Non-invasive Brain-Computer Interface in Humans. J. Neural Eng. 5, 214–220 (2008)
11. Li, W., Jaramillo, C., Li, Y.: A Brain Computer Interface Based Humanoid Robot Control System. In: IASTED International Conference on Robotics, Pittsburgh, pp. 390–396 (2011)
12. Li, W., Jaramillo, C., Li, Y.: Development of Mind Control System for Humanoid Robot through a Brain Computer Interface. In: 2nd International Conference on Intelligent System Design and Engineering Application (ISDEA), pp. 679–682. IEEE Press, Hainan (2012)
13. Ma, Z., Gao, S.: P300-Based Brain-Computer Interface: Effect of Stimulus Intensity on Performance. J. Tsinghua Univ. Nat. Sci. Ed. 48, 415–418 (2008)
14. Gonsalvez, C.J., Polich, J.: P300 Amplitude is Determined by Target-to-Target Interval. Psychophysiology 39, 388–396 (2002)
15. Polich, J., Ellerson, P.C., Cohen, J.: P300, Stimulus Intensity, Modality, and Probability. Int. J. Psychophysiol. 23, 55–62 (1996)

Preliminary Experiments on Soft Manipulating of Tendon-Sheath-Driven Compliant Joints

Wang Kai and Wang Xingsong[*]

School of Mechanical Engineering, Southeast University
Nanjing, Jiangsu Province, China
seuwangkai@126.com, xswang@seu.edu.cn

Abstract. This paper introduced a kind of Tendon-Sheath-actuated compliant joints which fixed on some long and slender robot for searching and inspection interior of confined space safely. The actuator discussed in this paper is comprised of the three modules of the clamping pair, bending pair and rotational pair. Each joint is compliant structure and has embedded sensors as part of its structure. The validity of the design is evaluated by finite element analysis. The robot is actuated by Tendon-Sheath and utilizes strain gages to measure both the force and position. In order to realize soft and stable contact with human body, a position based impedance controller is adopted. The experimental results of the compliant gripper show that the position based impedance controller is capable to realize accurate force tracking.

Keywords: Tendon-sheath actuation, compliant joint, impedance control, robot.

1   Introduction

Human had undergone many disasters in the last decades, such as Japan earthquake, mine accidents, etc. Only a small part of the people can survive from these disasters and most of the survivors are surface victims that can be easily found by rescue workers. However, the interior is where most of the trapped people located. So most of them can't be found timely, investigate its reason, the disaster ruins make it difficult for people to effectively search and rescue. All this shows that, it is necessary to develop a recue robot with slender and soft body, Fig 1. The rescue robot is mainly composed of an actuator and a long flexible body. In order to realize the safety of the rescue, the soft manipulating control strategy was proposed which covers compliant structure design, force feedback, coordinated control et al.

Many researchers have studied the characteristics of the tendon-sheath transmission system and developed many tendon driven robots previously. M. Kaneko et al. [1] first proposed a static model of a single tendon system and found a hysteresis characteristic which is similar to the backlash behaviour by using Coulomb friction model. G. Palli et al. used tendons to drive a robot hand named UBH-IV [2]. They proposed a   model-based optimal law to control the tendon tension [3] [4].

---

[*] Corresponding author.

J.-H. Kim et al. (eds.), *Robot Intelligence Technology and Applications 2*,   503
Advances in Intelligent Systems and Computing 274,
DOI: 10.1007/978-3-319-05582-4_44, © Springer International Publishing Switzerland 2014

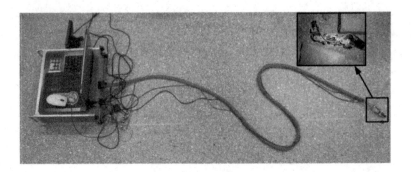

Fig. 1. The whole type of the rescue robot

Kitagawa et al. developed a small diameter active hose for search and life-prolongation of victimsunder ruins [5]. Hirayama et al. developed a search robot with duplex mechanism that is realized by connecting two manipulators in parallel [6].

As is known to all, compliant mechanisms are able to produce movements and transmit force due to the elastic deformation in single parts. The compliant mechanism has developed rapidly in the last decades, Larry L. Howell first proposed this concept [7], and pointed out that the material with high yield strength to Young Modulus ratio allows greater deformation. The compliant mechanisms have many advantages such as, no backlash and friction relative to the rigid articulated mechanisms [8].

The problem of controlling robot manipulators in contact with objects in their workspace is of central importance in many applications. In order to implement compliant control during the rescue, a hybrid position and force control strategy was proposed by [12].Based on this concept, position control was used to control the actuator's position until it touches object. After touching, control system was switched from position control to force control. Position based impedance control is a position controller nested within a force feedback loop. The combination of tendon-sheath transmission, compliant joints and impedance control will improve the safety and success probability in rescue actions.

This paper is organized as follows. In section 2, compliant actuator design will be introduced in detail. Section 3 describes an experimental setup about the tendon – sheath-driven compliant gripper and the control experiment of the gripper is presented. Finally, section 4 concludes the paper and talk about the future work.

2   Compliant Joints Design

A typical terminal actuator is composed of bending joint, rotary joint and the gripper. A long and slender robot which is actuated by tendon-sheath transmission system for searching and rescue was developed in our previous research [9]. But the drawbacks of the robot are obvious, such as the backlash of the transmission system and the dimensions of the robot is too large, etc. All these drawbacks affected the proper functions of the rescue robot; compliant mechanism is applied to redesign every joint which could overcome the above disadvantages. The 3D model of the actuator is shown in Fig.2. The actuator can be implemented in holding, bending and rotation movement. In Fig.3, the photographs of the actuator were shown in different postures.

To enable the actuator be used in rescue after earthquake and detection within the narrow space, two preconditions were considered [9]:

(1) Diameter less than 40 mm. In an actual disaster site, the shapes of ruins is various, so it is difficult for conventional robot to go through in its space, usually only small dimensions robot can across the barrier. Therefore, diameter less than 40 mm is desired for searching.

(2) Large deformation and high fatigue strength. The small voids channel could not be straight through. Thus, the robot has to change its motion direction to pass smoothly. At the same time, each joint must have high fatigue strength which determines the service life of the rescue robot.

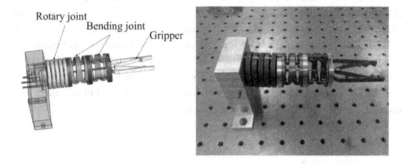

Fig. 2. 3D model of the actuator

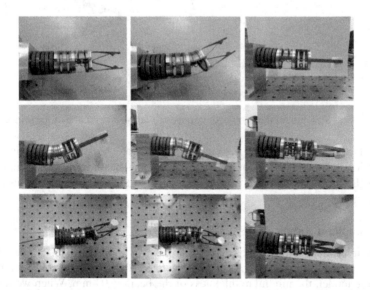

Fig. 3. Photographs of the Tendon-Sheath-Driven compliant actuator

2.1   Compliant Gripper

The structure of the compliant gripper is shown in Fig. 4 (a). One tendon is fixed on the beam of the gripper. When tendon is pulled, the gripper will clamp quickly and hold the object. On the contrary, when the tendon is relaxed, the gripper will automatically open under the action of elastic potential energy.

Because the deformation of the gripper is nonlinear, finite element method (FEM) was used to simulate the Von Mises of the gripper. We fixed the bottom rectangular block, and applied tensile on the beam. The result is shown in Fig 4(b). In the model, the minimum thickness of the beam is only 0.5mm. By using FFEPlus method, and choose beryllium bronze as the material which yield strength to Young Modulus ratio is 9.2, we have got a very good result.

As mentioned above, the compliant gripper must have large deformation. But due to its complex structure, it is impossible to use mathematical equations to describe the relationship between the displacement of the tendon and the gripper. In the next section some experiments are conducted to find out the relationship between them.

To design the compliant gripper structure, many methods have been tried, such as pseudo-rigid body method, structure matrix method and structural optimization method [10] [11]. Finally, due to the various reasons we adopt structural optimization method, by giving boundary motion conditions and using optimization algorithms to generate a reasonable model and the structure parameter.

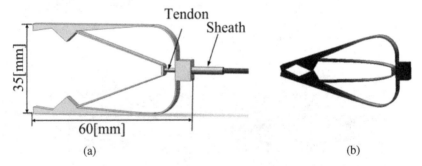

Fig. 4. (a) Structure of the compliant gripper. (b) Stress distribution of the gripper.

In the Prior works, our group has conducted the characteristic of the tendon-sheath transmission system. The advantage of using tendon-sheath transmission is that the shape of the transmission route can be arbitrary and the actuator can far away from the drive motor.

2.2   Compliant Rotary Joint

In a typical terminal actuator, the rotary joint plays an important role. The structure of the rotary joint is shown in Fig. 5. The diameter of the compliant rotary joint is 40 mm. In the model, the minimum thickness of the beam is 0.4mm. When we fixed the outer ring in three holes, the torque acting on the center hole makes the inner ring and the outer ring rotate relatively.

Compliant joints are limited to a finite range of rotation, while the rigid counterparts rotate infinitely. In fact, the rotation angle of this compliant joint is limited with the restrictions of the permissible stress and strain of the material. During the deformation, if the yield stress is reached, elastic deformation becomes plastic, after which, the behavior of the joint is changed and is difficult to restore.

Similar to the gripper, we use beryllium bronze as the raw material to processing the joint. Simulation result shows that the maximal rotation angle is 15.47deg in a single direction. The simulation result is shown in Fig. 5. This cannot satisfy our application requirements, therefor, several joints were connected in series. The thickness of each joint is 4 mm, when five rotary joints were connected in series the total thickness becomes 30 mm and the structure of the whole rotary joint is shown in Fig.6.

Fig. 5. The structure of the rotary joint and stress distribution of the joint

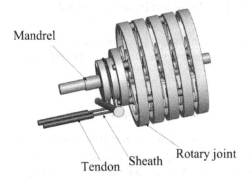

Fig. 6. The structure of the whole rotary joint

2.3   Compliant Bending Joint

Usually the terminal actuator must have a bending joint. In our actuator, we designed a two axes bending joint. The structure graphing is shown in Fig. 7. The diameter and the length of the bending joint is 40mm and 25mm, respectively .We processing it with beryllium bronze and the simulation work have been completed. The ideal of the bending joint structure has been inspired from compliant C-shape mechanism [8].

From the Fig. 7 we can see that the C-shape structure allows greater deformation and the preliminary test demonstrate that the bending angle can achieve $\pm45^{0}$. Next, we will post some strain gauges on the C-shape structure where the minimal thickness is 0.5 mm to do some characteristic experiments.

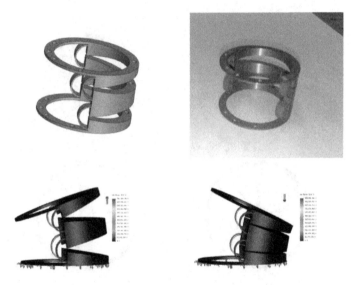

Fig. 7. Structure of the bending joint and stress distribution of the limit position

3   Experimental Results of Gripper Control

To study further about the performance of the compliant joints, some experiments will be done. At present we have conducted the displacement and force experiments of the gripper.

3.1   The Structure of the Compliant Gripper Sensor

In order to realize the safety operation, we must fix sensors on the actuator. Four high-precision resistance strain gauges are symmetrically arranged on the beams of the elastic body, which are connected into a double-arm bridge with the character of temperature compensation and linear output.

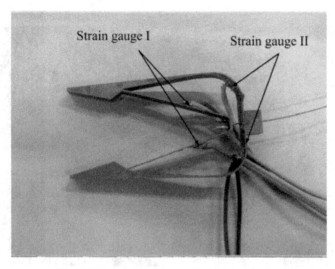

Fig. 8. Compliant gripper with strain gauges

The compliant gripper with strain gauge is shown in Fig.8. There are divided into two groups. Strain gauge I is used to measure the displacement of the gripper and strain gauge II is to measure the clamping force.

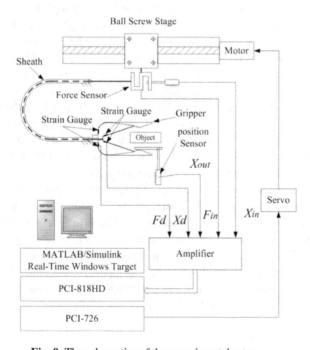

Fig. 9. The schematics of the experimental setup

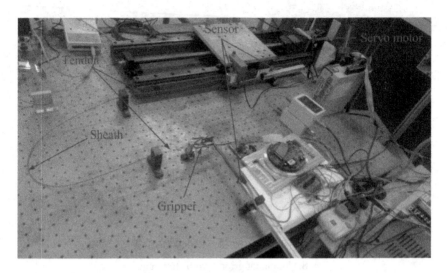

Fig. 10. Photo of the experiment setup

3.2   Displacement Experiment

In this experiment, we use tendon-sheath to drive the compliant gripper. The whole system is shown in Fig.9.Where the sheath is fixed to a plate at two ends; the tendon is attached to the gripper at its distal end that acts as a load, and to a ball screw stage at the proximal end that pulls the tendon. The stage is controlled by a servo motor; in the proximal end, force and displacement are measured with CTL-YZ and KTS-75 sensors, respectively; to get the displacement of the gripper we used a new method to measure the size of the opening and closing. Finally all signals of the sensors are acquired and processed via a data acquisition card (PCI-818HD) with Real-Time Windows Target of MATLAB. The tendon diameter is 0.5mm, uncoated stainless steel 6×7 ropes. The stainless steel sheath is made from 0.5 mm diameter wire wrapped into a close packed spring with an inner diameter 1.2mm.

The static displacement characteristic of compliant gripper is shown in Fig.11. In these tests, the shape of the sheath is straight and the elongation of the tendon is neglected because the pulling force is small. From Fig 11(a), we can know that the displacement of the gripper and the strain in the place where strain gauge I pasted are substantially linear. Thus we can use the voltage signal of the strain gauge I to represent the displacement of the compliant gripper. At the same time, the relationship between the ball screw stage and the gripper's displacement is also linear. A short horizontal line represents that the tendon is not fully tight.

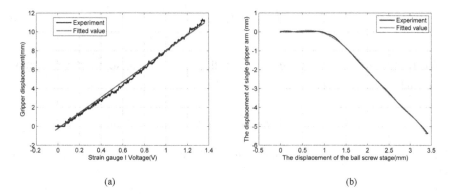

Fig. 11. Static displacement characteristic (a) the relationship of strain gauges. (b) The real displacement between ball screw stage and gripper.

3.3 Position Tracking Control

The goal of this experiment is to let the gripper's displacement follow a desired position commands. A position control system was proposed, as shown in Fig 13(a). When the ball screw stage moves at a low velocity, a phenomenon called stick-slip is present, which causing chattering in the position signal. The phenomenon is represented in Fig.14 (a) without considering the system lag. To overcome these issues, a new position control system was proposed in Fig 13(b). In the experiment, the bending   angle of the sheath is almost 180 degree, thus, the characteristic of the tendon-sheath transmission must be considered.

The applications of tendon-sheath mechanism can reduce the size and weight of the actuator, but friction between the tendon and the sheath which introduces some nonlinear phenomenon and lowers the performance of the system.

In [11], L. Chen *et.al.* considered the tendon-sheath transmission characteristic, and derived the relationship between input force and output force:

$$T(s) = \begin{cases} T_{in}e^{-\mu sign(\dot{s})\int_0^s \kappa(s)ds} & 0 \leq s \leq L_1 \\ keeps\ original\ value & L_1 \leq s \leq L \end{cases} \tag{1}$$

Where μ and $\kappa(s)$ are the Coulomb friction coefficient and the curvature. And L_1 is the length that divides the tendon into two parts. For a change of T_{in} ,the tension change appears only in the part $0 \leq s \leq L_1$, the rest of tension keeps its original value. The relationship between T_{in} and T_{out} can be written as:

$$\begin{cases} T_{out} = T(L) = T_{in} e^{-\mu sign(\dot{s}) \int_0^L \kappa(s) ds} \\ x_{out} = x_{in} - \delta(L) \end{cases} \tag{2}$$

Where x_{in} and x_{out} are the displacement of the input and output, δ is the elongation of the tendon and can be got as follows:

$$\delta(L) = \frac{T_{in}}{EA} L_{\mu\theta}, L_{\mu\theta} = \int_0^L e^{-\mu.sign(\dot{s}) \int_0^s \kappa(s) ds} ds$$

The simulation results of tendon-sheath transmission characteristics were shown in Fig 12.

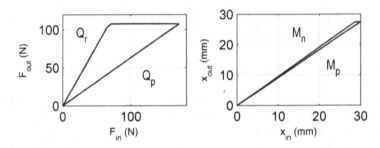

Fig. 12. Tendon- sheath transmission characteristic; (a) Force transmission characteristic, (b) Displacement transmission characteristic

In order to eliminate the nonlinear of the tendon-sheath transmission characteristic, the inverse compensation control was introduced. Before we conduct the experiment, the parameters of the tendon-sheath were obtained by prior identification. Based on the inverse backlash model, the inverse force and position models of single tendon-sheath transmission were proposed as [13]:

$$F_{in}(t) = \begin{cases} \dfrac{F_d(t)}{Q_p} & \dot{F}_d(t) > 0 \\ \dfrac{F_d(t)}{Q_r} & \dot{F}_d(t) < 0 \\ F_{in}(t^-) & \dot{F}_d(t) = 0 \end{cases} \tag{3}$$

$$x_{in}(t) = \begin{cases} \dfrac{x_d(t)}{M_p} & \dot{x}_d(t) > 0 \\ \dfrac{x_d(t)}{M_r} & \dot{x}_d(t) < 0 \\ x_{in}(t^-) & \dot{x}_d(t) = 0 \end{cases} \tag{4}$$

Where F_d and x_d are desired force and displacement, F_{in} and x_{in} are the actual inputs of the force and displacement. And Q_p, Q_r, M_p and M_r are the slope coefficient and explained in Fig. 12.

In the experiment, the requirement is to follow a sinusoid signal. With the feed forward control, the hysteresis nonlinearity is reduced. It is also found that the output position always lags behind the desired trajectory when the system is under no compensation. The addition of inverse model control can reduce the tracking error obviously. The experiment result is shown in Fig. 14(b).

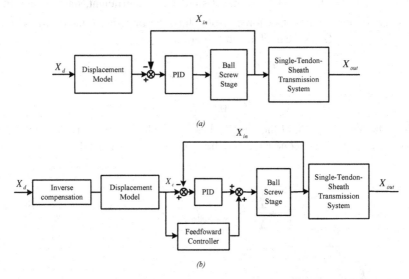

Fig. 13. The structure of the position control model. (a) Position control model without compensation. (b) Position control model with feed forward and inverse compensation.

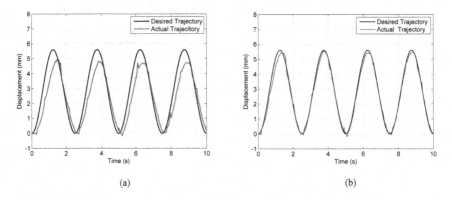

Fig. 14. The results of position tracking control: (a) Results without feed forward and compensation, (b) Results with feed forward and compensation

3.4   Position Based Impedance Control

The impedance control can absorb the reaction force from the object during the contact. Moreover, the impedance control is also able to control the position of the actuator. The equation of the desired impedance of the system is described as [12]:

$$M\ddot{x} + B\dot{x} + K(x - x_r) = f_e \tag{5}$$

Where x, $\dot{x}$ and $\ddot{x}$ are the position, velocity and acceleration. And x_r is the reference position. And, M, B, K and f_e are the desired mass coefficient, damper coefficient, spring coefficient and contact force error. Thus, the impedance model transfer function $G_x(s)$ and $G_F(s)$ can obtained from (5).

$$G_x(s) = \frac{K}{MS^2 + BS + K} \tag{6}$$

$$G_F(s) = \frac{1}{MS^2 + BS + K} \tag{7}$$

The schematic diagram of the position based impedance control is shown in Fig. 15. some experiments were conducted and the result is shown in Fig. 16.

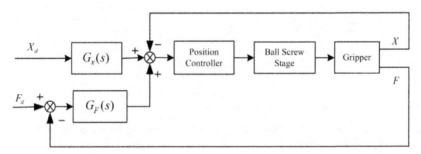

Fig. 15. The schematic diagram of impedance control

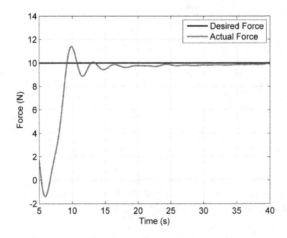

Fig. 16. Force response of the position based impedance control

4   Conclusion and Future Work

This paper presents a tendon-sheath driven actuator. Unlike other actuators this one was composed of compliant joints with different structures. Three compliant joints were introduced detailed especially the compliant gripper. It was showed that compliant mechanisms had many advantages such as, no backlash and friction relative to the rigid articulated mechanisms. Finally, preliminary displacement and force experiment of the gripper was carried out to realize the precious position and force tracking. In order to make a rescue robot, the coordinated control of the robot is under studied to realize the precise control.

Acknowledgement. This work was supported partly by the China Nation Nature Science Foundation under grant 50875044 and 51175078.

References

1. Kaneko, M., Yamashita, T., Tanie, K.: Basic considerations on transmission characteristics for tendon drive robots. In: Proc. Int. Conf. Adv. Robot., vol. 1, pp. 827–832 (1991)
2. Palli, G., Borghesan, G., Melchiorri, C.: Modeling, identification, and control of tendon-based actuation systems. IEEE Trans. Robot. 28(2), 277–290 (2012)
3. Palli, G., Melchiorri, C.: Model and control of Tendon-Sheath Transmission Systems. In: IEEE ICRA 2006, Orlando, Florida, pp. 988–993 (2006)
4. Palli, G., Melchiorri, C.: Optimal control of tendon-sheath transimission systems. Presented at the Int. Fed. Autom. Control Symp. Robot Control, Bologna, Italy (2006)
5. Kitagawa, A., Tsukagoshi, H., Igarashi, M.: Development of Small Diameter Active Hose-II for Search and Life-prolongation of Victims under Debris. Journal of Robotics and Mechatronics 15(5), 474–481 (2003)
6. Hirayama, A., Kazuyuki, I.: Development of rescue mainpulator to search narrow space for victims. Artif Life Robotics 13, 331–335 (2008)
7. Howell, L.L.: Compliant mechanisms. Wiley, New York (2001)
8. Rotinat-Libersa, C., Solano, B.: Compliant Building Blocks For the Development of New Portable Robotized Instruments for Minimally Invasive Surgery. In: The Fourth IEEE RAS/EMBS International Conference on Biomedical Robotics and Biomechantronics, Roma, Italy, June 24-27 (2012)
9. Chen, L., Wang, X.S., Lu, J., Xu, W.L.: Design and Preliminary Experiments of a Robot for Searching Interior of Rubble. In: Proc. IEEE Int. Conf. Syst. Man. Cybern., Anchorage, USA, pp. 1956–1961 (2011)
10. Trease, B., Moon, Y., Kota, S.: Design of Large-Displacement Compliant Joints. Journal of Mechanical Design, Transaction of the ASME 127, 788–798 (2005)
11. Chen, L., Wang, X.S.: Tendon-sheath actuated robots and transmission system. In: Proc. IEEE Int. Conf. Mech. Autom., Changchun, China, pp. 3173–3178 (2009)
12. Heinrichs, B., Sepehri, N.: A Limiation of Position Based Impedance Control in Static Force Regulation. In: Theory and Experiments, IEEE Conf. on Robotics and Automation, pp. 2165–2170 (1999)
13. Chen, L., Wang, X.S., Xu, W.L.: Inverse Transmission Model and Compensation Control of a Single-Tendon-Sheath Actuator. IEEE Transactions on Industry Electronics 61(3), 1424–1433 (2014)

Maintaining Model Consistency during In-Flight Adaptation in a Flapping-Wing Micro Air Vehicle

John C. Gallagher[1], Laura R. Humphrey[2], and Eric T. Matson[3]

[1] Dept. of Computer Science and Engineering
Wright State University, Dayton OH, USA
john.gallagher@wright.edu
[2] Aerospace Systems Directorate
Air Force Research Laboratory, WPAFB, OH, USA
Laura.Humphrey@wpafb.af.mil
[3] Dept. of Computer and Information Technology
Purdue University, West Lafayette, IN, USA
ematson@purdue.edu

Abstract. Machine-learning and soft computation methods are often used to adapt and modify control systems for robotic, aerospace, and other electromechanical systems. Most often, those who use such methods of self-adaptation focus on issues related to efficacy of the solutions produced and efficiency of the computational methods harnessed to create them. Considered far less often are the effects self-adaptation on Verification and Validation (V&V) of the systems in which they are used. Simply observing that a broken robotic or aerospace system seems to have been repaired is often not enough. Since self-adaptation can severely distort the relationships among system components, many V&V methods can quickly become useless. This paper will focus on a method by which one can interleave machine-learning and model consistency checks to not only improve system performance, but also to identify how those improvements modify the relationship between the system and its underlying model. Armed with such knowledge, it becomes possible to update the underlying model to maintain consistency between the real and modeled systems. We will focus on a specific application of this idea to maintaining model consistency for a simulated Flapping-Wing Micro Air Vehicle that uses machine learning to compensate for wing damage incurred while in flight. We will demonstrate that our method can detect the nature of the wing damage and update the underlying vehicle model to better reflect the operation of the system after learning. The paper will conclude with a discussion of potential future applications, including generalizing the technique to other vehicles and automating the generation of model consistency-testing hypotheses.

Keywords: Verification and Validation, Flapping-Wing Micro Air Vehicles, Evolvable and Adaptive Systems.

J.-H. Kim et al. (eds.), *Robot Intelligence Technology and Applications 2*,
Advances in Intelligent Systems and Computing 274,
DOI: 10.1007/978-3-319-05582-4_45, © Springer International Publishing Switzerland 2014

1   Introduction

Optimization methods drawn from a variety of traditions are often used to tune and/or improve controllers for a wide variety of electro-mechanical systems. The most aggressive users of such techniques will attempt "self-repair" of controllers while the controlled system is in regular service. In such cases, the goal of optimization is often recast into intent to recover at least adequate system performance after some unexpected damage to the system. Often, persons attempting to recover system performance via online optimization focus on one or both of the following items:

1. They are often concerned with showing that the modified and presumably recovered systems have returned to producing adequate behavior.
2. They are often concerned with showing that their methods of controller adaptation can produce those adaptations efficiently – or at least quickly enough to prevent further damage and/or risk to the system's mission.

Less often considered are issues related to how self-adaptation and self-healing impact Verification and Validation (V&V) efforts. Lack of consideration of how one explains how the newly adapted system functions and how one can provide assurances that it functions correctly can lead to costly, cascading faults in the future. For example, simply knowing that a Flapping-Wing aircraft has relearned an ability to hover in place after wing damage reveals nothing about how those same fixes might, perhaps catastrophically, impact the safety of other more aggressive maneuvers. Further, simply knowing that the same example Flapping-Wing aircraft has relearned an ability to hover reveals nothing about where the original fault occurred and provides no guidance to assess if the automatically generated fix consumed inappropriate resources that should have been held in reserve to help compensate for more serious problems that might arise later.

Foundational to any comprehensive V&V effort applied to self-adapting robotic and/or aircraft systems is an ability to maintain consistency between continuously adapting systems and the models describing those systems. Maintaining model-to-system consistency can be difficult because many aspects of system operation may not be directly observable by instrumentation inside the system. Additional and significant difficulty can be introduced by the very nature of comprehensive self-adaptation.

This paper will introduce a technique we will call *Evolutionary Model Consistency Checking (EMCC)* in which adaptive learning is used to simultaneously produce fault-correcting solutions as well as testable hypotheses that can be used to identify areas of inconsistency between the adapted system and its model. Once those inconsistencies are identified, they would be corrected so that the system's model is as consistent as possible with the adapted system. This technique will be applied to a simulation of a Flapping-Wing Micro Air Vehicle. We will show that using minimal resources and without the use of special sensors that directly detect faults, it is possible to determine the location of faults and modify the vehicle's model appropriately. The paper will conclude with a brief discussion of the implications of EMCC and possible extensions to the method.

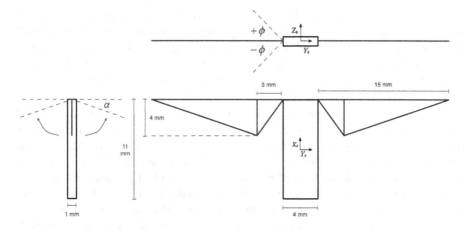

Fig. 1. Orthographic View of Insect-Scale Flapping-Wing Vehicle

2   Background

2.1   The Flapping-Wing Micro Air Vehicle (FW-MAV)

This work employs a variant of the Harvard RoboFly [1] that has been previously used in other works related to FW-MAV control and adaptation. Full descriptions of system parameters, kinematics and dynamics, control, and controller adaptation can be found in [2] – [9]. For the purposes of this paper, we will summarize only those aspects of the vehicle required to make clear our demonstration of EMCC.

The simulated vehicle (Figure 1) is modeled as a 15x4x1mm rectangular prism with two 15mm wings extending from each side. In work reported here, we will consider the system to be constrained to move only "up and down" so that only control of altitude is required. This constraint is similar to that imposed on the "first lift off" of the 2008 RoboFly due to the power and ground wires that extended through the vehicle along the longest axis of the body.

The 15mm lines extending from each side of the body in the top view portion of Figure 1 represent wing span spars that can be independently moved to angles ϕ_L and ϕ_R for the left and right wings respectively. The triangular wing planforms (see the front view in lower right hand part of Figure 1) hang down from the support spars to which they are passively hinged. As the wing spars rotate through their respective ϕ ranges, dynamic air pressure lifts the triangular plan form wings to an angle α under the plane of the spars (see the side view in the lower left hand part of Figure 1). The moving wings produce lift and drag forces that one can resolve into net body forces and torques using the kinematics and dynamics models derived in previous work.

For this and previous work, we adopted *cycle-averaged* control methods. In *cycle-averaged* control, one computes the forces and torques one wishes to apply to the body and then computes specific wing motions that would generate those forces and torques *on average* over the course of a full wing beat cycle. Most commonly used are *split-cycle-cosines* in which each wing is, at the beginning of its wing beat cycle, provided with a wing beat frequency (ω) and a shape parameter (δ) that determines the specific motion of the wing, which is presumed to be a cosine with frequency ω and with the "bottom" of the cosine trough shifted forward or back by δ radians per second. For this work, in which we are only controlling hover altitude, pure cosine actuation ($\delta_{left} = \delta_{right} = 0$) with synchronized motion of the wings ($\omega_{left} = \omega_{right}$) is sufficient and will be used here.

High-level control of this vehicle would be provided by a control law that computes desired body forces and torques followed by a mapping of the desired forces and torques to wing beat frequency (ω) and shape parameters (δ), which are given to the appropriate wing for the next wing beat (Figure 2). It is also presumed that there would be some form of control allocation applied for situations in which there were more than one position and/or pose states of the vehicle that required correction. For the experiments in this paper, no allocation is required, $\delta_{left} = \delta_{right} = 0$, $\omega_{left} = \omega_{right}$, torques are neglected, and altitude x is the only vehicle state under consideration. Given these assumptions, a SISO controller for the vehicle only needs to provide a wing flap frequency for the wings, as shown in Figure 2.

2.2   Controller Self-Repair after Wing Damage

The SISO control law given in Figure 2 critically depends on accurate modeling of the relationship between wing motions and produced cycle-averaged forces and torques. Equation (1), which is extracted from the SISO Altitude Command Tracking Controller shown in Figure 2, is an algebraic manipulation of just such a model that provides the flapping frequency both wings should be given if one wishes a particular upward cycle-averaged force. The derivation presumes that the wings are being actuated in a synchronized manner, that the wing tips are actually sweeping out the desired cosine trajectories, and that the wing membranes are themselves not damaged. Of course, if any of these assumptions are not true, then the force model is incorrect and the vehicle will not be controlled appropriately.

$$\omega_{F_x}(t) = \sqrt{\frac{2F_{x_{des}}}{\rho I_A C_L(\alpha)}} \tag{1}$$

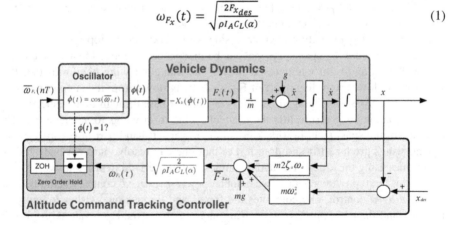

Fig. 2. Altitude Command Tracking Controller

In previous work [5] - [9], we employed a variety of Evolvable and Adaptive Hardware (EAH) [10] methods to learn new periodic, but non-cosine, wing motion functions that would restore the relationship given in Equation (1) in vehicles that had suffered some damage to one or both of their wings. This approach was distinct from neuromorphic approaches previously attempted in [11] – [13]. In all cases, one or more wings received simulated damage in the form of either a restriction on its rotation or a percentage loss in the generation of lift and drag forces. The EAH techniques would, while the vehicle was still attempting to fly its normal flight trajectory, assemble two new wing motion functions from a library of pre-computed basis functions stored in the learning engine. The salient observations from that previous work are that the best EAH methods could in minutes find wing motion patterns that restored the *required* flapping frequency to force relationship using only observations of ongoing whole-vehicle position error. An example of learned left and right wing motion functions to fix hovering errors induced by a specific single-wing force deficit fault is shown in Figure 3.

3   Evolutionary Model Consistency Checking

3.1   Conditions and Constraints

As implied in the first section, our previous work largely concerned itself with demonstrating efficacy and efficiency of the adaptation. Never before have we explicitly considered the implications of our previous successes for Verification and Validation (V&V) efforts. Such consideration is critical. The EA may, for example, easily solve the proximal problem of restoring appropriate hovering behavior. Without further analysis, however, one is in no position to predict how the evolved fixes might affect other flight modes or how those fixes might implicitly constrain the system's ability to adapt to future faults. Let us now consider the issue of maintaining consistency between the adapting system and its own model of itself.

We will presume that prior to self-repair, there existed an accurate mathematical model of the system. We will also presume that any kind of self-adaptation or repair, especially if it crosses abstraction layers [14], is potentially very disruptive to that model's ability to guide V&V efforts. As a prequel to more comprehensive V&V for self-modifying air and ground vehicles, we will here concern ourselves with the following two problems:

   a. Detecting and diagnosing the specific fault that triggered the need to self-repair
   b. Updating the model of the vehicle to reflect the fault and its repair

We will accept as a constraint that any solution to (a) and (b) above must not use any sensors or instrumentation that is not already on the vehicle in support of normal navigation and control. In other words, our solutions will not require the addition of special sensory systems that directly detect faults. We will accept as a further constraint that any techniques used to determine answers to (a) and (b) above must be usable onboard the vehicle while it is conducting normal missions and not require significant lengths of time or large disruptions to the vehicle's mission.

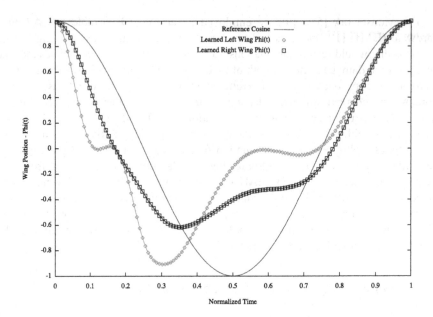

Fig. 3. Example Learned Left and Right Wing Motion Functions Compared to a Reference Cosine

3.2   Evolutionary Model Consistency Checking – The Concept

If one is not permitted to directly detect fault conditions via the use of specialized sensors added to the vehicle, one might reason that the only way fault conditions can be detected is by comparing operation of the actual, damaged vehicle to a model of an undamaged vehicle and then reasoning about where the faults must lie. Assuming this can be done, it in principle would become possible to update the model after self-repair so that the model, and any future reasoning about the system based on that model, remains consistent with the actual operation of the adapted vehicle. A major problem is, of course, in determining what tests uncover the most useful information about where the system-to-model inconsistencies lie. Another major problem is in collecting evidence that supports or refutes those hypotheses. Especially if one does not want to take the system out of normal service, evaluation opportunities might be exceedingly rare.

We propose to address these problems by leveraging the same Evolutionary Algorithm (EA) techniques previously employed to repair the system. In addition to compensating for faults, we now employ EAs to simultaneously produce easily testable hypotheses that determine where in the vehicle the fault occurred and where and how the internal vehicle model should be updated to reflect the adaptations that occurred. The idea is to leverage the time and information collected during the learning of the solution to, in parallel, create testable hypotheses about where the faults occurred. If *Evolutionary Model Consistency Checking (EMCC)* can be made to work, the cost of learning a repair solution and identifying the specific fault and updating the model would not be significantly greater than the cost of just learning the repair solution alone.   Although in the future we anticipate automating the generation of *EMCC* plans, in this work we will manually generate and demonstrate a sample plan as a proof-of-concept.   After examining this proof-of-concept, we will return to discussion of more general approaches.

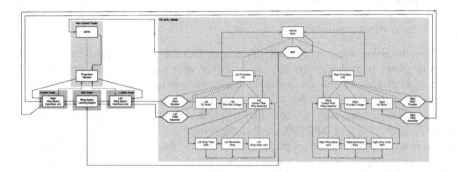

Fig. 4. Block-Level Model of FW-MAV, Control, and Learning Systems

3.3   A Sample of Evolutionary Model Consistency Checking

Figure 4 presents a complete block-level model of the FW-MAV, its control systems, and its EA learning engine. The block-level model defines the relationships among various models of system components at varying levels of abstraction, with less abstract models being "lower" in the hierarchy. Note that computational entities are shown at a height corresponding to the level of physical device abstraction at which they operate. It is presumed that the blocks of this hierarchical relationship model each contain a mathematical model that relates inputs and outputs and that these block models are constructed using a suitable formalism. For sake of brevity, we will only expand model blocks as necessary to make our core arguments. From examining the block relationships (Figure 4) and the model implicit in the ACTC (Equation (1)), we can draw the following conclusions:

1.   If the vehicle can hover at a commanded altitude with precision, the sum of the upward forces produced by both wings together *must* be a close approximation of 0.0005892 N at a flapping frequency of 755.061 radians/sec (120.1718 Hz). This is a consequence of the exact physical parameters of the vehicle (mass, moment of inertia of the wings, etc.) and environmental properties (air density, gravity, etc.) along with the specific form of the ACTC. If the actual relationship between upward force and flapping frequency is not as it is defined above, the vehicle will hover, but not at its commanded altitude.

2.   The total upward force produced is the sum of the forces produced by the right and left wings. This is a consequence of the construction of the vehicle and can be seen both in the block level diagram (Figure 4) and in the allocation assumptions implicit in the derivation of the ACTC.

3. In a vehicle with undamaged wings both driven with the same wing motion functions (cosine or not), exactly one half of the upward force applied to the vehicle comes from each wing. This is also a consequence of the construction of the vehicle and can be in the allocation assumptions implicit in the derivation of the ACTC and in the nature of the wings. Identical wings should produce identical forces (modulo orientation in some cases) for the same wingtip motions.

From the above, we can conclude that a wing motion pattern applied symmetrically to both wings that causes the vehicle to hover with precision must produce about 0.0005892 N and if neither wing is damaged, each wing will produce about half of that force value (0.0002946 N). If one wing is damaged, but the vehicle still hovers with precision, the sum of the forces produced by the wings must still be approximately $F_{trim} = 0.000592\ N$, but the wings will not be producing symmetric forces even though they are supplied with symmetric motion command profiles. The key to detecting if at least one wing is damaged is implicit in the just given conclusion. The first stage of EMCC proceeds as follows:

Step 1: Evolve a single wing motion function that when used for both wings, restores correct, precise vehicle hovering in the real FW-MAV. This evolution takes place on the real vehicle using techniques previously referenced. Note that if one or more wings are broken, Step 1 should restore correct flight behavior. At this point we will not yet know which wing(s) are broken, nor will we know how to update the model of the vehicle to reflect the implications of the newly learned wing motion strategies, but at least correct flight behavior will be restored. Note the fact that we're learning a symmetric wing motion pattern is key to what will happen in subsequent steps.

Step 2: Using the wing motion pattern learned in Step 1, run a simulation of an undamaged wing using that motion pattern to determine the upward force an undamaged wing would produce with that motion. This step is done in simulation using a wing model available from within the model hierarchy of Figure 4. Ideally, the simulation is run onboard the vehicle, which is carrying its own model for internal reference and local updating. If that force multiplied by 2 equals F_{trim} within some user selected sensitivity bound, then conclude that neither wing is damaged and that any improvements made to hover performance were due to correcting some factor common to both wings.

Steps one and two together inform us if one or more of the wings is broken by evolving a wing motion solution that, in addition to fixing any problems with hover precision, creates a testable hypothesis about model-to-vehicle consistency. In this case, we simply use the internal vehicle model to test if our initial assumption (in the model) of undamaged wings is consistent with the vehicle's actual behavior. Imposing a wing symmetry condition on the evolved flapping motions provides us with this model consistency checking opportunity. If we are issuing symmetric commands and not producing lift forces consistent with the assumption of symmetry of wing structure, then it is safe to conclude that the wings are, in fact, significantly different from one another.

For purposes of this proof-of-concept, we will at this point assume that only one of the two wings is damaged and leave the problem of checking for dual wing damage for a later work. Even assuming, however, that only one wing is damaged, we are still left with the significant problem of determining which of the two is damaged and in determining how much cycle-averaged lift force is being produced by each wing. We need to know both of these items to successfully update the internal model to reflect the operation of the vehicle after damage and subsequent repairs discovered in Step 1. We can predict the forces being generated by each wing as follows:

Step 3: The undamaged wing MUST produce the force predicted by the model of an undamaged wing. If the vehicle hovers with precision, which it does at the end of Step 1, then the damaged wing MUST produce a force of 0.0005892 N minus the predicted undamaged wing force. We will refer to the predicted force of the undamaged wing as F_A and the predicted force of the broken wing as F_B. To summarize:

$$F_{trim} = 0.0005892\ N \tag{2}$$

$$F_A = Model\ Predicted\ Force\ using\ Learned\ Wing\ Motion\ Pattern \tag{3}$$

$$F_B = F_{trim} - F_A \tag{4}$$

Assuming that only one wing is damaged, at the end of Step 3 we know that the undamaged wing produces a force of F_A and the damaged wing produces a force of F_B when both wings are driven with the wing motion pattern learned in Step 1. We do not yet know which of the two wings is damaged. In the next step, we will use the model of the vehicle that is presumably carried in the vehicle to generate a wing motion test pattern that when briefly applied to the real vehicle will tell us directly which of the two wings is actually damaged.

Step 4: Using the model of an undamaged wing extracted from the internal vehicle model, use an EA to internally evolve a wing motion pattern that would generate F_B on an undamaged wing. This evolution is conducted entirely in simulation, in faster than real time, on board the vehicle. We will designate the wing motion pattern evolved in Step 1 as pattern O and the wing motion pattern evolved in Step 4 as pattern B. We will denote using pattern O on the left wing and pattern B on the right wing as OB. We will denote using pattern B on the left wing and pattern O on the right wing as BO. Using pattern O on both wings will be denoted as OO. When using wing motion OO, the ACTC should enable the FW-MAV to hover at a desired position with high precision. Using wing motion OB or BO, however, will cause hovering at other, incorrect, altitudes. If one uses pattern OB and the LEFT wing is broken, then the left (broken) wing will produce a force of F_B and the right (unbroken) wing will also produce a force of F_B for a total cycle-averaged body lift force of $2F_B$. An isomorphic argument can be made for pattern BO when the RIGHT wing is damaged. Using the body dynamics model, an assumption of unbroken wings, and wing motion pattern B driving each wing, one can predict the altitude that the vehicle will trim at when O is applied to the broken wing and B is applied to the non-damaged wing in the real vehicle.

> To decide which wing is broken, use an internal vehicle simulation to
> compute the predicted hover height for a force of $F_{body\_lift} = 2F_B$.
> For brief time periods, control the real vehicle with OB and then again
> with BO supplied to the wings. Whichever pairing (OB or BO)
> produces a hover height closest to that predicted for a lift force of $2F_B$
> is considered to be the defining pair for the wing damage status of the
> vehicle. If it is OB, then the LEFT wing is damaged. If it is BO, then
> the RIGHT wing is damaged.

This experiment is conducted on the real vehicle and should require no more
than 10 to 20 seconds of flight time beyond that required to learn the O and B
wing motion patterns

3.4   Method Summary

The example of EMCC just given interleaves real system learning, model learning,
and real system experiments to not only repair system faults, but to simultaneously
leverage the learning already being conducted to repair the system to also generate
testable hypotheses that can be used to determine where the faults occurred. In this
specific case, with little more than 30 seconds of flight time in addition to the time
needed to find the OO solution on the real vehicle, it is possible to diagnose the nature
of the fault and to extract the knowledge required to later update the vehicle's own
model of itself. That extraction of information, notably, is done with no need for
special sensors beyond the normal navigational systems one presumes would be on
board to support simple path following. Naturally, there are some potential
drawbacks and room for improvement. We will return to those issues in the
conclusion section of this paper. First, however, we will present experimental
validation of the method just presented.

4   Experimental Verification

Although the given EMCC plan is sound in concept, several items related to practical
implementation required experimental evaluation for the following reasons:

1. It was unclear if it were even possible to evolve symmetric wing motion solutions
   that effectively corrected for wing damage as this had never before been
   attempted. In the past, we only ever evolved asymmetric wing motion solutions
   on the belief that different wings needed different motions. Without the
   capability to evolve symmetric wing motion solutions that in fact restored correct
   vehicle behavior, the EMCC procedure from Section 3 would not be usable in
   practice.
2. Assuming that symmetric wing motion solutions are evolvable, we needed to
   determine if the decision procedure outlined in Section 3 was able to detect and
   identify wing faults without direct knowledge of wing damage and/or force
   production abilities.

3. Even assuming that perfect detection of fault conditions is in principle possible, it was unknown if the differences in altitude one would observe in fault determinations Step 4 would be resolvable within the precision of reasonable altitude sensors a vehicle might carry on board. If the distances between target altitudes are less than the precision of the sensors, the provided EMCC technique would be, in practice, without value.

Each of the above concerns was evaluated via simulation-based experiment as described in this section.

4.1   Experiment Methods

We ran ten thousand independent evolutionary learning runs on the vehicle referenced in Section 2. For each of these runs, one wing was selected at random to receive damage. Wing damage was modeled as a force deficit in the range of zero to twenty-five percent selected from a uniform random distribution. This means that although technically all experiments would suffer from one damaged wing, some experiments would have wing damage so slight that any reasonable decision procedure would judge them as undamaged. We, of course, expected our method to detect those cases automatically.

Each vehicle was expected to hover at a set altitude of 1 meter above the ground. Symmetric wing motion patters for each vehicle were evolved in-flight using the ICGA evolutionary algorithm as described in [9]. Learning parameters of the ICGA were set as follows: Wing Flaps per Evaluation = 150; Independent Fitness Islands = 16; Simulated Population Size = 64; Immigrant Probability = 0.125; Hypermutate Probability = 0.125. These learning parameters were chosen conservatively to help ensure convergence on what we perceived to be a difficult problem of learning wing motions under a stringent symmetry constraint. Since optimizing learning speed is not the main issue of this paper, we will address search optimization in another work.

Each of the 10,000 test vehicles were allowed to run until they learned to consistently hover within 0.0001 meters of the commanded target height of one meter. This assumes we have altitude sensors capable of at least 0.1 mm precision, or in other words, sensors capable of precision of about 0.01 of the length of the long axis of the body. We will examine the effects of using sensors of less precision later in this paper.

4.2   Experiment Results

We will organize results with respect to the questions laid out at the beginning of Section 4:

1. Successful symmetric wing motion solutions were evolved for all 10,000 test cases. The mean flight time required to return vehicle hover performance to within 0.0001 meters was about 113 minutes with a 75th percentile of about 147 minutes. Although we feel that a learn time of about two hours of flight time is

unacceptable and although we have achieved 10 minutes or less in previous experiments with asymmetric wing functions, it should be noted that none of our learning algorithms have yet been tuned for compatibility with the symmetry constraint and extended learning times were expected. For purposes of this paper, we only need show that symmetric wing motion solutions exist and can be found with regularity. Optimization of the evolutionary learning will occur in subsequent work.

2. Using a percent delta of 0.005 in step 1 of the decision procedure, 715 vehicles were determined to have "no damage," which should be interpreted as force deficits so slight as to be judged not worthy of repair. 9285 vehicles were judged to possess significant wing faults. For all of those 9285 vehicles, the EMCC decision procedure correctly identified the faulty wing and correctly assigned the F_A and F_B forces to the correct wings. This was determined by direct comparison of the simulated "real" vehicle and the predictions made, without direct knowledge of "real" vehicle internal state, by the EMCC procedure.

3. Direct examination of the experimental data allows us to put bounds on the amount of precision required to resolve various levels of wing damage. These results are summarized in Table 1. Note that with the sensor precision of 0.1 mm, we would be able to reliably detect wing faults of over 2% using the given EMCC procedure. The specificity of the technique naturally drops off with reductions in altitude sensor precision. However, even when we reduce to a sensor only capable of resolving one tenth of the body length, we can still reliably detect force faults of more then 6.8%. This demonstrates that we can conduct meaningful EMCC fault analysis on this vehicle without requiring navigational sensors of greater precision than we would need for basic navigation.

Table 1. Levels of Force Deficit Resolvable at Different Altitude Sensor Precisions

Sensor Precision (Meters)	Approximate Resolvable Force Deficit	Approximate Percent of Normal Force Produced
0.0001	2.00%	98.00%
0.005	5.00%	95.00%
0.001	6.80%	93.20%

5   Conclusions and Discussion

In this paper, we introduced the concept of Evolutionary Model Consistency Checking and demonstrated the use of this technique for determining the location and nature of wing faults in a simulated flapping-wing micro air vehicle. We demonstrated that it is possible to interleave evolutionary learning and hypothesis generation to assist in detecting faults and presumably, in updating the models of specific vehicles to more accurately describe the actual condition of the vehicles they describe. The success of this proof-of-concept work leaves us with a number of intriguing questions related to both the specific FW-MAV application and the general applicability of EMCC.

Regarding the specific FW-MAV application, there are a number of open issues. First, the nearly two hours of flight time needed to recover hover are likely unacceptable for practical situations. We believe that the extended learning times are due to a lack of tuning of the ICGA to this specific problem and that this issue can be overcome. Second, our proof-of-concept does not extend to determining the fault conditions of wings when more than one wing breaks simultaneously. Again, we believe that the EMCC concept can detect these types of faults; we simply have not yet developed the necessary decision procedures. In subsequent work, we intend to push the technique and attempt to formulate a more comprehensive EMCC approach that will enhance the decision procedures discussed in this paper. Finally, we have not yet attempted any of this in a real vehicle. A test vehicle is currently under construction and we should soon be in a position to run both evolution and EMCC tests on real hardware.

Regarding the general applicability of EMCC to other problems, we believe there exists one very large open issue. The decision procedure presented in this paper was generated manually by persons well-acquainted with the vehicle's capabilities and limitations. For more complicated systems, it seems unlikely that human intuition will be able to generate EMCC plans. Further progress will require specific study of methods to traverse hierarchical models of systems and automatically generate specific EMCC plans. Study of automated the generation of effective EMCC plans is an intended area of future study.

Acknowledgements. This material is based upon work supported by the National Science Foundation under Grant Numbers CNS-1239196, CNS-1239171, and CNS-1239229. Additional support was provided by the AFRL SFFP program. This work was cleared for public release on 27 August 2013 with case number 88ABW-2013-3879.

References

1. Wood, R.: The first takeoff of a biologically-inspired at-scale robotic insect. IEEE Trans. on Robotics 24(11), 341–347 (2008)
2. Doman, D., Oppenheimer, M., Sigthorsson, D.: Dynamics and control of a minimally actuated biomimetic vehicle: Part I – aerodynamic model. In: Proceedings of the AIAA Guidance, Navigation, and Control Conference (2009)
3. Oppenheimer, M., Doman, D., Sigthorsson, D.: Dynamics and control of a minimally actuated biomimetic vehicle: Part II – control. In: Proceedings of the AIAA Guidance, Navigation, and Control Conference (2009)
4. Doman, D., Oppenheimer, M., Bolender, M., Sigthorsson, D.: Altitude control of a single degree of freedom flapping wing micro air vehicle. In: Proceedings of the AIAA Guidance, Navigation, and Control Conference (2009)

5. Gallagher, J., Doman, D., Oppenheimer, M.: Practical in-flight altitude control learning in a flapping-wing micro air vehicle (submitted)
6. Gallagher, J., Oppenheimer, M.: An improved evolvable oscillator for all flight mode control of an insect-scale flapping-wing micro air vehicle. In: Proceedings of the 2011 IEEE Congress on Evolutionary Computation (2011)
7. Gallagher, J., Oppenheimer, M.: Cross-layer learning in an evolvable oscillator for in-flight control adaptation of a flapping-wing micro air vehicle. In: The 45th Asilomar Conference on Signals, Systems, and Computers (2011)
8. Gallagher, J., Doman, D., Oppenheimer, M.: The technology of the gaps: An evolvable hardware synthesized oscillator for the control of a flapping-wing micro air vehicle. IEEE Trans. on Evolutionary Computation 16(6), 753–768 (2012)
9. Gallagher, J.C.: An islands-of-fitness compact genetic algorithm approach to improving learning time in swarms of flapping-wing micro air vehicles. In: Kim, J.-H., Matson, E., Myung, H., Xu, P. (eds.) Robot Intelligence Technology and Applications. AISC, vol. 208, pp. 855–862. Springer, Heidelberg (2013)
10. Greenwood, G., Tyrrell, A.: Introduction to Evolvable Hardware: A Practical Guide for Designing Self-Adaptive Systems. IEEE Press (2005)
11. Boddhu, S.K., Gallagher, J.: Evolved neuromorphic flight control for a flapping-wing mechanical insect. In: Proc. of the 2008 Congress on Evolutionary Computation, Hong Kong. IEEE Press (2008)
12. Boddhu, S.K., Gallagher, J.: Evolving non-autonomous neuromorphic flight control for a flapping-wing mechanical insect. In: Proc. of the 2009 IEEE Workshop on Evolvable and Adaptive Hardware (2009)
13. Boddhu, S.K., Gallagher, J.: Evolving neuromorphic flight control for a flapping-wing mechanical insect. International Journal of Intelligent Computing and Cybernetics (2010)
14. Gallagher, J.: Hierarchical decomposition considered inconvenient: self-adaptation across abstraction layers. In: Proceedings of the 2012 SPIE Defense, Security, and Sensing Conference, Baltimore, MA (2012)

Model Checking of a Flapping-Wing Mirco-Air-Vehicle Trajectory Tracking Controller Subject to Disturbances

James Goppert[1], John C. Gallagher[2],
Inseok Hwang[1], and Eric T. Matson[1]

[1] Purdue University, West Lafayette IN, USA
{jgoppert,ihwang,ematson}@purdue.edu
[2] Wright State University, Dayton OH, USA
john.gallagher@wright.edu

Abstract. This paper proposes a model checking method for a trajectory tracking controller for a flapping wing micro-air-vehicle (MAV) under disturbance. Due to the coupling of the continuous vehicle dynamics and the discrete guidance laws, the system is a hybrid system. Existing hybrid model checkers approximate the model by partitioning the continuous state space into invariant regions (flow pipes) through the use of reachable set computations. There are currently no efficient methods for accounting for unknown disturbances to the system. Neglecting disturbances for the trajectory tracking problem underestimates the reachable set and can fail to detect when the system would reach an unsafe condition. For linear systems, we propose the use of the H-infinity norm to augment the flow pipes and account for disturbances. We show that dynamic inversion can be coupled with our method to address the non-linearities in the flapping-wing control system.

Keywords: Verification and Validation, Model Checking, Hybrid System, Flapping-Wing Micro Air Vehicles, Disturbance, H-infinity, Flow Pipe.

1 Introduction

Verification and Validation (V&V) of a flapping-wing micro air vehicle (MAV) control system is not a trivial task due to the coupled discrete and continuous dynamics. The interacting discrete state (or mode) transitions and continuous dynamics can be modeled as a hybrid system and thus a hybrid model checker must be used for V&V of a flapping-wing MAV. If the system is fully autonomous and the mode transitions are state dependent, then the hybrid system model can be approximated using an approximate quotient transition system (AQTS) that partitions the state space. In AQTSs, the continuous states are handled by computing invariants (or flow pipes) for each discrete mode. The transition between flow pipes is governed by state dependent guards. The AQTS creates a finite state machine representation of the original hybrid system so that standard

J.-H. Kim et al. (eds.), *Robot Intelligence Technology and Applications 2*,
Advances in Intelligent Systems and Computing 274,
DOI: 10.1007/978-3-319-05582-4_46, © Springer International Publishing Switzerland 2014

model checkers can be used to verify its properties [1]. If the invariant partitions are represented using convex polyhedrons, then the approximation is called a polyhedral invariant hybrid automaton (PIHA) [2]. The CheckMate tool implements a PIHA based model checker and is integrated with the Matlab Simulink development environment [3]. Unfortunately, CheckMate cannot be efficiently extended to systems with unknown disturbances [4]. The d/dt model checker uses the concept of orthogonal polyhedra and is capable of handling unknown input/disturbances that are bounded by a convex polyhedron but at an increased computational cost [4, 5]. HyTech is one of the first hybrid model checkers but only applies to systems that can be represented by $\dot{x} = Ax < b$, which does not account for unknown disturbances [6].

We choose to employ the PIHA model due to its speed and the popularity of CheckMate; however, for the problem of the flapping-wing MAV there are several significant differences between systems typically modeled by PIHAs. First, the flapping-wing MAV is regulated around a nominal trajectory. Second, the vehicle is subject to disturbances. If we naively apply PIHA analysis without accounting for disturbances, the problem rapidly becomes trivial since the trajectory approaches the nominal trajectory. Ignoring the disturbances therefore dramatically underestimates the reachable set of the system, and model checking does not identify errors that may exist in the actual flight control system.

In this paper, we propose augmenting the computed flow pipes of existing PIHA methods for linear systems with the H-infinity norm of the disturbance to the output transfer function. The H-infinity norm is often used in robust control to account for system disturbances and represents the steady-state bound of the worst case sinusoidal disturbance to the system. While this method is not strictly conservative, since transients and non-sinusoidal inputs could exceed the H-infinity norm, the H-infinity norm still gives an adequate bound for practical disturbances and can be used to identify previously undetected errors in hybrid systems. Our objective is to develop an efficient real time model checking algorithm for adaptive systems.

2   Polyhedral Invariant Hybrid Automaton (PIHA) Based Model Checking

In the section, we discuss the existing methods for V&V of PIHAs [2, 1]. This knowledge is fundamental for understanding the usefulness of the method and how we extended the existing methods to account for disturbances in Section 3.

2.1   Polyhedral Invariant Hybrid Automaton (PIHA)

The main distinguishing property of a PIHA is that it uses convex polyhedra to describe the invariant sets and transitions. We use the definitions of a PIHA from [1]:

Definition 1. *A PIHA is a tuple $H = (X, X_0, F, E, I, G)$ where*

- $X = X_C \times X_D$, *where $X_C \subseteq R^n$ is the continuous state space and X_D is a finite set of discrete locations.*
- F *is a function that assigns to each discrete location $u \in X_D$ a vector field $f_u(\cdot)$ on X_C.*
- $I : X_D \mapsto 2^{X_C}$ *assigns $u \in X_D$ an invariant set of the form $I(u) \subseteq X_C$ where $I(u)$ is a non degenerate convex polyhedron.*
- $E \subseteq X_D \times X_D$ *is a set of discrete transitions.*
- $G : E \mapsto 2^{X_C}$ *assigns to $e = (u, u') \in E$ a guard set that is a union of faces of $I(u)$.*
- $X_0 \subseteq X$ *is the set of initial states of the form $X_0 = \cup_i (P_i, u_i)$ where each $P_i \subseteq I(u_i)$ is a polytope and $u_i \in U$; here, the notation (P, u) means the set $\{(x, u) \in X \mid x \in P\}$.*
- I, G, *and E must satisfy the following covariance requirements:*
  1. *for each u, $\partial I(u) = \cup_{e \in E \mid e = (u, u')} G(e)$ for some $u' \in X_D$, that is the guards for u cover the faces of the invariant for u.*
  2. *for all $e = (u, u') \in E$, $G(e) \subseteq I(u')$, that is events do not lead to transitions that violate invariants.*

2.2 Approximate Quotient Transition Systems (AQTS)

A standard approach for V&V of hybrid systems is to create a finite state bisimulation [1]. Bisimulations simulate the original system (have the same state histories given the same input) and can be simulated by the original system. Bisimulations can be created through partitioning the continuous state space into discrete regions and this is known as a quotient transition system (QTS). Unfortunately, it has been shown that finite state bisimulations only exist for hybrid systems with trivial continuous dynamics [7]. However, a QTS is a simulation of the system, meaning that any universal specifications (i.e. specifications that must be true for all possible trajectories) that are true for the QTS are also true for the hybrid system. Universal specifications include specifications such as safety and reachability [1].

In order to construct a QTS, for all partitions the reachable set of the continuous states must be computed. Consequently constructing a precise QTS is only possible for simple systems such as those with clock dynamics [7]. For more complicated continuous dynamics, it is only possible to construct an approximate quotient transition system (AQTS). An AQTS can either over-approximate the reachable set of the continuous system and be used to show that certain states are not reachable (safety), or under-approximate the reachable set and show that certain states are reachable (reachability).

2.3 Computation of Flow Pipes

The reachable set for a convex partition of an autonomous nonlinear systems can be computed by propagating the boundaries of the convex partition. For linear systems, this can be accomplished more efficiently by using a cached propagation shape for the reachable set that can be scaled as time progresses [1].

In Figure 1, we show how the hybrid system can be converted to an AQTS that is capable of verifying safety and reachability specifications. A flapping wing system moves from the square initial set X_0 at waypoint 1, $W1$, to flow pipe 1, $T1$, to waypoint 2, $W2$, to flow pipe 2, $T2$, to waypoint 3, $W3$, to flow pipe 3, $T3$, and finally to waypoint 4, $W4$. From the initial set, several trajectories are propagated using Monte-Carlo simulation (shown in blue). The existing flow pipe computation would enclose the trajectories originating from X_0. As can be seen, for the tracking problem shown in Figure 1(a), the blue trajectories rapidly collapse to the green reference trajectory. The red outer bounds shown in the plot are computed by our method and represent the bounds considering unknown disturbances which we discuss in Section 3.

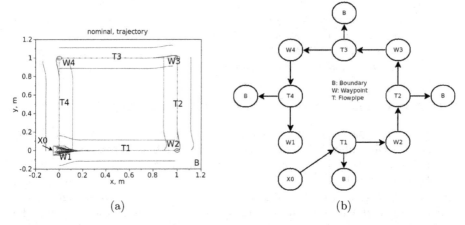

(a)   (b)

Fig. 1. The generation of an AQTS from computed flow pipes. Note that the green line is the nominal trajectory, the blue lines are Monte-Carlo simulations starting from the initial set X_0, and the red bounds are constructed using the H-infinity norm which we discuss in Section 3.

3   H-infinity Norm Flow Pipe Augmentation

The H-infinity norm is the maximum singular value of a linear system. It can be computed rapidly and can be precomputed for each mode of a linear hybrid system before computing the reachable set of the system. The core idea of this paper is to use the H-infinity norm to augment the computed flow pipe for the AQTS. We have not yet implemented a model checker with this modification, but we do verify the concept through Monte-Carlo simulation. For linear systems, the computation of the flow pipes can also be done efficiently, so computation of the H-infinity norm augmented flow pipes in real time for adapting systems should be feasible.

In Figure 2, a computational flow pipe is shown for the set X_0. The H-infinity norm is used to augment the bound based upon the magnitude of the H-infinity norm for the current face of the flow pipe.

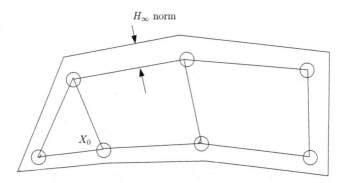

Fig. 2. The initial starting triangular set X_0 is propagated using standard reachability techniques to create the inner flow pipe with boundaries indicated by circles. The H-infinity norm augments the existing flow pipe to account for disturbances.

4   Flapping Wing Dynamics and Dynamic Inversion Based Control Law

In this section, our goal is to formulate the trajectory tracking control problem so that it can be verified through the use of the H-infinity augmented flow pipe based model checking method described in Section 3. A precise model of the flapping wing aerodynamics would be difficult to analyze. Therefore, we analyze a cycle average force and moment model [8]. The vehicle analyzed in this paper is shown in Figure 3. The flapping wing vehicle is mounted rigidly on top of a disc that is hovering on an air table. In this way, we can design the control system without concern for lifting the weight of the vehicle. The air table also limits the motion of the vehicle by keeping it level on the surface of the table.

For this flapping wing system, control is obtained through modulation of the flapping cycle. The flapping cycle has two parameters for each wing, the flapping frequency and the flapping delta shift which moves the trough of the function. The effect of the delta shift parameter on the flapping cycle is shown in Figure 4.

We assume that a wing flap cycle produces:

– A force in the body x direction proportional to the average dynamic pressure of the wing (dependent on the flapping frequency and the distance from the axis of rotation to the center of the wing).
– A force in the body y direction proportional to the average dynamic pressure of the wing (dependent on the flapping frequency and the distance from the axis of rotation to the center of the wing). It is assumed that this force has less magnitude since it is not aligned with the lifting direction of the wing.

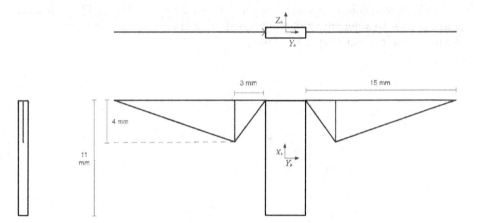

Fig. 3. The flapping wing vehicle analyzed in this paper

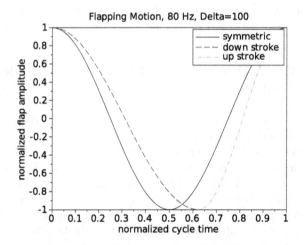

Fig. 4. Flapping wing cycle modulation through variation of the trough position of the stroke

- A force in the body z direction proportional to the product of the delta-shift of the flapping cycle and the average dynamic pressure of the wing.

Therefore we can derive the following aerodynamic force and moment model.

$$F_{aero} = \rho l_f^2 s/2 \begin{bmatrix} C_x(\omega_R^2 + \omega_L^2) \\ C_y(\omega_R^2 - \omega_L^2) \\ C_z(\delta_R\omega_R^2 + \delta_L\omega_L^2) \end{bmatrix} \tag{1}$$

$$M_{aero} = \rho l_f^2 l_w s/2 \begin{bmatrix} C_z(\delta_R\omega_R^2 - \delta_L\omega_L^2) \\ 0 \\ C_x(\omega_R^2 - \omega_L^2) \end{bmatrix} \tag{2}$$

where ρ is the air density, l_f is the distance from the flapping axis of rotation to the center of the wings, l_m is the moment arm of the wings, s is the reference area of the wing, (ω_R, ω_L) are the angular velocity of the (right,left) wings, (δ_R, δ_L) are the flapping cycle shift parameters for the (right,left) wings, and (C_x, C_y, C_z) are the (x, y, z) aerodynamic force coefficients.

Due to the air table, there are reaction moments in y and z that keep the flapping wing vehicle level and a reaction force in x, the normal force from the air table. We neglect friction forces and aerodynamic drag due to the low speed of the vehicle. We obtain the following equations of motion for the system:

$$m \begin{bmatrix} \ddot{x} \\ \ddot{y} \\ \ddot{z} \end{bmatrix} = \begin{bmatrix} 0 & \sin(\phi) & \cos(\phi) \\ 0 & -\cos(\phi) & \sin(\phi) \\ 1 & 0 & 0 \end{bmatrix} \left(F_{aero} + \begin{bmatrix} R_{Fx} - mg \\ 0 \\ 0 \end{bmatrix} \right) + \begin{bmatrix} d_x \\ d_y \\ d_z \end{bmatrix} \quad (3)$$

$$J \begin{bmatrix} \ddot{\phi} \\ \ddot{\theta} \\ \ddot{\psi} \end{bmatrix} = \begin{bmatrix} 0 & \sin(\phi) & \cos(\phi) \\ 0 & -\cos(\phi) & \sin(\phi) \\ 1 & 0 & 0 \end{bmatrix} \left(M_{aero} + \begin{bmatrix} 0 \\ R_{My} \\ R_{Mz} \end{bmatrix} \right) + \begin{bmatrix} d_\phi \\ d_\theta \\ d_\psi \end{bmatrix} \quad (4)$$

where ϕ is the roll angle of the MAV (here, the roll angle is about the vertical axis on the air table), R_{F_x} is the reaction force in the body x direction (the vertical axis on the table), R_{M_y} is the reaction moment about the y axis (side axis on the table), R_{M_z} is the reaction moment about the z axis (forward axis on the table), d_x, d_y, d_z are force disturbances, and d_ϕ, d_θ, d_ψ are moment disturbances.

For the trajectory tracking controller we need a model to track. We choose a simple system where the reference trajectory is specified by the velocity commands as shown in (5). The heading of the vehicle on the air table, ϕ, is chosen so that the vehicle regulates its heading in the direction of its velocity vector. This choice was made to keep the primary accelerations in the direction of the body z axis, pointing forward on the air table, where the wings have the most control authority since they are most aligned with the direction of the lift force from the wings. The ability of the wings to generate a side force through disparity of flapping frequency has less control authority than variation of the delta shifts in the flapping pattern to generate a forward force. This enables the model to match the desired linear system dynamics over a wider range of the flight envelope.

$$\begin{bmatrix} \dot{x}_r \\ \dot{y}_r \\ \dot{\phi}_r \end{bmatrix} = \begin{bmatrix} V_{rx} \\ V_{ry} \\ atan2(V_{ry}, V_{rx}) \end{bmatrix} \quad (5)$$

where $atan2$ is the quadrant accurate inverse tangent function, (x_r, y_r) are the Cartesian coordinates of the reference trajectory, (V_{rx}, V_{ry}) are the reference trajectory's velocities, and ϕ_r is roll angle in body or heading on the air table.

We now set the trajectory tracking error dynamics through dynamic inversion. We are only concerned with the states that are not constrained by the air table (x, y, z, ϕ). We invert the dynamics so that they match the model:

$$
\begin{bmatrix} \ddot{e}_x \\ \ddot{e}_y \\ \ddot{e}_z \\ \ddot{e}_\phi \end{bmatrix} + 2\zeta\omega_n \begin{bmatrix} \dot{e}_x \\ \dot{e}_y \\ \dot{e}_z \\ \dot{e}_\phi \end{bmatrix} + \omega_n^2 \begin{bmatrix} e_x \\ e_y \\ e_z \\ e_\phi \end{bmatrix} = \begin{bmatrix} d_x \\ d_y \\ d_z \\ d_\phi \end{bmatrix}
\tag{6}
$$

where $e_x = x_r - x$, $e_y = y_r - y$, $e_z = z_r - z$, w_n is the natural frequency of the 2nd order system and ζ is the damping ratio. The dynamics for each state are not coupled, so it is not necessary to have the same natural frequency or damping ratio, but we choose this for simplicity.

We now can compute the force and moments as a function of the tracking errors in order to match the above model. We can neglect F_x because the flapping wing vehicle is constrained to remain on the surface of the air table since the lift is less than the weight. In the equations, the value F_x sets the steady state flapping frequencies.

$$
\begin{bmatrix} M_{bx} \\ F_{by} \\ F_{bz} \end{bmatrix} = \begin{bmatrix} 0 & 0 & 1 \\ \sin(\phi) & -\cos(\phi) & 0 \\ \cos(\phi) & \sin(\phi) & 0 \end{bmatrix} \begin{bmatrix} -\omega_n^2 e_x - 2\zeta\omega_n \dot{e}_x \\ -\omega_n^2 e_y - 2\zeta\omega_n \dot{e}_y \\ -\omega_n^2 e_\phi - 2\zeta\omega_n \dot{e}_\phi \end{bmatrix}
\tag{7}
$$

Because there are 4 control variables and 4 degrees of freedom, there is a unique mapping of the controls to the desired forces and moments.

$$
\begin{bmatrix} \omega_R^2 \\ \omega_L^2 \\ \delta_R \\ \delta_L \end{bmatrix} = \frac{1}{\rho l_f^2} \begin{bmatrix} \frac{F_{bx}}{C_x} + \frac{F_{by}}{C_y} \\ \frac{F_{bx}}{C_x} - \frac{F_{by}}{C_y} \\ \frac{1}{\omega_R^2} \left(\frac{F_{bz}}{C_z} + \frac{M_{bx}}{C_z l_w} \right) \\ \frac{1}{\omega_L^2} \left(\frac{F_{bz}}{C_z} - \frac{M_{bx}}{C_z l_w} \right) \end{bmatrix}
\tag{8}
$$

Note in (8) that the ω_R^2 and ω_L^2 used in computation of δ_R and δ_L could be expressed instead as explicit functions of the desired forces and moments as given by the first two elements of (8).

In Figure 5, we plot the delta shifts and flapping frequencies required to generate the forces and moments experienced during a simulation of the flapping wing trajectory tracking control system for both the nominal and expected disturbance case. Both the delta shift and flapping frequencies are in reasonable ranges. The spikes in delta shift and flapping frequency occur when the vehicle is at a new waypoint and beginning to move in a new direction. Dynamic inversion based control is susceptible to modeling error, but these modeling errors can also be included as unknown disturbances to the system. Therefore, dynamic inversion integrates well with the proposed H-infinity based flow pipe augmentation model checking method we propose.

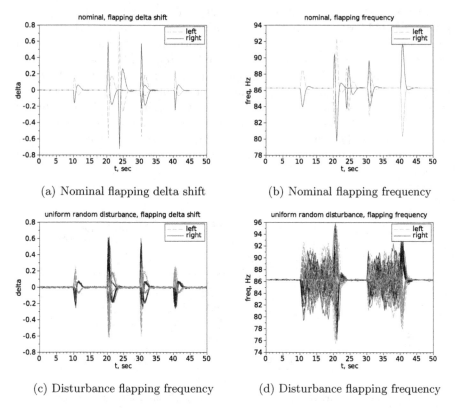

(a) Nominal flapping delta shift

(b) Nominal flapping frequency

(c) Disturbance flapping frequency

(d) Disturbance flapping frequency

Fig. 5. The time history of the delta shift and flapping frequency required for the dynamic inversion based control for both the nominal and uniform random disturbance case. Uniform random noise was used since it is bounded. Bounded Gaussian noise could also be used.

5   Flapping Wing Controller Model Checking

Now that we have successfully inverted the dynamics of the flapping wing trajectory tracking control system, we can use a PIHA model of the system with H-infinity norm augmented flow pipes to check the control system in the presence of bounded disturbances. We will use Computational Tree Logic (CTL) to express the universal properties we wish to verify [9]. CTL is a branching time logic which can represent non-deterministic transitions and is used by model checkers such as NuSMV[10]. For example, we wish to verify the property *AF AG W*1 (always end at waypoint 1) in the presence of bounded disturbances of magnitude 0.4 Newtons of force and 0.4 Newton-meters of torque. Disturbances that are H-infinity worst case frequency sinusoids and disturbances that are uniform random numbers sampled at 20 Hz will be tested with the given magnitudes. For the Monte-Carlo test, the phase of the H-infinity sinusoids will be sampled from a uniform distribution.

In Figure 6, the nominal trajectory of the vehicle is shown. The H-infinity norm augmented flow pipes are shown as the red lines in both the trajectory and position error plots. Note that the H-infinity norm is a constant that is added to the reachable set from the initial state (the blue region). For these plots both the x and y errors are plotted on the position error plot and that is why there are two sets of bounds. At the beginning of the simulation, the trajectories are widely dispersed but they quickly converge to the reference trajectory due to the tracking control system. The bounds could be improved by recomputing the flow pipes with no disturbance given the initial set from the previous flow pipe with disturbance. This would be implemented in a final version of our model checker.

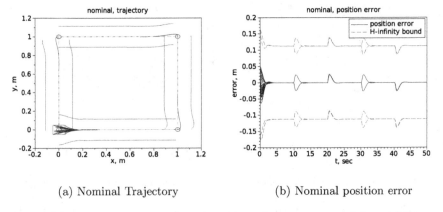

(a) Nominal Trajectory   (b) Nominal position error

Fig. 6. A Monte-Carlo simulation with no disturbance propagated from initial set x_0 (the square at (0,0))

For the existing AQTS representing the trajectory tracking control system as shown in Figure 1, we plot Monte-Carlo simulations in Figure 7. The uniform and H-infinity disturbances are both able to cause the system to violate the specification. This occurs because the MAV transitions to the next flow pipe only if the current state is within a set radius of the waypoint. When the disturbance level is large enough, it becomes possible for the vehicle not to pass inside this radius and the vehicle fails to return to $W1$.

If we no longer use a transition guard that depends on the distance of the vehicle from the waypoint, and we now use the along track distance of the vehicle, a more robust control system can be obtained. The along track distance is the distance between the previous waypoint and the projection of the current vehicle position onto the line between the waypoints. The new guidance law will transition to the next waypoint when the along track distance is greater than the distance between the waypoints, meaning the vehicle has passed the waypoint. In Figure 8, the modification to the AQTS is shown. Note that in the modified system it is possible that the vehicle will not be within the defined radius, but it

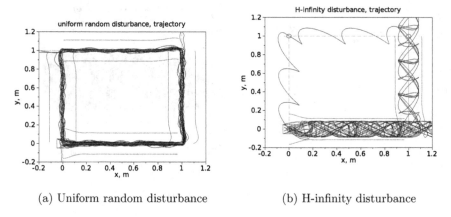

(a) Uniform random disturbance   (b) H-infinity disturbance

Fig. 7. A Monte-Carlo simulation showing failures of the radius based waypoint transition guard

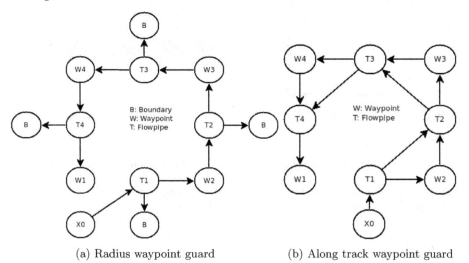

(a) Radius waypoint guard   (b) Along track waypoint guard

Fig. 8. The AQTSs representing the flapping wing system with the original radius based waypoint guard and the along track based waypoint guard

will always reach $W1$. Finally, we check the new modified AQTS with the along track guidance law using Monte-Carlo simulations with H-infinity and uniform random disturbances in Figure 9. We see that there are no trajectories that enter the boundary and all trajectories return to $W1$. We see that for the new guidance law the specfication $AF\ AG\ W1$ (always end at waypoint 1) is satisfied.

The guidance logic that we have improved in this simple example would be obvious to any autopilot designer, but the complications that disturbances have on more intricate components of a guidance system can be more difficult to discover and having an automated model checking program to verify these systems in the presence of large disturbances will improve the safety of such systems.

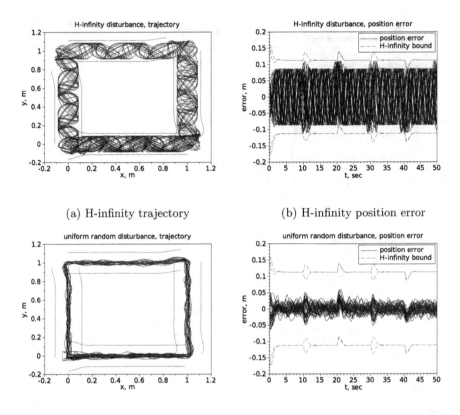

(a) H-infinity trajectory

(b) H-infinity position error

(c) Uniform random disturbance trajectory

(d) Uniform random position error

Fig. 9. A Monte-Carlo simulation showing that the along track based waypoint transition guard is resilient in the presence of disturbances

6   Conclusion

We have proposed a method for verification and validation (V&V) of a flapping-wing micro air vehicle controller (MAV). We have extended the existing algorithms for V&V of Polyhedral Invariant Hybrid Automatons (PIHAs) to account for bounded disturbances in linear hybrid systems using the H-infinity norm. While the H-infinity norm does not strictly bound all possible disturbances, because it only bounds the worst case steady state sinusoidal disturbance, it does successfully bound disturbances expected in practice. If large unbounded disturbances are expected, it will be necessary to consider probabilistic model checking and the probability density function of the regulated system around the nominal trajectory. In order for the H-infinity norm bound to be applicable, the system dynamics must be linear over the flight envelope; however, we have shown how our method can be coupled with dynamic inversion based controllers to extend our method to nonlinear hybrid systems. The H-infinity norm of the system can be computed efficiently and only requires updating when the linear system model changes. Coupled with the efficient reachable set computations for linear systems, this makes our approach attractive for runtime-assurance of adaptive systems.

Acknowledgements. This material is based upon work supported by the National Science Foundation under Grant Numbers CNS-1239196, CNS-1239171, and CNS-1239229. Additional support was provided by the AFRL SFFP program.

References

[1] Chutinan, A., Krogh, B.H.: Computational techniques for hybrid system verification. IEEE Transactions on Automatic Control 48(1), 64–75 (2003)

[2] Chutinan, A., Krogh, B.H.: Verification of polyhedral-invariant hybrid automata using polygonal flow pipe approximations. In: Vaandrager, F.W., van Schuppen, J.H. (eds.) HSCC 1999. LNCS, vol. 1569, pp. 76–90. Springer, Heidelberg (1999)

[3] Silva, B.I., Richeson, K., Krogh, B., Chutinan, A.: Modeling and verifying hybrid dynamic systems using checkmate. In: Proceedings of 4th International Conference on Automation of Mixed Processes, pp. 323–328 (2000)

[4] Asarin, E., Dang, T., Maler, O.: The d/dt tool for verification of hybrid systems. In: Brinksma, E., Larsen, K.G. (eds.) CAV 2002. LNCS, vol. 2404, pp. 365–370. Springer, Heidelberg (2002)

[5] Silva, B.I., Stursberg, O., Krogh, B.H., Engell, S.: An assessment of the current status of algorithmic approaches to the verification of hybrid systems. In: Proceedings of the 40th IEEE Conference on Decision and Control, vol. 3, pp. 2867–2874. IEEE (2001)

[6] Henzinger, T.A., Ho, P.-H., Wong-Toi, H.: Hytech: A model checker for hybrid systems. In: Grumberg, O. (ed.) CAV 1997. LNCS, vol. 1254, pp. 460–463. Springer, Heidelberg (1997)

[7] Alur, R., Courcoubetis, C., Halbwachs, N., Henzinger, T.A., Ho, P.H., Nicollin, X., Olivero, A., Sifakis, J., Yovine, S.: The algorithmic analysis of hybrid systems. Theoretical Computer Science 138(1), 3–34 (1995)

[8] Doman, D.B., Oppenheimer, M.W., Sigthorsson, D.O.: Wingbeat shape modulation for flapping-wing micro-air-vehicle control during hover. Journal of Guidance, Control, and Dynamics 33(3), 724–739 (2010)

[9] Baier, C., Katoen, J.P., et al.: Principles of model checking, vol. 26202649. MIT Press, Cambridge (2008)

[10] Cimatti, A., Clarke, E., Giunchiglia, E., Giunchiglia, F., Pistore, M., Roveri, M., Sebastiani, R., Tacchella, A.: NuSMV 2: An openSource tool for symbolic model checking. In: Brinksma, E., Larsen, K.G. (eds.) CAV 2002. LNCS, vol. 2404, pp. 359–364. Springer, Heidelberg (2002)

Design Constraints of a Minimally Actuated Four Bar Linkage Flapping-Wing Micro Air Vehicle

Benjamin M. Perseghetti[1], Jesse A. Roll[2], and John C. Gallagher[1]

[1] Dept. of Computer Science and Engineering
Wright State University, Dayton OH, USA
{perseghetti.2,john.gallagher}@wright.edu
[2] Dept. of Mechanical Engineering
Purdue University, West Lafayette, IN, USA
jroll@purdue.edu

Abstract. This paper documents and discusses the design of a low-cost Flapping-Wing Micro Air Vehicle (FW-MAV) designed to be easy to fabricate using readily available materials and equipment. Basic theory of operation as well as the rationale underlying various design decisions will be provided. Using this paper, it should be possible for readers to construct their own devices quickly and at little expense.

Keywords: Flapping-Wing Micro Air Vehicles, Quick Fabrication.

1 Introduction

The construction and control of Flapping-Wing Micro Air Vehicles (FW-MAVs) requires broad knowledge drawn from a wide variety of disciplines including biology, mechanics, electrodynamics, materials science and control theory. As with most small-sized air vehicles, FW-MAVs must be optimized for minimal weight and maximal lift. The components most responsible for weight are the actuators (motors and related components) and the power systems (batteries and related components). One strategy, therefore, to eliminate weight is adopt minimally actuated designs in which the number of actively driven degrees of freedom is minimized. By using split-cycle constant-period frequency modulation (SCCPFM) [1], [2], two actuators (and a bob-weight) can provide moments capable of effecting all six degrees of vehicle body pose and position.

SCCPFM works by concatenating two half cosine waves (upstroke (ϕ_U) and downstroke (ϕ_D) of the wing) such that the overall period is unchanged, however, the relative periods of the two half cycles can vary. When the upstroke and downstroke are symmetric equation (1) and equation (2) hold true and the wing oscillates with the fundamental wingbeat frequency (ω) [3],[4].

$$\phi_U = \cos(\omega t), \quad 0 < t < \frac{\pi}{\omega} \tag{1}$$

$$\phi_D = \cos(\omega t), \quad \frac{\pi}{\omega} < t < \frac{2\pi}{\omega} \tag{2}$$

To create an asymmetric upstroke and downstroke the value of δ is changed such that δ≠0 and equations (3) through (6) are implemented.

$$\phi_U = \cos[(\omega - \delta)t], \quad 0 < t < \frac{\pi}{\omega - \delta} \tag{3}$$

$$\phi_D = \cos[(\omega + \sigma)t + \xi], \quad \frac{\pi}{\omega - \delta} < t < \frac{2\pi}{\omega} \tag{4}$$

$$\sigma = \frac{\delta\omega}{\omega - 2\delta} \tag{5}$$

$$\xi = \frac{-2\pi\delta}{\omega - 2\delta} \tag{6}$$

For δ > 0 the upstroke has a smaller frequency and downstroke larger frequency than the fundamental wingbeat frequency (δ = 0) (Figure 1). For δ < 0 the upstroke has a larger frequency and downstroke smaller frequency than the fundamental wingbeat frequency (δ = 0) (Figure 2). The resultant of having different upstroke and downstroke frequencies is a non-zero cycle-averaged drag caused by differing dynamic pressures. This allows the vehicle to roll as well as translate forward and backwards.

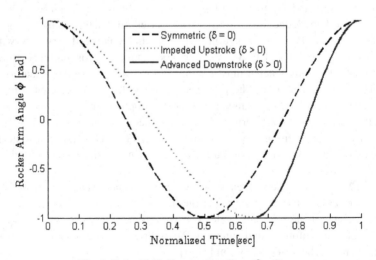

Fig. 1. Delta Shift Greater than Zero (δ > 0)

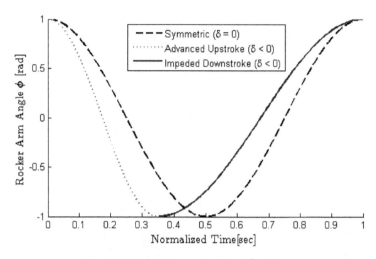

Fig. 2. Delta Shift Less than Zero ($\delta < 0$)

A four-bar linkage driven by a DC motor provides a relatively simple mechanism for driving the wing of a FW-MAV. A four-bar linkage for this application would be comprised of a crank arm (l_2), a coupler arm (l_3), and a rocker arm (l_4). The distance between the centers of rotation for the crank and rocker arms is referred to as the line of centers (l_1) (Figure 3). Although the four-bar linkage provides a mean to transform rotary motion of a DC motor into the required flapping/sweeping motions of a wing, the mechanism does introduce some inconvenience, primarily that the four-bar linkage is temporally asymmetric. This means that when the crank arm is given a fixed angular velocity ($\dot{\theta}$) the rocker arm angular velocity ($\dot{\phi}$) is asymmetric between upstroke and downstroke (Figure 4). This is caused by the non-proportional crank arm angle (θ) to rocker arm angle (ϕ) (Figure 5).

The temporal asymmetry is not insurmountable, although it does require kinematic inversion to remedy the latent temporal asymmetry [5]. Via kinematic inversion, one can compute the DC motor shaft positions and speeds that would be required to produce a smooth cosine wave motion at the tip of the rocker arm (where the wing would be mounted). The pairings of position and speed could be stored into a lookup table that the motor controller would consult as the DC motor was spinning. The motor would be commanded to rotate at the speed specified by its current shaft angular position, and thereby produce cosine motion at the end of the rocker arm. Naturally, the initiation of the controller is dependent upon the rocker arms initial angular position [6]. If the FW-MAV starts with the rocker arm in the wrong angular position then the look-up table will be shifted resulting in the improper alignment of the cosine-formed SCCPFM upstroke and downstroke. If a sensor is used to align the rocker arm with a known rocker arm angle the look-up table can be properly initiated.

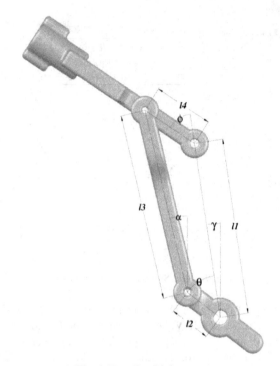

Fig. 3. Four Bar Linkage

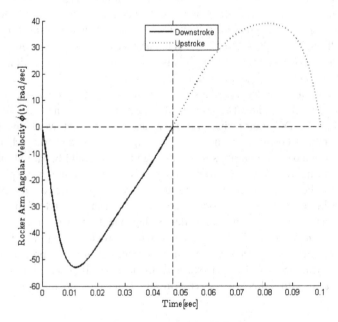

Fig. 4. Rocker Arm Angular Velocity

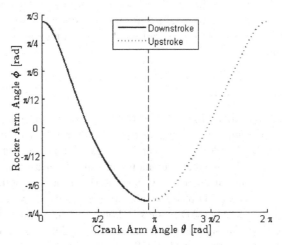

Fig. 5. Rocker Arm vs. Crank Arm

2 Design Considerations

2.1 Four Bar Linkage FW-MAV Construction Geometry

When designing a four bar linkage FW-MAV mechanism (Figure 3) the midpoint of the rocker arm (l_4) stroke, which houses the passively rotating wing, should be orthogonal to the plane of bi-lateral symmetry of the base frame. This allows for equal stroke amplitude on both sides and avoids pre-biased forward or backward moments. To determine the angle (γ) adjacent to the line of centers (l_1) between the crank arm (l_2) and rocker arm rotation centers relative to the plane of frame bi-lateral symmetry, the maximal (ϕ_{max}) and minimal (ϕ_{min}) rocker arm angles relative to the line of centers must first be determined by the law of cosines in Equation (6) and Equation (7) respectively. From there Equation (8) can be used to determine the angle (γ) that aligns the midpoint of the rocker arm stroke orthogonal to the plane of bi-lateral symmetry of the base frame.

$$\phi_{max} = \mathrm{acos}\left(\frac{l_1{}^2+l_4{}^2-(l_3+l_2)^2}{2l_1l_4}\right) \tag{6}$$

$$\phi_{min} = acos\left(\frac{l_1{}^2+l_4{}^2-(l_3-l_2)^2}{2l_1l_4}\right) \tag{7}$$

$$\gamma = \frac{\phi_{max}+\phi_{min}-\pi}{2} \tag{8}$$

The base frame design needs to incorporate a sensor to allow for the determination of relative rocker arm stroke location. The sensor type combined with the choice of actuators (motors) will determine the overall width and spacing requirements between the chiral rocker arms.

2.2   Construction Constraints

Choosing a material to design a four bar FW-MAV plays a major role in construction constraints. The use of aluminum and titanium provides for a desired high strength-to-weight ratio, however, a complex non-planar design can dramatically increase the cost of construction and result in an excessive amount of time for machining (assuming the precision required can be machined). A particularly attractive option is to take advantage of fast prototyping 3D printers. This allows for the direct actualization of complex non-planar FW-MAV CAD models in hours instead of potentially weeks or months. Due to the lower strength-to-weight ratio of 3D printing materials the design has to be thickened to accommodate for deformable material loading. Furthermore, the four-bar linkage should be designed in a way to assure it will stay planar during operation.

3   Fast Prototyping

3.1   Available 3D Printing Technologies

While there are many types of 3D printers, not all are equally useful for the construction of FW-MAVs. Some typical types of 3D printers are photopolymerization printers, lamination printers, extrusion deposition printers, and granular material binding printers. Photopolymerization printers allow for the highest level of precision and leverage stereolithography to build parts layer by layer using a large potential variety of photopolymer materials. This style of printer is highly effective at creating structures that require support material to be printed and can provide for uniform and smooth surfaces. Lamination printers are relatively low precision and provide significantly less selection of materials with high strength-to-weight ratios. Extrusion deposition printers rely on Fused Deposition Modeling (FDM) where a material is melted and placed in horizontally adjacent lines from an extrusion nozzle to form layers. FDM has relatively low precision (since it is limited to the diameter of the nozzle head), rougher surfaces, and longer build times than photopolymerization. However, it has an excellent selection of high strength-to-weight materials and is relatively inexpensive. Granular material binding uses Selective Laser Sintering (SLS) in which granulated material is bound together by a heat from a high energy laser. This, in some applications, allows for high precision but also results in a longer build time. Granular material binding printing is also capable of printing metal parts but is significantly more expensive than photopolymerization printing.

3.2 Material Selection

When selecting the material for the 3D printer of choice it's imperative to know the full material properties. Knowledge of materials response to temperature, water, oil, impact, compression, strain, stress as well as the longevity of material properties (strength, hardness, deformation, conductivity or lack thereof) is critical. It is crucial to avoid materials that deform at temperatures at which the FW-MAV is designed to operate. If a material is overly soft it could wear away quickly from operational friction and if it is too hard it could break from impact. If a material volumetrically changes as it absorbs or dissipates water and/or oil, then cleaning and lubricating as well as heat and humidity can have a significant effect on the operation of the FW-MAV. Furthermore, if the material is ferrous it severely limits the effectiveness of Hall effect and related magnetic sensors packages that might be required to identify the rocker arm's relative angular position.

4   Design Implementation

4.1   BMP-MAV5

The BMP-MAV5 (Figure 6) is a minimally actuated FW-MAV that is designed to operate while mounted on a circular puck supported from below on a cushion of air. Although not a free-flying vehicle, the BMP-MAV5 can nevertheless be used to, without risk of crashing, analyze the generation of aero forces required to roll the vehicle and to traversing forward and backward along the table's surface. The BMP-MAV5 uses the lengths of linkages in table 1. It was also designed to allow for passive wing rotation angle of attack selection by insertion of small gauge stainless steel wires in the selected holes at the end of the rocker arm and are highlighted green in Figure 7. This allows for the selection of multiple angles of attack and is part of a rapid wing replacement design.

Instead of using ball bearings which have significant weight and diametric area, a highly polished micro brass sleeve-bearing (OD 1.2mm ID 1.0mm) was used with spring strength stainless steel rods as the shafts to allow the articulating joints to move. Brass sleeve-bearings are commonly used in naval and high frequency or high dynamic loading applications. Since the stainless steel is exceedingly hard and both surfaces are highly polished the coefficient of friction is small enough to not introduce a major dynamic effect. The brass sleeve-bearings are also used in the end of the rocker arm as part of the rapid wing replacement design and are highlighted yellow in Figure 7. This allows for the carbon fiber rod of the wing to easily passively rotate while the end inside the rocker arm is capped to keep the wing from falling out. In order to keep the four-bar linkage in plane the coupler connects to the rocker arm from the top as well as the bottom side. This type of dual sided in-plane support was also used on the base frame to articulate with the top and bottom side of the rocker arm's center of rotation.

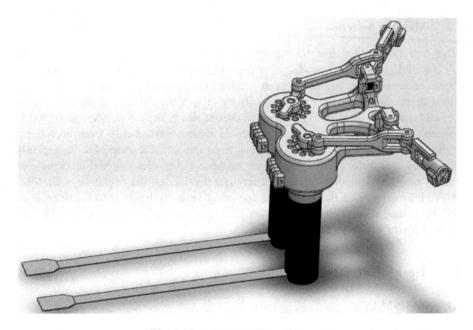

Fig. 6. BMP-MAV5 Without Wings

Table 1. Linkage Lengths

Linkage	Length (mm)
Line of Centers (l_1)	24.2000
Crank Arm (l_2)	5.7625
Coupler Arm (l_3)	25.7500
Rocker Arm (l_4)	8.3750

The motors are attached to the base frame with a locking sleeve so as to make the motor component modular. Various sizes and types of motors can be used including motors as large as the 10mm OD Micromo 1028M006BIEM3-1024 to however small of a motor desired. The Micromo 1028M006BIEM3-1024 motor was chosen due to its high torque, efficiency, encoder resolution, and compact design. In order to determine the initial position of the rocker arm angle a Honeywell SS30AT latching bipolar Hall effect sensor was inserted between the upper and lower base frame rocker arm support guides. The SS30AT can be carefully slid forward or backwards in the grooved slot to alter the magnitude of the magnetic flux at the surface of the sensor and is highlighted red in Figure 7. The 0.75mm OD neodymium magnets are located in the end of the rocker arm and are radially pointing outwards from the center of rotation. The magnet housing holes correspond to the endpoints of the upstroke, downstroke and midstroke and are highlighted blue in Figure 7. The magnetic wire connecting the SS30AT latching bipolar hall effect sensor is ran through the base frame as to avoid getting entangled with the four bar linkage.

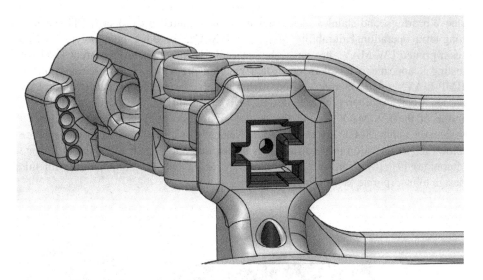

Fig. 7. Rocker Arm and Base Frame Components

The material selected for the BMP-MAV5 was Stratasys RGD525 white. This material has low water absorption, elongation at break, and Izod notched impact. Stratasys RGD525 white also has high tensile strength, modulus of elasticity, flexural strength, flexural modulus, Rockwell hardness, and heat deflection. Furthermore, since it is a non-ferrous material it doesn't interact with the magnets or Hall effect sensors. All parts were printed on an Objet Eden 350V and cleaned using a lye solution to remove the support material. Cyanoacrylate was used to adhere the Alliance Model Works TW006 brass sleeve-bearings to the RGD525 parts as well as the stainless steel rods. The entire vehicle can be printed and assembled within a day if necessary. To ensure the vehicle doesn't overheat during operation to the point of material deformation, four Vishay NTC thermistors (NTCLE305E4502SB) were placed on the vehicle. One was placed on each motor at the top side of the motor housing mount and one was placed next to the rocker arm center of rotation. This allows for the device to be immediately shut down if the temperature of the motor or rotating rocker arm exceeds the deformation threshold temperature. The crank arm and coupler arm for the BMP-MAV5 are identical for both sides. However, due to magnet locations the rocker arms are chiral.

5   Conclusions and Discussion

The culmination of modular design and fast prototyping (3D printing) greatly increased the production and testing rate of the FW-MAVs. The use of Hall effect sensors (figure 8) requires a significant amount of testing to determine the appropriate biased angle for insertion of the magnets in the end of the rocker arm. However, the Hall effect sensors do an outstanding job of providing reliable and expedient feedback for wing location. Some reaming of the holes used for housing the magnets, brass sleeve bearings, and stainless steel rods might be required for proper fit. The overall long-term operational durability of the BMP-MAV5 or other photopolymer based resin printed FW-MAVs has yet to be determined. Although, from current levels of testing it appears to maintain consistent material properties. Future variants of the BMP-MAV5 might include a free flight model constructed primarily of molded carbon fiber. Another potential design revision might include designing the base frame of the FW-MAV around the power supply and control hardware such as a system on a chip (SoC). The control of the vehicle is inspired by EAH methods which allows for assimilation of modified parts during operation (such as a wing or linkage alteration) [7], [8]. Hardware controllers are being built in conjunction with this design to incorporate those principals [9].

Fig. 8. BMP-MAV5 Printed In Stratasys FullCure 720 Material

Acknowledgements. This material is based upon work supported by the National Science Foundation under Grant Numbers CNS-1239196, CNS-1239171, and CNS-1239229.

References

1. Doman, D., Oppenheimer, M., Sigthorsson, D.: Dynamics and control of a minimally actuated biomimetic vehicle: Part I – aerodynamic model. In: Proceedings of the AIAA Guidance, Navigation, and Control Conference (2009)

2. Oppenheimer, M., Doman, D., Sigthorsson, D.: Dynamics and control of a minimally actuated biomimetic vehicle: Part II – control. In: Proceedings of the AIAA Guidance, Navigation, and Control Conference (2009)
3. Doman, D., Oppenheimer, M., Sigthorsson, D.: Wingbeat Shape Modulation for Flapping-Wing Micro-Air Vehicle Control During Hover. Journal of Guidance, Control and Dynamics 33(3), 724–739 (2010)
4. Oppenheimer, M., Doman, D., Sigthorsson, D.: Dynamics and Control of a Biomimetic Vehicle Using Biased Wingbeat Forcing Functions. Journal of Guidance, Control and Dynamics 34(1), 204–217 (2010)
5. Doman, D., Tang, C., Regisford, S.: Modeling Interactions Between Flexible Flapping-Wing Spars, Mechanisms, and Drive Motors. Journal of Guidance, Control, and Dynamics 34(5), 1457–1473 (2011)
6. Oppenheimer, M., Sigthorsson, D., Weintraub, I., Doman, D., Perseghetti, B.: Wing Velocity Control System for Testing Body Motion Control Methods for Flapping Wing MAVs. In: 51st AIAA Aerospace Sciences Meeting including the New Horizons Forum and Aerospace Exposition, January 7-10 (2013)
7. Boddhu, S., Gallagher, J.: Evolved Neuromorphic Flight Control for a Flapping-Wing Mechanical Insect. In: IEEE World Congress on Computational Intelligence (2008)
8. Boddhu, S., Gallagher, J.: Evolving Non-Autonomous Neuromorphic Flight Control for a Flapping-Wing Mechanical Insect. In: Proceedings of IEEE workshop on Evolvable and Adaptive Hardware (2009)
9. Boddhu, S., Botha, H., Perseghetti, B., Gallagher, J.: Improved Control System for Analyzing and Validating Motion Controllers for Flapping Wing Vehicles. In: Proceedings of 2nd International Conference on Robot Intelligence Technology and Applications (2013)

Improved Control System for Analyzing and Validating Motion Controllers for Flapping Wing Vehicles

Sanjay K. Boddhu, Hermanus V. Botha, Ben M. Perseghetti, and John C. Gallagher

Department of Computer Science & Engineering, Wright State University, Dayton, USA
{boddhu.2,botha.2,perseghetti.2,john.gallagher}@wright.edu

Abstract. In previous work, the viability of split-cycle constant-period frequency modulation for controlling two degrees of freedom of flapping wing micro air vehicle has been demonstrated. Though the proposed wing control system was made compact and self-sufficient to be deployed on the vehicle, it was not built for on-the-fly configurability of all the split-cycle control's parameters. Further the system had limited external communication capabilities that rendered it inappropriate for its integration into a higher level research framework to analyze and validate motion controllers in flapping vehicles. In this paper, an improved control system has been proposed that could addresses the on-the-fly configurability issue and provide an improved external communication capabilities, hence the wing control system could be seamlessly integrated in a research framework for analyzing and validating motion controllers for flapping wing vehicles.

Keywords: Flapping Flight Control, Split-Cycle Control, PID Control, Gumstix Control.

1 Introduction

Since last decade, there exists ever growing interest in study of flapping wing flight [1] [6] [5] and engineering challenges associated with realizing the flapping wing based vehicles at different sizes [1] [7]. One of the pioneering research activities in the field [1][7], that is aimed at implementing a self-sustainable flapping wing vehicle, had provided the research community with a prototype flapping wing vehicle [7], with an control capability, that can achieve two degrees of freedom control of the vehicle, via manipulation of single actuator in each wing. The prototype of flapping wing aircraft, shown in figure 1, was built using four-bar linkage assembly, where a single direct current (BLDC) motor's rotary motion is transformed to flapping motion of the respective wing. The control system was designed to manipulate the motion of flapping wings (thru BLDC), remotely, such that differential upstroke and down stroke velocities can be generated over the course of a wing beat cycle. The drag generated thru the designed control system was tested to demonstrate the ability to control two degrees of freedom, namely, roll and horizontal translation of the vehicle that corroborated with previous studies mentioned in [2] [3].

J.-H. Kim et al. (eds.), *Robot Intelligence Technology and Applications 2*, 557
Advances in Intelligent Systems and Computing 274,
DOI: 10.1007/978-3-319-05582-4_48, © Springer International Publishing Switzerland 2014

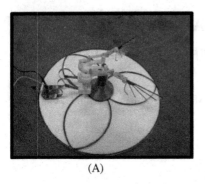
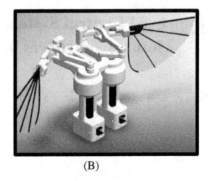

(A)   (B)

Fig. 1. Vehicle's four-bar linkage assembly. This mechanism can allow two degree control with using only one actuator per wing as described in [7]. Figure 1(A) show a mounted vehicle built using the mechanism and figure 1(B) shows the CAD simulation model of the mechanism.

Though the designed control system provided a mechanism to configure most of the parameters of the wing motion (during vehicle's flight), which are flapping rate and stroke active periods, it provided limited configurability of the shape parameter of each stroke phase, limiting the controllers to adhere to pre-determined shape burnt into the control system. This limited ability to configure the shape of the wing strokes, was justifiable for the project's vision [7] in the context, as it had effectively reduce the communication overhead of the system, making it as an ideal deployable system for specific vehicle structure and mission requirements. But, it was demonstrated in varied research efforts [6] [5], that the configurability of the shape parameter of the wing's trajectory would provide interesting insights in to the flapping wing flight mechanism.

Further, the user of the control system was limited to use a radio frequency transmission mechanism to communicate the flapping parameters of the wing, which required less intuitive user-side implementation of custom modulation techniques to encode all the control parameters on one channel. Hence, the existing control system in its current state [7], cannot be employed as a general purpose system that would aid in research efforts essential to promote advanced understanding and implications of the flapping wing flight mechanism.

Thus, in an effort to provide a more general purpose control system that could be most amicable for advanced flapping wing research, this paper proposes an improved control system that would readily address the above two issues of shape configurability and intuitive user control communication. In this vein, the second section of the paper provides a brief overview of the split-cycle frequency modulation theory, followed by which the proposed improved control system is described along with the implementation and validation details. Lastly, the paper ends with a discussion section that would outline current implications of the improved control system for the interested research community and also provides insights into the possible future improvements.

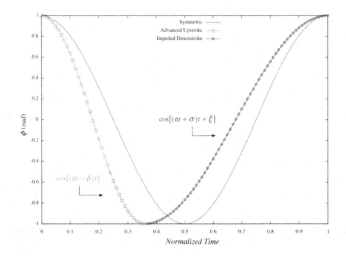

Fig. 2. A Sample Split Cycle Cosine

2 Split-Cycle Frequency Modulation Theory

A split-cycle cosine model is used to generate force from the wings. The model is intended to produce a frequency wave that maintains a constant period. The complete cycle is a combination of two different waves that maintain a similar frequency. Without the split-cycle modulation or the symmetric strokes the force generated from the wings during an upstroke would negate the force generated from a downstroke.

The wingbeat motion is dictated by the shape of a single split-cycle motion for both upstrokes and downstrokes, a sample cosine shape is shown in figure 2. The two different cosine waves that define the full split-cycle are respectively responsible for the upstroke and downstroke. For example, as shown in figure 2, to produce aerodynamic forces, a wing must advance one stroke phase velocity and impede the other stroke phase to complete the wingbeat cycle. The split-cycle cosine wave that describes this wingbeat motion contains a half wave cosine (1 radian to -1 radian) that reaches this point before the first half of the period is complete. The second section of the split-cycle cosine contains another half wave cosine that completes the cycle period.

Thus, in brief, a split-cycle wing motion control can be characterized three parameters, namely wing beat frequency, differential stroke period or ratio (between upstroke and downstroke period) and wing trajectory shape. Using the four-bar linkage assembly with single BLDC motor based actuator setup, these parameters get mapped to motor parameters thru the inverse kinematics mathematical model described in [7] , using which the base wing shape parameter is mapped into a commutation table, which is a time indexed position and velocity of the wing.

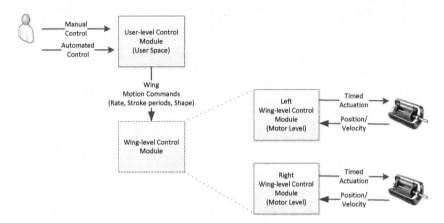

Fig. 3. Conceptual Overview Diagram of Improved Adaptive Control System

3   Improved Adaptive Control System

The proposed in-flight improved adaptive control system consists primarily of two conceptual blocks, namely wing-level control module and user-level control module as shown in figure 3. As one can perceive, the user-level control module, is responsible of providing an intuitive communication interface for a manual user or any automated system to provide wing motion parameters, like wing beat rate, stroke periods and base shape, in a feed-forward manner to the vehicle with appropriate acknowledgment mechanism, during an active flapping flight. Further, the wing-level control module conceptually is composed of two sub wing-level control modules, each for left and right wing motors. Each of these wing-level control modules are responsible of providing appropriate parameterized motor level feedback control mechanism to the accurately actuate the motor's motion in conjunction with the four-bar linkage mechanism guaranteeing the user provide wing motion parameters.

The next two sub sections provide the design and implementations details of the wing-level control module and user-level control module.

3.1   Wing-Level Control Module

The wing-level control module is the lowest level control module, whose objective is to produce user desired wing stroke motion, satisfying the split-cycle parameters and producing apt traversal of the upstroke and downstroke periods with varying velocities. Further, this module should also support configurability of split-cycle parameters during the active flight motion of the vehicle.

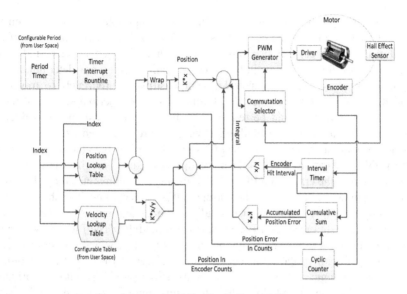

Fig. 4. Motor-level Feedback Control Flow

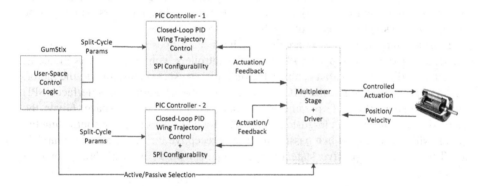

Fig. 5. Wing-level Control Multiplexing Architecture

In brief, the first challenge is solved by employing a proportional, derivative, and integral (PID) control, which has shown to effectively perform actual wing trajectory tracking, using the encoder , fitted to the BLDC motor, to detect the crank position and velocity [7]. The PID control tracks the actual wing trajectory against and reference trajectory that is stored in time indexed position/velocity tables, called commutation tables. These commutation tables define the shape of the wing motion and are designed to be configurable between the wing beats. Figure 4 provides an overview of the wing-level control actuation module for fixed split-cycle parameters, which are wing beat rate, downstroke and upstroke periods and the commutation tables. As seen in figure 4, the encoder fitted to the motor, provides pulses that indicate position of the motor shaft. These encoder pulses are used as position measurements via a counter and the rate of encoder pulse transitions is used as the velocity. These measured position and velocity are used to generate deviations (cumulative errors) against the stored expected position and velocity values in lookup-tables as per the given timer index. Here, the timer period determines the wing beat rate, which can be configured at end of the wing beat and also based on the user configured stroke periods, the increment values for the index into the commutation tables are determined. These increments might differ from upstroke to downstroke appropriately to traverse the complete shape of the wing trajectory. Further, due to the periodic nature of the control system, wrapping would be required on the computed position and velocity errors. The generated errors are multiplied with the pre-computed gains for the given vehicle assembly and the subsequent PID signal is used to tune the duty-cycle of the pulse width modulation (PWM) generator that is used to energize the motor's coil. Based on sign of the error from PID control, the motor is commutated by choosing the appropriate motor coil, based on the current hall sensors reading. The above described closed-loop trajectory has been successfully implemented on a PIC-based microcontroller, which is equipped with PWM generation modules. The PIC software control logic used timer interrupts and level-change interrupts to read hall sensors and encoder signal aptly with in the wing beat period. Besides the PID and interrupt-driven sensor feedback logic, the motors are driven by driver circuitry, built by H-bridges, as described in [7]. The current closed-loop trajectory implemented on the PIC controller is tightly coupled due to the time critical nature of the PID control logic during high wingbeat control and the configurability of the split-cycle parameters would not be possible with a single PIC controlling the each motor of the wing.

Thus, a control multiplexing architecture has been proposed, as shown in figure 5, where each motor is equipped with two PIC controllers with similar control logic as described earlier along with a controllable multiplexing stage connecting to the motor. Besides the control logic, the PIC controllers use Serial Peripheral Interface (SPI) logic to receive split-cycle parameters commanded from the user space. Given the time-critical nature of the closed-loop control, only one PIC is actively controlling the motor, while other PIC is in a passive mode and receptive for user space configurability. The active and passive state transition among two PICs is explained in next section.

3.2   User-Level Control Module

The primary focus of this module is to provide discrete (manual) and continuous (automated) control of vehicle's motion for different experiments. Thus, this module should at very least provide a user with the ability to program the vehicle's high-level controller as well as record experimental data. A small and lightweight computer-on-module named Gumstix [8] that can run a Linux operating system has been chosen as a platform to implement this user-level control module. Further, for the purpose of experimentation, the vehicle must remain free of unnecessary obstructions such as electrical cabling. A viable option is a wireless communication interface. The chosen Gumstix module provides built in wifi that allows a user to remotely log into the Linux system and transfer code that would likely have been written on another Linux system. The intent of use of this module in the design is to provide accessibility and

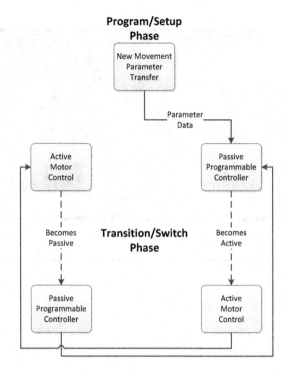

Fig. 6. Flow chart of Passive/Active Program and Transition

modularity, towards the aim for providing an intuitive interface for integration into a higher research framework. The implemented module on Gumstix, provides a user with the option to program the device on-the-fly which allows for ease of use and simplicity when performing and recording experiments.

This proposed Gumstix-based User space control module, would be useful for a user program that is designed to control vehicle movements that require wingbeat patterns that change multiple times. The controller can actively communicate the split-cycle parameters for each wing of the vehicle independently and intuitively to the low-level controllers, using the multiplexing architecture shown in figure 5. The Gumstix platform has been built with TI OMAP-3730 application processor [8], which contains two accessible SPI busses. Vehicle control parameters are transferred to the wing-level from the user-level controllers via SPI. Using the dual stage active-passive low-level controller, new parameters are transferred to the low-level controller without interrupting the actual wingbeat control. The low-level active PIC is actively controlling a flapping wing while a passive PIC is available for programming with new parameters. The high-level controller takes advantage of this fact, as shown in figure 5, to select the passive PIC and to transfer new movement parameters that will dictate the next wingbeat pattern. Once a user program is satisfied that new parameters have been uploaded successfully, the high-level controller transfers active control of a flapping wing from the active PIC to the passive PIC. The previous

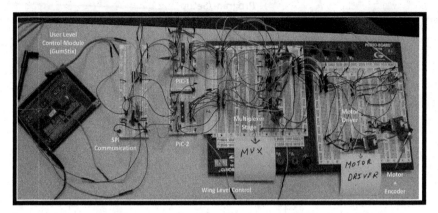

Fig. 7. Bread Boarded Implementation of the Proposed Adaptive Control System

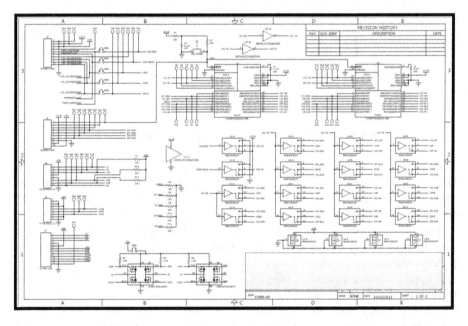

Fig. 8. Schematic design of the proposed control system

active PIC now becomes the passive PIC and the previous passive PIC becomes the active PIC. The high-level controller can now repeat the process with a different wingbeat pattern. The cycle nature of the implemented multiplexing logic is shown as a flow in figure 6. The Gumstix platform is programmable with various different types of Linux systems. However, the Yocto project [8] provides a set of tools which allow users to design their own highly configurable Linux distributions. The Yocto project allows users to overwrite and/or multiplex pins from the module processor for different uses.

Table 1. Electronic Components used in building Wing-Level Control Module and User-Level Control Module. (The above table is missing Gumstix employed in the User-Level Control Module)

Device	Package	Description
52745-1290	SMT	FFC/FPC-to-Board
C-EUC0201	C0201	CAPACITOR-0.1uF
MAX333ACWP	SO20L	ANALOG SWITCH
74LS04D	SO14	Hex Invertor
WM1389CT-ND	MOLEX-503480-1000	Connector
OR1037CT-ND	XF2W-0815-1A	Connector
ERJ-1GEF1002C	SMT	RESISTOR - 10K
ERJ-1GEF1501C	SMT	RESISTOR - 1.5K
ERJ-1GEF4700C	SMT	RESISTOR - 470 Ohm
B3U-3000P	SW1020CT-ND	Manul Switch
PIC24FJ64GA102	QFN28-ML_6X6MM	PIC MCU
ZXMHC6A07N8DICT-ND	8-SOIC	H-Bridge
C1608X5R1A105K	SMT	Capacitor-1µF
C0603X5R0J105M	1µF	Capacitor
21020-0101	Flat Flexible Cable	MCU Programming

The above User-space control logic has been implemented in C language using native libraries and an intuitive simple API has been developed as below for each wing:

SetWingBeat(float *frequency***)**
SetWingStrokePeriodRatio(double *StrokePeriodRatio***)**
SetWingStrokeBaseShape(double[] *CommutationTable***)**

Using the above API calls the users can develop more automated vehicle motion analyzing experiments as well validate the existing vehicle controllers impeccably. Further, the extensible nature of the Gumstix, provides a possibility of seamless integration with other experiment automation frameworks.

3.3   Implementation

The above described Wing-level and User-level modules have been implemented on respective platforms, towards an aim of creating a stand-alone control module on a compact single Printed Circuit Board (PCB). At the time of writing this paper, the whole proposed control system has been implemented and tested on a bread boarded implementation shown in figure 7. The implementation was successfully end-to-end tested thru remote login capabilities of the User level control module. Currently, a compact PCB design is being worked on based on the schematic shown in figure 8. Further, the table 1 provides the list of the electronic parts employed to implement the proposed control system.

4   Conclusion

In this paper, an improved and general purpose control system has been proposed and scrutinized that the proposed control system would readily address two issues, of shape configurability and intuitive user control communication, which existed in previous control system. Further, the proposed design has been discussed in detail and implementation details have been provided. The functionality of the implemented control system has been verified and it was discussed that the system can be employed to validate the vehicle controllers in a manual mode and also can be integrated into a higher-level research framework in an automated mode.

Further, the benefit of using Gumstix as the platform, to implement User-control module would have other benefits of designing various experiments that require varied external sensor integration and real-time in-flight data analysis. An example of such an experimental usage of this design has already been under development at author's facility, where an array of external cameras signals would be used to track the vehicle in real-time to develop complex vehicle maneuvering controllers. The experiment has been designed to relay camera tracking information thru wireless medium to the onboard Gumstix module, where authors can develop sensor fusion algorithms to learn vehicle controllers in an automated fashion. Additionally, the presented improved control system has been envisioned to effectively control next generation flapping wing micro air vehicles like one mentioned in [9]. Finally, the control system can be employed to validate the Evolvable and Adaptive Hardware based controllers mentioned in [10]-[11].

Acknowledgements. This material is based upon work supported by the National Science Foundation under Grant Numbers CNS-1239196, CNS-1239171, and CNS-1239229.

References

1. Doman, D.B., Oppenheimer, M.W., Sigthorsson, D.O.: Wingbeat Shape Modulation for Flapping-Wing Micro-Air Vehicle Control during Hover. Journal of Guidance, Control and Dynamics 33(3), 724–739 (2010)
2. Anderson, M.L.: Design and Control of Flapping Wing Micro Air Vehicles. Ph.D. thesis, Air Force Institute of Technology (2011)
3. Finio, B.M., Shang, J.K., Wood, R.J.: Body Torque Modulation for a Microrobotic Fly. In: 2010 IEEE International Conference on Robotics and Automation (2010)
4. Oppenheimer, M.W., Doman, D.B., Sigthorsson, D.O.: Dynamics and Control of a Biomimetic Vehicle Using Biased Wingbeat Forcing Functions. Journal of Guidance, Control and Dynamics 34(1), 204–217 (2010)

5. Gallagher, J.C., Oppenheimer, M.: An improved evolvable oscillator for all flight mode control of an insect-scale flapping-wing micro air vehicle. In: Proceedings of the 2010 Congress on Evolutionary Computation. IEEE Press (2011)
6. Boddhu, S.K., Gallagher, J.C.: Evolving neuromorphic flight control for a flapping-wing mechanical insect. International Journal of Intelligent Computing and Cybernetics 3(1), 94–116 (2010)
7. Oppenheimer, M.W., Sigthorsson, D.O., Weintraub, I.E., Doman, D.B., Perseghetti, B.: Wing Velocity Control System for Testing Body Motion Control Methods for Flapping Wing MAVs. In: 51st AIAA Aerospace Sciences Meeting including the New Horizons Forum and Aerospace Exposition (2013)
8. Gumstix development tools, https://www.gumstix.com/
9. Perseghetti, B.M., Roll, J.A., Gallagher, J.C.: Design Constraints of a Minimally Actuated Four Bar Linkage Flapping-Wing Micro Air Vehicle. In: For the Proceedings of the 2nd International Conference on Robot Intelligence Technology and Applications (2013)
10. Boddhu, S.K., Gallagher, J.C.: Evolved Neuromorphic Flight Control for a Flapping-Wing Mechanical Insect. In: The Proc. of 2008 Congress on Evolutionary Computation, Honk Gong. IEEE Press (2008)
11. Boddhu, S.K., Gallagher, J.C.: Evolving Non-Autonomous Neuromorphic Flight Control for a Flapping-Wing Mechanical Insect. In: For the Proceedings of 2009, IEEE Workshop on Evolvable and Adaptive Hardware (2009)

Adding Adaptable Stiffness Joints to CPG-Based Dynamic Bipedal Walking Generates Human-Like Gaits

Yan Huang, Yue Gao, Baojun Chen, Qining Wang*, and Long Wang

Intelligent Control Laboratory, College of Engineering,
Peking University, Beijing 100871, China
qiningwang@pku.edu.cn

Abstract. In this paper, we propose a seven-link passivity-based dynamic walking model, in order to further understand the principles of real human walking and provide guidance in building bipedal robots. The model includes an upper body, two thighs, two shanks, flat feet and compliant joints. A bio-inspired central pattern generator (CPG)-based control method is applied to the proposed model. In addition, we add adaptable joint stiffness to the motion control. To validate the effectiveness of the proposed bipedal walking model, we carried out simulations and human walking experiments. Experimental results indicate that human-like walking gaits with different speeds and walking pattern transitions can be realized in the proposed locomotor system.

Keywords: Passive dynamic walking, adaptable stiffness joints, central pattern generators, walking speed control, walking pattern transition.

1 Introduction

Stable bipedal walking is one of the most important components of humanoid robot design, which can help us better understand human natural walking. Compared with bipedal walking based on trajectory-control methods [1], which are commonly applied in industrial robots, passive dynamic walking pioneered by McGeer [2] shows more natural gaits and more efficient motions [3]. In order to understand motion characteristics of passive walkers with more natural anthropomorphic features, researchers have added the upper body [4], knee joints [5], feet with different shapes [6, 7] and compliant ankle joints [8–10]. Although passivity-based walkers can achieve higher energetic efficiency, they have limitations of practical uses [11].

Variable joint stiffness attracted increasing attention in passivity-based walking community, to improve the motion adaptability and versatility. In human walking gaits, joint stiffness plays as an energy-conserving mechanism [12]. The elastic elements of muscles absorb and release mechanical energy alternatively in each step and the adaptivity and efficiency can be improved [9]. Several studies analyzed the effects of the ankle stiffness on motion characteristics of flat-foot dynamic walkers [9, 13]. Owaki *et al.* added hip torsional spring and leg springs to a two-link passive bipedal model and obtain multiple gaits with different stiffness [14]. Our previous studies indicated the

---

* Corresponding author.

J.-H. Kim et al. (eds.), *Robot Intelligence Technology and Applications 2*, 569
Advances in Intelligent Systems and Computing 274,
DOI: 10.1007/978-3-319-05582-4_49, © Springer International Publishing Switzerland 2014

important effects of ankle compliance on gait selection [10] and realized speed and step length control of bipedal walkers with adaptable joint stiffness [15]. However, few efforts have been made in systematic control methods in variable-joint-stiffness bipeds and the comparison between the control performance and human natural gaits.

A variety of central pattern generator (CPG) models have been designed and applied to locomotion control of passivity-based bipedal walkers [16–18]. Several studies have reported evidences of the existence of CPG in vertebrates [19, 20]. Previous studies indicated that the biologically inspired CPG-based control methods could enhance robustness against perturbations, improve efficiency, and modulate complex motion behaviors by receiving only a few input signals [21].

In this paper, we propose a human-like passivity-based dynamic walking model with adaptable joint stiffness and CPGs for motion control. The model consists of an upper body, two thighs, two shanks and flat feet. The CPG models reduce the control parameters and simplify the control structure in a natural way. In most existing studies on CPG-controlled bipedal walking, the higher center generates only one driven signal for adjusting the basic rhythm of joint torques and controlling walking behavior [16–18]. The novelty of this study is introducing real-time stiffness control to the CPG-based control system, which shows great resemblance with natural human walking. Through simulations, the locomotion with different motion cycles is studied and the effects of control parameters on walking performance are investigated. In addition, we measure the kinematic data of human normal walking and gait transitions by the 3D human motion capture system. Comparison of the proposed bipedal walking with human gaits shows that our locomotor system can produce human-like gaits and gait transitions. The proposed system may help understand the principles of human normal walking and provide guidance in building efficient and practical bipedal robots.

The rest of this paper is organized as follows. In Section 2, we describe the mechanical and control systems of the seven-link bipedal walking model and depict the protocol of human walking experiments. Section 3 shows the results from both simulation model and human experiments. We conclude in Section 4.

2   Method

2.1   Bipedal Walking Model

We developed a seven-link bipedal walking model, which consists of an upper body, two thighs, two shanks and two flat feet. Each leg includes a hip joint, a knee joint and an ankle joint. The joint stiffness is modeled as a torsional spring at each joint. Thus the control parameters of the mechanical system are all the equilibrium positions and spring constants, which determine the torque and stiffness of each joint. The proposed bipedal walker travels forward on level ground. Fig. 1 shows the structure and the related variables of the bipedal model. A kinematic coupling has been added at the hip to keep the upper body midway between the two thighs. The knee joint is released in push-off and locked when the shank swings to the direction same to the thigh.

Different from a lot of existing passivity-based bipedal walking models with round feet or point feet, flat-foot walkers have the ability of standing stably and more complex walking phase sequences [10]. When the flat foot strikes the ground, there are two

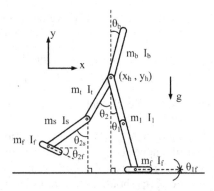

Fig. 1. The passivity-based dynamic bipedal walking model with compliant joints. To simplify the motion, we make the assumptions: 1) the upper body, the legs and the feet are modeled as rigid sticks without flexible deformation; 2) the mass of each part is averagely distributed among the corresponding stick; 3) the friction between the walker and the ground is sufficient. Thus the flat feet do not deform or slip; 4) all strikes are modeled as instantaneous, fully inelastic impacts where no slip and no bounce occurs.

impulses, "heel-strike" and "foot-strike", representing the initial impact of the heel and the following impact as the whole foot contacts the ground, respectively.

The Equation of Motion (EoM) of the proposed bipedal walking model can be obtained by using Lagrange's equation of the first kind. We suppose that the x-axis is along the forward direction while the y-axis is vertical to the ground upwards, as indicated in Fig. 1. The configuration of the walker is defined by the position of the hip joint and the angle of each stick. Thus the posture of the model can be arranged in a generalized vector $\mathbf{q} = (x_h, y_h, \theta_1, \theta_2, \theta_b, \theta_{2s}, \theta_{1f}, \theta_{2f})'$. The superscript $'$ means the transposed matrix (the same in the following paragraphs). The positive directions of all the angles are counter-clockwise. Note that the dimension of the generalized vector in different phases may be different. When the knee joint of the swing leg is locked, the freedom of the shank is reduced and the angle θ_{2s} is not included in the generalized coordinates. Consequently, the dimensions of the mass matrix and the generalized active force are also reduced in some phases.

The model can be defined by the Euclidean coordinates $\mathbf{r}$, which can be described by the x-coordinate and y-coordinate of the center of mass (CoM) of each stick and the corresponding directions. The walker can also be described by the generalized coordinates $\mathbf{q}$ as mentioned before:

$$\mathbf{q} = (x_h, y_h, \theta_1, \theta_2, \theta_b, \theta_{2s}, \theta_{1f}, \theta_{2f})' \tag{1}$$

We defined matrix J as follows:

$$J = d\mathbf{r}/d\mathbf{q} \tag{2}$$

Thus the Jacobian matrix J transfers the generalized velocity $\dot{\mathbf{q}}$ into the velocity of the euclidean coordinates $\dot{\mathbf{x}}$. The mass matrix in rectangular coordinate $\mathbf{r}$ is defined as:

$$M = diag(m_l, m_l, I_l, m_t, m_t, I_t, m_b, m_b, I_b,$$
$$m_s, m_s, I_s, m_f, m_f, I_f, m_f, m_f, I_f) \tag{3}$$

where m-components are the masses of each stick, while I-components are the moments of inertia, as shown in Fig. 1(a).

The constraint function is marked as $\xi(\mathbf{q})$, which is used to maintain foot contact with ground, the direction of the upper body and locking at the stretched knee joint. Note that $\xi(\mathbf{q})$ in different walking phases may be different since the constraint conditions change. Each component of $\xi(\mathbf{q})$ should keep zero to satisfy the constraint conditions. Each element of the constraint function corresponds to the generalized constrain force.

Then we can obtain the EoM as follows:

$$M_q \ddot{\mathbf{q}} = \mathbf{F_q} + \boldsymbol{\Phi}' \mathbf{F_c} \tag{4}$$

$$\xi(\mathbf{q}) = \mathbf{0} \tag{5}$$

where $\boldsymbol{\Phi} = \frac{\partial \xi}{\partial \mathbf{q}}$. $\mathbf{F_c}$ is the constraint force vector. M_q is the mass matrix in the generalized coordinates:

$$M_q = J' M J \tag{6}$$

$\mathbf{F_q}$ is the active external force in the generalized coordinates:

$$\mathbf{F_q} = J' \mathbf{F} - J' M \frac{\partial J}{\partial \mathbf{q}} \dot{\mathbf{q}} \dot{\mathbf{q}} \tag{7}$$

where $\mathbf{F}$ is the active external force vector in the Euclidean coordinates. For the walking model in this paper, $\mathbf{F}$ includes gravitation, the damping torques, and the joint torques generated by the torsional springs.

Equation (5) can be transformed to the followed equation:

$$\boldsymbol{\Phi} \ddot{\mathbf{q}} = -\frac{\partial(\boldsymbol{\Phi}\dot{\mathbf{q}})}{\partial \mathbf{q}} \dot{\mathbf{q}} \tag{8}$$

Then the EoM in matrix format can be obtained from Equation (4) and Equation (8):

$$\begin{bmatrix} M_q & -\boldsymbol{\Phi}' \\ \boldsymbol{\Phi} & 0 \end{bmatrix} \begin{bmatrix} \ddot{\mathbf{q}} \\ \mathbf{F_c} \end{bmatrix} = \begin{bmatrix} \mathbf{F_q} \\ -\frac{\partial(\boldsymbol{\Phi}\dot{\mathbf{q}})}{\partial \mathbf{q}} \dot{\mathbf{q}} \end{bmatrix} \tag{9}$$

The equation of the strike moment can be obtained by integration of Equation (4):

$$M_q \dot{\mathbf{q}}^+ = M_q \dot{\mathbf{q}}^- + \boldsymbol{\Phi}' \Lambda_c \tag{10}$$

where $\dot{\mathbf{q}}^+$ and $\dot{\mathbf{q}}^-$ are the generalized velocities just after and just before the strike, respectively. Here, Λ_c is the impulse acted on the walker which is defined as follows:

$$\Lambda_c = \lim_{t^- \to t^+} \int_{t^-}^{t^+} \mathbf{F_c} dt \tag{11}$$

Since the strike is modeled as a fully inelastic impact, the walker satisfies the constraint function $\xi(\mathbf{q})$. Thus the motion is constrained by the followed equation after the strike:

$$\frac{\partial \xi}{\partial \mathbf{q}} \dot{\mathbf{q}}^+ = \mathbf{0} \tag{12}$$

Then the equation of strike in matrix format can be derived from Equation (10) and Equation (12):

$$\begin{bmatrix} M_q & -\boldsymbol{\Phi}' \\ \boldsymbol{\Phi} & 0 \end{bmatrix} \begin{bmatrix} \dot{\mathbf{q}}^+ \\ \Lambda_c \end{bmatrix} = \begin{bmatrix} M_q \dot{\mathbf{q}}^- \\ \mathbf{0} \end{bmatrix} \tag{13}$$

2.2   CPG-Based Control Method

Central Pattern Generators (CPGs) are seemed as neural circuits which can produce coordinated patterns of high-dimensional rhythmic output signals while receiving only simple, low-dimensional, input signals [21]. Thus bio-inspired CPG-based control methods are very suitable for controlling bipedal robots with adaptable stiffness and walking pattern transitions. In this paper, we introduce real-time stiffness control to CPG. The CPG model controls not only the joint torque but also the joint stiffness, which is different from most existing studies on CPG-controlled bipedal walking [16–18]. Thus the natural dynamics of our model can be controlled by adjusting joint stiffness.

The input of the control system in this study is the desired walking pattern while the outputs (i.e. the commands sent to musculo-skeletal system) are joint torque and joint stiffness. The control system receives feedbacks from the motion states of the walker

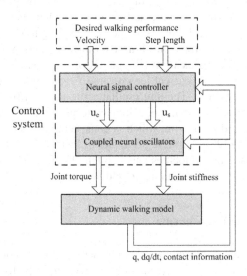

Fig. 2. The diagram of the control scheme. The control system receives the expected walking performance and sends commands as joint torque and joint stiffness to the mechanical system. The sensory feedback is from the motion states of the walker to both the neural signal controller and the coupled neural oscillator.

and the interaction between the mechanical system and environment. The architecture is shown in Fig. 2. The performance of different walking patterns is evaluated by velocity and step frequency, which is equivalent to velocity and step length, since frequency is the ratio of velocity to step length. Different from most previous studies which controlling only the speed, the CPG model in this paper is expected to control both velocity and step length simultaneously. Thus the walking behavior can be modulated over a wide range by controlling natural dynamics.

The control system consists of the neural signal controller and the coupled neural oscillators. The neural signal controller generates appropriate signals u_e and u_s according to the desired walking pattern and the actual walking performance. The signals, for setting the level of activity of the neural coupled oscillators, can be compared to the stimulation from the brain activating the spinal cord of many vertebrate animals [22]. The two parameters u_e and u_s are responsible for adjusting the equilibrium positions and the stiffness of each joint, respectively. The coupled neural oscillators receive input signals u_e and u_s and output rhythmic patterns of joint torques and joint stiffness, to generate periodic stable gaits. The control system contains twelve unit oscillators, associated with walking phase-dependent sensory feedback from the motion states (i.e. the generalized coordinates and velocities) and foot contact information. Each joint is controlled by two unit oscillators, producing the equilibrium position and stiffness respectively.

Inter-limb coordination of the two legs are established between the hip unit oscillators on the contralateral side. Inhibitory connection of equilibrium positions results in phase difference between hip angles and thus form periodic motion, while the stiffness of the two hip joints are positively correlated by the coupling. Intra-limb coordination makes the stiffness of ipsilateral joints increase or decrease proportionally. Derivative feedbacks of hip and ankle angles are added to the coupled neural oscillators for decreasing time delay effects and preventing the limb moving too fast to maintain stable walking. The unit oscillator for controlling equilibrium position of the knee joint of the swing leg receives feedback from the amount of foot clearance. The knee torque of the swing leg adapts to the current leg posture to avoid foot scuffing by as low energy consumption as possible. The unit oscillator for ankle stiffness of the stance leg receives sensory feedback from the ankle joint angle and angular velocity. The stiffness increases adaptively in dorsiflexion, which is consistent with the general tendency of human normal walking [23]. All these principles of feedback mentioned above are appropriate for different gaits, velocities and step lengths. Thus flexible walking pattern transitions can be realized by just tuning u_e and u_s.

2.3   Human Walking Experiments

For the validation of the proposed model, we observed human motion at constant and varied velocities and recorded the kinematic data. Five subjects (five males, 24.4 ± 3.0 years old, $1.7m \pm 0.07m$ height, $69.60kg \pm 3.97kg$ weight) were asked to walk uniformly at the natural speed, a smaller speed and a larger speed respectively. Then they performed walking pattern transitions from a slow gait to a fast gait in three manners as following:

1) Method 1: The subjects increase their walking speeds mainly by adjusting the step lengths, while the step frequency has only a small change;

2) Method 2: The speed transitions are realized in the self selected way;

3) Method 3: The subjects increase their speeds mainly by adjusting the step frequency, while the step length has only a small change.

We placed notice lines on the ground to help the subjects adjust the step lengths. The human motion data were obtained by Codamotion (Charnwood Dynamics Ltd.). The speed, step length and joint trajectories were represented by the average values and the standard deviations over all the subjects, as shown in the following section.

3   Results

In this section, we display both the simulation results and human motion results. The walking performance of the proposed model and human motion are compared.

3.1   Steady-State Walking

Different stable walking patterns are obtained by adjusting joint torques and joint stiffness of our model. Fig. 3 and Fig. 4 indicate the joint trajectories of cyclic motion with

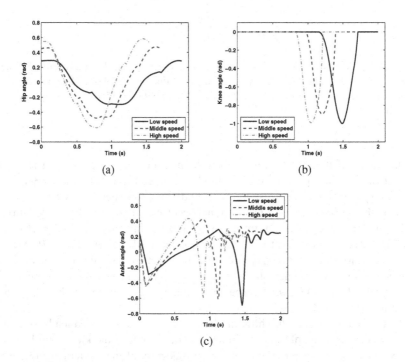

Fig. 3. Joint trajectories of different speeds of the proposed model. (a), (b) and (c) are hip angle, knee angle and ankle angle respectively. The blue solid line, the red dashed line and the green dot-dashed line represent the motions at $0.46m/s$, $0.83m/s$ and $1.15m/s$ respectively.

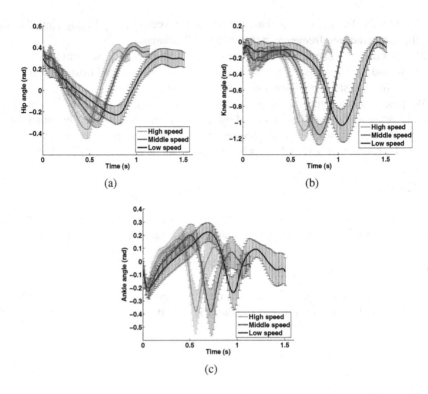

Fig. 4. Joint trajectories of different speeds of human walking. (a), (b) and (c) are hip angle, knee angle and ankle angle respectively. The blue solid line, the red dashed line and the green dot-dashed line represent the motions at $0.61 \pm 0.15m/s$, $0.98 \pm 0.15m/s$ and $1.40 \pm 0.29m/s$ respectively.

different speeds of the proposed model and human motion, respectively. The speeds of the selected motion patterns of the simulation model are $0.46m/s$, $0.83m/s$ and $1.15m/s$, respectively. The corresponding Froude numbers (defined as $Fr = V/\sqrt{gl}$, where V is the velocity, g is the gravitational acceleration and l is the leg length) are $0.16, 0.30$ and 0.41 respectively as the leg length is $0.8m$. In human motion experiments, the walking speeds of different walking patterns are $0.61m/s \pm 0.15m/s$, $0.98m/s \pm 0.15m/s$ and $1.40m/s \pm 0.29m/s$, respectively (represented as the mean value of different subjects $\pm$ the standard deviation). The corresponding Froude numbers are $0.21 \pm 0.05, 0.33 \pm 0.04$ and 0.47 ± 0.09, respectively, which are a little larger than the speeds of the simulation model.

Both the two types of results show similar tendencies. Larger speed leads to smaller step period and larger amplitude of hip angle. The flexion of the knee joint becomes earlier with increasing speed. The ankle angle trajectory of the model performs more obvious oscillation in swing phase than that of human motion, for the joint stiffness is modeled as a torsional spring in the proposed model. In general, the joint trajectories of our model are close to those of human motion, which demonstrates that our model

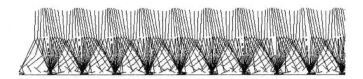

Fig. 5. The stickgram of stable walking at $0.83m/s$ of the bipedal walking model with adaptable stiffness joints and the CPG-based control method

can generate natural bipedal walking performance and reflect the motion characteristics over different speeds. Fig. 5 shows the walking trajectory of the proposed bipedal model at a speed of $0.83m/s$.

3.2   Walking Speed Transition

Since the proposed bipedal model has adaptable joint stiffness and controllable joint torques, real-time walking pattern transitions can be achieved. Adding adaptable joint stiffness to the CPG-based model makes it possible to control speed and step length (or step frequency) independently. Therefore, we show walking speed control with different step frequency behaviors.

In simulation, the speed is changed from $0.56m/s$ to $1.0m/s$ by adjusting control parameters on-line. The initial step frequency is $1.06Hz$ (thus the initial step length is $0.53m$). We applied three different methods to the speed transition. The target step frequencies are $1.12Hz$, $1.20Hz$ and $1.26Hz$ in method 1, method 2 and method 3, respectively. Therefore, the corresponding target step lengths are $0.89m$, $0.83m$ and $0.79m$ respectively. These three methods are accordance with the three pattern transition methods in human motion experiments mentioned in the previous section. The portions of step length variation and step frequency variation are different in different methods. Fig. 6(a) shows the speed transitions of the three methods of the simulation model. All the methods have acceptable control precision since the final speeds are close to the desired value. The rise time of speed variation in method 1 is shortest, which indicates that walking performance is more sensitive to step length variation than to step frequency variation. Adding joint stiffness control can improve the control accuracy. Fig. 6(c) represents the changes of step lengths of the bipedal model in the three methods. One can find that the step lengths of different methods achieve different steady-state values with almost the same speed change trend. Thus adding adaptable stiffness can realize more precise walking pattern control and obtain multiple speed transition manners. Fig. 7 shows the walking trajectory of speed transition in method 2.

The speed transitions in human locomotion also show similar trends. In different methods, the subjects achieve almost the same ultimate speeds while obviously different step lengths. When the change of step length plays a primary role, the final step length will reach a relatively large value (as shown in method 1 of Fig. 6(d)). Contrarily, if the speed transition is caused mainly by the change of step frequency, the final step length will stay at a low level and the subject will increase the speed mainly by increasing the step frequency. Comparison of hip angle trajectories of the proposed model and human

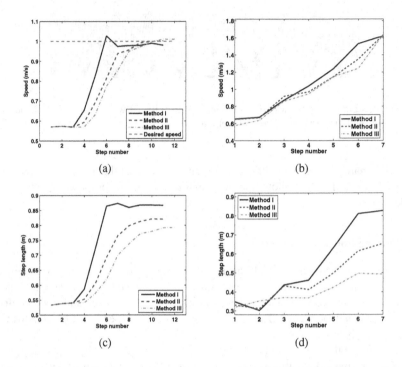

Fig. 6. Comparison of speed transitions of the proposed bipedal model and human walking. (a): speed variation of the model. The gray dashed line represents the desired speed. (b): speed variation of human walking. (c): step length variation of the model. (d): step length variation of human walking.

Fig. 7. The stickgram of speed transition of bipedal walking model with adaptable stiffness joints and the CPG-based control method. The initial speed is $0.56m/s$ and the ultimate speed is $1.0m/s$. The step length is changed from $0.53m$ to $0.83m$.

locomotion is illustrated in Fig. 8. The hip angle curve of method 3 has a small rise in amplitude and a large increment in frequency, while method 1 has the largest final amplitude, which corresponds to the largest final step length. Similar results can be observed in human locomotion. Comparison of the results from the simulation model and from human locomotion experiments indicates that the proposed model can explain different manners of speed transitions in human walking.

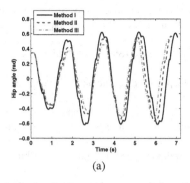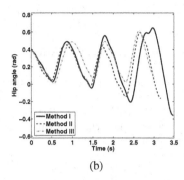

(a)   (b)

Fig. 8. Comparison of hip angle trajectories during speed transitions of the proposed bipedal model and human walking. (a): hip angle trajectories in three methods of the bipedal model. (b): Hip angle trajectories in three methods of human walking.

4   Conclusion

In this paper, we established a passivity-based dynamic bipedal walking model with an upper body, compliant knee and ankle joints, and flat feet. A bio-inspired CPG-based method is applied to the motion control. In addition, we add adaptable joint stiffness to the locomotor system. To validate the effectiveness and the natural performance of the model, we carried out simulation experiments and human motion experiments. Simulation results show that stable motion cycles with different walking speeds can be achieved in the proposed model. Comparison of the results from simulations and from human motion experiments indicates that human-like walking pattern transitions and multiple speed control methods can be realized by adding joint stiffness control. The model can reproduce natural bipedal locomotion, help better understand the principles of real human walking, and provide a new solution of building efficient bipedal robots with natural gaits.

To extend the study in this paper, we intend to improve the method to raise the control accuracy, and apply the idea of the proposed bipedal model to a physical prototype in the future.

Acknowledgements. This work has been funded by National Natural Science Foundation of China (No. 61005082, 61020106005) and the 985 Project of Peking University (No. 3J0865600).

References

1. Hirai, K., Hirose, M., Haikawa, Y., Takenaka, T.: The development of the Honda Humanoid robot. In: Proc. of the IEEE Int. Conf. Robotics and Automation, Leuven, Belgium, pp. 1321–1326 (1998)
2. McGeer, T.: Passive dynamic walking. Int. J. Robot. Res. 9, 68–82 (1990)
3. Collins, S., Ruina, A., Tedrake, R., Wisse, M.: Efficient bipedal robots based on passive-dynamic walkers. Science 307, 1082–1085 (2005)

4. Wisse, M., Hobbelen, D.G.E., Schwab, A.L.: Adding an upper body to passive dynamic walking robots by means of a bisecting hip mechanism. IEEE Trans. Robot. 23(1), 112–123 (2007)
5. Borzova, E., Hurmuzlu, Y.: Passively walking five-link robot. Automatica 40, 621–629 (2004)
6. Kwan, M., Hubbard, M.: Optimal foot shape for a passive dynamic biped. J. Theor. Biol. 248, 331–339 (2007)
7. Wang, Q., Huang, Y., Zhu, J., Wang, L., Lv, D.: Effects of foot shape on energetic efficiency and dynamic stability of passive dynamic biped with upper body. Int. J. Humanoid Robotics 7(2), 295–313 (2010)
8. Hobbelen, D.G.E., Wisse, M.: Controlling the walking speed in limit cycle walking. Int. J. Robot. Res. 27(9), 989–1005 (2008)
9. Wang, Q., Huang, Y., Wang, L.: Passive dynamic walking with flat feet and ankle compliance. Robotica 28, 413–425 (2010)
10. Huang, Y., Wang, Q., Chen, B., Xie, G., Wang, L.: Modeling and gait selection of passivity-based seven-link bipeds with dynamic series of walking phases. Robotica 30, 39–51 (2012)
11. Vanderborght, B., Van Ham, R., Verrelst, B., Van Damme, M., Lefeber, D.: Overview of the lucy project: dynamic stabilization of a biped powered by pneumatic artificial muscles. Adv. Robotics 22(10), 1027–1051 (2008)
12. Ishikawa, M., Komi, P.V., Grey, M.J., Lepola, V., Bruggemann, G.: Muscle-tendon interaction and elastic energy usage in human walking. J. Appl. Physiol. 99, 603–608 (2005)
13. Hobbelen, D.G.E., Wisse, M.: Ankle actuation for limit cycle walkers. Int. J. Robot. Res. 27(6), 709–735 (2008)
14. Owaki, D., Osuka, K., Ishiguro, A.: On the embodiment that enables passive dynamic bipedal running. In: Proc. of the IEEE Int. Conf. Robotics and Automation, Pasadena, CA, USA, pp. 341–346 (2008)
15. Huang, Y., Vanderborght, B., Van Ham, R., Wang, Q., Van Damme, M., Xie, G., Lefeber, D.: Step length and velocity control of a dynamic bipedal walking robot with adaptable compliant joints. IEEE-ASME Trans. Mechatron. 18, 598–611 (2013)
16. Taga, G., Yamaguchi, Y., Shimizu, H.: Self-organized control of bipedal locomotion by neural oscillators in unpredictable environment. Biol. Cybern. 65, 147–159 (1991)
17. Verdaasdonk, B.W., Koopman, H.F.J.M., van der Helm, F.C.T.: Energy efficient walking with central pattern generators: from passive dynamic walking to biologically inspired control. Biol. Cybern. 101, 49–61 (2009)
18. Owaki, D., Kano, T., Tero, A., Akiyama, M., Ishiguro, A.: Minimalist CPG model for inter- and intra-limb coordination in bipedal locomotion. In: Lee, S., Cho, H., Yoon, K.-J., Lee, J. (eds.) Intelligent Autonomous Systems 12. AISC, vol. 194, pp. 493–502. Springer, Heidelberg (2013)
19. Amemiya, M., Yamaguchi, T.: Fictive locomotion of the forelimb evoked by stimulation of the mesencephalic locomotor region in the decerebrate cat. Neurosci. Lett. 50, 91–96 (1984)
20. Cazalets, J.R., Borde, M., Clarac, F.: Localization and organization of the central pattern generator for hindlimb locomotion in newborn rat. J. Neurosci. 15, 4943–4951 (1995)
21. Ijspeert, A.J.: Central pattern generators for locomotion control in animals and robots: a review. Neural Netw. 21(4), 642–653 (2008)
22. Grillner, S., Georgopoulos, A.P., Jordan, L.M.: Selection and initiation of motor behavior. In: Stein, P.S.G., Grillner, S., Selverston, A., Stuart, D.G. (eds.) Neurons, Networks and Motor Behavior. MIT Press (1997)
23. Frigo, C., Crenna, P., Jensen, L.M.: Moment-angle relationship at lower limb joints during human walking at different velocities. J. Electromyogr. Kines. 6, 177–190 (1996)

Hand-Eye Calibration and Inverse Kinematics of Robot Arm Using Neural Network

Haiyan Wu, Walter Tizzano, Thomas Timm Andersen,
Nils Axel Andersen, and Ole Ravn

Automation and Control, Department of Electrical Engineering,
Technical University of Denmark, Elektrovej, 2800, Kgs. Lyngby
{hwua,ttan,naa,or}@elektro.dtu.dk, waltertizzano@gmail.com

Abstract. Traditional technologies for solving hand-eye calibration and inverse kinematics are cumbersome and time consuming due to the high nonlinearity in the models. An alternative to the traditional approaches is the artificial neural network inspired by the remarkable abilities of the animals in different tasks. This paper describes the theory and implementation of neural networks for hand-eye calibration and inverse kinematics of a six degrees of freedom robot arm equipped with a stereo vision system. The feedforward neural network and the network training with error propagation algorithm are applied. The proposed approaches are validated in experiments. The results indicate that the hand-eye calibration with simple neural network outperforms the conventional method. Meanwhile, the neural network exhibits a promising performance in solving inverse kinematics.

Keywords: Neural Network, Calibration, Inverse Kinematics, Robot Arm.

1 Introduction

Applying robot arm for dexterous tasks, e.g. object gasping and hanging, challenges both machine vision and control fields. These tasks often requires high-speed sensor feedback as well as fast and precise control of robot arm. Thanks to the advances of camera technology and image processing methodology, visual information has been widely used in various tasks, e.g. for object recognition, 3D localization, scene reconstruction, motion detection and so on [1–3]. In order to utilize the visual feedback in the control loop, it often requires a calibration between the vision system and the robot arm, e.g. for a 'eye to hand' setup, see Fig.1. The hand-eye calibration gives a mapping of 2D image coordinates $[u,v]$ to 3D Cartesian coordinates $[x,y,z]$. Moreover, the control tasks with robot arm are usually performed in Cartesian space with visual feedback providing 3D object information. However, the robot arm is manipulated in joint space. Therefore, it involves a mapping from Cartesian space $[x,y,z]$ to joint space $[q_1,q_2,\ldots,q_n]$. This paper focuses on these two fundamental issues

J.-H. Kim et al. (eds.), *Robot Intelligence Technology and Applications 2,*
Advances in Intelligent Systems and Computing 274,
DOI: 10.1007/978-3-319-05582-4_50, © Springer International Publishing Switzerland 2014

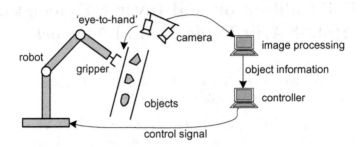

Fig. 1. Eye-to-hand setup

in robot arm control, namely the hand-eye calibration issue and the inverse kinematics issue, which will be studied in a 6-DOF (degrees of freedom) robot arm control system equipped with a stereo vision system.

The hand-eye calibration problem has fascinated many researchers over the last 30 years. It is traditionally dealt with through triangulation methods [4–6]. However, the triangulation method usually requires camera calibration, calibration from camera frame to gripper frame or robot base frame, and stereo triangulation (if a stereo vision system is applied). Therefore, the calibration procedure is cumbersome and time consuming. Besides, the triangulation involves camera intrinsic and extrinsic parameters. As a result, the performance of the traditional calibration method is susceptible to these parameters. Regarding the inverse kinematics problem, there are several techniques to simplify it and they can be in general grouped into two categories: analytical solution and geometrical solution [7]. For manipulators with higher DOF or with non-spherical wrist, the inverse kinematics problem becomes complex due to the high non-linearities in the kinematics model, and thus, it is difficult to find a closed-form solution. Inspired by animals' noteworthy abilities in a very wide range of different tasks with their remarkable analogous in biology, the artificial counterpart-neural network is used in robotics for several different purposes [9–14]. In [9] neural network estimators are trained offline for non-linear state estimation assuming non-Gaussian distributions of the states and the disturbances; neural networks can also be used to solve the camera calibration and 3D reconstruction problems [10,11]; the fast inverse kinematics of a 3-DOF planar redundant robot is achieved through neural network in [13]; in order to overcome the problems of singularities and uncertainties in arm configurations, artificial neural network is applied for the motion control of a 6-DOF robot manipulator in [14].

The benefits of neural network with regard to low computational cost and high efficiency inspires neural network based solutions for hand-eye calibration and inverse kinematics of a 6-DOF robot arm presented in this paper. The main purpose of this work is to implement the neural network for a real robot arm equipped with a stereo vision system. Different from some existing works, the robot arm used in this work has an offset wrist which poses difficulties for solving inverse kinematics in the traditional analytical or geometrical way. Data samples from real robot arm is utilized for performance analysis. Furthermore, this work is implemented as an initial step towards a realtime vision-guided robotic manipulation system based on neural network.

The remainder of this paper is organized as follows: The overall platform consisting of a robot arm and a stereo vision system is described in Section 2. In Section 3, the feedforward neural network is briefly introduced. Then, the training of neural networks for calibration and inverse kinematics is presented. In Section 4, the experimental validation and performance comparison are discussed. Finally, conclusions are given in Section 5.

2   System Overview

As mentioned in the previous section, in the present work the neural network is applied for a 6-DOF robot arm ('hand') with a stereo vision system ('eye'). In this section, these two hardware components will be introduced.

2.1   6-DOF Robot Arm

Universal robot UR5 is a 6-DOF robot arm with relatively lightweight (18.4 kg), see Fig. 2. It consists of six revolute joints which allows a sphere workspace with a diameter of approximately 170 cm. The Movements close to its boundary should be avoided considering the singularity of the arm. The arm is equipped with a graphic interface PolyScope, which allows users to move the robot in a user-friendly environment through a touch screen. In this work, the connection to the robot controller is realized at script level using TCP/IP socket. Once the

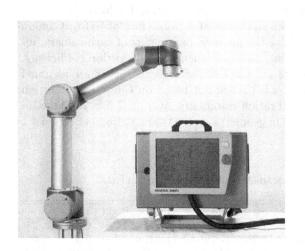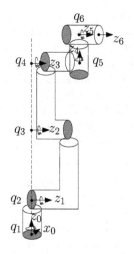

Fig. 2. UR5 robot and its joint coordinate system. *Left*: UR5 robot and PolyScope GUI [15]; *Right*: joint coordinate system.

Table 1. Denavit-Hartenberg parameters of UR5

Joint	a_i [m]	α_i [rad]	d_i [m]	θ_i [rad]
1	0	$pi/2$	0.089	q_1
2	-0.425	0	0	q_2
3	-0.392	0	0	q_3
4	0	$pi/2$	0.109	q_4
5	0	$-pi/2$	0.095	q_5
6	0	0	0.082	q_6

TCP/IP connection is established, the robot broadcasts data packets with a rate of 125 Hz.

The UR5 robot arm has a non-spherical wrist, as shown in Fig. 2. The D-H (Denavit-Hartenberg) parameters of the robot arm is given in Table 1. With the D-H parameters the forward kinematics can be calculated through the equation given below

$$T_6^0 = T_1^0 T_2^1 T_3^2 T_4^3 T_5^4 T_6^5, \tag{1}$$

where T_j^i denotes the homogeneous transformation from frame i to frame j, and

$$T_i^{i-1} = \begin{bmatrix} \cos(\theta_i) & -\sin(\theta_i)\cos(\alpha_i) & \sin(\theta_i)\sin(\alpha_i) & a_i\cos(\theta_i) \\ \sin(\theta_i) & \cos(\theta_i)\cos(\alpha_i) & -\cos(\theta_i)\sin(\alpha_i) & a_i\sin(\theta_i) \\ 0 & \sin(\alpha_i) & \cos(\alpha_i) & d_i \\ 0 & 0 & 0 & 1 \end{bmatrix}, \quad i = 1, 2, \ldots, 6.$$

2.2 Stereo Vision System

The stereo vision system consists of two Guppy Cameras F-036. Each camera has a resolution of 752×480 pixels and runs at a frame rate of 64 fps. Camera calibration is required to determine the intrinsic parameters of each camera, including focal length, principle point, skew coefficients and distortion coefficients. In this work the skew coefficients are not calculated as the pixels are assumed to be rectangular. The calibration is carried out based on Camera Calibration Toolbox in Matlab [16]. The calibration results are given in Table 2. These intrinsic parameters will be applied in geometric stereo triangulation in Section 4.1 for performance comparison.

Table 2. Intrinsic parameters of left and right cameras

Parameter	Left camera	Right camera
Focal length [pixel]	[683.89±2.49, 685.71±2.51]	[674.03±2.44, 676.18±2.55]
Principle point [pixel]	[307.76±3.26, 236.63±3.08]	[327.05±3.78, 247.40±2.88]
Distortion coef.	[-0.24±0.010, 0.222±0.046]	[-0.210±0.008, 0.168±0.034]

3   Neural Network for a 6-DoF Robot Arm

3.1   Feedforward Neural Network

According to the different ways of arranging neurons in a network, there are several different architectures of neural network, e.g. *feedforward neural network, radial basis function network, recurrent neural network,* etc. [8,17]. In this section, the feedforward neural network is introduced.

In feedforward neural network all the neurons are divided into layers, as shown in Fig.3. The neurons in one layer are not connected with each other. Moreover, each neuron in a layer receives inputs from all the neurons of the previous layer and has an output connected to all the neurons of the following layer. The first layer is called *input layer* which receives input from external environment, while the last layer is called *output layer* which transmits the results of the network. The layers in between are called *hidden layers* which are invisible to the external environment. The number of neurons in the input/output layer is related with the number of variables that are dealt with. The number of neurons in hidden layers is decided by the complexity of the problem. Generally speaking, a more complex problem requires more hidden layers and neurons. In most cases, one hidden layer will be sufficient [18].

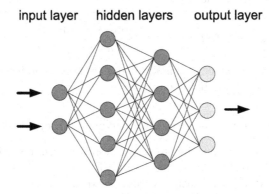

Fig. 3. An example of feedforward neural network

In order to determine the connection weights among the neurons, a network training strategy needs to be carefully selected. In this work, the well-known *error back propagation* algorithm is applied. The back propagation algorithm is a gradient descent procedure that finds the set of weights minimizing the output error [19].

3.2   Training of Neural Network for Hand-Eye Calibration

In hand-eye calibration, the 2D information is used to retrieve the 3D information with regard to a reference frame. In this work, the aim of hand-eye calibration is to determine the 3D location of an object in the robot base frame, based on

the 2D coordinates in two cameras. In order to build the neural network, the
network training is required to set the weights of the connections between the
neurons. Therefore, a training sample is demanded which is acquired through
the approach described below:

1. First, attach the object to the end-effector of the robot arm;
2. Move the robot arm to random positions within its workspace;
3. Detect the coordinates of the object in image space;
4. Finally, save both the 3D information and 2D information in a file.

With the collected sample data the network is trained based on error back propagation algorithm. After the training, the neural network for hand-eye calibration
is built, which transforms 2D image coordinates $[u, v]$ to 3D Cartesian coordinates $[x, y, z]$. Then, the performance of the neural network is evaluated with
another set of samples. The *cumulative distribution function* (CDF) and *root
mean square error* (RMSE) are selected for results analysis. The CDF describes
the probability of a variable X being less than or equal to x

$$CDF_X(x) = P(X \leq x). \tag{2}$$

The RMSE is often used as a measure of the difference between the predicted
values and the observed values, defined as

$$RMSE = \sqrt{\frac{\sum_{i=1}^{n}(\text{obs}_i - \text{pre}_i)^2}{n}}, \tag{3}$$

where obs_i is observed value, pre_i is predicted value at time instant i, and n is
the number of observed/predicted data.

3.3   Training of Neural Network for Inverse Kinematics

Through hand-eye calibration, the object position is determined with regard to a
reference base frame, e.g. the robot base frame. For tasks such as object tracking
and grasping the inverse kinematics is required to convert the 3D Cartesian coordinates $[x, y, z]$ to joint coordinates $[q_1, q_2, \ldots, q_n]$. It has to be mentioned that
in this work the orientation of the object $[\alpha, \beta, \gamma]$ is also considered for inverse
kinematics. The orientation information can be obtained through several meth-
ods, e.g. through image feature matching algorithm utilized in [20] and particle
filter based on one position sensor and one IMU described in [21]. Therefore,
the neural network for inverse kinematics has six inputs $[x, y, z, \alpha, \beta, \gamma]$. For a
6-DOF robot-arm, the output of the network is the corresponding six joint values $[q_1, q_2, \ldots, q_6]$.

Similar to the neural network training in Section 3.2, a set of sample data is
required. It is obtained through the following approach:

1. Move the end-effector of the robot arm to random positions and orientations
   within its workspace;
2. Save the joint angles, end-effector positions and end-effector orientations.

The step 1 and step 2 are repeated several times in order to better approximate the uniform distribution of the sample data points. Furthermore, another sample of data is needed for performance evaluation. The CDF and RMSE defined by Eqs. (2)(3) are also applied for results analysis.

4   Experiments

In this part, experiments are carried out to evaluate the performances of the neural networks for hand-eye calibration and inverse kinematics. The experimental setup is shown in Fig. 4. A table-tennis ball is chosen as the object, which is mounted on the end-effector of the robot arm. Two Guppy Cameras F-036 are used to build the stereo vision system. Each camera works at 30 fps@640×480 pixels. Before stereo triangulation, image processing algorithm is applied to extract the ball in the 2D image space. First, frames from each camera are converted from RGB colour space to HSV colour space, see Fig. 5(a)(b). Then, the coordinate of the ball is determined based on the colour, contour and radius information. Fig. 5(c) shows an example of the image processing result. After the image processing, the ball positions in both images from left and right cameras are found and saved to a file for neural network training.

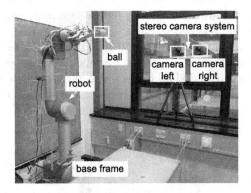

Fig. 4. Experimental setup

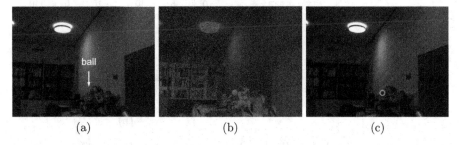

(a)   (b)   (c)

Fig. 5. Image processing for determining the 2D coordinate of a ball in the image space. (a): RGB image; (b) HSV image converted from RGB image; (c): ball detected (highlighted by green-circle) in image space.

4.1 Experiment I: Neural Network for Hand-Eye Calibration

In this experiment, neural network is trained for hand-eye calibration, as introduced in Section 3.2. The ball was randomly moved to 2000 different positions by the robot arm. Then, the 2D ball coordinates $[u_l, v_l]$, $[u_r, v_r]$ in the left and right cameras were saved to a file. Besides, the 3D ball position $[x, y, z]$ with regard to the robot base frame was also recorded. The network was trained using 80% of the sample, while the remaining 20% was utilized to test the performance.

A network with 4 input neurons, 3 output neurons and 2 hidden layers with 8 neurons each was heuristically selected. The results are in shown in Fig. 6, which gives the CDF (*left*) and histogram (*right*) of the RMSE and the mean absolute error $mean(|x|, |y|, |z|)$ along different axis. Here, the RMSE describes the absolute distance between the measured 3D ball position and the predicted 3D ball position. The results are also summarized in Table 3. It is noticed that the error is not homogeneously distributed along the different directions: the error on the x-direction is considerably bigger. The mean absolute error on the x-axis is $mean_{|x|} = 1.81$ mm. It is due to the fact that the x-direction of the base frame is roughly parallel to the optical axis of the cameras (see Fig. 4), which is more susceptible to the estimation uncertainty.

Moreover, the performance of the neural network based approach is compared with the one of the standard geometrical triangulation approach. As shown in table 3, the hand-eye calibration error is reduced dramatically by the proposed approach. For example, the RMSE of the proposed approach is 2.65 mm, while that of the geometrical approach is 19.10 mm. In addition, a deduction of approximate 80% of the mean absolute error on axis x, y, z is gained through the neural network approach. Based on the experimental results it is concluded that hand-eye calibration with the neural network can achieve a good performance and thus, can be implemented for different applications such as object tracking and grasping.

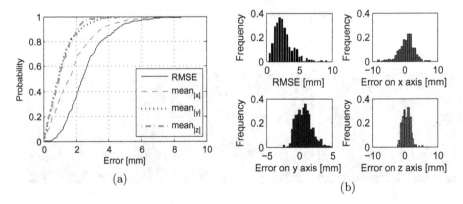

(a)

(b)

Fig. 6. Cumulative distribution (left) and histogram (right) of the calibration error based on the neural network approach. The neural network has 4 input neurons, 3 output neurons and 2 hidden layers with 8 neurons each.

Table 3. Results of the neural network for hand-eye calibration and against that of the geometrical triangulation

Parameter	Neural Network	Geometric Triang.
RMSE	2.65 (-86.1%)	19.10
95th percentile	5.29	32.42
99.5th percentile	8.44	41.41
Mean absolute error $[\|x\|, \|y\|, \|z\|]$	[1.81, 1.10, 1.07] $(-86.7\%, -88.0\%, -78.8\%)$	[13.56, 9.193, 5.039]
95th percentile $[\|x\|, \|y\|, \|z\|]$	[4.49, 3.01, 2.38]	[27.46, 21.74, 11.61]
99.5th percentile $[\|x\|, \|y\|, \|z\|]$	[8.02, 4.20, 4.56]	[38.70, 25.75, 16.09]

*All the values are expressed in [mm].

4.2 Experiment II: Neural Network for Inverse Kinematics

In this experiment, the neural network is trained for solving inverse kinematics as proposed in Section 3.3. The robot arm is moved randomly in its workspace. A sample of data is recorded, including the six joint values $[q_1, q_2, \ldots, q_6]$ and the position and the orientation of the end-effector $[x, y, z, \alpha, \beta, \gamma]$. Similarly to the experiment in Section 4.1, 80% of the sample data is used for training the neural network, while the remaining 20% is utilized for performance evaluation.

A network with 6 input neurons, 6 output neurons and 1 hidden layer with 12 neurons was trained to solve the inverse kinematics problem. The experimental results are shown in Fig. 7 and Table 4. The maximum mean error of the joints is 8.57×10^{-3} rad with a standard deviation of 10.95×10^{-3} rad. In general

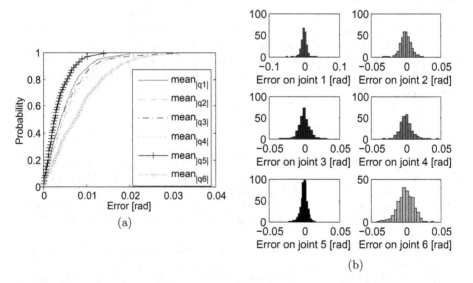

Fig. 7. Cumulative distribution (left) and histogram (right) of the error of inverse kinematics based on the neural network approach. The neural network has six input neurons, six output neurons and one hidden layer with 12 neurons.

Table 4. Results of the neural network for inverse kinematics

Joint	1	2	3	4	5	6
RMSE	5.11	5.73	5.50	6.16	3.42	8.57
Std. deviation	7.02	7.54	7.63	8.14	4.53	10.95
95th percentile	13.03	14.46	14.29	15.95	9.01	21.29
99.5th percentile	28.18	27.29	30.91	29.99	15.34	36.73

*All the values are expressed in $[10^{-3}\,\text{rad}]$.

the error is smaller than 0.04 rad. It is also noticed that the neural network gives comparatively more imprecise results on the last joint, which needs to be further investigated in the future.

5 Conclusion

In order to overcome the high complexity and high computational cost of traditional technologies, the neural networks are applied for hand-eye calibration and inverse kinematics in this paper. The feedforward neural network is selected and trained through the well-known error back propagation algorithm. The hand-eye calibration and inverse kinematics problems solved through neural network are investigated in a 6-DOF robot-arm control system equipped with a stereo vision system. The neural network is trained to transform the 2D image coordinates from the stereo vision system to the 3D Cartesian coordinates. Furthermore, the 3D coordinates combined with orientation information is converted to joint coordinates through inverse kinematics based also on neural network.

The experimental results indicate a significant improvement (a 80% deduction in the mean error) in the performance of the hand-eye calibration compared to the traditional method. A promising performance of inverse kinematics (mean error smaller than 9×10^{-3} rad) is achieved through neural network based approach. The proposed approaches can be directly extended to different hand-eye setups and robot-arms.

References

[1] Bay, H., Tuytelaars, T., Van Gool, L.: SURF: Speeded up robust features. In: Leonardis, A., Bischof, H., Pinz, A. (eds.) ECCV 2006, Part I. LNCS, vol. 3951, pp. 404–417. Springer, Heidelberg (2006)
[2] Wu, H., Zou, K., Zhang, T., Borst, A., Kühnlenz, K.: Insect-inspired high-speed motion vision system for robot control. Biological Cybernetics 106(8-9), 453–463 (2012)
[3] Smisek, J., Jancosek, M., Pajdla, T.: 3d with kinect. In: Consumer Depth Cameras for Computer Vision, pp. 3–25. Springer (2013)
[4] Horaud, R., Dornaika, F.: Hand-eye calibration. The International Journal of Robotics Research 14(3), 195–210 (1995)
[5] Daniilidis, K.: Hand-eye calibration using dual quaternions. The International Journal of Robotics Research 18(3), 286–298 (1999)

[6] Strobl, K.H., Hirzinger, G.: Optimal hand-eye calibration. In: The Proceedings of IEEE/RSJ International Conference on Intelligent Robots and Systems, pp. 4647–4653. IEEE (2006)

[7] Spong, M.W., Hutchinson, S., Vidyasagar, M.: Robot modeling and control. John Wiley & Sons, New York (2006)

[8] Norgaard, M.: Neural networks for modelling and control of dynamic systems: A practitioner's handbook. Springer (2000)

[9] Bayramoglu, E., Andersen, N.A., Ravn, O., Poulsen, N.K.: Pre-trained neural networks used for non-linear state estimation. In: The Proceedings of the 10th International Conference on Machine Learning and Applications and Workshops (ICMLA), vol. 1, pp. 304–310. IEEE (2011)

[10] Memon, Q., Khan, S.: Camera calibration and three-dimensional world reconstruction of stereo-vision using neural networks. International Journal of Systems Science 32(9), 1155–1159 (2001)

[11] Ahmed, M.T., Hemayed, E.E., Farag, A.A.: Neurocalibration: a neural network that can tell camera calibration parameters. In: The Proceedings of the Seventh IEEE International Conference on Computer Vision, pp. 463–468. IEEE (1999)

[12] Tejomurtula, S., Kak, S.: Inverse kinematics in robotics using neural networks. Information Sciences 116(2), 147–164 (1999)

[13] Mayorga, R.V., Sanongboon, P.: Inverse kinematics and geometrically bounded singularities prevention of redundant manipulators: An artificial neural network approach. Robotics and Autonomous Systems 53(3), 164–176 (2005)

[14] Hasan, A.T., Ismail, N., Hamouda, A.M.S., Aris, I., Marhaban, M.H., Al-Assadi, H.: Artificial neural network-based kinematics jacobian solution for serial manipulator passing through singular configurations. Advances in Engineering Software 41(2), 359–367 (2010)

[15] http://www.universal-robots.com/

[16] Bouguet, J.-Y.: Camera calibration toolbox for matlab (2004)

[17] Haykin, S.: Neural networks: a comprehensive foundation. Prentice Hall PTR (1994)

[18] Heaton, J.: Introduction to neural networks with Java. Heaton Research St. Louis 200 (2005)

[19] Rumelhart, D.E., Hinton, G.E., Williams, R.J.: Learning representations by back-propagating errors. Cognitive Modeling 1, 213 (2002)

[20] Wu, H., Lou, L., Chen, C.-C., Hirche, S., Kolja, K.: Cloud-based networked visual servo control. IEEE Transactions on Industrial Electronics 60(2), 554–566 (2012)

[21] Won, S.-H., Melek, W.W., Golnaraghi, F., et al.: A kalman/particle filter-based position and orientation estimation method using a position sensor/inertial measurement unit hybrid system. IEEE Transactions on Industrial Electronics 57(5), 1787–1798 (2010)

An Evolutionary Feature Selection Algorithm for Classification of Human Activities

Si-Jung Ryu and Jong-Hwan Kim

Department of Electrical Engineering, KAIST,
335 Gwahangno, Yuseong-gu, Daejeon 305-701, Republic of Korea
{sjryu,johkim}@rit.kaist.ac.kr

Abstract. This paper proposes an evolutionary feature selection algorithm to classify human activities. Feature selection is one of the key issues in machine learning, along with classification when some parts of features are not available or have redundant information. It enhances learning accuracy by selecting essential features and eliminating non-essential features. In the proposed algorithm, a feature selection algorithm integrated with an evolutionary algorithm (EA) is developed. We use the wrapper approach, which repeatedly calls the learning algorithm to evaluate the effectiveness of the selected features. Quantum-inspired evolutionary algorithm (QEA) is utilized as an evolutionary algorithm and multi-layer perceptron (MLP) is used as a classifier. The proposed algorithm is applied to classification of the human activities using smartphone sensors.

1 Introduction

Real life data have noise which makes subsequent machine learning processes difficult. The task of the classifier could be simplified by eliminating features that are seemed to be redundant for classification. The maintenance of only necessary features could reduce size of the dataset and subsequently allow more comprehensible analysis of the data.

In the feature selection problem, there are two big approaches. The first approach is reducing the dimensionality of the feature set, referred to *feature extraction* [1–4]. It is thought to create new features based on transformations or combinations of the original feature set. Popular dimension reduction algorithms include linear discriminant analysis (LDA), principal component analysis (PCA), locality preserving projection (LPP), neighborhood preserving embedding (NPE), graph optimization for dimensionality reduction with sparsity constraints (GODRSC).

The second approach is selecting essential features. To deal with this approach, *wrapper* and *filter* methods are commonly integrated to select essential features. Filter method selects a subset of features as a preprocessing step, and then learning algorithm is executed. Wrapper method uses a learning algorithm in the feature selection step to evaluate the performance of the feature subset [5, 6].

J.-H. Kim et al. (eds.), *Robot Intelligence Technology and Applications 2,*
Advances in Intelligent Systems and Computing 274,
DOI: 10.1007/978-3-319-05582-4_51, © Springer International Publishing Switzerland 2014

Such two methods are combined with various mathematical algorithm including mutual information [7], fuzzy-rough set [8], and local-learning [9].

In this paper, wrapper feature selection based on evolutionary algorithm is introduced. Evolutionary algorithm (EA) generates a subset of the features considering both classification accuracy and size of the selected features subset. To demonstrate the effectiveness of the proposed algorithm, the classification of human activities using smartphone sensors is carried out.

The rest of this paper is organized as follow. In Section 2, the quantum-inspired evolutionary algorithm (QEA) is introduced. Section 3 the details of the evolutionary feature selection algorithm is presented. The experimental results are discussed in Section 4 and concluding remarks follow in Section 5.

2   Quantum-inspired Evolutionary Algorithm (QEA)

Building block of classical digital computer is represented by two binary states, '0' or '1', which is a finite set of discrete and stable state. In contrast, QEA utilizes a novel representation, called a Q-bit representation [10], for the probabilistic representation that is based on the concept of qubits in quantum computing [11]. Quantum system enables the superposition of such state as follows:

$$\alpha|0\rangle + \beta|1\rangle \tag{1}$$

where α and β are the complex numbers satisfying $|\alpha|^2 + |\beta|^2 = 1$.

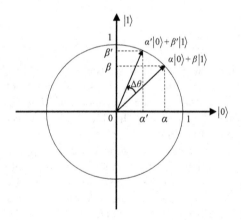

Fig. 1. Qubit described in two dimensional space

Qubit is shown in Fig. 1, which can be illustrated as a unit vector on the two dimensional space as follows:

$$q_{ji}^t = \begin{bmatrix} \alpha_{ji}^t \\ \beta_{ji}^t \end{bmatrix} \tag{2}$$

$$\mathbf{q}_j^t = \begin{bmatrix} \alpha_{j1}^t & \alpha_{j2}^t & \cdots & \alpha_{jm}^t \\ \beta_{j1}^t & \beta_{j2}^t & \cdots & \beta_{jm}^t \end{bmatrix} \tag{3}$$

where m is the string length of Q-bit individual, and $j = 1, 2, ..., n$ for the population size n. The population of Q-bit individuals at generation t is represented as $Q(t) = \{\mathbf{q}_1^t, \mathbf{q}_2^t, ..., \mathbf{q}_n^t\}$.

Since Q-bit individual represents the linear superposition of all possible states probabilistically, diverse individuals are generated during the evolutionary process. The procedure of QEA and the overall structure for single-objective optimization problems are described in [10].

3 Evolutionary Feature Selection Algorithm

3.1 Feature Generation

The characteristics of sensor signals can be obtained by extracting features. We extract five features including mean, var, rms, MAD, and IQR from the signals of the triaxial accelerometer and gyroscope. Each feature indicates average, variance, root mean square, mean absolute deviation, and interquartile range, respectively. The values of the features are defined as:

$$\text{Mean} = \frac{1}{L} \sum_{i=1}^{L} x_i \qquad\qquad \text{Var} = \frac{1}{L-1} \sum_{i=1}^{L} (x_i - m)$$

$$\text{rms} = \frac{1}{L} \sum_{i=1}^{L} x_i^2 \qquad\qquad \text{MAD} = \frac{1}{L} \sum_{i=1}^{L} |x_i - m| \tag{4}$$

where L is the length of the signals, and m is mean value. IQR represents the dispersion of the data and eliminates the influence of outliers in the data. The features are extracted from each axis of the triaxial accelerometer and gyroscope, thus an initial set of the features has 30 elements.

3.2 Feature Selection Based on Evolutionary Algorithm

Feature selection aims at finding a subset of the features that has the most discriminative information from the original feature set because most of data set from the real life has redundancy. Due to such redundancy, the dimensionality of the data set increases and the subsequent learning processes could have poor performance. In addition, it makes the learning process slow down. Among the feature selection algorithms, we use a randomized approach that could avoid the possibility of local optima problem compared to the other feature selection methods based on mathematical formula. In this paper, a evolutionary algorithm (EA) is used as an operator for the feature selection.

In the classification problem using a feature selection algorithm, there are two approaches mainly used; the *wrapper* approach depicted in Fig. 2 and the *filter* approach depicted in Fig. 3. The wrapper approach uses an actual classification algorithm to find a subset of features, while the filter approach extracts undesirable features out of the feature set before the classification process. Filter method is computationally efficient, but has poor performance compared to wrapper method. Hence, wrapper method is utilized to integrate evolutionary algorithm with learning algorithm.

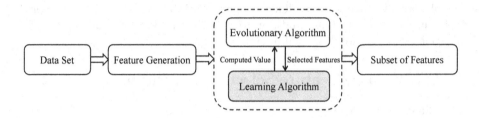

Fig. 2. Wrapper method for the feature selection

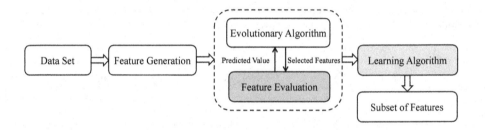

Fig. 3. Filter method for the feature selection

In this paper, quantum-inspired evolutionary algorithm (QEA) is adopted as an evolutionary algorithm. QEA uses probabilistic binary string which is called Q-bit individual defined as eq. 3. Each feature is linked to each corresponding Q-bit represented in eq. 2. Then, population $\mathbf{B}_j^t$ is generated by observing Q-bit individual, which is binary string. If an element of the population $\mathbf{B}_j^t$ has a value of '1', corresponding feature is selected, otherwise is not selected. A feature subset consisting of the selected features is forwarded to learning algorithm. Overall structure is depicted in Fig. 4.

3.3   Fitness Function

The fitness function evaluates a subset of features designated by the feature selection algorithm, providing classification accuracy and computational complexity. For considering both aspects, the fitness function that contains classification accuracy with size of feature subset is proposed. The proposed fitness function is as follows:

$$\text{fitness function} = \alpha f_1 + \beta \frac{1}{f_2} \tag{5}$$

where α, β are parameters that indicate weights of two objectives. The first term f_1 corresponds to the classification accuracy, and second term f_2 corresponds to the number of the selected features. In our experiments, the value of α, β set to 0.6, 0.4.

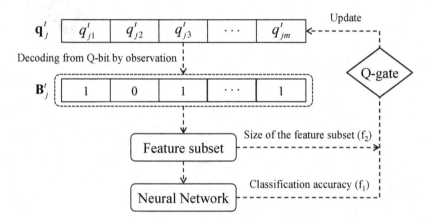

Fig. 4. Feature selection with QEA

4   Experimental Result

4.1   Experimental Setup

The proposed algorithm was applied to human activity classification problem using acceleration and gyroscope data in a smartphone. We classified four activities: *sitting in the chair, walking straightly, running straightly, jumping.* The acceleration and gyroscope signals in a smartphone were transmitted to the computer through Bluetooth, and the dataset was obtained from a subject who did activities with a smartphone. Each activity was repeated 10 times. Overall data flow of the hardware platform is depicted in Fig. 5.

4.2   Results

The proposed algorithm was able to find the optimized subset of the features. In the evolutionary feature selection algorithm, the maximum generation was set to 1000. The classification accuracy was highest when 19 features were used. Four human activities were classified with 95% accuracy when the optimized subset of the features was used. The results show that a large number of features did not always guarantee the highest classification accuracy. In the classification

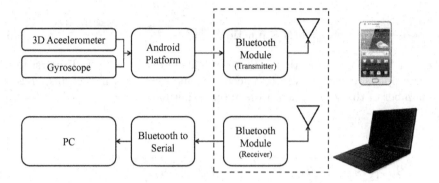

Fig. 5. Overall structure of the hardware platform

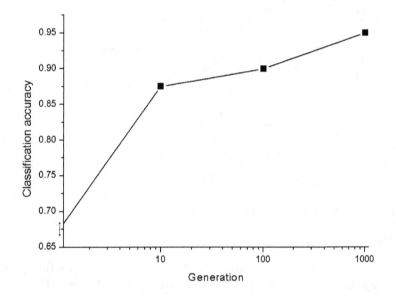

Fig. 6. The classification accuracy according to generation

problem, the retention of all features is not suitable for classifying data because some features might not have discriminative information to distinguish different data.

5   Conclusions

This paper proposed an evolutionary feature selection algorithm for classification of human activities. As an evolutionary algorithm, quantum-inspired evolutionary algorithm (QEA) was applied to feature selection, and multi-layer perceptron (MLP) was used as a classifier. The proposed algorithm was designed to enhance the classification accuracy and to reduce the size of the feature set. To validate the effectiveness of the proposed algorithm, we carried out the classification of human activities. As a result, we could obtain the subset of the features considering not only learning accuracy but also the size of the feature set, and classify the four human activities with high accuracy.

Acknowledgement. This research was supported by the MEST (The Ministry of Education, Science and Technology), Korea, under the Mid-career Researcher Program, supervised by the NRF (National Research Foundation)(2009-0080432, Robust Unified Navigation Technology of Humanoid Robot Using Gaze Control, Posture Learning and Footstep Planning).

This research was also supported by the MOTIE (The Ministry of Trade, Industry and Energy), Korea, under the Human Resources Development Program for Convergence Robot Specialists support program supervised by the NIPA (National IT Industry Promotion Agency)(H1502-13-1001, Research Center for Robot Intelligence Technology).

References

1. Song, L., Smola, A., Gretton, A., Bedo, J., Borgwardt, K.: Feature selection via dependence maximization. The Journal of Machine Learning Research 98888, 1393–1434 (2012)
2. Jeong, Y.S., Kang, I.H., Jeong, M.K., Kong, D.: A New Feature Selection Method for One-Class Classification Problems. IEEE Transactions on Systems, Man, and Cybernetics, Part C: Applications and Reviews 42(6), 1500–1509 (2012)
3. Kouchaki, S., Boostani, R., Parsaei, H.: A new feature selection method for classification of EMG signals. In: 2012 16th CSI International Symposium on Artificial Intelligence and Signal Processing (AISP), pp. 585–590. IEEE (May 2012)
4. Wang, J.S., Chuang, F.C.: An Accelerometer-Based Digital Pen With a Trajectory Recognition Algorithm for Handwritten Digit and Gesture Recognition. IEEE Transactions on Industrial Electronics 59(7), 2998–3007 (2012)
5. Kamath, U., Compton, J., Islamaj Dogan, R., De Jong, K., Shehu, A.: An Evolutionary Algorithm Approach for Feature Generation from Sequence Data and its Application to DNA Splice Site Prediction. IEEE/ACM Transactions on Computational Biology and Bioinformatics (TCBB) 9(5), 1387–1398 (2012)
6. Bermejo, P., de la Ossa, L., Gamez, J.A., Puerta, J.M.: Fast wrapper feature subset selection in high-dimensional datasets by means of filter re-ranking. Knowledge-Based Systems 25(1), 35–44 (2012)
7. Estevez, P.A., Tesmer, M., Perez, C.A., Zurada, J.M.: Normalized mutual information feature selection. IEEE Transactions on Neural Networks 20(2), 189–201 (2009)
8. Jensen, R., Shen, Q.: New approaches to fuzzy-rough feature selection. IEEE Transactions on Fuzzy Systems 17(4), 824–838 (2009)
9. Sun, Y., Todorovic, S., Goodison, S.: Local-learning-based feature selection for high-dimensional data analysis. IEEE Transactions on Pattern Analysis and Machine Intelligence 32(9), 1610–1626 (2010)

10. Han, K.-H., Kim, J.-H.: Quantum-inspired evolutionary algorithm for a class of combinatorial optimization. IEEE Trans. Evol. Computat. 6(6), 580–593 (2002)
11. Hey, T.: Quantum computing: an introduction. Computing and Control Eng. J. 10(3), 105–112 (1999)

Automatic Linear Robot Control Synthesis Using Genetic Programming

Tiberiu S. Letia and Octavian Cuibus

Technical University of Cluj Napoca, Cluj Napoca, RO 400114, Romania
Tiberiu.Letia@aut.utcluj.ro, ocuibus@yahoo.com

Abstract. An automatic controller synthesis method for a single axe linear robot is considered. The robot motions are modeled by a Delay Time Petri Net (DTPN). The search refers to automatically finding a controller modeled by a Time Petri Net (TPN) that fulfills some specified requirements. The controller model is synthesized using a Genetic Programming (GP) method. The mapping between TPN model and the tree representation of individual genotypes is performed using a formal language named here TPNL (Time Petri Net based Language). This language is suited for formal description of the controller behavior traits like sequential, concurrent, selection, loop or input/output. The use of control traits guaranties the construction of individuals that are capable and useful to control the robot moves. To diminish the search durations, besides the usual genetic operators like mutation, permutation and crossover, a new atrophy operator was introduced.

Keywords: linear robot, control synthesis, genetic programming, Petri nets, formal language, control traits.

1   Introduction

Robot control generally concerns path planning, decision making and motion control, depending on the problems the applications solve. Robots are enhanced with sensors to get the environment structure or to observe its behavior. Robots have tasks to fulfill and they have to react to environment or other system participant behaviors. Some applications use robots in a Flexible Manufacturing System (FMS) that is capable to process parts performing different activities according to their specified technologies. A FMS can process concurrently different parts involving different sequences of activities. Some types of parts could be manufactured once, and others repeatedly. Flexibility means to change as quickly as possible the scheduling and accept the processing of new parts involving other technologies; meantime the previous demands are not terminated. In such applications, the robots react to the environment demands and changes.

The structure of the considered FMS composed of one single axe linear robot (R) that moves parts to a set of machine tools (positioned in the places p_1, p_2, p_3 and p_4) performing different activities is represented in Fig. 1. The number of places is reduced for presentation purposes. The job of the robot is to precisely

J.-H. Kim et al. (eds.), *Robot Intelligence Technology and Applications 2*,
Advances in Intelligent Systems and Computing 274,
DOI: 10.1007/978-3-319-05582-4_52, © Springer International Publishing Switzerland 2014

set a part in the processing place and then the control system can start the processing activity. Until an activity is finished, the robot can move other parts to different places. The robot moves from one place to another could have different durations depending on the distance it has to cover. The positioning of the robot in front of a place or the demand for a new task can be signaled to the robot control system. The robot obeys to *left, right* and *halt* control signals.

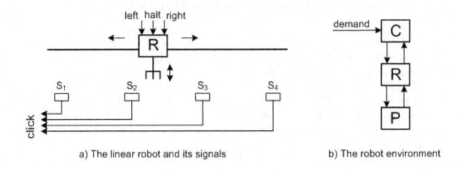

a) The linear robot and its signals   b) The robot environment

Fig. 1. The application architecture

The controller (C) receives the demands to move parts from one place to another and controls the robot moves taking into account the position signaled by the plant (P).

The current research problem is to synthesize automatically the controller behavior using the robot and plant model and some performance evaluations.

2   Related Works

The current application consisting of a single axe linear robot and some machine tools is transformed into a Discrete Event System (DES). An automatic synthesis control method for a kind of DES is based on bipartite directed graphs that yield the feasible control trajectories and their corresponding states [1]. This is combined with the supervisory synthesis.

The design of logic controllers for event-driven systems that relies on intuitive methods leads to control codes that require extensive verification and are hard to maintain and modify, and may even fail at times. Supervisory control theory provides a formal approach to logic control synthesis. This is used to derive a supervisor that enforces the specifications offering maximum flexibility [2].

2.1   Petri Nets

Some methods of synthesis use Petri nets to prevent the entrance into forbidden state and construct maximally permissive controllers [3].

More complex Petri nets models enhanced with time are introduced like Delay Time Petri Nets (DTPNs) [4] or Time Petri Nets (TPNs) [5]. Reduction methods are used to get more simple models and analysis methods based on reachability serve to model behavior verification.

In the current research relative to the TPN the following semantics are used. The TPN controller models are deterministic and fulfill the following assumptions:

1. The TPN has no conflicts or free elections (see Figure 2).
2. If more than one timed transition is executable from the same marking, the transition with the shortest delay is first chosen for execution;
3. If the TPN model has conflicts or free elections, the following semantics are accepted: the order of transitions chosen for execution is given by their index;
4. The system works with reserved tokens. A transition that started the execution cannot be cancelled by another one with a shorter delay;
5. The places of the input and output interface sets (P_C and P_R in Figure 7) are used exclusively only for input or output operations respectively.
6. The transitions correspond to actions and their executions have no durations.

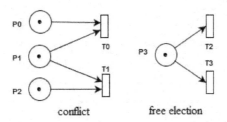

Fig. 2. Conflict and free election TPN structures

2.2 Genetic Programming

The problem of automatic control of a robot model in an arbitrary two-dimensional environment can be obtained based on behavior evolution with the use of GP [6]. This allows different behaviors suited to the environment and user requirements. One of the difficult problem of the GP is the uncontrolled growth of the program size (i. e. bloat). This is usually directly linked to the genotype dimension. Many methods to control the bloat are proposed [7]. An efficient bloat control mechanism is based on examining each function node in the programs. The nodes without contribution are removed before the creation of offspring [8].

3   Robot and Plant Model

The control synthesis requires (is based on) the robot and the plant model. This was constructed taking into account their specifications. The DTPN model is presented in Figure 3. The delays are not represented to diminish the figure complexity for the presentation and understanding purposes. There can be seen the control signals *left*, *halt* or *right* and the sensor signals activated when the robot reaches a place p_1, p_2,p_3 or p_4. These places model and store the robot position; meantime the upper part of Figure 3 describes the robot behavior when it receives the control signals.

Two plant constructions can be used. One construction signals the reach of each place separately, and another signals when any place is reached. From another point of view, the plant states can be accessible or not. The two constructions involve different efforts for the control synthesis.

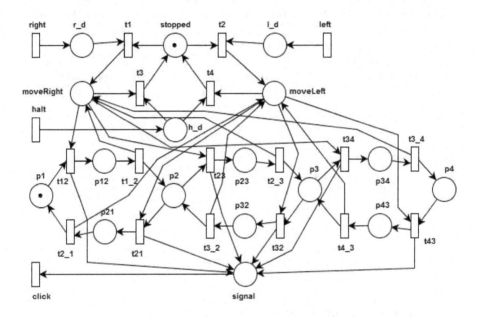

Fig. 3. The robot and the plant DTPN model

4   Control Synthesis

To get the controller behavior the GP was used [9]. The controller behavior is modeled by a TPN. So, the controller solution is a GP individual. As has been mentioned above, GP methods code the individual genotype using a tree structure. The mapping between TPN model and its GP genome is performed by TPNL.

GP operators, selected based on performance evaluation functions, act on genotypes. The performance evaluations use the robot-plant model, the controller TPN model and some parameters that increase the search speed.

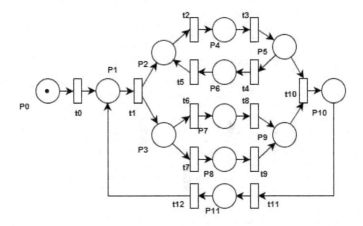

Fig. 4. Example of a TPN

4.1 Time Petri Net Based Language (TPNL)

The role of TPNL is to transform TPN models into string expressions that can be transformed later into the Lisp-S expressions required by the GP algorithms. The transformation of the TPNL expressions into Lisp-S expressions can be performed automatically by changing the operator positions. TPNL uses the following operator symbols:

- '*' for sequence,
- '+' for selection,
- '&' for concurrency and
- '#' for the loop composition.

Considering the example in Figure 4, the following relations describe:

- $t_2 * t_3$ the sequence of t_2 and t_3,
- $(t_2 * t_3) \# (t_4 * t_5)$ the loop of two sequences,
- $(t_6 * t_8) + (t_7 * t_9)$ the selection between two sequences and
- $((t_2 * t_3)\#(t_4 * t_5))\&((t_6 * t_8) + (t_7 * t_9))$ the concurrent composition.

The sequence generated by the entire TPN can be described by the expression:

$$\sigma = t_0 * (((((t_2 * t_3)\#(t_4 * t_5))\&((t_6 * t_8) + (t_7 * t_9))))\#(t_{11} * t_{12})) \qquad (1)$$

The timings (delays) are not represented in figure and not given in the previous formula for simplification reasons.

TPNL can describe the relation between a controller and a plant. For the model presented in Figure 5 containing a plant and its controller, the descriptions using TPNL are:

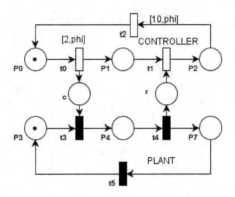

Fig. 5. Controller - plant example

$$\sigma_{controller} = (t_0[2,c] * t_1[r,\phi]) \# t_2[10,\phi] \tag{2}$$

$$\sigma_{plant} = (t_3[c,\phi] * t_4[\phi,r]) \# t_5[\phi,\phi] \tag{3}$$

The first argument of the transition symbol $t_i[a,b]$ represents a (time) delay or an event signaled by an input channel a. The second argument corresponds to an event signaled by the current component through the output channel b. The ϕ symbol is used to specify an inexistent channel (i. e. the lack of an input event, time or output event).

GP is used here to guide the search of the controller solution through a huge space. TPNL can be used to perform the bijective mapping between genotypes and individuals. For example, for the TPN controller model presented in Figure 5, the TPNL expression can be transformed into the following Lisp-S expression:

$$\sigma_{controller} = \#(*t_0[2,c], t_1[r,\phi]), t_2[10,\phi] \tag{4}$$

The tree representation of the above genotype is given in Figure 6. The TPNL operators become nodes; meantime the transition arguments become leaves.

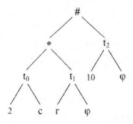

Fig. 6. The simple controller genotype tree representation

4.2   Controller Model Synthesis

Figure 7 depicts the general interfaces between the (robot and) plant and its controller. In the general case of a discrete event system, the plant is modeled by a DTPN and the controller by a TPN. The notations in the figure are:

- P_C denotes the set of control places
- T_C denotes the set of controlled transitions
- T_R denotes the set of reaction transitions and
- P_R the set of reactive places.

For the current problem they are:

- $T_C = \{\text{right, left, halt}\}$
- $T_R = \{\text{click}\}$

Another application variant includes the robot position signals, too.

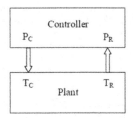

Fig. 7. Controller - plant (the robot is included too) interfaces

The synthesis problem refers to the conceiving of a TPN that has the interface sets P_C and P_R and that maximizes some given performance criteria for the given plant model.

The analysis of the robot-plant model presented in Figure 3 shows that it is a system that has memory. Many discrete event plants provide memory traits. The construction of the plant-controller interfaces can be done to provide the plant current state or not. The current research focuses on the following types of controllers.

First type controller (FTC) uses an interface that provides the plant internal state. This leads to a controller that receives a demand, reacts accordingly and does not (need to) store the plant final state to be able to perform the next request.

Second type controller (STC) does not use an interface that provides the plant internal state, but it is enhanced with a structure that models the plant state. STC is started simultaneously with the plant having specified the plant initial value. STC receives a demand and at the end of fulfilling the requirement, the controller has to store the plant observed state for use in the next demands.

Third type controller (TTC) does not have any information about the plant structure and its initial state value. TTC has to be constructed such that it maintains the plant state information and gets the plant initial state. TTC receives demands, performs their requirements and stores the plant state for future use.

The GP algorithm involves the following activities:

1. the random creation of the initial population
2. the individual evaluations
3. the random individual selection for reproduction
4. the creation of the offspring using randomly genetic operators (mutation, crossover and permutation)
5. the selection of the solution (the individual that won the competition)

Steps 2-4 are executed until a stopping criterion is fulfilled.

The three GP operators act on the genotype trees. The crossover uses two selected individuals (parents), it chooses in each parent a subtree and interchanges them.

Figure 8 shows the TPN models of two selected parents. The TPNL descriptions are:

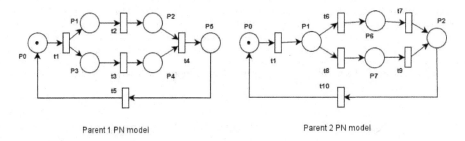

Parent 1 PN model   Parent 2 PN model

Fig. 8. Two parents PN models

$$\sigma_1 = (t_2 \& t_3) \# t_5 \tag{5}$$

$$\sigma_2 = (t_1 * ((t_6 * t_7) + (t_8 * t_9))) \# t_{10} \tag{6}$$

Transforming the TPNL descriptions into Lisp expressions gives:

$$\sigma_{1Lisp} = \#(\& t_2, t_3), t_5 \tag{7}$$

$$\sigma_{2Lisp} = \#(* t_1, (+(* t_6, t_7), (* t_8, t_9))), t_{10} \tag{8}$$

It is supposed the random crossover chooses the subtree t_2 from Parent 1 and the subtree $(+(* t_6, t_7), (* t_8, t_9))$ from Parent 2 for spring construction. Figure 9 represents the two parent trees and the selected subtrees for crossover.

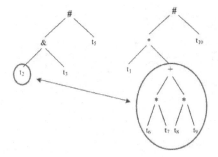

Fig. 9. Parent trees and crossover marking

Performing the crossover, the children described by σ_3 and σ_4 are obtained:

$$\sigma_{3Lisp} = \#(\&(+(*t_6, t_7), (*t_8, t_9)), t_3), t_5 \qquad (9)$$

$$\sigma_{4Lisp} = \#(*t_1, t_2), t_{10} \qquad (10)$$

$$\sigma_3 = (((t_6 * t_7) + (t_8 * t_9))\&t_3)\#t_5 \qquad (11)$$

$$\sigma_4 = (t_1 * t_2)\#t_{10} \qquad (12)$$

The children trees are given in Figure 10 and their TPN models in Figure 11.

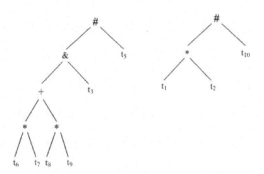

Fig. 10. Crossover resulted children trees

The mutation genetic operator acts on a node changing the TPNL operator, or on a leaf changing the input channel, the delay duration or the output channel. The TPN model presented in Figure 12 is chosen to show the operator mutation.

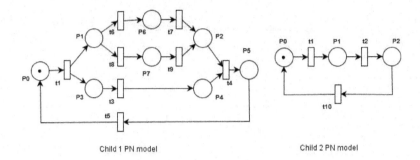

Fig. 11. Children PN models

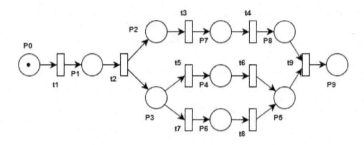

Fig. 12. PN model for operator mutation

The TPNL description of TPN model chosen for operator mutation is:

$$\sigma = t_1 * ((t_3 * t_4)\&((t_5 * t_6) + (t_7 * t_8)))) \tag{13}$$

The transformation into Lisp expression provides:

$$\sigma_{Lisp} = *(*t_1, (\&(*t_3, t_4), (+(*t_5, t_6), (*t_7, t_8))))) \tag{14}$$

The corresponding tree is drawn in Figure 13. The node '&' was chosen randomly to be replaced with the operator #.

The Lisp expression of the mutated individual model is:

$$\sigma_{newLisp} = *(*t_1, (\#(*t_3, t_4), (+(*t_5, t_6), (*t_7, t_8))))) \tag{15}$$

The mutated individual model can be transformed into:

$$\sigma_{new} = t_1 * ((t_3 * t_4)\#((t_5 * t_6) + (t_7 * t_8)))) \tag{16}$$

Figure 14 presents the TPN model obtained after operator mutation.

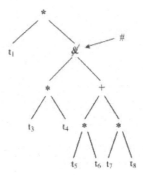

Fig. 13. The tree representation for operator mutation

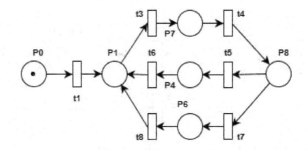

Fig. 14. PN model of operator mutation result

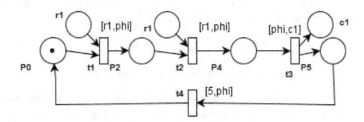

Fig. 15. PN model of leaf mutation

Figure 15 presents a TPN model that has to suffer mutations on transition arguments. The TPNL description and Lisp expression are:

$$\sigma = (t_1[r_1, \phi] * t_2[r_1, \phi] * t_3[\phi, c_1]) \# t_4[5, \phi] \tag{17}$$

$$\sigma_{Lisp} = \#(*(*t_1[r_1, \phi], t_2[r_1, \phi]), t_3[\phi, c_1]), t_4[5, \phi] \tag{18}$$

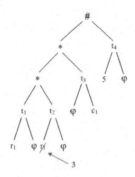

Fig. 16. Tree representation of argument mutation model

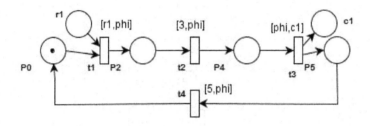

Fig. 17. PN model of argument mutation result

The mutation of the first argument of transition t_2 is supposed to change the input channel r_1 into a 3 time units delay. That is described by:

$$\sigma_{newLisp} = \#(*(*t_1[r_1, \phi], t_2[3, \phi]), t_3[\phi, c_1]), t_4[5, \phi] \tag{19}$$

$$\sigma_{new} = (t_1[r_1, \phi] * t_2[3, \phi] * t_3[\phi, c_1])\#t_4[5, \phi] \tag{20}$$

The mutation is represented on the tree in Figure 16. The result obtained after first argument mutation of the transition t_2 is drawn in Figure 17. Similar results are obtained if the mutation genetic operator acts on the second arguments, but in this case the control signal channel is changed.

The permutation changes within the same individual one node to another node or one leaf to another leaf. The TPN model used to apply the permutation is the Child 1 in Figure 11. It is supposed that the permutation interchanges the operators '&' and '+' that leads to the σ_5 expressions and consequently the tree presented in Figure 18.

$$\sigma_{5Lisp} = \#(+(\&(*t_6, t_7), (*t_8, t_9)), t_3), t_5 \tag{21}$$

$$\sigma_5 = ((t_6 * t_7)\&(t_8 * t_9)) + t_3)\#t_5 \tag{22}$$

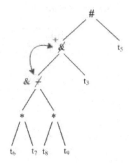

Fig. 18. Permutation representation on the tree

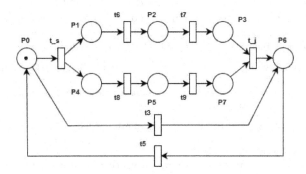

Fig. 19. Resulted PN model after permutation

After the permutation, the TPN model in Figure 19 is obtained.

The pure random creation of an individual could lead to many individuals that will not survive (aborted being useless) due to the fact that they do not have at least the compulsory control traits. For this reason some control traits are conceived and they have specified probabilities to be used for initial individual creation or by mutation operator during the reproduction phase. These are introduced in the so called trait pool.

In the current research the used trait pool is presented in Table 1. There t_i represents transitions and $\sigma_i(i = 1, 2)$ correspond to sequences of transitions. Different probabilities are assigned for trait use depending on the used model of the plant.

4.3   Controller Behavior Evaluation

According to GP the fitness function plays an important role for the individual selection and this guides indirectly the search for the desired solution. The evaluation is achieved using the concurrent simulation of the (robot plus) plant

Table 1. Trait pool

Trait	Coding	Probability
input	$t_i[a, \phi]$	12
output	$t_i[\phi, b]$	12
input-output	$t_i[a, b]$	12
loop	$\sigma_1 \# \sigma_2$	10
concurrency	$\sigma_1 \& \sigma_2$	10
selection	$\sigma_1 + \sigma_2$	10
memory	$(t_i[r, s] * t_j[\phi, \phi]) \# (t_l[l, s] * t_k[\phi, \phi])$	10
read (state 0)	$t_1[0, c_3] * t_2[r_0, \phi]$	5

and controller models. In the current case, the controller receives a specified test sequence (demands) containing the demanded movements.

An example of a demanded sequence used for controller evaluation is:

$$demand = p_3[0, \phi] * p_1[5, \phi] * p_2[4, \phi] * p_4[3, \phi] * p_1[6, \phi] * p_3[4, \phi] *$$
$$p_2[2, \phi] * p_1[3, \phi] * p_4[0, \phi] * p_1[7, \phi] * p_3[4, \phi] * p_2[3, \phi] * p_3[3, \phi] \quad (23)$$

The durations of demands should be chosen such that the robot can perform them under right controller supervision.

Two criteria can be used for individual evaluations:

- individual behavior evaluation
  - robot reaches or not the demanded state in a specified duration
  - number of transitions executed by plant or controller to reach the destinations (i. e. the time to fulfill a demand)
  - robustness relative to plant reaction delays
  - controller reaction delays
- individual structure evaluation
  - number of transitions
  - number of channels used for the interfaces plant-controller or plant-supervisor
  - number of parallel threads of execution for controller implementation
  - number of loops etc.

In the current research they are expressed using the following formulas:

$$Criterion_1 = \alpha_1 \cdot targetReached - \alpha_2 \cdot transitionExecuted - \alpha_3 \cdot duration \quad (24)$$

where:

- targetReached means that when the controller signals the end event, the plant reached the target,
- transitionExecuted counts how many transitions were executed to reach the target,
- duration counts the time (clock tic) from the start until the end signal of each demand

$$Criterion_2 = -\alpha_4 \cdot transitionNo - \alpha_5 \cdot channelNo - \alpha_6 \cdot threadNo - \alpha_7 \cdot loopNo \tag{25}$$

where:

- transitionNo represents the number of transitions used for individual construction,
- channelNo corresponds to the number of channels used by the controller,
- threadNo counts the number of threads and
- loopNo counts the number of loops.
- $\alpha_i, i = 1, \ldots, 7$ are weight coefficients.

4.4   Increasing the Search Speed

The search guiding can be achieved using mono or multi-objective functions. These are used for the individual selection. On the other hand, the probabilities of individual selection for applying the genetic operators also influence the search directions and speeds. Different probabilities can be assigned to perform the individual initial construction or modifications during the evolution.

In the case of GP, it can be assigned probabilities for operator selections and constraints for the number of the operators of specified types that are accepted in any individual during the construction of the initial population or the mutation operation.

The two criteria proposed to guide the solution search are used for individual selection, but the constrained numbers of the GP operators are involved too.

Besides the usual GP operators a new operator called *atrophy* was introduced due to the fact that during the individual evaluations some transitions are not used. The atrophy operator is applied (unlike the usual GP operators) after the individual selection and it has the role to remove from the individual the transitions that are not involved in any behavior demand. This removal further affects the individual genotype.

To avoid the unlimited increase of an individual (that leads to bloat) and the excessive use of some TPNL operators besides the choosing probabilities, some constraints have been assigned as can be seen in Table 2. On the other hand, GP operators can have different selection probabilities depending on the criterion used for individual selection as presented in Table 3. The tests were performed for a population of 2000 individuals and 300 generations. Table 4 contains the parameter values of the evaluation criteria. Their value can increase or decrease the search speed too.

5   Tests and Results

The controller synthesis involves the construction of the environment and the individuals, followed by the phenotype evaluations.

The environment was obtained by the implementation of the robot DTPN model. This requires two matrices with dimensions equal to the number of places

Table 2. Operator probabilities and constraints

Operator	Probablity	Constraint
*	40	50
+	20	20
&	5	50
#	10	10

Table 3. Genetic operator numbers used for the different criteria on a population of 2000 individuals

Criterion	Mutation	Crossover	Permutation
$Criterion_1$	285	850	150
$Criterion_2$	190	450	75

and the number of transitions. Two delay vectors were used, one with the dimensions equal to the number of transitions and the other equal to the number of places.

The individual constructions were achieved by the implementation of the TPN models that require two matrices with dimensions equal to the number of places and the number of transitions (like for the DTPN model) and a delay vector with the dimension equal to the number of transitions.

The phenotype evaluation was performed in simultaneous simulation of the environment, individuals and their interactions.

Two separate tests were performed on a personal computer with a 2.6GHz dual core processor:

1. The considered plant has an additional structure that can output the exact state of the plant (i. e. the robot position). The read command is denoted c3. The generated controller has 49 transitions and is made up of traits and individual transitions, connected with operators. The performance function evaluates the behavior of the (plant + controller) system in 16 different scenarios (the plant is in state i, the command is to go to state j, i,j=1..4). The solution is generated in 300 generations and it took around 9.5 hours to run. The resulted Lisp-S expression of the solution is (+(+(+(*(*(*(t1,r5,c3)(t2,0,c1))(+(+(+(t3,r0,c3)(*(t4,1,c3)(t5,r2,fi))) (t6,r1,fi))(+(t7,1,c3)(t8,r0,fi))))(*(*(t9,r6,c3)(t10,1,c3))(+(+(*(*(*(* (t11,r0,fi)(t12,0,c0))(*(t13,1,c3)(t14,r1,fi)))(t15,0,c2))(t16,r1,fi)) (*(*(*(t17,r2,fi)(t18,0,c1))(*(t19,2,c3)(t20,r1,fi)))(t21,0,c2)))(*(*(*(*

Table 4. The value of the coefficients of the evaluation functions

Coeficient	α_1	α_2	α_3	α_4	α_5	α_6	α_7
Value	350	80	10	20	10	10	10

(t22,r3,fi)(t23,0,c1))(*(t24,1,c3)(t25,r2,fi)))(*(t26,1,c3)(t27,r1,fi)))
(t28,0,c2)))))(*(*(t29,r7,c3)(t30,1,c3))(+(+(+(*(*(*(*(t31,r0,fi)
(t32,0,c0))(*(t33,1,c3)(t34,r1,fi)))(*(t35,1,c3)(t36,r2,fi)))(t37,0,c2))
(*(*(*(t38,r1,fi)(t39,0,c0))(*(t40,1,c3)(t41,r2,fi)))(t42,0,c2)))(t43,r1,fi))
(*(*(*(t44,r3,fi)(t45,0,c1))(*(t46,1,c3)(t47,r2,fi)))(t48,0,c2)))))(t49,0,c0))

2. The plant is without the additional structure, the controller is forced to track the state of the plant. Using this strategy, no result was generated in 300 generations. However, if we add the additional structure (memory) to the trait pool (such as to let the GP algorithm include it in the controller), the generated solution contains 39 transitions and is generated in 12 hours running time. Here, P1,..., P4 represent the places where the state of the plant is being stored (these places have been given as input channels for the rest of the controller). The best solution is

(+(+(+(*(t1,cmd1,fi)(+(+(+(t2,P1,stop)(*(t3,P2,left)(t4,P1,stop)))
(*(*(t5,P3,left)(t6,P2,fi))(t7,P1,stop)))(*(*(t8,P4,left)(t9,P3,fi))(t10,P2,fi))
(t11,P1,stop))))(*(t12,cmd2,fi)(+(+(+(*(t13,P1,right)(t14,P2,stop))
(t15,P2,stop))(*(t16,P3,left)(t17,P2,stop)))(*(*(t18,P4,left)(t19,P3,fi))
(t20,P2,stop)))))(*(t21,cmd3,fi)(+(+(+(*(*(t22,P1,right)(t23,P2,fi))
(t24,P3,stop))(*(t25,P2,right)(t26,P3,stop)))(t27,P3,stop))(*(t28,P4,left)
(t,P3,stop)))))(*(t29,cmd4,fi)(+(+(+(*(*(*(t30,P1,right)(t31,P2,fi))(t32,P3,fi))
(t33,P4,stop))(*(*(t34,P2,right)(t35,P3,fi))(t36,P4,stop)))(*(t37,P3,right)
(t38,P4,stop)))(t39,P4,stop))))

6   Conclusions

The newly introduced control traits diminish significantly the solution search duration. A control designer usually knows the main traits of the expected controller. But the probabilities of the trait appearances are unknown. These are problem dependent.

The proposed method leads to a TPN that is equivalent to a set of interconnected state machines which can be easily programmed. The obtained controller can be implemented using a single or multi threading execution.

The TPN solution can be easily implemented on a FPGA (Field Programmable Gate Array) or on a micro-controller using their programmable languages. The execution durations of the corresponding programs are very short relative to the robot temporal behavior, so it can be stated that a real-time controller has been obtained.

The proposed synthesis method can be successfully applied to any discrete event system. The controller designer should focus its efforts on specifying the controlled part of the application and the control performance evaluation, instead of finding a control algorithm that fulfills the specification.

References

1. Kapkovic, F.: Automatic control synthesis for agents and their cooperation in MAS. Computing and Informatics 29, 1045–1071 (2010)

2. Chandra, V., Zhongdong, H., Kumar, R.: Automated control synthesis for an assembly line using discrete event system control theory. IEEE Transactions on Systems, Man, and Cybernetics, Part C: Applications and Reviews 33(2), 284–289 (2003)
3. Dideban, A., Alla, R.: Controller synthesis by Petri nets modeling. In: Proc. of the 3rd International Workshop on Verification and Evaluation of Computers and Communication Systems (2010)
4. Juan, E.Y.T., Tsai, J.P., Murata, T., Zhou, Y.: Reduction methods for real-time systems using delay time Petri nets. IEEE Transactions on Software and Engineering 27(5), 422–448 (2001)
5. Wang, J., Deng, Y., Xu, G.: Reachability Analysis of Real-Time Systems Using Time Petri Nets. IEEE Transactions on Systems, Man, and Cybernetics, Part B: Cybernetics 30(5), 725–736 (2000)
6. Paic-Antunovic, L., Jakobovic, D.: Evolution of automatic robot control with genetic programming. In: Proceedings of the 35th International Convention MIPRO, pp. 817–822 (2012) ISBN: 978-1-4673-2577-6
7. Alfaro-Cid, E., Merelo, J.J., Fernndez de Vega, F., Esparcia-Alczar, A.I., Sharman, K.: Bloat Control Operators and Diversity in Genetic Programming: A Comparative Study. Evolutionary Computing 18(2), 305–320 (2010)
8. Song, A., Chen, D., Zhang, M.: Bloat control in genetic programming by evaluating contribution nodes. In: GECCO 2009: Proceedings of the Genetic and Evolutionary Computation Conference, pp. 1893–1894 (2009)
9. Letia, T.S., Hulea, M., Cuibus, O.: Controller synthesis method for discrete event systems. In: IEEE International Conference on Automation Quality and Testing Robotics (AQTR), pp. 85–90 (2012), doi:10.1109/AQTR.2012.6237680

Estimation of Stimuli Timing to Evaluate Chemical Plume Tracing Behavior of the Silk Moth

Jouh Yeong Chew, Kotaro Kishi, Yohei Kinowaki, and Daisuke Kurabayashi

Dept. of Mechanical and Control Engineering, Tokyo Institute of Technology,
2-12-1 Ookayama, Meguro-ku, 152-8552 Tokyo, Japan
{jychew,kotaro.glx914,kinowaki.y,dkura}@irs.ctrl.titech.ac.jp

Abstract. Insects serve as ideal models for replicating the adaptability of biological systems in Chemical Plume Tracing (CPT) because they perform efficient olfactory tracking. In this paper, we propose to evaluate the CPT behavior of the silk moth (*Bombyx Mori*) from the perspective of machine learning. We use a classification approach consisting of the Gaussian Mixture Model with Expectation Maximization (GMMEM) and the Echo State Networks (ESN) to identify the initial motion phases upon stimulation. The former method classifies the locomotion observation consisting of the linear and angular velocity into Gaussian density components which represent different elemental motions. Then, these motions are used as training data for the ESN to estimate the initial motion phases upon stimulation which represents the stimuli timing. The same procedure is implemented on different moths and cross-evaluation is done among the moths in the sample to evaluate their behavior singularity. This method achieves decent estimation accuracy and serves as a feasible approach to complement the conventional neurophysiology analysis of insects' behavior. The results also suggest the presence of CPT behavior singularity for silk moths.

Keywords: biomimetic, recognition, learning and adaptive systems.

1   Introduction

Olfactory sensing and localization is an important field of study which is useful to detect and track chemical sources in events of natural or industrial disasters, terrorist attacks and security checking. The significance of this study is increasing due to higher threats from natural disasters and more sophisticated chemical weapons. Some of the previous studies include olfactory detection of animals [1], static or dynamic sensor arrays [2-3] and chemical plume tracing (CPT) mobile robots and their bio-inspired version [4-11].

However, traditional olfactory sensing methods are less effective in countering the increasing threats. For example, using animals for sensing and tracking is not suitable in hazardous environment when there is chemical plume discharge in factories or public places. Fixed sensor arrays could be a better option but its efficiency is limited due to the static positions of sensor nodes which do not mimic the highly dynamic and

J.-H. Kim et al. (eds.), *Robot Intelligence Technology and Applications 2*,   619
Advances in Intelligent Systems and Computing 274,
DOI: 10.1007/978-3-319-05582-4_53, © Springer International Publishing Switzerland 2014

complex chemical plume flow in the air. Although the effectiveness of sensor arrays can be improved by having a larger number of sensor nodes, this will inflate the installation cost. Thus, mobile robots are more feasible for CPT and previous studies include localization and tracking of odor sources using single and multiple agents [4-7], multi-sensory tracking using vision and olfactory sensors [8] and bio-inspired CPT mobile robots mimicking adaptability and efficiency of biological CPT [9-11].

However, biological systems are known to exhibit singularity and behavior variance which complicates the bio-mimetic task. Previous study suggested the qualitative CPT behavior of the silk moth (*Bombyx Mori*) upon stimulation which consists of a sequence of motions [10]. Nonetheless, this information is inadequate for CPT behavioral analysis. Thus, it is important to measure the CPT behavioral data and decompose it into elemental motions without prior information. This provides a basis for quantitative analysis of its CPT behavior by facilitating comparison among different sets of data. For this purpose, machine learning seems to be an ideal approach.

We achieve a breakthrough by using the machine learning methods to evaluate the CPT behavior of the silk moth. This quantitative approach estimates the stimuli timing of the silk moth and uses the results to evaluate its CPT behavior. We suggest the presence of behavior singularity. Despite that, the classifier is able to achieve decent accuracy in detecting the presence of stimuli in the atmosphere by observing the CPT locomotion of the silk moth.

In this paper, the current section describes the motivation and related works. Section Two describes the problem settlement which includes the objectives, experimental subject and the measurement system. Section Three discusses about the methodology and system structure. The results and discussion are presented in section Four and the last section summarizes the current work and proposes further verification methods.

2   Problem Settlement

In this section, we identify and describe two problems which we are going to solve in this paper. First, we use a quantitative approach to estimate the stimuli timing of the silk moth. This is done by using the supervised classification method to identify the initial motion phases; the straight walking and looping pattern. The training data for the motion patterns is obtained by decomposing the locomotion data without prior information by using the unsupervised classification method. The stimuli timing estimation is the first step for quantitative analysis of the silk moth's CPT behavior.

Secondly, we compare the behavioral data under different situations. This facilitates the understanding of behavior singularity by evaluating the fitness of the behavioral data from a moth to the classification filter of another moth. This is done by training the moths with data from different moths. Like most biological systems, silk moths could exhibit behavioral singularity which should be taken into consideration in bio-mimetic.

2.1   The Silk Moth

We use the male silk moth as the experimental subject. It exhibits good olfactory sensing and serves as an ideal model for replicating efficient biological CPT on artificial systems. We illustrate its physical appearance in Fig.1(a). It detects the pheromone using a pair of black-colored antennas on its head. It also possesses a pair of wing but it does not fly and exhibits only the walking movement using its six legs. Thus, it is relatively easy to measure its behavioral data.

Apart from these physical properties, it does not exhibit olfaction for food or homing. It responds only to the pheromone released by the female silk moth for mating purpose. Moreover, it exhibits no voluntary movement in which it does not move unless stimulated by the pheromone. This presents a clear relationship in between the olfaction input and behavioral output. In addition, extensive studies have been conducted to reveal its qualitative olfactory behavior.

Previous study suggested its CPT behavior which consists of a sequence of programmed motions as shown in Fig.1(b) where t denotes the time [10]. The moth is in an initial idle state when given the olfactory stimuli at t. Then, it starts to move in a straight walking pattern. This is followed by the turn left and right walking pattern before ending with the looping pattern which does not have a definite duration. We focus on the straight walking pattern and consider the rest of the motions as looping because we are interested only in the initial motion phases upon stimulation to detect the stimuli timing.

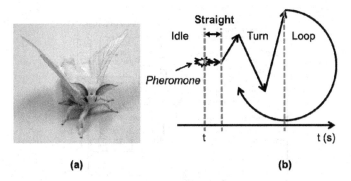

Fig. 1. (a) Silk moth (*Bombyx Mori*) (b) Locomotion behavior upon stimulation

2.2   The Tethered System

The Tethered system is an experimental system which is built to observe the CPT behavior of the silk moth in a virtual pheromone field. The system measures and records the x- and y-axis linear and angular velocities which are represented by $\hat{v}_x, \hat{v}_y$ and $\hat{\omega}$ for classification and stimuli time estimation. The observation is simplified further to obtain v and ω using (1) and (2) respectively because the locomotion of the moth is considered as a two degree of freedom non-holonomic system.

$$v = \sqrt{\left(\hat{v}_x^{\ 2} + \hat{v}_y^{\ 2}\right)} \tag{1}$$

$$\omega = \left|\hat{\omega}\right| \tag{2}$$

Referring to Fig.2, the moth is tethered on the experimental system with its legs on the Styrofoam sphere. The stimulation is given using the *Bombykol*, which is the synthetic pheromone of the female silk moth. The system consists of three main subsystems which are the pheromone stimulation, measurement and exhaust system. The pheromone stimulation system on the left consists of several components. The arrangement of cotton, charcoal and distilled water is used to purify and deodorize the air from the compressor. The delivery tube for *Bombykol* has its outlet leading to the tethered moth's antennas to deliver the pheromone. The stimulation is controlled by the virtual pheromone field on the personal computer (PC) which activates the solenoid valve based on its behavioral data measured by the pair of 2D optical sensors.

This is where the measurement system starts. The Styrofoam sphere is floated by the air blowing from an electrical fan beneath it. Then, the optical sensors measure the rotation of the floating sphere which is produced by the moth's movement. This effectively completes a loop in between the measurement and stimulation system. On the other hand, the exhaust system consists of only the enclosed space and the ventilation fan which removes the excess pheromone.

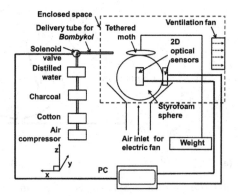

Fig. 2. Tethered system

3   Methodology

In this section, we describe the approach to estimate the stimuli timing and to evaluate the CPT behavior of the silk moth. The vertical system structure illustrated in Fig.3 summarizes the methods starting from data recording to behavioral analysis. The top layer of the structure measures $\hat{v}_x$, $\hat{v}_y$ and $\hat{\omega}$ of the silk moth using the Tethered system and converts the values into v and ω. This layer flows to the second or the third layer.

The gray arrows indicate one-off process using unsupervised classification to obtain the training data for supervised classification in the third layer. The Gaussian Mixture Model, which is estimated by the Expectation Maximization algorithm (GMMEM), is used for unsupervised classification whereas the Echo State Networks (ESN) is used for the supervised classification. On the other hand, the solid black arrows indicate the normal process flow. The final layer evaluates the fitness of the behavioral data from a moth to the classification filter of another moth, which we define as behavioral singularity.

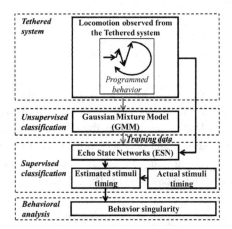

Fig. 3. The system structure

3.1   The Gaussian Mixture Model with Expectation Maximization (GMMEM)

In the case of unsupervised classification, the GMMEM is often used to represent sub-populations in observation data. Application of GMM is broad and it has been used extensively for classification of vibrational spectroscopy data [12], seismic activities [13], phonetic [14] and images [15-17]. In this paper, the GMMEM is used to decompose the behavioral data into elemental motions because it enables classification without prior information, except for the expected number of components. In our case, this is known based on the number of qualitative motions [10]. Furthermore, the straight walking pattern is brief and the moths exhibit behavioral variances among different individuals which make it difficult to identify these motions. The segmentation of the locomotion data into elemental motions is necessary to acquire the training data for the supervised classification.

The number of expected Gaussian components is estimated based on the number of motions which formed the qualitative behavior described in Section 2.1. A multivariate Gaussian distribution is parameterized by a mean vector μ and a covariance matrix Σ in (3). The variable κ is the dimensionality of the observation space. The vector $\mathbf{x}$ is the observation at the time-step n which consists v and ω. The combination of multivariate Gaussian density components forms a GMM represented by $G(\mathbf{x})$ in (4) where m is the total number of mixture components and c_i is the mixing coefficients for the ith component. The probabilistic model is represented in (5) where the Gaussians parameters are given by $\lambda = (c_1,..c_m, \mu_1,..\mu_m, \Sigma_1,..\Sigma_m)$.

$$N(\mathbf{x};\mu,\Sigma) = \frac{\exp(-\frac{1}{2}(\mathbf{x}-\mu)'\Sigma^{-1}(\mathbf{x}-\mu))}{\sqrt{(2\pi)^\kappa |\Sigma|}} \tag{3}$$

$$G(\mathbf{x}) = \sum_{i=1}^{m} c_i N(\mathbf{x};\mu_i,\Sigma_i) \tag{4}$$

$$p(\mathbf{x}|\lambda) = \sum_{i=1}^{m} c_i p_i(\mathbf{x}|\mu_i,\Sigma_i) \tag{5}$$

The observation $\mathbf{x}$ is also assumed to be incomplete due to missing values for simplifying the parameter estimation procedure. The missing values are represented by an unobserved dataset $y = \{y_i\}_{i=1}^{n}$ where y_i is the GMM component which generates the ith observation. The observed and unobserved datasets are considered as complete observation and $\lambda^g = (c_1^g,..c_m^g, \mu_1^g,..\mu_m^g, \Sigma_1^g,..\Sigma_m^g)$ represents its GMM parameters. Equation (6) and (7) are obtained based on the Bayes' rule [18].

$$p(y_i|\mathbf{x}_i,\lambda^g) = \frac{c_{yi}^g p_{yi}(\mathbf{x}_i|\lambda_{yi}^g)}{\sum_{k=1}^{m} c_k^g p_k(\mathbf{x}_i|\lambda_k^g)} \tag{6}$$

$$p(y|\mathbf{x},\lambda^g) = \prod_{i=1}^{n} p(y_i|\mathbf{x}_i,\lambda^g) \tag{7}$$

The EM algorithm uses the maximum likelihood to iteratively estimate the parameters of the distribution of an incomplete data set. The missing values could be caused by actual missing values or assumption of missing parameters to simplify the likelihood function. The procedure consists of two steps starting with finding the expected value of the complete log likelihood followed by maximizing the expected value. These two steps are performed simultaneously until the likelihood function converges to a local maximum using (8) to (10). In this case, we maximize the mixing coefficients c, mean vectors μ and covariance matrices Σ for $l \in 1,...,m$. With the component parameters known, the probabilistic assignment of each observation $\mathbf{x}$ to one of the m components is calculated. Then, these components are categorized further to eliminate components which are descriptive of the same motion and used as the training data for supervised ESN classification.

$$c_l^{new} = \frac{1}{n} \sum_{i=1}^{n} p(l|\mathbf{x}_i,\lambda^g) \tag{8}$$

$$\mu_l^{new} = \frac{\sum_{i=1}^{n} \mathbf{x_i} \, p(l \mid \mathbf{x}_i, \lambda^g)}{\sum_{i=1}^{n} p(l \mid \mathbf{x}_i, \lambda^g)} \tag{9}$$

$$\Sigma_l^{new} = \frac{\sum_{i=1}^{n} p(l \mid \mathbf{x}_i, \lambda^g)(\mathbf{x}_i - \mu_l^{new})(\mathbf{x}_i - \mu_l^{new})^T}{\sum_{i=1}^{n} p(l \mid \mathbf{x}_i, \lambda^g)} \tag{10}$$

3.2   Echo State Networks (ESN)

The ESN is used for supervised classification to estimate the stimuli timing. For this purpose, the system is trained to identify the walking patterns using the categorized locomotion data from the GMMEM. This procedure is necessary to compare the CPT behavior among the moths. Then, the estimation output is compared with the actual stimuli input to calculate the accuracy, which is used to analyze the CPT behavior. The ESN is one of the methods for Reservoir Computing (RC) which facilitates the training of Recurrent Neural Networks (RNN) by training only the connections in between the internal and the output layer [19]. It is suitable for tasks with high nonlinearity such as the biological system. This method has been successfully applied in various fields such as financial prediction [20], speech recognition [21] and engineering tasks [22-23] with good robustness [22, 24].

An ESN model is made up of the input, internal and output layers as illustrated in Fig.4. The solid black lines are fixed connections whereas the grey and dotted lines are optional and trainable connections, respectively. The basic idea is to reduce the connections to be trained by fixing the weights between the elements in the internal layers and compensate the dynamics using a large number of internal elements with sparse and random connections which fulfill the Echo State Property [25]. Transposed vectors $\mathbf{q(t)}$, $\mathbf{r(t)}$ and $\mathbf{s(t)}$ are state variables and matrices $\mathbf{W_{in}}$, $\mathbf{W}$ and $\mathbf{W_{out}}$ are connection weights for the input, internal and output layers, respectively.

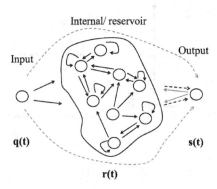

Fig. 4. Topology of the ESN model

In our case, we have two inputs v and ω. Matrix **M** consists of reservoir states which are calculated using the Leaky Integrator in (11) where u = tanh and a is the leaking rate valued in between 0 and 1. The connectivity is set at 10% based on the trial and error basis. Ridge regression in (12) is used for training the output weights **W**^{out} where matrix **D** contains the teachers' signal. The output is calculated using (13) where β is the ridge parameter. We train the system to identify the continuous straight walking and looping motion, which is shown in Fig.5. Stimulus time is identified based on the initial time t of this continuous motion when the stimulus is given. Subsequently, the stimuli time is grouped as a set of t and compared to the actual stimuli time for the experimental system.

We use 1,000 elements for the reservoir size whereas the time-step is 0.05 s each. An initialization period of 500 steps is used to enhance the effect of the training signal. There are two outputs and the training parameters are given below.
Training duration = 100 steps

Input, $\mathbf{q(t)} = \begin{bmatrix} v(t) \\ \omega(t) \end{bmatrix}$

Training output, $\mathbf{d(t)} = \begin{bmatrix} 11 \\ 10 \end{bmatrix}^T = \begin{bmatrix} Straight \\ Looping \end{bmatrix}^T$

$$r(t+1) = (1-a)r(t) + u(\mathbf{W}^{in}q(t+1) + \mathbf{W}r(t)) \tag{11}$$

$$\mathbf{W}^{out} = \left(\mathbf{M}^T\mathbf{M} + \beta^2\mathbf{I}\right)^{-1}\mathbf{M}^T\mathbf{D} \tag{12}$$

$$s(t+1) = u^{out}(\mathbf{W}^{out}(q(t+1), r(t+1), s(t))) \tag{13}$$

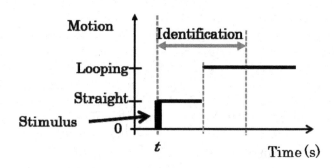

Fig. 5. Identification of the stimuli time

4   Results and Discussion

4.1   Estimation of the Stimuli Timing

The accuracy of the stimuli timing estimation is evaluated using the True Positive and Precision parameters of the confusion matrix [26]. They are represented by *TP* and *P* respectively. The reason for selecting only these two parameters is to obtain an evaluation result which is independent of the true negative cases. This is because the high number of true negative cases in the evaluation causes a bias in the estimation accuracy. The parameters are represented by Equation (14) and (15) where $Z_{correct}$, Z_{actual} and Z_{total} denote the correct predicted positive cases, total actual positive cases and total predicted positive cases respectively.

$$TP = \frac{Z_{correct}}{Z_{actual}} \tag{14}$$

$$P = \frac{Z_{correct}}{Z_{total}} \tag{15}$$

The stimuli timing estimation using the ESN is done by training the system with the moth's own data and also by using the data from other moths. Table 1 presents the cross-checking table which shows the mean and standard deviation for three trials. The training and behavioral data for moth A are denoted by A_{train} and A respectively. The same notion applies to the other moths. The estimation accuracy is calculated for all the five moths in the sample and Fig.6 illustrates the estimation results when fitting the behavioral data from moth C to the training data from moth D. The *x*-axis represents the time-steps whereas the *y*-axis represents the stimulation. The value 0 represents no stimulation whereas the value 1 represents the otherwise. The correct estimations are shown by the black solid lines whereas the actual answers are represented by the gray dashed lines.

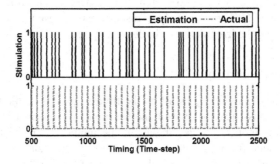

Fig. 6. Estimation of stimuli timing

Table 1. Cross-checking estimation accuracies for five moths

		Atrain		Btrain		Ctrain		Dtrain		Etrain	
		μ	σ	μ	σ	μ	σ	μ	σ	μ	σ
A	TP	72.5	1.4	77.5	0.0	55.0	4.3	63.3	3.8	56.7	2.9
	P	77.0	2.8	59.6	0.0	70.2	8.4	70.4	6.4	54.4	3.6
	Zcorrect	29	0.6	31	0.0	22	1.7	25	1.5	23	1.2
	Ztotal	38	0.6	52	0.0	31	1.2	36	1.0	42	0.6
B	TP	56.7	2.9	92.5	1.5	54.2	1.4	72.5	0.0	74.2	6.3
	P	68.7	7.5	64.9	2.2	72.2	1.9	65.9	0.0	59.7	6.6
	Zcorrect	23	1.2	37	0.6	22	0.6	29	0.0	30	2.5
	Ztotal	33	1.7	57	1.0	30	0.0	44	0.0	50	1.2
C	TP	72.5	2.5	50.8	1.5	77.5	1.5	61.7	1.4	46.7	3.8
	P	55.1	4.2	46.9	2.0	60.8	2.4	59.7	3.7	53.9	5.4
	Zcorrect	29	1.0	20	0.6	31	0.6	25	0.6	19	1.5
	Ztotal	53	2.1	43	0.6	51	1.0	41	1.5	35	0.6
D	TP	43.3	1.5	65.0	0.0	40.0	0.0	75.0	2.5	50.8	3.8
	P	74.3	4.4	51.3	0.6	75.0	2.1	88.2	5.7	50.8	7.7
	Zcorrect	17	0.6	26	0.0	16	0.0	30	1.0	20	1.5
	Ztotal	23	0.6	51	0.6	21	0.6	34	1.0	40	2.7
E	TP	10.0	0.0	65.8	1.4	5.8	1.4	45.0	0.0	75.0	0.0
	P	66.7	0.0	51.0	2.3	70.0	35.6	61.4	2.5	61.2	0.0
	Zcorrect	4	0.0	26	0.6	2	0.6	18	0.0	30	0.0
	Ztotal	6	0.0	52	1.2	3	0.6	29	1.2	49	0.0

Excluding the possible effect of behavioral singularity, we focus on the diagonal cases which are the estimation accuracies when the moths are trained by their own data. It seems that the classification system is able achieve an average *TP* and *P* of more than 70% and 60% respectively. This estimation result is in line with our objective considering the irregularity of biological systems.

4.2 Behavioral Singularity

We define behavioral singularity again as the fitness analysis of the behavioral data from a moth to the classification filter of another moth. We calculate only the average accuracy for all the moths. The accuracies of the moths which are trained by their own data are excluded from the calculations. Each row represents the average fitness of the behavioral data of a moth to the classification filters of different moths in the sample.

Figure 7 shows the average estimation accuracies for all the five moths in the sample. The bars with light gray represent the average accuracies when the moths are trained by their own data. They are plotted based on the values from the diagonal cases in Table 1. On the other hand, the bars with deep gray represent the average accuracies for the four cases in each row of Table 1, which is excluding the diagonal cases. This indicates the average fitness accuracies for each moth when trained by the data from different moths in the sample. The results seem to suggest that a moth always exhibits a better fitness to its own filter and the average fitness to other filters has a relatively lower performance. In other words, the moths exhibit varying degree of behavioral singularity.

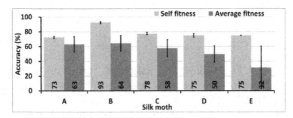

Fig. 7. Estimation accuracies when the moths are trained by their own data and the average accuracies when trained by data from other moths

4.3   Discussion

The preliminary result suggests that the classification approach is able to achieve a decent accuracy of above 70% in the estimation of the stimuli timing for each moth, provided that they are trained using their own behavioral data. It seems to be reasonable to exclude the cases when the moths are trained by the data from different moths because of their behavior singularity. The accuracy of the results seems to be acceptable, considering behavioral variance of biological systems. In addition, the results of the behavior evaluation using the estimation accuracy of the stimuli timing is in line with our hypothesis that the silk moth exhibited behavioral singularity like most of the other biological systems.

In addition, the use of the behavioral data and machine learning algorithms for behavioral analysis serves to complement the conventional neurophysiology method which is not a trivial task. This is because the neuron signals and behavioral data have to be measured simultaneously for comparison. However, the micro-electrode used to capture the neuron signal is connected to the neuron using the suction method which is delicate. Any excessive motion from the insect could cause unstable connection, thus leading to inaccurate or missing data.

5   Conclusion

As the conclusion, this work provides a quantitative evaluation for the CPT behavior of the silk moth by estimating its stimuli timing. The results seem to suggest the presence of a varying degree of CPT behavioral singularity. However, these are only preliminary results in the evaluation of the silk moth's behavior using the machine learning algorithms. Further verification methods are necessary to verify the consistencies in between the experimental results and previous works. The future works include increasing the number of samples to evaluate further the CPT behavior and verification from the perspective of neurophysiology.

Acknowledgements. This study was partially supported by Grants-in-Aid for Scientific Research, MEXT Japan 25420212. The experiments in this presentation were approved based on Article 22 clause 2 of the safety management rule of Tokyo Institute of Technology.

References

1. Cablk, M.E., Sagebiel, J.C., Heaton, J.S., Valentin, C.: Olfaction-based detection distance: A quantitative analysis of how far away dogs recognize tortoise odor and follow it to source. Sensors 8(4), 2208–2222 (2008)
2. Becher, C., Kaul, P., Mitrovics, J., Warmer, J.: The detection of evaporating hazardous material released from moving sources using a gas sensor network. Sens. Actuators, B 146(2), 513–520 (2010)
3. Trincavelli, M., Coradeschi, S., Loutfi, A.: Classification of odors with mobile robots based on transient response. In: IEEE/RSJ Int. Conf. Intelligent Robots and Systems, pp. 4110–4115 (2008)
4. Li, J.G., Meng, Q.H., Wang, Y., Zeng, M.: Odor source localization using a mobile robot in outdoor airflow environments with a particle filter algorithm. Auton. Robot. 30(3), 281–292 (2011)
5. Ishida, H., Tanaka, H., Taniguchi, H., Moriizumi, T.: Mobile robot navigation using vision and olfaction to search for a gas/odor source. Auton. Robot. 20(3), 231–238 (2006)
6. Li, J.G., Yang, J., Cui, S.G., Geng, L.H.: Speed limitation of a mobile robot and methodology of tracing odor plume in airflow environments. Procedia Eng. 15, 1041–1045 (2011)
7. Liu, Z.Z.: Odor source localization using multiple plume-tracking mobile robots, Ph.D. dissertation, Dept. Mech. Eng., Univ. Adelaide, Australia (2010)
8. Marjovi, A., Nunes, J., Sousa, P., Faria, R., Marques, L.: An olfactory-based robot swarm navigation method. In: IEEE Int. Conf. Robotics and Automation, pp. 4958–4963 (2010)
9. Harvey, D.J., Lu, T.F., Keller, M.A.: Comparing insect-inspired chemical plume tracking algorithms using a mobile robot. IEEE Trans. Robot. 24(2), 307–317 (2008)
10. Takashima, A., Minegishi, R., Kurabayashi, D., Kanzaki, R.: Construction of a brain-machine hybrid system to analyze adaptive behavior of silkworm moth. In: IEEE/RSJ International Conference on Intelligent Robot and Systems, pp. 2389–2394 (2010)
11. Ando, N., Emoto, S., Kanzaki, R.: Odor-tracking capability of a silkmoth driving a mobile robot with turning bias and time delay. Bioinspir. Biomim. 8(1), 1–14 (2013)
12. Jacques, J., Bouveyron, C., Girard, S., Devos, O., Duponchel, L., Ruckebusch, C.: Gaussian mixture models for the classification of high dimensional vibrational spectroscopy data. J. Chemometr. 24(11-12), 719–727 (2010)
13. Kuyuk, H.S., Yildirim, E., Dogan, E., Horasan, G.: Application of k-means and Gaussian mixture model for classification of seismic activities in Istanbul. Nonlinear Proc. Geoph. 19, 411–419 (2012)

14. Chang, H.A., Glass, J.R.: Hierarchical large-margin Gaussian mixture models for phonetic classification. In: IEEE Workshop on Automatic Speech Recognition and Understanding, pp. 272–277 (2007)
15. Ari, C., Aksoy, S.: Unsupervised classification of remotely sensed images using Gaussian mixture models and particle swarm optimization. In: IEEE Int. Geoscience and Remote Sensing Symposium, pp. 1859–1862 (2010)
16. Nacereddine, N., Tabbone, S., Ziou, D., Hamami, L.: Asymmetric generalized Gaussian mixture models and EM algorithm for image segmentation. In: The 20th Int. Conf. Pattern Recognition, pp. 4557–4560 (2010)
17. Peñalver, A., Escolano, F., Sáez, J.M.: Color image segmentation through unsupervised Gaussian mixture models. In: Sichman, J.S., Coelho, H., Rezende, S.O. (eds.) IBERAMIA 2006 and SBIA 2006. LNCS (LNAI), vol. 4140, pp. 149–158. Springer, Heidelberg (2006)
18. Bilmes, J.A.: A gentle tutorial of the EM algorithm and its application to parameter estimation for Gaussian Mixture and Hidden Markov Models. Technical Report TR-97-021, International Computer Science Institute, California (1998)
19. Jaeger, H., Haas, H.: Harnessing nonlinearity: Predicting chaotic systems and saving energy in wireless communication. Science 304(5667), 78–80 (2004)
20. Lin, X., Yang, Z., Song, Y.: Intelligent stock trading system based on improved technical analysis and Echo State Networks. Expert Syst. Appl. 38(9), 11347–11354 (2011)
21. Skowronski, M.D., Harris, J.G.: Noise-robust automatic speech recognition using a predictive Echo State Network. IEEE Audio, Speech, Language Process. 15(5), 1724–1730 (2007)
22. Xing, K., Wang, Y., Zhu, Q., Zhou, H.: Modeling and control of McKibben artificial muscle enhanced with echo state networks. Control Eng. Pract. 20, 477–488 (2012)
23. Antonelo, E.A., Schrauwen, B., Campenhout, J.: Generative modeling of autonomous robots and their environments using reservoir computing. Neural Process. Lett. 26(3), 233–249 (2007)
24. Hermans, M., Schrauwen, B.: Memory in linear recurrent neural networks in continuous time. Neural Networks 23(3), 341–355 (2010)
25. Jaeger, H.: A tutorial on training recurrent neural networks, covering BPPT, RTRL, EKF and the "echo state network" approach. GMD Report 159, German National Research Center for Information Technology, Germany (2002)
26. Kubat, M., Holte, R., Matwin, S.: Machine learning for the detection of oil spills in satellite radar images. Machine Learning 30, 195–215 (1998)

Vibration Occurrence Estimation and Avoidance for Vision Inspection System

Kap-Ho Seo, Yongsik Park, Sungjo Yun, Sungho Park, and Jeong Woo Park

Applied Technology Division, Korea Institute of Robot and Convergence,
San 31, Hyoja-dong, Nam-gu, Pohang, Gyeongbuk, 790-784, Korea
neoworld@kiro.re.kr

Abstract. Disturbance / vibration reduction is critical in many applications using machine vision. The off-focusing or blurring error caused by vibration degrades its performance. Instead of going with the more familiar approach like vibration absorber, a real-time disturbance estimation and avoidance is proposed.

Instantaneous motion due to the disturbance is sensed by an accelerometer inertial measurement unit (IMU). Modeling of periodic vibration is done to provide better performance. According to its modeling, the algorithm for vibration avoidance was described.

Keywords: vibration occurrence estimation, vibration avoidance, vision inspection system.

1 Introduction

Typical application of machine vision system in the automated production is quality control, measurement or classification of moving parts, placed on conveyor belts. Different technical issues (lighting problems, vibrations near camera or conveyor belt, etc.) can lead to noisy images and to wrong classifications or faulty measurements by the vision inspection system.

The machine vision in automated manufacturing is used for two-dimensional object classification, optical character recognition, code reading, part measurement and inspection. Normally, all the inspection algorithms are applied on a single image and special lighting in accordance with the required image analyses is used. The optimal lighting conditions assure that there will be no additional noise added to the image, except the one from the camera sensor and the images will be sharp and not blurred. The camera is firmly placed on special mounting and high quality lenses are used.

Vibratory motion of machines and mechanical structures is a source of noise. Consequently vibration analysis and control is an important discipline in engineering. Vibrations of machine parts, illumination or camera system can lead to blurred images. This results in blurred edges in the picture and will have a negative effect on the precision of measurement applications or completely prevent finding the edges.

J.-H. Kim et al. (eds.), *Robot Intelligence Technology and Applications 2,* 633
Advances in Intelligent Systems and Computing 274,
DOI: 10.1007/978-3-319-05582-4_54, © Springer International Publishing Switzerland 2014

Object or camera movement during exposure generates blur. Removing this motion blur is challenging due to estimation of the quantity of motion, which can vary over the time. The motion blur can cause the entire automatic visual inspection to fail especially if the measurements are performed on the the contour of the objects. Additional techniques must be applied if there are possibilities of mechanical camera shaking in high speed vision inspections.

Slight effects can possibly be reduced by means of a shorter exposure time, but cannot completely be avoided. Stronger vibrations must be avoided by all means. Vibration bumpers can avoid the response of a system. They should protect the whole system from vibration, decoupling only the camera system is not sufficient, as the inspection system would swing over inspected parts.

In this paper, the inspection is limited to the semi-automated factory environment, where most of the machinary operates in periodical cycle. Therefore, it the vibration from heavy machine like press occurs periodically, it can be estimated by the proposed algorithm. Consequently, iIt means that the start of image capture can be controlled.

In this research, algorithm for vibration occurrence estimation and avoidance is proposed. Section 2 describes the configuration of testing environment, and the vibration is modelled by mathmatical form. Section 3 shows the proposed algorithm and experimental results, respectively. The efficiency of the developed system is evaluated by the simulation using real vibration data. Concluding remarks are discussed in section 4.

2   System Description

The vision inspection system should operate in the extreme environment where large vertical vibrations are made by several press machines as shown in figure 1-(a). It is simplified into the vibration noise sources and a visual inspection system as figure 1-(b). In visual inspection system, the stable setup is very important to gather an image from camera. Without consideration on periodic vibration, the motion blurred image causes the faulty decision on the correct product. The example of motion blurred image is shown in figure 1-(c). It is hard to determine the physical properties (size, position, and so on) from blurred image.

2.1   Modeling of Vibration

Generally, A simple static equation for vibration is:

$$m\ddot{x} + c\dot{x} + kx = 0. \tag{1}$$

Dividing through by m, the dimensionless parameters ω and ζ:

$$\ddot{x} + 2\zeta\omega_n\dot{x} + \omega_n^2 x = 0 \tag{2}$$

where ω_n represents the undamped natural frequency, and ζ is the viscous damping ratio. For the purpose of this example, we assume the underdamped case ($\zeta < 1$). The solution to this equation is

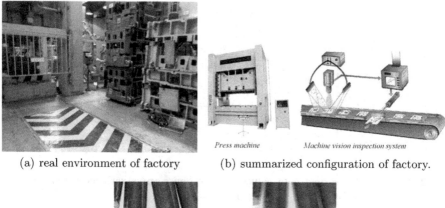

(a) real environment of factory

Press machine   Machine vision inspection system

(b) summarized configuration of factory.

Before   After

(c) motion blur

Fig. 1. Failure problem of machine vision inspection system from the nearby vibratory system. Press machinary generates periodic vibration. It results in faulty inspection.

$$x(t) = Ae^{-\zeta\omega_n t}sin(\omega_d t + \phi) \tag{3}$$

where ω_d is the damped natural frequency, equal to $\omega_n\sqrt{1-\zeta^2}$.

This equation is more useful if all of the terms as functions of parameters ω_n and ζ are written

$$x(t) = \sqrt{\frac{(v_0 + \zeta\omega_n x_0)^2 + (x_0\omega_n\sqrt{1-\zeta^2})^2}{(\omega_n\sqrt{1-\zeta^2})^2}}e^{-\zeta\omega_n t}$$
$$\times sin[(\omega_n\sqrt{1-\zeta^2})t + tan^{-1}(\frac{x_0\omega_n\sqrt{1-\zeta^2}}{v_0 + \zeta\omega_n x_0})].$$

$$\tag{4}$$

While this equation admittedly looks intimidating, note that it only depends on four quantities: x_0, v_0, ω_n, and ζ. Note the similarity between the parameters identified here and the ones relevant to the undamped case; the only difference is the addition of the viscous damping coefficient.

2.2   Occurrence Function

It assume that vibration signal occurs periodically during inspecting. Therefore, periodic function can express the periodic vibration signal as shown in figure 3.

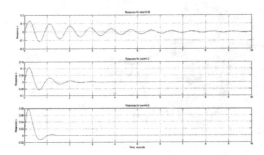

Fig. 2. Response for various zeta values. The response varys with increasing viscous damping coefficient for $x_0 = 0$, $v_0 = 1$, and $\omega_n = 7$.

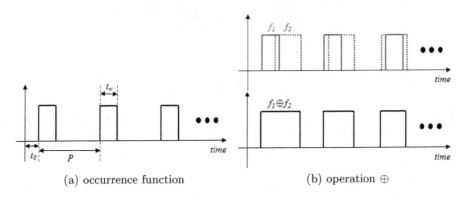

(a) occurrence function   (b) operation $\oplus$

Fig. 3. Graphical representation of an occurrence function and its operation $\oplus$

Before exact modelling of signal, the occurrence of interested signal should be identified. Therefore, the occurrence function was defined as equation (5). This function has the logical meaning.

$$f(t_{P,t_0,t_w}) = u(t - nP - t_0) - u(t - nP - t_0 - t_w). \tag{5}$$

where P is the period of vibration signal, t_0 is time offset, and t_w is the duration of vibration over threshold level. Occurrence function, f means that significant vibration signal occurs.

In this paper, it means that a captured image during occurrence can not be used for inspection, since it contain lots of motion noise. Therefore, visual inspection system should gather an image when the alarm from occurrence function disappears.

Vibration in real environment with several press machines can be combination of function 5 as

$$V_{total} = f_1 \oplus f_2 \oplus \cdots \oplus f_n. \tag{6}$$

Since the operation $\oplus$ of occurrence function is logical combination, maximized V covers whole time for testing signal. In break time of V, the camera is allowed to capture an image for inspection. Therefore, the break time should be maximized, through finding minizied coverage of each function f.

3   Vibration Occurrence Estimation and Avoidance

The standard GA process [1] is used for finding the proposed occurrence function. First, a population of chromosomes is creadted. Second, the chromosomes are evaluated by a defined fitness function. Third, some of the chromosomes are selected for performing genetic operations. Forth, genetic operations of crossover and mutation are performed. The prodeced offspring replace their parents in the initial population. In this reproduction process, only the selected parents in the third step will be replaced by their corresponding offspring. This GA process repeats until a user-defined criterion is reached. In this paper, some characteristic configuration of chromosome, operation, and fitness function is described in detail. Since other procedure is similar to that of standard genetic algorithm, that is not explained in this paper.

3.1   GA Approach to Occurrence Function Estimation

Determinating the n occurrence functions is same as finding its $3n$ parameters as equation (5). The number of necessary occurrence function, n, can be intuitively known from real signal.

As mentioned in previous section, in break time of V, the camera is allowed to capture an image for inspection. Therefore, the two following objectives should be obtained together. The one goal is maximization of the break time enough for capturing an image. Another is to reduce the motion blur in an captured image, which means to avoid the periodic vibration sequences. In order to match both goals, the exact matching between the occurrence function from an original signal and the calculated signal, is required. Then, the problem is changed to find a best solution to minimize mis-coverage of total occurrence function V_{total} to the original signal.

For this purpose, the parameters of each function are determined through the genetic algorithm (GA). Each gene is encoded with real value within its bound as shown in figure 4-(a).

3.2   Genetic Operations

The genetic operations are to generate some new chromosomes (offspring) from their parents after the selection process. They include the crossover and the mutation operations.

The crossover operation is mainly for exchanging information from the two parents, chromosomes P_1 and P_2, obtained in the selection process.

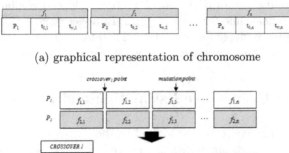

(a) graphical representation of chromosome

(b) operators (crossover, mutation)

Fig. 4. Configuration of genetic algorithm. the crossover and mutation is operated from the selected parents. The chromosome is described in (a).

The mutation operation is to change the genes of the chromosomes. Consequently, the features of the chromosomes inherited from their parents can be changed.

The two parents generate four chromosomes: two by crossover 1, one by crossover 2, and one by mutation. The each operator is described as follows:

- Crossover 1: one-point crossover. The genes is exchanged on the randomly selected point. The new offspring by crossover 1 is generated after exchanging their chromosome string.
- Crossover 2: average value with two parent. The selected parents share their value. Then, it generates the middle points in order to converge the better solution between two parents.
- Mutation: random value on randomly selected control point. This value is randomly selected between predefined upper and lower bounds in each value.

The graphical representation of operators is described in figure 4-(b).

3.3   Evaluation

Each chromosome in the population will be evaluated by a defined fitness function. The optimization problem is defined to be solved to minimize a fitness. The better chromosomes will return lower values in this process.

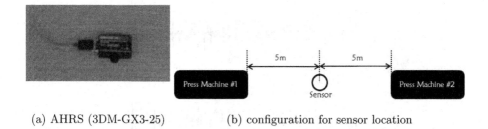

(a) AHRS (3DM-GX3-25)   (b) configuration for sensor location

Fig. 5. Measurement condition and location of sensor. Attitude Heading Reference System (AHRS) was used for measuring vertial vibration signal using acceleration according to gravity direction.

Fitness function for genetic algorithm is designed as following:

$$fitness = \frac{card((V_{total} \oplus f_d)^c)}{card((V_{total} \oplus f_d))} \tag{7}$$

where V_{total} is the combination of occurrence functions found from GA, f_d is the occurrence function from real signal, f^c means the complementary of occurrence function f, which is calculated the difference between the whole testing time for testing and the occurrence timee. $card_i(.)$ is the cardinality (i.e., occupancy) on through the whole testing time.

Through the whole procedure (operation, evaluation, selection), the better solution is obtained.

3.4   Simulation Result Using Real Vibration Data

In order to verify the effectiveness of proposed algorithm, simple system for the vibration measurement and image gathering was prepared. Attitude Heading Reference System (AHRS) was used for measuring vertial vibration signal using acceleration according to gravity direction as shown in figure 5.

Vibration signal for 100 seconds was used for the proposed algorithm. The raw signal and its occurrence is plotted in figure 6-(a). Since raw signal usually has a noise, the user-defined threshold was determined experimentally. The occurrence data is used for calculating a fitness value as f_d in equation 7.

According to the proposed algorithm in the previous section, the combination of occurrence functon, V_{total}, was obtained. After 10,000 generations on GA procedure, V_{total} was calculated as shown in figure 6-(b). Comparison between original signal and calculated occurrence shows a good result for function estimation.

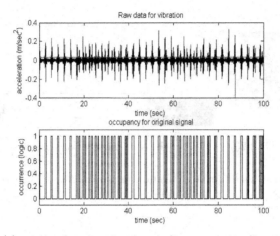

(a) raw acceleration signal according to gravity direction
and its occurrence

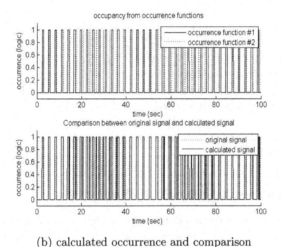

(b) calculated occurrence and comparison

Fig. 6. Signal processing for occurrence estimation. (a) raw acceleration signal according to gravity direction and its occurrence (b) comparison between original signal and calculated occurrence.

4   Conclusion

In order to reduce the effect from disturbance such as vibration, the algorithm for vibration occurrence estimation and avoidance was proposed. Through the proposed algorithm, Instantaneous motion due to the disturbance is sensed by an accelerometer inertial measurement unit (IMU). The effectiveness was proven with experiment in real environment.

Acknowledgment. This research was partially supported by

Ministry of Knowledge Economy (MKE), Korea Institute for Advancement of Technology (KIAT) and Daekyung Leading Industry Office through the Leading Industry Development for Economic Region,

Ministry of Culture, Sports and Tourism(MCST) and Korea Creative Content Agency(KOCCA) in the Culture Technology(CT) Research & Development Program (Immersive Game Contents CT Co-Research Center),

Ministry of Knowledge Economy (MKE), Korea Institute for Advancement of Technology (KIAT) and Biomedical IT Convergence Center.

References

1. Holland, J.H.: Adaptation in Natural and Artificial Systems. Univ. Michigan Press, Ann Arbor (1975)
2. Seo, K., Shin, J., Kim, W., Lee, J.: Real-Time Object Tracking and Segmentation Using Adaptive Color Snake Model. International Journal of Control, Automation, and Systems 4(2), 236–246 (2006)
3. Seo, K., Lee, J.: Object Tracking and Recognition Based on CPCA. In: Proc. 2005 CACS, CDROM (2005)
4. Seo, K.: Unified Framework for Object Tracking and Recognition Based on Condensation Principal Component Analysis in a Structured Environment. Ph. D. Thesis, Dept. EECS, KAIST (2009)

Part III
Emerging Applications of Robot Intelligence Technology

Peter Xu

The intelligence technology enables robots to think, feel and interact like their biological counterparts, which have been witnessed by many robotic systems such as those exploring on mars, operating on patients, driving on roads, caring the elderly in retirement villages and performing dance on stages. This chapter consists of seven groups of 30 papers, presenting some latest emerging robot intelligence technologies and applications. They are Computer Vision, Humanoid Robots, Assistive/Rehabilitative/Medical Robots, Underwater Vehicles, Aerial/Wheeled/Legged Robots, Heterogeneous Robotic System and Smart Actuators and Sensors.

Computer Vision has a group of 8 papers dedicated to its applications in rehabilitation of paralyzed lower extremity, soccer robots, self-driving vehicles, structural health monitoring, object recognitions and human visual processing intelligence, where a variety of intelligence technologies other than computer visions also play a important role, such as biomechatronics, simulation technology, embedded computing and processing, and cloud computing.

1) Computer Vision System as Part of Bio-mechatronic Rehabilitation Simulator
2) FPGA Implementation of Global Vision for Robot Soccer as a Smart Camera
3) Obstacle & Lane Detection and Local Path Planning for IGVC Robotic Vehicles using Stereo Vision
4) Adaptive Regions of Interest based on HSV Histograms for Lane Marks Detection
5) One-way ViSP (Visually Servoed Paired Structured Light) for 6-DOF Structural Displacement Measurement
6) Target-driven Visual Words Representation via Conditional Random Field and Sparse Coding
7) Cloud-based Object Recognition: A System Proposal
8) Automatic Salient Object Detection Using Principal Component Analysis

In the group of **Humanoid Robots,** there are 4 papers talking about the design, programming and control of humanoid robots for dancing, bowling, traffic guiding and soccer playing, respectively. A number challenges in developing such a special purposed humanoid robot are dealt with, involving light-weighted construction, steady coordination of walking, localization of pins and ball and throwing angle decision making, integration of speeding camera and radar sensor, and fast localization of soccer.

1) Design of Child-sized Humanoid Robot for Performance
2) Bowling with the DARwIn-Op Humanoid Robot
3) Design and Control of a Humanoid Robot for Traffic Guidance
4) Trigonometry Technique for Ball Prediction in Robot Soccer

Assistive/Rehabilitative/Medical Robots of 5 papers deals with robotic intelligence in the applications of powered prostheses, wearable jaw and hand exoskeletons, computer assisted minimally invasive surgery and facial paralysis analysis. Novel flexible sensors, actuators, touchless interface, surface electromyography (sEMG), chaos theory can be seen in these applications.

1) Classifier Selection for Locomotion Mode Recognition Using Wearable Capacitive Sensing System
2) Development of Jaw Exoskeleton for Rehabilitation of Temporomandibular Disorders
3) Towards a Touchless Master Console for Natural Interactions in Sterilized and Cognitive Robotic Surgery Environments
4) Flexible and Wearable Hand Exoskeleton and Its Application to Computer Mouse
5) Application of the Chaos Theory in the Analysis of EMG on Patients with Facial Paralysis

Underwater Vehicle contains 4 papers that are related one to another. They present design, software architecture, localization, navigation of an underwater vehicle for effective inspection of underwater structures, in which the localization and navigation is achieved by artificial landmarks through fusion of sensors such as Doppler velocity log, inertial sensors and electronic compass; the objects are detected and recognized through use of acoustic images; and the software architecture is implemented by a middleware.

1) Design, Implementation, and Experiment of an Underwater Robot for Effective Inspection of Underwater Structures
2) Issues in Software Architectures for Intelligent Underwater Robots
3) Estimation of Vehicle Pose Using Artificial Landmarks for Navigation of an Underwater Vehicle
4) A New Approach of Detection and Recognition for Artificial Landmarks from Noisy Acoustic Images

In the group of **Aerial/Wheeled/Legged Robots,** the first paper presents a flapping wing micro aerial robot with a focus on the design, implementation and fabrication of the circuitry used for split-cycle constant period wind beat modulation; the second paper discusses the dynamic balance of a single wheeled vehicle with two fly wheels through neural network control; the third paper deals with the dynamics and simulation of a legged robot; and the last paper is about the kinematics and dynamics of a 4DOF Delta parallel manipulator.

1) Implementation of Split-Cycle Control for Micro Aerial Vehicles
2) Neural Network Control for the Balancing Performance of a Single-wheel Transportation Vehicle: Gyrocycle
3) Dynamic Simulation of a Sagittal Biped System
4) Analysis of Kinematics and Dynamics of 4DOF Delta Parallel Robot

Heterogeneous Robotic Systems has 2 papers in its group. One paper talks about the development of an easy-to-use middleware for the purpose of integrating different heterogeneous robotic systems, where the notion of efficient structural data exchange is defined. The other paper takes on the challenge that a heterogeneous network of humans, robots and agents constantly change their requirements for mobile provider.

1) Axon: a Middleware for Robotic Platforms in an Experimental Environment
2) Enhancing Wi-Fi Signal Strength of a Dynamic Heterogeneous System Using a Mobile Robot Provider

Smart Actuators and Sensors is the last group of 3 papers. The first paper deals with an IPMC (ionic polymer metal composite) actuator where an adaptive neuro-fuzzy inference system is used for its trajectory tracking control. The second paper uses a combination of MEMS inertia sensors (accelerometer and gyroscope) for the fall detection of a wireless excavator control to prevent any unintentional fall of the physical excavator. The third paper seeks to harness the solar energy and store it in hydrogen to power an industrial mobile robot platform.

1) Adaptive Neuro-Fuzzy Control for Ionic Polymer Metal Composite Actuators
2) Fall Detection Interface of Remote Excavator Control System
3) Solar Energy as an Alternative Energy Source to Power Mobile Robots

Computer Vision System as Part of Bio-mechatronic Rehabilitation Simulator

Sergey M. Sokolov, Alexander K. Platonov, Andrey A. Boguslavskiy,
and Oleg V. Triphonov

Keldysh Institute for Applied Mathematics RAS, Moscow, Russia
sokolsm@list.ru, platonov@keldysh.ru, anbg@mail.ru, tob@imail.ru

Abstract. There is a description of the first stage of researches aimed at equipping with the vision system ("VS") for the rehabilitation simulator being developed for the rehabilitation of spinal patients who suffer paraplegia. Based on the VS information it is supposed to develop some software tools to solve the issue of mechatronics of the bio-training simulator in order to create artificial motions of the paralyzed legs imitating natural legged movement. There's discussed a hardware- and software of the system, methods of data-acquisition concerning the leg movements of a healthy person. There're given results of the first experiments aimed at acquiring data needed for the analysis of the geometrical parameters of the leg movement. There've been received some accuracy assessments to define angles and angular velocities in the hip, knee joint, and ankle joint.

This research has been performed with the support of RFBR grants No. 11-01-12060-ofi-m-2011, No 12-08-12030-ofi-m-2012.

Keywords: bio-mechatronic rehabilitation simulator, vision system, geometrical parameters of walking, artificial leg movement.

1 Introduction

One of the main problems of medicine is the rehabilitation of spinal patients with paraplegia, i.e. legs motionlessness caused by a backbone trauma. At the rehabilitation simulator when studying the spinal cat in the Physiology Institute n.a. I.P. Pavlov (hereinafter "IPh") there've come a famous experimental result, i.e. after making some special effects the spinal cat began to walk almost independently. This result opened some hope for the possibility to cure people of a similar serious illness. As a note, the effect of stepping generators recovery in the medulla spinalis of the spinal cat was obtained by IPh expert Mr. O.A. Nikitin, Biological Sciences PhD, who not only applied the regular cat's walking at the simulator, but, which is also important, made long-time massaging of the cat's hind feet on his knees while making forced movement of the cat's feet to the beat of walking [1].

With these experiments IPh showed that the key thing of the possible Paraplegics rehabilitation is a poorly-studied problem of organizing afferent (centripetal) signals coming to the cerebrospinal axis during the rehabilitation walking on the simulator.

These afferent signals are formed with the nerve endings of a foot and musculoskeletal system of an animal and a human being at the moment of positioning legs, and the following supporting and shifting of legs. Meantime, the formation of these afferent signals with the simulator is needed not only to rehabilitate the stepping generators in the medulla spinalis, but, which is also important, to prevent any atrophy of inactive muscles and dying of the nerve cells related to the muscles during the *apoptosis*, i.e. termination of physiological processes in the muscles and nerve endings in case of their inactivity.

Researches concerning the formation of afferent and stimulating signals were arranged by IPh n.a. I.P.Pavlov together with Physical Training Academy (Velikie Luki town) and Institute of Medical-and-Biological Problems at the Russian Science Academy (hereinafter "IMBP") . The works were also done in the USA with the participation of the Russian scientists. These researches have acute need for the development of mechatronic means of managing and synchronizing the movement processes with the processes of forming some stimulating effects on the nerves of the back bone and return afferent signals within the multilevel nerve system for the motion synthesis of a human being.

In terms of applying these results in mid-nineties there appeared an acute issue concerning the creation of a specific bio-mechatronic simulator for a human being which would differ from a usual treadmill-type apparatus and would be equipped as a medical rehabilitation simulator with effector and sensor control elements. This simulator is needed to study the processes of docking the management of high-power engines of legs drive components with the low return signals concerning their physiological condition.

The important place in the system of "sensing"/information supply for the mechatronic simulator is given to the Vision System (VS). VS permits to non-connately register movements of both the parts of the patient's body and parts of the simulator. VS solves the important issue of verifying the movement models of the simulator, and mainly the control of correspondence of the patient's feet movements in the simulator with the natural movements when walking.

One of known systems for registration of parts of a body movements is the system using magnetic markers. There is a positive experience of use of electromagnetic systems of type Spatial Tracking System (Fastrack™ Polhemus), Flock of Birds, MiniBirds and TrakStar (Ascension Technology Corporation). Such system is successfully applied to control of hands movements [2-4]. Application of similar system has a number of the restrictions which are not allowing effectively of it to use as a part of the projected rehabilitation simulator. To restrictions are: a working zone (a range of controllable movements) less than 1 m; necessity of removal from a working zone of all metal subjects which essentially influence accuracy of work. Overcoming the specified restrictions of system on the basis of magnetic markers, the vision system should provide accuracy of movements registration not worse, than ±3~5 mm (average parameter of electromagnetic system).

On the first stage of the project execution there has been selected the following way for the development/formation of VS as part of the simulator. The software is based on the frame for the stereo-system design (fig.1-2) and tracking algorithms for the small targets [5, 6]. The hardware is formed under COTS technology with account of the world and own experience in VS development.

2   Packaging of the VS Hardware

VS tracking of the human movements has been used for a long time [7]. Most of works are known for the research of the efficiency of the sportsmen's movements and for the creation of various effects in the cinema. With the recent growth of the capacities of computer technologies and hardware support for the collection and primary processing of the visual data, there have been created compact devices for a wide range of users. These devices allow in ordinary conditions to rehabilitate the general kinematics of the observed person's action [8]. Almost in all enlisted systems the kinematic model of a human leg is rather sluggish, mainly in the tibio-tarsic. Here, as a rule, there's defined one point [8] or two pints. At the same time, anatomic features of the leg constitution and their detailed observation say that the actual legs movement is definitely more complicated and demands a more appropriate model when rendering. Moreover, the accuracy of feet movements' determination shall be very high with regard to the range of observed movements. With the foot size about 1000mm and space movements within one step cycle (two steps) of about 1600mm, it is necessary to fix shifting of the concerned objects with millimeter accuracy. VS characters demand a range of the described area with one vision field (VS more accurately register shifting of the objects within one vision field, which do not need "linking" of various vision fields). Thus, qualitative estimations of the VS characteristics are being formed in order to be applied as a part of the bio-mechatronic simulator.

The basic thing is the high-resolution cameras with the following characteristics, i.e.:

- resolution 1600x1200, color model RGB, 8 bit/color channel;
- external synchronization (allows to arrange a stereo pair to watch dynamic objects);
- digital output Gigabit Ethernet;
- progressive scanning (access to the random fragments of raster);
- frequency of visual data input from 15 to 200 hz.

These resolution characteristics allow to receive the accuracy of the space resolution when registering:

- person's movements when walking up to 3mm;
- movements of the patient's feet and those of mechanical parts in the rehabilitation simulator up to 1mm.

Light-emitting diodes (fig. 3, 4) are applied to mark the interesting parts of the patient's body. This non-expensive thing definitely accelerates and increases the effectiveness of the development, mainly on the stage of forming an experimental model of the rehabilitation simulator.

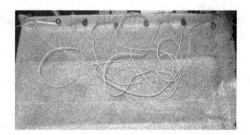

Fig. 1. Light-emitting diodes (without fixing devices)

In order to provide verification of, first, kinematic, then dynamical schemes/ models of the observed objects, it is necessary to separate from the visual data some invariant peculiarities, i.e. characteristic points, to be reference points for the establishment of mathematic models for the movement of concerned objects. In previously uncertain conditions of the breadboard model of the rehabilitation simulator, when it's hard to unambiguously determine objects in the video-camera vision field while the system adjustment for the recognition of a new object demands significant efforts and time, the location of the known objects in the video-system vision field substantively simplifies the task of determining the concerned objects.
Requirements for the markers:

- to provide fixture on the human body with ±1 mm accuracy;
- to provide repeatability for the attaching point
- to clearly indicate (observation on all movement stages, simplicity in terms of separation on brightness and geometry) the concerned place of the human body or details of the rehabilitation simulator
- Fixing convenience (not to impede moving, other equipment)

The structure of the applied marker is shown on fig. 2. The marker is done of two parts, i.e.: cone-shaped part - the base with the incorporated piston which provides the attachment to the human skin or to the smooth surface of the dummy, and a cylindrical element with a fixed light-emitting diode and battery.

To collect and process the visual data a common laptop with the following characteristics is used: CPU Intel Core i5 3.4 GHz, RAM 4 Gb.

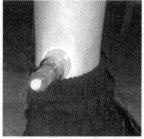

Fig. 2. Structure of the luminous marker (left: overall view, right: attachment to the human leg)

Fig. 3. Location of markers on the human being on the side. Four markers with numbers 3 – 6 are fixed on the leg.

3   Collection and Processing of the Visual Data

The task of detecting the moving marker icons on the images may be referred to the tasks of tracking small targets. The shape of a small target may change with the time due to the change of orientation of object with regard to the visual detector, and, as a result, of discretization errors at the interface of the icon when making a digital image. As a whole, the shape is not a reliable character of a small object. This definite restriction complicates the application of a target model on the image in the tracking system.

Circular markers on the images are applied quite frequently [9, 10, 11]. In the work [12] ref. the analysis for the accuracy of positioning the location of markers when comparing the markers presented as circular icons and chess squares it's noted that the angles of chess squares are less liable for moving during projective transformations and nonlinear distortion deformations compared with the circular markers. This effect becomes important when we have distortion deformations of the camera lens [9]. The markers' layout of supporting points of the human foot or dummy as chess squares is difficult for execution and inconvenient in practice. Thus, as already noted, for the above-described researches we used markers, which form of icons is similar to the circular form on the image (fig. 2, 3). In [5, 6] we review the structure of algorithms for the detection and tracking of a small-size target on the sequence of images.

Similar to the approach described in [5, 6], the algorithm of markers' tracking on the sequence of images is divided into two parts, i.e. initial detection of images and tracking these icons.

Algorithm of initial detection is based on the processing of the image within a square sliding window, which dimensions are several times bigger than the dimensions of the marker (4-5 times bigger). In every position of the window there's performed a thresholding with the separation of an image of the brightest connected component. Coordinates of the center of gravity of the components, which dimensions and form are consistent with the marker icon, are noted as candidates for the tracking.

After the initial processing of the image, on the following images of the sequence there's provided some tracking of the icons of the images. In the position of the predicted location of every marker there's preformed a thresholding which is analogous to the initial detection algorithm. For the account of the time reduction by excluding the image processing with the sliding window, the time needed for the markers detection is decreased by 10-15 times [5, 6].

4   Experiments with the Pilot System

The task of the experiments on this stage of researches was to determine the system capacities in terms of velocity and accuracy of recording the movements of the concerned objects. As such objects we used some marked parts of the human legs and parts of the dummy simulating the positioning of the patient's feet in the rehabilitation simulator.

The following kinematic scheme of the human foot (on the first stage it's flat) (fig. 4) is proposed. In the performed researches the experiments were arranged within traditional models (within 3-6 markers).

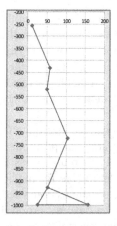

Fig. 4. Example of a flat pivoted model of a human foot which takes into account the peculiarities of the ankle-joint anatomy

Based on the VS measurements there're calculated values of the angles and angle velocities in the pivoted foot model (fig. 4). These data are calculated for the markers located in the places corresponding to the model of the human feet joints, i.e.: hip, knee-joint, ankle-joint (3, 4, 5, 7 on fig. 3 and 4).

The experiment controlled the movements of a person walking 5km/h (~ 1,4 m/s). There were tracked movements within four steps, i.e. two cycles of walking in the straight and opposite directions.

Movements of the dummy's legs in the rehabilitation simulator (fig. 5) were done "manually" one-by-one with each degree of movement. The velocity of movements approximately (measured with the second-counter) corresponded to the velocities of legs walking 5km/h.

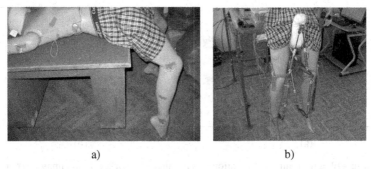

a)   b)

Fig. 5. Dummy with the markers as part of a model of the bio-mechatronic simulator. a) markers fixed on the dummy; b) the dummy as part of a model of the bio-mechatronic simulator.

4.1   Determination of Kinematic Characteristics of a Human Walking

Fig. 6-9 show examples of determining the location of markers on the human leg. σ_x, σ_y – mean square deviations of the measured quantities along the axis Ox and Oy respectively. These characteristics of the measurement accuracy are obtained as a result of comparing the measured coordinates of the markers on the picture with the results of the smoothing averaging filtration with the time window 1.5 mm/pixel. The scaling coefficient is about 1.5 mm/pixel.

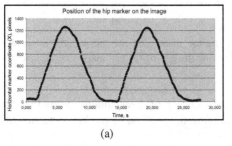
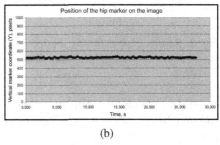

(a)   (b)

Fig. 6. Results of determining the location of the hip marker on the image ($\sigma_x \approx 3.1$ pixel, $\sigma_y \approx 0.8$ pixel). (a) Horizontal marker coordinate on the image; (b) vertical marker coordinate on the image.

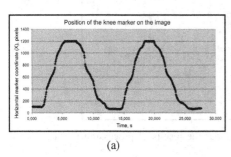
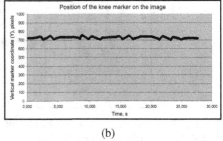

(a)   (b)

Fig. 7. Results of determining the location of the knee marker on the image. ($\sigma_x \approx 4.4$ pixel, $\sigma_y \approx 1.2$ pixel). (a) Horizontal marker coordinate on the image; (b) vertical marker coordinate on the image.

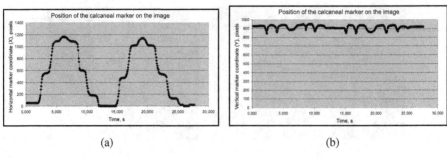

(a)   (b)

Fig. 8. Results of determining the location of the calcaneal marker on the image. ($\sigma_x \approx 6.6$ pixel, $\sigma_y \approx 3.7$ pixel). (a) Horizontal marker coordinate on the image; (b) vertical marker coordinate on the image.

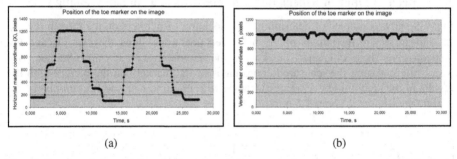

(a)   (b)

Fig. 9. Results of determining the location of the toe marker on the image. ($\sigma_x \approx 9.4$ pixel, $\sigma_y \approx 3.8$ pixel). (a) Horizontal marker coordinate on the image; (b) vertical marker coordinate on the image.

On the example of chain "hip – knee" the accuracy of determination is $\sigma \approx 14$ mm and may be increased at the account of changing the method of markers attachment (see p.2).

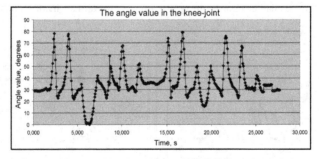

Fig. 10. The value of angle in the knee-joint under the results of processing the location of markers in the human walking process

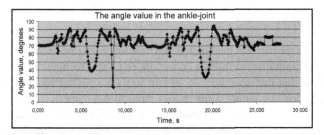

Fig. 11. The value of angle in the ankle-joint under the results of processing the location of markers in the human walking process

4.2   Determination of Kinematic Characteristics of the Dummy's Movements

Fig. 12-13 show examples of determining the location of markers on the dummy's leg. σ_x, σ_y mean square deviations of the measured quantities along the axis Ox and Oy respectively. These characteristics of measurements accuracy are obtained as a result of comparing the measured coordinates of the markers on the image, with the results of the smoothing averaging filtration with the time window 0.2 s. The scaling coefficient is about 1.5 mm/pixel.

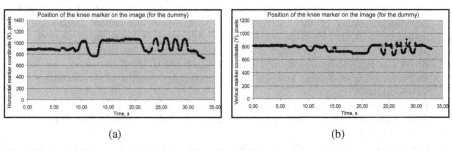

(a)   (b)

Fig. 12. Results of recording the location of the knee marker on the image. ($\sigma_x \approx 3.6$ pixel, $\sigma_y \approx 6.2$ pixel). (a) Horizontal marker coordinate on the image; (b) vertical marker coordinate on the image.

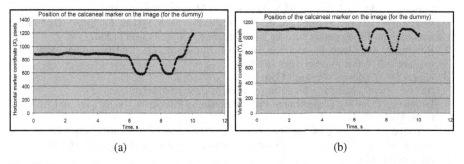

(a)   (b)

Fig. 13. Results of recording the location of the calcaneal marker on the image. ($\sigma_x \approx 3.2$ pixel, $\sigma_y \approx 3.6$ pixel). (a) Horizontal marker coordinate on the image; (b) vertical marker coordinate on the image.

5   Conclusion

The work describes the VS packaging and researches to be included into the bio-mechatronic rehabilitation simulator, done under COTS technology. The performed researches show that the reviewed VS permit to record the movements of the human legs with high accuracy within a flat model, both in the natural walking and in walking simulation as provided by the simulator. Meantime, there's provided convenience of the non-contact registration and cost-effectiveness (for the account of unification of various VS components).

The obtained accuracies permit to apply VS as a part of control system for the simulator when doing natural human legs' movements.

The future of the just-described studies will be the development of a system for recording space movements of the concerned objects (transfer to 3D model). On the following stages, when recording the location of the whole equipment and conditions for the surveillance over the rehabilitation simulator space, it's possible to replace the specific markers with different clothes elements or parts of equipment.

References

1. Okhotsimsky, D., Nikitin, O., Gerasimenko, Y., Serbenjuk, N., Mitutsova, L., Delchev, K., Vitkov, V., Platonov, A., Yaroshevsky, V.: A biomechanical stimulator for scientific-experimental study of the regeneration of spinal cord locomotion capabilities after traumatic break. In: Proc. of the Int. Conf. Advanced Problems in Mechanics 2005, Minisymposium on Biomechanics, St. Petersburg, Russia, pp. 394–400 (2006)
2. Biryukova, E.V., Roby-Brami, A., Frolov, A.A., Mokhtari, M.: Kinematics of human arm reconstructed from Spatial Tracking System recordings. Journal of Biomechanics 33(8), 985–995 (2000)
3. Prokopenko, R.A., Frolov, A.A., Biryukova, E.V., Roby-Brami, A.: Assessment of the accuracy of a human arm model with seven degrees of freedom. Journal of Biomechanics 34, 177–185 (2001)
4. Platonov, A.K., Frolov, A.A., Birjukova, E.V., Prjanichnikov, V.E., Yemelyanov, S.N.: Biomechanics methods in the training apparatus of a human hand. Preprint KIAM RAS, No 82 (2012)
5. Boguslavsky, A.A., Sokolov, S.M.: Algorithms for tracking a small-sized target as part of VS. In: Materials of the Scientific School-Conference on Mobile Robots and Mechatronic Systems, pp.81-90. MSU n.a. M.V. Lomonosov, Moscow, Russia (2004)
6. Boguslavsky, A.A., Sokolov, S.M.: The realtime Vision System for small-sized target tracking. Int. J. Computing Science and Mathematics 1(1), 115–127 (2007)
7. Rosenhahn, B., Klette, R., Metaxas, D. (eds.): Human Motion. Understanding, Modelling, Capture and Animation. Springer (2008)
8. Kinect for Windows, Microsoft Corp.,
   http://www.microsoft.com/en-us/kinectforwindows/
9. Luhmann, T., et al.: Close Range Photogrammetry: Principles, Techniques and Applications. Wiley (2007)
10. Heikkilä, J., Silvén, O.: A four-step camera calibration procedure with implicit image correction. In: Proc. IEEE Conf. on Computer Vision and Pattern Recognition (CVPR 1997), pp. 1106–1112 (1997)
11. Wohler, C.: 3D Computer Vision. Springer (2009)
12. Mallon, J., Whelan, P.F.: Which pattern? Biasing aspects of planar calibration patterns and detection methods. Pattern Recognition Letters 28(8), 921–930 (2006)

FPGA Implementation of Global Vision
for Robot Soccer as a Smart Camera

Miguel Contreras, Donald G. Bailey, and Gourab Sen Gupta

School of Engineering and Advanced Technology
Massey University, Palmerston North, New Zealand
{M.Contreras,D.G.Bailey,G.SenGupta}@massey.ac.nz

Abstract. An FPGA-based smart camera is being investigated to improve the processing speed and latency of image processing in a robot soccer environment. By moving the processing to a hardware environment, latency is reduced and the frame rate increased by processing the data as it is steamed from the camera. The algorithm used to track and recognise robots consists of a pipeline of separate processing blocks linked together by synchronisation signals. Processing of the robots location and orientation starts while the image is being captured so that all the robot data is available before the image has been fully captured. The latency of the current implementation is 4 rows, with the algorithm fitting onto a small FPGA.

1   Introduction

The goal of this paper is to improve the accuracy of the robot position and orientation data in a robot soccer environment, by increasing the resolution and frame rate as outlined in a previous paper [1]. Because of the rapidly changing nature of robot soccer, information needs to be processed as quickly as possible. The longer it takes to capture and process the information, the more inaccurate that information becomes. Therefore the accuracy can be improved by increasing the frame rate of the camera and reducing the latency of the processing. This can be achieved by implementing the image processing within an FPGA-based smart camera. By processing streamed data instead of captured frames it is possible to reduce the latency. By also increasing the frame rate it is possible to gather more data and thereby improve the ability to control the robots in the fast changing environment.

The idea of using FPGAs as a platform for a smart camera is not a new one. In fact researchers have used them in various ways in their smart camera implementations. Broers *et al.* [2] outlines a method of using an FPGA-based smart camera as the global vision system for a robot soccer team competing in the RoboCup. Even though they demonstrated that it is possible to use a smart camera to implement the image processing and produce positioning data in real time, their system required multiple FPGAs.

Dias *et al.* [3] describes a more modular approach, where interface cards are connected to the FPGA to perform different tasks, such as Firewire communication, memory modules, as well as the image sensor. This approach does use a single FPGA, however it needs multiple daughter boards in order to function correctly. This can take up a lot of space and add weight, which can be problematic for a mobile application.

This paper describes an implementation of a smart camera to function as the global vision system for a robot soccer team. To accomplish this, a single off the shelf FPGA development board (Terasic DE0 Nano, although a DE0 was used for initial prototyping because it has a VGA output) was used with a low cost image sensor (Terasic D5M CMOS camera module) to recognise and track the position and orientation of the robot players. The FPGA then transmits the robot data to a separate computer for strategy processing and robot control.

2   The Algorithm

The algorithm is an implementation of that proposed in [1]. It is split up into separate blocks, each handling a particular task or filter, shown in Fig. 1. The first part of the algorithm is completed using pipelined processing of streamed data. This allows the image processing to begin as soon as the first pixel is received, removing the need to buffer the frame into memory.

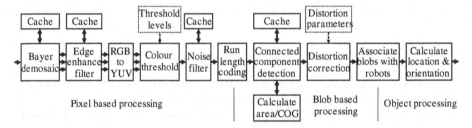

Fig. 1. The block diagram of the algorithm (from [1])

The camera streams 12-bit raw pixel data to the Bayer filter, which derives separate RGB colour channels. The edge enhancement filter removes colour bleeding introduced during the Bayer interpolation process; this allows for more accurate centre of gravity calculation. The signal is then converted into a YUV-like colour space and thresholded to detect the colours associated with the ball and colour patches. A noise filter is used to remove single pixel wide noise and artefacts of the Bayer filter, especially around edges and lines. The filtered signal is then passed through a single pass connected component analysis algorithm which groups the detected pixels together into blobs associated with different colours on top of the robots. The centre of gravity is then extracted from each blob and this is used to recognise and label the separate robots and the ball.

Data flow between blocks in the pipeline is controlled by synchronisation signals. These indicate whether the current streamed pixel is in an active region (valid pixel) or in the horizontal or vertical blanking region (invalid pixel).

2.1   Camera

The D5M digital image sensor can acquire 15 frames per second at its full resolution of 2592×1944. Because it is a CMOS sensor, it allows for windowed readout where a smaller resolution can be captured from the same sensor thus increasing the frame rate. By reducing the resolution to 640×480 it is possible to capture up to 127 frames per second.

Unfortunately the maximum camera height of 2 metres makes the reduced resolution unsuitable because the field of view cannot cover the entire playing area. There are two ways to correct this: The first is by using a wider angle lens, and the second is by implementing another feature of the camera called skipping.

The default lens that comes with the D5M camera has a focal length of 7.12 mm. To see the entire 1.5 m × 1.3 m board, a lens with a focal length of 1.5 mm or lower would be needed. Unfortunately, such a wide angle lens introduces a large amount of barrel distortion. The distortion from a 1.5 mm lens would be difficult to correct so this method on its own is unsuitable.

Skipping is a feature where only every 2nd, 3rd or 4th pixel is read out from the image sensor, effectively producing a smaller resolution from a larger resolution. This makes it possible to output a resolution of 640×480 but sample pixels from a 1280×960 or 2560×1920 area. The disadvantage of using skipping is that it introduces subsampling artefacts onto the image; the greater the skipping factor is, the greater the subsampling artefacts are. This is complicated further with Bayer pattern sensors because of the clustering resulting from the Bayer mask as illustrated in Fig. 2. To completely see the field at 640×480, 4× skipping would be required, however this will also add a lot of area outside of the board and will require a more complex Bayer filter to account for the skipping. A compromise was to use 2× skipping with a 3 mm focal length lens.

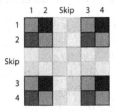

Fig. 2. 2× skipping with a Bayer pattern (from [10])

2.2   Bayer Interpolation

The D5M camera uses a single chip sensor to capture a colour image with a Bayer pattern. Each pixel only captures one of the RGB components as illustrated in Fig. 2, so the missing components must be recovered through interpolation (the difference between the raw image and the interpolated image is shown in Fig. 3).

A bilinear interpolation filter provides reasonable quality interpolation at relatively low cost [6]. For simple Bayer pattern demosaicing, bilinear interpolation simply averages adjacent pixels to fill in the missing values. However because of the skipping introduced by the camera the available pixels are no longer evenly spaced. An altered bilinear interpolation was created to adjust for the skipping and give evenly spaced output pixels.

Fig. 3. Left: Raw image captured from the camera, Right: after Bayer interpolation

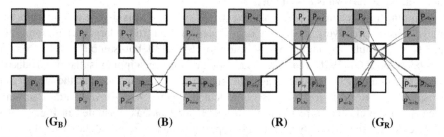

Fig. 4. Pixel locations used to interpolate the RGB values for Green on a blue row (G_B), Blue (B), Red (R), Green on a red row (G_R)

Referring to the pixel locations in Fig. 4, the equations are

$$G_B : R = (3P_{+y} + P_{-y})/4$$
$$G_B : G = P$$
$$G_B : G = (3P_{+x} + P_{-x})/4$$
$$B : R = (3(P_{+x+y} + P_{-x+y}) + P_{+x-y} + P_{-x-y})/8$$
$$B : G = (P_{+x} + P_{-x})/2$$
$$B : B = (3P + P_{+2x})/4$$
$$R : R = (3P + P_{+2y})/4 \tag{1}$$
$$R : G = (P_{+y} + P_{-y})/2$$
$$R : B = (3(P_{+x-y} + P_{+x+y}) + P_{-x-y} + P_{-x+y})/8$$
$$G_R : R = (3(P_{-x} + P_{+x}) + P_{-x+2y} + P_{+x+2y})/8$$
$$G_R : G = (2P + P_{+x+y})/3$$
$$G_R : B = (3(P_{-y} + P_{+y}) + P_{+2x-y} + P_{+2x+y})/8$$

The altered bilinear interpolation requires a 4×4 window instead of the 3×3 window used by standard bilinear interpolation. This requires 3 row buffers to form the window, and adds a 2 row latency to the algorithm.

To implement the bilinear interpolation as a hardware solution it is important to first optimise the equations so that they can be performed efficiently in real time. Multiplication by 3/8 and 5/8 can be implemented by an addition and a bit shift. More problematic is the division by 3 required for G_R:G. There are a few ways this can be implemented: The first is to change the divisor to a power of 2, such as multiplication by 5/16 or 85/256. Another method would be to change which green pixels the bilinear interpolation samples from, giving

$$G_R : G = (P_{-x-y} + P_{+x+y})/2 \qquad (2)$$

However this method makes ¼ of the pixels in the image redundant and therefore ¼ of the image information is effectively lost. Both methods reduce the logic resources and the computational time needed to complete the task, however the second method produced a better quality result, making it the preferred method.

2.3   Colour Edge Enhancement

The colour edge enhancement filter outlined by Gribbon *et al.* [4] was used to reduce the blur introduced by area sampling and bilinear interpolation. The filter utilises a 3×3 window to reduce blurring along both horizontal and vertical edges. The window is comprised or 2 row buffers and adds a single row of latency to the algorithm.

 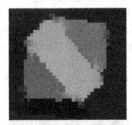

Fig. 5. Left: Blurred image from sampling and bilinear interpolation. Right: After edge enhancement filter.

2.4   YUV Transform

Lighting is very important with all image processing projects. Ideally the light should be spread evenly across the entire area to avoid forming shadows or altering colour values. The RGB colour space is very susceptible to changes in light level. This leads to wide thresholds and results in possible overlap of the threshold values. One solution to this problem is to convert the image into a colour space that is less susceptible to changes in light levels. The YUV colour space maps most of any light level changes onto the Y component leaving the other components for matching the colour. A simplified YUV transform only using powers of 2 is [7], as shown in the equation.

$$\begin{bmatrix} Y' \\ U' \\ V' \end{bmatrix} = \begin{bmatrix} \frac{1}{4} & \frac{1}{2} & \frac{1}{4} \\ \frac{1}{4} & -\frac{1}{2} & \frac{1}{4} \\ \frac{1}{2} & 0 & -\frac{1}{2} \end{bmatrix} \begin{bmatrix} R \\ G \\ B \end{bmatrix} \tag{3}$$

Because of its simplicity this would only add a single clock cycle delay onto the total algorithm, and allow for simple forward and inverse transformations requiring only additions or subtractions (and shifts).

2.5   Colour Thresholding

The first step is to threshold the image to separate colours from the background. All of the pixels for a particular colour are clustered together in YUV space. Each colour is defined by a rectangular box in YUV, delineated by minimum and maximum thresholds for each component. A colour label is then assigned to each pixel that falls within the threshold limits. This process introduces one clock cycle of latency to the algorithm and can be performed directly on the streamed data.

2.6   Noise Filtering

Next the image is filtered to remove isolated pixels from the thresholded image as these can be interpreted as separate blobs. This is performed by implementing a morphological filter to detect single pixel wide noise in either the horizontal or vertical direction. This noise is then cancelled out by changing the colour label to equal the surrounding pixels (see Fig. 6). This allows for most of the noise to be cancelled.

This filter is made up of a 3×3 window which adds another row of latency to the algorithm.

Fig. 6. Left: Labelled image before filtering. Right: After morphological filtering.

2.7   Connected Component Labelling

The purpose of the connected component labelling is to group similar adjacent pixels together based on colour. Typically a two pass algorithm is used [9; 8; 5]. The first pass labels pixels into initial groups and the second pass re-labels touching groups with a common label. This allows concave shapes to be labelled as a single group.

However since all of the shapes within robot soccer are convex, the second pass is therefore redundant in this application.

A 4 point detection grid can be used to search for adjacent pixels in the streamed data, as shown by Fig. 7(a).

(a)   (b)

Fig. 7. Left: 4 point detection grid and Right: 2 point detection grid for connected component labelling

Because the shapes are convex, there is no reason why a group should be linked by a single diagonal connection. Therefore the simpler 2 point detection grid shown in Fig. 7(b) can be used. The previously labelled row is required for the pixel above. To avoid having to re-label a row when a connection with the row above is established, a whole group of pixels is saved into the row buffer at once using run length coding whenever a background pixel is encountered. This reduces the amount of memory used and minimises the number of accesses per frame.

During labelling, the algorithm also accumulates the total number of pixels in the group, and the sum of the X and Y coordinates. When a blob is completed (by detecting that it does not continue onto the next row) this information is used to calculate the centre of gravity of each blob. This processing adds one row of latency to the algorithm for detecting blob completion.

2.8   Blob Processing

The final process is to group the blobs into individual robots and calculate their location and orientation. This stage has not yet been implemented however the first step is to calculate the centre of gravity for each blob using data collected during the connected component labelling algorithm (the number of pixels, N, within the blob and the sum of the coordinates that each pixel is located at). The equation for calculating the centre of gravity is

$$COG = \left(\frac{\sum x}{N}, \frac{\sum y}{N} \right) \tag{4}$$

A search window approach is used to find blobs with centres of gravity within close proximity. A robot is recognised and labelled once all of the blobs associated with the robot have been recovered.

3 Results

With the camera fully working we are able to see some very promising results. With the current implementation, up to but not including the centre of gravity, we are utilising 3610 LUTs which is only 29% of the total DE0's logic units, and 23% of the FPGA's memory blocks.

This design is capable of operating at 127 frames per second, which is the maximum that this camera is capable of for this resolution. However in order to display the output on a computer screen for debugging purposes it is necessary to limit the frame rate to 85 frames per second, as this is the maximum frame rate the LCD display allows. For the final implementation a display will not be necessary, therefore the frame rate of the camera can be increased to operate at the full 127 frames per second.

In total there are 4 rows of latency added to the algorithm. This means that the data for a robot is available 4 rows after the last pixel for the robot is read out from the camera. Therefore, it is possible to have located and processed all the robots before the frame finishes capturing.

To test the accuracy of the algorithm the blob data was captured from each frame and analysed over a period of 20 seconds. The differences in blob size and position between frames were compared with robots at different areas of the playing field. With the robot in the centre of the playing field (where the light is brightest) the standard deviation for the x and y coordinate is 0.15 mm with a max error of 2.4 mm. A robot was placed in one of the corners furthest away from the centre of the camera and light. The standard deviation for the x and y coordinates in this location was 0.15 mm with a max error of 2.3 mm.

4 Conclusion

In conclusion this paper has described an algorithm that accurately detects the positions of robots in real time, with only 4 rows of latency. Even though at this stage the project is a work in progress it is already possible to operate the camera at 85 frames per second and achieve 2.4 mm accuracy in a worst case scenario. Future study on this project will include automatic setup of the camera and its thresholds as well as implementing a distortion correction calibration to adjust for the added parallax error introduced by the lens.

Acknowledgements. This research has been supported in part by a grant from the Massey University Research Fund (11/0191).

References

[1] Bailey, D.G., Gupta, G.S., Contreras, M.: Intelligent camera for object identification and tracking. In: Kim, J.-H., Matson, E., Myung, H., Xu, P. (eds.) Robot Intelligence Technology and Applications. AISC, vol. 208, pp. 1003–1013. Springer, Heidelberg (2013)

[2] Broers, H., Caarls, W., Jonker, P., Kleihorst, R.: Architecture study for smart cameras. In: Proceedings of the EOS Conference on Industrial Imaging and Machine Vision, Munich, Germany, pp. 39–49 (2005)

[3] Dias, F., Berry, F., Serot, J., Marmoiton, F.: Hardware, design and implementation issues on a FPGA-based smart camera. In: First ACM/IEEE International Conference on Distributed Smart Cameras (ICDSC 2007), Vienna, Austria, pp. 20–26 (2007)

[4] Gribbon, K.T., Bailey, D.G., Johnston, C.T.: Colour edge enhancement. In: Image and Vision Computing New Zealand (IVCNZ 2004), Akaroa, NZ, pp. 291–296 (2004)

[5] He, L., Chao, Y., Suzuki, K., Wu, K.: Fast connected-component labeling. Pattern Recognition 42(9), 1977–1987 (2009)

[6] Jean, R.: Demosaicing with the Bayer pattern. University of North Carolina (2010)

[7] Johnston, C.T., Bailey, D.G., Gribbon, K.T.: Optimisation of a colour segmentation and tracking algorithm for real-time FPGA implementation. In: Image and Vision Computing New Zealand (IVCNZ 2005), Dunedin, NZ, pp. 422–427 (2005)

[8] Park, J., Looney, C., Chen, H.: Fast connected component labelling algorithm using a divide and conquer technique. In: 15th International Conference on Computers and their Applications; New Orleans, Louisiana, USA, pp. 373–376 (2000)

[9] Rosenfeld, A., Pfaltz, J.: Sequential operations in digital picture processing. Journal of the Association for Computing Machinery 13(4), 471–494 (1966)

[10] Terasic: TRDB-D5M 5 Mega Pixel Digital Camera Development Kit. vol. Version 1.2. Terasic Techologies (2010)

Obstacle & Lane Detection and Local Path Planning for IGVC Robotic Vehicles Using Stereo Vision

Christopher Kawatsu, Jiaxing Li, and C.J. Chung

Department of Mathematics and Computer Science, Lawrence Technological University,
21000 West Ten Mile Road, Southfield, MI, 48075-1058, USA
{ckawatsu,jli,cchung}@ltu.edu

Abstract. A robotic car, also known as a driverless or self-driving car, is an autonomous vehicle capable of sensing its environment and navigating itself without human input. Robotic cars exist mainly as prototypes and demonstration systems, but are likely to become more widespread in the near future. Usually, autonomous vehicles sense their surroundings for lane detection and obstacle avoidance with radar, 2D LIDAR, 3D LIDAR, and camera sensors.

This article describes our approach in developing an affordable stereo vision system using two ordinary webcams and OpenCV (Open Source Computer Vision) library. We applied our stereo vision system to Vulture 2 and iWheels robotic platforms to enter IGVC (Intelligent Ground Vehicle Competition) 2012 and 2013. The results show that stereo vision based navigation is a promising and affordable mechanism to develop robot cars.

Keywords: lane following, obstacle avoidance, stereo vision.

1   Introduction

The Intelligent Ground Vehicle Competition (IGVC) is an annual international robotics competition for teams of undergraduate and graduate students since 1993. Teams design and build an autonomous ground vehicle capable of completing several difficult challenges. The competition is well suited to senior design ourses as well as extracurricular design projects [1] [2].

In IGVC, the fully autonomous ground vehicle must navigate an outdoor obstacle course. The course consists of a lane painted with white or yellow lines on a grassy field (See Fig. 1). Obstacles come in a wide variety of shapes and colors including barrels, cones, saw horses, and posts (See Fig. 2 and Fig. 3).

J.-H. Kim et al. (eds.), *Robot Intelligence Technology and Applications 2*,   667
Advances in Intelligent Systems and Computing 274,
DOI: 10.1007/978-3-319-05582-4_57, © Springer International Publishing Switzerland 2014

Fig. 1. Example of IGVC Lane

Fig. 2. Example of the IGVC obstacle course

Fig. 3. Example of the IGVC obstacle course

Lawrence Technological University (LTU) computer science students have been participating in the competition since 2003. Early LTU IGVC teams used cameras to detect obstacles and lines since all obstacles were orange barrels. After more diverse obstacles were introduced, we adopted laser rangefinders to detect obstacles and continued to use cameras to detect boundary lines. The Viper robot used in the 2008 and 2009 competitions is shown in Fig. 2, Fig. 3, and Fig. 4.

Fig. 4. Viper robot vehicle with a camera and a SICK laser rangefinder

Using a camera and a laser rangefinder was a popular solution, but we experienced four problems. First, the laser scanner was too expensive. In 2008, it was over $7,000, almost half of the total cost to build the robot. Second, the power consumption of the laser rangefinder was very high. Third, the complexity of combining camera data and rangefinder data was non-trivial. Fourth, the weight of the laser rangefinder was heavy. To solve the problems, we introduced another robot called Culture Shock which used a stereo vision camera instead of a laser rangefinder. The stereo vision camera (Fig. 5) was purchased from a company, Videre, together with software libraries.

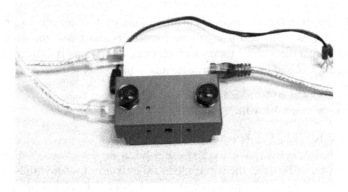

Fig. 5. Videre stereo vision camera used for Culture Shock robotic vehicle

In 2012, when we tried to re-use the Videre camera on a new platform called Vulture, we found the Videre company was not supporting the maintenance of software libraries any more. Due to the errors and poor performance of the old libraries, we had to abandon the purchased stereo cameras, and decided to make our own stereo vision system using two low cost webcams. This article describes how we made our own stereo vision system, how we detect obstacles and lanes, and how we navigate in the obstacle course using stereovision.

2   Introduction to Stereovision Algorithms

Stereovision relies on the same property as human depth perception: two cameras positioned next to each other will see slightly different images. Differences in an object between the two images depend on the distance of the object to the cameras. We used existing algorithms implemented in the OpenCV library [2].

Computing (x, y, z) coordinates for pixels in stereo camera images can be broken into several sub problems. First intrinsic and extrinsic camera parameters must be found. Finding intrinsic and extrinsic camera parameters is a well-known problem [3] [4] [5] and beyond the scope of this paper.

The camera parameters from the previous step can be used to rectify the images, this ensures that pixels that correspond to the same (x, y, z) location will be in the same row in both camera images. Image rectification reduces the complexity of the next step to searching a single row of the image, rather than the entire image.

After rectifying the images, the problem is finding the matching pixels within the same row of the left and right images. This process produces a disparity image which contains the number of pixels between matching points in the left and right images. There are two methods of block matching implemented in the OpenCV library, Block Matching (BM) [3] and Semi Global Matching (SGM) [4]. The BM algorithm has faster performance compared to the SGM algorithm; however, the SGM algorithm produces higher quality disparity images. We chose to use the SGM algorithm since it produces much better disparity images for the types of obstacles present in IGVC. Many other stereo vision algorithms exist see [8] and [9] for a more comprehensive survey.

Once the disparity image has been calculated, the next problem is to find (x, y, z) points for each pixel. The extrinsic camera parameters along with the disparity value can be used to find the z value for each pixel. Once the z value is known, the x and y values can be calculated using the intrinsic camera parameters.

3   LTU Stereo Vision System

A standard webcam that costs only $49 each was chosen for the development of our own stereo vision system. We use extrusion beam as the base in order to easily change the distance between the two cameras. The fixtures to mount the cameras and connect to the extrusion beam were purchased from a local hardware store (see Fig. 6). Our stereo vision system was developed with Emgu CV, which a C# wrapper around the OpenCV library.

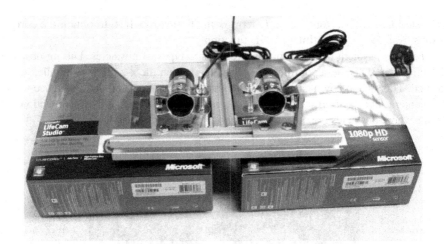

Fig. 6. LTU stereo vision camera developed with two webcams and Emgu CV

The stereo vision processing described in Section 3 is accomplished using Emgu CV library functions. First, the cameras must be calibrated using a series of chessboard images. The location of the chessboard corners for the left and right camera images is found to the nearest pixel and further refined using a sub pixel search function. The locations of the chessboard corners are then used to calculate camera parameters for the left and right cameras in addition to the transformation which relates the position of the two cameras. Once these calibration parameters are known the next step is to rectify the images. Rectifying the images guarantees that corresponding pixels in the left and right images will be in the same pixel row. An example rectified image for an IGVC obstacle can be seen in Fig. 7. Once the images are rectified, a Semi Global Matching (SGM) algorithm [2] is used to find the location difference between corresponding pixels in the left and right rectified images. An example disparity image is shown in Figure 8. The three dimensional coordinates can be determined using the disparity image and Emgu's ReprojectImageTo3D function. This function relies on the following matrix

$$Q = \begin{bmatrix} 1 & 0 & 0 & -C_x \\ 0 & 1 & 0 & -C_y \\ 0 & 0 & 0 & f \\ 0 & 0 & -1/T_x & (C_x - C'_x)/T_x \end{bmatrix}$$

where (C_x, C_y) is the principle pixel in the left rectified image, f is the focal length of the left camera in pixels, C'_x is the x coordinate of the principle pixel in the right camera and T_x is the spacing between the two cameras. The value of the Q matrix is obtained during the calibration process; however, some of the values must be modified in order for the function to work. The disparity image in Emgu actually contains the difference in pixels multiplied by 16 so the $-1/T_x$ term must be divided by 16. Additionally, the $1/T_x$ term has a sign error which causes nonlinear error in the resulting

3D coordinates. Therefore the $-1/T_x$ term is multiplied by $-1/16$ to obtain the correct results from the ReprojectImageTo3D function.

The vision system distinguishes between three types of obstacles. During obstacle detection the 3D camera coordinates are rotated so that the x-z plane is parallel to the ground. This is done by measuring the height of the cameras and the z distance to the left camera. For each obstacle type the vision system provides a list of rotated (x, z) points where obstacles are present.

Fig. 7. Left Rectified Image

Fig. 8. Disparity Image

4   Obstacle and Lane Detection

Obstructions are detected by finding all points where the rotated y distance from the plane is greater than 10 inches. For each pixel in the disparity image, 3D coordinates are found and rotated. The (x, z) coordinates are then passed to the local path planning.

Lane detection is done using the color version of the left rectified image. The brightness of the blue values is adjusted by subtracting 60 from all blue values. The contrast of the blue values is adjusted using

$$B' = B \cdot contrast + 255 \cdot (1 - contrast) \cdot 0.5$$

where contrast = 200, B' is the contrast adjusted blue value, and B is the original value. These adjustments cause white lines painted on green grass to appear blue. The pixels with B' > 150 are considered to be candidates for lines. If more than half of the pixels in a 4 by 4 pixel block have B' > 150 the block is considered to be a line.

The 3D coordinates for each pixel are found using the depth image from SGM . The coordinates are rotated so x-z is parallel to the ground and the (x, z) coordinates are passed to the local and global grids. Pixels in the left rectified image identified as lines can be seen in Fig. 9. Note that part of the obstacle is detected as a line. These points are later discarded because the y value for the pixel is too large.

Flags, other IGVC obstacles, are detected by finding points 6 inches above the x-z plane. A simple threshold value in the red and blue channel along with a maximum height is used to distinguish red and blue flags from other obstacles. No contrast or brightness adjustments are done for flag detection.

Fig. 9. Rectified image after contrast adjustment. The white areas of the obstacle are incorrectly identified as lines; however, these points can be removed based on the y (height from ground) value at each pixel.

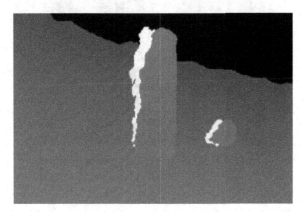

Fig. 10. Pixels colored by z distance from the camera, green is closer, red is farther away. White areas have no disparity value and black areas are greater than 5 meters away.

5   Local Path Planning

The path planner receives (x, z) coordinates for obstructions, lines, and flags from the vision system. It tests a set of rectangles to see if they contain obstacles. Each rectangle starts at the front of the robot and extends about 5 meters ahead. The rectangles are placed at five different angles relative to the robot's heading (straight, hard left, soft left, hard right, and soft right). The rectangle containing no obstacles with a heading which is closest to the next GPS waypoint is chosen as the desired driving direction. In the case where all rectangles contain obstacles, the robot will turn in place until an obstacle free rectangle appears.

6   Results and Conclusion

The LTU stereovision system and a simple local path planning algorithm described in Section 5 was tested on the Vulture robot (see Fig. 11) in 2012 and the iWheels robot (see Fig. 12) in 2013. The system was able to accomplish obstacle detection as well as lane detection tasks successfully. Vulture won 5th place and iWheels won 6th place in autonomous navigation challenge category. Our robot was the only robot that is solely dependent on cameras in the competition with around 50 University teams. Future work includes testing of our robot platform with improved stereovision system (1) at higher speeds (IGVC competition does not allow driving faster than 10 miles per hour.) and (2) at night with headlights.

Fig. 11. vuLTUre – IGVC 2012 robotic platform

Fig. 12. iWheels – IGVC 2013 robotic platform

References

[1] Theisen, B.L., Nguyen, D.: The 14th Annual Intelligent Ground Vehicle Competition: intelligent teams creating itelligent ground robots. In: Intelligent Robots and Computer Vision XXIV: Algorithms, Boston, MA (2006)

[2] Kosinski, A., Nguyen, D., Theisen, B.: Intelligent Ground Vehicle Competition, http://www.igvc.org/ (accessed August 19, 2013)

[3] Bradski, G.: The OpenCV Library. Dr. Dobb's Journal of Software Tools (2000)

[4] Zhang, Z.: Flexible camera calibration by viewing a plane from unknown orientations. In: The Proceedings of the Seventh IEEE International Conference on Computer Vision (1999)

[5] Clarke, T.A., Fryer, J.G.: The development of camera calibration methods and models. The Photogrammetric Record 16(91), 51–66 (1998)

[6] Wang, F.-Y.: A simple and analytical procedure for calibrating extrinsic camera parameters. IEEE Transactions on Robotics and Automation 20(1), 121–124 (2004)

[7] Konolige, K.: Small Vision Systems: Hardware and Implementation. In: Robotics Research, pp. 203-212 (1998)

[8] Hirschmuller, H.: Accurate and Efficient Stereo Processing by Semi-Global Matching and Mutual. In: Proceedings of the 2005 IEEE Computer Society Conference on Computer Vision and Pattern Recognition (CVPR 2005), vol. 2 (2005)

[9] Scharstein, D., Szeliski, R.: Middlebury Stereo Vision Page, http://vision.middlebury.edu/stereo/ (accessed August 19, 2013)

[10] Scharstein, D., Szeliski, R.: A taxonomy and evaluation of dense two-frame stereo correspondence algorithms. International Journal of Computer Vision 47, 7–42 (2002)

Adaptive Regions of Interest Based on HSV Histograms for Lane Marks Detection

Vitor S. Bottazzi[1,*], Paulo V.K. Borges[2], Bela Stantic[1], and Jun Jo[1]

[1] School of Information and Communication Technology
Griffith University, Gold Coast
vitor.bottazzi@griffithuni.edu.au, {b.stantic,j.jo}@griffith.edu.au
[2] Autonomous Systems Laboratory, CSIRO ICT Centre
Brisbane, Australia
vini@ieee.org

Abstract. The lane detection is a vital component of autonomous vehicle systems. Although many different approaches have been proposed in the literature it is still a challenge to correctly identify road lane marks under abrupt light variations. In this work a vision-based ego-lane detection system is proposed with the capability of automatically adapting to abrupt lighting changes. The proposed method automatically adjusts the feature extraction and salient point tracking cues introduced by the GOLDIE (Geometric Overture for Lane Detection by Intersections Entirety) algorithm. The variance of the lighting conditions is measured using hue-saturation histogram and abrupt light changes on the road are detected based on the difference between histograms. Experimental comparison with previously proposed algorithms demonstrated that this method achieved efficient lane detection in the presence of shadows and headlights. In particular, the accuracy of the algorithm applied on the footage with highest light variation increased 12.5% on average. The overall detection rate increased 4%, which illustrated the applicability of the method.

1 Introduction

Automated lane detection is an important vehicle functionality when considering urban driving. It has developed the interest of the research community on both autonomous driving systems, and advanced driver assistance system (ADAS). Lane detection is commonly used to identify road boundaries, unwanted lane changes, and to help estimate the upcoming geometry of the road. This article proposes a histogram-based method that measures light conditions on the roads to reduce false positive lane detection caused by dynamic light conditions. Saliency histogram features were recently used to simplify existence detection using the probability distribution function (PDF) of the saliency maps [1]. In this case, the regions-of-interest (ROI) positioning for ego-lane detection was solved using an efficient prior triangle model [2] to approximate the road behaviour.

---

* Corresponding author.

J.-H. Kim et al. (eds.), *Robot Intelligence Technology and Applications 2*,
Advances in Intelligent Systems and Computing 274,
DOI: 10.1007/978-3-319-05582-4_58, © Springer International Publishing Switzerland 2014

As environmental lighting conditions make lane mark detection challenging, a light variance measurement method based in histograms is proposed to quantify light changes. The light variance ranges are used as reference to adjust relevant variables to improve overall ego-lane mark detection. This strategy measures the light change on a road surface to automatically tune object tracking improving the algorithm robustness. The light variance measurement is introduced to the image processing pipeline to adjust the ego-lane mark tracking according to dynamic light conditions. The variance of the light is measured using the hue-saturation histogram. Abrupt light changes on the road are detected based on the difference between histograms. First, the histograms of the whole frame and the road fragment of the frame are calculated. Then, the difference between the two histograms is computed using the chi-square [3] approach to set the minimum, maximum and average values to be used as references. Depending on the light variance, the number of corners tracked by the algorithm and the amount of pixels analysed is adjusted to adapt to it. This paper extends the GOLDIE [4] system by increasing robustness under dynamic light conditions. The GOLDIE approach introduced an efficient vision-based ego-lane detection system based on a simplified drivable corridor model. Experiments and comparisons with the previous algorithm illustrated the applicability of the method. Often expensive computer vision techniques are used to extract specific features from road images in order to identify road marks. As real-time performance is a mandatory requirement for applications that target autonomous vehicle navigation, the focus is to increase performance with minimum overhead, monitoring the computational cost attached to feature enhancements. The remainder of the paper is organised as follows. Section 2 presents the related work and some state-of-the-art lane detection algorithms. Section 3 describes the light variation estimation process and the relevant variables set, while section 4 shows the experiment setup with its practical considerations, results and comparison with the previous approach. Section 5 concludes the paper and presents future work.

2   Related Work

A large number of works has been proposed [5–9] on lane detection where dynamic light conditions strongly affect the accuracy of algorithms. Common approaches include detection based on texture [10], colour [9], appearance models [11] or edge points [7, 8]. Current literature still addresses histograms as a support method to enhance object detection [1], but there are no reference of using histograms to automatically tune up unsupervised algorithms for road lane tracking. Multiple models have been applied to describe lane mark configuration, such as piecewise linear segments [12], clothoids [13], parabola [7], hyperbola [8], splines [5], or snakes [12,14]. The lane tracking issue has been addressed in many research articles [6,12,15,16]. Significant research has focused on lasers [6,15] and cameras [7,9,17] to support autonomous vehicles on lane mark detection. Most robust solutions on autonomous vehicle systems depend on multiple sensors to provide a reliable road boundary reference for navigation.

3   Proposed Method

One of the main problems related to vision-based soultions is robustness. Unsupervised approaches have shown good results when applied to controlled scenarios where a priori knowledge about the road is assumed [18]. A recent analysis [19] has illustrated that the dynamic light on roads still challenges the most advanced algorithms. Nevertheless, approaches using contrast enhancement and brightness compensation still present limitations on vision-based algorithms [20]. Light variations over the road painted marks may reflect on reduced accuracy for lane detection. Therefore, the issue was addressed by measuring the light variance using the hue-saturation histogram comparison. This approach adapts the lane mark tracking on GOLDIE algorithm when abrupt light changes are detected. During the first step, the RBG image is converted to the HSV colour space so that hue and saturation can be analysed. Furthermore the two histograms generated using the whole image and the visible portion of the road are compared to identify minima and maxima light variance references. If the light variance exceeds a threshold, tuning is performed over the tracking variables aiming to increase algorithm's robustness.

This strategy consists of resizing the dynamic regions-of-interest (DROIs) [4] used for tracking while adapting the number of detected corners on the tracking algorithm according to light changes on the road. The motivation for using smaller image processing windows is to reduce sensibility to noise, reduce the computational effort and concurrently increase precision. The optimised ROI processing strategy native to GOLDIE, combined with the histograms calculation, adapts the image processing effort aiming for maximised robustness and throughput (see Fig 1). Depending on external light variance, ROIs are automatically resized and re-positioned using a prior triangle reference that represents the drivable corridor on the ego-lane. The following pseudocode summarises the automatic tuning algorithm based on light variance:

Data: Input frame
Result: Adaptive lane detection
while *valid input frame* **do**
    Check histograms variance;
    if *variance bigger than threshold* **then**
        Reduce tracking ROIs area to $1/4$ of its normal size;
        Reduce number of tracked corners;
    else
        Reset tracking ROIs size;
        Increase number of tracked corners;
    end
    Apply appearance segmentation cue;
    Apply tracking cue;
end

Algorithm 1. Simplified pipeline

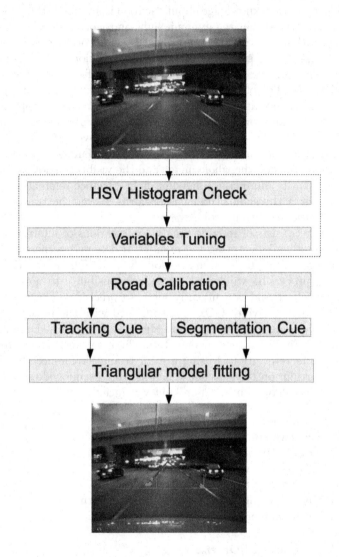

Fig. 1. Pipeline illustrating the light variance analysis before appearance-based segmentation and salient point tracking

The comparison method was chosen due to its similarity to the Mahalanobis [21] distance, which highlights outliers with abnormal large deviations. The Chi-Square method has also been used in online learning [3] to find correlations between images. It is applied here to detect abnormal light changes that might disrupt the ego-lane detection accuracy. Its property is relevant to this approach when looking for high sensitivity on road light changes, to adjust the ego-lane tracking ROIs in real-time.

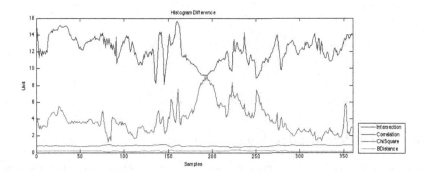

Fig. 2. Histogram comparison methods applied to one of the samples with predominantly dynamic light

The Chi-Square method (see Fig 2) is denoted by:

$$c(M, M') = \sum_{I} \frac{(M(I) - M'(I))^2}{M(I)} \tag{1}$$

where M is the full HSV frame and M' is the bottom half of the HSV image where road is visible.

3.1 Lane Mark Segmentation

The calibration step aims to optimally position the segmentation ROIs defined by the parallel segments shown in Fig. 3. Firstly, a predefined position at the centre of the bottom half of the image is determined. Secondly, the image segmentation is applied over the predefined position to calculate the distance in between the vanishing point and the bottom part of the image. This strategy aims to position the segmentation ROIs concentrating the feature extraction efforts on regions where road lane marks are visible.

From this point, the extended ego-lane segments are extracted from the new ROI and the distance between extended ego-lane segments is updated. This step aims to approximate the ego-lane size, setting up the triangle's baseline ($\overline{AC}$ in Fig. 3) as a reference.

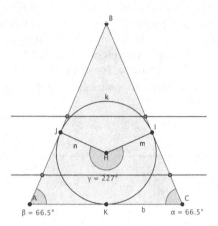

Fig. 3. Ego-road model

3.2   Lane Painting Tracking

Painted lane marks often become discontinuous, disappear or change colour depending on traffic rules (i.e. white-yellow-white). The algorithm uses the triangular prior model allied to Lucas-Kanade [22] tracking to validate the road lane references captured by the segmentation stage. Additionally, the algorithm considers the relative motion of the ego-lane mark references, within an expected range defined by road legislation. The Harris [23] corners are tracked across consecutive frames and selected following a linear approximation model based on Laplace expansion theorem. The points are classified according to their proximity to the prior triangle's reference using (2):

$$\det(M) = (-1)^{|r|} \sum_c (-1)^{|c|} \det S(M; r, c) \det S'(M; r, c) \qquad (2)$$

This schema calculates the determinant of a matrix to discard outlier points that are too far from the prior road model reference [4]. The GOLDIE method describes how to efficiently combine two independent image processing cues to enhance ego-lane mark detection (see Fig. 1). The image processing cues are respectively:

- Appearance-based approach for lane segmentation; and
- Harris' corner tracking, clustered according to lane-like configuration.

The appearance-based detection is executed in parallel with the painted mark tracking to validate the ego-lane mark reference position. Both references are compared and used to process the model fitting. A mathematical model was generalised to incrementally validate information extracted from road image samples, aiming to segment left and right lane marks. Furthermore, the vanishing point [7] detection is included as part of the test case, to restrict the amount of lines detected in the frame, and to increase the accuracy of the ego-lane mark detection. Finally, two different cues are combined to incrementally improve the detection results and to conclude the preliminary experiments.

4   Experiments

This section presents the results related to the adaptive tracking under dynamic light conditions. The experiments exposed a list of important facts to address. First, histogram-based approaches are capable of measuring light variances in a shot. Second, runtime and the sensibility to noise can be reduced using this approach. Furthermore, a salient point tracker based on a prior representation of the road can effectively guide DROIs. Finally, independent DROIs can keep track of the ego-lane marks, based on previous frame references, to accurately estimate lane mark position and automatically adapt themselves to different light conditions.

4.1   Practical Considerations

This research aims to extract all of the visual information from a single camera. Regular driving situations under different light conditions were used as a case study to create an intelligent algorithm capable of self-adaptation according to external light. The following events were identified as main causes of abrupt light change: post pole lights, shadows introduced by objects surrounding the road, road surface reflectiveness, car headlights and brake lights. According to previous experiments, most of the cited variables still introduce considerable noise to image processing pipelines making automatic tuning a very difficult task. Two different datasets were analysed in this experiment: the local dataset created in residential roads in Queensland, Australia and the external one published by [17]. The external dataset was used to support the test-driven development method providing incrementally challenging scenarios to test the robustness of this approach. The following videos were involved to compare how the algorithm performs on:

- a clear day with some clouds and predominant discontinuous road lines (sequence labeled S1C1);
- a cloudy day with predominant abnormal road painting and heavy traffic (sequence labeled S6C2); and
- a cloudy day during dawn time with heavy traffic and car light disturbances (sequence labeled S7C2).

The three sample videos used the same resolution 640×480 at 30 frames per second. The resolution of the local sample videos is 320×240 and the frame rate is 30 frames per second. The footage used for development provides many different situations where lane detection is challenged by highly dynamic light conditions. The local sample videos were recorded on a clear day with no clouds. The path was chosen due to the predominance of continuous line painting, the ideal scenario for road calibration purposes. As the ground truth definition is a time consuming task, the 4 samples used have the average duration between 5 and 20 seconds. The datasets were recoded at 30 frames per second, which gives a frame set from 150 to 600 frames per footage, totalling 1500 shots in total. The automatic calibration time increased considerably compared to the past experiment where parameters were fixed and non-dependent on light variance. For this reason, the initial frames where calibration is being processed were excluded from the lane detection statistics. The sample labeled S6C2, for example, took up to 20% of the available frames to automatically calibrate the algorithm to the road size. The false negatives analysis is not included in the statistics as the new version of the algorithm uses the past ego-lane marks' reference to reconstruct a driveable corridor, when no references are available nor detected. The extraction of a single lane mark through pure segmentation may also be regarded as a successful detection. By using the prior road model, the position of the ego-lane marks may be extrapolated incorporating the road size reference determined during the calibration stage.

4.2   Results

The results were analysed by comparing the previous GOLDIE algorithm with the GOLDIE algorithm enhanced by the new light checking feature. The results are divided in three parts: (i) single performance, (ii) overall performance and (iii) computational complexity verification. The light quantification approach combined to the GOLDIE approach improved the accuracy of the ego-lane mark detection in 50% of the samples (S6C2 and S7C2). The samples where light conditions are stable (less that 30% variance between minima and maxima) suffered a small reduction in accuracy of -3.5%. However, the samples where light conditions oscillated more than 30% demonstrated the improvement of 12.5% on average, which is considered relevant when compared with the previous results. Additionally, the overall performance of the system increased from 84% to 88%. The statistics were calculated using true positive indicators on a *per frame* basis, shown in Table 1.

Table 1. Ego-Lane detection in dynamic light

Sample data	Previous Approach	New Approach (CF)
Local Sample	98%	94% (-4%)
S1C1	91%	88% (-3%)
S6C2	82%	93% (+11%)
S7C2	63%	77% (+14%)

The fail rate increased in some of the footage with stable light conditions due to the new window resizing feature. It directly influences the painted mark search when dealing with discontinuous painting. However, the influence of uninteresting road painting such as speed limit indicators, bicycle lanes and turning signs

Fig. 4. Increased accuracy in presence of abnormal lane mark painting

marked on the road, was considerably reduced (see Fig. 4). The point tracking produced an excellent response to the automatic variable adjustments, increasing the overall accuracy by 4%.

4.3 Comparisons

Single appearance-based segmentation is strongly affected by intense light variation as confirmed on similar approaches [7–9]. This paper demonstrated the possibility of reducing the number of false true patterns mismatched by headlights of vehicles changing lanes, shadows, non-uniform pavement colour, among others variables. Additionally, the algorithm demonstrated increased accuracy when dealing with abnormal road painting, as seen in Fig 4. The algorithm was tested using the database sample published in [17], providing different scenarios that shall be extended to challenge the proposed approach towards future enhancements. The results were promising, achieving an overall tracking average of up to 84%. It is important to mention that the high rate of false positives detected on samples *S6C2* and *S7C2* were mainly caused by a long calibration effort at the start of the video due to discontinuous lane marks and floating road size. After road calibration, the algorithm showed improved detection results in highly dynamic light conditions. The computer used to test the lane detection algorithm is an ordinary Intel Core 2 Duo 1.4 GHz with 4GB of RAM. As with other state-of-the-art methods, the experiment showed the capability of identifying and reconstructing missing parts of the road lanes, with the enhancement of handling dynamic light conditions automatically. The new GOLDIE architecture including light variance measurement, road calibration and interactive cues still attends to the real-time demands of running the ego-lane detection algorithm in less than 0.04 second.

5   Conclusions

This paper introduced a new framework for automatic parameter calibration combining HSV histograms, colour-based lane segmentation and feature point tracking for lane mark position estimation. This solution does not require excessive computational effort compared to other state-of-the-art methods, as our method uses optimised and simplified geometric fitting models. The main benefit of the GOLDIE is that the system becomes more robust to lighting variations, different road colours, shadows across the road, breaks in line painting and line colour changes. The results verify that this method presents good performance under different road sizes and dynamic light conditions. However, one of the limitations of the algorithm is that it relies on well-painted road marks (or high contrast curbs) on the calibration stage to perform fast calibration. Nonetheless, GOLDIE performed efficiently. This new approach demonstrated how flexible the architecture is in being extended. It can be easily combined with semantic aspects of roads, information regarding camera angle and other sensorial modalities not limited to support for autonomous driving experience, but also ADAS. Future work includes homography, and to implement a sliding window that will move according to the speed of the car.

Acknowledgment. The authors would like to thank the Institute for Integrated and Intelligent Systems(IIIS) at Griffith University, and the Commonwealth Scientific and Industrial Research Organisation (CSIRO) for the valuable support with this research.

References

1. Scharfenberger, C., Waslander, S.L., Zelek, J.S., Clausi, D.A.: Existence detection of objects in images for robot vision using saliency histogram features. In: 2013 International Conference on Computer and Robot Vision (CRV), pp. 75–82. IEEE (2013)
2. Bottazzi, V., Jo, J.: Road lanes prediction algorithm based in trigonometry. In: International Robot Olympiad Committee, IROC, p. 14 (2010)
3. Felsberg, M., Larsson, F., Wiklund, J., Wadstromer, N., Ahlberg, J.: Online learning of correspondences between images. IEEE (2013)
4. Bottazzi, V., Borges, P.V.K., Jo, J.: A vision-based lane detection system combining appearance segmentation and tracking of salient points. In: 2013 IEEE Intelligent Vehicles Symposium (IV 2013). IEEE (2013)
5. Wang, Y., Shen, D.G., Teoh, E.K.: Lane detection using spline model, vol. 21, pp. 677–689 (June 2000)
6. Sehestedt, S., Kodagoda, S., Alempijevic, A., Dissanayake, G.: Efficient lane detection and tracking in urban environments. In: Proc. European Conf. Mobile Robots, pp. 126–131 (2007)
7. Zhou, S., Jiang, Y., Xi, J., Gong, J., Xiong, G., Chen, H.: A novel lane detection based on geometrical model and gabor filter. In: IEEE Intelligent Vehicles Symposium, pp. 59–64 (2010)
8. Assidiq, A., Khalifa, O., Islam, R., Khan, S.: Real time lane detection for autonomous vehicles. In: IEEE International Conference on Computer and Communication Engineering, ICCCE 2008., pp. 82–88 (2008)
9. Chiu, K.Y., Lin, S.F.: Lane detection using color-based segmentation. In: Proceedings of IEEE Intelligent Vehicles Symposium, pp. 706–711. IEEE (2005)
10. Bui, T.H., Nobuyama, E., Saitoh, T.: A texture-based local soft voting method for vanishing point detection from a single road image. IEICE Transactions on Information and Systems 96(3), 690–698 (2013)

11. Broadhurst, R.E., Stough, J., Pizer, S.M., Chaney, E.L.: A statistical appearance model based on intensity quantile histograms. In: 3rd IEEE International Symposium on Biomedical Imaging: Nano to Macro, pp. 422–425 (2006)
12. Huang, S.S., Chen, C.J., Hsiao, P.Y., Fu, L.C.: On-board vision system for lane recognition and front-vehicle detection to enhance driver's awareness. In: ICRA, pp. 2456–2461. IEEE (2004)
13. Danescu, R., Nedevschi, S., Meinecke, M., To, T.: Lane geometry estimation in urban environments using a stereovision system. In: IEEE Intelligent Transportation Systems Conference, ITSC 2007, pp. 271–276. IEEE (2007)
14. Kang, D.J., Choi, J.W., Kweon, I.S.: Finding and tracking road lanes using line-snakes. In: Proceedings of the 1996 IEEE Intelligent Vehicles Symposium, pp. 189–194 (1996)
15. Wijesoma, W.S., Kodagoda, K.R.S., Balasuriya, A.P.: Road-boundary detection and tracking using ladar sensing 20, 456–464 (2004)
16. Butdee, S., Suebsomran, A.: Automatic guided vehicle control by vision system. In: IEEE International Conference on Industrial Engineering and Engineering Management, IEEM 2009, pp. 694–697. IEEE (2009)
17. Borkar, A., Hayes, M., Smith, M.T.: A novel lane detection system with efficient ground truth generation. IEEE 13, 365–374 (2012)
18. Chen, W., Jian, L., Kuo, S.Y.: Video-based on-road driving safety system with lane detection and vehicle detection. In: 2012 12th International Conference on ITS Telecommunications (ITST), pp. 537–541. IEEE (2012)
19. Ding, H., Zou, B., Guo, K., Chen, C.: Comparison of several lane marking line recognition methods. In: 2013 Fourth International Conference on Intelligent Control and Information Processing (ICICIP), pp. 53–58. IEEE (2013)
20. Yoo, H., Yang, U., Sohn, K.: Gradient-enhancing conversion for illumination-robust lane detection. IEEE (2013)
21. Todeschini, R., Ballabio, D., Consonni, V., Sahigara, F., Filzmoser, P.: Locally centred mahalanobis distance: A new distance measure with salient features towards outlier detection. Analytica Chimica Acta (2013)
22. Bouguet, J.: Pyramidal implementation of the affine lucas kanade feature tracker description of the algorithm (2001)
23. Harris, C., Stephens, M.: A combined corner and edge detector. In: Alvey Vision Conference, Manchester, UK, vol. 15, p. 50 (1988)

One-Way ViSP (Visually Servoed Paired structured light) for 6-DOF Structural Displacement Measurement

Haemin Jeon[1], Wancheol Myeong[1], Youngjai Kim[2], and Hyun Myung[1,2]

[1] Dept. of Civil and Env. Engg., KAIST (Korea Advanced Institute of Science and Technology)
[2] Robotics Program, KAIST
291 Daehak-ro, Yuseong-gu, Daejeon 305-701,
Republic of Korea
{inhishand,wcmyeong,david-kim,hmyung}@kaist.ac.kr

Abstract. Structural health monitoring (SHM) of civil infrastructures has gained great attention since it is directly related to public safety. To estimate structural displacement, which is considered as one of important categories of SHM, a one-way type vision and laser-based 6-DOF displacement measurement system (one-way ViSP) is proposed. The system is composed of a transmitter with two laser pointers, a 1-D laser range finder (LRF), and a 2-DOF manipulator; and a receiver with a camera and a screen. The lasers and LRF project their parallel beams to the screen and the camera attached to the screen captures the image of the screen. The manipulator controls poses of the lasers and LRF for projected laser beams to be always inside the screen. By calculating positions of the laser beams and obtaining the distance from LRF, the relative displacement between two places can be estimated. The performance of the system has been verified from simulations.

Keywords: structural health monitoring (SHM), displacement measurement, laser, laser range finder (LRF), camera.

1 Introduction

As civil structures are directly related to public safety and exposed to various external loads such as traffic, wind, or wave loads, it is important to evaluate the structural condition at any moment. The structural displacement measurement is considered one of the important categories of structural health monitoring (SHM), and conventional sensors such as accelerometers, GPS, or Laser Doppler Vibrometers (LDVs) have been widely used [1]. However, the accelerometer is sensitive to temperature changes and a reconstructed displacement is neither accurate nor stable due to the signal drift, and the cost of GPS and LDV is very high.

To solve these problems, vision-based displacement measurement systems have been researched. Most of the vision-based systems should install an artificial target on the structure and a high resolution camera with a zoom lens captures the movement of the target from a far [2, 3]. Although the system can directly estimate the displacement with high accuracy, it is highly sensitive to environmental changes such as weather or lighting conditions. Also, it can measure only 2 or 3-DOF displacement.

Therefore, Myung *et al.* proposed a paired structured light (SL) system composed of two sides facing each other, each with one or two lasers, a camera, and a screen

J.-H. Kim et al. (eds.), *Robot Intelligence Technology and Applications 2*,
Advances in Intelligent Systems and Computing 274,
DOI: 10.1007/978-3-319-05582-4_59, © Springer International Publishing Switzerland 2014

[4–7]. The lasers project their parallel beams to the screen on the opposite side and the camera near the screen captures the image of the screen. As the distance between the camera and the screen is short, it is highly robust to the environmental changes. By calculating positions of the projected laser beams, the relative 6-DOF displacement can be estimated.

As the next version of the paired SL system, a visually servoed paired SL system (ViSP) was proposed to extend measurable displacement range [8, 9]. In other words, the newly added 2-DOF manipulators on each side of ViSP control the pose of lasers for the projected laser beams to be inside the screen all the time. The relative 6-DOF displacement can be estimated by calculating positions of the projected laser beams and rotation angles of the manipulators. For reducing computation time of the displacement estimation, an incremental displacement estimation (IDE) algorithm which updates the previously estimated displacement by calculating difference of the previous and the current observed data was proposed [10]. Also, a displacement estimation error back-propagation (DEEP) method for multiple modules of ViSP was developed to reduce a propagation error [11].

But it is sometimes not easy to install the lasers on both sides in this two-way type ViSP system. To mitigate this problem, a one-way type vision and laser-based structural displacement measurement system with minimal configuration, one-way ViSP in short, is proposed in this paper. In comparison with the former two-way type ViSP, the system has laser pointers in one side. The transmitter is composed of two lasers, a laser range finder (LRF), and a 2-DOF manipulator; and a receiver is composed of a screen and a camera. The lasers and LRF project their parallel beams to the screen on the receiver and the camera attached to the screen captures the image of the screen. By calculating positions of the three projected laser beams and obtaining the distance from LRF, the relative 6-DOF displacement between two places can be estimated. The performance of the system has been verified from simulations. The remainder of this paper is organized as follows. In Section 2, the design and kinematics of the one-way ViSP are introduced. The simulations are conducted in Section 3. In Section 4, conclusions and further research directions are discussed.

2   One-Way ViSP

2.1   Design of the One-Way ViSP

A design of the one-way ViSP is shown in figure 1 . As shown in the figure, the system consists of a receiver and a transmitter facing each other. The receiver is composed of a camera and a screen, and the transmitter is composed of lasers, a laser range finder (LRF), and a 2-DOF manipulator. The lasers and LRF project their parallel beams to the screen on the receiver and camera attached to the screen captures the image of the screen. Since the distance between the camera and the screen is short, usually less than 20 cm, it is robust to environmental changes such as weather or illumination. The manipulator controls the pose of the lasers and LRF for projected laser beams to be inside the screen all the time.

The procedure of the 6-DOF displacement estimation using the proposed system is shown in figure 2 . As shown in the figure, the camera on the receiver first captures

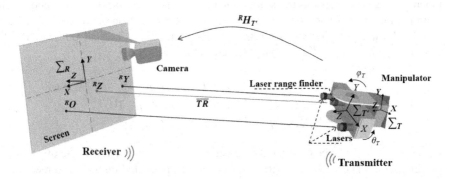

$^{R}H_{T'}$: Homogeneous Transformation matrix

Fig. 1. The schematic diagram of the one-way visually servoed paired structured light system (ViSP)

the image of the screen. After correcting the lens distortion of the captured image, a screen boundary and positions of the projected laser beams are detected. In case the laser beams get off the screen boundary, the rotation angles of the manipulator are calculated based on the current and the desired positions of the center of the laser beams. The relative translational and rotational displacement between two positions can be estimated by calculating positions of the projected laser beams, the measured distance from LRF, and the rotation angles of the manipulator.

2.2 Kinematics of the One-Way ViSP

The kinematics of the system is defined by a geometric relationship between the observed data $m = [^{R}O, {^{R}Y}, {^{R}Z}, \overline{TR}]^{T}$ and estimated displacement $p = [x, y, z, \theta, \phi, \psi]^{T}$. Here, ^{R}O and ^{R}Y, and ^{R}Z are the coordinate points of the projected laser beams on the screen from the lasers and LRF, respectively; $\overline{TR}$ is a measured distance from LRF; x, y, and z are translational displacements along the X, Y, and Z axes, respectively; and θ, ϕ, and

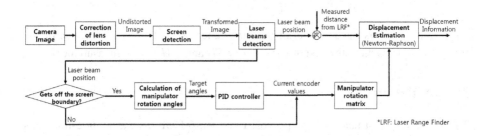

Fig. 2. Procedure of the 6-DOF displacement estimation using one-way ViSP

ψ are rotational displacements about the X, Y, and Z axes, respectively. To derive the kinematics, a homogeneous transformation matrix ${}^{R}H_T$ that transforms the coordinate from the manipulator on the transmitter ($\Sigma_{T'}$) to the screen on the receiver (Σ_R) can be used. By using the homogeneous transformation matrix and installed laser and LRF positions, ${}^{R}O$, ${}^{R}Y$, and ${}^{R}Z$ can be obtained as follows:

$$
{}^{R}O = {}^{R}H_T \cdot {}^{T}H_{T'} \cdot [L \quad 0 \quad Z_{TR} \quad 1]^T \tag{1}
$$

$$
{}^{R}Y = {}^{R}H_T \cdot {}^{T}H_{T'} \cdot [-L \quad 0 \quad Z_{TR} \quad 1]^T \tag{2}
$$

$$
{}^{R}Z = {}^{R}H_T \cdot {}^{T}H_{T'} \cdot [0 \quad L \quad Z_{TR} \quad 1]^T \tag{3}
$$

where Z_{TR} is a distance between the transmitter and receiver and L is the offset of the installed lasers and LRF. Since the laser beams are projected on the 2D screen, the z components of ${}^{R}O$, ${}^{R}Y$, and ${}^{R}Z$ should be constrained to zero. In a similar way, $\overline{TR}$ can be obtained as follows:

$$
\overline{TR} = \sqrt{{}^{R}Z_x^2 + ({}^{R}Z_y^2 - L) + Z_{TR}^2} \tag{4}
$$

where ${}^{R}Z_x$ and ${}^{R}Z_y$ denote the x and y components of ${}^{R}Z$, respectively. By putting the constraints together, the kinematics can be obtained as follows:

$$
M = [{}^{R}O_x \quad {}^{R}O_y \quad {}^{R}Y_x \quad {}^{R}Y_z \quad {}^{R}Z_x \quad {}^{R}Z_y \quad \overline{TR}]^T \tag{5}
$$

$$
= \begin{bmatrix} x - Lc_\phi s_\psi - (s_\phi(z + Lc_\psi s_\theta + Lc_\theta s_\phi s_\psi))/(c_\theta c_\phi) \\ y + L(c_\theta c_\psi - s_\theta s_\phi s_\psi) + (s_\theta(z + Lc_\psi s_\theta + Lc_\theta s_\phi s_\psi))/c_\theta \\ x - Lc_\phi c_\psi - (s_\phi(z - Ls_\theta s_\psi + Lc_\theta c_\psi s_\phi))/(c_\theta c_\phi) \\ y - L(c_\theta s_\psi + c_\psi s_\theta s_\phi) + (s_\theta(z - Ls_\theta s_\psi + Lc_\theta c_\psi s_\phi))/c_\theta \\ x + Lc_\phi c_\psi - (s_\phi(z + Ls_\theta s_\psi - Lc_\theta c_\psi s_\phi))/(c_\theta c_\phi) \\ y + L(c_\theta s_\psi + c_\psi s_\theta s_\phi) + (s_\theta(z + Ls_\theta s_\psi - Lc_\theta c_\psi s_\phi))/c_\theta \\ \sqrt{(z + L(c_\psi s_\theta + c_\theta s_\phi s_\psi))^2 + (x - Lc_\phi s_\psi)^2 + (y - L + L(c_\theta c_\psi - s_\theta s_\phi s_\psi))^2} \end{bmatrix}
$$

where s_θ and c_θ denote $\sin\theta$ and $\cos\theta$, respectively. Because it is not easy to directly calculate the 6-DOF displacement from the observation, Newton-Raphson method is employed. By using Newton-Raphson method, the displacement p can be estimated iteratively as follows:

$$
\hat{p}(k + 1) = \hat{p}(k) + J_p^\dagger(m(k) - \hat{m}(k)) \tag{6}
$$

where $J_p = \frac{\delta M}{\delta p}$ is the Jacobian of the kinematics, $J_p^\dagger$ is the pseudo-inverse of the Jacobian, $\hat{m}(k)$ is the estimated observation by M, and $m(k)$ is the measurement.

3   Simulations

To verify the performance of the proposed system, 6-DOF displacement estimations with and without random noise of the measurement have been conducted. In the

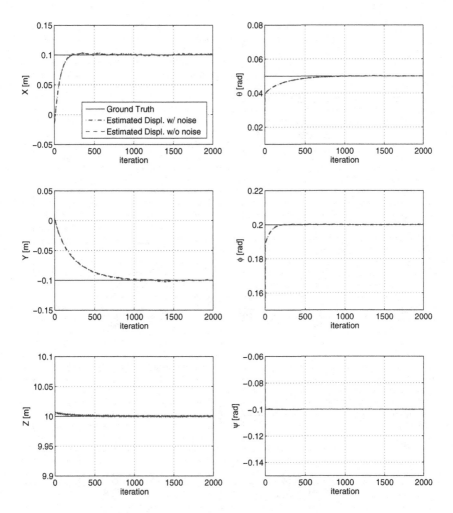

Fig. 3. Static displacement estimation using Newton-Raphson method with and without random uniform measurement noise. The true value is set to $p_1 = [0.1, -0.1, 10, 0.05, 0.2, -0.1]^T$.

following simulations, the initial displacement is set to $p_0 = [x_0, y_0, z_0, \theta_0, \phi_0, \psi_0]^T = [0, 0, 10, 0.001, 0.001, 0.001]^T$ where all units are meters and radians. The random uniform measurement noise of $[-0.4 \ 0.4]$ mm and $[-1.5 \ 1.5]$ mm are set for projected laser beams and LRF, respectively, considering the sensors' specification. The simulation results of static displacement estimation when the true values are set to $p_1 = [0.1, -0.1, 10, 0.05, 0.2, -0.1]^T$ and $p_2 = [0.2, -0.1, 10, 0.1, 0.1, -0.2]^T$, where all units are meters and radians, are shown in figures 3 and 4, respectively. As the results show, the estimated displacement from the proposed system converges to the true value for both cases. In the figures, the solid line represents the true value, the dashed dot line represents the estimated displacement with the measurement noise, and the dashed line represents the estimated displacement without the measurement noise.

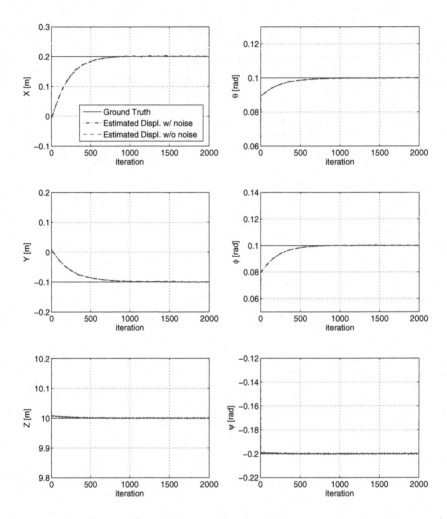

Fig. 4. Static displacement estimation using Newton-Raphson method with and without random uniform measurement noise. The true value is set to $p_2 = [0.2, -0.1, 10, 0.1, 0.1, -0.2]^T$.

4 Conclusion

In this paper, a one-way vision and laser-based structural displacement measurement system (one-way ViSP) is proposed to monitor the structural condition. The one-way ViSP is composed of a transmitter with two lasers, a laser range finder (LRF), a 2-DOF manipulator; and a receiver with a screen and a camera. The lasers and LRF from the transmitter project their parallel beams to the screen on the opposite side and the camera attached to the screen captures the image of the screen. By calculating the positions of the projected laser beams and using the measured distance from LRF, the 6-DOF relative displacement between two positions can be estimated. To verify the performance of the system, simulations with and without the uniform random noise of measurement data are conducted. The simulation results show that the estimated displacements from both cases converge to the true value. In the future, a prototype of the system will be built and applied to real civil structures. Moreover, the extended Kalman filtering (EKF) scheme will be applied to consider the uncertainties in the measurements.

Acknowledgment. This study was supported by the National Research Foundation of Korea (NRF) and funded by the Korea government (Grant No. 2009-0075397). This research was also supported by a grant (13SCIPA01) from Smart Civil Infrastructure Research Program funded by Ministry of Land, Infrastructure and Transport (MOLIT) of Korea government and Korea Agency for Infrastructure Technology Advancement (KAIA). Mr. Myeong is supported by Korea Minister of Ministry of Land, Transport and Maritime Affairs (MLTM) as U-City Master and Doctor Course Grant Program.

References

1. Balageas, D., Fritzen, C.P.: Structural Health Monitoring. John Wiley & Sons, Inc., New Jersey (2006)
2. Lee, J.J., Shinozuka, M.: Real-time displacement measurement of a flexible bridge using digital image processing techniques. Exp. Mech. 46(1), 105–114 (2006)
3. Park, J.W., Lee, J.J., Jung, H.J., Myung, H.: Vision-based displacement measurement method for high-rise building structures using partitioning approach. NDT & E Int. 43(7), 642–647 (2010)
4. Myung, H., Lee, S.M., Lee, B.J.: Paired structured light for structural health monitoring robot system. Struct. Health Monit. 10(1), 49–64 (2011)
5. Jeon, H., Bang, Y., Myung, H.: Structural inspection robot for displacement measurement. In: IEEE- Digital Ecosystem Science and Technology (DEST), Daejeon (2011)
6. Myung, H., Jung, J.D., Jeon, H.: Robotic SHM and model-based positioning system for monitoring and construction automation. Advances in Structural Engineering 15(6), 943–954 (2012)
7. Jeon, H., Bang, Y., Myung, H.: Design of a modular structural inspection robot. In: Int'l Workshop on Design in Civil and Env. Eng., Daejeon (2011)
8. Jeon, H., Bang, Y., Myung, H.: A paired visual servoing system for 6-DOF displacement measurement of structures. Smart Mater. Struct. 20(4), 45019 (2011)
9. Jeon, H., Bang, Y., Myung, H.: Measurement of structural displacement using visually servoed paired structured light system. In: Intl Symposium on Automation and Robotics in Construction, Seoul (2011)
10. Jeon, H., Shin, J.U., Myung, H.: Incremental displacement estimation of structures using paired structured light. Smart Struct. Syst. 9(3), 273–286 (2012)
11. Jeon, H., Shin, J.U., Myung, H.: The displacement estimation error back-propagation (DEEP) method for a multiple structural displacement monitoring system. Meas. Sci. Technol. 24(4), 045104 (2013)

Target-Driven Visual Words Representation via Conditional Random Field and Sparse Coding

Y.-H. Yoo and J.-H. Kim

Department of Electrical Engineering, KAIST
335 Gwahangno, Yuseong-gu, Daejeon, Republic of Korea
{yhyoo,johkim}@rit.kaist.ac.kr

Abstract. At any given moment, humans eye captures a large amount of information simultaneously. Among these information, human visual system is able to select specific information in which human is interested. In recent years, there have been trials for (system) experimental, computational and theoretical studies on imitating human visual system, which are commonly referred as sparse coding. When any visual stimuli are given, human visual system makes a minimal number of neurons activated efficiently. It increases the storage capacity in associative memories. A set of activated neurons and deactivated neurons are called sparse code and the process to make sparse code is called sparse coding. In this paper, the effectiveness of the proposed method is demonstrated for Graz-02 dataset. And visual words were visualized that were relevant to activated neurons as patch-level images and sparse coding. By displaying active neurons that are represented by visual words, sparse coding could be a solution to top-down visual object detection.

Keywords: Sparse coding, dictionary learning, CRF, max-margin approach, saliency map.

1 Introduction

At any given moment, human's eye captures a large amount of information simultaneously. Among these information, human visual system is enable to select specific information in which human is interested. A bag of model is similar method to imitate human visual system[1]. This model consists of codewords chosen from training samples. This bag of model is enable to classify scenes for arbitrary view. However, there is limit to target-oriented computer vision that is reflected human internal state such as object detection, segmentation and image classification.

Recently, sparse coding is in the limelight to resolve target-oriented computer vision task. In recent years, a combination of experimental, computational, and theoretical studies have pointed that information is represented by a relatively small number of simultaneously active neurons out of a large population, commonly referred as sparse coding[2]. For instance, assume that there are optic neurons that corresponds to their own color. When yellow signal comes to visual system, neurons taking charge of red and green signal

J.-H. Kim et al. (eds.), *Robot Intelligence Technology and Applications 2*,
Advances in Intelligent Systems and Computing 274,
DOI: 10.1007/978-3-319-05582-4_60, © Springer International Publishing Switzerland 2014

could be activated. But, human visual system gets only one neuron that take charge of yellow signal activated to handle this yellow signal more simply. As any stimuli is given, human visual system makes minimal neurons activated efficiently. It increases the storage capacity in associative memories. A set of activated neuron and de-activated neuron is called sparse code and the process to make sparse code is called sparse coding. In sparse coding, dictionary is a kind of function that is mapping external stimuli to sparse code and in computer vision, dictionary is called visual words. Sparse code's sparsity depends on how dictionary is defined. So, dictionary learning is essential process to represent sparse code efficiently.

Jimei Yang *et al.*[3] tried to find top-down visual saliency using joint CRF and dictionary learning. Conditional random field(CRF)[4] is modeled using sparse code extracted from image patches by SIFT[5] and its corresponding ground truth label. CRF's parameter and dictionary is derived by max-margin approach[6] and gradient descent algorithm[7]. This method showed top-down object detection is possible via joint CRF and dictionary learning.

In this paper, the effectiveness of the proposed method is demonstrated for Graz-02 dataset. And visual words were visualized that were relevant to activated neurons as patch-level images and sparse coding. The rest of this paper is organized as follows: Section 2 reviews related work such as sparse coding(dictionary learning), CRF and max-margin approach. Section 3 explains how to generate visual words by using CRF and sparse coding. Section. Experimental results are presented in section 4, and this paper is concluded with a future research plan in section 5.

2   Related Work

In this section, related work such as sparse coding, CRF, max-margin classification to demonstrate that top-down object detection is described.

2.1   Sparse Coding(Dictionary Learning)

At any given moment, high-dimensional information is captured by natural sensors: eyes, ears, etc. This information is largely redundant. However, a few information is necessary for human to generate physical process and make decision[8]. There are many dimension reduction methods such as PCA and ICA[9,10]. The goal of these methods is to reduce dimension by training given dataset. So, the number of basis is fixed as generated basis depend on their training dataset. However, an arbitrary stimuli is given to human, they respond their own way. Assume that each responce consists of factors and let each factor is basis. Then, the response could be expressed as linear combination of basis. This concept is called sensory coding, as first outlined by Barlow[11]. It is crucial for efficient dimensional reduction to select basis in observed data. Dictionary learning is promising research field to select basis efficiently and sparsity constraint is key issue to dictionary learning.

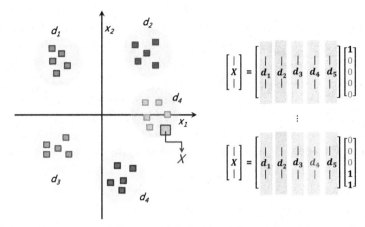

(a) Input vectors and their own labels.   (b) Parts of sparse representation.

Fig. 1. Sparse coding simple example

Let us denote the dictionary as D and sparse code as S. Then, a given signal X is represented as linear combination of a small number of signal[8].

$$X = Ds = \sum_{k=1}^{K} d_k s_k \tag{1}$$

Assume that input signal X is two dimension vector. In this space, there are many regions that represent their own features. By calculating Equation 1, there are many solution S as dictionary D is over-complete. The yellow signal X is represented as linear combination of features that are in red and green region. Also, this signal X is represented as feature in yellow region. Latter case is more sparse so that this signal X could be represented only one basis.

Like above example, there are many solutions to sparse codes to represent arbitrary X. The sparsest solution is usually derived by minimizing cost function[8,12],

$$\min \|s\|_1 \quad subject\ to\ \|x - Ds\|^2 \le \epsilon, \tag{2}$$

where $\|.\|$ denotes the l_p norm. As combinatorial l_0-norm minimization is impractical. L_1-norm, as the closest convex function to l_0-norm minimization is applied[13]. Equation 2 could be expressed one cost function by adding lagrange multiplier as

$$s = arg \min_s \|x - Ds\|^2 + \lambda \|s\|_1, \tag{3}$$

where λ is a parameter controlling the sparse penalty. The method to derive sparsest solution is proven by H. Lee *et al.*[12].

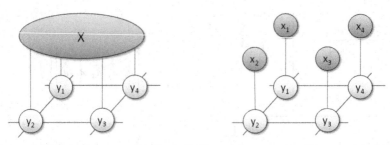

(a) General CRF models.   (b) Used CRF models assumed all latent variable i.i.d.

Fig. 2. CRF models

2.2   CRF

CRF(Conditional Random Field) is an undirected graphical model, or a kind of markov random field. Generally, $G = (V, E)$ is used for describing undirected graph model. Each vertex in V is indexed as y_i in Y. CRF is satisfied when each y_i in Y is conditioned by x_i in X and all y_i obey markov property as

$$P(y_i|x, y_{\backslash\{i\}}, w) = P(y_i|x, y_{n_i}, w) \tag{4}$$

where $y_{\backslash\{i\}}$ are vertices except y_i and w is CRF weight vector for defining probability model. General model is described in Fig 2.(a). But, in this paper, CRF model is described like Fig 2.(b) by assuming all nodes in X is i.i.d.

In the CRF model, probability could be expressed as

$$P(Y|X, W) = \frac{1}{Z} \exp -E(X, Y, W), \tag{5}$$

where Z is the normalized parameter and E is energy function. Thus, maximizing a posteriori probability is identical to minimizing the energy function. Energy function is consist of two energy term such as unary potential and pairwise potential. Unary potential is a node y_i's probability conditioned by x_i. And pairwize potential is probability between some nodes y_i in Y unrelated to X. This energy function could be expressed as follows. In this equation, CRF weight $w = [w_1; w_2]$ have to be trained and this is similar to most CRF models[3,6].

$$E(X, Y, W) = \sum_{i \in V} \psi(x_i, y_i, w_1) + \sum_{i,j \in E} \psi(y_i, y_j, w_2) = \sum_{i \in V} -y_i w_1 x_i + \sum_{i,j \in E} w_2 I(y_i, y_j),$$
$$\tag{6}$$

where $I(y_i, y_j)$ is one for different labels. Like above equation, enery minimization could be converted to max-margin classification by assuming energy function is linear to weight vector W.

2.3   Max-margin Approach

Given arbitrary input vector X and CRF weight vector W, the goal of inference is finding maximum a posteriori labeling y* such as $Y = arg\max_Y P(Y|X, W)$. Also, weight vector W have to be found for satisfying

$$P(Y^{(n)}|X^{(n)}, W) \geq P(Y|X^{(n)}, W) \tag{7}$$

for all Y not equal to $Y^{(n)}$ all n[3,6]. This equation could be expressed in term of energy.

$$E(Y^{(n)}, X^{(n)}, W) \leq E(Y, X^{(n)}, W) \tag{8}$$

Given $X^{(n)}$, energy of $Y^{(n)}$ is called ground truth energy that is always less than energy of Y except $Y^{(n)}$. To satisfy these conditions, many constraint equations are needed. To reduce the number of constraint equation, most violated energy closest to ground truth enery is chosen. Let us denote most violated energy $\hat{Y}^{(n)}$. Then, constraint equation is expressed as

$$E(\hat{Y}^{(n)}, X^{(n)}, W) - E(Y^{(n)}, X^{(n)}, W) \geq \gamma \tag{9}$$

where γ is margin for maximizing. This equation could be converted as

$$\min_w \frac{1}{2} \|W\|^2 \quad subject \ to \ E(\hat{Y}^{(n)}, X^{(n)}, W) - E(Y^n, X^{(n)}, W) \geq 1, \tag{10}$$

for $\forall Y \neq Y^{(n)}$ by transforming $\gamma = \frac{1}{\|W\|}$. This equation could be converted a cost function as

$$F^n(D, W) = \lambda \frac{\|W\|^2}{2} - E(\hat{Y}^{(n)}, X^n, W) + E(Y^n, X^n, W)). \tag{11}$$

where λ represent lagrangian multiplier.

3   Visual Words Generation via Sparse Coding

3.1   Model

In this section, the process how visual words are generated is explained using sparse coding, CRF and max-margin classification. Graz-02 dataset is used for training. In this dataset, each image size is 640 by 480. Each image is divided as 37 by 27 patches that are sampled 64 by 64 pixels by shifting 16 pixels. Feature vector x could be derived from each patches via SIFT[5]. By dictionary learning, each feature vector x is converted to sparse code that is corresponded to labeling y_i. By doing this, CRF model is generated like Fig. 3.

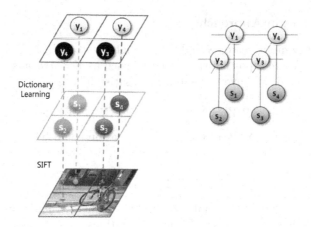

Fig. 3. (a) Procedure for generating CRF model. (b) simplicated CRF model.

3.2   Learning Algorithm

Given CRF model like Fig. 3, cost function could be defined as

$$F^n(D,W) = \lambda \frac{\|W\|^2}{2} - l^n(D,W) \tag{12}$$

where $l^n(D,W) \equiv < W, f(S^{(n)}, \hat{Y}^{(n)}) > - < W, f(S^{(n)}, Y^{(n)}) >$. Energy term could be transformed by inner product between CRF weight vector W and potential function $f(S,Y)$. To find visual words, dictionary D and CRF weight vector W have to be found for minimizing above cost function. These parameters could be updated by gradient descent method. The gradient descent with respect to W is

$$\frac{\partial F}{\partial W} = \lambda W - f(S^{(n)}, \hat{Y}^{(n)}) + f(S^{(n)}, Y^{(n)}). \tag{13}$$

As the cost function is not explicit function with respect to D, the gradient with respect to D has to be evaluated by chain rule,

$$\frac{\partial F^n}{\partial D} = \sum_{i \in V} (\frac{\partial F^n}{\partial s_i})^T \frac{\partial s_i}{\partial D}. \tag{14}$$

Gradient $F^{(n)}$ with respect to S is expressed as $(y_i^{(n)} - \hat{y}_i^{(n)})W$. And gradient s_i with respect to D is evaluated by using implicit differentiation on the fixed point equation[3,14],

$$\frac{\partial \tilde{s}}{\partial D} = (\tilde{D}^T \tilde{D})^{-1} (\frac{\partial \tilde{D}^T x}{\partial D} - \frac{\partial \tilde{D}^T \tilde{D}\tilde{s}}{\partial D}), \tag{15}$$

where symbol tilde means taking only non-zero elements. Using this gradient descent method, weight W and dictionary D could be derived efficently.

4   Experiment Results

Dictionary D is trained dependent on targets such as bycicle, car and person that are in Graz-02 dataset. Among visual words corresponding to non-zero element with respect to each target, visual words of high frequency are showed in Fig. 4. Actually, visual words are not images, but feature vectors. However, patch nearest to visual words is chosen to show visual words as images.

Using trained visual words, an arbitrary image whose target already defined is expressed as a combination of their own patches representing probabilities in Fig. 5. And their precision-curve is shown in Fig. 6. In this figure, baseline is the result before learning. As seen from Fig. 6, the result shows that sparse coding could improve the CRF model's performance more.

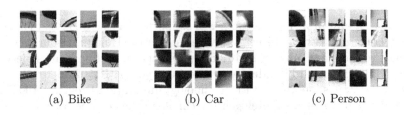

(a) Bike   (b) Car   (c) Person

Fig. 4. Trained visual words of high frequency

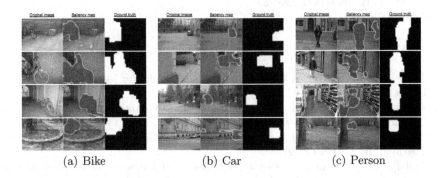

(a) Bike   (b) Car   (c) Person

Fig. 5. Saliency map for each target

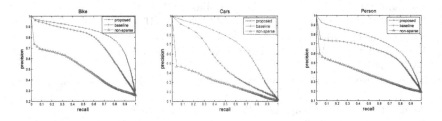

Fig. 6. Precision-recall curve

5 Conclusion

In this paper, target-driven visual words were displayed via sparse coding, CRF and max-margin approach. For that, Graz-02 dataset including bicycle, car and person is used. Using trained parameters by dictionary learning, saliency map was generated for any arbitrary image. A precision-recall curve was plotted to evaluate the proposed method and this result showed the performance was improved. Our future work includes generating dataset in accordance with any circumstance, in particular, indoor environment. Using this generated dataset, the saliency map will be generated in regard to any targets as well as bike, car and person. During generating the dataset, RGB-D sensor will be used for constructing 3D saliency map. And this map will be applied to pre-processing for cleaning-up a messy room using a robot.

Acknowledgements. This research was supported by the MOTIE (The Ministry of Trade, Industry and Energy), Korea, under the Human Resources Development Program for Convergence Robot Specialists support program supervised by the NIPA (National IT Industry Promotion Agency)(H1502-13-1001).

This work was supported by the Technology Innovation Program, 10045252, Development of robot task intelligence technology that can perform task more than 80% in inexperience situation through autonomous knowledge acquisition and adaptational knowledge application, funded By the Ministry of Trade, industry & Energy (MOTIE, Korea).

References

1. Fei-Fei, L., Perona, P.: A bayesian hierarchical model for learning natural scene categories. In: CVPR 2005 (2005)
2. Olshausen, B., Field, D.: Sparse coding of sensory Inputs. Current Opinion in Neurobiology 14(4), 481–487 (2004)
3. Yang, J., Yang, M.-H.: Top-down visual saliency via Joint CRF andDictionary Learning. In: CVPR (2012)
4. Laffety, J., McCallum, A., Pereira, F.: Conditional random fields: Probabilistic models for segmenting and labeling sequence. ACM (2001)
5. Lowe, D.: Distinctive image features from scale-invariant keypoints. IJCV 60(2), 91–110 (2004)
6. Szummer, M., Kohli, P., Hoiem, D.: Learning cRFs using graph cuts. In: Forsyth, D., Torr, P., Zisserman, A. (eds.) ECCV 2008, Part II. LNCS, vol. 5303, pp. 582–595. Springer, Heidelberg (2008)
7. Yang, J., Yu, K., Huang, T.: Supervised translation-invariant sparse coding. In: CVPR (2010)
8. Tosic, I., Frossard, P.: Dictionary Learning. IEEE Signal Processing Magazine (2011)
9. Jolliffe, I.: Principal Component Analysis. Springer, New York (1986)
10. Information-maximization approach to blind separation and blind deconvolution. Neural Comput. 7(6), 1129–1159 (1995)

11. Barlow, H.B.: Possible principles underlying the transformations of sensory messages. In: Rosenblith, W. (ed.) Sensory Communication, ch. 13, pp. 217–234. MIT Press, Cambridge (1961)
12. Lee, H., Battle, A., Raina, R., Ng, A.Y.: Efficient sparse coding algorithms. In: NIPS (2006)
13. Joachims, T., Finley, T., Yu, C.-N.: Cutting-plane training of structural svms. Machine Learning 77(1), 27–59 (2009)
14. Yang, J., Yu, K., Huang, T.: Supervised translation-invariant sparse coding. In: CVPR (2010)

Cloud-Based Object Recognition: A System Proposal

Daniel Lorencik, Martina Tarhanicova, and Peter Sincak

Department of Cybernetics and Artificial Intelligence, Technical University of Kosice,
Letna 9, Kosice, 04200 Slovak Republic
{daniel.lorencik,martina.tarhanicova,peter.sincak}@tuke.sk

Abstract. In this chapter, we will present a proposal for the cloud – based object recognition system. The system will extract the local features from the image and classify the object on the image using Membership Function ARTMAP (MF ARTMAP) or Gaussian Random Markov Field model. The feature extraction will be based on SIFT, SURF and ORB methods. Whole system will be built on the cloud architecture, to be readily available for the needs of the new emerging technological field of cloud robotics. Besides the system proposal, we specified research and technical goals for the following research.

Keywords: cloud computing, cloud robotics, object recognition, SIFT, SURF, ORB, MF ARTMAP, Gaussian Random Markov Field.

1   Introduction

Since the history of computers and computing began roughly 70 years ago, we have seen the large scale computers replaced by affordable personal computers. In the last years, we are witnesses to another notion – the personal computers shrank in size to tablets, netbooks and even smartphones, and the heavy computational and storage tasks are offloaded to the cloud. Also, the applications available on the cloud have high impact on the productivity as they allow for easy implementation on sharing data between several users, thereby promoting real-time collaboration and aggregation of crowd knowledge, example being the Google Apps suite [1] or Microsoft Office 365 [2].

With this knowledge in mind, it is possible to envision the similar system of application which will be available for use by robots. The obvious benefit is the possibility of creating small robots with greater longevity of battery life since the heavy computation is done elsewhere. These robots will not have to be highly sophisticated. Therefore, they can be cheap or can be created from available resources like smartphones combined with the wheeled chassis. More than that, the robots can benefit from the sharing of knowledge. This idea was presented by professor James Kuffner in his talk "Robots with their heads in the Clouds" [3]. The knowledge sharing in real time has a potential to influence the ability of robots to exist in the real world as the knowledge gained in the learning process is gained by all robots using the service. We will provide more detailed information on the cloud robotics in the next section.

With the availability of cloud computing and method of artificial intelligence provided as a service in the cloud environment, the idea of remote brain [4] resurfaces

J.-H. Kim et al. (eds.), *Robot Intelligence Technology and Applications 2*,
Advances in Intelligent Systems and Computing 274,
DOI: 10.1007/978-3-319-05582-4_61, © Springer International Publishing Switzerland 2014

again. It is true that the connection to the cloud is crucial; therefore it is a weak link in the chain, but with the available connection options via WiFi and availability of 3G and 4G networks, this is more of a technological problem.

The structure of the chapter is as follows: in the second section, we will provide an introduction to the cloud computing and define the cloud robotics. In this section, we will also present the projects of cloud robotics. In the third section, we will provide an overview of three methods (SIFT, SURF, ORB) for the feature extraction from the image, and two methods for classification (Membership Function ARTMAP and Gaussian Markov Random Field model) of the objects on the image based on extracted features. In the fourth section, we will propose a cloud based system for object recognition on image. Fifth section contains the conclusion of the chapter.

2   Cloud Robotics

The cloud computing can be viewed as grid computing with the added concepts from utility, service and distributed computing [5]. The relationship between different distributed computing systems is shown on Fig. 1:

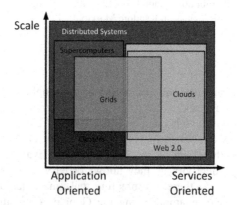

Fig. 1. Relationship between orientation and scale of different distributed computing systems

The cloud computing was defined by the National Institute of Standards and Technology as the "model for enabling ubiquitous, convenient, on-demand network access to a shared pool of configurable computing resources (e. g. networks, servers, storage, applications and services) that can be rapidly provisioned with minimal effort or service provider interaction." [6].

The clouds are provided in four deployment models:

- Private Cloud – cloud infrastructure is used exclusively by a single organization
- Community Cloud – cloud infrastructure is used by a group of consumers with shared concerns
- Public Cloud – cloud infrastructure is provided for public use

- Hybrid Cloud – combines at least two of the previous models with the clear distinction between models of cloud infrastructure but provides the possibility to port applications from one model to another
- Besides the deployment models, the cloud also provide three types of services:
- Infrastructure as a Service (IaaS) – the user has the ability to create and manage virtual machines depending on his/her individual needs. The administration of machines, networking and all settings are the responsibility of the user
- Platform as a Service (PaaS) – the user is provided with an access to the high-level integrated environment to build, test and deploy applications. Part of the required settings is managed by the platform itself (also the scaling is done automatically). However, this can present some restrictions on the use of programming language or tools.
- Software as a Service (SaaS) – the software or application is provided to the end users. Benefits are instant update of the application, and the minimal footprint of it on the user computer (usually is used from internet browser). Examples of cloud services are Google Apps [1], or storage oriented services with automatic synchronization like Dropbox [7] or SkyDrive [8].

Cloud robotics is based on the notions of "cloud" and combines the computational power of the computer cloud and the availability of internet-connected devices. Device can be any hardware that has the ability to connect to the internet, and can be programmed to use the cloud services. It can be virtually any robot that has wired or wireless connection, or it can be smartphone, small computer (NetDuino [9], Raspberry Pi [10]). Especially when using smartphones with connected actuators (Romo [11], SmartBot [12]), or low cost small computers like Raspberry Pi, it is possible to create affordable robots, for which the cloud robotics can provide the software needed. This software can be in the form of AI bricks [13].

Most of the projects in cloud robotics until now were focused on the task of creating the cloud robotics infrastructure. In the process, similar to the services in cloud computing, the Robot as a Service (RaaS) was defined [14]. RaaS has to have the features of Service oriented architecture, namely it has to be a service provider, service broker and service client. The RaaS makes available the actions it can perform, accepts connections to it, and is able to use other services as well.

As was said, there are several projects concerned with cloud robotics:

DAvinCi. It is a cloud-based framework for service robots [15], which allows several robots to communicate together and collaborate on the creation of the environment map using FastSLAM algorithm.

MyRobots.com. It is a web based project focused on "connecting all robots and intelligent devices to the Internet" [16]. It is promoted as a cloud service for robots, although currently only app store and basic monitoring service are available. It is possible to download the application for the device, and also upload user-created application. Monitoring service allows for remote monitoring of robot status, and it can send alerts if the robots encounter problem.

ASORO. It is an acronym for A*Star Social Robotics [17]. The main goal of the project is to create and promote social robots. From the cloud robotics point of view, this project is intriguing because all the robots created use the Unified Robotics Framework (UROF), which is essentially an operating system allowing to connect modules for robot functions. These are similar to the AI bricks already mentioned, and are used as needed for the tasks as path planning, task planning, navigation control and other.

RoboEarth. It is a project which goal has been to "create a World Wide Web" for robots [18]. RoboEarth is a collection of databases storing actions, objects data and environments data. These databases are shared amongst connected robots. Therefore if one robot has learned to identify certain object, all others robots gain this knowledge as well. The same is true for actions (or action recipes), which describe how to do tasks, and environments, which store information about the object and their locations. The data in databases are encoded in semantic language OWL, so it is possible to derive new knowledge from existing or to use the same approach to the similar action. Also, the actions are finely tuned with the use. The action recipes are composed of atomic actions, which are again similar to the notion of the AI bricks.

The proposal of the system is based upon knowledge gained from these projects and aims to an AI brick, which can be used in already available cloud robotics frameworks.

3   Image Processing and Object Classification

As our proposed system will provide a cloud based service for object classification on the image, in this section we will provide an overview of methods we will use. We will start with methods for extraction of local features from the image - SIFT, SURF and ORB, and continue with the classification methods based on Membership Function ARTMAP and Gaussian Markov Random Fields model.

Scale-invariant Feature Transform. It was described in detail in [19]. SIFT extracts local features from scale-space extrema called key points. Key points are identified as a minimum (or maximum) of difference of Gaussians occurring at multiple scales of image pyramid. Next, the unstable key points are removed, and the orientation is assigned. Computing from the image pyramid provide invariance to scale, assigning the orientation based on a peak in local histogram provides invariance to the rotation, and invariance to illumination is provided by thresholding the values of descriptor vector comprised of values of orientation histograms. The final descriptor has 128 dimensions. Experimental results from [19] suggest that SIFT can recognize even partially occluded objects, as for the object recognition only 3 descriptors are needed. It is invariant to scale, rotation and translation and partially to the affine translation (up to 50 degrees). Several improved method were proposed – Affine SIFT [20] to improve invariance to the tilt of the camera and Principal Component Analysis SIFT (PCA-SIFT) [21] which creates the descriptor vector using PCA.

Speeded Up Robust Transform. It was described in [22]. It is inspired by SIFT, uses three stages: detection, description and matching and is tailored to provide high speed and accuracy similar to SIFT. Detection is based on using basic Hessian matrix approximation and integral images. Interest points are blobs located at maxima of the determinant of Hessian matrix. The searching of interest points on different scales is done by gradually larger filters as opposed to the resampling of the image in the SIFT [22], [23]. Description is similar to the SIFT approach, and based on Haar wavelet responses and produces a 64 dimensional vector (there are modification with different length of descriptor vector – SURF-36 and SURF-128). In matching phase, the sign of Laplacian is used to determine if the interest point is a bright blob on a dark background or dark blob on a white background. Only the features with the same sign are compared, which leads to speed improvement over SIFT.

Oriented FAST and Rotated BRIEF. It is a new approach proposed in [24]. It uses modified edge detector FAST (Features from Accelerated Segment Test, [25], [26]) and modified feature point descriptor BRIEF (Binary Robust Independent Elemental Features, [27]). It uses FAST edge detector to detect corners, and to provide scale invariation, the FAST is run on the scale image pyramid. Orientation of the corner is found by use of intensity centroid, which assumes that the corner intensity is offset from the center of the corner. BRIEF is found to be equal in performance to SIFT and SURF, and since it uses binary strings as description vectors, it is computed faster [27]. As the BRIEF is not rotation-invariant, the corners found by oriented FAST are normalized in orientation and then the BRIEF descriptor is computed. Then the uncorrelated key points are selected and used to construct a Rotated BRIEF descriptor. ORB was designed to be faster than existing local feature detectors SIFT and SURF. Experiments described in [24] suggest it is magnitude faster than SURF and two magnitudes faster than SIFT with the similar recognition ability.

Membership Function ARTMAP. It is a classification tool [28], [29] based on Adaptive Resonance Theory [30], [31] and the theory of Fuzzy Sets. The knowledge representation is based on the hypothesis that input samples are in *fuzzy clusters* in feature space, which is the universe of fuzzy sets. Therefore, it is possible to calculate the membership value of each point from feature space to every fuzzy cluster defined in this space. The MF ARTMAP network consists of:

- The input layer of neuron, which normalizes input and maps the input samples to the comparison layer, and the number of neurons is the same as the number of dimensions in feature space;
- Comparison layer is n to m grid, where n is the dimension of the feature space, and m is the number of neurons in recognition layer; size of the layer is dynamically changed in the process of learning;
- Recognition layer contains neurons representing the fuzzy clusters in feature space; therefore the number of neurons can change in the learning process
- MapField layer which consists of neurons representing fuzzy classes. Here is computed the value of the membership of the sample to the fuzzy class

Learning algorithm of MF ARTMAP is divided into two steps: structure adaptation, where new fuzzy clusters are created (therefore changing recognition and comparison layer); and parameters adaptation, where parameters of membership function stored in connections between layers are changed.

Gaussian Markov Random Field Model (GMRF). It was investigated for the task of object classification in [32], [33] on texture data and in [34] the GMRF was investigated for the task of image classification. The use of GMRF is interesting because any data distribution can be approximated by Gaussian mixtures and there are many mathematical techniques to allow easy work with these data [35]. In the [32], the textures were modeled with the GMRF with parameters estimated from training samples observed at given angle as the GMRF is not invariant to scale and rotation. Classifier was based on modified Bayes rule consisting of obtaining the maximum likelihood estimate of the rotation and scale parameters for each class hypothesis, comparing the results and mapping the input sample to the class with the highest estimate. GMRF model of texture was parameterized, and the rotation and scale parameters have been a part of the model using spectral density of the GMRF. The results of the experiments show that this approach proved successful, and can be found in [32].

4   System Proposal for Cloud-Based Object Classification

Based on the study of the cloud robotics, we have identified a challenging research topic in this field – cloud-based system for object classification from image data.

This system contributes to the cloud robotics as a distributed vision system can be available for any device capable of connecting to the internet and able to use cloud service, will be based on the shared knowledge base and will fulfill the criteria for becoming and AI brick as defined in [13]. This system should be usable in existing cloud robotics frameworks like RoboEarth. The high level overview of the proposed system is shown on the Fig. 2.

The system will accept the image from the device in most commonly used image formats. Then, the image will be preprocessed, and features will be extracted using already mentioned local feature extracting methods. The final decision which method will be used will be done depending on the results of performance tests. By using the single feature extraction method, we will ensure that the feature space will be normalized for all users. The clustering and classification will be done in classification module which will use shared knowledge from all users. The classification service will then send back the result of classification which will consist of at least five most probable object classes. In case the object on the image will not be classified or classified wrongly, the user will have the option to offer a better result.

The main contribution of the system will be availability for various devices, and as a result of use of cloud computing platform should be available "everywhere and every time". Second main contribution will be the use of shared knowledge. This will allow for increased rate in building a knowledge base, and will provide a higher quality service with a higher number of users. The knowledge sharing, easy availability and implementation are the characteristics of proposed system.

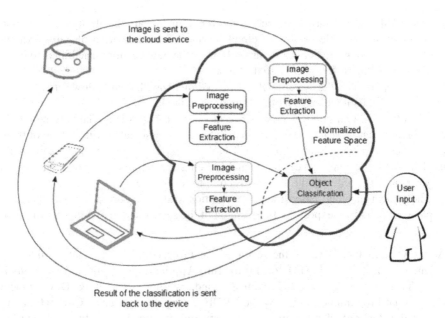

Fig. 2. High level overview of architecture of the proposed system. Object classification module is shared between users and knowledge is stored in the structure of classifier.

The challenge here will be to adapt the MF ARTMAP and GMRF model for the cloud architecture. These two methods will be compared to their stand-alone versions.

To test this system, we will also create a test to verify the robustness and performance of the system. We will use standard classification tests, as well as test the classification of standard household objects.

Most decisive test will be the comparison of the system which is open to general use versus the system open to only handful of expert teachers. Open system can benefit from the crowd knowledge, where every user of the system can also contribute to the learning process. Hypothesis stands that the system open to general use will be faster in training, will have more object classes and will provide more accurate results in the long-term use.

5   Conclusion

In this chapter, an overview of cloud robotics, local features extraction methods SIFT, SURF, ORB and classifiers Membership Function ARTMAP and Gaussian Markov Random Field model were presented.

Based on the knowledge gained, the cloud-based object classification system was proposed.

The system was proposed as an AI brick, and aims to provide powerful and easy to use object classification from the image data for existing and future robots. The advantages of providing this system as a cloud service will be availability, both geographical, and for various devices (only requirements are the ability to connect to the internet and be able to use the cloud services), instant sharing of gained knowledge between connected devices, easy rollout of the new version, scalability, reliability and offloading of heavy tasks to the cloud.

More importantly, the system will be created in a way that will allow for easy integration to the existing cloud robotics frameworks like RoboEarth.

From the research point of view, since the cloud robotics is a relatively new field of technology and research, there are many challenges associated with it. One of them is the question if the cloud robotics will have an impact on the method traditionally used in Artificial Intelligence, the example being the implementation of the neural network on the cloud with the shared structure for every user.

The project of cloud-based image recognition system is currently researched and implemented, and we expect the first version of the system in the first quarter of 2014.

Acknowledgments. Paper is the result of the Project implementation: **University Science Park TECHNICOM for Innovation Applications Supported by Knowledge Technology**, ITMS: **26220220182**, and by the Research & Development Operational Programme funded by the ERDF and by the **Center of Competence of knowledge technologies for product system innovation in industry and service**, with ITMS project number: **26220220155** for years 2012-2015.

References

[1] Google Apps, http://www.google.com/enterprise/apps/ (accessed June 5, 2013)
[2] Microsoft Office 365, http://office.microsoft.com/en-001/ (accessed July 24, 2013)
[3] Guizzo, E.: Robots With Their Heads in the Clouds. IEEE Spectrum (February 28, 2011)
[4] Inaba, M.: Remote-brained humanoid project. Advanced Robotics 11(6), 605–620 (1996)
[5] Foster, I.T., Zhao, Y., Raicu, I., Shiyong, L.: Cloud Computing and Grid Computing 360-Degree Compared. In: 2008 Grid Computing Environments Workshop, pp. 1–10 (2008)
[6] Mell, P., Grance, T.: The NIST Definition of Cloud Computing Recommendations of the National Institute of Standards and Technology. Nist Special Publication 145, 7 (2011)
[7] DropBox, http://www.dropbox.com/ (accessed June 3, 2013)
[8] SkyDrive, https://skydrive.live.com/ (accessed June 5, 2013)
[9] netduino, http://www.netduino.com/ (accessed July 22, 2013)
[10] Raspberry Pi, http://www.raspberrypi.org/ (accessed July 22, 2013)
[11] Romo, http://romotive.com/ (accessed July 25, 2013)
[12] SmartBot, http://www.overdriverobotics.com/SmartBot/ (Accessed July 25, 2013)
[13] Ferraté, T.: Cloud Robotics - new paradigm is near. Robotica Educativa y Personal (January 20, 2013)
[14] Chen, Y., Du, Z., García-Acosta, M.: Robot as a Service in Cloud Computing. In: 2010 Fifth IEEE International Symposium on Service Oriented System Engineering, pp. 151–158 (June 2010)

[15] Arumugam, R., Enti, V.R., Baskaran, K., Kumar, A.S.: DAvinCi: A cloud computing framework for service robots. In: 2010 IEEE International Conference on Robotics and Automation, pp. 3084–3089 (2010)

[16] MyRobots.com, http://myrobots.com (accessed June 8, 2013)

[17] Li, H.: A*Star Social Robotics, http://www.asoro.a-star.edu.sg/index.html (accessed June 13, 2013)

[18] RoboEarth Project, http://www.roboearth.org/ (accessed June 3, 2013)

[19] Lowe, D.G.: Object recognition from local scale-invariant features. In: Proceedings of the Seventh IEEE International Conference on Computer Vision, vol. 2, pp. 1150–1157 (1999)

[20] Yu, G., Morel, J.-M.: A Fully Affine Invariant Image Comparison Method. In: IEEE International Conference on Acoustics, Speech and Signal Processing, pp. 1597–1600 (2009)

[21] Ke, Y., Sukthankar, R.: PCA-SIFT: a more distinctive representation for local image descriptors. In: Proceedings of the 2004 IEEE Computer Society Conference on Computer Vision and Pattern Recognition, vol. 2, pp. 506–513 (2004)

[22] Bay, H., Ess, A., Tuytelaars, T., Van Gool, L.: Speeded-Up Robust Features (SURF). Computer Vision and Image Understanding 110(3), 346–359 (2008)

[23] Bay, H., Tuytelaars, T., Van Gool, L.: SURF: Speeded Up Robust Features. In: Leonardis, A., Bischof, H., Pinz, A. (eds.) ECCV 2006, Part I. LNCS, vol. 3951, pp. 404–417. Springer, Heidelberg (2006)

[24] Rublee, E., Rabaud, V., Konolige, K., Bradski, G.: ORB: an efficient alternative to SIFT or SURF. In: IEEE International Conference on Computer Vision, pp. 2564–2571 (2011)

[25] Rosten, E., Drummond, T.W.: Machine learning for high-speed corner detection. In: Leonardis, A., Bischof, H., Pinz, A. (eds.) ECCV 2006, Part I. LNCS, vol. 3951, pp. 430–443. Springer, Heidelberg (2006)

[26] Rosten, E., Porter, R., Drummond, T.: Faster and better: a machine learning approach to corner detection. IEEE Transactions on Pattern Analysis and Machine Intelligence 32(1), 105–119 (2010)

[27] Calonder, M., Lepetit, V., Strecha, C., Fua, P.: BRIEF: Binary Robust Independent Elementary Features. In: Daniilidis, K., Maragos, P., Paragios, N. (eds.) ECCV 2010, Part IV. LNCS, vol. 6314, pp. 778–792. Springer, Heidelberg (2010)

[28] Sinčák, P., Hric, M., Vaščák, J.: Membership Function-ARTMAP Neural Networks. TASK Quarterly 7(1), 43–52 (2003)

[29] Smolár, P.: Object Categorization using ART Neural Networks. Technical University of Kosice (2012)

[30] Carpenter, G.A., Grossberg, S.: The ART of adaptive pattern recognition by a self-organizing neural network. Computer 21(3), 77–88 (1988)

[31] Carpenter, G.A., Grossberg, S.: Adaptive Resonance Theory. MIT Press, Boston (2003)

[32] Cohen, F.S., Fan, Z., Patel, M.A.: Classification of rotated and scaled textured images using Gaussian Markov random field models. IEEE Transactions on Pattern Analysis and Machine Intelligence 13(2), 192–202 (1991)

[33] Rellier, G., Descombes, X., Falzon, F., Zerubia, J.: Texture feature analysis using a gauss-Markov model in hyperspectral image classification. IEEE Transactions on Geoscience and Remote Sensing 42(7), 1543–1551 (2004)

[34] Berthod, M., Kato, Z., Yu, S., Zerubia, J.: Bayesian image classification using Markov random fields. Image and Vision Computing 14(4), 285–295 (1996)

[35] Gopinath, R.A.: Maximum likelihood modeling with Gaussian distributions for classification. Proceedings of the 1998 IEEE International Conference on Acoustics, Speech and Signal Processing, ICASSP 1998 (Cat. No.98CH36181) 2(914), 661–664 (1998)

Automatic Salient Object Detection Using Principal Component Analysis

Hansang Lee, Jiwhan Kim, and Junmo Kim[*]

Dept. of Electrical Engineering, Korea Advanced Institute of Science and Technology,
291 Daehak-ro, Yuseong-gu, Daejeon 305-701, South Korea
junmo@ee.kaist.ac.kr

Abstract. In this paper, a novel method for salient object detection from natural images is proposed. In order to extract the object from an image which is visually attractive, non-redundancy is conceptually incorporated to define the saliency of image pixels by applying principal component analysis (PCA) to color components of an image. From the principal component images, seed pixels for object and background are extracted. Using these object and background seed pixels as training samples, linear discriminant analysis (LDA) is applied to image pixels so that the pixels are classified as object or background. Experiments on test images show that not only the performance of the proposed method is promising, but also it works competitively with state-of-the-art salient object detection methods.

Keywords: salient object detection, principal component analysis, linear discriminant analysis.

1 Introduction

Salient object detection is a task of detecting "salient" region from natural images, where the term "saliency" can be defined as visual attractiveness. Salient object detection is motivated by an assumption that in most cases an object in an image is visually attractive so that it can be distinguished from background as seen in Fig. 1. Since an automatic object extraction technique is required for recent advances in machine vision or image search in huge database, salient object detection is researched recently as a powerful tool for automatic object extraction [1]. Various methods for salient object detection have been proposed, while most of them are contrast-based [2, 3] such as center-surround difference [4], pixel-wise difference [5], and context-aware [6].

In this paper, two novel methods for automatic salient object detection are proposed. Unlike the recent works on salient object detection which mainly focused on the contrast or difference as saliency measure, statistical non-redundancy is incorporated as a saliency measure conceptually in the proposed method. To define non-redundancy in an image, the principal component analysis (PCA) is performed on RGB and CIE-LAB color components of an image to create prior maps. In the first

---

[*] Corresponding author.

J.-H. Kim et al. (eds.), *Robot Intelligence Technology and Applications 2*,
Advances in Intelligent Systems and Computing 274,
DOI: 10.1007/978-3-319-05582-4_62, © Springer International Publishing Switzerland 2014

original method, several prior maps are evaluated with the cluster density measure, which measures how so-called objects are displayed densely on each prior map, and the prior map with the best score is selected as a final saliency map. In the second improved method, object and background seed are extracted from the prior maps and the linear discriminant analysis (LDA) is performed on image pixels with the extracted seed as training set, so that the salient object is defined by the classified pixels.

Fig. 1. Examples of salient object detection from natural images: Natural images from MSRA dataset (top row) and their salient object detection results (bottom row), which are detected by the proposed method, are shown

The paper is organized as follows: In section 2, an original work with PCA-based salient object detection using cluster density evaluation is described. In section 3, an improved works with PCA-based salient object detection via linear discriminant analysis is described. In section 4, experimental results of two proposed methods and their analyses are provided. Finally, in section 5, the paper is concluded with further discussions.

2   PCA-Based Salient Object Detection Using Cluster Density Evaluation

The key assumption of the first method is that the information about non-redundancy is included in the color distribution of an image. For example, in a natural image showing a red flower on grass, the salient object, which is a flower in this case, can be distinguished from grass with color components of red and green. This assumption is similar to that of contrast-based salient object detection method which also assumes that an object has different color distribution with that of background. In the proposed method, however, the saliency of an image is extracted by not only the local contrast of color distribution, but also its global non-redundancy.

Outline of the first method is summarized in Fig. 2. For an image with the color components vector $\mathbf{R}, \mathbf{G}, \mathbf{B}, \mathbf{L}, \alpha, \beta$ for RGB and CIE-LAB color space, the principal component analysis (PCA) [7, 8] is performed on data matrix $[\mathbf{R}, \mathbf{G}, \mathbf{B}, \mathbf{L}, \alpha, \beta]^{T}$ to obtain the vectors of principal components $\mathbf{V} = [\mathbf{v}_1, \mathbf{v}_2, \mathbf{v}_3, \mathbf{v}_4, \mathbf{v}_5, \mathbf{v}_6]^{T}$. Since two principal components $\mathbf{v}_1, \mathbf{v}_2$ represent the two most major axis of color component distribution, it can be assumed that the two components either have the most of redundancy information of color component distribution. Thus, two projection images P_1, P_2 with the principal components $\mathbf{v}_1, \mathbf{v}_2$ are computed by

$$P_1^T = \mathbf{v}_1^T \left[\mathbf{R, G, B, L}, \alpha, \beta\right]^T , \tag{1}$$

$$P_2^T = \mathbf{v}_2^T \left[\mathbf{R, G, B, L}, \alpha, \beta\right]^T . \tag{2}$$

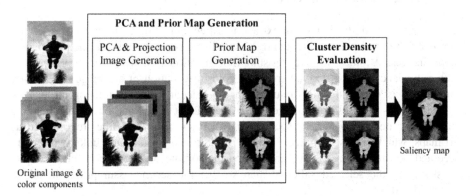

Fig. 2. Pipeline of PCA-based salient object detection using cluster density evaluation

To enhance the contrast of non-redundancy of an image, four saliency map candidates called prior maps are created by computing sums and differences of two projection images as follows:

$$S_1 = P_1 + P_2 \tag{3}$$

$$S_2 = P_1 - P_2 \tag{4}$$

$$S_3 = -P_1 + P_2 \tag{5}$$

$$S_4 = -P_1 - P_2 \tag{6}$$

In the first method, the saliency map is selected from the prior maps with the criteria based on the characteristics of saliency map. As a selection criteria, cluster density [9] is incorporated in the first method. Cluster density estimates how the clusters, detected objects in this case, are represented densely in an image. With the binary map created by thresholding the prior maps with Otsu's method, cluster density d is computed by the sum of average center-surround distances as follows:

$$d = \sum_W \frac{\displaystyle\sum_{(x,y)\in W} \sqrt{(x-x_0)^2 + (y-y_0)^2}}{|W|} \tag{7}$$

where W is a set of clusters for objects and (x_0, y_0) is a coordinate of a center of a cluster W. As d increases, it means that objects could be separated into several clusters or represented in jagged shape, whereas smaller d means that objects tend to be represented as single smooth cluster. Thus, with the cluster density scores calculated for four prior maps, the prior map with the lowest cluster density value is selected as a final saliency map, as seen in Fig. 2.

3   PCA-Based Salient Object Detection with Linear Discriminant Analysis

In the first method, the method of extracting salient object from an image by selecting optimal prior map which is created by simple linear combination of principal component images. Besides the key assumption mentioned above, which states that non-redundancy can be derived from the color distribution of an image, in the first method, it is unintentionally assumed in the selection step that the prior map is represented as nearly binary image so that the salient object and background can be easily separated. However, as seen in Fig. 2, prior maps are rather represented as tri-maps, which consist of object (a man), background (cloud), and unknown region, tree and sky, in this case. Thus, in the first method, noise, like tree in this example, is shown in the selected saliency map so that the second assumption should be reconsidered.

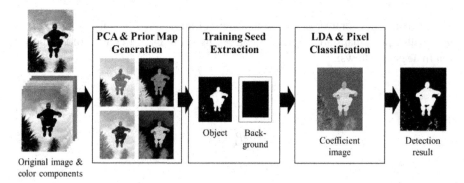

Fig. 3. Pipeline of PCA-based salient object detection with linear discriminant analysis

In the second improved method, it is assumed that at least the prior maps include the non-redundancy information of object and background. Thus, as seen in the Fig. 3, assured information of object and background, called training seed in the method, is extracted from the prior maps. First, Otsu's thresholding is performed on the selected prior map from the first method to extract the salient region and cut the edge region since objects tend to be located at the center of an image. With this center location assumption, the background training seed is defined on the edge region of an image as seen in Fig. 3.

With the training seed, the Fisher's linear discriminant analysis [10] is performed on each pixel of an image with its RGB and CIE-LAB color components, so that every pixel of an image is classified into object or background region with the information from training seed, as seen in Fig. 3.

4   Experimental Results

To evaluate the performance of the proposed methods, experiments are performed on the MSRA salient object database [5], which consists of 1,000 test images and their ground truth binary masks. On this dataset, the test results by the proposed methods are compared with those by Goferman *et al.* (CA) [6], Cheng *et al.* (RC) [2], Perazzi *et al.* (SF) [3], and Shen *et al.* (LR) [11].

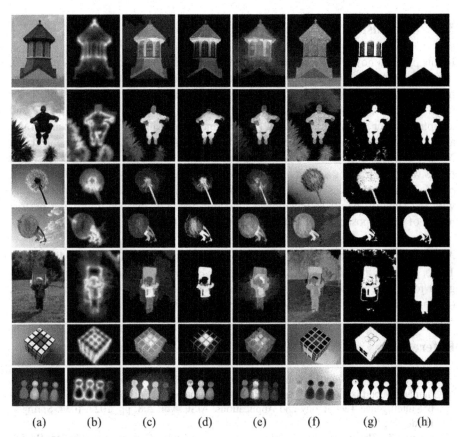

(a)	(b)	(c)	(d)	(e)	(f)	(g)	(h)

Fig. 4. Experimental results of the proposed methods and comparison with those of state-of-the-art methods: (a) Original images from MSRA dataset, salient object detection results of (b) CA [6], (c) RC [2], (d) SF [3], (e) LR [11], (f) PCA with cluster density evaluation, (g) PCA with LDA, (h) ground truth labels

Selected test images and their corresponding salient object detection results from mentioned methods are shown in Fig. 4. As shown in Fig. 4, while the detection results from the first method in column (f) suffer the noisy background such as tree in the second row image and forest behind the man in the fifth row image, the second method shows improved performance where those noises are avoided in the second results, so that the results seem similar to those of ground truth masks. Compared to the detection results of the state-of-the-art methods, the second proposed method also shows competitive performance. For example, in the flower image in the third row, the proposed method captures the blossom while others mostly miss. Furthermore, in the four toys image in the seventh row, the proposed method captures all four toys, while most of others fail to capture the rightmost blue toy, which has similar blue color with the background, with the help of considering non-redundancy in the color distribution, not just color contrast information.

5   Conclusion

In this paper, a novel method for fully automatic salient object detection is proposed. A non-redundancy driven saliency detection with PCA is introduced with two proposed methods, one original first and improved second. In the first method, prior maps are generated by simple linear combination of principal component images and an optimal prior map is selected as a final saliency map with the cluster density evaluation. In the second method, training seed including assured information of object and background is extracted from prior maps and Fisher's LDA is performed on image pixels with this training seed as training set. As a result, two proposed methods show promising performances and the second improved method performs competitively with the state-of-the-art salient object detection method. Further improving performance of the proposed method and extended analysis is remained as future works.

Acknowledgments. This research was supported by the MOTIE (The Ministry of Trade, Industry and Energy), Korea, under the Human Resources Development Program for Convergence Robot Specialists support program supervised by the NIPA (National IT Industry Promotion Agency)(H1502-13-1001).

References

1. Kang, S., Lee, H., Kim, J., Kim, J.: Automatic Image Segmentation Using Saliency Detection and Superpixel Graph Cuts. In: Kim, J.-H., Matson, E., Myung, H., Xu, P. (eds.) Robot Intelligence Technology and Applications. AISC, vol. 208, pp. 1023–1034. Springer, Heidelberg (2013)
2. Cheng, M., Zhang, G., Mitra, J., Huang, X., Hu, S.: Global Contrast Based Salient Region Detection. In: Proc. of the Conference on Computer Vision and Pattern Recognition (CVPR), pp. 409–416 (2011)

3. Perazzi, F., Krahenbuhl, P., Pritch, Y., Hornung, A.: Saliency Filters: Contrast Based Filtering for Salient Region Detection. In: Proc. of the Conference on Computer Vision and Pattern Recognition (CVPR) (2012)

4. Itti, L., Koch, C., Niebur, E.: A Model of Saliency-based Visual Attention for Rapid Scene Analysis. IEEE Trans. on Pattern Analysis and Machine Intelligence 20(11), 1254–1259 (1998)

5. Achanta, R., Hemami, S., Estrada, F., Susstrunk, S.: Frequency-tuned Salient Region Detection. In: Proc. of the Conf. on Computer Vision and Pattern Recognition (CVPR), pp. 1597–1604 (2009)

6. Goferman, S., Zelnik-Manor, L., Tal, A.: Context-aware Saliency Detection. In: Proc. of the Conf. on Computer Vision and Pattern Recognition (CVPR), pp. 2376–2383 (2010)

7. Pearson, K.: On Lines and Planes of Closest Fit to Systems of Points in Space. Philosophical Magazine 2(11), 559–572 (1901)

8. Hotelling, H.: Analysis of a Complex of Statistical Variables into Principal Components. J. Educational Psychology 24, 417–441 (1933)

9. Vu, C.T., Chandler, D.M.: An Algorithm for Detecting Multiple Salient Objects in Images via Adaptive Feature Selection. In: Proc. of the IEEE Intl. Conf. on Image Processing (ICIP), pp. 657–660 (2012)

10. Fisher, R.A.: The Use of Multiple Measurements in Taxonomic Problems. Annals of Eugenics 7(2), 179–188 (1936)

11. Shen, X., Wu, Y.: A Unified Approach to Salient Object Detection via Low Rank Matrix Recovery. In: Proc. of the Conference on Computer Vision and Pattern Recognition (CVPR) (2012)

Development of Child-Sized Humanoid Robot for Dance Performance

Young-Jae Ryoo

Department of Control Engineering and Robotics,
Mokpo National University
1666 Youngsan-ro, Cheonggye-myeon,
Muan-gun, Jeollanam-do, 534-729, Korea
yjryoo@mokpo.ac.kr

Abstract. In this paper, a development of a child-sized humanoid robot for dance performances is described. The mechanism of a humanoid robot of 1.10 meters tall was designed. By considering dance performance, the mechanism of the robot with lightweight was fabricated. The frame and links were designed by 3D CAD software and was manufactured by precision machining with aluminum and plastics. A motion editing tool to generate dance motions of a humanoid robot is used. The motion editing tool can generate performance motions composed of several still motions which are acquired from every joint while the robot plays. The motion editing tool generates the continuous motion interpolated between each still motion. The developed child-sized humanoid robot was experimented with dance performance. The robot was demonstrated the smooth performance.

Keywords: humanoid robot, motion editing tool, dance performance.

1   Introduction

While humanoid robots have excellent motility, it has a complex inverse kinematics to realize the motion like a human. And robot's stability has to be considered because the robot to stand by two feet is to be fallen and can lead to additional accident. Also, the hardware that controls the robot in real time has to recognize a variety of external environment. When the robot is lighter, the stability is increased and control is getting easier.

Humanoid robots have been concerned in robotics a lot of activity. So it is being studied as a variety of ways, and it became increasingly popular. When the humanoid robot plays the performance, the robot should deliver the robot's movement and the means of action as well as it looks like a human in order for the smooth communication and interaction with the human and humanoid robot. Therefore, the robot must be capable for smooth movement like a human.

In this paper, we designed the lightweight humanoid robot of 1.1 meters tall using 3D CAD software. In order to realize the motions similar to a human, joints was designed by using 9 servo motors in upper body and 12 servo motors in lower body.

J.-H. Kim et al. (eds.), *Robot Intelligence Technology and Applications 2*,
Advances in Intelligent Systems and Computing 274,
DOI: 10.1007/978-3-319-05582-4_63, © Springer International Publishing Switzerland 2014

To realize the natural robot's movement, the hardware was designed to control 21 servo motors in real time. We used a motion editing tools to generate smooth performance motions for a humanoid robot. The suggested motion editing tool terminates order to robot's torque at pause of each step and control with customized orders form the producer. Tool loads the saved movement values and saves those as one motion. Each still motion has pausing time. Between still motions there are movement time which connects movements as well as motion time which can control the entire motion time. This realizes the movement on the exact burst of music beat. By adjusting parameter, motion editing tool that robot can have synchronization according to beat timing was proposed.

2   Design of Humanoid Robot

2.1   Concept of Humanoid Robot

In this paper, we designed and manufactured a humanoid robot. In order to minimize unexpected shocks caused by accidents that the robot falls down, we will reduce the overall weight of it.

To realize the walking and movement like a human in the real world, it has been designed into the structure of the robot as shown in Figure 1.

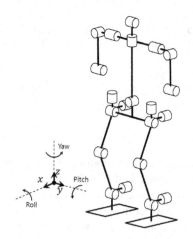

Fig. 1. Schematic structure of humanoid robot

The mechanism of the humanoid robot was designed by using the 3D design tool. In particular, the lower body was designed with human body proportions. Figure 2 shows the joint configuration and proportion of the lower body.

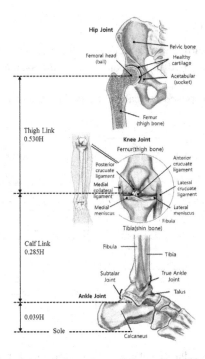

Fig. 2. Joint configuration and proportion of lower body

Also, we can predict the weight of the robot designed by applying the volume density of materials that will be used for frames and links when designing.

The mechanism of the humanoid robot designed by applying the proportion of the human is shown in Figure 3.

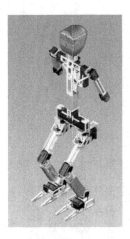

Fig. 3. Mechanism of designed humanoid robot

2.2   Design of Hardware

Electrical and electronic hardware required to realize the walking and movement of the humanoid robot is described in this paper. Also, the system for interaction with humans is required.

The humanoid robot proposed to realize it has the structure of the electrical and electronic hardware.

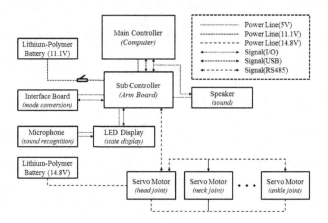

Fig. 4. Block diagram of electrical and electronic hardware

2.3   Development of Humanoid Robot

To configure the system for interaction between humans and humanoid robots, the humanoid robot was developed. Figure 5 is developed humanoid robot.

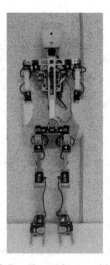

Fig. 5. Developed humanoid robot

The developed humanoid robot is possible for image processing using a camera, and it transfers the recorded video to the user in order to see the information through a wireless network. Also, by using the speaker and microphone, it can play or recognize the sound. In particular, it has been developed to facilitate the remote control and information sharing by using a wireless network. The specification of the robot is shown in Table 1.

Table 1. Specifications of the developed humanoid robot

Height(cm)		110
Weight(kg)		7.0
DOF	Head	2 DOF, MX-28R
	Arm	3×2 DOF, MX-28R
	Waist	1 DOF, EX-106+
	Leg	6×2 DOF, EX-106+
	Total	21 DOF
Sensor	Gyro	L3G4200D
	Accelerometer	LIS331DLH
Main Controller		PC(Ubuntu 9.10, C++)
Sub Controller		ARM core
Camera		Logitech C905
Power	Motors	Li-po 4000 4cell 14.8V
	FitPC2 and Sub	Li-po 2000 3cell 11.1V

3   Structure of Motion Editing Tool

In creating animation, there is a technic called 'morphing' which connects image to image minimizing glitch by finding corresponding points between plural pictures. Robot technology produces smooth movement of robot using 'motion editing tool' similar to morphing technic. In this paper, suggested motion editing tool terminates order to robot's torque at pause of each step and control with customized orders from the producer. Tool loads the saved movement values and saves those as one motion.

Fig.6 shows the motion structure of motion editing tool mentioned above. In this structure, still motion acts towards step. Meaning, each still motion is saved as each step in this tool. Numbers of still motions are connected to be one motion and one motion is saved as one page.

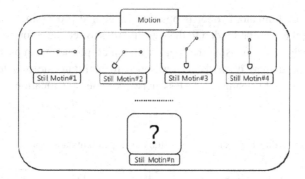

Fig. 6. Procedure to capture still motion at each step during a motion

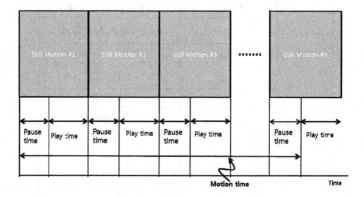

Fig. 7. Timing chart of motion editing tool

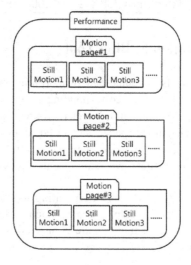

Fig. 8. Structure of performance and connection of motion page

Fig.7 shows the movement time graphs of still motion and motion page. Each still motion has pausing time. Between still motions there are movement time which connects movements as well as motion time which can control the entire motion time. This realizes the movement on the exact burst of music beat.

Fig.8 explains the structure and the behavior of the motion page. Numbers of still motion forms one motion, and one motion is saved as one page. Motion page can be customized for each acceleration time, page speed, repetition time and others. Customized motion pages are connected to be behavior and behavior becomes a form of dance or performance.

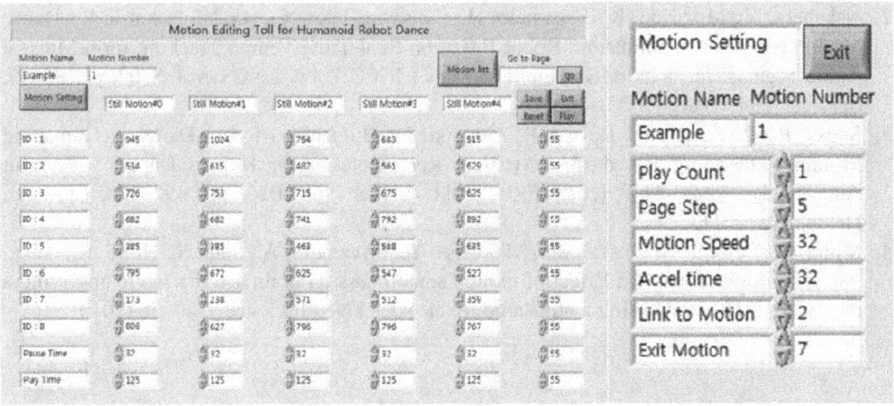

Fig. 9. User interface of motion editing tool for humanoid robot's dance performance

Fig.9 is the exemplified program which has the structure of movement editing tool that was explained above, showing the part where still motion 0 loads the current value of robot's articulation. This program makes still motion 1 to n, excluding still motion 0, as customized by the producer. In this user interface, dance and performance can be produced by filling up the time parameter related to each still motion, producing customization of repetition time, number of still motions, motion page numbers to be connected at motion setting window.

4   Conclusion

In this paper, we proposed the structure of the motion editing tools of the humanoid robot for the performance. Among various parameters of motion editing tools, we experimented on matching real motion after the flow of joint's waveform that is using various parameters for smooth movement, and output applied to developed robots. When proposed motion editing tools were applied, we confirmed the humanoid robot moved more smoothly and showed smooth move.

Acknowledgements. This research was financially supported by the Ministry of Education, Science Technology(MEST) and National Research Foundation of Korea(NRF) through the Human Resource Training Project for Regional Innovation (No. 2012026068).

References

1. Kim, C., Kim, D., Oh, Y.: Adaptation of human motion cap-ture data to humanoid robots for motion imitation using optimization. Integrated Computer-Aided Engineering, Bol. 13(4), 377–389 (2006)
2. Nakahara, N., Miyazaki, K., Sakamoto, H., Fujisawa, T.X., Nagata, N., Nakatsu, R.: Dance Motion control of a Humanoid Robot Based on Real-Time Tempo Tracking form Musical Audio Signals. In: Natkin, S., Dupire, J. (eds.) ICEC 2009. LNCS, vol. 5709, pp. 36–47. Springer, Heidelberg (2009)
3. Sousa, P., Oliveira, J.L., Reis, L.P., Gouyon, F.: Humanized Robot Dancing: Humanoid Motion Retargeting Based in a Metrical Representation of Human Dance Styles. In: Antunes, L., Pinto, H.S. (eds.) EPIA 2011. LNCS, vol. 7026, pp. 392–406. Springer, Heidelberg (2011)
4. Miura, N., Sugiura, M., Takahashi, M., Sassa, Y., Miyamoto, A., Sato, S., Horie, K., Naka-mura, K., Kawashima, R.: Effect of motion smoothness on brain activity while observing a dance: An fMRI study using a humanoid robot. Social Neuroscience 5, 40–58 (2010)

Bowling with the DARwIn-Op Humanoid Robot

Saltanat B. Tazhibayeva, Mukhtar Kuanyshbaiuly, Aibek Aldabergenov,
Ji Hyeon Hong, and Eric T. Matson

M2M Lab/RICE Research Center
Department of Computer and Information Technology
Purdue University, West Lafayette, USA
{stazhiba,mkuanysh,aaldaber,hong,ematson}@purdue.edu

Abstract. In this paper, we will describe our approach in building an application, which empowers the DARwIn-OP Humanoid robot to play a bowling game. The main difficulties of bowling, in both humans and robots, is steady walking control, vision processing to detect the pins and ball, precise localization of the ball and decision-making of angles to throw. The aim of this project is to contribute to better and more enjoyable robot and human interaction as well as to humanoid robot research area.

Keywords: bowling, entertaining robots, artificial intelligence, gripper hands.

1 Introduction

Nowadays, different many types of robots are having an increasing demand and interest. For example, robot types can be as robot-assistants, social robots, and humanoid robots. Among them, humanoid robots are becoming more and more common place in the area of artificially intelligent robots, and in the lives of humans.

Humanoid type robots are an excellent choice to interact with humans rather than computers or other types of robots, due to anthropomorphic similarity. Humanoid robots typically have very complex logic and tremendous potential to be a pervasive and common platfom for human/robot interaction, in the physical and mental environment. Physical similarity to humans gives the humanoid robots some ability to act and function in a human-like manner, which results to increase the bond between humans and robots and provides additional motivation for interaction.

This research has been done in order to increase the potential of this interaction between humanoid robots and humans. The ability to play games not only with desktop computers, but also physically standing against a robot and play with it, in normal environment, has a great potential for robotics to grow rapidly and is a very important aspect of robotics technology development.

The main purpose of our research is to extend abilities of DARwIn-OP [4] so that it could be able to play bowling autonomously and as well as to extend general capabilities of humanoid robots.

In section 2, background material is described, as section 3 covers the realization of the project in detailed. Section 4 draws overall conclusions to this research.

J.-H. Kim et al. (eds.), *Robot Intelligence Technology and Applications 2,*   733
Advances in Intelligent Systems and Computing 274,
DOI: 10.1007/978-3-319-05582-4_64, © Springer International Publishing Switzerland 2014

2   Background

There are several works on the subject of game playing artificially intelligent humano-id robots as well as in robot entertainment: a robot as a toy and educational tool Robota[14], [15] a pet type legged robot AIBO, a ping pong playing TOPIO [16], REEM-A playing chess [17]. In addition, the well known RoboCup Humanoid League [7] is a soccer game competition between different types of humanoid robots, grouped as teams, divided into different sub-leagues according to the size of the robots. This represents an arena for industries and researchers to demonstrate their humanoid ro-bot's capabilities. The way of solving the game challenges such as ball detection, posi-tioning, steady walking, keeping balance while kicking, contributes a lot to the field of robots and sensors.

For this project, the choice is to use one of the most popular and favorite worldwide games – bowling; because one can play it with no age limit. [6] Ten-pin bowling is a sports game, which includes physical exercise and competition for score, knocked down pins. In comparison with soccer, it comprises a new set of abilities such as grab-bing the bowl, aiming the target and, finally, throwing the ball. It demands complex movements from the upper parts of the body, whereas soccer from the lower side of the body. However, there are still minimal cases of humanoid robots bowling with human or with other robot agents.

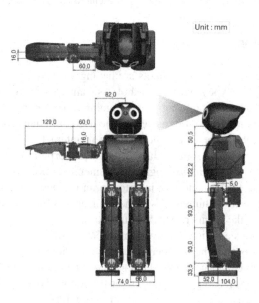

Fig. 1. DARwIn Parameters

We implemented our research project on DARwIn-OP (Dynamic Anthropomorphic Robot with Intelligence - Open Platform) humanoid robot developed by Virginia Tech in collaboration with Purdue and University of Pennsylvania. [4] and Robotis.

DARwIn, shown in Figure 1, is open source, has a relatively large user community, and is also easily extendible. It is convenient robot for the development of entertaining games, as well as, assistive applications.

3   Implementation

The bowling mode of DARwIn provides the same play-mode as the real bowling game. DARwIn firstly detects pins, then gets the bowl, grabs it and after all throws the bowl toward the pins. We constructed and adjusted our own bowling game-play area to our robot's proportions, refer to the Figure 1 for robot parameters, as close as it is possible, so that it would take place no more than 4 meters (the length of real bowling lane is about 18m) this was due to the DARwIn's vision system limitations [5], also would have mini ball with diameter about 100mm and mini pins set with about 200mm height each.

The main phases of our application and the process of implementation are divided into the following subcategories, which we will discuss further:

- Vision processing
- Replacing hands by grippers
- Keeping up the balance while walking
- Picking up the ball
- Aiming the target
- Throwing the ball

3.1   Vision Processing

In robotics, vision processing is utilized so that a robot could recognize an object in an environment. To recognize an object, there have been several researches in vision processing with various approaches for object recognition. There are various approaches of vision processing for object recognition, especially for ball detection: template-based [8], motion matching [9], morphological-feature-based [10], trajectory-based [11], and chromatic [12] approaches. DARwIn-OP, which is the application of the bowling context, is developed with the chromatic approach. The chromatic approach utilizes color-spaces and segmentation models to recognized objects based on their colors; according to Caleiro, António, and Armando, there are mainly five categories within the color-based color spaces for robotics applications [13]: RGB, YUV, HLS, HSV, and IHLS. Among these diverse color models, DARwIn-OP utilizes HSV (Hue, Saturation, and Value), which the authors chose as the best color space in real-time object recognition, to find a ball and distinguish different colors; vision processing in the DARwIn-OP is only color-based.

The colors that we utilized for this study were green, yellow, and red for a background of bowling pins, bowling pins, and a bowling ball, respectively. DARwIn-OP is set to recognize red color as a ball in default; we utilized the default color for recognition of a bowling ball. The reason of selecting red color for a ball was because its accuracy of recognizing a red ball was high enough during our experiment. Also, we used yellow for bowling pins and green for the background of the bowling pins; the reason of selecting these contrast colors was because it was important for the robot to recognize bowling pins and the background of the pins without any confusion in some degree of tolerance limits. By using the color contrast between yellow and green, we calculated the distance as well as the angel from DARwIn-OP to the targeted bowling pins, including left bowling pins after a first-round trial, in order for the robot to determine its position and the direction of throwing a ball.

During the experiment, a difficulty that we experienced was to find exact values needed for its color segmentation. In order to find the exact colors of a bowling ball, pins, and the background, we measure the exact values of the colors in HSV from the captured images taken from the vision system of robot. Figure 2 demonstrates an example of such image. In order to optimize the color segmentation for each object, an appropriate tolerance value was also needed to set to find each of the objects in some ranges of its color. Through multiple tests with various values, this difficulty could be overcome.

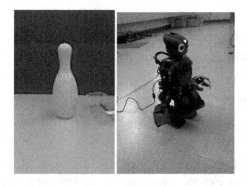

Fig. 2. Bowling pin and background from DARwIn's vision system and the DARwIn with gripper hands

3.2   Replacing Hands by Grippers

First of all, it took some time to have the gripper hand delivered and start working on the physical and mechanical aspects of the project. During the replacement of the hands the team found out that there 2 extra motors were needed for the full functioning of a gripper hand. The team had to detach the motors from the other Darwin robot's joint motors and use them for the building of the gripper hand. Figure 3 shows gripper hand scheme from TellDarwin project [1].

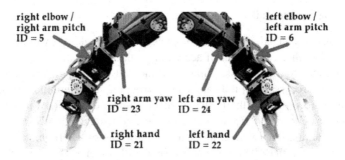

right elbow /
right arm pitch
ID = 5

left elbow /
left arm pitch
ID = 6

right arm yaw   left arm yaw
ID = 23   ID = 24

right hand   left hand
ID = 21   ID = 22

Fig. 3. 5DOF DARwIn's gripper hand

There were many problems and disadvantages even during the replacement of the gripper hands. Several screwdrivers were used and got broken during the building of the gripper hand, because of the very little screws in the robot's hands. And interestingly enough, there were not enough information about how to build a gripper hand, so we got an approximate idea from the Figure 3 shown above.

One of the major aspects of replacement of hands was to imitate the human hand for Darwin's grippers. It was very difficult to use grippers to keep a ball in the grippers. The lager balls were not even considered to be picked up because the grippers didn't suit at all. Also the main wrist motor of the gripper hand tended to easily shut down because of the pressure made by the soft ball on the inner sides of the grippers. The problem was fixed when the angle of the motor was changed and so that pressure could be decreased.

With the successful replacements of hands to gripper hands and successfully configuring the every motor of Darwin it was easier for us to proceed to the next step of our project.

3.3   Keeping the Balance When Walking

As it was mentioned previously, our team faced one major problem after replacing Darwin's one hand with a gripper hand. When running the walk program, Darwin would tilt to the right and fall after a few steps. In order to counterbalance Darwin's heavy right hand, we used "Walking Tuner" which was installed with the Darwin's framework. We used the web interface of Walking Tuner so that we could test the new settings immediately and change settings as needed. The very first setting that we changed was Darwin's walking speed. We decreased the robot's default walking speed by about 15%. The next setting we changed was Swing left/right setting. By default, Darwin would swing to left and right equally, and because the right hand was heavier, the robot would fall or go slightly in the right direction. We changed that setting so that Darwin would swing less to the right. The last setting was Hip pitch offset. By default, Darwin's body is slightly tilted forward, which sometimes caused the robot to fall forward because of the right hand. We changed the Hip pitch offset of the robot so that Darwin's body would be perpendicular to the floor. With these settings, we managed to get a straight walking Darwin with the replaced right hand.

3.4   Picking Up the Ball

One of the most challenging parts of our project was to make Darwin accurately find and track the ball then position itself according to the ball position and grab it with its gripper hand. Due to the processor limits and the ball tracking system formula errors, the accuracy of processing all the data coming from the high definition camera wasn't good enough. According to the technical specifications of the Darwin Ball Tracking works through color-based feature extraction methods detection and because of not consistent lightening environment it results on error rate of determining non-occluded tennis ball with less than 10% error at a distance of 5 meters. [5] As it was claimed before, the ability of Darwin accurately track the ball was a big issue for the ROBO-Cup teams participating in Darwin playing soccer tournament. However we decided on further continuation of using its functions of tracking and following the ball. We slightly changed previously done working code, so that whenever it finds the ball it would go and stand exactly from the right from the position of the ball. The difficulties that we had encountered during this task were not only to position but also not to lose ball while following it. Because in some cases whenever it finds the ball after standing on the proper position from it, the shoulder of DARwin might block the camera that is directed to the ball. Also a big issue is that robot perceives the world in 2D images, so we could not detect the exact distance between moving object and the ball. The path planning was also difficult because even if robot goes towards the ball, refer to Figure 4 (a), and stands in front of it, it's becoming more difficult for the robot then to move to the right for the precise distance, because in that case body parts of the robot blocks the image of the ball and robot gets distorted. So we decided on division of the path to the ball by some short intervals so that if it is away from the ball, then it would go forward first, and only when it gets closer to the ball it would shift it self left from the ball to the additional adjustment parameter θ' shown in Figure 4 (b) [3]. This was done through the measurements of head angle and tilt angle. By these two parameters we could abstractly detect the farness of the ball as well as its position whether it is from right side or from left side.

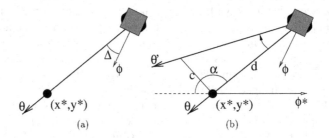

Fig. 4. (a) The parameters used to reach a target configuration (x^*, y^*) without specified target orientation. (b) The adjustment of θ to θ' to reach a target configuration of the form (x^*, y^*, ϕ^*).

Other problem was that the color detecting algorithm worked by counting the center color of a mass, if it detected some color not referring to the ball but red color of environment it would obtain false position to the ball.

3.5   Aiming the Target

Other problem was that the color detecting algorithm worked by counting the center color of a mass, if it detected some color not referring to the ball but red color of environment it would obtain false position to the ball.

After picking up the ball, DARwIn shoots at the bowling pins. There are two possible ways of throwing the bowl in bowling, one with rotation and the other is without [2]. However robot capabilities are restricted and not the same as humans' e.g. it is considered that a human arm has 7 DOF(Degrees of freedom), there are shoulder, yaw, roll, pitch, 3 for wrist, additionally we have fingers by which we cling the bowl and hold it fixed, whereas our DARwIn has only 5 DOF gripper hand, which mean that it's too complicated for him to throw the ball with rotation. Therefore we chose the first option for our game logic where the robot will throw the ball directly without rotation.

In bowling, there are numerous ways of shooting the pins, however in order to get the maximum score we chose to throw the ball toward the center of the group at the front pin. In our case DARwIn is able to utilize its vision system, which can detect a target point as the center of the color intensity in our case of yellow. Now having two central points of the background and pins DARwIn will be able to align himself computing formula (1) of line equation between given two points: center of the background (x_1, y_1) and center for pins (x_2, y_2).

$$y = ax + b = \frac{y_2 - y_1}{x_2 - x_1} x + \frac{x_2 y_1 - x_1 y_2}{x_2 - x_1} \tag{1}$$

$$\theta = \frac{180}{\pi} \tan^{-1} \left(\frac{y_2 - y_1}{x_2 - x_1} \right) \tag{2}$$

Other problem was that the color detecting algorithm worked by counting the center color of a mass, if it detected some color not referring to the ball but red color of environment it would obtain false position to the ball.

Robot alternately will shift to the right or two the left in order to get to that line and turn itself to the angle θ^* towards the pins' central point. The big issue has occurred while robot was trying to turn and to shift to the sides simultaneously, so that he couldn't maintain balance fast enough. As result it fell down. For that purpose we divided task by two and applied semaphores programming technique with lockers, so that he couldn't start turning until shift is done and vice versa, until proper adjustment end error rate range wasn't satisfied.

3.6   Throwing the Ball

Basically as it was explained in the previous part about picking up the target ball, Darwin comes up from the left side of the ball to pick up and aims at the target. After this process, the Darwin places his body at the certain angle and throws the ball towards the pins. The whole motion of throwing the ball is divided up into several movements to be recorded to the software to memorize it and play at the right time, shown in Fig. 5. The main software used for the recording of every single movement was RoboPlus [18]. This particular software gave us an ability to control the speed of every single movement and also to link these movements to other operations.

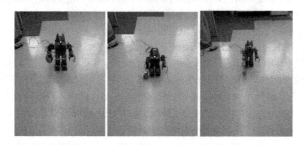

Fig. 5. Throwing the ball

Regarding the movements, in order to throw the ball Darwin needs to make a back swing which requires bringing the ball back closer to his shoulder height, slowly. Further movements are bringing the ball from backswing to a downward arch in a highest speed, and releasing it at the very right moment. As you see according to the description of the movements above, there were 3 main movements: backswing, throw and release.

One of the difficulties in this scenario was to accurately adjust the angles of every motor and adjust the speed of every movement in order the ball could hit the target.

Preliminary results of the game gave us success rate of 90% of knocking down at least 1 pin, with the mean equal to 3.94 pins. On the other hand, the average bowling player would have almost the same success rate, which means that our robot may possibly compete against a human.

4   Conclusions

This bowling game could be installed on any type of humanoid robot, which could have gripper hand configuration and also would have vision system. Our application could be extended so that humans especially children and robots may play this game together one against other. We've encountered some difficulties during the process of this project, and our team solved almost all of the problems that we had. Certainly, there are still lots of room for improvement. The walking motion and vision processing features of Darwin could be fine-tuned to work with very little or no error.

The separate motions of the robot like picking up the ball and throwing the ball can also be configured to replicate the human's motions. Our team's next plans include further work on Darwin and sharing the source code of our application on the Internet.

References

[1] Krzyzaniak, M.: TellDarwin Reference. Idea Lab. University of Georgia Web (June 15, 2013), http://idealab.uga.edu/Projects/Darwin/Reference.html
[2] Pan, Z., Xu, W., Huang, J., Zhang, M., Shi, J.: Easybowling: a small bowling machine based on virtual simulation. Computers & Graphics 27, 231–238 (2003)
[3] Bowling, M., Veloso, M.: Motion control in dynamic multi-robot environments. In: Proceedings of the 1999 IEEE International Symposium on Computational Intelligence in Robotics and Automation, CIRA 1999. IEEE (1999)
[4] Inyong, H.A., et al.: Development of open humanoid platform DARwIn-OP. In: 2011 Proceedings of the SICE Annual Conference (SICE), pp. 2178–2181. IEEE (2011)
[5] Budden, D., Fenn, S., Walker, J., Mendes, A.: A novel approach to ball detection for humanoid robot soccer. In: Thielscher, M., Zhang, D. (eds.) AI 2012. LNCS, vol. 7691, pp. 827–838. Springer, Heidelberg (2012)
[6] Kosof, M.E.: Method of playing a bowling game. U.S. Patent No. 4,597,575 (July 1, 1986)
[7] Kitano, H., et al.: Robocup: The robot world cup initiative. In: Proceedings of the First International Conference on Autonomous Agents. ACM (1997)
[8] Yow, D., Yeo, B., Yeung, M.: Analysis and Presentation of Soccer Highlights from Digital Video. In: Second Asian Conference on Computer Vision, pp. 499–503 (1995)
[9] Ohno, Y., Miura, J., Shirai, Y.: Tracking players and estimation of the 3D position of a ball in soccer games. In: Proceedings of the 15th International Conference on Pattern Recognition, vol. 1. IEEE (2000)
[10] D'Orazio, T., Ancona, N., Cicirelli, G., Nitti, M.: A ball detection algorithm for real soccer image sequences. In: Proceedings of the 16th International Conference Pattern Recognition, vol. 1. IEEE (2002)
[11] Yu, X., Xu, C., Leong, H.W., Tian, Q., Tang, Q., Wan, K.W.: Trajectory-based ball detection and tracking with applications to semantic analysis of broadcast soccer video. In: Proceedings of the Eleventh ACM International Conference on Multimedia. ACM (2003)
[12] Gevers, T., Smeulders, A.W.: Color-based object recognition. Pattern Recognition 32(3), 453–464 (1999)
[13] Caleiro, P.M.R., Neves, A.J.R., Pinho, A.J.: Color-spaces and color segmentation for real-time object recognition in robotic applications. Electrónica e Telecomunicações, 940–945 (2013)
[14] Billard, A.: Robota: Clever toy and educational tool. Robotics and Autonomous Systems 42(3), 259–269 (2003)
[15] Fujita, M., Kitano, H.: Development of an autonomous quadruped robot for robot entertainment. Autonomous Robots 5(1), 7–18 (1998)
[16] Sun, Y., et al.: Balance motion generation for a humanoid robot playing table tennis. In: 2011 11th IEEE-RAS International Conference on Humanoid Robots (Humanoids). IEEE (2011)
[17] Faconti, D.: Technical description of REEM-A. In: RoboCup Humanoid League Team Descriptions, Atlanta, GA (July 2007)
[18] Robotis. RoboPlus.Robotis support site, Web (June 15, 2013), http://support.robotis.com/en/software/roboplus_main.htm

Design and Control of a Humanoid Robot for Traffic Guidance

Qingcong Wu and Xingsong Wang

College of Mechanical Engineering, Southeast University of China
No.2, Dongnandaxue Road, Jiangning District, Nanjing, Jiangsu Province, 211189, China
xswang@seu.edu.cn

Abstract. In this paper, a humanoid traffic guidance robot with 9 active degrees of freedom (DOF) is designed to relieve the traffic pressure and improve the working safety situation for the road traffic policemen and road maintenance workers. The proposed robot is able to perform the standard traffic command gestures such as turning left, turning right, slowing down and stopping according to the feedback signal of a radar sensor. Besides, there is a digital camera capturing the high resolution images of the vehicles having a speed higher than the preset limitation. The image information is recorded in a SD memory card and can be used to check the illegal driving history. The mechanical structure, the kinematics and the control system are described. On each robot joint a PID controller is implemented for trajectory tracking control. Several preliminary experiments have been implemented to verify the effectiveness of traffic conducting in the laboratory environment and realistic application.

Keywords: humanoid robot, traffic guidance, kinematics, trajectory planning.

1   Introduction

With the rapid increase in the number of motor vehicles, many road traffic problems are becoming more and more serious in all countries, such as traffic jams, environment pollution, and road safety. Traffic guidance, highway construction and maintenance are all extremity important, necessary and, however, dangerous jobs. According to epidemiological investigation, most of the traffic policemen and road maintenance workers have to stay in a sub-healthy or disease state due to long-term exposure to photochemical smog, overworking with high risk and strength, and continue tense state of mind brought by the huge psychological and physical pressure. Proper traffic control is critical to guarantee their safety and decrease traffic accidents [1, 2]. At present, there are several researches about the traffic guidance robot application. Farritor et al. [3, 4] designed a mobile safety marker system that used mobile robots to transport safety markers for highway construction and maintenance and serve as a visible barrier between traffic and work crews. Hamid et al. [5] proposed a simple humanoid robot for road traffic safety assurance in highway and digging places. This robot has 2 DOF at the upper arms and can only perform a few command gestures through a microcontroller.

J.-H. Kim et al. (eds.), *Robot Intelligence Technology and Applications 2,*   743
Advances in Intelligent Systems and Computing 274,
DOI: 10.1007/978-3-319-05582-4_65, © Springer International Publishing Switzerland 2014

For this special application, this paper presents a 9 DOF Humanoid Traffic Guidance Robot (HTGR) to assist or replace the traffic policemen to perform the standard traffic command gestures and, manage traffic under the poor working environment. Besides, it can be used in road repairing or constructing places as a flagger worker, informing the vehicles and pedestrians from the existence of danger. The reminder of this paper is organized as follow: Section 2 described the mechanical structure of our prototype. The kinematics model of the robot is built in section 3. Section 4 presents the control strategy of the robot system for vehicles detection and trajectories tracking. Section 5 and Section 6 presents the preliminary experiments of guidance gestures execution in the laboratory environment and the realistic application, respectively. Finally, a brief conclusion and future work are made in section 7.

2   Mechanical Design

HTGR consists of 9 actuated DOF from the head (horizontal rotation) to the left arm and right arm (shoulder flexion/extension, internal/external rotation, elbow pronation/supination, wrist flexion/extension). That is corresponding to the minimum number of joints needed to perform the traffic command gestures, such as, stopping, slowing down, turning right and turning left.

As shown in figure 1, the kinematic structure is chosen to be simple and similar to human anatomy. Meanwhile, the range of motions (ROM) of each joint is designed to meet the requirement of all kinds of gestures as well. The operating element of each joint is a brushless DC motor with Hall-effect sensors built-in to measure the position and velocity. There are two kinds of motors (BLDC-56PA77G for shoulder, BLDC-36PA71G for others) being used in HTGR, coupled with a planetary gearbox with reduction ratio between 25 and 70. The motor torque, the output power, the reduction, the final output torque and the ROM of each joint are shown in table 1. We utilized a piece of plastic body clothes (not shown in figure) with natural face to make the robot look like a real traffic police as much as possible and keep waterproof in any weather condition. A set of differential gear train is used to transfer the driving torque from the actuators placed inside the robot chest to the shoulder joints at each side. The motors for elbow pronation/supination and wrist flexion/extension are located inside the upper arm and the forearm, respectively. This kind of structure is convenient for the body clothes dressing and, moreover, helps to decrease the unbalanced torque and stabilize the whole robot device. A radar sensor (IPS-146) is placed in front of the robot chest to detect the moving vehicles and measure their speeds within a maximum range of 100 meters. A digital camera is placed above the radar sensor to capture the high resolution images of the vehicles having a speed higher than the preset limit. The posture of the camera can be adjusted to get a better view of the vehicles. A control system consisting of 10 microcontrollers is used to analyze the feedback information and send the appropriate control signals to the robot system. All the microcontrollers and devices are connected to each other by a Controller Area Net (CAN) bus network. The power supply of the system is a rechargeable 24v/48Ah lead acid battery with complementary recharging circuit, allowing HTGR to work 24 hours.

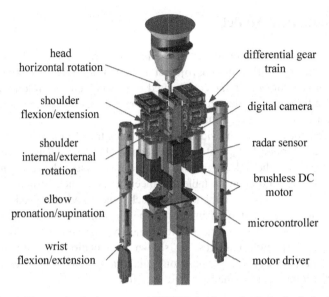

Fig. 1. The mechanical structure of HTGR (without plastic body clothes)

Table 1. Joints characteristics

	Shoulder flexion/ extension	Shoulder internal/ external	Elbow pronation/ supination	wrist flexion/ extension	head horizontal rotation
Range of motion (deg)	-30~135	-60~60	-90~90	0~90	-45~45
Motor torque (Nm)	0.5	0.5	0.12	0.12	0.12
Output power (W)	200	200	50	50	50
Reduction ratio	70	70	40	40	25
Output torque (Nm)	35	35	4.8	4.8	3

Table 2. D-H parameters of HTGR

Link i	θ_i*home (deg)	α_i (deg)	a_i (mm)	d_i (mm)
1	θ_1*90	90	0	0
2	θ_2*0	-90	0	d_2
3	θ_3*0	90	0	d_3
4	θ_4*90	90	a_4	0
5	θ_5*0	-	-	-
6	θ_6*180	90	0	0
7	θ_7*180	90	0	d_7
8	θ_8*0	90	0	- d_8
9	θ_9*90	90	a_9	0

3   Kinematical Model

In this study, the forward kinematics of the proposed robot is modeled based on the Denavit-Hartenberg (D-H) Notation [6], just as shown in figure 2. In order to get D-H parameters, it is assumed that the coordinated frames coincide with joint axes of rotation and have the same number in order. The forward kinematics can be used to analyze and adjust the command gestures of HTGR.

The general form of link individual homogeneous transformation matrices that relates frame {i} to frame {i-1} is shown in equation (1). Where θ_i is the joint angle, α_i is the link twist, a_i is the link length and d_i is the link offset. The resultant D-H parameters for the robot are given in table 2. Since the anatomical structure of the right arm and the left arm are symmetrical to each other, and the head movement is independent, we only take the kinematics of the left arm in consideration. The transformation matrix that relates frame {1} to frame {5} can be obtained by multiplying individual transformation matrices, as shown in equation (2). Where $s_i=\sin(\theta_i)$, $c_i=\cos(\theta_i)$. Thus, we can get the positions and orientations of the reference frame attach to wrist joint regard to the base reference frame.

$$
{}^{i-1}T_i = \begin{bmatrix} \cos\theta_i & -\cos\alpha_i\sin\theta_i & \sin\alpha_i\sin\theta_i & a_i\cos\theta_i \\ \sin\theta_i & \cos\alpha_i\cos\theta_i & -\sin\alpha_i\cos\theta_i & a_i\sin\theta_i \\ 0 & \sin\alpha_i & \cos\alpha_i & d_i \\ 0 & 0 & 0 & 1 \end{bmatrix}
\tag{1}
$$

$$
{}^1T_5 = {}^1T_2\,{}^2T_3\,{}^3T_4\,{}^4T_5 = \begin{bmatrix} r_{11} & r_{12} & r_{13} & p_x \\ r_{21} & r_{22} & r_{23} & p_y \\ r_{31} & r_{32} & r_{33} & p_z \\ 0 & 0 & 0 & 1 \end{bmatrix}
\tag{2}
$$

$$r_{11} = -c_4(s_1s_3 - c_1c_2c_3) - c_1s_2s_4$$
$$r_{21} = c_4(c_1s_3 + c_2c_3s_1) - s_1s_2s_4$$
$$r_{31} = c_2s_4 + c_3c_4s_2$$
$$r_{12} = c_3s_1 + c_1c_2s_3$$
$$r_{22} = c_2s_1s_3 - c_1c_3$$
$$r_{32} = s_2s_3$$
$$r_{13} = c_1c_4s_2 - s_4(s_1s_3 - c_1c_2c_3)$$
$$r_{23} = s_4(c_1s_3 + c_2c_3s_1) + c_4s_1s_2$$
$$r_{33} = c_3s_2s_4 - c_2c_4$$
$$p_x = 125s_1 + 100c_4(s_1s_3 - c_1c_2c_3) + 500c_1s_2 + 100c_1s_2s_4$$
$$p_y = 500s_1s_2 + 100s_1s_2s - 125c_1 - 100c_4(c_1s_3 + c_2c_3s_1)_4$$
$$p_z = -100c_2s_4 - 500c_2 - 100c_3c_4s_2$$

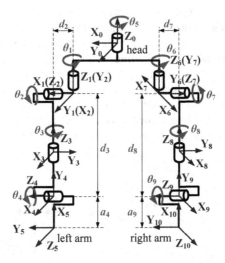

Fig. 2. D-H link coordinate system for HTGR

4   Control System Design

The humanoid traffic guidance robot can be used in many applications with different control strategies, ranging from assisting traffic policemen in managing traffic to working as a road safety flagger in the construction place. The whole control system of HTGR can be divided into ten subsystems, and they can communicate to each other by a CAN bus network [7, 8]. The subsystem mainly consists of a microcontroller to process data and a transceiver to transmit and receive messages. Particularly, one of the subsystem works as the host processor to receive the velocity feedback signal from the outer radar sensor and give the corresponding commands to other subsystems (lower controllers), respectively. Besides, an image processing based data measurement unit is integrated in the host processor to capture the high resolution images of the vehicles having a speed more than a preset limitation. A SD memory card driven by the host processor is used to record the images information, allowing the traffic policemen to check the illegal driving history.

Each of the brushless DC motor is connected to a motor driver and controlled by a lower control unit independently. The lower controller generates the planning trajectories for different command gestures depending on the control strategy preset in the host processor. Moreover, it analyzes the feedback information from hall sensors and sends the appropriate control signals to the driver. A PID controller with a speed closed loop and a position closed loop is implemented for trajectory tracking control at each joint. The schematic diagram of the control system is shown in figure 3.

In this work, as a previous step, we describe a simple control strategy for the traffic management. First of all, the operator needs to select a specific command mode on the control panel, such as, stopping, slowing down, turning left or turning right. And then, the radar sensor detects and measures the velocity of the moving vehicles within the maximum range. If the velocity equals to zero, that means there is no moving object on the road, thus HTGR needs to stay in waiting state to save the power. Otherwise, if the radar sensor detects something moving, the host processor will send commands to the lower controllers. Then the lower controllers are required to generate the planning trajectories corresponding to the gesture selected before. Finally, the motors of each joint are driven to track the trajectories and implement traffic control functions. Meanwhile, if the speed is higher than limitation, several high resolution images will be captured and recorded. An over view of the control flow strategy is shown in figure 4. By the way, the functions of HTGR can be extended to other applications by adjusting the planning trajectories and updating the control program.

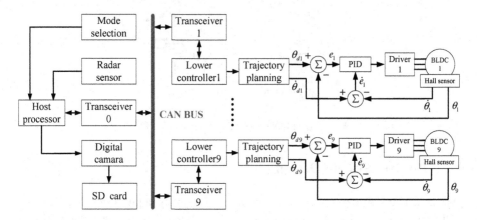

Fig. 3. Schematic diagram of the control system

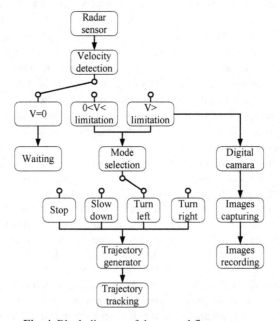

Fig. 4. Block diagram of the control flow strategy

5   Preliminary Experiments

Several preliminary experiments have been implemented to verify the feasibility of performing different traffic control gestures in the laboratory environment, just as shown in figure 5. To execute the guidance missions, the experimenter is required to walk in front of HTGR at different speed to imitate the movement of vehicles. The radar sensor measures the walking speed of the experimenter with a sample time of 24 GHz. And then, the control system controls the robot base on the speed information. The planning trajectories for different command gestures are shown in table 3. Since the gestures of turning left and turning right are symmetrical to each other, the latter is neglected in this paper. We make use of cubic polynomials functions to generate smooth motion trajectories for each command procedures [9]. The movement periods of each gesture are set to 2 seconds for the stopping command, and 3 seconds for others, which are similar to the realistic traffic police.

According to the experimental results, we can see that the robot can easily detect the movement within the effective range and complete the preset action. However, the outside disturbances such as fallen leaves and floating garbage may lead to the incorrect operation of the robot. Thus, an improved control algorithm should be proposed to filter the disturbances. There is about 0.2 seconds delay from motion recognition to action execution, but it is acceptable to the traffic control application.

Table 3. Planning trajectories for different gestures in a performance period

(deg)	Stopping		Slowing down		Turning left	
θ_1	90	$0 \leq t \leq 2$	90	$0 \leq t \leq 3$	90	$0 \leq t \leq 1$
					$60 (t-1)^3 - 90 (t-1)^2 + 90$	$1 \leq t \leq 2$
					$90 (t-2)^2 - 60 (t-2)^3 + 60$	$2 \leq t \leq 3$
θ_2	$270 t^3 - 405 t^2$	$0 \leq t \leq 1$	0	$0 \leq t \leq 3$	0	$0 \leq t \leq 1$
	$405 (t-1)^2 - 270 (t-1)^3 - 135$	$1 \leq t \leq 2$			$90 (t-1)^3 - 135(t-1)^2$	$1 \leq t \leq 2$
					$135(t-2)^2 - 90(t-2)^3 - 45$	$2 \leq t \leq 3$
θ_3	$180 t^3 - 270 t^2$	$0 \leq t \leq 1$	0	$0 \leq t \leq 3$	0	$0 \leq t \leq 3$
	$270(t-1)^2 - 180(t-1)^3 - 90$	$1 \leq t \leq 2$				
θ_4	$90 t^3 - 135 t^2 + 90$	$0 \leq t \leq 1$	90	$0 \leq t \leq 3$	0	$0 \leq t \leq 3$
	$135(t-1)^2 - 90(t-1)^3 + 45$	$1 \leq t \leq 2$				
θ_5	0	$0 \leq t \leq 2$	0	$0 \leq t \leq 3$	0	$0 \leq t \leq 1$
					$90 (t-1)^3 - 135 (t-1)^2$	$1 \leq t \leq 2$
					$135 (t-2)^2 - 90 (t-2)^3 - 45$	$2 \leq t \leq 3$
θ_6	180	$0 \leq t \leq 2$	180	$0 \leq t \leq 3$	180	$0 \leq t \leq 3$
θ_7	180	$0 \leq t \leq 2$	$180 t^3 - 270 t^2 + 180$	$0 \leq t \leq 1$	$180 t^3 - 270 t^2 + 180$	$0 \leq t \leq 1$
			$135(t-1)^2 - 90(t-1)^3 + 90$	$1 \leq t \leq 2$	90	$1 \leq t \leq 2$
			$135(t-2)^2 - 90(t-2)^3 + 135$	$2 \leq t \leq 3$	$270(t-2)^2 - 180(t-2)^3 + 90$	$2 \leq t \leq 3$
θ_8	0	$0 \leq t \leq 2$	$-180 t^3 + 270 t^2$	$0 \leq t \leq 1$	$-180 t^3 + 270 t^2$	$0 \leq t \leq 1$
			90	$1 \leq t \leq 2$	90	$1 \leq t \leq 2$
			$180(t-2)^3 - 270(t-2)^2 + 90$	$2 \leq t \leq 3$	$180(t-2)^3 - 270(t-2)^2 + 90$	$2 \leq t \leq 3$
θ_9	90	$0 \leq t \leq 2$	90	$0 \leq t \leq 3$	$180 t^3 - 270 t^2 + 90$	$0 \leq t \leq 1$
					0	$1 \leq t \leq 2$
					$270(t-2)^2 - 180(t-2)^2$	$2 \leq t \leq 3$

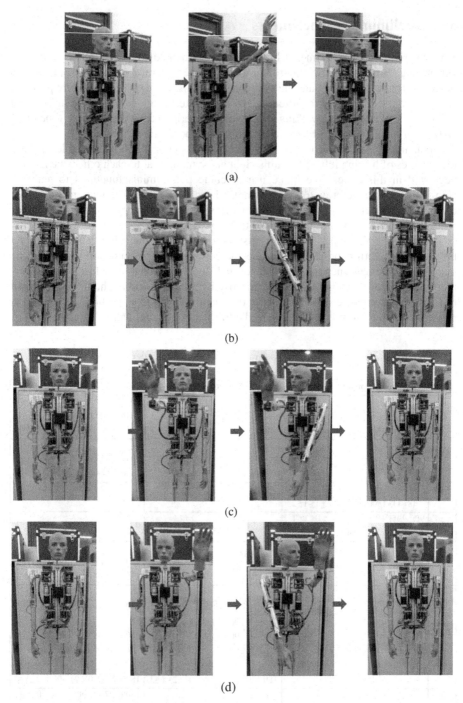

Fig. 5. Experiments of different traffic control gestures. (a) stopping; (b) slowing down; (c) turning left; (d) turning right.

6   Realistic Application

In order to verify the effectiveness of HTGR when managing the traffic in realistic application environment, we place the proposed robot along the roadside and command it to work in the slowing down mode to decrease the average velocity of vehicles. The predefined maximum limitation velocity is set as 100km/h in the experiments. We respectively calculate and analyze the distribution characteristics of vehicles velocities before and after the installation of HTGR. In the former situation, we just use the radar sensor to measure the velocities of 200 vehicles randomly, sending no instruction to the drivers. Analogously, the vehicles detected in the latter situation are as many as the former. However, the only difference is that if the car is detected to be overspeed, HTGR will perform the slowing down gesture, warning the driver to decrease the driving velocity.

The statistics results of the experiments are shown in figure 6. For ease of comparison, the range of vehicles speed can be basically divided into five parts (A~E). From the results we can see that, with the help of HTGR, the proportion of speeding vehicle is reduced from 57% to 13% and the average velocity is reduced by 16.8%. This means that the traffic robot is effective to force the drivers to complying with the desired driving command and, thus, helps to relieve the working pressure of traffic policemen and improve road traffic safety situation.

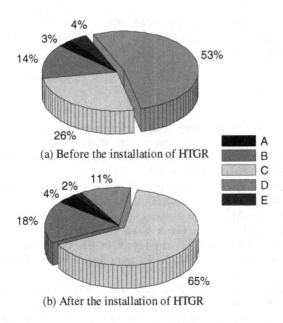

(a) Before the installation of HTGR

(b) After the installation of HTGR

Fig. 6. Distribution of vehicles velocities before and after the installation of HTGR (A: v<60km/h; B: 60 km/h<v<80km/h; C: 80 km/h<v<100km/h; D: 100 km/h<v<120km/h; E: v >120 km/h)

7   Conclusion and Future Work

In this paper, we propose a 9 DOF humanoid robot named HTGR to manage traffic at the crossroads and work as a road safety worker in road repairing or constructing places. HTGR can perform the standard traffic command gestures and capture the images of the vehicles having a speed higher than limitation base on the signals of a radar sensor. The mechanical structure and the kinematics of the robot are designed to be similar to human anatomy. A CAN bus network is employed in the control system to connect the host processor and the lower controllers. On each joint a PID controller is applied to ensure that the robot can track the planning trajectories correctly. The planning motion trajectories are all described in cubic polynomials functions. Several preliminary experiments have been carried out in the laboratory environment and realistic application to verify the effectiveness of traffic management.

Future works will include the improvement of structure design and control algorithm. Firstly, the mass distribution of HTGR needs to be improved to reduce the mass of the robot arms and the upper body. It will help to reduce the power consumption and enhance the overall stability. And then, to lower the cost of production, the mechanical structure is supposed to be simplified according to different applications. At last, we are motivated to develop an improved control algorithm to make HTGR more intelligent for complex traffic conditions.

Acknowledgements. This research has been supported by Ordinary University Research Innovation Project in Jiangsu Province, China (CXZZ13_0085) and the China Nation Nature Science Foundation under grant 50875044 and 51175078.

References

1. Ha, T.J., Nemeth, A.: Detailed study of accident experience in construction and maintenance zones. Transportation Research Record 1509, 38–45 (1995)
2. Tang, M.D., Zhen, Y.Y.: The effect study of traffic policemen healthy by traffic contaminated of x city. China Public Health 16(8), 71–713 (2000)
3. Shen, X., Dumpert, J., Farritor, S.: Design and control of robotic highway safety markers. IEEE/ASME Transactions on Mechatronics 10(5), 51–520 (2005)
4. Farritor, S., Rentschler, M.: Robotic highway safety markers. In: Proceedings of ASME International Mechanical Engineering Congress, pp. 17–22 (2002)
5. Abdi, H., Abdi, M.: Humanoid traffic control robot. In: UK. GB2449836 (2008)
6. Craig, J.: Introduction to Robotics-Mechanics and Control, 3rd edn. Pearson Prentice Hall (2005)
7. Tindell, K., Hansson, H., Wellings, A.: Analysing real-time communications: controller area network (CAN). In: Real-Time Systems Symposium, pp. 259–263 (1994)
8. Ran, P., Wang, B., Wang, W.: The design of communication convertor based on CAN bus. In: IEEE International Conference on Industrial Technology, pp. 1–5. IEEE (2008)
9. Kozlowski, K.: Robot motion and control: recent developments. LNCIS, vol. 335. Springer, Heidelberg (2006)

Trigonometry Technique for Ball Prediction in Robot Soccer

Muhammad Nuruddin Sudin, Siti Norul Huda Sheikh Abdullah,
Mohammad Faidzul Nasrudin, and Shahnorbanun Sahran

Pattern Recognition Research Group,
Center for Artificial Intelligence Technology
Faculty of Technology and Science Information
National University of Malaysia, 43600 Bangi, Selangor, Malaysia
nudin64@gmail.com,
{mimi,mfn,shah}@ftsm.ukm.my

Abstract. The main challenge in a robot soccer competition is to estimate the best robot's position according to two aspects: the ball and other robots positions given by visualize system and the game strategies. The ultimate aim is that to assign the right robot to the right position at the right time to win the ball for attacking or defending. Most of the time, the movements of the robots are determined by the position of the ball. The paper presents a ball position prediction technique based on trigonometry. We demonstrate the precision of the predicted ball position from the proposed technique and compare the precision result with those obtained from several existing techniques.

Keywords: robot soccer, attacking strategies, ball prediction, trigonometry.

1 Introduction

Robot soccer competition has been introduced since 1992. The objective of the competition is to provide a testing platform for artificial intelligence techniques in a realistic and dynamic ambience. In middle league robot soccer, a match will consist of 2 teams with five autonomous two-wheel robots in each team. Positions of the ball and all robots are detected by the vision system and send to a computer for strategy computation. Then, the decision made by the strategy algorithm will be transmitted to each robot for an appropriate action. The fast and nearly random movement of the ball and other robots makes the ball prediction very handy. Attacking robots need the ball-prediction ability to estimate the best shooting position as well as to control the ball.

In this paper, we demonstrate a trigonometry-based ball prediction technique using several current techniques as the comparisons. In Section 2 we will present the background to various ball prediction techniques. In Section 3 and Section 4, we will elaborate on the experiments and results respectively. We shall conclude in Section 5.

J.-H. Kim et al. (eds.), *Robot Intelligence Technology and Applications 2*,
Advances in Intelligent Systems and Computing 274,
DOI: 10.1007/978-3-319-05582-4_66, © Springer International Publishing Switzerland 2014

2   Background

Various techniques have been used to predict the randomly moving ball in robot soccer. The basic idea is that to locate the right intercept position for the robots to move. A position on the field is given by two numbers, the first tells where it is on the x-axis and the second which tells where it is on the y-axis. These are the current ball prediction techniques that have been used in robot soccer.

2.1   Ball Prediction (Merlin's Techniques)

Merlin's ball prediction technique was designed by the Merlin Corp. This technique is based on distance computation and very straightforward. It compose of two equations defined as (1) and (2)

$$dx = ball_{x_t} - ball_{x_{t-1}}$$
$$dy = ball_{y_t} - ball_{y_{t-1}} \tag{1}$$

The first equation is to find the distance, dx and dy, between two robot's positions at time, t and $t-1$. $ball_x$ and $ball_y$ represents the ball x-coordinate and y-coordinate correspondingly. t and $t-1$ denotes the ball's current and last position respectively. The time t is relatively related to the images captured in frame per second (fps). The second equation is then to add the distances to the current position of the ball, defined as:

$$ball_{x_P} = ball_{x_t} + dx$$
$$ball_{y_P} = ball_{y_t} + dy \tag{2}$$

This techniques uses distance differences (dx and dy) between ball's current positions acquired frame with the previous one to calculate the ball's predicted position. The technique has been known as fast and suitable for all types of vision camera regardless of all speeds of frame capturing. However, this technique is said to be failed in slow moving ball prediction.

2.2   Real-Time Collision Prediction Model (Hong Liu)

Real-time collision prediction model is a technique presented by Hong Liu et al in 2008[1]. This technique is used in robot soccer and has been applied at Peking University, Beijing, China. In general, this technique predict the time of collision between the robot from our own team with the ball, and with the other robots in the team. However, the central aspect is to predict the time of collision between the ball with the robot from our own team. It takes into account the real-time factors which are the robot position, the surface of friction and the inertia of the robots. The main objective of this technique is to determine the periods for collision to be combined with other functions as follow up. This technique is defined by the following vectors:

$$C(C_0, C_1, C_2, C_3, C_4, C_5, C_6) \tag{3}$$

In the original equation, the C represent the predicted time before collided where C_0 is the predicted time before collided with ball and C_1 to C_6 each represents the predicted time before collided with another robot in the same team. Thus, the vector for the collision with the robots alone can be defined just as $C(C_0)$. This vector is associated with another vector and replaced by other external factors aforementioned (position, surface friction and the robot's movement). The vector is defined as:

$$A(X_{(t)}, Y_{(t)}, V_{X(t)}, V_{Y(t)}, M_{(t)}, W_{(t)}) \tag{4}$$

In this equation, A represents a robot and $(X_{(t)}, Y_{(t)})$, $(V_{x(t)}, V_{y(t)})$ respectively represents the location and velocity of a robot in the time of t. While $M_{(t)}, W_{(t)}$ are the weight scope and monument in time, t. Keep in mind that every movement is taken in the form of axial coordination system (x and y), so in order to find the difference of coordination between the robot and the ball, the equation is as follows:

$$\Delta X = |X_{(t)} - X_{0(t)}|,$$
$$\Delta Y = |Y_{(t)} - Y_{0(t)}| \tag{5}$$

Based on the above equation, the difference of x and y axis were obtained from the x and y coordinates of the robot and the ball. The differences will then be calculated in the modulus to obtain an absolute value. The difference of velocity can be obtained in the same manner.

$$T_m = \frac{(D_m - D_R)}{(2 \times V_M)} + T_t \tag{6}$$

In all equations mentioned above, V_M is the largest velocity of the robot and T_t is the periods for changing direction to avoid collisions. However, In cases involving collision between the robot and the ball, T_t is the period for activating other functions such as showing the ball into opponent's goal, gripping or changing the incoming ball's direction.

$$D_m = max\{|X_{(t)} - X_{0(t)}|, |Y_{(t)} - Y_{0(t)}|\} \tag{7}$$

Taking into account for directional changes, the predicted collision time is defined as the equation follows:

$$C_0 = T_m / P_t \tag{8}$$

One of the advantages of this technique is that it can be used with other functions. In addition, it gives the robot ample time to take action without affecting its movement. Nevertheless, the coordination prediction technique is more popular than the collision prediction technique. Besides, this program needs to be used with other complementary programs such as velocity and network functions, which will result as a longer program and burden the computer.

3   Proposed Trigonometry Method

This technique computes linear motion as in (10) that considers displacement, initial and final velocity, and the momentum of the ball. These elements are part of basic elements in physics. There are four fundamental of constructing the linear motion. According to Hong Liu *et al* [5], the movement of the robot or the ball in the field is influenced by the external factor such as friction and inertia force. In this technique, there are two phases in the calculation process which are initial phase and prediction phase.

3.1   Path Conditions

There are several conditions to be specified during the initial phase in order to determine the ball's direction. They are illustrated in Figure 1.

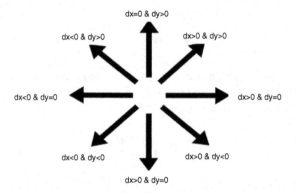

Fig. 1. Conditions for dx and dy to determine the ball's direction

3.2   Initial Phase

In the initial phase, let's assume that the ball is in stationary. Thus, the value for initial velocity of the ball (u) is zero. The objective of this phase is to find the current velocity (y) of the ball to be used in the next phase. The gap between a frame to another is one second, hence the value of t is 1. The distance of the path taken by the ball is determined by values that are preprogrammed in the Merlin Vision System. *Currentball* variable locates the ball in the x-axis and y-axis from the current frame while *lastball* variable locates the ball from the previous frame. The following equations are used to find the distance of the ball:

$$dx = currentball.pos.x - lastball.pos.x$$
$$dy = currentball.pos.y - lastball.pos.y \qquad (9)$$

However, the dx and dy values are not the accurate displacement of the ball. Displacement of the ball is the hypotenuse value obtained from the dx and dy values. The figure 2 below describes the relationship between dx, dy and the hypotenuse.

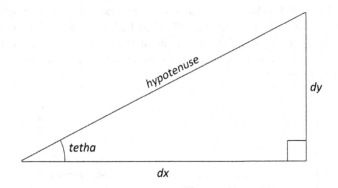

Fig. 2. Relationship of dx, dy and hypotenuse

In a right triangle, the hypotenuse, s, can be calculated using the dx and dy. Later, s can be used to obtain acceleration, a, in:

$$a = 2(s - ut)/t^2 \tag{10}$$

Where:

$s=$ hypotenuse	$u =$ initial velocity
$t =$ time	$a =$ acceleration

To complete the initial phase, the equation:

$$v = u + at \tag{11}$$

Where:

$v =$ final velocity	$u =$ initial velocity
$a =$ acceleration	$t =$ time.

3.3   Prediction Phase

In the second phase namely prediction phase, the main objective is to find the final displacement (d_2) of the moving ball using the velocity obtained in the first phase. By assuming that the ball is travelling at constant acceleration (a_2), the value for initial velocity (u_2) in this phase is same as the value of final velocity in the initial phase. Thus, $u_2 = v$ and the acceleration, $a_2 = a$. This prediction estimates the location of the ball in the next frame by having second time interval between each frames. By using the equation:

$$d_2 = u_2 t + \frac{1}{2} a_2(t)^2 \tag{12}$$

The d_2 value, will be obtained. By assuming that the ball is in a smooth-surfaced field and there is no obstacle that will affect the ball's movement, the angle formed will the same as the angle obtained in initial phase. By using the trigonometry equation, the predicted location will be obtained. The figure 3 below shows the processes in both phases:

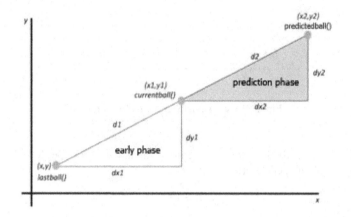

Fig. 3. Overview for the trigonometry technique

Table 1 shows the implementation steps to find the predicted coordinate of the ball by using the trigonometry technique.

Table 1. Implementation of the Trigonometry Technique

Implementation Steps of Trigonometry Technique
1. Determine the final location of the ball in y-axis and x-axis in the variables.
2. Determine the current location of the ball in y-axis and x-axis in the variables.
3. Calculate the distance difference between final and current location of the ball in both axes.
4. Based on the Figure 3, verify the required conditions for direction of the ball.
5. Calculate the absolute value of the differences to find the vector.
6. Applying Pythagoras theorem in both vectors, find the hypotenuse value.
7. Using trigonometry equation, find the angle formed.
8. Using equation $a = 2(s - ut)/t^2$ and $v = u + at$, to find the value of a and v.
9. Replace the value of a into a_2 and v into u_2.
10. Use the value of a_2 and u_2 in $s = ut + 1/2a(t^2)$ to find d_2.
11. Use the value of d_2, that is the *hypotenuse2*, and the angle formed earlier into trigonometry equation to find the new difference values for x and y.
12. Add the difference value obtained to the current ball location according to the conditions stated.

4   Result and Analysis

During prediction testing of our method, four of the eight directions of the ball are chosen and each direction was repeated five times. The similar situations are then tested using the existing Jun Jo's programming to be compared with the first test. Execution time is taken to measure the speed taken to run a task by robot. Beside that a testing in real time environment also carried out.

4.1   The Evaluation Technical Proposed Result

The testing phase was conducted in the simulator and real-time environment robot soccer. Four directions are chosen randomly and that are towards north, north-west, north-east and east. The ball was toasted toward these directions. This technique is tested with Jun Jo's programming for the comparison. For the prediction techniques testing, four of the eight directions of the ball are chosen and for each direction tested was repeated five times. The similar situations are then tested using Jun Jo's programming. In a testing in real time environment, for each direction, the tests are repeated 25 times.

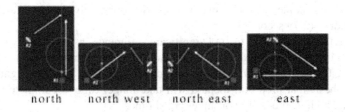

north   north west   north east   east

Fig. 4. Overview of testing estimation techniques

As in figure 4, R2 is a robot that predicts the position of the moving ball and R1 as a ball pusher.

4.2   The Evaluation Technical Proposed Result

Both of the proposed and Jun Jo's methods are evaluated using the effectiveness percentage for the robot soccer to predict the ball defined as shown in the equation follows:

$$effectiveness\ percentage = \frac{total\ succesfull}{total\ trial} \times 100 \qquad (13)$$

After the percentage for each direction is obtained, the overall average is used to evaluate the effectiveness of this technique as the whole.

4.3   Assessment through Simulations

Table below shows the result obtained from the simulator testing:

Table 2. The proposed and Jun Jo's methods results for simulator testing in all tested directions

Direction	Effectiveness, %		Average Time, *s*	
	Proposed	Jun Jo	Proposed	Jun Jo
North	60	40	3.38	2.62
North East	100	0	2.9	2.36
North West	80	80	2.58	3.18
East	40	40	2.9	3.0
Average	70	40	2.94	2.79

As aforementioned, the four directions chosen were tested using proposed ball prediction and Jun Jo's techniques, five times each and then were evaluated using equation 13. Based on table 2, this technique shows 70% more effective than Jun Jo (40%). This is because, this technique consider more factor than Jun Jo's technique. It also caused trigonometry need more time (2.94 seconds) to complete task then Jun Jo's techniques (2.79 seconds).

After performed a T-test, to find significant difference in time taken, the value of t = 0.958521. So, according to the t sig/probability table with df= (40-2) = 28, t must be at least 2.048 to reach p<0.05. Hence, the difference is not statically significant. An early hypothesis that proposed method will take longer time to complete a task was rejected and both methods are consumes equal time.

4.4   Real-Time Evaluation

Table below shows the result that has been gain by real time evaluation:

Table 3. The proposed results for real time testing in all tested directions versus two types of ball speed

Direction	Effectiveness, %		Average Time, *s*	
	Slow	Fast	Slow	Fast
North	56	76	2.48	2.76
North East	88	76	1.29	3.27
North West	76	72	1.67	2.79
South East	88	80	2.23	2.86
Average	77	76	1.92	2.92

A total of 25 tests were conducted on the robot in real-time ball in that same direction during the test in the simulation. Robot been used and the engine that run the strategy was from the Merlin. The gain value is 0.5.

In real time environment tested, proposed technique shows a good effectiveness in slow or fast moving ball. The difference in time is because of the ball speed. The greater the speed, longer the distant ball path to intercept.

4.5   The Results of Comparison

After the results obtained, these results should be compared to measure the performance achieved from the proposed techniques presented. Table 4 shows the comparison between the proposed ball prediction technique and Jun Jo's technique in simulation and real-time environment testing.

Table 4. Comparison of estimation techniques ball

Prediction Technique	Overall Effectiveness, %	Overall Time Average, s
Proposed	70	2.94
Jun Jo's Technique	40	2.79
Real Time (slow)	77	1.92
Real Time (fast)	76	2.92

4.6   Discussion

All the techniques had been developed and well tested. Our prediction technique is also successfully applied in robot soccer simulator and provided good predictions.

Based on Table 4, the overall effectiveness obtained in the robot soccer simulator may not be equivalent to a dynamic real-time environment. However, improvements can be made from time to time. The calculation for prediction may be simplified further to shorten execution time.

5   Conclusion

The ball prediction using trigonometric technique is capable to provide quite good prediction. By taking the proposed factors into account, the prediction calculation for the movement of the ball becomes more accurate. However, it will take a longer time to execute that might cause the movement of the ball to be less efficient.

The weakness of this prediction technique is that it need to compute for every frames. This technique also does not take the rotation of the ball speed into account that can cause changes in direction. In addition, this technique is not suitable for robot soccer that moves slowly.

In the future, this technique should be made more dynamic so that the movement of the ball will be more accurately predicted.

Acknowledgements. This project research is fully funded by GGPM-ICT-119-2010 project entitled "Logo and Text Detection using vision guided approach", PTS-2011-047 entitled "Empowering Robot Soccer Teaching and Learning in FTSM" and from Centre for Research and Instrumentation (CRIM) PIP number UKM-UP-PIP-2012. It is part of robot soccer research development in UKM.

References

1. Xian, L.: Artificial intelligence and modern sports education technology. In: 2010 International Conference on Artificial Intelligence and Education (ICAIE), pp. 772–776 (2010)
2. Brunette, E.S., Flemmer, R.C., Flemmer, C.L.: A review of artificial intelligence. In: 4th International Conference on Autonomous Robots and Agents, ICARA 2009, pp. 385–392 (2009)
3. Nakashima, T., Takatani, M., Ishibuchi, H., Nii, M.: The Effect of Using Match History on the Evolution of RoboCup Soccer Team Strategies. In: 2006 IEEE Symposium on Computational Intelligence and Games, pp. 22–24, 60–66 (2006)
4. Davids, A.: Urban search and rescue robots: from tragedy to technology. IEEE Intelligent Systems 17(2), 81–83 (2002)
5. Hongbin, Z., Keming, C., Peng, W.: A New Real-Time Collision Prediction Model for Soccer Robots. National Lab. on Machine Perception, Peking University, 100871, China (2008)
6. Ishii, T.K.: Linear coefficients of equation of motion of a drift electron. Proceedings of the IEEE 71(8), 1010–1012 (1983)
7. Joshi, A.: Analysis of Bang-bang CDR circuits with equations of linear motion. In: 15th IEEE International Conference on Electronics, Circuits and Systems, ICECS 2008, pp. 1143–1146 (2008)
8. Miabot, P.B.: v2 User Manual, Merlin Systems Corp. Ltd. 2002-2004 (2004)
9. Jo, J., Smith, R., Rodwell, A., Clarke, M., Truesdell, S.: 3D Robot Soccer Simulator. The School of Information Technology, Griffith University, Australia

Classifier Selection for Locomotion Mode Recognition Using Wearable Capacitive Sensing Systems

Yi Song[1,3], Yating Zhu[2], Enhao Zheng[1,3], Fei Tao[2], and Qining Wang[1,3,*]

[1] Intelligent Control Laboratory, College of Engineering,
Peking University, Beijing 100871, China
[2] School of Automation Science and Electrical Engineering,
Beihang University, Beijing 100191, China
[3] Beijing Engineering Research Center of Intelligent Rehabilitation Engineering,
Beijing 100871, China
qiningwang@pku.edu.cn

Abstract. Capacitive sensing has been proven valid for locomotion mode recognition as an alternative of popular electromyography based methods in the control of powered prostheses. In this paper, we analyze the characteristics of the capacitive signals and extract suitable feature sets to improve the recognition accuracy. Then the classification results of different classifiers are compared and one optimal classifier which can offer highest accuracy within a reasonable time limit is selected. Experimental results show that the recognition accuracy of the wearable capacitive sensing system has been improved by using the selected classifier.

Keywords: Capacitive sensing, classifier selection, locomotion mode recognition, wearable systems, lower-limb prostheses.

1 Introduction

Capacitive sensing is a technology that allows detection and tracking of conductive as well as non-conductive objects [1]. This concept can be used in the sensing of human daily movements [2]. Rekimoto proposed a capacitance based approach of detecting human gestures and the prototype was implemented with a wristwatch [3]. Cheng *et al.* presented a concept for using capacitance sensing in human activity recognition and designed a prototype [4]. Our recent studies indicated that the capacitive sensing is a promising solution for locomotion mode classification as an alternative of electromyography (EMG) based systems in the control of powered prostheses [5, 6] and recognition of human normal gaits [7].

Compared with other body activity recognition, studies of lower-limb movement recognition have to pay more attentions to the accuracy, due to high demands of the control of lower-limb prostheses to guarantee human safety.

---

* Corresponding author.

J.-H. Kim et al. (eds.), *Robot Intelligence Technology and Applications 2,*
Advances in Intelligent Systems and Computing 274,
DOI: 10.1007/978-3-319-05582-4_67, © Springer International Publishing Switzerland 2014

Therefore, it is necessary to select an optimal classifier for the specific locomotion mode recognition problem with higher accuracy and better robustness. Several studies have attempted various classifiers, including linear discriminant analysis (LDA), quadratic discriminant analysis (QDA), support vector machine (SVM) and multiple binary classifier (MBC), to improve the recognition performance of EMG based systems [8–10]. In [9], a multi-class SVM with one-against-one (OAO) scheme and C-Support Vectors Classification (CSVC) were used to identify different locomotion intent. A Naive Bayes classifier with sequential forward search can locate and elucidate the primary factors contributing to the discrimination between several types of terrain [10]. However, no effort has been made on classifier selection for locomotion mode recognition using capacitive sensing.

In this paper, based on our former study [5], we analyze the characteristics of the capacitive signals and extract suitable feature sets to improve the recognition accuracy. A capacitive sensing system, which is made up of sensing bands, a signal processing circuit, a gait event detection module and a computer to receive data streams, is presented to provide capacitance signals during human gaits. Then we select an optimal classifier for locomotion mode recognition with the wearable capacitive sensing system through the comparison among various types of classifiers. Experimental results show that recognition accuracy of the capacitive sensing system has been improved.

The rest of the paper is organized as follows. In section 2, we introduce the prototype of the capacitive sensing system. In section 3, we select some promising algorithms for later comparison according to their performance. Section 4 shows classification results. We conclude in section 5.

2   Capacitive Sensing System

For human body capacitance measurement, the electrodes of the capacitor can be divided into two parts, the transmitter electrode and receiver electrode (see Fig. 1). If the transmitter electrode is stimulated with a wave signal at a frequency of several hundred kilohertz, the receiver electrode will receive the wave. The magnitude of the signal is proportional to the frequency and the voltage of the transmitted signal. The impedance Z of the capacitor is a function of the capacitance:

$$Z = (j\omega C_{body})^{-1}, \qquad (1)$$

where ω is the frequency of the driven signal and it remains constant in the system. The resistor R in series and the capacitance C_{body} makes an impedance matching circuit. The voltage across the resistor will vary with C_{body}. Thus, the signal of the receiver electrode can be extracted as the capacitance signal. The raw signal is a sinusoid wave with constant frequency ($100kHz$) and varying amplitude, which is unsuited for direct sampling. Thus, the signal needs to be converted to root mean square (RMS) voltage before the analog-to-digital converter (ADC). We can detect the human motion by fixing the electrodes on the human body. The muscle deformation and contact conditions can be reflected by the change of capacitance. With proper regulation, the magnitude of the signals

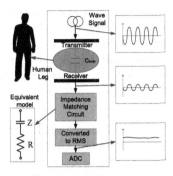

Fig. 1. Design concept of the proposed capacitance based sensing system for human locomotion recognition. The magnitude of the signal on the receiver electrode is influenced by the body movement. The signal was converted to RMS value before the ADC.

on receiving electrodes will vary with the human movement. By measuring the voltage, we can get the motion information accordingly.

Based on the design concept mentioned above, the proposed capacitive sensing system is made up of four parts: sensing bands, a signal processing circuit, a gait event detection module and a computer to receive data streams. In the capacitive sensing system, two separate arrays of capacitors were implemented on the sensing bands to record the shape changes of the thigh and the shank respectively [5]. Seventeen receiver electrodes made from copper sheets are mounted on the inner surface (seven for the shank and ten for the thigh). Specific implementation of sensing bands can be found in [5]. The muscular cross section of the thigh changes visibly according with lower-limb movement. The shape change of lower limb can slightly change the relative positions between the electrodes and skin. This tiny change will have a strong effect on the capacitance signals.

3   Classifier Selection

3.1   Capacitance Signal Analysis

The analysis of capacitance signal takes place in three parts: the data from capacitive sensors is first filtered; then features that characterized the capacitance signal from each channel are extracted; finally these features are classified into corresponding locomotive modes.

Data Source. The capacitance signal used in this paper were collected from previous experiments reported by Chen et al. [5]. The data of six locomotion modes, including standing, sitting, stair ascent, stair descent, obstacle clearance and normal level-ground walking, were collected from eight able-bodied subjects. Two capacitive sensing bands were worn on the thigh and shank of the experiment subjects respectively and five channels out of ten on each band were recorded for later data analysis.

The experiment consisted of seven blocks. All six locomotion modes were tested in each block. The numbers of trials for every mode were designed to keep the number of steps approximately the same across locomotive modes, which ensured the balance of the data source.

Feature Extraction. The capacitance signals can reflect the quasi-cyclic changes of leg shapes caused by muscle contractions during locomotive movements. According to [5], the continuous data stream within one gait cycle is segmented with the optimal parameters including $190ms$ of phase size, $150ms$ of window size and $20ms$ of window increment.

The extracted features should be capable of presenting the characteristics of the capacitance signals for different locomotive modes. Since locomotion mode recognition (LMR) is a problem in the real-time application, computation load and operation time are important concerns. Time-domain features, instead of frequency-domain features, are extracted from the raw capacitive signals, such as mean, variance, standard deviation, median, maximum, minimum, range, root mean square, integration, correlation, zero-crossings and magnitude.

Feature Selection. In the previous work [5], a forward selection method was employed to select the feature set. Five features, including mean, maximum, minimum RMS and standard deviation, were selected out of the above-mentioned feature array.

It is obvious that fewer features bring lower cost and higher efficiency. Therefore, the optimal feature set for any classifier should be as small as possible. We will examine a short list of only five features. Because the number of features is small, in the following research work we will use the exhaustion method to search for the optimal feature set. Therefore, 31 ($2^5 - 1$) combination of 5 features will be tested for every classifier. The feature set that provides the highest accuracy will be selected as the optimal.

3.2   Brief Review of Related Classifiers

Pattern classification is to classify objects into a number of classes. In spite of almost 50 years of research, design of a general purpose machine pattern recognizer remains an elusive goal. The well-known "No Free Lunch" theorem [11] indicates that there does not exist a pattern classification method that is inherently superior to any other, or even to random guessing without using additional information. It is the type of problem, prior information, and the amount of training samples that determine the form of classifier to apply.

In this paper, we are going to find the most suitable classifier for the specific LMR problem. First of all, LMR is a statistical pattern recognition problem. For the statistical pattern recognition [12], a pattern is represented by a set of n features, viewed as an n-dimensional feature vector. The pattern classification system is operated in two modes: training and testing. In the training mode, the feature extraction/selection module finds the appropriate features for representing the input patterns and the classifier is trained to partition the feature space. In the testing mode, the trained classifier assigns the input pattern to one of the pattern classes under consideration based on the measured features.

In addition, LMR belongs to supervised classification, i.e., given a set of objects, each of which belongs to a known class, and each of which has a known vector of variables, our aim is to construct a rule which will allow us to assign future objects to a class, given only the vectors of variables describing the future objects. Many methods for constructing such rules have been developed.

3.3   Classifier Candidates

The most intuitive method is based on the concept of similarity: patterns that are similar should be assigned to the same. One of the simplest classifiers is the Rote classifier, which memorizes the entire training data and performs classification only if the features of the test object exactly match the attributes of one of the training objects.

A more sophisticated approach, k-nearest neighbor (kNN) classification, finds a group of k objects in the training set that are closest to the test object, and bases the assignment of a label on the predominance of a particular class in this neighborhood. However, kNN classifier is a lazy learner, i.e., the model is not built explicitly. Thus, building the model is cheap, but classifying unknown objects is relatively expensive since it requires the computation of the k-nearest neighbors of the object to be labeled. This, in general, requires computing the distance of the unlabeled object to all the objects in the labeled set, which can be expensive particularly for large training sets. Obviously that is not a good solution suitable for the LMR problem and we will skip the kNN classifier in the following comparison.

Another category of classifiers is to construct decision boundaries directly by optimizing certain error criterion. The LDA classifier is a classical example of this type of classifier is Fisher's linear discriminant, which we have tested in the previous work, and the result will be compared with other classifiers in this paper.

Naive Bayes. Based on the probabilistic approach, the Naive Bayes classifier (with the 0/1 loss function) assigns a pattern to the class with the maximum posterior probability [13]. This method is well-known for its extreme simplicity, permitting easy estimation and straightforward interpretation via the weights of evidence. And, particularly important, it often does surprisingly well: It may not be the best possible classifier in any given application, but it can usually be relied on to be robust and to do quite well.

We label the class by i, and define $P(i \mid x)$ to be the probability that an object with feature vector $x = (x_1, ..., x_n)$ belongs to class i, $f(x \mid i)$ to be the conditional distribution of x for class i objects, $P(i)$ to be the probability that an object will belong to class i (the prior probability of class i), and $f(x)$ to be the overall mixture distribution of i classes:

$$f(x) = \sum_{i=1}^{n} f(x \mid i)P(i) \tag{2}$$

A simple application of Bayes theorem yields:

$$P(i \mid x) = f(x \mid i)P(i)/f(x) \tag{3}$$

To obtain an estimate of $P(i \mid x)$ from this, we need to estimate each of the $P(i)$ and each of the $f(x \mid i)$. The core of the naive Bayes method lies in the method for estimating the $f(x \mid i)$, which assumes that the components of x are independent within each class, so that

$$f(x|i) = \prod_{j=1}^{n} f(x_j|i) \tag{4}$$

This assumption is highly unrealistic since the majority of features are related in complicated ways. However, various factors may come into play which means that the assumption is not as detrimental as it might seem [14]. Thus, we use the Naive Bayes classifier as one of our classifier candidates.

C4.5. The decision tree classifier is a series of questions systematically arranged so that each question queries an attribute and branches based on the value of the attribute. At the leaves of the tree are placed predictions of the class variable.

C4.5 algorithm, designed by Quinlan as a descendant of ID3 approach to inducing the decision trees [15], can be divided to two steps. The first step is to construct iteratively a decision tree using training set, and to prune the result tree; the second step is to classify the input pattern using the pruned tree. For the input pattern, its value is tested in a proper order from the root node of the decision tree to one of leaf node, finally the class of the input pattern can be found [16].

The entropy of a class random variable that takes on c values with probabilities $p_1, p_2, ..., p_c$ is given by:

$$\sum_{i=1}^{c} -p_i log_2 p_i \tag{5}$$

For a given random variable, say attribute a, the improvement in entropy is calculated as:

$$Gain(a) = Entropy(a \text{ } in \text{ } D) - \sum_{v} \frac{|D_v|}{|D|} Entropy(a \text{ } in \text{ } D_v) \tag{6}$$

where v is the set of possible values, D denotes the entire dataset, D_v is the subset of the dataset for which attribute a has that value, and the notation $|\cdot|$ denotes the size of a data set.

The default splitting criterion is actually the gain ratio for the attribute a, which is defined as:

$$GainRatio(a) = \frac{Gain(a)}{Entropy(a)} \qquad (7)$$

In addition to inducing trees, C4.5 is capable of pruning tree to avoid overfitting the data. Since C4.5 is often used as a benchmark, it is one of our classifier candidates.

Support Vector Machine. Support vector machine (SVM), including support vector classifier (SVC) and support vector regressor (SVR), are among the most robust and accurate methods in all well-known data mining algorithms. SVM, which was originally developed by Vapnik in the 1990s, has a sound theoretical foundation rooted in statistical learning theory, requires only as few as a dozen examples for training, and is often insensitive to the number of dimensions [17].

Margin maximization is the initial motivation of the SVM algorithm. Consequently, SVM (SVC) usually places more focus on the separability between the classes of samples, instead of the prior data distribution information within classes.

To solve linearly inseparable problems, there are two approaches for SVC: soft margin and kernel trick. The soft margin slackens the constraints in the original input space and allows some errors to exist. However, when the problem is heavily linearly inseparable and the misclassified error is too high, the soft margin is unworkable. The kernel trick maps the data to a high-dimension feature space implicitly by the kernel function in order to make the inseparable problem separable. In practice, we often integrate them to exert the different advantages of the two techniques and solve the linearly inseparable problem more efficiently. As a result, the corresponding dual form for the constrained optimization problem in the kernel soft margin SVC is as follows:

$$\max_\alpha W(\alpha) = \sum_{i=1}^{n} \alpha_i - \frac{1}{2} \sum_{i=1}^{n} \sum_{j=1}^{n} \alpha_i \alpha_j y_i y_j K(x_i, x_j)$$
$$s.t. \sum_{i=1}^{n} \alpha_i y_i = 0$$
$$0 \le \alpha_i \le C, i = 1, ..., n \qquad (8)$$

Following the Lagrange multipliers method, we can obtain the optimal classifier:

$$f(x) = \sum_{i=1}^{n} \alpha_i^* y_i K(x_i, x) + b^* \qquad (9)$$

where $b^* = 1 - \sum_{i=1}^{n} \alpha_i^* y_i K(x_i, x_s)$, for a positive support vector $y_s = +1$.

The kernel function selection is critical to SVM classifier. The most popular ones includes linear: $K(x_i, x_j) = x_i \cdot x_j$, polynomial: $K(x_i, x_j) = (\gamma x_i \cdot x_j + r)^d$, RBF: $K(x_i, x_j) = exp(-\gamma \parallel x_i - x_j \parallel^2)$ and sigmoid: $K(x_i, x_j) = tanh(\gamma x_i \cdot x_j + r)$. Here, $r, \gamma, d > 0$ are the kernel parameters. Generally, RBF is suggested for use. To our surprise, the linear function performs quite well for our specific LMR problem.

For multi-class problem, there are alternative schemes for SVM: one-against-all (OAA) or one-against-one (OAO). In contrast, OAA scheme is simple to implement and fast running, while OAO scheme is more accurate in most cases. According to the importance of accuracy in the LMR problem, we use OAA scheme in our SVM classifier.

This paper employs library for support vector machines (LIBSVM) [18] as the core of our SVM classifier with C-SVC model, linear kernel function and OAO scheme.

Back Propagation. Artificial neutral networks (ANNs) are structured according to human brain. ANNs consist of groups of processing elements called neurons. Neurons are connected into a network in the way that the output of each neuron represents the input for one or more other neurons. According to its direction, the connection between neurons can be either one-directional or bi-directional, and due to its intensity the connection can be excitatory or inhibitory. The ANNs can be classified into feed-forward networks and recurrent networks according to the connection way of neurons.

The most commonly used family of neural networks for pattern classification is feed-forward network [19]. Back Propagation (BP) algorithm, as a typical feed-forward network classifier, is good at nonlinear mapping, generalization and fault tolerance. In our BP classifier, the input neurons are as many as the features. However, when the feature number is only one, the BP classifier can hardly converge. Therefore, at least two features are used. The number of hidden neurons varies according to feature numbers, 12 for 2 features, 13 for 3 features, 14 for 4 features and 15 for 5 features. Since there are 6 locomotive modes in our LMR problem, 3 output neurons are needed. The learning rates of input layer to hidden layer and hidden layer to input layer are both set to 0.7. The learning error is calculated as

$$E(\vec{w}) = \frac{1}{2} \sum_{d \in D} \sum_{k \in outputs} (t_{kd} - o_{kd})^2 \qquad (10)$$

where $outputs$ is the set of output neurons, t_{kd} and o_{kd} are output values relative to training sample d and k-th output neuron.

The loop ends if the learning error is under 0.01. The sigmoid function, which is a smooth differentiable function with thresholds, is used in the gradient descent algorithm.

3.4   Performance Evaluation

The ultimate object of the LMR problem is to control the powered lower-limb prostheses. The solution should be as accurate and fast as possible. Therefore, accuracy and time are both significant in the classifier selection. In addition, the data amount for training, the robustness against bad data and the computing complexity are also in our concerns.

Accuracy. The classification accuracy is defined as the rate of correct classification to all data in a test data set and can be calculated as:

$$Accuracy = \frac{N_{correct}}{N_{total}} \times 100\% \tag{11}$$

where $N_{correct}$ is the number of correct test events and N_{total} is total number of test events.

We will use 7-fold cross validation for calculating classification error. Six blocks out of seven will be used as the training data set and the rest one as the test data set. We will repeat this procedure for seven times and calculate the average classification accuracy of each subject.

Time Limit. As previously mentioned, pattern classification consists of two phases: training and testing. For LMR problem, the training process can be offline with low requirement of time limit, which can take minutes. On the contrary, the testing has to be online and continual with high time limit requirement, which can only take milliseconds.

Miscellaneous. The influence of training data amount is different for various classifiers. Some are sensitive, others are not. We prefer the latter. It is obvious that the shorter the experiment takes, the better the subject feels.

Although classifier ensemble is a solution for higher accuracy, it means more computation load and takes more operation time. Single classifier is strongly recommended.

4 Experimental Results

Four candidate classifiers (NB, C4.5, SVM and BP) are examined for feature selection separately. It is not surprising that the optimal feature sets of different classifier are not the same. The details are shown in Table 1.

The optimal feature set of the NB classifier includes mean, minimum and standard deviation, whereas it includes mean, maximum, minimum and RMS for SVM, mean, maximum, RMS and standard deviation for C4.5, mean, maximum, minimum, RMS and standard deviation for BP.

The highest accuracy is 78.18% for the NB classifier, 93.75% for SVM , 82.76% for C4.5 and 79.49% for BP. It is obvious that SVM outperforms other three classifiers. Besides, the result is also better than those of all three classifiers in [5], viz., linear discriminant analysis (LDA), quadratic discriminant analysis (QDA) and LR (linear regression).

Since SVM is selected, more work is done on the classifier. First, we try to optimize the penalty parameter c in the SVM model of each subject separately, which is set as its default value 1 in the former experiment. By this means, better classification results are achieved. The average classification accuracy of SVM increases to 94.0%.

Table 1. Optimal feature set selection for NB, SVM, C4.5 and BP

Mean	Max	Min	RMS	STD	NB	SVM	C4.5	BP
★					72.06%	90.42%	75.62%	
	★				67.94%	85.20%	71.72%	
		★			74.13%	87.07%	77.42%	
			★		71.16%	90.20%	74.67%	
				★	53.14%	66.66%	67.12%	
★	★				73.29%	90.42%	76.11%	71.02%
★		★			75.09%	90.42%	78.16%	67.53%
★			★		71.98%	90.42%	79.30%	59.40%
★				★	77.65%	90.42%	75.32%	73.15%
	★	★			75.26%	85.20%	75.68%	66.57%
	★		★		72.33%	85.20%	78.45%	70.96%
	★			★	75.80%	85.20%	81.37%	60.67%
		★	★		74.43%	87.07%	81.27%	68.13%
		★		★	76.84%	87.07%	82.01%	44.72%
			★	★	77.14%	90.20%	81.18%	71.42%
★	★	★			75.87%	90.42%	78.61%	72.16%
★	★		★		73.25%	90.42%	75.95%	69.85%
★	★			★	77.65%	90.42%	78.27%	73.50%
★		★	★		74.73%	90.42%	79.04%	71.85%
★		★		★	78.18%	90.42%	81.54%	73.29%
★			★	★	77.70%	90.42%	81.78%	75.70%
	★	★	★		75.13%	85.20%	82.00%	71.57%
	★	★		★	77.41%	85.20%	81.42%	75.05%
	★		★	★	76.67%	85.20%	81.61%	76.11%
		★	★	★	77.55%	87.07%	81.95%	75.78%
	★	★	★	★	77.54%	93.44%	82.39%	79.44%
★		★	★	★	77.96%	93.30%	82.25%	79.09%
★	★		★	★	77.18%	93.41%	82.76%	78.77%
★	★	★		★	78.07%	93.43%	81.96%	78.69%
★	★	★	★		75.40%	93.75%	78.49%	73.44%
★	★	★	★	★	77.73%	93.55%	81.65%	79.49%

Secondly, we build the confusion matrixes of SVM for four gait phases of all subjects as shown in Fig. 2 The average classification accuracies of SVM are $92.8 \pm 1.0\%$, $94.4 \pm 0.9\%$, $95.6 \pm 0.7\%$, $93.2 \pm 1.7\%$ for the gait phase of Pre-FC, Post-FC, Pre-FO and Post-FO respectively. More details about the error distribution can be found. Higher off-diagonal numbers indicate higher error rates.

Thirdly, we measure the modeling and classifying time. The experiments are done on a computer with an Intel Core $i7 - 2600$ 3.4 GHz CPU and Windows 7 operating system. The average modeling time is 1.2 seconds, while the average classifying time is 0.15 second. Both are well below the time limits we have mentioned before.

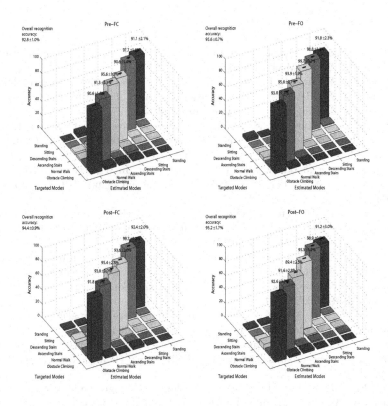

Fig. 2. SVM confusion matrixes of all gait phases

5   Conclusion

This paper aims at selecting an optimal classifier for LMR with capacitance signals. Various classifiers are introduced and four typical classification algorithms are realized with C++ program language. According to the results, SVM outperforms other classifiers and is considered as the first choice. In the future, more research on the SVM classifier will be done. Meanwhile, we will improve the wearable capacitive sensing system and design new experiments with more locomotive modes and fewer training data. It is promising to find a solution for the locomotion modes recognition problem with capacitive sensing in the practical applications of powered prostheses.

Acknowledgment. This work was supported by the National Natural Science Foundation of China (No. 61005082, 61020106005) and the National Key Technology Research and Development Program (No. 2011BAK16B00).

References

1. Luo, R.C.: Sensor technologies and microsensor issues for mechatronics systems. IEEE/ASME Trans. Mechatronics 1(1), 39–49 (1996)
2. Wimmer, R., Kranz, M., Boring, S., Schmidt, A.: A capacitive sensing toolkit for pervasive activity detection and recognition. In: Proc. of the Fifth IEEE Int. Conf. Pervasive Computing and Communications (2007)
3. Rekimoto, J.: Gesturewrist and gesturepad: unobtrusive wearable interaction devices. In: Proc. of the 5th Int. Symp. Wearable Computers, pp. 21–27 (2001)
4. Cheng, J., Amft, O., Lukowicz, P.: Active capacitive sensing: exploring a new wearable sensing modality for activity recognition. In: Floréen, P., Krüger, A., Spasojevic, M. (eds.) Pervasive 2010. LNCS, vol. 6030, pp. 319–336. Springer, Heidelberg (2010)
5. Chen, B., Zheng, E., Fan, X., Liang, T., Wang, Q., Wei, K., Wang, L.: Locomotion mode classification using a wearable capacitive sensing system. IEEE Trans. Neur. Sys. Reh. Eng. 21(5), 744–755 (2013)
6. Zheng, E., Wang, L., Wei, K., Wang, Q.: Non-contact capacitance sensing for continuous locomotion mode recognition: design specifications and experiments with an amputee. In: Proc. of the 13th International Conference on Rehabilitation Robotics, Seattle, USA (2013)
7. Zheng, E., Chen, B., Wei, K., Wang, Q.: Lower limb wearable capacitive sensing and its applications to recognizing human gaits. Sensors 13(10), 13334–13355 (2013)
8. Huang, H., Kuiken, T.A., Lipschutz, R.D.: A strategy for identifying locomotion modes using surface electromyography. IEEE Trans. Biomed. Eng. 56(1), 65–72 (2009)
9. Zhang, F., Fang, Z., Liu, M., Huang, H.: Preliminary design of a terrain recognition system. In: Proc. of the Annual International Conference of the IEEE Engineering in Medicine and Biology Society, pp. 5452–5455 (2011)
10. Farrell, M.T., Herr, H.: A method to determine the optimal features for control of a powered lower-limb prostheses. In: Proc. of the Annual International Conference of the IEEE Engineering in Medicine and Biology Society, pp. 6041–6046 (2011)
11. Wolpert, D.H., Macready, W.G.: No free lunch theorems for optimization. IEEE Trans. Evolutionary Computation 1, 67–82 (1997)
12. Jain, A.K., Duin, R.P.W., Mao, J.: Statistical pattern recognition: a review. IEEE Trans. Pattern Analysis and Machine Intelligence 22(1), 4–37 (2000)
13. Webb, A.R.: Statistical pattern recognition, 2nd edn., pp. 50–75. John Wiley & Sons, Ltd. (2002)
14. Hand, D.J., Yu, K.: Idiot's Bayes-not so stupid after all? Int. Statistical Rev. 69, 385–398 (2001)
15. Quinlan, J.R.: Induction of decision trees. Machine Learning 1(1), 81–106 (1986)
16. Sethi, I.K., Sarvarayudu, G.P.R.: Hierarchical classifier design using mutual information. IEEE Trans. Pattern Recognition and Machine Intelligence 1, 194–201 (1979)
17. Vapnik, V.: Statistical learning theory. John Wiley & Sons, Ltd. (1998)
18. Chang, C.-C., Lin, C.-J.: LIBSVM: A library for support vector machines, http://www.csie.ntu.edu.tw/cjlin/libsvm
19. Jain, A.K., Mao, J., Mohiuddin, K.M.: Artificial neural networks: a tutorial. Computer 29(3), 31–44 (1996)

Development of Jaw Exoskeleton for Rehabilitation of Temporomandibular Disorders

Xiaoyun Wang[1,*], Johan Potgieter[1], Peter Xu[2], and Olaf Diegel[1]

[1] School of Engineering & Advanced Technology, Massey University,
Private Bag 102 904, North Shore City 0745, New Zealand
{X.Wang,J.Potgieter,O.Diegel}@massey.ac.nz
[2] Department of Mechanical Engineering, University of Auckland, New Zealand
P.Xu@auckland.ac.nz

Abstract. A jaw exoskeleton is proposed in this paper to assist the exercise for the purpose of practical rehabilitation of temporomandibular disorder (TMD). The jaw, attached to the skull by muscles and pivoted at the condyle via the temporomandibular joint (TMJ), can be simplified as moving in the two-dimensional sagittal plane. Based on the in-vivo recording the jaw movement from the healthy subject, the motion pattern is justified to be the primary specification to design the jaw exoskeleton. A planar four-bar linkage is synthesized to reproduce the specified normal jaw motion in terms of incisor and condyle trajectory on the coupler point to meet the kinematic specification. Adjustable lengths of the links are used to achieve a group of trajectories of any possibility.

Keywords: TMD, rehabilitation, four-bar linkage.

1 Introduction

The temporomandibular disorder (TMD), which refers to a collective term generally encompassing a group of dysfunctions in the masticatory system causing mandibular hypomobility and muscle stiffness, severely compromises the accomplishment of the masticatory function [1]; it considerably impairs speech and oral hygiene and brings discomforts and inconveniences to a large number of patients [2]. Studies of the prevalence of the TMD reported different values of the proportion in the general population in terms of their samples, but the TMD sufferers commonly occupied a considerate percentage [3].

The pathology of TMD remains to be unclear yet, but more than one diagnosis is likely to blame for mandibular hypomobility that can be classified into musculoskeletal and neuromuscular causes, depending on the dysfunction that takes place. The jaw mobility loss due to musculoskeletal reasons usually occurs on the joint, accompanied with joint ankylosis, resulting from external wound [4]. The neuromuscular reasons on the other often associate with neurologic abnormalities in the area of central nervous system (CNS) and brainstem related to masticatory muscles which

---

* Corresponding author.

J.-H. Kim et al. (eds.), *Robot Intelligence Technology and Applications 2*,
Advances in Intelligent Systems and Computing 274,
DOI: 10.1007/978-3-319-05582-4_68, © Springer International Publishing Switzerland 2014

lose the functional control from the upper level and then go into spasm and trismus [5].

Apart from the medication that traditionally has been adopted in treatment, such as botulinum toxin for alleviating trismus following stroke, the physical exercise has gained more support of usage in an appropriate way to improve the muscular conditions, in light of the definite effectiveness to relieve the pain and restore oral function [6, 7]. The physical training that is task-oriented in repetitive delivery can prevent the adhesion formation in the surrounding tissue and enhance local metabolism in the joint in the musculoskeletal level [8], while on the other hand can stimulate healthy cortical cells to establish control of the disabled parts of body that lost original connection from impaired one in the central nervous system through the reorganization process, achieving rehabilitative effect in neuromuscular level [9].

Significant efficacy the exercise has been revealed in restoration of the jaw opening range and the pain mitigation amid clinical trials, bringing with the enhancement for quality of life in terms of the functional recovery of the mouth [10]; since the manual training which is labor-intensive appears to be hardly upon implementation, the robotic device for jaw exercise which is uncommonly found has been called in development. TheraBite plus WY-5 dental robot is barely the representative devices in the existence designed to provide stretching on the masticatory muscles via external forces; the former is strictly a hand-held man-powered device that traces anatomical-resemblance trajectories on the incisor, while the other takes advantage of a six-DOF (degree of freedom) parallel mechanism to achieve a complete reproduction of the range of motion as human [11, 12].

While the existing devices were dedicated to their own objectives, the simple structures or the full range of motion, respectively; this paper presents the description of the mandibular movement on the sagittal plane from investigation of the workspace of the temporomandibular joint (TMJ), which establishes the justification of the conceived device in delivering the jaw movement, trading-off the structure to shrink the weight and the range of motion to meet every individual. Though the jaw movement has been acquired on specific points in numerous studies via three-dimensional (3D) recording, unfortunately no substantial data were available for analysis of the TMJ motion pattern in exactly numerical representation. The live recording is hereby preferably carried out to derive the normal movement by correlating the trajectories of two points on the jaw.

The concept of the jaw exoskeleton is then proposed in this paper to accommodate the therapeutic requirement such as the long duration and repetitive session, which is implemented by a simple linkage that fully exhibits the process of the development.

2   Justification of the Jaw Movement

TMJ is the two-sided articulation surrounded between the fossa of the temporal bone and the head of the mandible, forming the superior and the inferior bound of TMJ, respectively [13]; the translation of the mandible against the articular fossa at the condyles plus the rotation composes of six-DOF mandibular movement in 3D space.

Basically the jaw movement on the condylar point does not stick along an invariant trajectory. The workspace of the jaw exoskeleton movement relies on the chief functional implementation of the jaw, which does not have to be identical with the complete one.

Ostry, Vatikiotis-Bateson [14] indicated in their studies that the sagittal plane motion is independently varying from other DOFs, as the movement necessarily critical for daily functionality; additionally, the restricted mouth opening range patients mostly suffer from impairing the oral functions implicates the importance of the sagittal movement. The sagittal plane is appreciably referred to be the primary functional plane for the mandibular movement, within which is hereby to be focused on the movement delivery by the jaw exoskeleton.

Each cycle of mandible open-close may have the ability to exhibit a little variation in the normal unloaded motion when evaluated by referencing the incisor, the trajectories however can be found to fall within a small narrow band on the sagittal plane. In fact, during a single cycle, masticatory muscles can be considered in recruitment of performing a definite jaw motion, which is regulated by the neurologic system for specific purpose, behaving to possess one DOF featured by a pair of the trajectories on two points according to the analysis. Additionally from the perspective of the therapy provision, the exercise is preferably taken with the periodical stretch training in a certain fixed manner for a relative long duration, as one protocol before switching to another in order to establish the control in the neuromuscular level or maintain the joint ROM in the musculoskeletal level, instead of keeping altering the pattern in one motion cycle.

Considering the device is expected to drive the mandible repetitively to follow the same trajectories as those at enabled state, a pair of trajectories that pilots the jaw to move along a series of exact position is specified as the first paramount requirement to replicate the jaw movement orientated to a single exercise session by the exoskeleton; and both trajectories on the incisal point (IP) and condylar point (CP) that lay into their own family of trajectories are herein selected to construct a typical pair for each individual usage whose relative correlation is established based on the inherent motion rhythm. A vast range of training choices in terms of the guiding traces can be achieved by configuring up adjustable links that are broadly applicable in the wearable exoskeletons to adapt the population.

2.1   Live Recording

Focused on the sagittal plane, the CP may be involved into three different types of movement, i.e. translation along the fossa and vertically away from the fossa, and the rotation; the travelling distance on the jaw facilitates to reveal the inherent pattern which is likely to exist in the way of the correlations inside TMJ concerning the jaw movement. The relationship among them that has been investigated in the literature is given in mean values which are so averaged across the whole trajectory on the IP with respect to that on the CP that only outlining a general mandibular movement trace; but each proposed ratio can be finely valued as distributed into separate phases covering the overall travelling distance. Theoretically, with a set of definite values of the correlation, the mandible can be exactly positioned in space.

Live recordings capable to provide the trajectories under the synchronous time frame are herein preferable to correlate the instantaneous position. Tracking the jaw movement on the incisor takes advantage of an Articulograph AG500 on the subject, which can record tracking positions of 0.1mm precision within a 300mm-radius cubic space for inside the mouth measurement. Since the condylar point is not supported in non-invasive accessibility directly, as an alternative way, two more points on the molars that could define a spatial position of a rigid-body plus the incisor are recruited in trajectory tracking for further computation of the joint variables.

The subject[1] has a full dentition and did not reported any history of TMD symptoms from a questionnaire-based survey; the size of the jaw was obtained by calculating the relative distances recorded on the incisor, the left and the right ear (assumed to be both sides of condyles) respectively by the machine; the length was processed into the same tip frame used in this context, where the anterior-posterior length was 89 mm and the superior-inferior length was 45 mm. In recording trials, the subject was sitting in the chair while maintaining an upright head position, while being ordered to relax the jaw and open the mouth naturally as usual till up the limit. This session was repeated for three times; the mean ratio is then averaged over the values of the corresponding item; note the statistical analysis in terms of standard deviation on the same subject is not carried out.

The recorded data present in numerical form are given in absolute coordinates of the moving sensors under the AG machine frame, and should be converted into the same representation under the subject frame before illustrating the composed trajectories. For word limit, the overall transformation is not unfolded here.

2.2   Movement Pattern

A string of positions of the IP and CP that assemble each trajectory can be matched together to unveil the inherent rhythm in a habitual depression-elevation, and they are the other essential component to sketch the continuous movement of the jaw on the sagittal plane. The TMJ rhythm, which should be built on inspection of a considerably large pool of participants, is not elaborately derived in this paper; instead, the typical trajectories on the CP and IP either of which respectively is representative of the movement of the jaw and the TMJ, are acquired through different in-vivo measurements, as shown in Fig. 1, where the lines linking the trajectories represents position under the same time frame.

3   The Mechanism Design

The four-bar linkage as the simplest mechanism intrinsically provides one-DOF repetitive movement to fulfill the replication of the trajectories on a coupler point, so the four-bar linkage is given an attempt to deliver the movement. A family of the IP trajectories is attainable by tracing the coupler point of the linkage with the adjustable

---

[1] Informed consent was obtained from the subject.

length of the links since the single-DOF four-bar linkage with fixed length of the links only lead to one set trajectory. The choice of the four-bar linkage implicitly carries two important pieces of information: manually adjustable links can be settled in more than one link in the linkage to enhance applicability for individual user; and the IP trajectory generated by the single-DOF linkage is totally independent of the control algorithm, based on which variable rehabilitation regimes are established via regulating the frequency, the duration and the resistance.

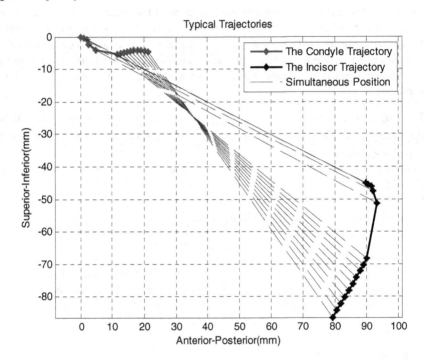

Fig. 1. Typical trajectories of CP and IP

The mandible for training is placed to contact with the linkage exactly on the coupler to do the rigid-body guidance movement. The accurate position of the links and the orientation of the linkage are precisely calculated in the process of approximate synthesis based on the numerical coordinates of the trajectories, which set the basis for motion generation of the four-bar linkage. To adapt the synthesis process, the CP at the rest position establishes the origin of the coordinate frame for this application, and the size of the jaw for initializing the IP coordinates applies the one referred in section, whose details can be acquired in great precision. Accordingly the initial position of the incisor can be written in the format of $[89.51, -45.01]$ with respect to the condyle of $[0,0]$.

The $[x_{Pj}, y_{Pj}]$ and θ_{1j} as known parameters are denoted as the coordinates of the coupler in a natural coordinate system and each corresponding rotated angle with respect to the initial position, respectively. The coordinates of the condyle and incisor that are travelling along each trajectory are respectively denoted as $[CP\_x_j, CP\_y_j]$ and $[IP\_x_j, IP\_y_j]$ at each position $[j]$, sequentially starting from the closed state of the jaw. The IP coordinates are invoked in the optimization equations to replace $[x_{Pj}, y_{Pj}]$; and the incremental rotation angles each step of which is denoted as $q_{j1} = q_j - q_1$, calculated from the $q_j = \frac{IP\_y_j - CP\_y_j}{IP\_x_j - CP\_x_j}$ and $q_1 = \frac{IP\_y_1 - CP\_y_1}{IP\_x_1 - CP\_x_1}$ are invoked to substitute all of the corresponding θ_{1j}.

Typical performances of the mechanism in terms of the mechanical system rely on its kinematic parameters, which form the constraints directly or indirectly with respect to the synthesis variables. The constraints the mechanism is subject to encircle a set of ranges for each variable in seeking a desirable combination that satisfies the requirements, among them the commonly referred ones are listed up in the aspects of the construction.

1. The length of each link L_{crank}, $L_{coupler}$ is restricted within a range of [5,20] for the crank, and [5,60] for the other, and the relation in terms of the length $L_{coupler} > L_{crank}$;
2. Rotational angle of the crank α over all prescribed points is limited within [0,180];
3. θ_{1j} is arranged in sequence of the points concerning j;
4. Structural error $|\varepsilon|$ is specified within a range.

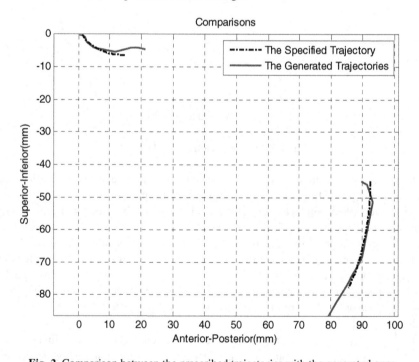

Fig. 2. Comparison between the prescribed trajectories with the generated ones

The dimensioning of all links in the linkage is given as follows: Crank *10mm*, Ground *47mm*, Follower *16mm*, Coupler *30mm*, and the coupler, *32mm*. The original pair of trajectories derived from the in-vivo recording is then compared with the one generated subject to the four-bar linkage which comes from those of two points on the coupler of the linkage maintaining relative position, as shown in Fig. 2. It can be seen that the achieved trajectory is a little scaled-down but basically matches the pattern point by point. The sum of squared residual at this solution is calculated amid the optimization.

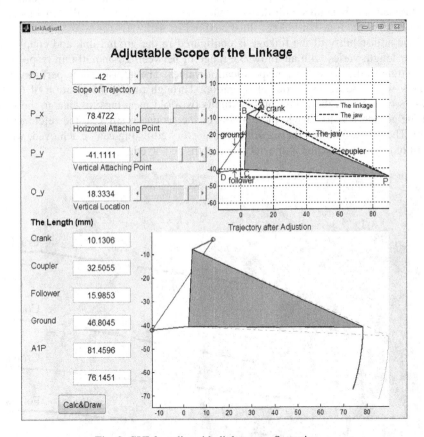

Fig. 3. GUI for adjustable linkage configuration

The family of the IP trajectories can be generated within a range of each link, apart from the crank which is difficult to implement in practice, the other links especially the coupler point that attaches to the jaw are configured adjustable within a bound, the ground is given 20mm range, while the attaching part on the coupler is given 30mm on both the anterior-posterior direction and superior-inferior direction. A GUI program is coded in MATLAB illustrated in Fig. 3, allowing tunable wearing position in several places for prospective users as a chart reference. It is noticed that the range of the motion on the incisor can not only be adjusted, but the shape of mouth moving can also be regulated, which proves the adjustable linkage is able to replicate a group of IP trajectories. Therefore, the attachment to the jaw in the device can also be adjusted through tuning the position of the attachment.

Given the aforementioned dimensions, the mechanical design of the linkage is implemented into being. In the course of designing, the basic requirements about the device's wearability and adjustability are referred to. For this first prototype, the parts of aluminum are made as compact and light as possible. For the adjusting screws and the guide-bolts as well as for the pins, stainless steel is chosen. The bushings are made out of acetal. The linkage contains simple links, which of adjacencies are connected by the pins and bushings with flanged outer ring on the joints, and the final mechanical design of the four-bar linkage illustrated in Fig. 4.

The adjustability of the linkage is configured in the ground link and coupler link whose length varies with an allowable range of between 20 and 30 mm respectively, according to the above design specification values. The adjustment is performed manually when the device is out of operation. Through tuning the position of the two blocks on the slot, the adjustability exhibits two-fold meanings: to change the length of the ground and to change the position of the ground relative to the reference system. Therefore, not only the size-related trajectory adjusting can be achieved, will the shift of the trajectory due to the varying lengths be solved.

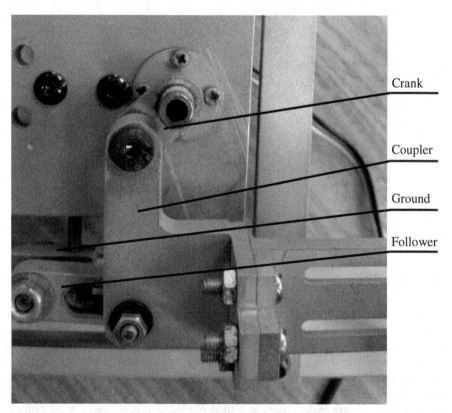

Fig. 4. Assembled four-bar linkage

4   Discussion and Conclusion

The jaw exoskeleton was firstly proposed in this paper with the intention of TMD treatment based on a comprehensive literature review of the rationales of the rehabilitation. Live recording of the jaw movement were taken on the healthy subjects intended to derive the TMJ motion pattern. The preliminary study was only focused on the primary motion of the jaw as a planar movement, forming the design specification in the aspect of the kinematics, which was fulfilled with a simple mechanism in order to shrink the weight. The fully described process of the mechanism design and the construction of the prototype reflected the concept of the adjustability and the applicability across a wide range of users. The GUI of configuration of the adjustable links and the corresponding trajectories on the incisor were plotted for the convenience adjusting.

The mandibular movement is rather complicated, on one hand owing to the particular construction of the TMJ. So far, the jaw movement has not been investigated by introducing the concept of motion rhythm to disclose the inherent pattern relative to the TMJ, which may be explained by large variation existing individually; but a general pattern is likely to exist amid a certain percentage of the population. The pattern has not been derived in this paper, since a typical pair of trajectories is adequate to define the exact instantaneous position of the jaw during a normal depression-elevation, and the authors considered a rather small pool of samples was too speculative to define the anatomical concept in biology.

For safety consideration, the generated trajectories on both concerned points on the mandible can meet the specified set, even in different configurations to adapt sizes and shapes, which can be achieved tuning the adjustable links. However, the potential detriment to the wearer was not given an outlook in terms of the inside TMJ, which is the most vulnerable spot. The original motion pattern amid adjusting the links to fit the trajectory on IP is likely to be altered from the original correlation between the IP and the CP that was specified through approximate synthesis, possibly leading to a hidden point to pose any impairment on the TMJ. The performance of the four-bar linkage based device was not given in detail, especially the control of the device to accomplish the training protocol, which is expected with a full statement in the future research.

Acknowledgement. The work presented in this paper was supported by New Zealand Foundation of Research, Science and Technology, under the contract MAUX0809. The authors would like to express their gratitude to the IFNHH, Massey University for kindly assisting the in-vivo measurement with the equipment.

References

1. Blasberg, B., Greenberg, M.: Temporomandibular Disorders. In: Burket's Oral Medicine Diagnosis and Treatment, pp. 271–306 (2003)
2. Costello, B., Edwards, S.: Pediatric mandibular hypomobility: current management and controversies. Oral and Maxillofacial Surgery Clinics of North America 17(4), 455–466 (2005)

3. Lehman-Grimes, S.P.: A review of temporomandibular disorder and an analysis of mandibular motion, The University of Tennessee (2005)
4. Fonseca, R.J.: Oral and Maxillofacial Surgery 2: Orthognathic Surgery, vol. 2. Elsevier Science Health Science Division (2000)
5. Tveteras, K., Kristensen, S.: REVIEW The aetiology and pathogenesis of trismus. Clinical Otolaryngology 11(5), 383–387 (1986)
6. Feine, J.S., Widmer, C.G., Lund, J.P.: Physical therapy: A critique. Oral Surgery, Oral Medicine, Oral Pathology, Oral Radiology, and Endodontics 83(1), 123–127 (1997)
7. McNeely, M.L., Armijo Olivo, S., Magee, D.J.: A Systematic Review of the Effectiveness of Physical Therapy Interventions for Temporomandibular Disorders. Physical Therapy 86(5), 710–725 (2006)
8. O'Driscoll, S.W., Giori, N.J.: Continuous passive motion (CPM): theory and principles of clinical application. Journal of Rehabilitation Research and Development 37(2), 179–188 (2000)
9. Krakauer, J.W.: Motor learning: its relevance to stroke recovery and neurorehabilitation. Current Opinion in Neurology 19(1), 84–90 (2006)
10. Cohen, E., et al.: Early use of a mechanical stretching device to improve mandibular mobility after composite resection: a pilot study. Archives of Physical Medicine and Rehabilitation 86(7), 1416–1419 (2005)
11. Takanobu, H., et al.: Mouth opening and closing training with 6-DOF parallel robot. In: Proceedings of the IEEE International Conference on Robotics and Automation, ICRA 2000 (2000)
12. Lin, C., Kuo, Y., Lo, L.: Design, Manufacture and Clinical Evaluation of a New TMJ Exerciser. Biomedical Engineering Applications Basis Communications 17(3), 135 (2005)
13. Drake, R., Vogl, A.W., Mitchell, A.W.: Gray's anatomy for students. Churchill Livingstone (2009)
14. Ostry, D., Vatikiotis-Bateson, E., Gribble, P.: An examination of the degrees of freedom of human jaw motion in speech and mastication. Journal of Speech, Language and Hearing Research 40(6), 1341–1351 (1997)

Towards a Touchless Master Console for Natural Interactions in Sterilized and Cognitive Robotic Surgery Environments

Cai Lin Chua[1], Hongliang Ren[1,*], Wei Zhang[2]

[1] Department of Biomedical Engineering
National University of Singapore, Singapore
[2] School of Control Science and Engineering
Shandong University, Jinan, China
ren@nus.edu.sg

Abstract. With the increasing benefits of minimally invasive surgery (MIS), computer assisted surgical systems are widely used to handle the challenges of sterilization, augmented reality, intracorporeal accessibility, dexterity in confined space and precision demanded in MIS. In conventional surgical environments, the sterilized objects in the surgical space are manipulated either manually or by tele-operated master-and-slave system. One of the challenges faced in this kind of manipulation is the physical controls can present risks of non-sterile contact for surgeons in a highly sterilized operating room. Gesture recognition has been widely researched as a Natural User Interface (NUI) for human-computer and human-robot interactions. This paper focuses on assessing the feasibility of using skeletal tracking as a touchless control interface for the master consoles of master-and-slave surgical robotic systems. A demonstration system that uses one RGB-D sensor is set up for the purpose of the assessment.

Keywords: computer assisted surgery, touchless interface, cognitive control, sterility, RGBD.

1 Introduction

Currently, minimally invasive surgery (MIS) is gaining popularity over traditional open surgery due to the medical benefits it brings to the patients, such as minimized trauma experienced and faster recovery [1]. Due to the nature of MIS where only small incisions are made on the patients, small and delicate surgical instruments have to be used in MIS. There have been demands for natural control, sterility, precise manipulation during physical clinician-machine interactions. Computer assisted surgical systems are emerging for precision [2], safety and efficacy [3] in a confined space inaccessible by conventional means. The commonly used tele-operated master and slave surgical robot [3], such as the *da Vinci* Surgical System®, used physical master console to control the instruments mounted on the slave robot.

---

* Corresponding author.

J.-H. Kim et al. (eds.), *Robot Intelligence Technology and Applications 2,*
Advances in Intelligent Systems and Computing 274,
DOI: 10.1007/978-3-319-05582-4_69, © Springer International Publishing Switzerland 2014

Improvements to the master and slave surgical robot systems have been made over the years. There are demands for natural control methods that do not restrict the surgeon's movements and can provide opportunity for sterility, and for intuitive systems that make user training rapid [3]. Today, the commercially available surgical robots require the surgeons to operate the master console through sophisticated physical buttons and controllers [4]. These touch-based interfaces may cause disruption to the surgeon's workflow and cause the surgeon to get in contact with non-sterile objects [5]. Thus, there is an area of interest in exploring touchless input methods for the master console, since touchless interfaces are able to provide the needed control features in the sterilized environment [6]. Gesture recognition, especially, is recognized as a possible touchless input modality for surgical robots [4].

Gesture recognition has been studied for various uses under surgical settings, such as image visualization in surgery [6], supervision of an operating room [7], control of the operating room lights [5], and alternatives to traditional menu interaction in medical robotics [4]. The application of gesture recognition in those researches has yield promising results. In general, advantages of using human body gestures for human-computer interaction include easier learning of the system for novices [6], more favorable user experience [4], and less cognitive effort needed to remember intuitive gestures [4].

In this paper, the application of gesture recognition as a touchless control modality is extended to the master console. It can be utilized for manipulating 3D images remotely, controlling the slave robot, or pose adjustments of overhead lights or more specific manipulations. For demonstration purposes, gesture recognition will be achieved through the skeletal tracking capability of the RGBD sensor.

2   Materials and Methods

The demonstration is set up according to Fig. 1 and makes use of the hardware and software components listed in Table 1.

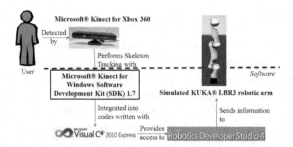

Fig. 1. The hardware and software components of the demonstration system

Table 1. Details of the main hardware and software components used in the demonstration system

Component	Details
Desktop computer	Windows 7 OS, 64-bit, with Intel® Core™ i5-3570 CPU processor at 3.40 GHz
Microsoft® Kinect RGB-D sensor	RGB-D camera, launched in 2010 [8], with an input stream rate of 30 frames per second [9]
Microsoft® Kinect for Windows Software Developement Kit (SDK)	Version 1.4, released in March 2013 [10]
Microsoft® Robotic Developer Studio (RDS)	Version 4, released in June 2012 [11]

2.1   RGB-D Sensor

The Kinect sensor contains a RGB colour and a depth camera that provide colour and depth information respectively [8]. Both colour and depth information are streamed at a rate of 30 frames per second (FPS) [9]. The depth sensor consists of an infrared (IR) laser projector and a monochrome complementary metal oxide semiconductor (CMOS) IR camera [8]. The depth-sensing technology is based on structured light principle [8], which makes depth-sensing possible by coding the scene with IR dots emitted from an IR projector and deciphering the reflected coded light using an IR camera [12].

The Kinect sensor contains a skeletal tracking engine that is able to identify and actively track 20 different joint points of up to two users at any one time [13]. The Kinect sensor's skeletal tracking method uses single depth image with no temporal data to predict the 3D positions of body joints [14]. Using an object recognition approach, the skeletal tracking method first represents the human pose depth data as an intermediate body parts representation, and finally as a per-pixel classification [14]. This skeletal tracking method predicts joint positions accurately with an overall of 0.731 mean Average Precision (mAP) [14]. The x, y, and z 3D spatial coordinates of each tracked joint point can then be accessed.

The elbow angle is calculated between the shoulder, elbow, and wrist joint points of a user as tracked by the Kinect sensor. The x and y coordinates of these three specific joint points are retrieved, and the shoulder-to-elbow, elbow-to-wrist and shoulder-to-wrist 2D vectors are obtained using vector subtraction. The magnitude of each 2D vector is calculated, and using Cosine Rule, the elbow angle, θ, is obtained from the magnitude of shoulder-to-wrist vector, c, magnitude of shoulder-to-elbow vector, a, and magnitude of elbow-to-wrist vector, b:

$$c^2 = a + b - 2ab\cos\theta \tag{1}$$

2.2   Kinect for Windows SDK

Kinect for Windows SDK allows access to the colour, depth and skeleton data collected by the Kinect sensor. The SDK uses a smoothing method, based on a combination of double exponential smoothing, jitter removal, and overshoot control, to remove the inherent noises present in raw joint position data obtained from the skeletal tracking system [15]. Holt double exponential smoothing method involves a second exponential filter that accounts for trends in the data received, while jitter removal limits changes in the output of each frame in order to dampen the spikes present in the input [15]. Equation (2) shows a more commonly used formulation of the Holt double exponential smoothing filter, where γ is the weight given to the last two filtered outputs $\hat{X}_n$ and $\hat{X}_{n-1}$ in calculating the trend, and $\hat{X}_n$ is the raw input at time t [15].

$$\text{Trend: } b_n = \gamma(\hat{X}_n - \hat{X}_{n-1}) + (1 - \gamma)b_{n-1} \tag{2a}$$

$$\text{Filtered Output: } \hat{X}_n = \alpha\hat{X}_n + (1-\alpha)(\hat{X}_{n-1} + b_{n-1}) \tag{2b}$$

The extent of smoothing applied to the skeleton joint points can be changed by changing the value of the following five parameters [16] shown in Table 2.

Table 2. The five parameters that control the smoothing applied to the tracked skeleton joints

Name of Parameter	Description of Parameter
Smoothing	Smoothing value that ranges from 0.0 to 1.0 meters, with higher values giving smoother output but at higher latency.
Correction	Correction value that ranges from 0.0 to 1.0 meters, with lower values correcting toward raw values more slowly and therefore, giving a smoother output.
Prediction	The number of frames, ranging from 0 onward, to predict into.
JitterRadius	Jitter-reduction radius, ranging from 0 meters onward, that controls the intensity of jitter removal applied to the raw data.
MaxDeviationRadius	The maximum radius, ranging from 0 meters onward, that a filtered position can deviate from the raw data.

As surgical robotics places utmost emphasis on precision [2] and safety [3], it is undesirable for the calculated elbow angle of a stationary pose to vary frequently over time. Thus, the skeleton tracked must be smoothed with tolerable latency to keep such variations to a minimum. In order to ensure optimal performance for the demonstration setup, the values shown in Table 3 are used.

Table 3. Values used for each parameter in the demonstration setup

Name of Parameter	Value used for Demonstration System
Smoothing	0.75
Correction	0.25
Prediction	0.50
JitterRadius	0.10
MaxDeviationRadius	0.02

In addition, the average calculated elbow angle over 60 consecutive frames, instead of the calculated elbow angle from each frame, will be used as input to the simulated arm. This would ensure that any minimal variations of calculated elbow angle present could be further smoothed out over time.

2.3   Robotic Developer Studio

RDS includes Visual Simulation Environment (VSE) that supports simulations for robots without the need for the actual robotic hardware [17] [18]. For the demonstration system, the simulated KUKA® LBR3 robotic arm, a six-degrees-of-freedom (DOF) robotic arm, is used. However, for simplicity, only one of the joints, namely Joint 3, will be used for the demonstration. Joint 3 is automatically initialized to be free only for twisting and has spring coefficient and damper coefficient of 500 000 and 100 000 respectively.

In order to achieve a more intuitive control of the simulated robotic arm, mapping of elbow angle to position of robotic arm is done (Fig. 2). Particularly, since the tracked skeleton is mirrored by default [19], different calculations for the mapped angle, θ_{LBR3}, have to be done for the calculated elbow angles, θ_{elbow}, from the different arms tracked. Equation (3a) is used for the right arm whereas (3b) is used for the left arm. In this way, the simulated arm will mimic the user's arm positions, thereby allowing for a more intuitive control.

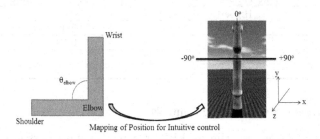

Fig. 2. Overall idea of the mapping of user's arm position to simulated robotic arm position using θ_{elbow}, the calculated elbow angle from the tracked user

$$\theta_{LBR3} = \theta_{elbow} - 90° \tag{3a}$$

$$\theta_{LBR3} = - (\theta_{elbow} - 90°) \tag{3b}$$

During initial trials of the demonstration system when the elbow angle is mapped to the input angle for LBR3 robotic arm in a 1° to 1° manner, it is found that the 1° angle changes will result in uncontrolled swings of the simulated arm. This behavior is extremely dangerous and an elbow angle segmentation and manual angle correspondence method, as described below, is implemented to solve this problem.

The largest possible elbow angle that can be obtained from either arm is 180°, and according to (3), this implies that the range of possible mapped angle θ_{LBR3} is [-90°, 90°]. This range of mapped angle θ_{LBR3} is segmented into groups of 10°, centering at 0°. Each group is then corresponded to a specific angle to be input to the simulated robotic arm (Fig. 3). As this method inevitably restraints the number of possible positions of the robotic arm, a fine tuning option is added to change the extent of the restraint in position as needed (Table 4).

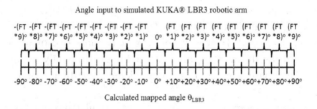

Fig. 3. Segmentation of calculated mapped angle θ_{LBR3} into groups of 10o, centering at 0°, and correspondence of each group to a specific angle that will serve as the input angle for Joint 3 of the simulated robotic arm. FT refers to the Fine Tuning value applied (Table 4).

Table 4. Fine Tuning (FT) values set for the simulation and their associated fine tuning level names. FT values are used in the manual angle correspondence step

Fine Tuning Level	Fine Tuning (FT) value
Highest	0.01
High	0.05
Medium	0.10
Low	0.15
Lowest	0.20

2.4   Start and Stop Tracking Controls

For total control over the simulated LBR3 robotic arm, it is necessary to implement a method that recognizes when a user wants to start moving the robotic arm and when the user wants to stop moving it. For the demonstration system, pose recognition is implemented as the start and stop controls. A user posing with either the left or right arm in a 90° position at shoulder level, and the other hand stretched downward such that the palm is below the hip level (Fig. 4a), will signal to the simulation program to start tracking that particular raised arm. If the user raises both arms in 90° positions at shoulder level, forming a psi (ψ) shape (Fig. 4b), the simulated robotic arm will stop responding to any further movements of the user's arm, until the user sends the start signal again.

Due to the similar conditions in defining the start and stop poses, it is necessary to include a three seconds time lag between the detection of stop pose and the check for start pose. This time lag is included as an added safety measure in case the stop pose is mistaken by the program to be one of the start poses.

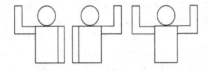

Fig. 4. (a. left) The two possible poses to signal the simulation program to start tracking the raised arm. (b. right) The stop pose that signals the simulation program to stop tracking the arm.

3   Results and Discussion

3.1   The Demonstration System

The demonstration system was tested under a controlled environment such that only the subject, and no other people, is present in the field of view of the Kinect sensor's depth camera. Under such an environment, the demonstration system is shown to be able to interpret and respond to the start and stop poses of the subject correctly as expected. In addition, the direction of movement of the simulated robotic arm mimics the direction of movement of the subject's arm, giving an intuitive control of the robotic arm movement. The different fine tuning levels applied also show visible differences in the extent of the movement of the simulated LBR3 arm, thereby showing the possibility of a control for fine tuning the movement of the robotic arm. This fine tuning ability will be significant for achieving more precise positioning of the tip of the robotic arm.

3.2   Assessment of Demonstration System

1) Variation Changes with Skeleton Smoothing

For the dependency experiment, an one-tailed two-sample F-test for variances is used to test the null hypothesis H_0, the two variances have no significant difference, against the alternative hypothesis H_1, that the variance for the second execution is significantly smaller than for the first execution of the program. The results of the dependency are tabulated in Table 5.

Table 5. Results of one-tailed two-sample F-test, at significance level of 0.05, for the dependency experiment

Skeleton Smoothing Parameters used		p-value	95% Confidence Interval
First Execution	*Second Execution*		
Default	Default	0.1716	[0.8898, Infinity]
Improved	Improved	1.854e-14	[2.8048, Infinity]

The results of the dependency experiment shows that at a significance level of 0.05, the variance of the second data set collected does not differ significantly from the first data set collected when the first module executed uses default skeleton smoothing parameters. Thus, for the experiment in investigating whether the improved skeleton smoothing parameters do significantly decrease the variances in the calculated elbow angle, the module with default skeleton smoothing parameters is executed first before the module with improved skeleton smoothing parameters.

The one-tailed two-sample F-test is again used to analyze the data collected for the main experiment, with the same null and alternative hypotheses as used in the dependency experiment. The two-sample F-test gives a p-value of less than 2.2e-16, with a 95% confidence interval of [7.1684, Infinity]. This concludes that the module with improved skeleton smoothing parameters gives calculated elbow angle values that have significantly lower variance than the module with default skeleton smoothing parameters does. This means that the joint position values and calculated elbow angle output from the module with improved skeleton smoothing parameters are significantly smoother. Since the calculated elbow angle is used to control the arm of a surgical robot, as demonstrated in the demonstration system, there is thus a need to use improved skeleton smoothing parameters in order to ensure safety if such skeletal tracking-based system is to be used under actual surgical settings.

2) Accuracy of Calculated Elbow Angle

The mean calculated elbow angle from the Kinect sensor, the ground truth elbow angle value determined by the Polaris® optical tracking system, and the percentage error are presented in Table 6.

Table 6. Results of the accuracy test conducted on the mean elbow angle obtained from the skeletal tracking of one subject using the Kinect sensor

Arm	Position	Mean Elbow Angle (°)		Percentage Error (%)
		Calculated	*Ground Truth*	
Left	180°	177.7626	177.8610	0.0553
	90°	98.3626	97.0572	1.3450
Right	180°	178.8193	179.9638	0.6360
	90°	107.7010	100.0878	7.6065

From the results, it can be seen that for this particular subject, the percentage errors for the $180°$ arm position are generally much lower than that for the $90°$ arm position. The higher percentage errors observed in $90°$ arm position may be due to the limited ability of the skeletal tracking algorithm developed by Microsoft® in determining the precise position of the shoulder joint point from the depth images. If this is so, it would imply that the accuracy of determination of the various joint points varies with different poses and positions of the subject. However, further work has to be done in order to test this hypothesis as well as to find out the cause of the greater percentage errors observed for the $90°$ arm positions in this experiment.

4   Conclusion

A demonstration system for evaluating the feasibility and potential of using skeletal tracking as a touchless control for a simulated robotic arm has been set up. In general, based on the qualitative observations of the usage of the demonstration system, the response of the demonstration system is shown to be highly uniform across the three different volunteer subjects. The demonstration system also behaves accordingly to the tasks defined in the module, such as the criteria for recognizing the subject's start and stop poses. More stable response of the simulated arm has been achieved by introducing the segmentation and correspondence method. However, the quantitative results from the demonstration system show a need for further improvements in the data filtering and skeleton joints determination. These improvements are needed in order for such a touchless control interface to be accurate and precise enough to meet the demands in surgical environments.

Overall, the demonstration system presented showed that the skeletal tracking ability of the Kinect sensor can be utilized to control the movement of a simulated robotic arm in an intuitive manner. As the presented master-console is touchless, it is possible to control the robotic arm's movements without a need for physically interactive master-device that can potentially present non-sterility in the OR that demands sterility. Although the control methods, accuracy and precision of this demonstration system need to be further improved to meet the demands in MIS, the demonstration system has shown an intuitive and touchless console for in a sterile and cognitive surgical environment.

Acknowledgment. This research is supported by Singapore Academic Research Fund, under Grants R397000139133, R397000157112 and NUS Teaching Enhancement Grant C397000039001, awarded to Hongliang REN. Ms. Cai Lin Chua carried out in-part as the Undergraduate Research Opportunities Programme (UROP) initiated by the Faculty of Engineering. We would like to thank Dr Alberto Corrias, Mr Aneel Kumar, Mr Matthew Tham, Mr Wei Liu and Mr Andy Lim for their guidance, advice and logistical help given throughout the project.

References

[1] Fiorini, P.: Robotic surgery: Past results and current developments. In: Advanced Technologies for Enhanced Quality of Life, AT-EQUAL 2009, p. 15 (2009)

[2] Mack, M.: Minimally invasive and robotic surgery. Journal of the American Medical Association 285, 568–572 (2001)

[3] Simorov, A., Otte, R., Kopietz, C., Oleynikov, D.: Review of surgical robotics user interface: what is the best way to control robotic surgery? 26(8), 2117–2125 (2012), http://dx.doi.org/10.1007/s00464-012-2182-y

[4] Staub, C., Can, S., Knoll, A., Nitsch, V., Karl, I., Farber, B.: Implementation and evaluation of a gesture-based input method in robotic surgery. In: 2011 IEEE International Workshop on Haptic Audio Visual Environments and Games (HAVE), pp. 1–7 (2011)

[5] Hartmann, F., Schlaefer, A.: Feasibility of touch-less control of operating room lights 8(2), 259–268 (2013), http://dx.doi.org/10.1007/s11548-012-0778-2

[6] Ruppert, G., Reis, L., Amorim, P., Moraes, T., Silva, J.: Touchless gesture user interface for interactive image visualization in urological surgery 30(5), 687–691 (2012), http://dx.doi.org/10.1007/s00345-012-0879-0

[7] Monnich, H., Nicolai, P., Beyl, T., Raczkowsky, J., Worn, H.: A supervision system for the intuitive usage of a telemanipulated surgical robotic setup. In: 2011 IEEE International Conference on Robotics and Biomimetics (ROBIO), pp. 449–454 (2011)

[8] Zhang, Z.: Microsoft kinect sensor and its effect. IEEE MultiMedia 19(2), 4–10 (2012)

[9] Microsoft Developer Network (MSDN) Library kinect for windows sensor components and specifications (2013), http://msdn.microsoft.com/en-us/library/jj131033.aspx (last accessed September 14, 2013)

[10] Microsoft Developer Network (MSDN) Library what's new (2013), http://msdn.microsoft.com/en-us/library/jj663803.aspx#SDK_1pt7 (last accessed September 14, 2013)

[11] Microsoft Download Center microsoft robotics developer studio 4 (2012), http://www.microsoft.com/en-us/download/details.aspx?id=29081 (last modified June 3, 2012)

[12] PrimeSense Ltd. our full 3d sensing solution (2013), http://www.primesense.com/solutions/technology/ (last accessed September 14, 2013)

[13] Microsoft Developer Network (MSDN) Library skeletal tracking (2013), http://msdn.microsoft.com/en-us/library/hh973074.aspx (last accessed September 14, 2013)

[14] Shotton, J., Fitzgibbon, A., Cook, M., Sharp, T., Finocchio, M., Moore, R., Kipman, A., Blake, A.: Real-time human pose recognition in parts from single depth images. In: 2011 IEEE Conference on Computer Vision and Pattern Recognition (CVPR), pp. 1297–1304 (2011)

[15] Azimi, M.: Skeletal joint smoothing white paper, Advanced Technology Group (ATG), Tech. Rep., http://msdn.microsoft.com/en-us/library/jj131429.aspx

[16] Microsoft Developer Network (MSDN) Library transformsmoothparameters properties (2013), http://msdn.microsoft.com/en-us/library/microsoft.kinect.transformsmoothparameters_properties.aspx (last accessed September 14, 2013)

[17] Morgan, S.: Overview of Robotics and Microsoft Robotics Studio. In: Programming Microsoft® Robotics Studio, pp. 1–18. Microsoft Press, United States of America (2008)

[18] Microsoft Developer Network (MSDN) Library simulated articulated entities (2013), http://msdn.microsoft.com/en-us/library/dd939191.aspx (last accessed September 14, 2013)

[19] Microsoft Developer Network (MSDN) Library coordinate spaces (2013), http://msdn.microsoft.com/en-us/library/hh973078.aspx (last accessed September 14, 2013)

[20] Webb, J., Ashley, J.: Kinect Skeletons. In: Beginning Kinect Programming with the Microsoft Kinect SDK. Apress Media LLC, New York (2012)

[21] Ryden, F.: Tech to the future: Making a "kinection" with haptic interaction. IEEE Potentials 31(3), 34–36 (2012)

Flexible and Wearable Hand Exoskeleton and Its Application to Computer Mouse

Woo-Young Go and Jong-Hwan Kim

Department of Electrical Engineering, KAIST
335 Gwahangno, Yuseong-gu, Daejeon, Republic of Korea
{wygo,johkim}@rit.kaist.ac.kr

Abstract. This paper proposes a flexible and wearable hand exoskeleton which can be used as a computer mouse. The hand exoskeleton is developed based on a new concept of wearable mouse. The wearable mouse, which consists of flexible bend sensor, accelerometer and bluetooth, is designed for comfortable and supple usage. To demonstrate the effectiveness of the proposed wearable mouse, experiments are carried out for mouse operation consisting of click, cursor movement and wireless communication. The experimental results show that our wearable mouse is more accurate than a standard mouse.

Keywords: wearable mouse, hand exoskeleton, computer pointing devices, cursor positioning, perceptual user interface.

1 Introduction

For most people who use computer, a mouse is assumed to be a standard device widely used. The concept of standard mouse is simple and useful to click and move the mouse point. However, there exist several inconvenient disadvantages. First of all, the standard mouse need bigger place than certain place for hand to grab. Secondly, it is under space constraints on the transparent or uneven space. Finally, It is impossible for our hand to use the keyboard and the mouse at the same time.

A lot of wearable or exoskeleton applications have been developed for its convenient reasons. Therefore, recently, in most of wearable or exoskeleton applications, those are on walking assistance for people with mobility problems or military purpose. Many types of wearable mouse were developed such as wearable wrist support for computer mouse [1], wearable computer pointing device [2] and wearable inertial mouse [3]. Although there have been various researches on wearable mouse, they are not convenient and have several problem when put to use.

The previous mouses have several problems. Thus, in this paper, wearable and flexible mouse is developed to solve those problems. The proposed mouse consists of flexible sensor to click, accelerometer to move the pointer and bluetooth for wireless communication with computer. On account of improved accuracy

(a) Board overview (b) Bluetooth (c) Accelerometer (d) Flexible sensor (e) ATMEGA8

Fig. 1. Wearable mosue main hardware components

and delicacy of this mouse, it can be applied in the domain of game, art and construction design.

This paper is organized as follows. Section 2 describes hardware design. In Section 3 and Section 4, performance property and technology of wearable mouse are described. An experiment result is discussed in Section 5. Finally, conclusion and further work are described in Section 6.

2   Hardware Design

2.1   Wearable Mouse Overview

The wearable mouse is composed of bi-directional flexible bend sensor, three axis low-g micromachined accelerometer, WLAN-bluetooth chip antenna and ATMEGA8-16PU in Fig. 1. The two bi-directional flexible bend sensors are placed on the forefinger and middle finger so that their movement is detected. It decreases resistance when it is bent or flexed in either direction. The use of three axis Low-g micromachined accelerometer allows the wearable mouse to sense movement of mouse pointer. WLAN-bluetooth chip antenna(w51-BF-RS) which is embedded in bluetooth module is used for wireless communication between the wearable mouse and the computer in order to give and take signals without USB. ATMEGA8-16PU which is 8-bit AVR with 8K bytes in-system programmable flash is used to program.

Fig. 2. Overall structure

2.2   Overall Structure Design

The mouse has three main functions, i.e., click, cursor movement and communication with computer. It is composed of buttons, photo sensor, and rotary encoder. However, it is under space constraints on the transparent or uneven space and is impossible for our right hand to use keyboard and mouse at the same time. Thats why, in this paper, wearable and flexible mouse is developed to solve those problems. In order for wearable mouse to have these essential mouse functions, flexible bend sensor, accelerometer and WLAN-bluetooth chip are used in Fig. 2. Flexible bend sensor is in charge of click instead of button of standard mouse. Accelerometer is used for cursor movement up and down instead of photo sensor of standard mouse. USB receiver is not used because WLAN-bluetooth chip can be operated for wireless communication with computer.

3   Performance Property

Table 1. Flexible sensor performance methods

	Flexible sensor performance methods
Right click	While both forefinger and middle fingers with flexible sensors are bent, bend and spread out middle finger
Left click	While both forefinger and middle fingers with flexible sensors are bent, bend and spread out forefinger
Scroll wheel up	While both forefinger and middle finger with flexible sensors are spread out, slightly incline right part of hand
Scroll wheel down	While both forefinger and middle finger with flexible sensors are spread out, slightly incline left part of hand

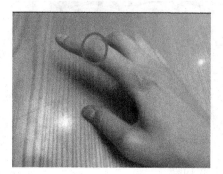 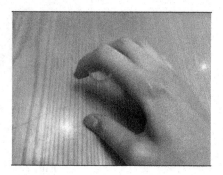

Fig. 3. Left click: like standard mouses left click, sensing whether left flexible sensor is bent or not is in charge of left click

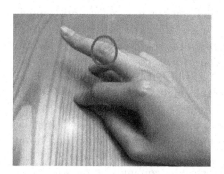 

Fig. 4. Right click: like standard mouses right click, sensing whether right flexible sensor is bent or not is in charge of right click

Fig. 5. Drag: like drag ofstandard mouse, sensing whether both flexible sensors are bent or not is in charge of drag

Table 2. Accelerometer sensor performance methods

Accelator sensor performance methods	
Cursor movement up	Hand move up
Cursor movement down	Hand move down

4 Technology of Wearable Mouse

Table 3. Advantages of flexible and wearable mouse

Problem of standard mouse	Strong point of wearable mouse
Portable problem	Wearable mouse has effective size and shape because it is attached on the hand
Space constraints	Transparent and uneven spaces do not affect the performance of the wearable mouse
Effectiveness problem	Wearable mouse is more convenient because it is possible for our hand to use keyboard and mouse at the same time
Inconvenience of touch pad and joystick	Wearable mouse is more familiar and practical by using only hand movement
Obvious fact	Wearable mouse is break from the normal design idea

5 Experimental Results

5.1 Experimental Setting

An experiment was carried out with both standard mouse and wearable mouse independently, in order to determine the accuracy of each device. For this purpose, the mouse tracker program was used to observe both trajectory pattern. The mouse tracker program was run 30 times for each device. And the average trajectory pattern result can be observed in Fig. 6. The experiment setting for the standard mouse was comprised with a commercially available laser mouse with its corresponding mouse pad.

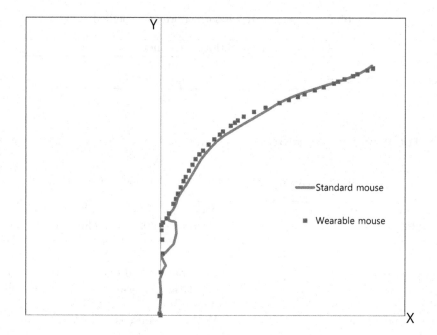

Fig. 6. Y versus X mean tragectory

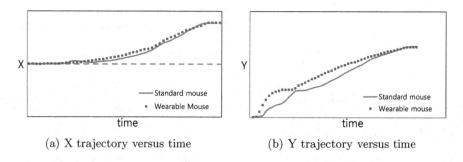

(a) X trajectory versus time   (b) Y trajectory versus time

Fig. 7. X and Y time-course

5.2   Experimental Results

One of the drawbacks of this setting is the existence of friction between the mouse and mouse pad and between the users arm and the desk, causing slight inaccuracies in the mouse trajectory. Even though during the experiment, both devices reached the goal point successfully, a big difference could be observed between both trajectory patterns. This phenomenon is due to the existence of friction, previously explained, in the standard mouse. Furthermore, for a standard mouse to operate correctly, it has to be in permanent contact with a smooth surface which cause friction, leading to accuracy errors if it is momentarily lifted from

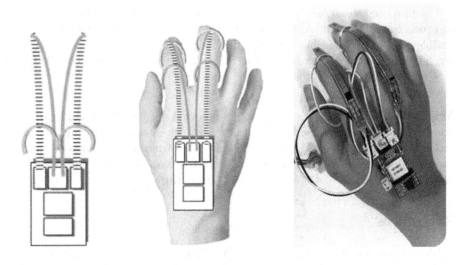

Fig. 8. Finished product

such a surface. However, when the wearable mouse is used, there are less accuracy errors compared to the standard mouse. In case of using wearable mouse, an accelerometer is used when the cursor move up and down. Thats why an accuracy of wearable mouse in the mouse trajectory is better because there do not exist friction between the mouse and mouse pad and between the users arm and the desk. The friction in y-axis is bigger than the friction in x-axis because the right arm, used to move the mouse, lies on y-axis in Fig. 7. Therefore, there is less x-axis error. On the other hand, the y-axis trajectory of standard mouse oscillates severely. This result shows us that wearable mouse with accelerometer instead of photo sensor has more accuracy. And click using flexible bend sensor is operated successfully.

6   Conclusion and Further Work

This paper proposes a wearable and flexible hand exoskeleton with a new concept and design, which can be used as a wearable computer mouse in our daily life. In experiments, the proposed device was used as a computer mouse and the results were compared with a standard mouse. While the standard mouse is simple and useful to click and move mouse point, there exist several inconvenient disadvantages such as limited space and non-operation on transparent or uneven surface and impossibility to use keyboard and mouse at the same time. In the case of using a wearable mouse, most of the problems in using a standard mouse were solved. Moreover, the experiments demonstrated that a wearable mouse had better accuracy than a standard mouse.

The proposed wearable device can be applied in domain of various game, art and construction design. Although there are a lot of merits of wearable mouse, flexible sensor is a little bit long and the board on the hand is also a little bit big. Thus, our further work is to reduce board's size and to replace flexible bend type with a more comfortable type.

Acknowledgements. This work was supported by the Technology Innovation Program, 10045252, Development of robot task intelligence technology that can perform task more than 80% in inexperience situation through autonomous knowledge acquisition and adaptational knowledge application, funded By the Ministry of Trade, industry & Energy (MOTIE, Korea).

References

1. Kim, D., Hilliges, O., Izadi, S., Butler, A., Chen, J., OikonomidiS, I., Olivier, P.: Digits: freehand 3D interactions anywhere using a wrist-worn gloveless sensor. In: User Interface Software and Technology, pp. 167–176 (2012)
2. Mackenzie, I.S., Tatu, K., Tatu, K., Silfverberg, M.: Accuracy measures for evaluating computer pointing devices. In: Proceedings of the SIGCHI Conference on Human Factors in Computing Systems, vol. 1, pp. 9–16 (2001)
3. Raya, R., Roa, J.O., Ceres, R., Pons, J.L.: Wearable inertial mouse for children with physical and cognitive impairments. Sensors and Actuators A 162, 248–259 (2009)
4. Jonathan, B.F., Nalini, A.: MouseTracker: Software for studying real-time mental processing using a computer mouse-tracking method. Behavior Research Methods 42, 226–241 (2010)
5. Lee, D.-H., Han, J.H., Kim, J.H.: Market-based Multiagent Framework for Balanced Task Allocation. In: Robot Intelligence Technology and Applications, vol. 1 (2012)
6. Hong, Y.-D., Kim, J.H.: Walking Pattern Generation on Inclined and Uneven Terrains for Humanoid Robots. In: Robot Intelligence Technology and Applications, vol. 1 (2012)
7. Kurata, T.: The Hand Mouse: GMM hand-color classification and mean shift tracking. In: Proceedings of the IEEE ICCV Workshop on, pp. 119–124 (2001)
8. Tu, J., Tao, H., Huang, T.: Face as mouse through visual face tracking. Computer Vision and Image Understanding 108, 35–40 (2007)

Application of the Chaos Theory in the Analysis of EMG on Patients with Facial Paralysis[*]

Anbin Xiong[1,2], Xingang Zhao[1], Jianda Han[1], and Guangjun Liu[1,3]

[1] State Key Laboratory of Robotics, Shenyang Institute of Automation (SIA),
Chinese Academy of Sciences (CAS), Shenyang, Liaoning, 110016, China
{xiongab,zhaoxingang,jdhan}@sia.cn
[2] University of Chinese Academy of Sciences (CAS), Beijing, 100049, China
[3] Department of Aerospace Engineering, Ryerson University. Toronto, Canada
gjliu@ryerson.ca

Abstract. Surface electromyography (sEMG) has been widely applied to disease diagnosis, pathologic analysis and rehabilitation evaluation. It is the nonlinear summation of the electrical activity of the motor units in a muscle and can reflect the state of neuromuscular function. Traditional linear and statistical analysis methods have some significant limitations due to the short-term stationary and lower signal-noise ratio of sEMG. In this paper, we introduce chaotic analysis into the field of sEMG process to investigate the hidden nonlinear characteristics of sEMG of patients with facial paralysis. sEMG on the bilateral masseter, levator labii superioris and frontalis of the 21 patients is recorded. Chaotic analysis is employed to extract new features, including correlation dimension, Lyapunov exponent, approximate entropy and so on. We discover the maximum Lyapunov exponents are all greater than 0, indicating that sEMG is a chaotic signal; correlation dimensions of sEMG on healthy sides are all smaller than that of diseased sides; and inversely, the approximate entropies of healthy sides are all greater than that of diseased sides. Consequently, chaotic analysis can provide a new insight into the complexity of the EMG and may be a vital indicator of diagnosis and recovery assessment of facial paralysis.

Keywords: sEMG, chaotic analysis, features extraction, facial paralysis.

1 Introduction

Surface electromyography (sEMG) is a technique to record the electrical activity produced by skeletal muscles and can be readily measured on the skin to provide an assessment to human neuromuscular system [1]. It is safe, noninvasive, real-time and convenient to be implemented. The signals can be used to detect medical abnormalities, activation level, recruitments of motor units and to analyze the biomechanics of human or animal movements [2-4,34].

---

[*] This work is supported by National Natural Science Foundation of China: 61273355, 61273356.

sEMG is a kind of weak signal ranging between 50 µV and 10 mV with a bandwidth between 50 and 500Hz. It is non-stationary and easy to be influenced by environment noise, which brings difficulties to the signal process. Conventionally, sEMG is regarded as the function of time and analyzed in time [5-8], frequency [9,10] and time-frequency[11-13] domains respectively to obtain features such as the Integral of Absolute Value(IAV), Median Frequency(MDF) and Wavelet Transform Coefficients(WTF) etc..

On the other hand, researcher employed chaotic theory to analyze EMG. Chaos theory is a field of study in mathematics. It refers to a determined but cannot be predicted system state and features local randomness and global stability [15].

Chen [14] utilized Fuzzy Entropy and Band Spectrum Entropy to analyze sEMG. Fuzzy Entropy not only owns higher relative consistency and less dependence on data length, but also achieves freer parameter selections and stronger robustness to noise; Band Spectrum Entropy can characterize changes in the sEMG complexity and its frequency shift during muscle fatigue and shows better reliability when estimating muscle fatigue compared with traditional EMG features. Zhang et al. [27] put forward standard multi-scale Entropy and the intrinsic mode entropy (IMEn) to apply to different patterns of spontaneous EMG. Significant differences were observed (p< 0.001) from any two of the spontaneous EMG patterns, while such difference may not be observed from the single-scale entropy analysis. The aforementioned researches have extracted sEMG's chaotic features to validate its chaotic characteristics but have not applied to pathological diagnosis in clinic.

Furthermore, Erfanian et al. [21] found that the state space of the muscle exhibits a strange attractor with a non-integer correlation dimension which increases with muscle fatigue. Bodruzzaman et al. extracted fractal dimension [18], Hurst's rescaled-range, Housdorff-Besicovich fractal dimension [19] and correlation dimension [20] respectively, to analyze sEMG acquired from the patients with neuromuscular diseases. The results indicated that chaotic features of sEMG are effective indicators for diseases diagnosis. Small et al. [23] carried out investigations into the construction of phase portraits, correlation dimension analysis, and dominant Lyapunov exponents of sEMG during the isometric contraction of trapezius. The results do not support the hypothesis that the EMG is chaotic. Lei et al. [24] proposed symplectic geometry method to estimate embedding dimension of reconstructed attractor to overcome the deficiency of Singular Value Decomposition (SVD) method and results showed that correlation dimension of sEMG is 6. Padmanabhan et al.[25] recorded sEMG signals from 10 muscles in the leg during walking. Maximum voluntary contraction (MVC) were also obtained and pre-processed using wavelet based denoising techniques. The results indicated that EMG signals were chaotic with a dimension between 2 and 3 for walking and 3 and 4 for MVC data. Liu et al. [26] introduced the Recurrence Quantification Analysis (RQA) strategy to the analysis of electrical stimulation evoked surface EMG, and found that the Percent Determination increases along with stimulation intensity.

To sum up, the above mentioned papers are all published before 2005; and the number of papers about the chaotic analysis decreased gradually in the past decade. What hindered this analysis method's development is that the correlation dimension of sEMG hasn't been definitely determined and its advantage hasn't been highlighted compared with traditional linear analysis methods.

In this paper, we introduce chaotic analysis into the field of sEMG process to investigate the hidden nonlinear characteristics of sEMG on the patients of facial paralysis. sEMG on the bilateral masseter, levator labii superioris and frontalis of the 21 patients is recorded. Chaotic analysis is employed to extract new features. The rest of the paper is organized as follows: chaotic theory will be introduced in Section 2; Section 3 describes the experiment procedure; results are presented in Section 4 in detail; and finally Section 5 draws the conclusion.

2   Chaotic Theory Analysis Method

Simple nonlinear system can produce certain simple deterministic behaviors as well as unstable, seemingly random, non-deterministic behaviors, which is also called chaos. Chaos is a complex motion which is always restricted in limited areas, extremely sensitive to initial values, long-term unpredictable, with track never repeated, fractal dimension and strange attractors. Amount of study demonstrated that Center Nervous System (CNS) generates chaotic firing patterns of action potentials; and a variety of physiological potential nerve signals, including EEG, ECG and EMG, etc., originated from CNS, have shown some degree of chaotic behavior, too[16-28]. Some nonlinear analysis methods are applied to the analysis of EMG.

The main analysis methods of chaos theory cover numerical analysis, statistics analysis and analytical analysis. Numerical analysis includes phase plane, power spectrum, Poincare sections, cell-to-cell mapping method; statistic analysis refers Lyapunov exponent, fractal dimension, metric entropy etc.; analytical analysis has renormalization group, KAM theory, Melnikov method , smale horseshoe maps, symbolic dynamics systems [29]. In this paper, we will combine the numerical analysis and statistical analysis to reconstruct the phase space of sEMG and to analyze the correlation dimension, Lyapunov exponent and approximate entropy. Specific methods are described as follows:

2.1   The Phase Space Reconstruction Based on Time Delay

A time delay τ is introduced to reconstruct a m-dimensional phase space $\{X_i\}$ for raw sEMG sequence $X(t) = \{x(1), x(2), ..., x(M)\}$, where $i = 1, 2, ..., m$, $X_i = \{x(i), x(i+\tau), ...x(i+(m-1)\tau)\}$ is the vector represented the ith point in phase space; and m is the embedding dimension. Consequently, the sEMG is reconstructed to form a m-dimensional space:

$$\begin{bmatrix} x(1) & x(2) & ... & x(M-(m-1)\tau) \\ x(\tau+1) & x(\tau+2) & ... & x(M-(m-2)\tau) \\ ... & ... & ... & ... \\ x((m-1)\tau+1) & x((m+1)\tau+2) & ... & x(M) \end{bmatrix} \quad (1)$$

2.2   The Determination of Time Delay and Embedding Dimension

The determination of time delay τ and the embedding dimension m is of great importance. In this paper, τ is determined according to the auto-regression function $R(\tau)$

$$R(\tau) = \frac{\frac{1}{N}\sum_{i=1}^{N}[x(i+\tau)-\bar{x}][x(i)-\bar{x}]}{\frac{1}{N}\sum_{i=1}^{N}[x(i)-\bar{x}]^2} \tag{2}$$

where $\bar{x}$ is the mean of sEMG sequence $X(t)$; $N = M -(m-1)\tau$ is the total number of the points in the phase space. The optimal value of τ is the time when $R(\tau)$ decrease below $1/e$.

According to Takens Theorem, $m \geq 2D_2 +1$, where D_2 is the correlation dimension and can be obtained with G-P algorithm[30].

$$D_2 = \lim_{I \to 0}\frac{\log C(I)}{\log(I)} \tag{3}$$

where $C(I)=\dfrac{1}{N(N-1)}\sum_{i=1,i\neq j}^{N}\sum_{j=1}^{N}H(I-\|X_i - X_j\|)$, X_i and X_j are any points in phase space; $\|\cdot\|$ is the euclidean distance between two points; I is the user-defined threshold value; and $H(\cdot)$ is Heaviside function.

But for the signal compounded with noise, G-P algorithm is time-consuming or D_2 is hard to converge. Thus, researchers proposed False Nearest Neighbor (FNN) method [31] and Cao method [32] to calculate m.

For each point X_i in phase space, there is surely a nearest point $X^{NN}(i)$; and the distance between them is $R_m(i)=\|X(i)-X^{NN}(i)\|$. When the dimension of phase space increases from m to $m+1$, the distance become $R_{m+1}^2(i) = R_m^2(i)+\|x(i+\tau m)-x^{NN}(i+\tau m)\|$. If $R_{m+1}(i)$ is much greater than $R_m(i)$, we could consider that the nearest neighbor point is false because the distance in higher dimension decreases when projected into lower dimension.

We define

$$a_2(i,m)=\frac{\|X_{m+1}(i)-X_{m+1}^{NN}(i)\|}{\|X_m(i)-X_m^{NN}(i)\|} \tag{4}$$

where $X_m(i)$, $X_m^{NN}(i)$ are the ith point and its nearest point in m-dimensional space and $X_{m+1}(i)$, $X_{m+1}^{NN}(i)$ are in $m+1$-dimensional space accordingly.

We define $E(m) = \dfrac{1}{N-m\tau} \sum\limits_{i=1}^{N-m\tau} a_2(i,m)$, $E1(m) = E(m+1)/E(m) \cdot E1(m)$ will con-

verge after m increases to a certain value m_0, which is the embedding dimension.

2.3 Lyapunov Exponent(LE)

In chaotic system, two phase trajectories, which are close to each other initially, will diverge exponentially with time. Lyapunov exponent is a quantitative indicator to describe the divergence rate of phase trajectories.

X_0 and Y_0 are points in phase space and the distance between them is $d_0 = \|X_0 - Y_0\|$. After the time of $i\Delta t$, the distance become

$d_i = \|X_i - Y_i\| = \dfrac{\|X_i - Y_i\|}{\|X_{i-1} - Y_{i-1}\|} \cdot \dfrac{\|X_{i-1} - Y_{i-1}\|}{\|X_{i-2} - Y_{i-2}\|} \cdots \dfrac{\|X_1 - Y_1\|}{\|X_0 - Y_0\|} d_0$. We define $e^{\lambda_{i-1}} = \dfrac{\|X_i - Y_i\|}{\|X_{i-1} - Y_{i-1}\|}$, and

hence $\dfrac{d_i}{d_0} = e^{\lambda_{i-1}} \cdot e^{\lambda_{i-2}} \cdot e^{\lambda_{i-3}} \ldots e^{\lambda_0} = e^{\lambda_{i-1} + \lambda_{i-2} + \lambda_{i-3} + \ldots + \lambda_0} = e^{n\lambda}$. Consequently, the average di-

vergence rate of signal, namely Lyapunov exponent is

$$\lambda = \lim_{i \to \infty} \frac{1}{i} \ln(\frac{d_i}{d_0}) \tag{5}$$

In a m-dimensional space, there are several Lyapunov exponents $\boldsymbol{\lambda} = \lambda_j, (j=1,2,\ldots,m)$. If the maximum Lyapunov exponent is greater than 0, the system is a chaotic system.

2.4 Approximate Entropy(ApEn)

Approximate entropy is a measurement to gauge the complexity of signals; specifically, it is the conditional probability that the vector continue to maintain its similarity when its dimension increases from m into $m+1$. The calculation procedure is as follows:

1. To reconstruct a m-dimensional vector with the original sEMG sequences $X(t) = \{x(1), x(2), \ldots, x(M)\}$, namely, $\chi_i = \{x(i), x(i+1), \ldots, x(i+(m-1)\tau\}$, where The L_∞ distance between χ_i and $\chi_j (i, j=1,2,\ldots,M-m+1, j \neq i)$ is defined as:

$$d_{ij} = \max_k \|x(i+k) - x(j+k)\|, k = 0,1,2,\ldots,m-1 \tag{6}$$

where $x(i+k)$ is the component of χ_i.

2. Define a threshold r, to caculate the total number that $d_{ij} < r$ for each i and the ratio between the number and $M - m + 1$, which is defined as $C_i^m(r)$:

$$C_i^m(r) = \frac{1}{M-m+1} num\{d_{ij} < r\}, i = 1, 2, ..., M - m + 1 \tag{7}$$

3. Define

$$\Phi^m(r) = \frac{1}{M-m+1} \sum_{i=1}^{M-m+1} \ln C_i^m(r) \tag{8}$$

4. To increase the dimension of the vector from m into $m+1$ and repeat the above steps to obtain $C_i^{m+1}(r)$ and $\Phi^{m+1}(r)$

5. Define approximate entropy(ApEn)

$$ApEn(m, r, M) = \Phi^m(r) - \Phi^{m+1}(r) \tag{9}$$

Approximate entropy using a smaller amount of data points can arrive at more robust estimation and has good anti-jamming capability, especially for occasional transient strong interference.

We will analyze sEMG with the above chaotic features to verify its chaotic characteristics and implement a comparison between the healthy and diseased sides of the patients of facial paralysis. This will contribute to the in-depth understanding of the activity patterns of the neuromuscular system and to establish a more scientific and reasonable evaluation of muscle function with the non-invasive techniques.

3   Experiment

3.1   Participants

Twenty-one participants, including thirteen males and eight females, were randomly selected from a pool of patients treated in Shenyang Traditional Chinese Medicine Hospital (STCMH), Shenyang, China, with an average weight of 60±9.45 kg, height of 1.65±0.07m, and age between 22 and 64 years old (44.2±4.84 years), participated in the tests. They were clinically examined with regard to inability to control facial muscles and received no other treatments before the tests. All participants signed the Term of Consent and were informed of the purpose and potential risks of the study before their written voluntary consent.

3.2   Procedure

In this study, the acupoints were selected by an experienced clinician of STCMH as Yingxiang (LI20), Sibai (ST2), Quanliao (SI18), Dicang (ST4). Each of the twenty participants was asked to puncture for 30 minutes every day. The needles were stainless, with a diameter of 0.2mm and length of 30mm. The depth, angle and manipulation of needle insertion depended on the participants' symptoms and the discretion of the clinician.

Before the acupuncture therapy, the sEMG electrodes (see Fig.1) were attached on the bilateral masseter (MS), levator labii superioris (LLS) and frontalis (FT) of both healthy and diseased sides (see Fig.2), and sEMG was measured while the patient was asked to do the actions of raising eyebrows (RE), cheek-bulging (CB), pouting (PT), grinning (GN) and nose-wrinkling (NW). Each movement lasted approximately for 2 seconds and was repeated 3 times with an interval of 3 seconds in each session.

Then the patient was treated by acupuncture. After the acupuncture, the electrodes were placed again on the same muscles, and sEMG was measured while the patient's doing the same actions as those before the acupuncture. Fig.3 shows the complete sEMG sampling procedure.

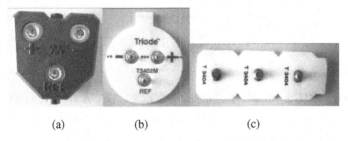

(a)   (b)   (c)

Fig. 1. The sEMG pre-amplifier and electrodes

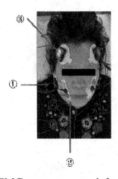

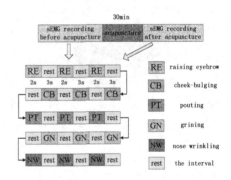

Fig. 2. sEMG measurement, left: diseased side, right: health side, and ①-masseter, ②-levator labii superioris, ③-frontalis

Fig. 3. The procedure of sEMG sampling in the experiments

In all the experiments, we used the sEMG electrodes and pre-amplifier made by FlexComp (Thought Technology Co. Ltd.®, Canada). All data was sampled with a frequency of 2kHz (0.5ms), and then digitally filtered by a bandpass filter of 2~500Hz. A notch filter of 50Hz was also applied on the filtered data before it was further analyzed. All the participants received the acupuncture therapy 7 days per week and the sEMG recording was conducted with a 3 or 4 days interval.

4   Results

As illustrated in Fig. 4, it is typical 6-channel sEMG acquired on bilateral masseter (MS), levator labii superioris (LLS) and frontalis (FT).

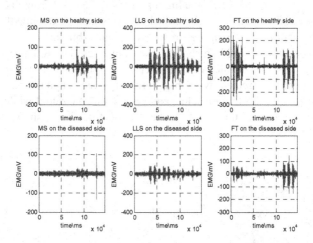

Fig. 4. The raw EMG acquired on the bilateral MS, LLS and FT on the patients suffering facial paralysis

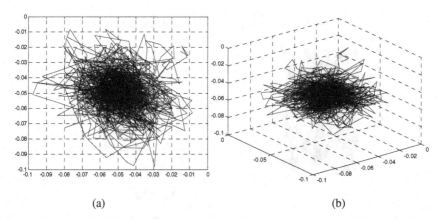

(a)   (b)

Fig. 5. (a) The Poincaré plot of EMG with embedding dimension of 2; (b) the Poincaré plot of EMG with embedding dimension of 3

According to Section 2.2, the reconstruction of phase space cannot be accomplished until the time delay τ and embedding dimension m are determined.

Firstly, we calculate the auto-correlation function of sEMG and find the first point on the function curve below $1/e$ is between 3 and 4. Consequently, the time delay is determined as 4.

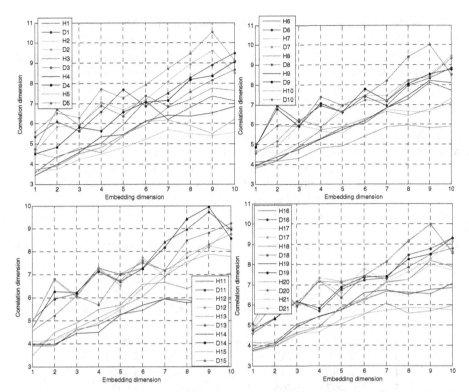

Fig. 6. The relationship between the correlation and embedding dimension. H means healthy side and D diseased side. The curves of the healthy sides are all below the diseased ones.

Secondly, the phase loci in 2 and 3 dimensional phase space are plotted in Fig.5. We cannot observe a clear attractor, but we cannot conclude that EMG is non-chaotic, because the attractor may hide in the hidden structure of the low-dimensional space. Therefore, we calculate the embedding dimension m utilizing G-P algorithm. The result is shown in Fig.6.

Fig. 6 shows that, with the increase of the embedding dimension, correlation dimension increases accordingly without convergence due to the noise compounded in sEMG. That goes against the determination of correlation dimension but we still find an interesting conclusion: The curves of the healthy sides are all below the diseased sides (except for the 1st and 12th patients), i.e. the correlation dimensions of the diseased sides are higher than that of healthy side, which tallies with the conclusions in [21,22] that correlation dimension of sEMG increases with muscle fatigue, indicating that patients with facial paralysis may result from muscle weakness and paralysis caused by the reduction of the recruited motor units number. The statistical data is listed in Table 1.

In addition, we use Cao method (introduced in Section 2.2) to calculate the correlation dimension of sEMG. We find when the embedding dimension increases to 6, the correlation dimension of sEMG no longer increases apparently for 21 patients with facial paralysis, indicating the embedding dimension of sEMG locates between 6 and 7, and there is no significant difference between healthy and diseased sides

According to Eq. (5) in Section 2.3, we can obtain the Lyapunov exponents of sEMG on 21 patients' bilateral muscles and the results are shown in Fig. 8. With the increase of embedding dimension, the maximum Lyapunov exponents converge above 0, indicating that sEMG is chaotic signal, and no significant difference between the healthy and diseased sides is observed.

As shown in Fig. 9, with the embedding dimension increases, the ApEn of sEMG remain stable approximately, indicating the probability of generating new model stays unchanged. Furthermore, except for 1st, 13th and 19th patients, ApEn of the healthy sides of the patients are all greater than that of the diseased sides, which means that sEMG on healthy sides is with a greater complexity than that of diseased sides. The statistical results are shown in Table 2.

Table 1. The statistics comparison of the correlation dimension of the EMG on the healthy and diseased sides

Patients	Healthy side			Diseased side		
Number	MAX	MIN	AVG	MAX	MIN	AVG
1	7.356	3.565	5.821	9.500	5.344	7.240
2	5.706	3.636	4.825	9.602	4.600	7.236
3	7.758	3.719	5.774	8.682	4.741	6.925
4	6.846	3.439	5.505	9.061	4.499	6.721
5	6.007	3.765	5.124	10.03	4.845	7.490
6	8.117	4.107	6.027	8.835	4.851	7.093
7	7.325	3.849	5.819	9.449	4.667	6.928
8	5.954	3.813	5.092	10.014	5.032	7.477
9	8.196	3.956	6.075	8.766	4.883	7.155
10	7.074	3.921	5.775	9.336	4.563	6.884
11	5.965	3.901	5.143	9.716	5.060	7.479
12	8.253	4.103	6.055	9.039	4.985	7.144
13	7.070	3.500	5.756	9.235	4.572	6.984
14	5.945	3.97	5.116	9.957	4.71	7.440
15	7.884	3.932	5.967	8.753	4.981	7.141
16	7.064	3.862	5.735	9.296	4.608	6.958
17	5.871	3.958	5.127	9.948	4.635	7.459
18	8.176	4.120	6.046	8.800	5.066	7.160
19	6.858	3.750	5.735	9.312	4.76	6.984
20	6.084	3.707	5.096	10.013	4.81	7.452
21	7.999	4.044	6.018	9.136	4.895	7.153

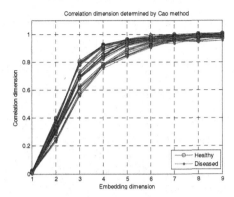

Fig. 7. The correlation dimension of sEMG calculated used Cao method for 21 patients with facial paralysis, which converges after the embedding dimension increase to 6

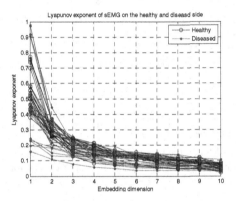

Fig. 8. The maximum Lyapunov exponents of the EMG on the healthy and diseased sides of 21 participants, which are all greater than 0

Studies have shown that, chaotic features of sEMG vary with muscle fatigue during the isometric contraction [21,22,33]. Some researchers suggest these change implied the complexity of EMG is reduced during muscle fatigue and sEMG signal tends to be regular. The facial paralysis caused by the dysfunction of facial muscle seems to be similar to muscle fatigue rooted in the reduction of motor units number. After searching in related database, we failed to find any literature which is research directly and demonstrate substantially on this hypothesis. In this paper, our analysis confirmed the hypothesis. It not only provides an important support to the feasibility study that muscle function can be evaluated based on the complexity of sEMG signals, but also brings a new idea to the pathological analysis methodology.

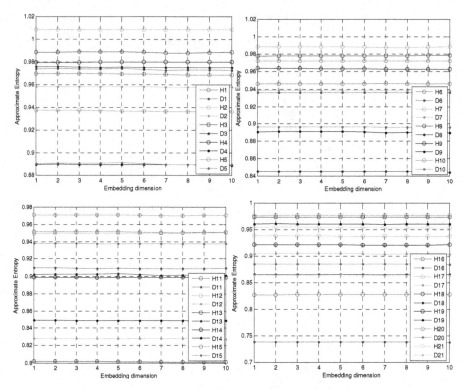

Fig. 9. The ApEn of the EMG on the healthy and diseased sides for all 21 patients; the Apen of the healthy side is greater than that of the diseased side except for the 1st, 13th and 19th patients

Table 2. The statistic comparison of the ApEn of the EMG on the healthy and diseased sides. The ApEn of the healthy side is greater than that of the diseased side except for the 1st , 13th and 19th patient

Patients number	Healthy side	Diseased side	Patient number	Healthy side	Diseased side
1	0.969	0.973	12	0.948	0.825
2	1.008	0.958	13	0.802	0.902
3	0.988	0.975	14	0.899	0.849
4	0.980	0.889	15	0.972	0.938
5	0.938	0.891	16	0.831	0.737
6	0.947	0.938	17	0.936	0.914
8	0.978	0.843	18	0.975	0.864
9	0.963	0.890	19	0.930	0.962
10	0.975	0.898	20	0.981	0.882
11	0.952	0.910	21	0.964	0.863

5   Conclusion

In this study, we acquire sEMG on the bilateral masseter, levator labii superioris and frontalis of 21 patients with early facial paralysis. Chaotic analysis is employed to extract new features. A time delay is introduced to describe the signal in a multi-dimensional phase space; correlation dimension and Lyapunov exponent is obtained to verify the chaotic characters of EMG signals; approximate entropy is calculated to quantify the amount of regularity and the unpredictability of EMG; and a comparison is implemented between the above features of healthy and diseased sides. We find that the maximum Lyapunov exponents are all greater than 0, indicating that sEMG is a chaotic signal; correlation dimensions of sEMG on healthy sides are all smaller than that of diseased sides; and inversely, the approximate entropyies of healthy sides are all greater than that of diseased sides. This method can provide a new insight into the complexity of the EMG and may be a vital indicator of diagnosis and recovery assessment of facial paralysis.

References

1. Tang, X.F.: Clinical electromyography. The Joint Publishing House of Beijing University of Science and Technology & China Union Medical University, Beijing (1995)
2. Bilodeau, M., Schindler-Ivens, S., Williams, D.M., et al.: EMG frequency content changes with increasing force and during fatigue in the quadriceps femoris muscle of men and women. Journal of Electromyography and Kinesiology 13(1), 83–92 (2003)
3. Chan, A.D.C., Englehart, K.B.: Continuous classification of myoelectric signals for powered prostheses using Gaussian mixture models. In: IEEE 25th Annual International Conference, vol. 3, pp. 2841–2844 (2003)
4. Frigo, C., Ferrarin, M., Frasson, W., et al.: EMG signals detection and processing for on-line control of functional electrical stimulation. Journal of Electromyography and Kinesiology 10(5), 351–360 (2000)
5. Graupe, D., Cline, W.K.: Function separation of EMG signals via ARMA identification methods for prosthesis control purposes. IEEE Trans. on Syst., Man, Cyber., SMC-5(2), 252–259 (1975)
6. Chen, X., Zhang, X., Zhao, Z.Y., et al.: Multiple Hand Gesture Recognition based on Surface EMG Signal. In: 1st International Conference on Bioinformatics and Biomedical Engineering, pp. 506–509 (2007)
7. Horiuchi, Y., Kishi, T., Gonzalez, J., et al.: A Study on Classification of Upper Limb Motions from Around-Shoulder Muscle Activities. In: IEEE 11th International Conference on Rehabilitation Robotics, pp. 311–315 (2009)
8. Doud, J.R., Walsh, J.M.: Muscle fatigue and muscle length interaction: effect on the EMG frequency components. Electromyography and Clinical Neurophysiology 35(6), 331–339 (1995)
9. Mix, D.F., Olejniczak, K.J.: Elements of Wavelets for Engineers and Scientists. John Wiley & Sons, Inc. (2003)
10. Saito, N., Coifman, R.R.: Local discriminant bases and their applications. J. Math. Imag. Vis. 5(4), 337–358 (1995)
11. Boostani, R., Moradi, M.H.: Evaluation of the forearm EMG signal features for the control of a prosthetic hand. Physiological Measurement 24, 309–319 (2003)

12. Englehart, K., Hudgins, B., Parker, P.A.: A Wavelet-Based Continuous Classification Scheme for Multifunction Myoelectric Control. IEEE Trans on Biomedical Engineering 48(3), 302–311 (2001)
13. Chu, J.U., Moon, I., Lee, Y.J., et al.: A Supervised Feature-Projection-Based Real-Time EMG Pattern Recognition for Multifunction Myoelectric Hand Control. IEEE/ASME Trans. on Mechatronics 12(3), 282–290 (2007)
14. Chen, W.T.: A Study of Feature Extraction from sEMG Singal Based on Entropy, A Dissertation for the Degree of Doctor of Philosophy, Biomedical Engineering. Shanghai Jiao Tong University (2008)
15. Ott, E.: Chaos in Dynamical Systems, 2nd edn. Cambridge University Press, Cambridge (September 9, 2002)
16. Han, C.X.: Research on the Acupuncture Neural Electrical Signals Conduction and Effect, A Dissertation for the Degree of Doctor of Philosophy, Control Science and Engineering. Tianjin University (2010)
17. Chen, W.T., Wang, Z.Z., Ren, X.M.: Characterization of Surface EMG Signals Using Improved Approximate Entropy. Journal of Zhejiang University Science B 7, 844–848 (2006)
18. Bodruzzaman, M., Zein-Sabatto, S., Marpaka, D., et al.: Neural network-based classification of electromyographic (EMG) signal during dynamic muscle contraction. In: IEEE Southeastcon 1992, April 12-15, pp. 99–102 (1992)
19. Bodruzzaman, M., Cadzow, J., Shiavi, R., et al.: Hurst's rescaled-range (R/S) analysis and fractal dimension of electromyographic (EMG) signal. In: IEEE Southeastcon 1991, April 7-10, pp. 1121–1123 (1991)
20. Bodruzzaman, M., Devgan, S., Kari, S.: Chaotic classification of electromyographic (EMG) signals via correlation dimension measurement. In: IEEE Southeastcon 1992, pp. 95–98 (1992)
21. Erfanian, A., Chizeck, H.J., Hashemi, R.M.: Chaotic activity during electrical stimulation of paralyzed muscle. In: IEEE 18th Annual International Conference, vol. 4, pp. 1756–1757 (1997)
22. Ehtiati, T., Kinsner, W., Moussavi, Z.K.: Multifractal characterization of the electromyogram signals in presence of fatigue. In: IEEE Canadian Conference on Electrical and Computer Engineering, May 24-28, vol. 2, pp. 866–869 (1998)
23. Small, G.J., Jones, N.B., Fothergill, J.C., et al.: Chaos as a possible model of electromyographic activity. In: IEEE Intl. Conf. on Simulation, pp. 27–34 (September 1998)
24. Lei, M., Wang, Z.Z., Feng, Z.J.: The application of symplectic geometry on nonlinear dynamics analysis of the experimental data. In: 14th International Conference on Digital Signal, pp. 1137–1140 (2002)
25. Padmanabhan, P., Puthusserypady, S.: Nonlinear analysis of EMG signals - a chaotic approach. In: IEEE Conf. on Eng. Med. Biol. Soc., vol. 1, pp. 608–611 (2004)
26. Liu, C., Wang, X.: Recurrence quantification analysis of electrically evoked surface EMG signal. In: Conf. on Eng. Med. Biol. Soc., vol. 5, pp. 4572–4575 (2005)
27. Zhang, X., Chen, X., Barkhaus, P.E., et al.: Multiscale Entropy Analysis of Different Spontaneous Motor Unit Discharge Patterns. IEEE Journal of Biomedical and Health Informatics 17(2), 470–476 (2013)
28. Gallez, D., Babloyantz, A.: Predictability of human EEG: a dynamical approach. Biol. Cvbern. 64(5), 381–391 (1991)
29. Ashwin, P.: Nonlinear dynamics: Synchronization from chaos. Nature 422(6930), 384–385 (2003)
30. Grassberger, P., Procaccia, I.: Characterization of Strange Attractors. Phys. Rev. Lett. 50, 346–349 (1983)

31. Abarbanel, H.D.I., Brown, R., Sidorowich, J.J., et al.: The analysis of observed chaotic data in physical systems. Rev. Mod. Phys. 65, 1331–1392 (1993)
32. Cao, L.Y.: Practical method for determining the minimum embedding dimension of a scalar time series. Physica D: Nonlinear Phenomena 110(1-2), 43–50 (1997)
33. Felici, F., Rosponi, A., Sbriccoli, P., et al.: Linear and non-linear analysis of surface electromyograms in weightlifters. European Journal of Applied Physiology 84(4), 337–342 (2001)
34. Luo, Z.Z., Yang, G.Y.: Prosthetic Hand Fuzzy Control Based on Touch and Myoelectric Signal. Robot 28(2), 224–228 (2006)

Design, Implementation, and Experiment of an Underwater Robot for Effective Inspection of Underwater Structures

Seokyong Kim, Hyun-Taek Choi, Jung-Won Lee, and Yeongjun Lee

Korea Institute of Ocean Science and Technology
32 1312 Beon-gil, Yuseong-daero, Yoseong-gu, Daejeon 305-343, Korea
{seokyong,htchoiphd,jwkitty,leeyeongjun}@kiost.ac

Abstract. This paper describes development of a specialized underwater robot for effective inspection of underwater structures. Among various inspection methods of underwater structures, using underwater robots becomes popular. Unfortunately, most underwater robots are not specialized for the inspection purpose. The inspection using traditional method is more inefficient and more inconvenient than the inspection by divers. To overcome this problem, functions to be specialized in the underwater inspection is conceived and these functions are implemented in the developed robot. The type of the developed underwater robot is ROV which is possible to communicate large amount of data in real-time and to supply power efficiently. Moreover, the performance of visual sensor is improved, because most inspection methods rely on visual information. Operational algorithms of the robot developed for stable and convenient operations. The performance of the developed robot is verified in a tank.

Keywords: underwater, structure, inspection, ROV, camcorder, assistance, operation.

1 Introduction

Many underwater structures have been damaged by long term usage. Underwater structures in the ocean are more severely damaged by salt water. These damaged structures may cause a disaster induced from a collapse. To prevent these accidents in advance, the inspection is important to confirm the state of the structures.

The general inspection of underwater structures is a visual inspection by divers. However, this method is not suitable for long time working, and the divers could be exposed to hazardous situations. An inspection using underwater robots can be an alternative. The underwater robots are applied in a various field of working underwater, including wreckage or munitions recovery, underwater sampling, maintenance and construction of underwater structure, surveying and inspection of underwater, and mining to name a few[1, 2]. For this purpose, this paper proposes a specialized underwater robot for inspection task in the underwater environment. The developed underwater robot is a ROV type with visual sensors which can provide high quality performance in the underwater environment.

J.-H. Kim et al. (eds.), *Robot Intelligence Technology and Applications 2*,
Advances in Intelligent Systems and Computing 274,
DOI: 10.1007/978-3-319-05582-4_72, © Springer International Publishing Switzerland 2014

The type of ROV is suitable for underwater inspection robot. First, the size of the robot can be minimized because it does not need battery and the main processor also can be separated from the robot. The minimized size of the robot helps to perform the inspection tasks efficiently. Second, it can perform long time operation in underwater environment. Last, it can provide a stable telecommunication for real time visual inspection by using a tether. In fact, the tether might be a major drawback of ROV. The tether might disturb the operation of the robot or be tangled when the robot is operated within a complex structure. For these cases, an expert is needed to operate the ROV. Despite of these drawbacks of ROV, it has several advantages to perform underwater inspection as aforementioned. For this reason, we developed the inspection robot as ROV type to be specialized in the underwater inspection.

The developed robot is equipped with a camcorder for the visual inspection. Traditional underwater inspection robots used underwater cams. In the developed system, the camcorder replaces traditional underwater cam for improvement of inspection performance. In the case of using camcorder, HD image can be obtained in underwater. Moreover, robot is able to use function of camcorder that is auto focusing, zoom in, zoom out and, image stabilization. Using the camcorder as a visual sensor, the developed robot can perform more effective inspection tasks.

The remainder of this paper is organized as follows. Section 2 describes design for specialized underwater inspection and implement of design. Section 3 shows performance of robot to require function of underwater inspection, and conclusion and future directions follow in Section 4.

2   Design and Implementation

2.1   Design of Conception

Figure 1 shows design and main frame of the developed underwater inspection robot. The robot should be easily controlled to maintain the stable pose of the robot and to provide stable images of the inspected structures. For this purpose, the main frame of the robot is designed to be appropriate for the stable control of the robot pose. By arranging a buoyancy material at the top of the main frame, the center of buoyancy could be located above the center of gravity. This structure is possible to maintain stable pose in term of kinematics. And if necessary, it is possible to control posture by thrusts. Other components of the robot including main pressure vessel are arranged at the bottom of the main frame.

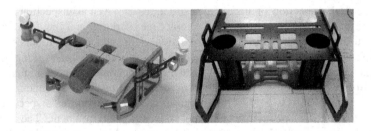

Fig. 1. 3D cad design of robot (right) and Main robot Frame (left)

Figure 2 describes the configuration of thrusters and the corresponding DOF (Degree Of Freedom) of the robot. The thrusters are arranged to restrict the pitching motion of the robot which is not necessary for the inspection tasks. The robot can successfully move with 3 translational motions and 2 rotational motions without the pitching motion using the arranged 6 thrusters [3-6].

Fig. 2. Arranged thruster (right) five DOF (left)

Figure 3 shows the camcorder model and the manufactured pressure vessel which are installed at the front of the developed robot. As mentioned, the developed robot uses visual inspection method. Therefore, the performance of visual sensor is crucial to execute the inspection tasks successfully. Traditional underwater cam has a low picture quality and insufficient function to obtain image effectively in underwater environment. Especially almost underwater cam shows low performance, because it is impossible to control image focus. For this reason, the camcorder is used to replace underwater cam in the developed system. A pressure vessel is made to use the camcorder in underwater environment. This vessel was verified through inner pressure test of 150m (15bar), because working depth of the robot is 100m (10bar). By locating the camcorder within the pressure vessel, it can be used to perform the inspection tasks successfully in underwater environments.

Fig. 3. Camcorder solid model (right) and camcorder pressure vessel (left)

2.2 Design and Implementation of System

The developed robot is equipped with various sensors. The equipped sensors are mainly classified by the sensors for localization and them of health monitoring as shown in Figure 4. The sensors for localization purpose are composed of AHRS (Attitude and Heading Reference System) to measure pose of the robot, DVL (Doppler

Velocity Log) to measure the movement, and a pressure sensor to measure the depth. The sensors for health monitor are composed a temperature sensor to measure heat inside vessel, and a humidity sensor to detect water leakage.

Localization sensor   Robot health sensor

Fig. 4. Second version of lamp actuator

Figure 5 shows a communication diagram of the developed robot. The robot is equipped with a sub-processor which collects and transmits sensor data to the main processor and receives control commands from the main processor. The communication of internal sensors and the sub-processor in the robot is based on a serial communication. The serial communication, this method uses each communication port. This method is inefficient to maintain each communication port. For this reason, to combine communication in this system is advantage and efficient. This role is assumed serial device server. This device combines communication ports to convert from serial to LAN. Obtained image to use camcorder is transmitted by optical fiber communication to generally use image transmission. The tether of ROV is designed for this communication method to include power cable, LAN cable, and optical fiber cable

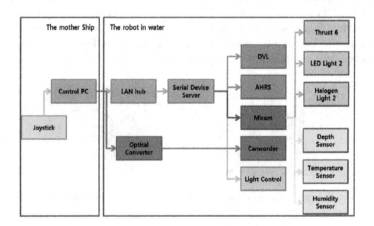

Fig. 5. Communication of ROV system

Figure 6 shows a software structure to operate the developed system. The software structure consists of several separate classes corresponding to each function. It is divided into a communication class, a mission class and a GUI. The communication class collects sensor data and transmits the order of operator by the serial communication. The mission class estimates position data of the robot to use collected sensor data. This localization Class is especially important because we have to know where the inspection image is obtained [6]. It makes command to operate the robot by the order of operator. The GUI performs a general role to represent state of the robot and to transmit the order of operator to the mission class.

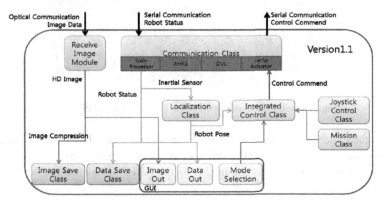

Fig. 6. Structure of system software

Figure 7 shows a pattern board, a LAN hub, and a serial device server. The pattern board is designed to connect devices of the robot to serial device server. This board includes the sub-processor to control thrusters and to collect data of health monitor sensors. Moreover, it performs to control supplying power to control relay switch in the board. This board is located in main pressure vessel.

Fig. 7. Robot pattern board and LAN hub, serial device server

3   Experimental Results

Experiments were conducted in a basin to verify the performance of the developed robot. Figure 8 shows the basin where the experiments were conducted. The experiment tested the performance of heading and position keeping algorithms when disturbances were affected to the robot. The pose keeping function is essential to improve the efficient of inspection because the robot should keep the heading and position toward target objects [7].

Fig. 8. Environment of experiments (water pool, KIOST in Korea)

For the experiment, the pose of the robot was acquired based on the localization sensors, and a conventional PID control based pose keeping algorithm was used. The heading data of the robot was obtained from the AHRS sensor and the position data of the robot was calculated by integrating velocity data from the DVL sensor.

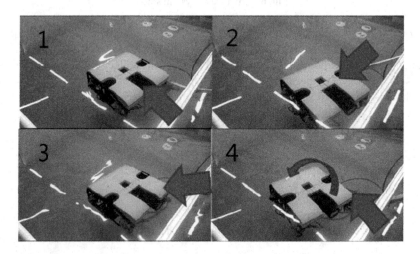

Fig. 9. Disturbance test to confirm performance of keeping algorithm (Backward, Right, Backward & Right, and Rotational displacement)

The disturbance was generated by applying forces to a desired direction. Figure 9 shows the directions of the applied disturbance to confirm the performance of keeping algorithm. Single direction disturbances, which are a backward disturbance (-x direction) and a right direction disturbance (+y direction), were applied first. Then, disturbances of right-backward direction and rotational displacement were applied to confirm the performance of the pose keeping algorithm under the complex disturbances.

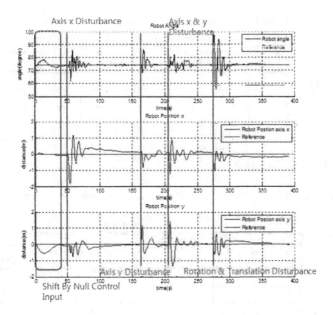

Fig. 10. Motion of robot by heading and position keeping

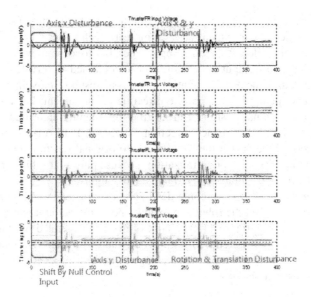

Fig. 11. Thruster input voltage by heading and position keeping

Table 1. Movement of the robot by thruster direction

		Movement of the robot									
		Front	Back	Right	Left	FR	FL	BR	BL	RR	RL
Thruster	FR	+	-	-	+	0	+	-	0	+	-
	TR	+	-	+	-	+	0	0	-	+	-
	FR	-	+	-	-	-	0	0	+	+	-
	TR	-	+	+	+	0	-	+	0	+	-

(FR: Front & Right, TR: Tale & Right, FL: Front & Left, TL: Tale & Left,
RR: Rotate Right, RL: Rotate Left)

Table 2. Direction of thruster and disturbance in experiment of keeping test

		Direction of disturbance			
		Back(axis x(-))	Right(axis y(+))	Back & Right	Rotate right & Translation
Thruster	FR	+	+	+	-
	TR	+	-	Small	Small
	FR	-	-	Small	-
	TR	-	+	-	-

Table 1 show to move the robot by direction to operate thruster. Table 2 represent direction of thruster in experiment of keeping test. Each controller that is composed axis x, y, and rotation decides control input when the robot is moved. These control input are decided control input of each thruster to combined by thruster position. Each thruster is operated by the decided control input. As a result of movement, we confirm that the thrusters operate in the opposite direction of movement. For this reason, the robot is return to original position and heading by keeping algorithm. So, thrusters have thrust in reversed disturbance direction as table 2. Performance of the keeping algorithm is good through that the robot move to original position and heading. The performance of algorithm is not optimization, because purpose of this experiment is to confirm that the robot is suitable moved by keeping algorithm.

These experiments are to overcome dead zone of thruster by null control input to supply 1V to thruster. But the robot is moved by not same output to supply same control input to thrusters as figure 11. The control input is generated by this movement. So the null control input is reduced until dead zone of thruster. To maintain control input voltage by null control input is advantage to rapidly effect control input by PID controller to thruster in the robot.

4   Conclusion

This paper described development of an underwater robot to be specialized in the inspection of underwater structures. ROV was selected as type of the underwater robot, because it has excellent working time by stable supplying power and enables to transmit high quality images by a tether. Moreover, a camcorder was used as a visual sensor instead of a traditional underwater cam. The camcorder is possible to effectively obtain image in underwater by auto focusing and definition of high performance. The developed robot was tested by experiment to confirm the dynamic performance in basin.

Future research will be to verify performance of visual sensor with camcorder by experiment. And algorithm will be developed to assist inspection of underwater structures. Algorithms for efficient control of the robot will be developed. To keep depth of the robot in the water and to verify tidal current is will be made. Additional, sensor of the robot is processed to improve performance. Algorithm to estimate the robot pose to interpolate error to obtain incorrect result of AHRS by magnetic field of structures will be developed to use image of structures.

Acknowledgements. This work has been done by "Development of technologies for an underwater robot based on artificial intelligence for highly sophisticated missions." funded by Korea Institute of Ocean Science & Technology (KIOST) and "Development of an autonomous swimming technology with less than 1.0m position error for underwater robot operating in man-made structural environment" funded by Ministry of Trade, Industry & Energy (MOTIE).

This research was done under the projects, "development of technologies for an underwater robot based on artificial intelligence for highly sophisticated missions" sponsored by the Korea Institute of Ocean Science and Technology, Korea and "development of inspection equipment technology for harbor facilities" sponsored by the Ministry of Oceans and Fisheries, Korea.

References

1. Yuh, J.: Underwater Robotic Vehicles: Design and Control. TSI Press, Albuquerque (1995)
2. Choi, S.K., Yuh, J.: Underwater Vehicle Technology. TSI Press, Albuquerque (1995)
3. Fossen, T.I.: Marine Control Systems: Guidance, Navigation, and Control of Ships, Rigs, and Underwater Vehicle, Marine Cybernetics, Trondheim (2002)
4. Corke, P., Detweiker, C., Dunbabin, M., Hamilton, M., Rus, D., Vasilescu, I.: Experiments with Underwater Robot Localization and Tracking. In: 2007 IEEE International Conference on Robotics and Automation, Roma, Italy, pp. 4556–4561, 10-14 (2007)
5. Negahdaripour, S., Xu, X.: Mosaic-Based Positioning and Improved Motion-Estimation Methods for Automatic Navigation of Submersible Vehicles. IEEE Journal of Oceanic Engineering 27(1), 79–99 (2002)
6. Tripp, S.T.: Autonomous Underwater Vehicles (AUV'S): A Look at Coast Guard Needs to Close Performance Gaps and Enhance Current Mission Performance. Transportation Security White Papers (2001)
7. Batlle, J., Ridao, P., Garcia, R., Carreras, M., Cufi, X., El-Fakdi, A., Ribas, D., Nicosevici, T., Batlle, E., Oliver, G., Ortiz, A., Antich, J.: Automation for the Maritime Industries. In: Aranda, J., Armada, M.A., de la Cruz, J.M. (eds.) URIS: Underwater Robotic Inteligent System, ch. 11, 1st edn. (2004)

Issues in Software Architectures for Intelligent Underwater Robots

Hyun-Taek Choi[1] and Joono Sur[2]

[1] Korea Insitute of Ocean Science and Technology,
32 1312 Beon-gil, Yuseong-daero, Yoseong-gu, Daejeon 305-343, Korea
htchoiphd@kiost.ac
[2] Samsung Thasles
259, Gongdan-Dong, Gumi-City, Korea
joono.sur@samsung.com

Abstract. Recently, as increasing demands of taking care of complex missions, robot software architecture has been focused on the provision of an effective development environment. In this paper, we describe some issues regarding software architecture for an underwater robot, in particular, a middleware. It has been popular because it provides a well-structured, unified and proven environment for a development. First, we summarized concept and requirements for a robot middleware with some names of well-known service robot middleware. Then, additional requirements for an underwater robot are addressed and recent cases of underwater robot software architecture are presented. Actually, there are limited time and budget for a development of software architecture, unless it is a goal. Considering this, two practical approaches are proposed; (1) using an open source software platform like MOOS-IvP, (2) using a simple but effective software structure, proposed for implementing mid-level intelligent algorithms. Under this streatgy, yShark2, a test-bed underwater robot developed by KIOST, is operated for developing various algorithms such as sensing, decision making, and controlling, and ready to move to MOOS-IvP.

Keywords: underwater robot, software architecture, robot middleware.

1 Introduction

Recently, great progresses in robot technologies have been made from industrial robots to intelligent service robots. With state-of-the-art technologies in materials, sensors, and robotic intelligence, underwater robots have been being developed for various kinds of applications, and their missions are getting more and more complicated when compared to conventional ones such as a hydrographic survey which is a simple and one-way data collecting.

As increasing demands of taking care of complexity mainly due to missions in unstructured, unknown and dynamic environments, lots of software components from the drivers of each hardware to high level learning algorithms should be integrated along with a framework of software. So, many researchers have been very interested in software architectures as a key issue in designing robot systems, because it is more than set of components like software library and it provides a powerful development environment in terms of abstraction and modularity [1].

J.-H. Kim et al. (eds.), *Robot Intelligence Technology and Applications 2*,
Advances in Intelligent Systems and Computing 274,
DOI: 10.1007/978-3-319-05582-4_73, © Springer International Publishing Switzerland 2014

Many researchers of underwater robots have proposed their own software architectures which are showing how to design their idea for autonomous capabilities, even though these results have not covered issues of implementation and of rigorous proof for its stability and performance [2]. Since almost all recently developed architectures have been classified as a hybrid, grouping of architecture might not be an interesting issue anymore. Our interest already moved to a generalized architecture which is associated with standards and compatibility [1]. We believe that the right choice of a software architecture will greatly facilitate specification, implementation and validation of robot performance as a backbone of robot systems.

Even though it is obvious that there is no clear definition of software architecture for a robot [3], and even though it is indeed difficult to evaluate software architectures directly, we could describe how good it is developed by several criteria including predictability, reactivity, robustness, modularity, extendibility, generality, and standardization [1]. However, it is too tough to implement an architecture based on the above criteria from scratch. Instead of this, a middleware as the common lower part of a software architecture has been popular in robotics society. It consists of lots of software components as shown Fig. 1, where an appropriate real-time operating system (RTOS) should be selected separately.

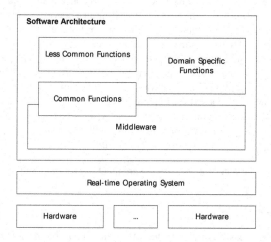

Fig. 1. Typical software architecture for a robot

Section 2 in this paper describes the concept and requirements of a robot middleware briefly and then discusses additional requirements for an underwater robot middleware in section 3. Then, as an example study, 2 candidates for an underwater robot, yShark2 are considered, then conclusion is made.

2   Robot Middleware

A middleware should be easy to use, robust against changing, reliable for any faults, easy to maintain, efficiently flexible, and support for heterogeneous and distributed hardwares, where what we want to get is minimizing the time and cost for a development by providing a well-structured and proven middleware with some often-needed and value added functions involved in the middleware. In the literatures [4-12], there are many demanded characteristics to realize above expectations. Requirements of robot middleware could be summarized as followings[3, 6]: (1) Simplifying the development process using higher-level abstractions with simplified interfaces, (2) Support communication and interoperability among robotic modules, (3) Providing efficient utilization of available resources like CPU, networks, (4) Providing heterogeneity abstractions which can hide the complexity of the low-level communication and the heterogeneity of the modules, (5) Supporting integration with other systems based on real-time interaction with other systems, (6) Offering often-needed robot services which allow us to save lots of efforts for implementation of existing algorithms, (7) Providing automatic recourse discovery and configuration because external devices can be dynamically available or unavailable for a robot's use, (8) Supporting embedded components and low-resource-devices. With these requirements, a robot middleware can be designed for distributed environments without any dependency with hardware and/or operating system and unpredictable environments which require reactiveness and robustness using time and event driven tasks. To implement a middleware, standards such as the Common Object Request Broker Architecture (CORBA), the Internet Communications Engine (ICE), and the Open Dynamics Engine (ODE) are applied [8], which brings more advantages. Basically it saves lots of effort from implementing and evaluating of a core engine from scratch, and some of these have additional functionalities such as standardized interface and security. Security aspects such as authentication, authorization and secure communication become essential for collaborations of multiple robot or classified applications [8].

As a user of middleware, issues when considering making an appropriate choice are more interesting. For examples, we should handle overheads imposed when moving to a new environment in terms of difficulties and the volume of tasks we have to do at the beginning stage. It is definitely based on available human resources in your group and long-term plans of the development. Specific backgrounds of your group or lack of development time should be another variable to your decision.

So far many research groups have proposed their own middleware for their robots such as MIRO (Middleware for Robot), ORCA (Open Robot Control Architecture), UPnP (Universal Plug and Play), RT (Robot-Technology), the Player & Stage, OPRoS (The Open Platform for Robotic Services), ROS (Robot Operating System), OROCOS (Open Robot Control Software), ERSP, MRDS (Microsoft Robotics Developers Studio), MOOS (The Mission Oriented Operating Suite), Webots, etc.

Actually, not like in other applications, middleware in the robotics doesn't seem to be firmly rooted. If it satisfied above key characteristics, we might not need so many middlewares, because we could evaluate the effectiveness of a middleware by how much it is used. We could find many insufficient and incomplete functions in current any middlewares, but we believe that there will be improvement to achieve our goals through several iterations of try and test with extensive feedback from the community [7].

3   Underwater Robot Middleware

Basically, an underwater robot middleware has almost the same requirements as a robot, but the weighting of requirements and point of view of requirements is little different. Some points we are considering are follows:

Limitation of resources: many functions of recent robots are built based on the high-speed network which allows low cost and low computing power hardwares in a robot and powerful server for remote intelligence away from the robot. But because of the low bandwidth of an underwater wireless communication, an underwater robot has to take care of the whole mission within computing resources inside the robot. Meanwhile, recent developed middlewares have a strong capability to handle multimedia data, which might bring unnecessary load to an underwater robot.

Unstable sensor information: Many sensors in robot provide stable and deterministic information. Unfortunately, underwater sensors based on sonar would   not stable and accurate enough to directly use data for robot control systems. Therefore, assisting algorithm such as well-tuned filtering algorithms and sensor fusion algorithms are needed and much consideration in a middleware is demanded for these algorithms.

Lack of accessibility: Unlike robots out of the water, an underwater robot doesn't allow any access by an electrical contact for debugging because of water-proof pressure vessels. This is one of the hardest things during a development process. A middleware should provide an independent channel for supporting debugging process.

A couple of recently developed software architectures for underwater robots are; T-Rex (Teleo-Reactive EXecutive) was developed for a specific underwater robot by the Monterey Bay Aquarium Research Institute. It is a goal-oriented hybrid executive with an embedded automated planning adaptive execution using agents. You can find more details in [9, 14, and chapter 3 in 16]. Huxley was developed by Bluefine Robotics originally for its fleet of AUVs (Autonomous Underwater Vehicle). Since it is for their wide range of commercial products, the most fundamental requirement was the flexibility in terms of robustness and reliability of other products, extensibility for new hardware, maintainability and testability [10]. The LCM system was designed as an alternative for implementing interprocess communication (IPC) for real-time robotics in the marine environment. This architecture was verified by two marine robotics applications [11, 15]

4   Practical Approaches

When you start to develop an underwater robot, basically two options could be taken into consideration for a system software; building new software architecture for your own robot or using existing middleware through some degree of customization. A development of software architecture itself is a tremendous work. As one can guess, fancy concepts are one thing and their implementations are the other thing, which are failure-prone works in particular for a robot system. The reasons are mainly due to (1) so many inherently distributed and heterogeneous components which are going to be changed to improve performance, (2) the limited and restricted resources available on a robot platform, and (3) high rate of system failures from incorporating complicated algorithms and interacting with the environment, causing exceptional situations [3, 7]. Therefore, before you make a decision to build own software architecture, you have to figure out allowable time, available budget and human resources for your project, and most importantly the purpose of your project.

Currently, MOOS-IvP would be the best candidate because it is an open source software, there is relatively much information on the website and case examples for AUVs and USVs. MOOS–IvP is the only software architecture which has been used by several AUVs in different groups including the Bluefin 21-inch UUV, the Hydroid REMUS-100 and REMUS-600 UUVs, the Ocean Server Iver2 UUV, the Ocean Explorer 21-inch UUV, autonomous kayaks from Robotic Marine Systems and SARA Inc, and two larger USVs from the NATO Underwater Research Center in La Spezia Italy. [12, 14]. It consists of two distinct open source software projects. The Mission Oriented Operating Suite (MOOS) is, as a core middleware, a suite of libraries and executables developed by the Mobile Robotics Group at the University of Oxford, and it provides inter-process communication using a publish-subscribe model, as a star topology where all messages go through a central MOOS server. IvP (Interval Programming) refers to a multi-objective optimization method used by the IvP Helm for arbitrating between competing behaviors in its behavior-based architecture. MOOS provides modules for navigation, control and data logging, tools for mission replay from log files and communication debugging, and provides driver modules for some sensors widely used with AUVs [11-13]. However, even customization of an existing software architecture is not trivial work.

In some case, we only need a simple but effective solution to test a robot instead of a formal software architecture. Since AUVs have a tele-operation function for testing and/or debugging, the simple software architecture of Fig. 2 could be considered as another alternative. This architecture has one real-time thread with several non real-time threads for controlling actuators and devices installed on the robot using tele-operation mode or low-level autonomous mode. Packet of ① conveys three types of commands from an operator; commands for tele-operation, commands for RT-thread itself and functions in the selected tasks, and commands for devices of the robot, where device control commands run in a non-RT thread in ② triggered by the RT-thread, typically every 1 sec. All sensor data including tele-operation commands and results of any sensor fusion algorithms are independently saved in common memory, ③ and thruster commands, i.e. control input are also saved here. Basically a robot is in one of two modes, sleep and run, where the sleep mode is doing the only health monitoring with basic sensors, the run mode runs a selected task in ④. This architecture has a manual task coordinator and each task is designed for specific purposes and it can easily be synthesized using functions in function pool, ⑤. Of course, default task should be tele-operation task. The RT-thread, ⑥ is carrying out three main functions; doing any critical device control commands first, DA converting for control input of previous sampling time saved in the common memory, and processing of algorithms using sensor input saved in the common memory. Since this RT-thread runs based on a basic sampling time, we could easily implement any multi sampling time structures for various latency time of sensors and load distributing structures for enhancing the usage of computing resources by just calling functions at multiples of the basic sampling time. This architecture has been applied to yShark2 developed by Korea Institue of Ocean Science and Technology (KIOST) as a test platform for developing algorithms. It could be a good starting point for research purpose. Fig. 3 shows the internal structure of yShark2 and its subsystems including pressure vessels, sensors and thrusters. yShark2 is about 2.0m long, 0.4m diameter, and weighs 80kg. There are two pressure vessels; one for two lithium polymer battery packs, the other for a computer control system and sensor processing units where three PC104s work for main control, vision and acoustic signal processing. There are two cameras with two LED lights, 8 ch ranging sonar, AHRS (Attitude and Heading Reference System), RLG(Ring Laser Gyro) and DVL(Doppler Velocity Log). As a key feature of yShark2, it has a reconfigurable thruster structure. The thruster configuration of Fig. 3 is good for an omni-directional motion. But, you can change thruster configurations along with your research objectives. We just finished basic and mid-level tests using the proposed architecture, and we are going to move to MOOS-IvP for higher level of algorithms.

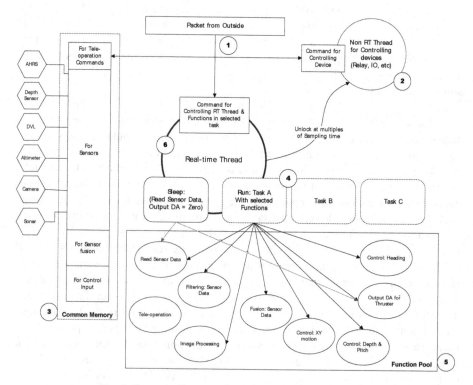

Fig. 2. Example diagram of a simple software architecture

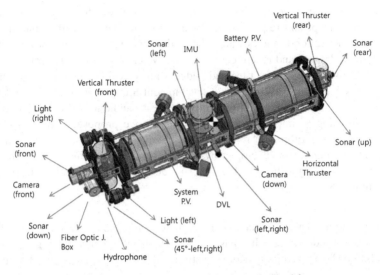

Fig. 3. Internal strcture and devices of yShark2

Fig. 4. yShark2 during pool testing

5   Conclusion

This paper describes some issues regarding software architecture for an underwater robot. A middleware as a part of robot software architecture has been popular because it provides well-structured, unified and proven environment for a development. We summarized concept and requirements for a robot middleware with some names of well-known service robot middleware. Then, additional requirements for an underwater robot are addressed and recent cases of underwater robot software architecture are presented. Considering limited time and budget of a typical development, two practical approaches for an underwater robot development are proposed; (1) using MOOS-IvP which is an open source and currently the best candidate for common software platform, (2) using a simple but effective structure we proposed, which is possibly applied to verify algorithms for mid-level intelligence. Under this strategy, yShark2, a test-bed underwater robot developed by KIOST as shown in Fig. 4, is operated for developing various algorithms such as sensing, decision making, and controlling, and ready to move to MOOS-IvP.

Acknowledgment. This research was done under the project, "development of technologies for an underwater robot based on artificial intelligence for highly sophisticated missions" sponsored by Korea Institute of Ocean Science and Technology, Korea.

References

1. Lin, C., Feng, X., Li, Y., Liu, K.: Toward a generalized architecture for unmanned underwater vehicles. In: 2011 IEEE International Conference on Robotics and Automation, pp. 2368–2373 (May 2011)
2. Yuh, J.: Design and Control of Autonomous Underwater Robots: A Survey. Autonomous Robots 8(1), 7–24 (2000)
3. Mohamed, N., Al-Jaroodi, J., Jawhar, I.: Middleware for Robotics: A Survey. In: 2008 IEEE Conference on Robotics, Automation and Mechatronics, pp. 736–742 (2008)
4. Valavanis, K.P., Gracanin, D., Matijasevic, M., Kolluru, R., Demetriou, G.A.: Control architectures for autonomous underwater vehicles. IEEE Control Systems 17(6), 48–64 (1997)
5. Ridao, P., Yuh, J., Batlle, J., Sugihara, K.: On AUV control architecture. In: Proceedings of the 2000 IEEE/RSJ International Conference on Intelligent Robots and Systems, pp. 855–860 (2000)
6. Namoshe, M., Tlale, N.S., Kumile, C.M., Bright, G.: "Open middleware for robotics. In: 15th International Conference on Mechatronics and Machine Vision in Practice, pp. 189–194 (2008)
7. Smart, W.: Is a Common Middleware for Robotics Possible? In: Proceedings of the IROS 2007 Workshop on Measures and Proceduresfor the Evaluation of Robot Architectures and Middleware (October 2007)
8. Elkady, A., Sobh, T.: Robotics middleware: a comprehensive literature survey and attribute-based bibliography. Journal of Robotics (2012)
9. McGann, C., Py, F., Rajan, K., Thomas, H., Henthorn, R., McEwen, R.: T-REX: A Model-Based Architecture for AUV Control. In: 3rd Workshop on Planning and Plan Execution for Real-World Systems (2007)
10. Goldberg, D.: Huxley: a flexible robot control architecture for autonomous underwater vehicles. In: OCEANS 2011. IEEE (2011)

11. Bingham, B.S., Walls, J.M., Eustice, R.M.: Development of a flexible command and control software architecture for marine robotic applications. Marine Technology Society Journal 45(3), 25–36 (2011)
12. http://oceanai.mit.edu/moos-ivp/pmwiki/pmwiki.php
13. Benjamin, M.R., Schmidt, H., Newman, P.M., Leonard, J.J.: Nested autonomy for unmanned marine vehicles with MOOS - IvP. Journal of Field Robotics 27(6), 834–875 (2010)
14. Fernández-Perdomo, E., Cabrera-Góomez, J., Domínguez-Brito, A.C., Hernández-Sosa, D.: Mission specification in underwater robotics. Journal of Physical Agents 4(1), 25–34 (2010)
15. Monferrer, A., Bonyuet, D.: Cooperative robot teleoperation through virtual reality interfaces. In: Proceedings of the Sixth International Conference on Information Visualisation. IEEE (2002)
16. Seto, M.L. (ed.): Marine Robot Autonomy. Springer (2013)

Estimation of Vehicle Pose Using Artificial Landmarks for Navigation of an Underwater Vehicle

Tae Gyun Kim[1], Hyun-Taek Choi[1,*], and Nak Yong Ko[2]

[1] Korea Institute of Ocean Science and Technology, Korea
{tgkim,htchoiphd}@kiost.ac
[2] Dept. Control, Instrumentation and Robot Eng., Chosun University, Korea
nyko@chosun.ac.kr

Abstract. This paper describes a localization method to localize a mobile vehicle in underwater environment. Particle filter based localization method is implemented, which is based on Bayesian filter to deal with non-linear system. This method comprises prediction step and correction step to estimate the pose of the vehicle. The prediction step is achieved by a motion model of the vehicle with inertial sensor data acquired from Doppler Velocity Log, inertia sensors, and electronic compass. In the correction step, the pose of the vehicle is updated using range and bearing information of externally fixed landmarks in the vehicle work space. The performance of the proposed localization method is verified by experiment in a tank environment using four artificial landmarks. In the experiment, the motion information of the underwater vehicle is used as surge and yaw velocities obtained from DVL and AHRS sensors. The landmark information is acquired from the artificial landmarks using an image sensor. The experimental result shows that the proposed method successfully estimated the pose of the vehicle.

Keywords: Underwater robot, navigation, localization, particle filter, artificial landmarks.

1 Introduction

Autonomous vehicles working in underwater environment essentially need navigation functions. One of the functions is localization which estimates the location and orientation of the vehicle by using motion and/or environmental information[1]. Localization is available for geometric map building, exploration, environment monitoring, and object manipulation in underwater environment[2-3].

A variety of localization methods for underwater vehicles have been researched. Dead-reckoning method has long been used for localization. The method estimates vehicle pose by using vehicle motion information received from DVL(Doppler Velocity Log), AHRS(Attitude Heading Reference System), IMU(Inertial Measurement Unit), etc. Unfortunately, it yields poor result of localization when used long time without correction due to accumulation of error for vehicle motion.

---

* Corresponding author.

J.-H. Kim et al. (eds.), *Robot Intelligence Technology and Applications 2,*
Advances in Intelligent Systems and Computing 274,
DOI: 10.1007/978-3-319-05582-4_74, © Springer International Publishing Switzerland 2014

One of practical methods for localization of a vehicle is sensor fusion of dead-reckoning and GPS information. This method estimates pose of a vehicle through dead-reckoning when working in underwater and GPS information is used to correct accumulated pose error. It spends time and power because the vehicle should move to the surface to receive GPS information. Moreover, this method is impossible to estimate pose of a vehicle in case when dead-reckoning is not worked in underwater.

Other methods using range and bearing information by installing an acoustic beacon system such as USBL[5], SBL[6], and LBL[7] in working area of a vehicle also have been developed. They estimate location using trilateration or triangulation along with least squares method[8]. Location error for the methods is not accumulated compared with dead-reckoning method since they received information instantaneously from the acoustic beacon system. However, the work space of the vehicle is limited as around the acoustic sensors, because accurate pose cannot be acquired outside of the detectable region of the sensor system[9].

The proposed method in the paper fuses dead-reckoning and range-bearing information which is received from artificial landmarks. It is based on Bayes theorem[10]. Generally, Bayes theorem is implemented by using particle filter[11-13] and Kalman filter methods[14-15]. The proposed method is developed using particle filter to cope with the non-linearity of the motion model of the vehicle and the measurement model. Moreover, the particle filter can provide reliable localization result even under the sensor errors because of multi-hypothesis concept. The proposed localization method uses underwater image information of artificial landmarks received from an image sensor. The bearing and range information is calculated by the received image information[16]. Thus, environmental information is given as the location of artificial landmarks. By fusing the dead-reckoning and landmark information, the proposed method can give a reliable localization result in underwater environment.

The remainder of this paper is organized as follows. Particle filter method to fuse motion of a vehicle and range and bearing received from artificial landmarks is described in the section 2. Section 3 depicts the experiment and comparison of proposed method with dead-reckoning method in a water tank. Section 4 concludes the paper with suggestions of the future research plans.

2   Estimation of Vehicle Pose by Particle Filter

The proposed method for localization of an underwater vehicle is particle filter method fusing motion and external environment information. The method is able to consider uncertainties for vehicle motion and external sensor information. Estimated pose of vehicle from the method is calculated by vehicle motion and external sensor information at time t and previous estimated pose at time t-1. Figure 1 shows a simple example of the estimation result for a vehicle pose by particle filter method. Estimated pose of a vehicle is represented by distribution of particles as shown the figure 1. Circles indicate the range and/or bearing measurement data of the robot from the external landmarks.

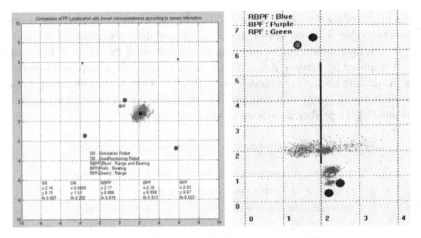

Fig. 1. An example of localization using particle filter

Table 1 depicts pseudo code of the localization method. The procedure is repeated at every time step using the estimated result from the previous time step. Line 3 illustrates the prediction step which applies *motion model*(·) to predict the pose of particles. This model uses previous estimated pose X_{t-1} at time t-1, along with vehicle motion u_t. Predicted particles are corrected by the process described at line 4~7. These lines illustrate the correction step. In line 4, *Sensor model*(·) obtains respective belief for predicted particles. This model uses measurement information z_t received from external sensors and predicted particles in *Motion model*(·), while external environment information $E=(E_{1,x}, E_{1,y}, E_{1,z}, ..., E_{n,x}, E_{n,y}, E_{n,z})$ which is locations for landmarks was already given. *Resampling*(·) procedure in line 7 corrects the distribution of particles. The distribution of particles is decided by calculating the beliefs in *Sensor model*(·).

Table 1. Pseudo code for particle filter

Localization PF(X_{t-1}, u_t, z_t, E)

1.	$\overline{X}_t = X_t = \phi$
2.	*for i = 1 to M do*
3.	$x_t^{[i]} = Motion\ model(u_t, x_{t-1}^{[i]})$
4.	$\omega_t^{[i]} = Sensor\ model(z_t, x_t^{[i]}, E)$
5.	*endfor*
6.	*for i = 1 to M do*
7.	$x_t^{[i]} = Resampling(\{(x_t^{[j]}, \omega_t^{[j]})j = 1, \cdots, M\})$
8.	*endfor*
9.	*return X_t*

The proposed method uses range and bearing information received from external landmarks. Table 2 shows sensor model and resampling procedure of particle filter method to utilize measurement information. As shown in the Table 2, belief of particles is acquired by *Feature-Based Sensor model*(·) and *Resampling*(·) is performed using the acquired belief. The models are conducted when at least one measurement is received at time t. If any measurement information is not received, the distribution is determined by the predicted particles in motion model.

Table 2. Pseudo code for sensor model and resampling

Procedure of Sensor model and Resampling

---

if there is even one detected landmark

$\bar{\mu}_{t,Belief}^{all} = 1$

Feature – Based Sensor model$(\bar{\mu}_t, z_t, c_t, m)$

$\bar{\mu}_t = Resampling(\bar{\mu}_{t,Belief})$

end if

return $\mu_t = \bar{\mu}_t$

Table 3 indicates detailed *Feature-Based Sensor model*(·) shown in the Table 2. The model calculates belief of predicted particles in *motion model*(·) of the Table 1 by utilizing received range and bearing. In this case, range and bearing information from landmarks are observed, and the range and bearing information between predicted particles and known lanmarks are calculated. The belief of predicted particles is obtained from the observed and the calculated range and bearing information.

Table 3. Pseudo code for sensor model of the proposed particle filter

---

Feature – Based Sensor model$(\bar{\mu}_t, z_t, c_t, m)$

---

for all particles $\bar{\mu}_t^k = (x_t^k \; y_t^k \; \theta_t^k \; Belief_t^k)^T$ *do*

  for all observed features $z_t^i = (r_t^i \; \phi_t^i)^T$ *do*

    $j = c_t^i$

    $q = (m_{j,x} - \bar{\mu}_{t,x}^k)^2 + (m_{j,y} - \bar{\mu}_{t,y}^k)^2$

    $\bar{z}_t^i = \begin{pmatrix} \bar{r}_t^i \\ \bar{\phi}_t^i \end{pmatrix} = \begin{pmatrix} \sqrt{q} \\ atan2(m_{j,y} - \bar{\mu}_{t,y}^k, m_{j,x} - \bar{\mu}_{t,x}^k) - \bar{\mu}_{t,\theta}^k \end{pmatrix}$

    $\bar{\mu}_{t,Belief}^k = \bar{\mu}_{t,Belief}^k \times prob(r_t^i - \bar{r}_t^i, \sigma_r) \times prob(\phi_t^i - \bar{\phi}_t^i, \sigma_\phi)$

  end for

end for

return $\bar{\mu}_{t,Belief}$

$prob(r_t^i - \bar{r}_t^i, \sigma_r)$ in the Table 3 is a weight value for uncertainty of received range information. The probability value is shown in equation (1). In (1), $p_{r,hit}$, $p_{r,short}$, $p_{r,max}$, and $p_{r,rand}$ are four probabilities of range information for measurement noise, unexpected objects, failure of detection, and unexplainable noise, respectively.

$$prob(r_t - \bar{r}_t, \sigma_r) = p_r(r_t / \mu_t, m) = \begin{pmatrix} z_{r,hit} \\ z_{r,short} \\ z_{r,max} \\ z_{r,rand} \end{pmatrix}^T \cdot \begin{pmatrix} p_{r,hit} \\ p_{r,short} \\ p_{r,max} \\ p_{r,rand} \end{pmatrix} \quad (1)$$

Equation 2 indicates a weight value of bearing information to expressed $prob(\phi_t^i - \bar{\phi}_t^i, \sigma_\phi)$ in the Table 2. The value has two uncertainties for received bearing information. The probabilities for the uncertainties indicate measurement noise $p_{\phi,hit}$ and unexplainable noise $p_{\phi,rand}$.

$$prob(\phi_t - \bar{\phi}_t, \sigma_\phi) = p_\phi(\phi_t / \mu_t, m) = \begin{pmatrix} z_{\phi,hit} \\ z_{\phi,rand} \end{pmatrix}^T \cdot \begin{pmatrix} p_{\phi,hit} \\ p_{\phi,rand} \end{pmatrix} \quad (2)$$

3   Experiment in a Tank

An experiment was carried out in a tank located at Korea Institute of Ocean Science and Technology[17] to verify the performance of the proposed particle filter method. Figure 2 shows the tank. The tank is 1.2 meters deep, 7.0 meters long, and 4.0 meters wide. Figure 3 indicates four artificial landmarks used for the experiment. They are cross, sphere, cylinder, and cone types. The size of the landmarks is maximum 0.2 meters height, 0.2 meters long, and 0.2 meters wide.

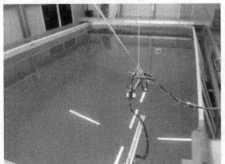

Fig. 2. Tank environment

Fig. 3. Artificial landmarks

Figure 4 shows the navigation system composed for the experiment. In the Figure 4(a), the overall navigation system is shown. The system consists of DVL[18] sensor, AHRS[19] sensor, image sensor[20], and lights. Figure 4(b) and (c) show DVL and AHRS sensor which measure motion information for the navigation system. Figure 4(d) shows image sensor which receives information of the landmarks.

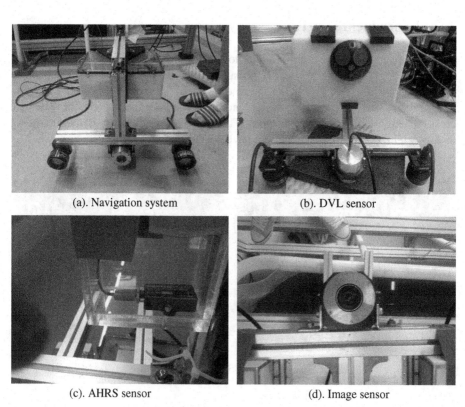

(a). Navigation system   (b). DVL sensor

(c). AHRS sensor   (d). Image sensor

Fig. 4. Navigation System Configuration

Figure 5 depicts the experimental condition and environment. Detail locations of the artificial landmarks are shown in Table 4. In the experiment, the navigation system moved back and forth between the two way points of wp1(2, 1.5)m and wp2(2, 5.5)m 9 times, repeatedly.

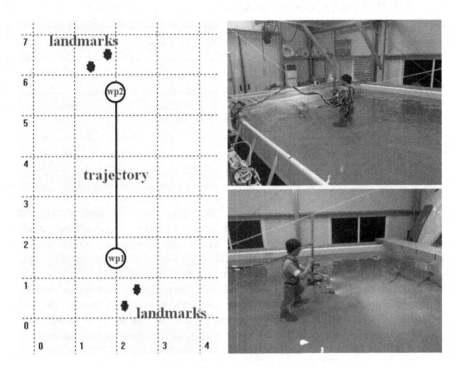

Fig. 5. Experiment environment

Table 4. Location of external landmarks

Landmarks		X(m)	Y(m)
1	Cross	1.8	6.5
2	Sphere	2.2	0.4
3	Cylinder	1.4	6.2
4	Cone	2.5	0.7

The motion information of the system was acquired as surge and yaw velocities received from DVL and AHRS sensors. This information was used for the prediction of particles in motion model of particle filter method. Range and bearing information of artificial landmarks was calculated by obtained image frame from the light and image sensor. The range and bearing information were used for calculating the belief of particles in sensor model.

In the experiment, 1000 particles were used for particle filter method and the initial poses of the particles were given as known initial pose of the navigation system.

The pose of the particles were updated by the proposed particle filter method with the rate 3Hz. Table 5 shows parameter values applied to motion and sensor model of the particle filter method and Table 6 shows weight parameter for the particles belief.

Table 5. Parameter for motion and sensor model of particle filter method

Motion model				Sensor model	
Surge velocity		Yaw velocity		Range	Bearing
α_{uu}	α_{ur}	α_{ru}	α_{rr}	σ_r(m)	σ_b(rad)
0.1	0.05	0.05	0.1	0.2	0.1

Table 6. Weight parameter for belief of particles

Range				Bearing	
z_{hit}	z_{short}	z_{max}	z_{rand}	z_{hit}	z_{rand}
0.9	0	0	0.1	0.9	0.1

Figure 6 shows the experimental result. The figure compares performance of the proposed particle filter method with dead-reckoning method. The coordinates marked by the PF represent the estimated locations by the proposed method and DR represents the location calculated by the dead-reckoning method. The estimated locations by PF are distributed between 1.567m~2.346m on x-axis and 1.232m~5.529m on y-axis. In case of DR, the calculated trajectory shows distribution which is 1.417m~2.808m and 0.602m~5.590m along x-axis and y-axis respectively. The experimental result shows that the proposed method is able to estimate pose of the vehicle by using information of landmarks when error of dead-reckoning is accumulated.

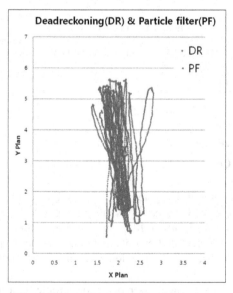

Fig. 6. Estimated location after 9 iterations

4   Conclusions

The paper describes particle filter method for the estimation of a vehicle in underwater environment. The proposed method fuses both dead-reckoning and landmark information. Especially, the sensor model for artificial landmarks considered the uncertainties of range information for measurement noise, unexpected objects, failure of detection, and unexplainable noise and bearing information for measurement noise and unexplainable noise. The experimental result in the tank environment shows that the proposed method can provide a reliable localization result even when the error of the dead-reckoing information is accumulated.

For further experiments, the performance of localization using proposed method will be compared with other filtering methods such as EKF and UKF. Moreover, the proposed particle filter method in this paper will be used to develop the new method for SLAM.

Acknowledgement. This work has been funded by Maritime and Ocean Engineering Research Institute(MOERI) in Korea Institute of Ocean Science & Technology (KIOST). The title of project is "Development of technologies for an underwater robot based on artificial intelligence for highly sophisticated missions".

References

1. Thrun, S., Burgard, W., Fox, D.: Probabilistic Robotics. MIT Press (2005)
2. Song, K.S., Chu, P.C.: Conceptual Design of Future Undersea Unmanned Vehicle (UUV) System for Mine Disposal. IEEE Systems Journal 6(4) (August 2012)
3. Moon, Y.S., Ko, N.Y., Sur, J., Lee, Y.O.: Development of Torpedo Type Mine Disposal Robot RUCUR MK-1. In: Proceedings of the Ninth International Symposium on Technology and the Mine Problem, pp. 17–21. Naval Postgraduate School, Monterey (2010)
4. Alameda, W.: Seadevil: A totally integrated inertial navigation solution. In: Proc. Underwater Intervention Symp., New Orleans, LA (2002)
5. Christ, R.D., Wernli Sr., R.L.: The ROV Manual: A User Guide for Observation Class Remotely Operated Vehicles, p. 320. Butterworth-Heinemann (August 2007)
6. Cheng, W.H.: Study mobile underwater positioning system with expendable and multi-functional bathythermographs. Ocean Engineering 32, 499–512 (2005)
7. Vaganay, J., Bellingham, J.G., Leonard, J.J.: Outlier rejection for autonomous acoustic navigation. In: Proceedings of the IEEE International Conference on Robotics and Automation, vol. 3, pp. 2174–2181 (April 1996)
8. Smith, J.O., Abel, J.S.: The Spherical Interpolation Method of Source Localization. IEEE Journal of Oceanic Engineering 12(1), 246–252 (1987)
9. Alcocer, A., Oliveira, P., Pascoal, A.: Underwater Acoustic Positioning Systems Based on Buoys with GPS. In: Proceedings of the Eighth European Conference on Underwater Acoustics, 8th ECUA, Carvoerio, Portugal (2006)
10. Wang, J., Gao, Q., Yu, Y., Wang, H., Jin, M.: Toward Robust Indoor Localization Based on Bayesian Filter Using Chirp-Spread-Spectrum Ranging. IEEE Transaction on Industrial Electronics 59(3), 1622–1629 (2012)

11. Donovan, G.T.: Position Error Correction for an Autonomous Underwater Vehicle Inertial Navigation System (INS) Using a Particle Filter. IEEE Journal of Oceanic Engineering 37(3) (July 2012)
12. Ko, N.Y., Kim, T.G., Noh, S.W.: Monte Carlo Localization of Underwater Robot Using Internal and External Information. In: 2011 IEEE Asia-Pacific Services Computing Conference Proceedings, pp. 410–15 (2011)
13. Maki, T., Matsuda, T., Sakamaki, T., Ura, T., Kojima, J.: Navigation Method for Underwater Vehicles Based on Mutual Acoustical Positioning With a Single Seafloor Station. IEEE Journal of Oceanic Engineering 38(1), 167–177 (2013)
14. Fallon, M.F., Papadopoulos, G., Leonard, J.J., Patrikalakis, N.M.: Cooperative AUV navigation using a single maneuvering surface craft. The International Journal of Robotics Research 29(12), 1461–1474 (2010)
15. Kim, T.G., Ko, N.Y.: Comparison and Analysis of Methods for Localization of a Mobile Robot. Journal of Korean Institute of Information Technology 11(1), 79–89 (2013)
16. Lee, D., Kim, G., Kim, D., Myung, H., Choi, H.: Vision-based object detection and tracking for autonomous navigation of underwater robots. Oceanic Engineering 48, 59–68 (2012)
17. http://www.kiost.ac/kordi_web/main
18. http://www.dvlnav.com/explorer.aspx
19. http://www.microstrain.com
20. http://www.cylod.com

A New Approach of Detection and Recognition for Artificial Landmarks from Noisy Acoustic Images

Yeongjun Lee, Tae Gyun Kim, and Hyun-Taek Choi

Korea Institute of Ocean Science and Technology, Korea
{leeyeongjun,tgkim,htchoiphd}@kiost.ac

Abstract. This paper presents a framework for underwater object detection and recognition using acoustic image from an imaging sonar. It is difficult to get a stable acoustic image from any type object because of characteristic of ultrasonic wave. To overcome the difficulties, the framework consists of the selection of candidate, the recognition, and tracking of identified object. In selection of candidate phase, we select candidate as possible objects using an initial image processing and get rid of noise or discontinuous object using a probability based method in series of images. The selected candidate is processed in adaptive local image processing and recognition using shape matrix recognition method. Identified object in previous phase is tracked without selection of candidate, and recognition phase. We perform two simple tests for the verification of each phase and whole framework operability.

Keywords: Underwater recognition framework, Imaging sonar, artificial landmark.

1   Introduction

In recent years, object recognition and position estimation using image processing have been researched in underwater robot fields. Although an optical camera is used for underwater image processing, it has a number of problems such as light, turgidity, floaters, electric power, etc. For this reason, an imaging sonar is employed by many researchers [1][2][3].

Imaging sonar is used to measure the direction and intensity of the echoes against the object. Unfortunately, the imaging sonar suffered from the characteristics of ultrasonic wave, acoustic image coordinate, and underwater condition even when a known target object is used [4][5]. These make it is difficult to detect and to recognize objects using imaging sonar in underwater environment. This paper proposes a framework for reliable object detection and recognition to cope with the defects of the imaging sonar.

1.1   Framework

The proposed method is expanded from our previous research [6]. A framework for object detection and recognition was proposed and a simple test was performed.

J.-H. Kim et al. (eds.), *Robot Intelligence Technology and Applications 2*,
Advances in Intelligent Systems and Computing 274,
DOI: 10.1007/978-3-319-05582-4_75, © Springer International Publishing Switzerland 2014

In this paper, we expand the algorithm to apply to the multi object detection and recognition. Fig.1 shows a proposed framework for the object detection and recognition in underwater environment. In the proposed method, we present a multiple selection of candidate and filtering method with in series of images for the multi object detection.

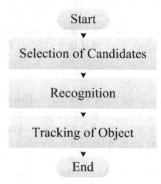

Fig. 1. A framework of detection and recognition method

1.2   Artificial Landmark

The proposed method uses artificial objects as target objects. Natural objects in underwater environment have non-fixed and complex shapes, which make it difficult to recognize them. Furthermore, the characteristics of the acoustic image are very unstable due to a speckle like a salt noise and unstable intensity in the same region [4]. For this, we develop an artificial landmark which is considering the characteristics of ultrasonic waves and imaging sonar [2], and the artificial landmark is used to verify the proposed object recognition algorithm.

2   Object Detection and Recognition Method

As shown in Figure 1, the proposed object recognition framework consists of three parts, which are the selection of candidates of target objects, the recognition of filtered candidates, and tracking of identified objects. The selection of candidates is specifically divided into two steps. First, we try to find likelihood candidates of objects from whole image. Then, a probability based method similar to particle filter is applied to filter out fake candidates in series of images. This makes a chance to use more efficient, forceful noise filtering. The filtered candidates are processed by local image processing in a small region, and then we use the shape matrix method to identify candidate's ID. After a target object is recognized, it is tracked by mean-shift tracking algorithms without performing the candidate selection and recognition procedures.

2.1   Selection of Multiple Likelihood Candidates

As mentioned, the proposed method used artificial landmarks as target objects. The candidates are obtained from the whole image according to the used landmark model. The used landmarks have circular shape with various inner fan shapes. Therefore, the likelihood candidates which could be matched with target objects are acquired as circular shapes in the image.

For the selection of likelihood candidates, a matching map is needed. It is calculated by comparing the landmark model and the extracted edge image using Hough circle transform (Fig.2-b). The matching map represents a probability that target

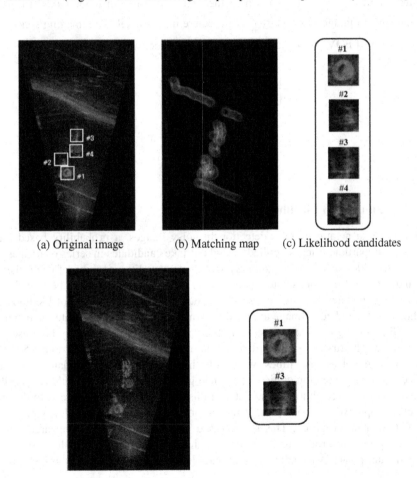

(a) Original image   (b) Matching map   (c) Likelihood candidates

(d) Filtering candidates   (e) Result (Selected Candidates)

Fig. 2. Process of selection of candidates

objects would be exist in the received acoustic image. The regions which have high probability in the matching map is considered as likelihood candidates (Fig.2-c). At most 5 regions are extracted as likelihood candidates from each image. . If all matching values are smaller than the number of edge pixels of landmark model, we decide that any target object does not exist in the current image.

An outer model of artificial landmark is calculated by the following equations. (1) is a distortion equation due to pitch (downward) of imaging sonar. (2) is a scale equation which consider the image scale from the range of the field of view (FOV). An outline of a landmark at X and Y axis is (x_{circle}, y_{circle}), and a distorted outline related to pitch angle is ($x_{distort}$, $y_{distort}$), where the unit is metric. An outer model of artificial landmark in image plane is (x_{img}, y_{img}), where unit is pixel. The imaging sonar's pitch and range of FOV are θ_D and $window\_length$, respectively.

$$\begin{bmatrix} x_{distort} \\ y_{distort} \end{bmatrix} = \begin{bmatrix} 1 & 0 \\ 0 & \cos(\theta_D) \end{bmatrix} \begin{bmatrix} x_{circle} \\ y_{circle} \end{bmatrix} \qquad (1)$$

$$\begin{bmatrix} x_{img} \\ y_{img} \end{bmatrix} = \begin{bmatrix} 1/(window\_length/512) & 0 \\ 0 & 1/(window\_length/512) \end{bmatrix} \begin{bmatrix} x_{distort} \\ y_{distort} \end{bmatrix} \qquad (2)$$

2.2   Filtering of Likelihood Candidates

To decrease a ratio of fake detection in noisy images, a probability based method similar to particle filter is applied to find out fake candidates in series of images. First, 1000 to 2000 particles are randomly scattered around the position of each likelihood candidate, and it is designated as a particle group (Fig.2-d). Then, the belief of each particle is updated by using consecutive images. Like the selection of likelihood candidate, extracted edge image and landmark model are used to calculate the belief.

All particle groups are considered three conditions because it may be noise or disappeared. The first, if a standard deviation of particle group is larger than 25% of size of outer model for five times, we decide that the likelihood candidate is not object. Second, it is the agglomeration of particle groups. The particles could be possible to detect the same likelihood candidate. In this case, we should be forced to delete particle groups except one. Third, if recognition of candidate is failed sequentially, we delete its particle group. This is processed when the object is disappeared in image. Filtered particle groups or likelihood candidates are to be the result of selection of candidates phase. A mean position of particle group is offered to recognition phase.

2.3   Recognition of Filtered Candidates

As a recognition phase, multiple trials for recognition allow using various pre-processing filters just for a small region of each likelihood candidate where pre-processing filters are prepared in case study in advance. Fig.3 is an example of pre-processing filter which is adapted in this research. The input image is filtered by the median and the bilateral filter. And then, filtered image is to be black-white image by otsu-thresholding method.

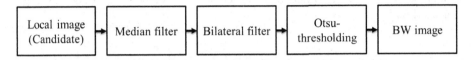

Fig. 3. Process of pre-processing filter

After each pre-processing, we use the shape matrix method for recognition. The artificial landmarks have an I.D separately. First, calculate a shape matrix at the candidate position in the image plane, then calculate a similarity between the given shape matrix of each landmark and obtained shape matrix [5]. If a value of similarity after pre-processing is over certain threshold, the candidate gets I.D. A matching function between shape matrix A (known) and B (obtained) is (3).

$$M(A,B)=1-\frac{1}{m\times n}\sum_{j=1}^{n}\sum_{i=1}^{m}\left|A_{ij}-B_{ij}\right| \tag{3}$$

2.4 Tracking of Identified Object

The identified objects (or landmark) are tracked without selection of candidates and recognition in series of images. The object tracking is achieved by general mean shift tracking method [8]. Instead of using Hue image, the proposed method performed mean shift tracking using intensity image because the imaging sonar provides only one channel image.

As shown in Fig.4, it shows the result of the tracking. Each tracking result indicates the target object properly, but it does not indicate on the center of object in Fig.4-b. This reason is the lack of the information which is one channel and unstable image.

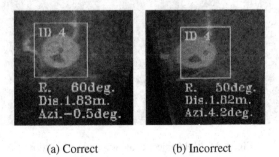

(a) Correct   (b) Incorrect

Fig. 4. Result of mean-shift tracking in single object

3   Experiment

To verify the proposed object recognition framework, an experiment was performed in a pool, and four artificial landmarks which have distinctive I.D. are used as target objects (Fig. 5).

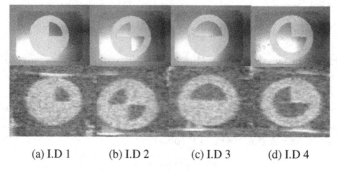

(a) I.D 1   (b) I.D 2   (c) I.D 3   (d) I.D 4

Fig. 5. An artificial landmarks

3.1   Multiple Object Detection and Recognition in Sequential Image

The first experiment was carried out to verify the performance of the object detection and recognition when the multiple objects appeared in successive images. In this experiment, only target objects were located in the pool.

The experimental result is shown in Fig.6. At the first image, ID.2 was recognized at the center of the image successfully. During tracking the recognized object (ID.2), another object appeared from upper right side of the subsequent images. Then, the new object was recognized as the landmark ID.3. At the last image, both the landmarks ID.2, and ID.3 are identified and tracked at the same time. It is verified that the object is not selected the same candidate repeatedly.

Fig. 6. Multiple object detection and recognition

3.2   Robust Object Detection and Recognition in Multi-shape Object

The second experiment was performed to verify the proposed method when unknown objects were present. The proposed algorithm is designed to detect circular shape of objects because the developed artificial landmark has a circle shape. So, we performed a test about the robustness of the detection algorithm in the presence of various shapes of objects in the pool.

Fig.7 shows that the various objects move from right to left in sequential images. The shapes of objects are star, rectangle, grid, etc. As a result, probability based filtering method worked well. All the unknown types of shapes except circle were filtered out. Even though some objects were not filtered out in the selection of candidates, it could be deleted in the recognition phase because of the similarity test of shape matrix.

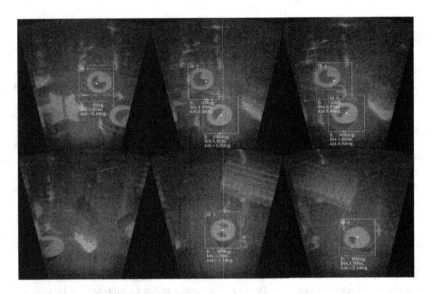

Fig. 7. Robust recognition in multi-shape object

4   Conclusion

This paper proposed the framework for underwater object recognition using imaging sonar. The acoustic image received from imaging sonar is unstable due to ultrasonic waves. So, it is difficult to detect and to recognize target objects. To solve this problem, we selected a number of likelihood candidates as high-score similarity object using Hough circle transform. Then, false likelihood candidates were filtered out by the probability based filtering method before recognition phase. The selected candidates after filtering process were used to recognize the target objects using shape matrix method. An identified object was tracked by the mean shift method. Using these processes, the proposed method can provide a reliable detection and recognition using imaging sonar in underwater environment.

We performed two simple tests in the pool for the verification the proposed method. As a future direction, we plan to run the proposed method in real-time and perform experiments in a real sea environment. And, we will try to extend proposed framework to recognize natural objects in the near future.

Acknowledgements. This work has been done by "Development of technologies for an underwater robot based on artificial intelligence for highly sophisticated missions." funded by Korea Institute of Ocean Science & Technology (KIOST) and "Development of an autonomous swimming technology with less than 1.0m position error for underwater robot operating in man-made structural environment" funded by Ministry of Trade, Industry & Energy (MOTIE).

References

1. Lee, Y., Choi, H.: A Comparative study on feature extraction methods for environment recognition using underwater acoustic image. In: The 8th Korea Robotics Society Annual Conference, South Korea (2013)
2. Lee, Y., Han, K., Choi, H.: Design and implementation of artificial landmark for underwater acoustic camera. In: Information and Control Symposium, South Korea (2012)
3. Johansson, H., Kaess, M., Englot, B., Hover, F., Leonard, J.: Imaging sonar-aided navigation for autonomous underwater harbor surveillance. In: IEEE/RSJ International Conference on Intelligent Robots and Systems (2010)
4. Lee, Y., Choi, H.: Design and implementation of artificial landmark for underwater acoustic camera. In: 2012 Information and Control Symposium, South Korea (2012)
5. Yu, S., Kim, T., Asada, A., Weatherwax, S., Collins, B., Yuh, J.: Development of high-resolution acoustic camera based real-time object recognition system by using autonomous underwater vehicle. In: Proceedings of MTS/IEEE OCEANS 2006 Boston Conference and Exhibition, Boston, USA, September 15-21 (2006)
6. Lee, Y., Kim, T., Choi, H.: Preliminary Study on A Framework for Imaging Sonar based Underwater Object Recognition. In: The 10th International Conference on Ubiquitous Robots and Ambient Intelligence, South Korea (2013)
7. Goshtasby, A.: Description and Discrimination of Planar Shapes Using Shape Matrices. IEEE Transactions Pattern Analysis and Machine Intelligence (1985)
8. Comaniciu, D., Ramesh, V., Meer, P.: Real-Time Tracking of Non-Rigid Objects using Mean Shift. In: IEEE Conference on Proceedings of Computer Vision and Pattern Recognition (2000)

Implementation of Split-Cycle Control for Micro Aerial Vehicles

Isaac E. Weintraub[1], David O. Sigthorsson[1],
Michael W. Oppenheimer[2], and David B. Doman[2]

[1] General Dynamics Information Technology,
2210 8th St., Air Force Research Laboratory, WPAFB, OH 45433-7531,
5100 Springfield St., Dayton, OH 45431, USA
{Isaac.Weintraub,David.Sigthorsson}@gdit.com
[2] Autonomous Control Branch, Power and Control Division, Aerospace Systems
Directorate, 2210 8th St., Air Force Research Laboratory, WPAFB, OH 45433-7531,
5100 Springfield St., Dayton, OH 45431, USA
{Michael.Oppenheimer,David.Doman}@wpafb.af.mil

Abstract. Flapping wing micro air vehicles have been of significant
research interest in recent years due to the flight capabilities of their
biological counterparts and their ability to hide in plain sight, inspiring
applications for military and civilian surveillance. This work introduces
the design, implementation, and fabrication of the circuitry used for split-
cycle constant-period wingbeat capable flapping wing micro air vehicle
platforms. Split-cycle constant-period modulation involves independent
control of the upstroke and downstroke wing velocity profiles to provide
the theoretical capability of manipulating five degrees of vehicle motion
freedom using only two actuators, namely, a brushless direct current motor for each wing. The control circuitry mainly consists of a control circuit board, a wireless receiver, three micro-controllers, and drivers. The
circuitry design is tested using a prototype vehicle mounted on an air-
table platform. A human operated transmitter relays split-cycle constant-
period commands to the vehicle to produce the desired vehicle motion.

1 Introduction

Flapping wing micro air vehicles (FWMAV) have been of significant interest in
recent years because of the attractive flight capabilities of their biological counterparts and their potential to hide in plain sight, for example for military and
civilian surveillance purposes. The development of flapping wing micro air vehicles involves a range of design criteria. Researchers have developed promising
platforms from insect sized piezoelectric actuated vehicles[1] to hummingbird
mimicking electromechanical systems[2], and a number of others[3–7]. This work
focuses on the design, implementation, and fabrication of the circuitry involved
with the development of split-cycle constant-period frequency modulation (SC-
CPFM) capable FWMAV platforms. SCCPFM involves independent control of
the upstroke and downstroke wing velocity profiles, providing the theoretical
capability of manipulating up to five degrees of freedom of FWMAV flight using

J.-H. Kim et al. (eds.), *Robot Intelligence Technology and Applications 2*,
Advances in Intelligent Systems and Computing 274,
DOI: 10.1007/978-3-319-05582-4_76, © Springer International Publishing Switzerland 2014

only two actuators[7–10]. Two brushless direct current (BLDC) motors drive two flapping wings and are controlled using amplified microcontrollers along with Hall effect sensors and encoders for position and velocity feedback. A human operated transmitter sends remote control commands which are interpreted using a microcontroller and used to manipulate the SCCPFM of the FWMAV wingbeat. The development of the needed circuitry involves the interconnectivity of the microcontrollers, motor driver circuitry, sensors, and remote control transceiver, along with the software involved with replicating and manipulating a desired variable velocity profile for the wing stroke. These features must be implemented under limiting mechanical and computational constraints, namely minimizing weight and size.

Figure 1 shows a rendering of the prototype FWMAV used in this work. The FWMAV platform consists of wings, a fuselage, a four-bar linkages connecting the wing spars (rockers) to the motor shafts (cranks) of two BLDC motors, and the associated encoders and Hall effect sensors. The (z, y) plane shown on the figure denotes the stroke-plane, the plane in which the wing spars lie during flapping, and the positive z-axis denotes the forward direction. The upstroke is defined as starting at the wings maximum positive position (forward of the origin of the body-axes) and moving to its most negative position (aft of the origin of the body-axes), while the downstroke is defined as starting at the wings most negative position and moving to its most positive position. The full range of wing motion is determined by the four-bar linkage and extends 1 rad in each direction away from the nominal position shown in Figure 1.

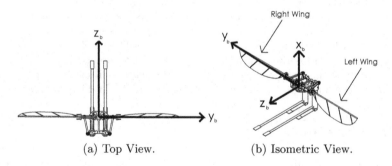

(a) Top View.   (b) Isometric View.

Fig. 1. Rendering of the Prototype FWMAV Fuselage and Motor Assembly

The general FWMAV architecture is presented by a flowchart in Figure 2. The bottom half of the flowchart corresponds to the electromechanical system shown in Figure 1 while the top half corresponds to the circuit board and wireless receiver module featured in this study. The wireless receiver module converts radio frequency signals from a human operated transmitter to pulse position modulated (PPM) signals fed to the main circuit board. The circuit board employs microcontrollers to map the PPM signals to SSCPFM commands fed to the embedded motor controller and motor driver circuitry. The motor control receives feedback from the motors using encoders and Hall effect sensors.

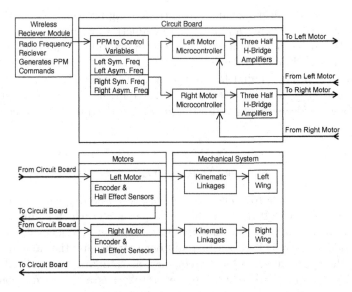

Fig. 2. FWMAV Architecture

This study is organized as follows. The concept of SCCPFM is introduced in Section 2. The electronic architecture employed to enable a human operator to control the movement of the FWMAV prototype using SCCPFM is described in Section 3 . Section 4 illustrates how the electronic architecture is incrementally designed and implemented. Experimental results verifying the primary function of the final circuit board design are given in Section 5. Finally, Section 6 provides concluding remarks.

2   Split-Cycle Constant-Period Frequency Modulation

The fundamental idea of split-cycle constant-period frequency modulation (SC-CPFM) is to, for a given period, piece together two half cosine waves prescribing the stroke angle such that the period remains unaltered while the half cycle durations may differ from each other. When no split-cycle modulation is applied, the two half cosine waves have the same frequency, namely that corresponding to the given fundamental period. When non-zero split-cycle modulation is applied, either the upstroke or the downstroke will be impeded, while the opposite stoke is advanced. The development of the split-cycle technique for the right wing is exactly the same as for the left wing. Therefore, only general expressions will be considered.

For symmetric flapping, which means the upstroke and downstroke utilize the same frequencies, the wing position is defined as

$$\phi_U = \cos(\omega t), \quad 0 < t < \pi/\omega$$
$$\phi_D = \cos(\omega t), \quad \pi/\omega \le t < 2\pi/\omega \tag{1}$$

where ϕ_U, ϕ_D are the upstroke and downstroke wing positions, respectively, for either the right or left wing and ω is the symmetric or fundamental wingbeat frequency. When flapping symmetrically, the magnitude of the wing velocities on the upstroke and downstroke are equal.

With SCCPFM, for the first half of the cycle (the upstroke), the frequency is altered from ω to $\omega - \delta$, while for the second half of the cycle (the downstroke), the frequency is altered from ω to $\omega + \sigma$. Thus, the position of the wing is defined by

$$\phi_U = \cos\left[(\omega - \delta)\,t\right], \qquad\qquad 0 < t < \pi/_{\omega-\delta}$$
$$\phi_D = \cos\left[(\omega + \sigma)\,t + \xi\right], \quad \pi/_{\omega-\delta} \le t < 2\pi/_{\omega} \qquad\qquad (2)$$

where δ is the split-cycle frequency, $\sigma = \delta\omega/(\omega-2\delta)$ is the frequency modification parameter, and $\xi = (-2\pi\delta)/(\omega-2\delta)$ is the phase parameter which maintains stroke angle continuity. If $\delta = 0$, the fundamental wingbeat frequency for both the upstroke and downstroke is prescribed and both strokes use the same frequency. For $\delta > 0$, the upstroke has a lower frequency, as compared to the fundamental frequency, and is impeded, while the downstroke has a higher frequency and is advanced. For $\delta < 0$, the upstroke has a higher frequency, as compared to the fundamental frequency, and is advanced, while the downstroke has a lower frequency and is impeded. By altering the frequencies of the upstroke and downstroke, non-zero cycle-averaged drag can be generated, resulting in the ability to translate fore, translate aft, and roll the vehicle. A nonzero split-cycle frequency yields a difference between the velocities of the upstroke and downstroke. This velocity difference results in different dynamic pressure on the wings and hence, different aerodynamic forces generated by the flapping wing.

Figures 3a and 3b show wingbeat patterns for $\delta > 0$ and $\delta < 0$ along with a symmetric wingbeat. In Figure 3a, the upstroke is impeded and the downstroke is advanced, while in Figure 3b, the upstroke is advanced and the downstroke is impeded. The split-cycle frequency, δ, along with ω for the right wing and left wing, are control parameters used to control the generated forces and moments on the vehicle.

3 Electronic Architecture

The electronic architecture is designed to implement SCCPFM on the prototype FWMAV platform, enabling a human operated remote control (RC) to dictate SCCPFM commands to the controlled FWMAV actuation in order to perform desired vehicle maneuvers. The actuator controller generates a position and velocity reference according to commands from the remote control receiver, dictating SCCPFM wing motion, measures the position and velocity of the motor shaft (via encoders), performs error integration, and implements PID control on the BLDC motors. Implicitly, the controller must also commute the BLDC motors in accordance to the rotor positions measured by Hall sensors. The implementation of these features needs be designed for simplicity in order to facilitate rapid computation on small and light electronics components (microprocessors).

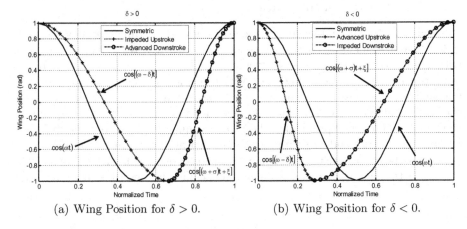

(a) Wing Position for $\delta > 0$.   (b) Wing Position for $\delta < 0$.

Fig. 3. Split Cycle Wing Positions.

Figure 4 shows the high level circuit architecture employed in order to implement remote controlled SCCPFM of the FWMAV actuation. Following from left to right in Figure 4, a wireless module receives commands from a human operated RC transmitter. The RC commands are translated into a pulse position modulated (PPM) signal by the wireless module. A microcontroller, PIC24FJ64GA102, senses the PPM signal and maps them to control variables dictating the reference SCCPFM wingbeat. The control variables are sent to the two motor controlling microcontrollers, PIC24FJ64GA102, using the serial peripheral interface (SPI) standard. These microcontrollers implement trajectory tracking for the motors, utilizing information from the encoders and Hall effect sensors for feedback. The motor controllers generate the reference prescribed by the control variables and employ the motor drivers to energize the motors according to a PID controller and the prescribed commutation sequence. The motor controllers account for inverting the kinematics of the four bar linkage and perform disturbance rejection. Pulse-width-modulation (PWM) is used to command the speed and torque of the motors.

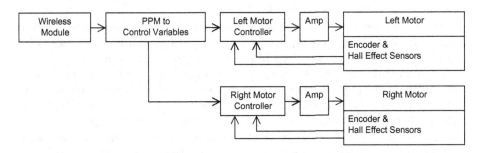

Fig. 4. Circuit Architecture

3.1   Remote Control Integration

The DelTang©RX31 remote control receiver, shown in Figure 5, provides a small (9.3 × 9.9mm) and light (0.21gram) package for a wireless module to interface with a human operated RC transmitter such as the Spektrum DX6i©. The RX31 outputs PPM signals, encoding the position of the user operated sticks on the transmitter. The position of the pulses are measured with an internal timer on the microcontroller and translated in software into four values corresponding to timing commands. Specifically, the timing commands indicate time duration of each half stroke and are supplied to the motor control microcontrollers on request at the end of each stroke, via SPI. The mapping from stick commands to SCCPFM timing commands is application specific and described in the results section.

Fig. 5. Deltang Wireless Receiver (Courtesy of www.deltang.co.uk)

3.2   Actuator Implementation

The two wing of the FWMAV prototype are actuated via a 4 bar linkage and two BLDC motors selected for small size, low weight, relatively high power and efficiency, and integration of encoders for position and velocity measurement. The BLDC motors are the Faulhaber© 0620C006BK1674. Each motor has an outside diameter of 6mm and contains an encoder and Hall effect sensor used to detect the motor shaft position and the commutation sequence. Each Faulhaber©package weighs 5g, which includes a motor, planetary gear head, magnetic encoder, and Hall effect sensor. Figure 6 shows two Faulhaber© 0620C006BK1674 motors installed in a FWMAV fuselage. The motor, gear-head, and electrical connections are labeled for clarity.

The cost of using the traditionally more efficient small BLDC motor over brushed motors is the increased complexity of the motor driving control. The (simplified) three motor windings must be energized in a given commutation sequence according to the rotor position in order to drive the motor shaft clockwise (CW) or counter-clockwise (CCW). The Hall effect sensors are used to obtain that position feedback and must be monitored by the motor control microcontroller. Figure 7 shows an example of the Hall sensor readings for different motor positions and the arrows on the three motor windings, connected in a star configuration, show how the motor windings need to be energized for CW rotation. This translates into computational cost on the microcontroller, which must

Fig. 6. BLDC Motor, Gear-head, Sensors, and Fuselage

be able to constantly monitor the Hall sensors. The field strength supplied by each winding is controlled by employing pulse width modulation (PWM) to the energizing of the windings at a frequency significantly higher than that of the rotational velocity of the motor, 16kHz in this study. The motor control microcontroller dictates the duty cycle of the PWM; the higher the duty cycle, the higher the torque of the motor, and vice versa.

The energization of the motor windings is performed by metal oxide semiconductor field effective transistors (MOSFET) connected in a standard three phase half bridge formation, with one MOSFET connecting each winding to the positive voltage supply and another MOSFET to ground (negative supply). The microcontrollers are connected to the gate terminals of the MOSFETs, with or without a pre-driver, and assign them to either be "on" or "off" according to the commutation, Hall effect sensor feedback, and the PWM duty cycle command. Labeling the windings as phases A, B, and C, Figure 8 shows the measured waveforms applied to the gate terminal of the MOSFETs as described in Table 1 while the motor control is in closed loop operation. Note that the controller is making online adjustments to the PWM duty cycle as it is adjusting to disturbances and changes in the reference velocity due to the inverse kinematics.

3.3   Actuator Control

The actuator controller must generate a position and velocity reference according to commands from the remote control receiver, dictating SCCPFM rocker

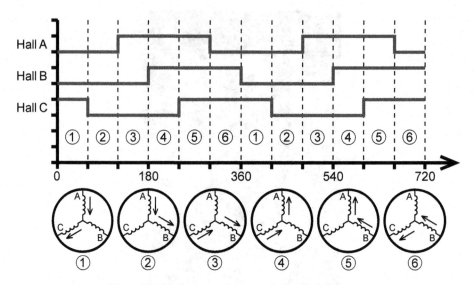

Fig. 7. Commutation Sequence of a BLDC Motor

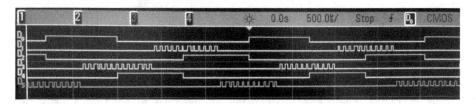

Fig. 8. BLDC Motor Commutation Signals Monitored by an Oscilloscope

Table 1. Description of The Commutation Monitoring Oscilloscope Connections

Signal	Description
D_0	The upper N-MOSFET connected to the A lead on BLDC motor
D_1	The lower P-MOSFET connected to the A lead on BLDC motor
D_2	The upper N-MOSFET connected to the B lead on BLDC motor
D_3	The lower P-MOSFET connected to the B lead on BLDC motor
D_4	The upper N-MOSFET connected to the C lead on BLDC motor
D_5	The lower P-MOSFET connected to the C lead on BLDC motor

motion, measure the position and velocity of the crank (via encoders), perform error integration, and implement PID control on the BLDC motors. The controller must also commute the BLDC motor in accordance with the rotor position measured by Hall effect sensors. The implementation of these features needs be designed for simplicity in order to facilitate rapid computation on small and light electronics components (microprocessors).

Figure 9 provides an overview of the control actuation. The encoder provides incremental position measurements via a counter and a timer is used to measure the duration between encoder transitions, which is converted into a velocity measurement. The timer information is also used for cumulative numerical integration of the position error. The measured position and velocity are compared to reference values generated using a period timer and lookup-tables. The lookup-tables are derived from the inverse kinematics of the crank-rocker mechanism to produce cosinusoidal half strokes when progressing between table entries at constant time intervals. The period timer is set according to the commanded lengths of time between table entries given via SPI communications with the receiver microcontroller. The increment values can be different for each half of the two tables in order to generate a split-cycle reference signal. The position and velocity errors are computed and the position error must be wrapped due to the periodic nature of the system. Gains are applied to the PID errors and the resultant PID signal is fed to a PWM generator to set the duty-cycle of the motor coil energization. Finally, Hall sensors provide feedback to control which motor coils are energized in order to commute the motor in the direction indicated by the sign of the PID control signal.

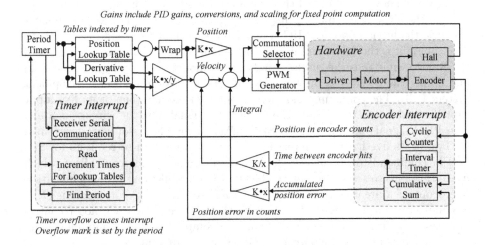

Fig. 9. Motor Control Block Diagram

4   Circuit Board Development

The circuit board design needs to accept commands from the wireless module and control the SCCPFM actuation of the FWMAV wings, providing the human operator with control over the motion of the FWMAV. The basic structure is to use three microcontrollers, one for communication and two for motor control, allowing for simplicity and scalability to a small size and lightweight designs, while maintaining the necessary computational capabilities. The three microcontrollers, software, and the motor driver circuitry were designed in an iterative process to allow for identifying challenges or faults in the design while gradually improving the size and weight of the circuit. The progression is divided into circuit board versions from the relatively crude original development stage Version 1, up to the latest design Version 4, which is under 1 square inch in size with minimal performance compromises.

4.1   Version 1

The first version of the control circuitry was designed mainly as a proof-of-concept using scalable electronics. Three MikroElektronika©LV 24-33 v6 development systems equipped with Microchip©PIC24FJ96GA010 16bit microcontrollers operating at a clock frequency of 32MHz with a phased-locked-loop (PLL) were chosen as they offered a versatile suite of computational abilities while allowing for scalability to smaller fully compatible Microchip©microcontrollers. The BLDC motor driver circuit was designed as a standard three phase half-bridge using power MOSFETs for switching with redundant current and voltage tolerances. The size and weight consideration for the drivers were not considered in this version but left for later iterations.

Figure 10 shows an image of the assembled Version 1 of the circuit implementation. The large green square printed circuit boards (PCB) are the development systems, the human operated RC transmitter is in the lower right corner, above it to the right is the driver circuit and inter-connection breadboard, to the left of the RC transmitter is the fuselage and motors mounted on a white plastic disk, and finally the receiver is on lower left of center beside the lower development system PCB. The communication microcontroller employs one of its five input captures and timers to measure the serial PPM signal from the wireless module. The PPM signals are mapped to SCCPFM commands in software written in C. Two of its five hardware interrupts are used to respond to requests for SC-CPFM commands from each of the two motor control microcontrollers, which use output pins to send information. The two SPI modules on the communication microcontroller are used to relay the SCCPFM commands to corresponding SPI modules on the motor control microcontrollers. Finally, the two motor control microcontrollers each use three of their five comparator/PWM modules along with three output pins to control the BLDC motor drivers.

This architecture implementation was used to successfully drive the BLDC motors and flap the wings with the commanded velocity profiles. Although this implementation was not flight ready, it was the first implementation of SCCPFM in flapping wing micro air vehicles.

4.2   Version 2

The development systems in Version 1 provided excessive versatility for this application and thus included numerous capabilities that could be eliminated to reduce size and weight. An incremental design step was taken by removing unnecessary electronics and hardware, doing away with the large development

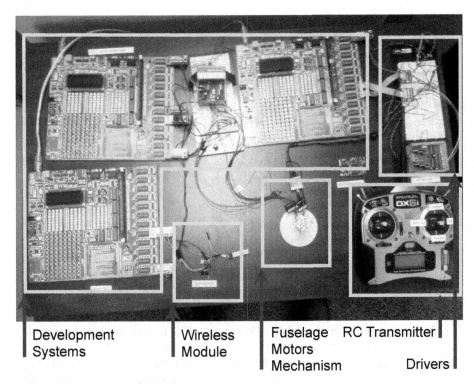

Development   Wireless   Fuselage   RC Transmitter
Systems   Module   Motors
                               Mechanism   Drivers

Fig. 10. Version 1 Circuit Board

systems. All control related hardware was implemented on one two-layer PCB, using all surface mount components except for a power switch, signal monitoring through-hole headers, and connector through-hole headers for convenience. All the driver circuitry was implemented on another two-layer PCB using surface mount driver components with less conservative power tolerances, thus greatly reducing size and weight. Breakout boards were used to connect the motors to the driver and control circuity. The resulting Version 2 of the development can be seen in Figure 11 with the part list given in Table 2. Viewing Figure 11, progressing left to right, the controller board is followed by the driver board, and then four breakout boards to allow connection to the BLDC motors, encoders, and Hall effect sensors.

The microcontrollers in Version 2 are small, 6×6mm, 28pin, PIC24FJ64GA102 which retain the required capabilities utilized in Version 1, namely their feature suite includes:

- Up to 16 MIPS Operation at 32MHz
- 16 Peripheral Pin Selection (PPS) Option
- 5.5V tolerant inputs
- 5 Timers
- 5 input Capture (IC) pins

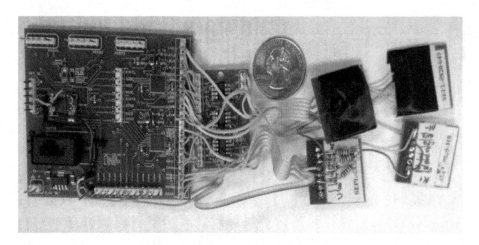

Fig. 11. Version 2 Circuit Board

- 5 Output Compare / Pulse Width Modulation (PWM) Outputs
- 2 Serial Peripheral Interface Buses

The PPS feature was not available in Version 1. This feature allows any module, such as PWM and IC, to be mapped to any of 16 pins using software only. This allows for considerable versatility in circuit layout design as well as allowing Version 2 to be used as a general development platform for later iterations, testing different pin layout designs and associated software.

This second development iteration, although not flight ready, provided a significant reduction in size and weight. Instead of a desktop crowded with electronics, a relatively small package with all required power, control, and amplification electronics was made available.

4.3   Version 3

Combining all communication, control, and driver electronics on one board allows significant reduction of size and weight. Version 3, shown in Figure 12, was the first iteration to implement this unification with Table 3 showing the part list. In the interest of simplicity and compatibility, the design and implementation of Version 2 were integrated with minimal alterations onto one four-layer PCB layout. The extra layers allowed for compacting the PCB without sacrificing signal integrity or power capacity. The integration included accurately spaced motor and sensor flexible flat cable (FFC) surface mount connectors on to the PCB for direct connection to the motors. Surface mount jumpers allow for configuration of the encoder sensors. This PCB still has a through-hole power switch for safety. A 14pin FFC surface mount connector provides access to programming pins via a breakout board with headers, not shown in the figure. Another 14pin FFC connector provided access to redundant microcontroller pins for debugging

Table 2. Parts List For Version 2

Part Number	Description	Package
PIC24FJ64GA102-I/ML	MCU	28-QFN
LP3875EMP-5.0/NOPB	5 Volt Regulator	3-SOT
AP1117Y33L-13	3.3 Volt Regulator	SMD
CRCW060310K0FKEA	10K Ohm Resitor	SMD
CRCW0603470RFKEA	470 Ohm Resistor	SMD
CRCW06031K50FKEA	1.5K Ohm Resistor	SMD
TMK107BJ104KA-T	0.1 μF Capacitor	SMD
TMK316B7106KL-TD	10 μF Capacitor	SMD
JMK316B7226ML-T	22 μF Capacitor	SMD
1N4148WS	Diode	SOD
B3U-1000P	Switch	SMT
L101011MS02Q	Slide Switch	THRU
22-03-2051	Header	THRU
22-03-2081	Header	THRU
22-03-2021	Header	THRU
22-03-2061	Header	THRU
22-03-2031	Header	THRU
ZXMHC6A07T8TA	H-Bridge	SM8

Fig. 12. Version 3 Circuit Board

purposes. Finally, the PCB was routed to a non-square custom shape to reduce size and weight. The weight of the PCB was still 16g, dominating the overall weight of the FWMAV prototype.

Table 3. Parts List For Version 3

Part Number	Description	Package
PIC24FJ64GA102-I/ML	MCU	28-QFN
LP3875EMP-5.0/NOPB	5 Volt Regulator	3-SOT
AP1117Y33L-13	3.3 Volt Regulator	SMD
CRCW060310K0FKEA	10K Ohm Resitor	SMD
CRCW0603470RFKEA	470 Ohm Resistor	SMD
CRCW06031K50FKEA	1.5K Ohm Resistor	SMD
TMK107BJ104KA-T	0.1 μF Capacitor	SMD
TMK316B7106KL-TD	10 μF Capacitor	SMD
JMK316B7226ML-T	22 μF Capacitor	SMD
1N4148WS	Diode	SOD
B3U-1000P	Switch	SMT
100SP1T1B4M2QE	Toggle Switch	THRU
22-03-2021	Header	THRU
ZXMHC6A07T8TA	H-Bridge	SM8
503480-0800	Flexible Cable Connector	SM8
14FLZ-RSM2-TB(LF)(SN)	Flexible Cable Connector	14-SMT

4.4 Version 4

The fourth development iteration represents the latest circuit board, which is finally at flight weight and shown in Figure 13 . All the components were selected as surface mount components. Each component on the Version 3 circuit board was reconsidered and smaller footprint replacements were chosen, if feasible. The required power for driver circuity and signal communication was calculated and components were selected to provide the required power, thus allowing for component size reduction. Resistors, capacitors, and power electronics were all sized appropriate for driving the motors. Thinner board thickness was selected to reduce weight, and the number of layers moved from four to six to fit the circuit layout on a smaller board than Version 3. Furthermore, both sides of the board are populated with electronic components in order to reduce the footprint of the board. Meticulous attention to weight savings was even taken into account when deciding all via's be through hole and that solder masking be done to remove excess solder on the board. A notch was removed in the corner of the circuit board and a mounting hole was placed in the middle to reduce the amount of weight even further. Reverse current protection was removed and a single reset button was used to reset all three microcontrollers simultaneously. The programming connector was reduced to 10pin by utilizing a combined reset signal for all three microcontrollers. Every effort was taken to reduce the size and weight of the board. Figure 13 shows the top and bottom of the fourth stage control board and Table 4 shows the parts list.

(a) Top of Version 4 Control Board.

(b) Bottom of Version 4 Control Board.

Fig. 13. Version 4 Control Board

Table 4. Parts List For Version 4

Part Number	Description	Package
PIC24FJ64GA102-I/ML	MCU	28-QFN
LP38690SD-5.0/NOPB	Voltage Regulator	6-LLP
LP38690SD-3.3/NOPB	Voltage Regulator	6-LLP
ZXMHC6A07N8TC	H-Bridge	8-SOIC
B3U-1000P	Switch	SMT
503480-1000	10p Connector for MCU Programming	SMT
XF2W-0815-1A	8p Connector for Motor and Encoder Connections	SMT
ERJ-1GEF1002C	10k Ohm Resistor	SMT
ERJ-1GEF1501C	1.5k Ohm Resistor	SMT
ERJ-1GEF4700C	470 Ohm Resistor	SMT
C0603X5R1A104K	0.1μF Capacitor	SMT
C1608X5R1A105K	1μF Capacitor	SMT
C0603X5R0J105M	1μF Capacitor	SMT
21020-0101	Flat Flexible Cable for MCU Programming	N/A

The Version 4 circuit board is under 1in^2 in size and weighs 1.713g, fully populated with all of its electronic components. Figure 13 demonstrates the board being approximately the size of a US quarter. The weight and size of this board is sufficiently small as to be considered smaller than the weight of the other components needed for the FWMAV, such as the motors (5g each), and fuselage (>3g). This comparison determines that Version 4 circuit board is flight weight, i.e., feasible to be part of a FWMAV of the size considered in this study.

5   Experimental Results

The basic functionality of each of the four versions of the circuit board development was tested using the prototype FWMAV seen in Figure 6. Visual inspection, internal circuitry, oscilloscopes, and high speed cameras were used to verify observed SCCPFM of the wing motion as commanded by the human operator of the RC transmitter. This allowed the progression of design to be advanced without sacrificing functionality for size and weight optimization.

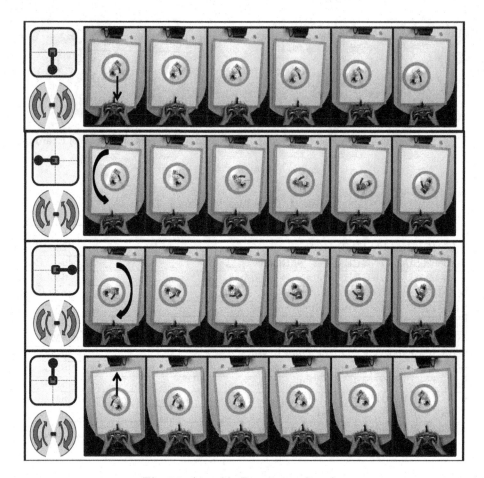

Fig. 14. Air-table Experiment Results

The version 2 of the circuit board was needed for the air table because when the tests were performed, Version 3 and 4 did not exist[7]. After developing Versions 3 and 4, the functionality was tested against Version 2 to ensure the boards were properly designed and manufactured. The FWMAV prototype equipped with the Version 2 circuit board, the wireless module, and a LiPo battery, were mounted on a disc and placed upon an air-table. The air-table provides the lift equal to weight condition and allows isolation of the degrees-of-freedom that can be controlled with split-cycle, namely, horizontal translation and roll. Figure 14 show the results of a test of the split-cycle control law on an air-table. The first column of the figure shows the commanded stick position on the RC transmitter. For example, the upper left hand corner shows the stick pulled back, which commands the vehicle to translate aft. The rows of six pictures, shown to the right of each stick position, show the time elapsed motion of the vehicle. The first picture in each row has an arrow to show the desired direction of translation or rotation. The amount of time elapsed during each set of six frames is 5 seconds. In the first row, the vehicle starts near the top of the air-table and moves backward, while exhibiting almost zero roll. The second row of plots shows the stick to the left, which commands a counter-clockwise rotation of the vehicle. The six frames in the second row show the vehicle rotating counter clockwise. The third row has the stick to the right, which commands a clockwise rotation of the vehicle, and the plots to the right show the vehicle rotating with a limited amount of translation. In the last row, the stick is forward, which commands a forward translation of the vehicle. The six frames to the right show the vehicle moving forward with nearly zero rotation.

The air-table experiment further confirms the operational abilities of the Version 2 circuit board, enabling a human operator to control the movement of a FWMAV prototype on an air-table. Similarly, the subsequent Version 3 and Version 4 showed comparable capabilities.

6 Conclusions

The incremental design approach taken in this work resulted in the implementation of a flight weight (1.713g), small ($<$1in^2), circuit board implementing human operated remote control of a prototype FWMAV utilizing split-cycle constant-period frequency modulation to realize commanded vehicle motions. This capability is a pivotal step towards an autonomous FWMAV with only two actuators capable of free flight, utilizing control over the wingbeat waveform to manipulate more than two degrees of freedom of vehicle motion with only two actuators. The use of a minimal number of actuators helps to reduce weight and allows for potential cost reduction in mass production. The engineering cost of using only two actuators lies in potentially needing better sensors and computational capacity when compared to FWMAVs not requiring control of wing motion at a sub-wingbeat period time scale. Future research needs to include more powerful motors in order to achieve higher flapping frequency, improved wing designs, and/or other means of improving aerodynamic lift and control capabilities to compensate for not only the weight of the current FWMAV but also future power sources. This will call for modifications of current designs in order to satisfy power and weight constraints. However, this work demonstrates a proof of concept that off-the-shelf components are likely to allow for such development provided that sufficient design effort is made to minimize all non-essential connections and structures.

References

1. Finio, B., Wood, R.: Open-loop roll, pitch and yaw torques for a robotic bee. In: 2012 IEEE/RSJ International Conference on Intelligent Robots and Systems, Piscataway, NJ, USA, pp. 113–119 (2012)
2. Keennon, M., Klingebiel, K., Won, H., Andriukov, A.: Development of the nano hummingbird: A tailless flapping Wing Micro Air Vehicle. AIAA-Paper-2012-0588 (2012)

3. Khan, Z., Agrawal, S.: Study of Biologically Inspired Flapping Mechanism for Micro Air Vehicles. AIAA Journal 49, 1354–1365 (2011)
4. Anderson, M.L.: Design and Control of Flapping Wing Micro Air Vehicles. Ph.D. thesis, Air Force Institute of Technology (2011)
5. Tan, X., Zhang, W., Ke, X., Chen, W., Zou, C., Liu, W., Cui, F., Wu, X., Li, H.: Development of flapping-wing micro air vehicle in Asia, pp. 3939–3942 (2012)
6. Hu, Z., Cheng, B., Deng, X.: Lift Generation and Flow Measurements of a Robotic Insect, AIAA-Paper-2011-1311 (2011)
7. Doman, D.B., Oppenheimer, M.W., Sigthorsson, D.O.: Wingbeat Shape Modulation for Flapping-Wing Micro-Air Vehicle Control During Hover. Journal of Guidance, Control and Dynamics 33(3), 724–739 (2010)
8. Oppenheimer, M.W., Doman, D.B., Sigthorsson, D.O.: Dynamics and Control of a Biomimetic Vehicle Using Biased Wingbeat Forcing Functions. Journal of Guidance, Control and Dynamics 34(1), 204–217 (2010)
9. Doman, D.B., Oppenheimer, M.W., Sigthorsson, D.O.: Dynamics and Control of a Minimally Actuated Biomimetic Vehicle: Part I - Aerodynamic Model, AIAA-Paper-2009-6160 (August 2009)
10. Oppenheimer, M.W., Doman, D.B., Sigthorsson, D.O.: Dynamics and Control of a Minimally Actuated Biomimetic Vehicle: Part II - Control, AIAA-Paper-2009-6161 (August 2009)

Neural Network Control for the Balancing Performance of a Single-Wheel Transportation Vehicle: Gyrocycle

Minsoo Ha and Seul Jung

Dept. of Mechatronics Engineering,
Chungnam National University
199 Daehak-ro, Yuseong-gu, Daejeon 305-764, Korea
jungs@cnu.ac.kr

Abstract. A single-wheel mobile robot has been developed for carrying a driver. Since a single-wheel mobile robot carries a human driver, the size and weight are larger compared with other sing-wheel mobile robots. To maximize the gyroscopic effects, Gyrocycle is designed to have two flywheels that need to be synchronized. In addition to the synchronization of two flywheels, Gyrocycle is tested for the robust balancing perfromance by unknown payloads. A neural network control method is used to control the balance. Experimental studies are conducted to verify the perfromance by the neural network controller.

Keywords: one-wheel mobile robot, balancing control, neural network control, synchronization, transportation vehicle.

1   Introduction

A balancing technique is one of intelligent characteristics for dynamical systems to achieve. The balancing technique plays an important role in many systems such as humanoid robots, two-wheel mobile robots, and single-wheel mobile robots. Specially, the balancing control technique is a major concern for one or two-wheel mobile robots to satisfy specifications.

Balancing control of two-wheel mobile robots has attracted many researchers since control becomes quite challenging. Various two-wheel robots have been presented. The goal of two-wheel mobile robots is to regulate the pitch angle by two wheels. Difficulties in control come when two wheel torques have to control three variables, position, heading angle and balancing angle. Segway and Joe are two well-known examples of two-wheel mobile robots [1,2]. Other two-wheel robots are also presented in the literature [3-7].

Control of a one-wheel robot is more challenging. To balance, instead of using two-wheel torques of two-wheel mobile robots, the gyroscopic effect induced by three actuators is used in a single-wheel robot. Gyroscopic effect has been used in many applications such as gyro sensors, camera stabilization systems, gimbals, and actuators. The principle of a single-wheel mobile robot is that the rotation of an inner

J.-H. Kim et al. (eds.), *Robot Intelligence Technology and Applications 2*,
Advances in Intelligent Systems and Computing 274,
DOI: 10.1007/978-3-319-05582-4_77, © Springer International Publishing Switzerland 2014

gimbal with a high spinning flywheel generates a gyroscopic force to turn the whole system in the yaw direction. This effect prevents the robot from falling down.

In the literature, research on a single-wheel robot is not as active as two-wheel robots. Gyrover is a well-known single-wheel robot using gyroscopic effects for balancing and driving. Gyrover has been developed and controlled for a long time and the successful control performance by Gyrover has been demonstrated [3]. In our research, Gyrobo has been designed and developed on the same concept of using the gyro effect [8]. Balancing and navigation control of Gyrobo have been successfully demonstrated.

In this paper, as an extension of Gyrobo, a single-wheel mobile vehicle is introduced for carrying a driver. Gyrocycle as a personal transportation vehicle is developoed and controlled. Control performance is presented for checking the feasibility of using a personal transportation vehicle. To achieve more accurate and robust balancing performance, a neural network control method is introduced. Experimental studies of balancing control of Gyrocycle are conducted to demonstrate the perfromance by the neural network controller.

2   Control Schemes

2.1   PD Control

For the balancing control of Gyrocycle, a key controller is the tilting angle of the gimbal system including flywheels. The flywheel is supposed to rotate at a high constant speed. Tilting the gimbal generates the gyroscopic motion to rotate the Gyrocycle in the yaw direction. Therefore, control algorithm is formed on the basis of controlling the tilting torques for the gimbal to satisfy the lean angle of Gyrocycle.

Since there are two flywheels to generate more power, the same PD controllers are used for both gimbals. From the measurement of the lean angle, β the lean angle error, e_β forms the PD control law for tiling the left gimbal, u_{TL}.

$$u_{TL} = k_{\beta D}(\dot{\beta}_d - \dot{\beta}) + k_{\beta P}(\beta_d - \beta) \tag{1}$$

where β is a balancing(lean) angle, $k_{\beta D}, k_{\beta P}$ are controller gains. Then control inputs are used for both gimbals. The control input for the right and left gimbals are same such as $u_{TL} = u_{TR}$. The PD control block diagram is described in Fig. 1. For the control of flywheel speed, the open loop control is used to generate the constant speed, $\dot{\gamma}$ of about 5,700 rpm.

2.2   Gain Scheduling Control Method

If we use PD controllers only for tilting gimbals, the gimbals tend to lean against one direction. This behavior leads to falling down of Gyrocycle. Different controller gains for different angles are obtained from experimental trials to prevent from leaning

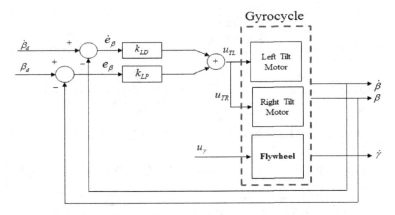

Fig. 1. PD control block diagram for the balancing control task

toward one direction. The gain scheduling method is purely empirical. This method makes gimbals back to zero angle position, not deviating away from the zero angle position.

The control input with an addition of the gain scheduling method is given by

$$u_{TL} = k_{\beta D}(\dot{\beta}_d - \dot{\beta}) + k_{\beta P}(\beta_d - \beta + \beta_\theta) \qquad (2)$$

where β_e is the added term from the gain scheduling method. If the lean angle is less than 1 degree, then there is no addition as listed in Table 1. Total 11 gains are classified with respect to the addition of two tilt angles. The PD control block diagram with the gain scheduling method is shown in Fig.2.

Table 1. Gains scheduling values

$(\theta_L + \theta_R)/2$(deg)	β_e (deg)
$10 < (\theta_L + \theta_R)/2$	-0.3
$6 < (\theta_L + \theta_R)/2 < 10$	-0.22
$4 < (\theta_L + \theta_R)/2 < 6$	-0.2
$2 < (\theta_L + \theta_R)/2 < 4$	-0.15
$1 < (\theta_L + \theta_R)/2 < 2$	-0.1
$-1 < (\theta_L + \theta_R)/2 < 1$	0
$-2 < (\theta_L + \theta_R)/2 < -1$	0.1
$-4 < (\theta_L + \theta_R)/2 < -2$	0.15
$-6 < (\theta_L + \theta_R)/2 < -4$	0.2
$-10 < (\theta_L + \theta_R)/2 < -6$	0.22
$(\theta_L + \theta_R)/2 < -10$	0.3

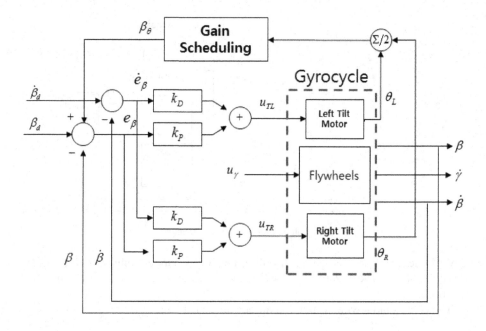

Fig. 2. PD control block diagram with gain scheduling for balancing control task

2.3   Neural Network Control Method

The neural network control method is known for its powerful adapting and leaning capabilities for controlling nonlinear systems. A nonlinear mapping capability provides powerful performances in pattern recognition as well as in the dynamical systems to compensate for uncertainties.

The reference compensation technique (RCT) is one of neural network control methods [9, 10]. The RCT scheme compensates at the reference level for uncertainties in system dynamics. The control input for the tilting control of the left gimbal is given as

$$
\begin{aligned}
u_{TL} &= k_{\beta D}(\dot{\beta}_d - \dot{\beta}) + k_{\beta P}(\beta_d - \beta + \beta_\theta + \beta_N) \\
&= k_{\beta D}(\dot{\beta}_d - \dot{\beta}) + k_{\beta P}(\beta_d - \beta + \beta_\theta) + k_{\beta P}\beta_N \\
&= v + k_{\beta P}\beta_N
\end{aligned}
\tag{3}
$$

where $k_{\beta D}, k_{\beta P}$ controller gains, β_N is the neural network output and v is the training signal.

The neural network compensates at the left gimbal only since the motion of the right gimbal is synchronized with that of the left gimbal. Then the training signal of the neural network is formed as a PD control output, v given in (3). The closed loop error equation can be obtained by equating (3) with the dynamic equation of Gyrocyle as

$$v = \tau_{TL} - k_{\beta P} \beta_N \tag{4}$$

where τ_{TL} is the dynamic equation.

The objective function is formed with a training signal as

$$E = \frac{1}{2} v^2 \tag{5}$$

The gradient is obtained in association with equation (4) as

$$\Delta w(t) = -\eta \frac{\partial E}{\partial w} = -\eta \frac{\partial E}{\partial v} \frac{\partial v}{\partial w}$$

$$= -\eta v \frac{\partial v}{\partial w} = \eta v k_{\beta P} \frac{\partial \beta_N}{\partial w} \tag{6}$$

The weights inside the neural network are updated at every sampling time as

$$w(t + 1) = w(t) + \eta \Delta w(t)$$

$$= w(t) + \eta v k_{\beta P} \frac{\partial \beta_N}{\partial w} \tag{7}$$

where η is the update rate.

The detailed control block description is shown in Fig. 3.

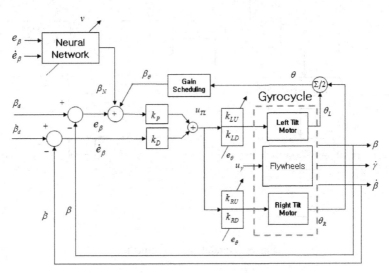

Fig. 3. Neural network control block diagram

3   Gyrocycle

Gyrocycle is designed to carry a human driver as a personal transportation vehicle. A human driver stands on the steps as shown in Fig. 4. Hardware specifications are described in Fig. 5. There are three motors, one for the flywheel spinning, another for the tilting of gimbal, the other for the driving. For the balancing performance, spinning and tilting motors are controlled. A DSP controls tilt and drive motors and an AVR controls the flywheel spin motor. A gyro sensor is used to detect the balancing angle.

Fig. 4. Gyrocycle

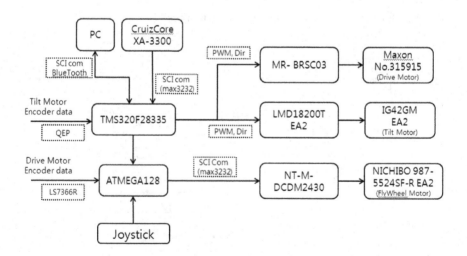

Fig. 5. Overall hardware structure

4   Experiment

4.1   Experimental Setup

The balancing performance of Gyrocycle is tested. The PD controller gains are set to $k_{\beta P} = 700$ and $k_{\beta D} = 65$. For the synchronization control, gains are selected as $k_{LU} = 1.2, k_{LD} = 0.8, k_{RU} = 1.2, k_{RD} = 1$ in Fig. 3. Those gains are selected by trial and error processes. The radial basis function (RBF) network is used as a neural network controller.

Without a neural network controller, the main controller can maintain balance. When payloads are placed on one of steps, however, Gyrocycle becomes unstable due to unbalance of the system. Thus, a neural network controller is added to the main controller to compensate for the unbalance due to the payload. While Gyrocycle is balancing, payloads are added. Payloads are about 350 g.

4.2   Experimental Results

Fig. 6 shows the actual balancing demonstration of Gyrocycle. Initially, Gyrocycle balances for a while. Then a payload is placed at 42 seconds and it balances for a while. Another payload is placed on the step at 118 seconds. Total weight of the payloads is about 0.35kg.

The balancing angle and the corresponding neural network output are shown in Fig. 7. After placing a payload at 42 seconds, the neural network out is changing severely. When the second payload is placed, the neural network out is also changing as shown in Fig. 7 (b).

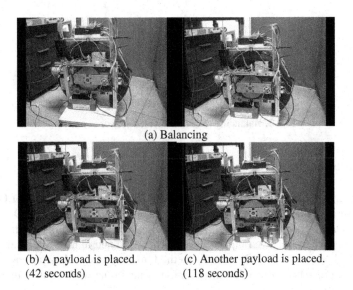

(a) Balancing

(b) A payload is placed.   (c) Another payload is placed.
(42 seconds)   (118 seconds)

Fig. 6. Balancing demonstration

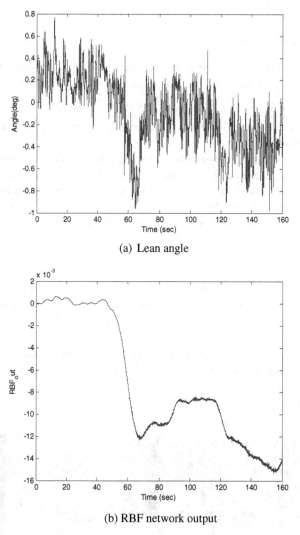

(a) Lean angle

(b) RBF network output

Fig. 7. Balancing results by neural network control

5   Conclusion

In this paper, we have demonstrated the balancing control performance of Gyrocycle by a neural network controller. Gyrocycle is primarily controlled by PD controllers with the gain scheduling method. In addition, synchronization control of two flywheels is performed. To compensate for payloads, a neural network control method is employed along with PD control method to improve balancing performance. Experimental results confirm that the addition of neural network control improves the balancing performance.

Acknowledgements. This research has been supported by Korea Research Foundation (KRF) through the general research program (2010-0024904) in Korea and the center for autonomous intelligent manipulation (AIM) for service robots of the MKE (The Ministry of Knowledge Economy), Korea, under the Human Resources Development Program for Convergence Robot Specialists support program supervised by the NIPA (National IT Industry Promotion Agency) (NIPA-2013-H1502-13-1001).

References

[1] http://www.segway.com
[2] Grasser, F., Darrigo, A., Colombi, S., Rufer, A.: JOE: A mobile, inverted pendulum. IEEE Trans. on Industrial Electronics 49(1), 107–114 (2002)
[3] Xu, Y., Au, K.W.: Stabilization and Path Following of a Single Wheel Robot. IEEE/ASME Transactions on Mechatronics 9(2), 407–419 (2004)
[4] Lee, H.J., Jung, S.: Balancing and navigation control of a mobile inverted pendulum robot using sensor fusion of low cost sensors. Mechatronics 22(1), 95–105 (2012)
[5] Lee, H.J., Jung, S.: Gyro sensor drift compensation by Kalman filter to control a mobile inverted pendulum robot system. In: IEEE International Conference on Industrial Technology, pp. 1026–1031 (2009)
[6] Pathak, K., Franch, J., Agrawal, S.: Velocity and position control of a wheeled inverted pendulum by partial feedback linearization. IEEE Trans. on Robotics 21, 505–513 (2005)
[7] Jeong, S.H.: Wheeled Inverted Pendulum Type Assistant Robot: Design Concept and Mobile Control. In: IEEE IROS, pp. 1932–1937 (2007)
[8] Kim, P.K., Park, J.H., Jung, S.: Experimental Studies of Balancing Control for a Disc-Typed Mobile Robot Using a Neural Controller: GYROBO. In: Multi-Conference on Systems and Control, ISIC 2010, pp. 1499–1503 (2010)
[9] Jung, S., Hsia, T.C.: Neural network Inverse Control Techniques for PD Controlled Robot Manipulator. ROBOTICA 19(3), 305–314 (2000)
[10] Jung, S., Cho, H.T.: Decoupled Neural Network Reference Compensation Technique for a PD Controlled Two Degrees of Freedom Inverted Pendulum. International Journal of Control, Automation, and Systems Engineering 2(1), 92–99 (2004)

Dynamic Simulation of a Sagittal Biped System

Riaan Stopforth and Glen Bright

Mechatronics and Robotics Research Group,
University of KwaZulu-Natal
Durban, South Africa
{stopforth,brightg}@ukzn.ac.za

Abstract. Controlling a system with chaotic nature provides the ability to control and maintain orbits of different periods which extends the functionality of the system to be flexible. A system with diverse dynamical behaviours can be achieved. Simulations of dynamic modelling of a sagittal biped system is investigated and results are obtained.

Keywords: Dynamic modeling, simulation, Biped.

1 Introduction

Passive walking bipeds were first investigated by McGeer in 1990 [1]. Passive bipeds are constructed such that when given an initial push, they will walk down a gentle slope. Potential energy is converted into kinetic energy while walking down a slope. Bipeds lose some of their energy due to collision in knee and heel strikes. Continuous walking is possible if the overall velocity of the biped remains constant. Based on previous research, McGeer has extended the passive walking model to one which has knees [2]. The research shows that even a model with passive hip and knees is able to walk down a gentle slope. Ikemata *et al.* have also constructed a passive walker with knees which walked 4010 steps on a tilted treadmill [3]. Figure 1 (a) [3] shows the photo of the walker used in these experiments. The design of this walker was similar to McGeer's kneed walking model. The design employed inner and outer paired legs which constrained the walkers to the sagittal plane.

Collins *et al.* have built the first three dimensional passive biped which consists of passive hips and knees, and arms as shown in figure 1 (b) [4]. This robot is capable of walking down a ramp with the correct initial push. It was designed to manoeuvre in all three planes of motion and is therefore a three-dimensional model [4].

Another three-dimensional passive walker was developed by Tedrake *et al.* [5]. This robot was designed with two straight legs joined by a passive hip as shown in figure 1 (c) [5]. The feet have curvatures both in the sagittal and frontal planes. Due to the feet curvatures, the robot oscillates laterally during its forward pitch when walking down a slope. These lateral oscillations generate ground clearance for the swinging foot while walking. The step of the robot is confined to short lengths since large steps will lead to large yaw rotations which will cause it to fall due to imbalance.

J.-H. Kim et al. (eds.), *Robot Intelligence Technology and Applications 2*,
Advances in Intelligent Systems and Computing 274,
DOI: 10.1007/978-3-319-05582-4_78, © Springer International Publishing Switzerland 2014

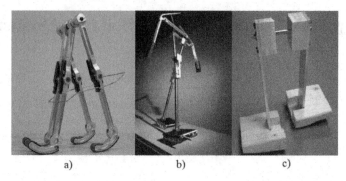

Fig. 1. a) The Nagoya Biped, b) The Cornell Biped, c) The MIT Biped

Model parameters of the biped are as defined in figure 2, figure 3 and figure 4. The sagittal plane model is of a triple linked inverted pendulum type. It has 3 and 6 DOF in configuration and phase space respectively. It was modelled on an inclined slope with an inertial reference frame as shown in figure 2.

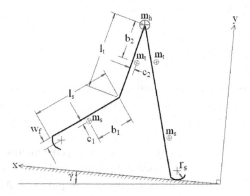

Fig. 2. Sagittal plane model and its inertial reference frame

The independent parameters of the model are shown in figure 2. The lower body of the biped was modelled using five segments. The segments consist of the hip, two thighs and two shanks. The segments were modelled as point masses (red crossed circles) and moments of inertias about Center of Mass (COM), which are not shown in the figure. The mass symbols are m_h, m_t, and m_s. The moment of inertia symbols are I_h, I_t and I_s. The subscripts h, t and s denotes hip, thigh and shank respectively. The position of the thigh and shank COMs are constrained by parameters b_2, c_2, and b_1, c_1 respectively. The leg segments lengths l_t and l_s are the thigh and shank lengths respectively. The radius of the roll-over foot is r_s and its centre is forward shifted with an offset of w_f. The angle of the slope is γ.

Figure 3 illustrates the angles used to describe the configuration of the model. The angles q_1, q_2 and q_3 and its respective velocities make up the dynamic variables (generalised coordinates) of the system. All the angles are measured from the normal of the ground to the respective limb, with clockwise being positive. The symbols u_1, u_2, u_3 and u_4 are joint torques acting on the stance foot, hip, swing knee and stance knee of the biped respectively. Clockwise torques has positive signs.

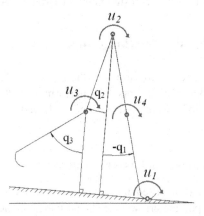

Fig. 3. Generalised coordinates and external torques of the kneed model

Figure 4 shows the dependant parameters which were derived from the parameters of figure 2. The symbols l_r, l_{mt} and l_{ms} represent lengths from the hip to foot radius centre, and to m_t, and from knee to m_s respectively. The symbol l_k is the length from the knee to the foot radius centre (not shown in the diagram). The symbol θ_{mt} is the angle between the line connecting m_h to m_t and thigh, and θ_{ms} is the angle between the line connecting the knee to m_s and the shank. The symbol ε_t is the angle between the line connecting m_h to the foot arc centre and thigh, and ε_k is the angle between the line connecting the knee to the foot arc centre and shank.

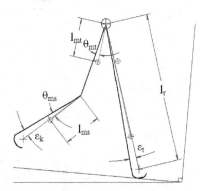

Fig. 4. Derived biped parameters

This paper presents the simulations of the sagittal biped's dynamics and stability. The simulation procedures of the modelling equations are outlined. The conditions and assumptions used in the simulation are declared. Ground clearance and basin of attraction size simulations (terminology explained in later sections) were used as stability quantifiers. The relationship between the two stability quantifiers and parameters were determined. Based on the trends of stability relations and anthropomorphic proportionality, a design parameter range was proposed.

2   Results of a Passive Walking Step

The path of locomotion and the dynamics of the biped model were obtained through simulations of its walking algorithms. To perform trajectory simulations, model conditions and assumptions were first defined followed by the simulation of the walking phases.

2.1   Simulation Conditions and Assumptions

Simulations of GC and BOA used the same model parameters but with different conditions. The model parameters were chosen such that the trends produced from simulations are amplified. Parameters used in simulations are listed in table 1. Model parameters were varied in simulations in order to determine the relationship between the varied parameters and stability quantifiers (GC and BOA). Two of the model's parameters were varied in each simulation. Variable parameters were accordingly paired to determine the possible relationship between the two. Six simulations have been conducted for each stability quantifier using six paired variable parameters. The paired variables are as shown in table 2.

Table 1. Simulation Parameters

Masses	[kg]	Lengths	[mm]
m_h	10	l_t	500
m_t	7	l_s	500
m_s	1	b_2	250
Moments of Inertia	[kg .mm^2]	c_2	0
I_h	3500	b_1	250
I_t	70000	c_1	15
I_s	30000	w_f	65
		r_s	130
		Slope Angle	[rad]
		γ	0.07

The moments of inertia of body segments were assumed to be proportional to the segment's mass. The relations with units of [kg. mm^2] were used in all simulations as shown in equation 1.

$$I_h = 3500m_h/10$$

$$I_t = 70\,000m_t/7 \qquad (1)$$

$$I_s = 30\,000m_s$$

Table 2. Variable Parameter Pairs of Simulations

Simulation No.	Variable Parameters	Range Minimum	Maximum	Units
1	m_h	1	10	[kg]
	m_t	1	10	
2	m_t	1	9	[kg]
	m_s	0.1	3	
3	l_t	200	500	[mm]
	l_s	200	500	
4	c_1	-50	65	[mm]
	c_2	-50	50	
5	b_1	100	400	[mm]
	b_2	100	400	
6	r_s	0	400	[mm]
	w_f	0	200	

Ground clearance was determined for the time span of the unlocked knee swing phase. Simulation of GC after knee strike was ignored because foot scuffing after knee strike is simply heel strike. Equation 1 was used in the calculation of the GC. The smallest local minimum of GC was recorded for every parameter variation. All negative GC was recorded as zero. The initial condition used for all GC simulation is shown in equation 2.

$$q_1 = -0.350 \text{ rad}$$

$$q_2 = -q_1 + 2\varepsilon_t \text{ rad}$$

$$q_3 = q_2 \qquad (2)$$

$$\dot{q}_1 = 1.600 \text{ rad/s}$$

$$\dot{q}_2 = 0.960 \text{ rad/s}$$

$$\dot{q}_3 = \dot{q}_2$$

Results from the simulations are dependent on initial conditions. A change in initial conditions, such as from a small step to a large step, will produce results that have the same trend pattern but different magnitudes of ground clearance height.

The BOA simulations categorise initial condition points as stable or unstable. In simulations, the initial conditions were categorised as stable if they do not fall within 50 steps. Note that all BOA simulations in this paper do not take foot scuffing into account in the results. The ranges of initial conditions are shown in equation 3.

Only two of the initial conditions are variables (q_1 and $\dot{q}_1$) where the others are dependants. The phase space of BOA in simulations is a 2D plane with axes of parameters q_1 and $\dot{q}_1$. The q_1 and $\dot{q}_1$ axes have 9 and 18 interval divisions respectively. This divides the BOA phase space into a grid with 162 cells. The simulated condition variables are located at the centre of each cell. The BOA is approximated as the area of cells, that produces stable walking. The BOA stability quantifier utilise the Cell-to-cell Mapping Method [8] which measures the number of stable cells expressed as a percentage of the total number of cells.

$$q_1 \in [-0.32, -0.15] \text{ rad}$$
$$q_2 = -q_1 + 2\varepsilon_t$$
$$q_3 = q_2 \tag{3}$$
$$\dot{q}_1 = \in [0.8, 0.16] \text{ rad/s}$$
$$q_2{}' = 0.6\dot{q}_1$$
$$\dot{q}_3 = \dot{q}_2$$

2.2   Simulation Results

Simulations were conducted on the range sweep of model parameters against the two stability quantifiers. The trend pattern or relationship between the model parameters and the quantifiers are presented.

Figure 5 shows the simulation plots where both the hip and thigh masses are varied. Figure 5 (a) shows that GC is directly proportional to m_t and inversely proportional to m_h. Maximum GC is obtained when both m_t and m_h are respectively maximised and minimised.   The changes in GC with 1 kg change in m_t or m_h mass are approximately 11 mm and 3 mm in height, respectively. The effect of varying m_h results in much smaller changes on GC when compared to m_t changes. Figure 5 (b) shows the direct proportionality of m_t and inverse proportionality of m_h to BOA size. Both plots in figure 5, show that stable walking is possible with low m_h but not a low m_t. For a m_t less than ±1.2 kg, no stable walking can be produced (figure 5 (b)). Large m_t mass produces larger swing thigh acceleration which causes large knee flexions hence larger GC.

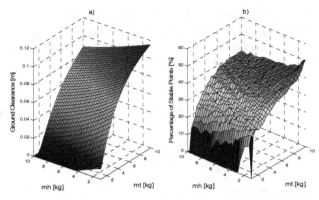

Fig. 5. Effects of hip and thigh mass variations - a) GC plot, b) BOA plot

Figure 6 shows are the simulations where both thigh and shank masses are varied. Figure 6 (a) shows that GC is directly proportional to m_t and inversely proportional to m_s. Maximum GC is obtained when both m_t and m_s are respectively maximised and minimised. Smaller m_s will result in slower acceleration of the swing shank. Slower shank accelerations cause larger knee flexions (greater GC) due to the difference in acceleration with the swing thigh. Figure 6 (b) shows that stable walking is possible when the ratio of m_s to m_t $\left(\frac{m_s}{m_t} \right)$ is as follows: the lowest and highest $\frac{m_s}{m_t}$ ratios are 0.10 and 0.86 respectively, with an average ratio of 0.67 (stable walks). For low m_t, the ratio of $\frac{m_s}{m_t}$ is low and vice versa. Larger BOA is possible if m_t is increased or m_s is decreased.

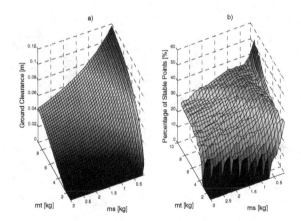

Fig. 6. Effects of thigh and shank mass variations - a) GC plot, b) BOA plot

Figure 7 shows the simulations where both the thigh and shank lengths are varied. Figure 7 (a) shows that GC is directly proportional to l_t and l_s up to lengths of ±0.46 m for both. When both parameters exceed ±0.46 m, GC starts to decrease. The above does not take the sudden rise of the GC spike ridge into account. The ridge with high

GC values results from parameters that do not generate local GC minimum during the unlocked knee swing phase. High GC peaks are initial-condition-dependant and may not appear in other conditions. Figure 7 (b) shows that BOA size is directly proportional to l_t and l_s lengths. The relationship between BOA size and increase in l_t or l_s length forms a saturation curve. The saturation value is about 40 % stable to simulated values.

Figure 7 (b) shows that no stable initial conditions exist if the following condition is met:

$$l_s + l_t < 0.7 \qquad (4)$$

where the units are in meters.

Figure 8 shows the simulations where both the b_2 and b_1 location lengths are varied. Figure 8 (a) shows that GC forms a saddle relation with b_2 and b_1. GC is the highest at $b_2 = 0.285$ m and $b_1 = 0.1$ m. This corresponds to m_t situated at about midpoint of the thigh and m_s being close to the knee. Knee flexion increases when m_s gets closer to the knee joint. The GC spike at $b_2 = 0.4$ m and $b_1 = 0.1$ m is due to dependence on initial condition. Figure 8 (b) shows that the BOA size is largest (40 % of simulated points) when b_2 and b_1 are about ±0.2 m. From the trends of figure 8, it is optimal to have b_2 and b_1 in the range of 40 % to 60 % of its segment length. This is to optimise both GC and BOA size without complicating the design process such as bringing the m_s COM close to the knee joint.

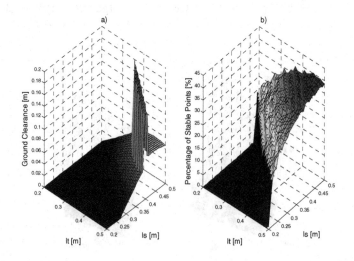

Fig. 7. Effects of thigh and shank length variations - a) GC plot, b) BOA plot

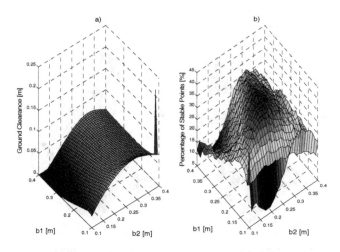

Fig. 8. Effects of b_2 and b_1 location variation - a) GC plot, b) BOA plot

Figure 9 shows the simulations where both the c_2 and c_1 location lengths are varied. Figure 9 (a) shows that GC is directly proportional to c_1 and inversely proportional to c_2. The largest GC is obtained when c_1 is small and c_2 is large. As in figure 17 (a), the ridge peak is due to the lack of local minimum of GC and its initial-condition-dependency. Positive c_2 values cause a rapid initial swing of the swing thigh which lasts for about a quarter of the time of the unlocked knee swing phase. The swing thigh slows down which leads to a smaller knee flexion and a shorter step. Negative c_2 values have the opposite effect of being: slow at beginning, faster at end, and having a larger knee flexion and a longer step. The effect of varying c_1 on GC is much smaller when compared to varying c_2. Figure 9 (b) shows that BOA size is optimised when c_2 is close to zero and c_1 is at its maximum. From the results of figure 9, having the mass centred at its geometric centre (c_2 and c_1 =0) will be design friendly without over-compensating for the GC and BOA size.

Figure 9 shows the simulations where the radius of the roll-over feet r_s and its centre offset w_f are varied. Figure 10 (a) shows that GC is highest when feet $r_s = 0$ m and $w_f = 0.113$ m (excluding the GC spike). The GC crest is situated according to the following equation:

$$w_f = -0.26r_s + 0.1231 \tag{5}$$

where the units of r_s and w_f are in meters.

The parameters that satisfy equation 5 have GC > 0.02 m. The GC crest positioning alters when initial condition and other model parameters are changed. GC is also high (GC > 0.05 m) in the region formed by $r_s \in [0; 0.144]$ m and $w_f \in [0.051; 0.159]$ m. GC spikes can also be found in figure 10 (a) due to the reason discussed previously. GC decreases until zero when r_s increases. Figure 10 (b)

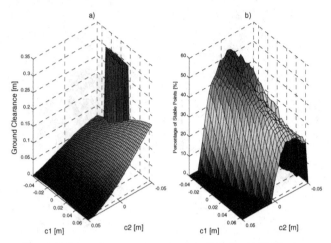

Fig. 9. Effects of c_2 vs. c_1 location - a) GC plot, b) BOA plot

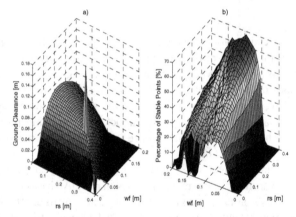

Fig. 10. Effects of w_f and r_s variation - a) GC plot, b) BOA plot

shows that as r_s increases, BOA size increase which agrees with the research findings of Wisse and van Frankenhuyzen [9]. Wisse and van Frankenhuyzen showed that increases in foot radius increases robustness (BOA size), but also increases the chance of foot scuffing. An increase in foot radius causes both the foot length and the chance of toe stubbing to increase. Similarly, smaller r_s results in larger GC but reduced BOA size. Both plots in figure 10, show that a horizontal shift in foot radius centre w_f improves stability. The optimal w_f offset is about 0.1 m for the simulated model. According to Franken, the w_f offset affects the acceleration of the stance leg as it changes the distance between foot contact and the total COM [10]. Figure 11 [10] illustrates the effects of w_f offset. For large w_f offset ($w_f \gg 0$), deceleration of the stance leg occurs due to gravitational force. Likewise if $w_f = 0$, acceleration of the stance leg occurs. Faster stance leg acceleration results in shorter steps as heel strike of the swing leg occurs quicker. Slower stance leg acceleration increases step length but may also cause backward falls if the w_f offset is too large. Figure 10 shows that w_f is bounded to a fixed range for stable walks.

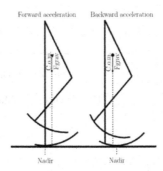

Fig. 11. Effects w_f offset on acceleration

2.3   Design Parameter Ranges

By integrating the results of GC and BOA simulations with human proportionality, a stable model with parameters of anthropometric proximity was achieved. The simulations were performed on a robust model which generated a relationship between the parameters, and GC and BOA size. The simulation trends can only be used as a guide as they are dependent on parameters and initial conditions. Actual biped parameters will not be exactly the same as the simulated model due to discrepancies. Design requirements such as the magnitude of moments of inertia, are difficult to achieve in the design. It is possible to change the dynamics of the design through small changes. Parameters that can be easily altered (manufacture-wise) will be selected as the stability tuners of the model. Typical parameters that alter stability greatly through small changes are l_t, l_s, r_s and w_f.

The design model was decided to be 0.6 m tall (hip height) as this height is the most suitable. It is large enough for operational handling while running experiments and small enough for large expensive parts not to be required. Although the simulated model is 1 m in hip height, the trend roughly applies to both models. Anthropometric parameters corresponding to the above model are listed in table 3. The range of stable design ranges are also shown in the table. The total mass of the design model ranges from 7.2 kg to 12.2 kg (lower body mass). When converting the design model to a full body model, the model's body mass index (BMI) ranges from 12.11 kg.m^{-2} to 20.52 kg.m^{-2}. The anthropometric model has a total lower body mass of 9.74 kg and a BMI of 16.39 kg.m^{-2}. Typical human adult BMIs are in the range of 20 kg.m^{-2} to 25 kg.m^{-2} [11]. The design model has a smaller BMI as it has less degrees of freedom than humans. Also, larger actuation force will be required for larger mass which leads to a larger actuator being required.

The selection of the design parameter range of table 3 was based on the results of previous simulations. The average of the design parameter range in the table is equal to the anthropometric model parameters. Large tolerance was selected for the hip mass (33 % of anthropometric m_h) because a parameter change in the hip does not alter the stability significantly according to figure 5. The hip is also where the actuator is located so mass flexibility is needed. The thigh design mass has a tolerance of approximately 20 % and a minimum mass of 1.8 kg. Simulation results of figure 5

Table 3. Anthropometric and Design Parameters

Masses	Anthropometric Parameters	Design Parameter Range	
	[kg]	Minimum [kg]	Maximum [kg]
m_h	2.98	2	4
m_t	2.1	1.8	2.6
m_s	1.28	0.8	1.5
Lengths	[mm]	Minimum [mm]	Maximum [mm]
l_t	277	250	300
l_s	323	300	325
b_2	-	$l_t/2 - 20$	$l_t/2 + 20$
c_2	-	-20	20
b_1	-	$l_s/2 - 20$	$l_s/2 + 20$
c_1	-	-20	40
w_f	9 = (0.008H_t) [12]	$r_s/2 - 20$	$r_s/2 + 20$
r_s	181 = (0.16H_t) [1]	$0.2\times(l_t + l_s) - 20$	$0.2\times(l_t + l_s) + 20$

and figure 6 shows that stability is improved with larger m_t but this requires more actuation torque for active control. Instead, m_t was bounded such that it optimised both stability and actuation power. The shank design mass has the opposite effect to thigh mass. It should be at a minimum without over-reducing the structure of the mechanical components. The body segments' moments of inertia range has not been stipulated due to the complexity of including this as part of the design requirement.

The segment lengths of the design are dependent on each other. Condition 4 regarding stability applies only to the simulated model but a similar condition will govern the design model. In general, the difference between the lengths of l_t and l_s should be less than 50 % of the longer segment. The proportionality between the leg lengths and masses should also be similar to the simulated model. The selection of l_t and l_s lengths will be very close to the anthropometric model as this selection does not influence stability and provides good anthropometric appeal. The COM locations b_2 and b_1 was proposed to be at midway of the segments which agrees with the results of figure 8 and the COM is also normally situated at the mid-point of uniform sections. Similarly COM locations c_2 and c_1 were selected to lie within the segments. Feet parameters r_s and w_f of the design model differs from the roll-over foot models proposed by other researchers. The feet design parameters are based on the trend of figure 10. Stable feet parameters are roughly located around $r_s = 0.2\times(l_t + l_s)$ and

$w_f = r_s/2$. This is $r_s = 0.106\ H_t$ and $w_f = 0.053\ H_t$ where the symbol H_t denotes total body height in meters. McGeer proposed $0.16\ H_t$ for r_s [1] and Hansen and Childress proposed $0.008\ H_t$ for w_f [12].

3   Conclusion

Simulation routines were extended to form GC and BOA stability quantifiers. These two robustness quantifiers determine the relationship between parameters and stability. Simulations of GC determine the occurrence of foot scuffing and sensitivity of walking to ground unevenness. GC simulations trends are initial-condition-dependent and fail to determine stability beyond a foot step. The BOA simulations determine the phase space size of initial conditions that produce long-term stability. Foot scuffing was ignored in the BOA simulations. This shortfall was overcome by combining trend patterns with GC simulations. A BOA size measurer was known to be a better long-term stability quantifier.

By combining the trend pattern of stability simulations and anthropometric proportionality, the design parameter range was determined. Certain parameter ranges are governed by essential stability trends.   Other parameters are based on anthropometric proportionality if they do not oppose the stability trend. The combined effects were aimed at generating stable anthropomorphic locomotion.

A lower body kneed biped was researched with design parameters to allow for the simulation of walking tests. It is an under actuated system whose dynamics are controlled through hip actuations.

References

1. McGeer, T.: Passive dynamic walking. International Journal of Robotics Research 9(2), 62–82 (1990)
2. McGeer, T.: Passive walking with knees. In: Proceedings of IEEE International Conference on Robotics and Automation, Los Alamitos, CA, pp. 1640–1645 (1990)
3. Ikemata, Y., Sano, A., Fujimoto, H.: A physical principle of gait generation and its stabilization derived from mechanism of fixed point. In: Proceedings of IEEE International Conference on Robotics and Automation, pp. 836–841 (2006)
4. Collins, S.H., Wisse, M., Ruina, A.: A three-dimensional passive-dynamic walking robot with two legs and knees. International Journal of Robotics Research 20(7), 607–615 (2001)
5. Tedrake, R., Zhang, T.W., Fong, M.-F., Seung, H.S.: Actuating a simple 3D passive dynamic walker. In: Proceedings of IEEE International Conference on Robotics and Automation, vol. 5, pp. 4656–4661 (2004)
6. Wisse, M., Atkeson, C.G., Kloimwieder, D.K.: Swing leg retraction helps biped walking stability. In: 5th IEEE-RAS International Conference on Humanoid Robots, pp. 295–300 (2005)
7. Schwab, A.L., Wisse, M.: Basin of attraction of the simplest walking model. In: Proceedings of ASME Design Engineering Technical Conference, DETC2001/VIB-21363, Pittsburgh, PA (2001)

8. Hsu, C.S.: Cell-to-cell mapping; a method of global analysis for nonlinear systems. Applied Mathematical Sciences, vol. 64. Springer, New York (1987)

9. Wisse, M., van Frankenhuyzen, J.: Design and construction of mike; a 2d autonomous biped based on passive dynamic walking. In: Proceeding of Conference on Adaptive Motion of Animals and Machines, Paper number WeP-I-1, Kyoto, Japan (2003)

10. Franken, M.: Ankle Actuation for Planar Bipedal Robots. unpublished MSc Report, University of Twente (2007)

11. Halls, S.B.: Formula for Body Mass Index (2008),
http://www.halls.md/body-mass-index/bmirefs.htm
(accessed September 16, 2010)

12. Hansen, A.H., Childress, D.S.: Effects of adding weight to the torso on roll-over characteristics of walking. Journal of Rehabilitation Research and Development 42(3), 381 (2005)

Analysis of Kinematics and Dynamics of 4-dof Delta Parallel Robot

Tuong Phuoc Tho and Nguyen Truong Thinh

Dept. of Mechatronics, HCMC University of Technical Education
1 Vo Van Ngan Street, Thu Duc District, Ho Chi Minh City, Viet Nam
tuongphuoctho@gmail.com, thinhnt@hcmute.edu.vn

Abstract. This paper will describe the kinematics and dynamics of parallel robot named Delta with 4 degree of freedom (dof). In this study, the model of Delta parallel kinematic robot 3 dofs combined with a mechanism, which is a kinematic chain with one dof of its links identified angle between the base and moving platform as the end-effector. The use of dynamics coupled with kinematics for the control of parallel robot has been gaining increasing popularity in recent years. Relationship between generalized and articular velocities is established, hence jacobian and inverse jacobian analyses are determines. The inverse formulas are generally shown simply and the direct formulas are also described. Besides, this paper deal with the direct and inverse dynamics to determine the relations between the generalized accelerations, velocities, coordinates of the end-effector and the articular forces based on simulation and control. Parallel robots have become the important machines to manufacturing using for various purposes in industry and life. The dynamic model of parallel robot with 4 dof is presented, and an adaptive control strategy based kinematic and dynamic models for this robot is described. Experiments were implemented to evaluate the responding of controlling system based on dynamics and kinematics controlling method for tracking desired trajectories. The results show that the use of the suitable control system based on kiematic and dynamic model can provide the high performance of the robot.

Keywords: Delta robot, Parallel robot, Kinematics, Dynamics.

1   Introduction

The parallel robot manipulators have taken great interest in many applications, such as handling, assembly and packaging tasks, high speed machining, various medical and space applications... It is understandable, because their static and dynamic characteristics are more favorable than the serial ones due to their closed loop kinematic chains. Parallel robots have been extensively studied by virtue of its high force-to-weight ratio, rigid and accuracy and widespread application. Parallel robots are closed-loop mechanisms presenting very good performances in terms of accuracy, regidity and abality to manipulate large loads. Many applications in the field of production automation, such as assembly and material handling, require machines capable of very high speeds and accelerations. The parallel robots are able to work on

J.-H. Kim et al. (eds.), *Robot Intelligence Technology and Applications 2,*
Advances in Intelligent Systems and Computing 274,
DOI: 10.1007/978-3-319-05582-4_79, © Springer International Publishing Switzerland 2014

some tasks with a much better performance. However, there are still several unanswered questions and few papers published studying robots with parallel architectures. This paper introduces a 4 d.o.f parallel robot dedicated to pick-and-place: Delta Parallel Robot. First a kinematics model of a Delta parallel robot is obtained using a generic geometrical formulation then the model is used for a workspace analysis. Delta robot has many advantages like operating required accurary, rigidity and manipulation of large loads. A parallel robot is a mechanism that has links that form closed kinematics chains. Because of this, Parallel mechanisms have many advantages compared to serial mechanisms, such as speed and accuracy. Generally, a parallel robot is made up of a mobile platform (end-effector) with n d.o.f, and a fixed base, linked together by at least two independent kinematics chains. Normally, each kinematics chain has a series of links connected by joints. Manipulators with *3* degrees prove extremely interesting for pick-and-place operations. Several prototypes have been suggested. The most famous robot with *3* dof is Delta. All the kinematic chains of this robot are *3* rotary actuators allowing to obtain 3 dof in translation and a dof in rotation. This paper introduces a 4-dof parallel manipulator architecture Delta dedicated with kinematics and dynamics analyses to pick-and-place and developed to perform high speed and acceleration. In this article we have discussed the inverse and direct kinematics solution as well as dynamics for the 4 dof Delta parallel robot. With this manipulator it is often difficult to determine the kinematics and dynamics analyses. Thus, this paper includes five seperated sections. The main properties of parallel robot are described in section 2 as well as focusing on kinematics and dynamics analyses, respectively. Experiment and discussions is established in Section 4. Finally, in Section 5 is shown the conclusion.

2   Background Kinematics and Jacobian Analysis

For kinematic studies, the kinematic description of a robot consists of two parts, one is pure Delta 3-dof robot model and the mechanism combined by passive prismatic joint and active revolute joint. Motion of the moving platform is generated by actuating the revolute joints which vary the angles of the legs, θ_i, $i = 1....4$. So, trajectory of the center point of moving platform is adjusted by using these variables. For modeling the Delta manipulator, a base reference frame O_0 $(O_0 x_0 y_0 z_0)$ is defined as shown in Fig. 1. A second frame O_P $(OP x_P y_P z_P)$ is attached to the center of the moving platform. The pose of the center point, OP, of moving platform is represented by the vector

$$\mathbf{x} = \{ x_P \quad y_P \quad z_P \quad \phi \} \tag{1}$$

where x_P, y_P, z_P are the cartesian positions of the point O_P relative to the frame O_0 and ϕ is the rotation angle, namely gripper's rotation angle, representing the orientation of frame O_P relative to the frame O_0 by rotations about the z_P axe. In this section, the description and kinematics of the parallel robot – 4 dof are shown in Fig.1. Generally, parallel robot is a closed loop manipulator is more difficult to calculate the kinematics. The moving plate always stays parallel to the base platform and its orientation around the axis perpendicular to the base plate is constantly zero. Thus, the parallelogram type joints (forearm) can be substituted by simple rods without changing the robot kinematic behaviour. The revolute joints (between the base plate and the upper arms and between the forearms and the travelling plate) are identically placed on a circle. Besides, a mechanism is a kinematic chain with one dof to form identified angle of the end-effector between the base and moving platform. The robot is driven by the actuated joints, the moving platform and all legs undergo constrained motions with respect to the base.

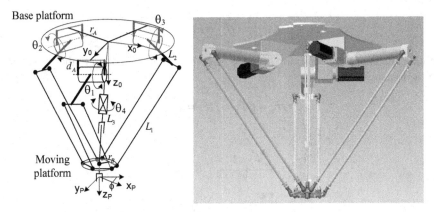

Fig. 1. Modelling of 4 dof Delta parallel robot

The modelling of Delta robot has the assumptions like as: φ_1, φ_2, φ_3 are the rotate angle of 3 link, d_A is the distance from the center of the base (origin) to the spin axis of the transmission, F_1; F_2; F_3 are the center of the spindle attached to the transmission, rA is the distance from the center stand on compared to the projection axis of the arm to stand on. And L_1, L_2 are the length of 2 links as describe in Fig. 2. Because, the inverse kinematics of Delta parallel robot is easier than direct kinematics (forward kinematics), so firstly the inverse kinematics is shown. The inverse kinematics of a parallel manipulator determines the θ_i angle of each actuated revolute joint given the (x,y,z) position of the travel plate in base-frame.

$$\theta_1 = \arctan\left(\frac{z_{j_1}}{y_{F_1} - y_{J_1}}\right) \tag{1}$$

The angle between the base and moving platform as the end-effector is identified as.

$$\theta_4 = \phi \tag{2}$$

Such algebraic simplicity follows from good choice of reference frame: joint F_1J_1 moving in YZ plane only, so we can completely omit X coordinate. To take this advantage for the remaining angles θ_2 and θ_3, we should use the symmetry of delta robot. First, let's rotate coordinate system in XY plane around Z-axis through angle of 120° counterclockwise. We've got a new reference frame X'Y'Z', and it this frame we can find angle θ_2, θ_3 using the same algorithm that we used to find θ_1.

$$\begin{cases} x_0^{'} = x.\cos\left(\pm120^\circ\right) + y.\sin\left(\pm120^\circ\right) \\ y_0^{'} = -x.\sin\left(\pm120^\circ\right) + y.\cos\left(\pm120^\circ\right) \\ z_0^{'} = z_0 \end{cases} \tag{3}$$

Now the three joint angles θ_1, θ_2 and θ_3 are given, and we need to find the coordinates (x_0, y_0, z_0) of end effector point E_0. Use of the vector translation of y-axis displacement, we have:

$$\overrightarrow{OJ_1^{'}} = \overrightarrow{OF_1} + \overrightarrow{F_1J_1} + \overrightarrow{J_1J_1^{'}} \tag{4}$$

With a length of the vector, the distance from the original quadrant to the swivel point of the transmission is:

$$OF_1 = OF_2 = OF_3 = \sqrt{r_A^{\ 2} + d_A^{\ 2}} \tag{5}$$

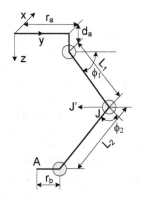

Fig. 2. Shows model simplification of the Delta parallel robot

Distance from center of the three spheres intersect at the center base

$$J_2J_2^{'} = J_2J_2^{'} = J_3J_3^{'} = r_B \tag{6}$$

Radius of the sphere is L_2, so:

$$F_1J_1 = L_2 \cos\varphi_1 \tag{7}$$

$$F_2J_2 = L_2 \cos\varphi_2 \tag{8}$$

$$F_3J_3 = L_2 \cos\varphi_3 \tag{9}$$

And (x, y, z) is the coordinates of sphere centers $J_1^{'}$, $J_2^{'}$, $J_3^{'}$. So the coordinate of $J_1^{'}$ is:

$$\begin{bmatrix} 0 & r + L_2\cos\varphi_1 & L_2\sin\varphi_1 + d_A \end{bmatrix}^T = \begin{bmatrix} x_1 & y_1 & z_1 \end{bmatrix}^T \tag{10}$$

Similarly we have the coordinates of J_2' and J_3' as follows:

$$J_2' = (x_2; y_2; z_2) = ((r + L_2 \cos \varphi_2) \cos 30^0; (r + L_2 \cos \varphi_2) \sin 30^0; L_2 \sin \varphi_2 + d_A) \quad (11)$$

$$J_3' = (x_3; y_3; z_3) = (-(r + L_2 \cos \varphi_3) \cos 30^0; (r + L_2 \cos \varphi_3) \sin 30^0; L_2 \sin \varphi_3 + d_A) \quad (12)$$

So the intersection of 3 spheres is:

$$\begin{cases} (x - x_1)^2 + (y - y_1)^2 + (z - z_1)^2 = L_1^2 \\ (x - x_2)^2 + (y - y_2)^2 + (z - z_2)^2 = L_1^2 \\ (x - x_3)^2 + (y - y_3)^2 + (z - z_3)^2 = L_1^2 \end{cases} \quad (13)$$

With help from computer this equation system can be solved. There will be two solutions that describe the two intersection points of the three spheres. Then the solution that is within the robots working area must be chosen. With the base frame {R} in this case it will lead to the solution with negative z coordinate. The Jacobian matrix that gives the relation between the revolute joint velocities and the velocity of the center point of moving platform, O_P, can be derived using the partial differentiation of the inverse geometric model of the manipulator.

3 Dynamic Analysis of 4 dof Delta Robot

One important step in design process of a robot is to understand the behaviour of the device as it moves around its workspace or doing a specific task. This behaviour is determined through the study of the dynamics of the mechanism, where the forces acting on the elements and torques required by the actuators can be determined. Consequently, each component must be optimized in dimensions and material to be used in the manufacturing processes. In section, the dynamics of Delta parallel robot is described based on Lagrangian formulation, which is based on calculus variations, states that a dynamic system can be express in terms of its kinetic and potential energy leading in an easy way the solution to the problem. In addition, it is considered a good option to be used for real-time control for parallel manipulators [4]. The Lagrange equations can be derived.

$$\frac{d}{dt}\left(\frac{\partial L}{\partial \dot{q}_j}\right) - \frac{\partial L}{\partial q_j} = Q_j + \sum_{i=1}^{k} \lambda \frac{\partial g_i}{\partial q_k} \quad (14)$$

Where L is the Lagrange function, where L = T - V, T is the total kinetic energy of the body, V is the total potential energy of the body, q is the k_{th} generalized coordinate, Q is a generalized external force, λ_i is the Lagrange multiplier and g_i is the constrain equation. By employing the formula above it is possible to determine the external forces of a body. However, friction forces are not constraints even though they play an important role in the dynamics analysis so they can be treated separately.

The Lagrange multipliers are derived as.

$$2\sum_{i=1}^{3}\lambda_i(p_x + h\cos\phi_i - r\cos\phi_i - a\cos\phi_i\cos\theta_{1i}) = (m_p + 3m_b)\ddot{p}_x \tag{15}$$

$$2\sum_{i=1}^{3}\lambda_i(p_y + h\sin\phi_i - r\sin\phi_i - a\sin\phi_i\cos\theta_{1i}) = (m_p + 3m_b)\ddot{p}_y \tag{16}$$

$$2\sum_{i=1}^{3}\lambda_i(p_z - a\sin\theta_{1i}) = (m_p + 3m_b)\ddot{p}_z + (m_p + 3m_b)g_c \tag{17}$$

When the Lagrangian multiplies are found the actuator torque can be determined as.

$$\tau_i = \left(\frac{1}{3}m_a a^2 + m_b a^2\right)\theta_{1i} + \left(\frac{1}{2}m_a + m_b\right)g_c a\cos\theta_{1i}$$
$$-2a\lambda_i\left[\left(p_x\cos\phi_i + p_y\sin\phi_i + h - r\right)\sin\theta_{1i} - p_z\cos\theta_{1i}\right] \tag{18}$$

The analytical inverse dynamics solutions for Delta parallel robot can be obtained from Eqs.(18).

4   Experiment and Discussions

Fig. 3. Delta parallel robot for experiments

To valid the analyses of kinematics and dynamics in previous section, an experimental setup was built to perform the control of Delta parallel robot. The specifications of Delta parallel robot is shown in Table 1.

Table 1. Specifications of Delta parallel robot

Parameters		Value
Upper robot arm m_a [kg]		1.1
Parallelogram m_b [kg]		0.9
Moving platform m_p [kg]		0.2
Upper arm length l_1 [mm]		250
Parallelogram length l_2 [mm]		480
No. of AC Servo motor		4
Motor power [W]		200
Encoder resolution [ppr]		1000
Maximum load capacity [kg]		1.0
Maximum moving platform velocity [m/s]		5.0
Position repeatability [mm]		0.2
Workspace	Diameter [mm]	1000
	Height [mm]	300

As shown in Fig. 3, the 4-DOF parallel robots is used to evaluate the accuracy, kinematic and dynamic modellings in practice. Four servo AC actuators are used to link a moving platform and a base. The geometrical parameters of the experimental parallel robots are summarized in Table 1.

This experimental implementation is built on the PC and Delta robot. The software for Delta parallel robot is implemented in Matlab using the kinematics and dynamics analyses from above solutions to control the moving platform. The proposed analyses are applied the Delta parallel robot for material drawing. The program is used to control the moving platform with predefined trajectory. We will apply the kinematics, Jacobi and dynamics to control suitable trajectory of parallel robot based on positions, velocities. In the section, some experimental results by kinematics - dynamics control are addressed. To demonstrate the capability controller, several responses were taken into account with several various trajectories. In these experiments, a pen attached to moving platform of Delta parallel robot is regulated following the several predefined paths including curves of circle, butterfly, flower, heart.

The first experimental results for controlling the moving platform with contour of flower are illustrated in Fig.4. A curve has the shape of a petalled flower and the polar equation of the rose is follows.

$$r = a \sin(n\theta) \qquad (21)$$

The drawing on paper reveals that the analysis results are almost near the desired ones shown in Fig.4(a). Compared desired contour, we can see that the very small differences between the desired and experimental values may be attributed to the following reasons: first, there is error of mechanical transmission and calculation of kinematics and dynamics of Delta robot. The improvements will bring better results for generating trajectories. And responding of three AC servo motors with time is shown in Fig.4(b).

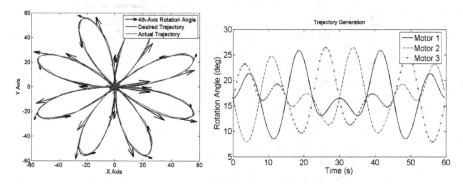

Fig. 4. Trajectory of gripper combine rotation angle (a) and responding of 3 motors for legs (b) with flower curve path

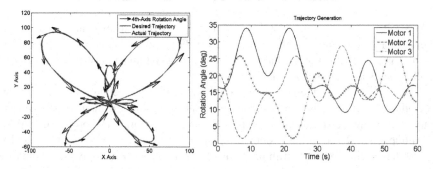

Fig. 5. Trajectory of gripper (a) and responding of 4 motors (b) with butterfly curve path

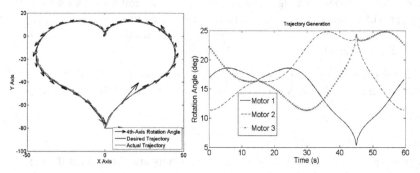

Fig. 6. Trajectory of moving platform (a) and responding of 3 motors (b) with heart curve path

Next, other responses for reference commands for butterfly contour are presented to evaluate the performance of the controller based kinematics and dynamics. The equation of butterfly curve is follows.

$$r = e^{\sin\theta} - 2\cos(4\theta) + \sin^5\left[\frac{(2\theta - \pi)}{24}\right] \tag{22}$$

Fig.5 shows output of responding trajectory and input responses for contour of butterfly. The control results for butterfly are good enough to track the perfect shape while moving path of pen has a little bit error.

Besides, we also generate the trajectory of heart curve with pole equation like as:

$$r = 2 - 2\sin\varphi + \frac{\sin\varphi\sqrt{|\cos\varphi|}}{\sin\varphi + \frac{7}{5}} \tag{23}$$

Fig.6(a) shows the actual time response signals and the command signals of parallel to a heart profile, and the time history of the controlled position output. The movement of moving platform followed the commanded signals quite well for long time. Present results show that the analyses of kinematics and dynamics can be successfully applied to the dynamic tracking of various contour profiles.

5   Conclusion

This paper is mainly concerned with kinematic and dynamic analyses as well as the application of solutions of kinematics and dynamics to modeling and control of parallel robots. A practical implementation is completed to evaluate the results of an designed controller for Delta manipulator control system. It can be said that, excepted results has been achieved for these cases. The inverse and forward kinematics and velocity equations have been derived. The results presented in the paper will be valuable for both the design and development of 4 dof Delta parallel robot for various applications. With the aid of computer, these equations with the design of this robot base on dynamic modeling and dynamic control in order to improve the behavior of the robot while reaching high acceleration. By fitting grippers or other tools to this small platform the delta robot can handle all sorts of items. Their design enables them to move both rapidly and accurately, and they are deployed for tasks varying from high-speed packaging to the assembly of miniature products.

Acknowledgments. This study was financially supported Ho Chi Minh city University of Technical Education (HCMUTE), Viet Nam.

References

1. Olsson, A.: Modeling and control of a Delta-3 dof robot (2009)
2. García, J.M.: Inverse-Forward Kinematics of a Delta Robot (2010)
3. Napole, M., Gutierrez, C.: Kinematics Analysis of a Delta Parallel Robot (2011)
4. Ha, S.M., Ngoc, P.V.B., Kim, H.S.: Dynamics Analysis of a Delta-type Parallel Robot. In: 2011 11th International Conference on Control, Automation and System (2012)
5. Ha, S.M., Ngoc, P.V.B., Kim, H.S.: Dynamics Analysis of a Delta-type Parallel Robot. In: 2011 11th International Conference on Control, Automation and Systems, pp. 855–857 (2011)

Axon: A Middleware for Robotic Platforms in an Experimental Environment

Michael Morckos[1] and Fakhreddine Karray[2]

[1] Voice Enabling Systems Technology
Waterloo Ontario, Canada
mmorckos@vestec.com
www.vestec.com
[2] Electrical and Computer Engineering, University of Waterloo
Waterloo Ontario, Canada
karray@uwaterloo.ca
www.uwaterloo.ca

Abstract. Major strides have been achieved recently in developing frameworks for multi-robot systems. The need to integrate different heterogeneous robotic systems has led to the emergence of robotic middleware design. The aim of the work presented in this paper is to develop an easy-to-use middleware that is able to effectively handle multiple robots and clients within an experimental environment. Unlike previous work in robotic middleware which models a robot as a network of components and provide low-level control, our proposed approach provides a high-level representation of robots. We also introduce the notion of efficient structured data exchange as an important aspect in robotic middleware research. We designed and developed our middleware using recent technologies. Moreover, two different robotic platforms, the PeopleBot mobile robot and the Cyton Alpha robotic arm, were used to test and evaluate the middleware's ability to integrate different types of robots. A series of performance measurement experiments were carried out to gauge the middleware's ability to handle multiple robots.

Keywords: Robotic middleware, distributed robot systems, multi-robot systems, mobile robot, robotic arm.

1 Introduction

A typical modern robot consists of a distributed system of hardware components including among other things, sensors and actuators. These components are controlled by sophisticated software ranging from low-level controllers and device drivers to high-level software libraries. There exists great diversity in robotic software and hardware components. To date, however, there is no global standard for multi-robot frameworks. Most of the vendor-provided robotic hardware and software are self-contained and proprietary, with minimal or nonexistent support for interoperability.

J.-H. Kim et al. (eds.), *Robot Intelligence Technology and Applications 2,*   911
Advances in Intelligent Systems and Computing 274,
DOI: 10.1007/978-3-319-05582-4_80, © Springer International Publishing Switzerland 2014

The evolution of middleware [1] has greatly simplified the task of building distributed and loosely coupled applications. In robotics, the purpose of middleware is to integrate several heterogeneous components into a single cohesive framework. However, distribution requirements extend beyond a single robot. The area of multi-robot control and collaborative task achievement among different robots has long been an important research topic. Many modern robotic middleware programs not only deal with the heterogeneity of components onboard a single robot, but also deal with network heterogeneity.

Numerous robotic middleware programs and frameworks have been introduced [2]. Current state-of-the-art include the Open Robot Control Software (OROCOS) project [3], Miro [4], Player/Stage [5], RT-Middleware [6], the Mobile and Autonomous Robotics Integration Environment (MARIE) [7], Orca [8], the Open Platform for Robotic Services (OPRoS) [9], and the Robot Operating System (ROS) [10]. Some other systems are also described in [11–13]. Many of the existing frameworks provide detailed representations of a robot as a bundle of hardware and software components, and provide tools for customized fine-grained control. While this is desirable in some applications, it can lead to unnecessary complexities in others. For instance, a researcher working on high-level applications such as collaborative robotic behavior may treat the robot as a (more-or-less) single entity that provides inputs for his/her application and can be acted upon in return. Such researchers may not be interested in lower-level details such as sonar ranging acquisition rate, and low level motor control signals [14, 15], to name a few.

In this work, we propose a robotic middleware that provides a high-level representation of a robot as a single entity within an experimental environment. The main goal is to provide a robust framework that achieves greater transparency for developers, and enables prototyping of robotic applications and algorithms in a straightforward manner. The middleware does not aim at replacing vendor-provided libraries and tools, but aims instead at introducing a generic-enough protocol that could accommodate heterogeneous robotic platforms. Lastly, we aim at introducing the notion of structured lightweight data and information exchange between multiple robots within a given environment as a novel component in robotic middleware research. The remainder of the paper is organized as follows: Sect. 2 presents an overview of our proposed robotic middleware. This includes the middleware architecture, used tools and technologies, as well as the middleware's components and how they work together. Section 3 describes the experimental setup used to test and evaluate the middleware in a real laboratory environment. Section 4 discusses the assessment metrics used to preliminarily evaluate the performance and scalability of the middleware. Section 5 presents a concise overview of two of the current state-of-the-art work in the area of robotic middleware, and provides a comparison between them and our work. Section 6 concludes the paper by summarizing the contributions of the presented work and discusses ideas for further research work.

2   Middleware Architecture

In this section we present our proposed robotic middleware Axon. Axon aims at eliminating the need of having prior experience in hardware or low-level intricate programming. It minimizes the time spent on setting up experiments/scenarios, and enables researchers to mainly focus on their research.

Axon recognizes a robot as a single entity that possess certain properties and has a number of services to offer to a human operator or to other entities. A typical flow of events in using the middleware can be as follows: the user, through an external client program, can browse through a list of all connected entities. The user can then "check out" one or more entities, and issues various tasks to them. After running some experiments and scenarios, the user can then release previously "checked-out" entities so that they become available to other users. Figure 1 illustrates an abstracted overview of the middleware architecture. The named components are discussed in subsequent subsections.

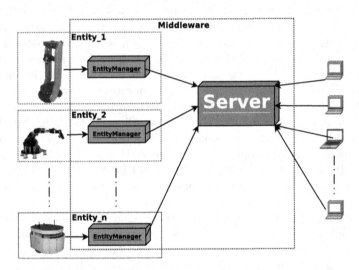

Fig. 1. Middleware Abstracted Overview

2.1   Information Base

In Axon, each entity in the environment is recognized by a unique *entity profile*. This XML-based profile contains information that uniquely defines an entity as an active actor in the environment. The entity profile is composed of two sub-profiles: the *basic profile* and the *services profile*.

The basic profile for an entity is made available to all clients, and it should help a user decide which entity best suits his/her needs. A typical basic profile stores the entity's name, ID, type, description, list of provided services, and current location in the environment.

The services profile contains all the details about the services offered by the entity. A service is defined as an executable task that can be performed by an entity upon request from a human operator, or from another entity (in a collaborative task). A typical service is defined using an ID, type, description, active flag, blocking flag, and child services list. In addition to attributes, each service might be defined using *control parameters* and *resources*. Control parameters are used to configure a service by providing input and runtime configurations, as well as communicating results and feedback to the requester. A resource represents external data or files that a service might require. Examples of resources are map files, images, or a batch of sensor readings.

2.2   Data Carrier

The notion of lightweight extensible data and information exchange was not much emphasized in robotic middleware design. On one hand, a large number of existing middleware and frameworks rely directly on the underlying network middleware such as the Common Object Request Broker Architecture (CORBA) or the Internet Communications Engine (ICE) to exchange data and information. Examples of these are work presented in [3], [4], [6], [8], [13], [16], [17], [18], and [19]. In addition, some of the middleware such as the one presented in [20] use custom-developed RPC mechanisms. Others use custom implemented low-level mechanisms, such as in [5], [10] and [21], or custom network protocols such as in [22]. We believe that data exchange mechanisms that are tightly coupled with the communications middleware does not allow for much extensibility.

On the other hand, in [23], robotic specifications can be written in XML and exchanged over the Simple Object Access Protocol (SOAP) protocol. In addition, [10] and [24] use XML-RPC over HTTP for data exchange. While XML is an ideal choice for extensible and structured information exchange, it is also considered to be verbose and heavyweight.

To have lightweight extensible information exchange, we relied on the Google Protocol Buffers (GPB) [25]. GPB is a collection of platform-and-language-independent mechanisms for serializing structured data. According to [25], "GPB is 20 to 100 times faster than XML and 3 to 10 times smaller." GPB is independent of the transport medium or protocol. Comparative evaluations presented in [26] and [27] illustrate that GPB compares favorably to XML and SOAP. In addition, GPB is used in the Robotics Service Bus (RSB) middleware [28] to assess its openness.

To create a type, one has to write the specification as a GPB message. A message is composed of uniquely numbered fields. The message file is then compiled into a simple data access class. In Axon, GPB is completely decoupled from other components and logic. The current version of Axon defines information and data exchange through six GPB message types: *Service*, *CtrlParam*, *Param*, *Resource*, *Basic Profile*, and *ByteStream*.

2.3   Network Middleware

We chose to rely on a robust general-purpose network middleware to provide the infrastructure that handles all communication between different entities and clients. The main reason behind this choice is to have a rapid development cycle and to focus on addressing the main issues of this work.

There are a number of high-profile network middleware technologies in use, such as CORBA, SOAP, and ICE [29]. The CORBA standard is still considered to be the de facto standard. Various work in robotic middleware are based on CORBA such as those presented in [4], [6], and [16–19].

However, based on a number of technical reports and developer experiences, CORBA suffers from numerous shortcomings [30]. Web-oriented technologies such as SOAP can introduce overhead in time-critical applications. Consequently, we chose ICE as the network middleware in Axon. ICE has an efficient component-based architecture. A great emphasis is put on eliminating network overhead and bottlenecks. Moreover, ICE is being used by a number of high-profile technology and defense companies [31]. In addition, ICE has been adopted in some robotic frameworks such as in [13], and it replaced CORBA in the OpenRTM middleware [32].

2.4   Entity Component

The entity component is the part of Axon that resides on the entity's onboard machine. This component is responsible for presenting a simple and extensible layer that can easily interface with custom-developed robotic applications. The entity component is composed of two modules: the **Entity Manager** (EM) and the **Entity Interfacer** (EI).

The EM manages the underlying entity and acts as its interface to the outside environment. In addition to providing complete transparency to developers, it also provides basic crucial safety features, such as a guaranteed stop signal that is dispatched instantly in cases of client requests or disconnections.

The core design of the EM is based on the state machine software design pattern [33]. The EM models the different states an entity can be in. Figure 2 illustrates the simplified UML state machine diagram of the EM. The UML class diagram of the EM is illustrated in Fig. 3.

The EI is the second module of the entity component. The EI provides a TCP socket-based communications link with the EM, as well as a simple and extensible API that the robotic application must interface with to be able to be seamlessly plugged in to the environment. Figure 4 illustrates the UML class diagram of the EI.

2.5   Server Component

The server component is the central hub of Axon. It is composed of two loosely coupled modules: the entity server, and the front-end server. The entity server keeps track of all connected entities and acts as a scalable lightweight registry

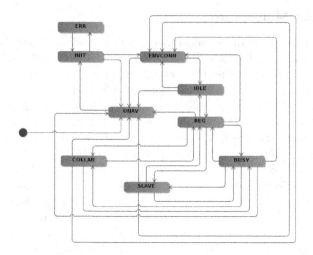

Fig. 2. Entity Manager Simplified UML State Machine Diagram

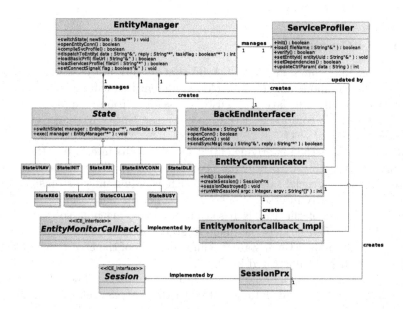

Fig. 3. Simplified Entity Manager UML Class Diagram

and message router for them. It also keeps the front-end server updated with the latest changes in connected entities' status.

The front-end server performs functionalities similar to those of the entity server, albeit for connected clients. This server can interface with different client programs, such as Web-based and desktop-based clients. In addition, it is responsible for processing and uploading of the *Global Services profile* (if available) to

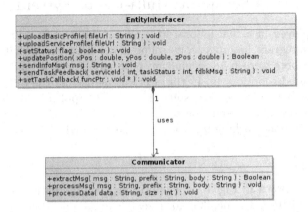

Fig. 4. Entity Interfacer UML Class Diagram

be downloaded by connected clients on request. This profile is a simple XML-based profile used to specify collaborative scenarios involving different robots.

2.6 Overall Architecture

Figure 5 illustrates the overall architecture of Axon, showing how the different components interface with each other. It also shows how a robotic application and a client program interface with the middleware.

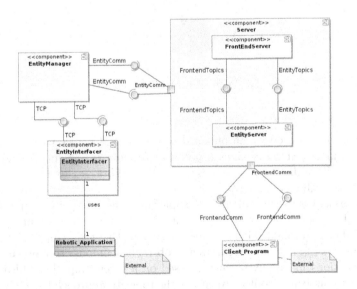

Fig. 5. Middleware UML Component Diagram

3   Experimental Setup for Multi-Robot System

Axon was deployed in a real laboratory environment to test and demonstrate its ability to integrate heterogeneous robot units. To achieve that, two different types of robots were used: a multipurpose mobile robot, and a stationary platform mounted robotic arm. In this section, we discuss the different robotic entities including their hardware, software, and the applications developed for each one. Moreover, the discussion will cover a collaborative scenario that involves the two robots working together to achieve a goal.

3.1   PeopleBot

PeopleBot [34], shown in Fig. 6, is a multi-purpose differential-drive robot developed by MobileRobots, Inc. [35], for research and applications involving HRI, cooperative robotics, and much more. PeopleBot comes with an array of sensors and range-finding devices for navigation and obstacle avoidance purposes. Moreover, PeopleBot is able to recognize objects and people, as well as track colors using a 120-degrees pan/tilt/zoom (PTZ) camera. In addition, PeopleBot is equipped with a 2-DOF gripper.

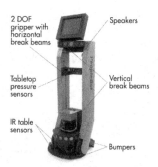

Fig. 6. PeopleBot Mobile Robot [34]

PeopleBot is powered by an onboard computer and comes with the Pioneer SDK [36]. Pioneer SDK provides a powerful framework for developing robotic applications. In addition, it provides a bundle of libraries for localization and navigation, as well as object tracking.

Three services were developed for PeopleBot. They range in complexity and utilize all the robot's hardware components. The first service is a manual control service enabling simple driving, turning, and operating the gripper. The second service employs intelligent navigation and obstacle avoidance to navigate to designated locations in an environment, based on a given map. The third service employs a sophisticated routine to enable the robot to approach a table and grab an object placed on the edge.

3.2   Cyton Alpha

Cyton Alpha, shown in Fig. 7, is a 7-DOF 1G robotic arm mimicking the human arm developed by Robai for the purposes of performing lightweight tasks, prototyping, and research [37]. Cyton Alpha features 7-DOF movement capability plus a single end effector gripper. Cyton Alpha comes with Actin-SE, a powerful cross-platform configurable software package which provides a rich set of APIs and tools for programmatic control of the arm [37].

Two services were developed for the Cyton Alpha arm: a service enabling manual control of the individual joints on the arm, and an intelligent service enabling the arm to pick an object from an elevated surface and place it accurately on a lower surface or a table edge. This is used to demonstrate collaboration with the PeopleBot mobile robot.

Fig. 7. Cyton Alpha Robotic Arm [37]

3.3   Collaborative Service

To test Axon, a collaborative service was created involving a PeopleBot robot and a Cyton Alpha arm. In this scenario, the two robots work together to transport

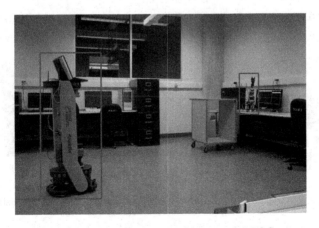

Fig. 8. PeopleBot and Cyton Alpha at Initial Setup

an object placed on an elevated surface at one side of a room to a table located at the opposite side. Figure 8 illustrates the initial setup in the CPAMI laboratory at the University of Waterloo.

All a human operator needs to do is designate a master entity and issue a task request for the service to it. The master entity will intelligently assign tasks to the most suitable entities and manage execution of the service. Figures 9, 10, and 11 illustrate the progress of the service.

(a) Picking from an Elevated Surface   (b) Placing on the Table Edge

Fig. 9. Cyton Alpha in Action

(a) Approaching the Table   (b) Picking Up the Object

Fig. 10. PeopleBot performing "Table Object Tracking and Grabbing"

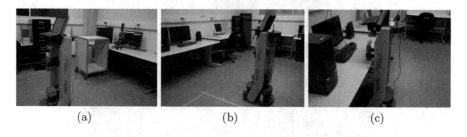

(a)   (b)   (c)

Fig. 11. PeopleBot Performing "Navigation". The Target is the Front of the Opposite Table.

4   Performance Evaluation

This section discusses preliminary evaluation of Axon. Axon was subjected to a series of performance tests that were tailored to match the needs of a robotic research environment. We do not intend to draw specific conclusions from these tests. Instead, we attempt to accurately measure the middleware performance to determine if the prototype has enough capability to be used readily.

4.1   Latency

Latency, in the networking and communications context, is the transmission delay between a source and a destination within a network. In a typical packet switching network, there are two types of latency measurements: one-way latency and two-way latency, also known as the round-trip time (RTT).

It is crucial to measure how long it takes a client to reach its checked-out entities, and how this time is affected by the number of connected entities. Thus, we were mainly interested in one-way latency in our evaluation. We carried out two different experiments. The first one consisted of a series of tests to evaluate the middleware performance under (what we think) normal laboratory conditions (a reasonable number of entities). The second experiment aims at roughly gauging the performance of the middleware while handling a large number of entities. This, we believe, might help investigate the prospect of deploying the middleware in multiple robotics laboratories.

The main measurement is the latency of transmitting batches of raw data and tasks. This simulates real interaction between a client and an entity, such as sending resources and requests for executing services.

4.2   Normal Operation

A typical WLAN router was used in this experiment. To simulate a data stream, the GPB `bytestream` message was used to create packets containing 1000 5-ASCII character strings each. All measurements are in milliseconds.

The experiment involved a single client and multiple entities, where the number of entities is increased from 1 up to 20 while recording the latency for each entity. For each run, an equal number of wired and wireless entities existed in the environment. Transmissions took place at random intervals. Figure 12 illustrates the plot of the average latency for data stream and task messages against the number of entities.

4.3   Load Testing

Tests in this experiment relied on the provided high speed LAN and WLAN in the campus building housing the laboratory. To simulate a data stream, we used packets containing 5000 5-ASCII character strings each. This size was chosen to create substantial network traffic.

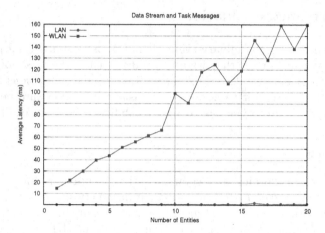

Fig. 12. Single Client - Multiple Entities Data Stream and Task Messages Average Latency Plot

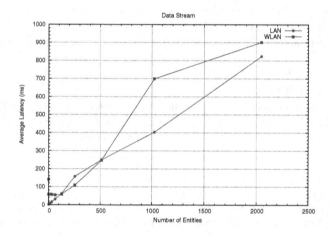

Fig. 13. Single Client - Multiple Entities Data Stream Average Latency Plot

The experiment involved exponentially increasing the number of entities from 1 up to 2048 while recording the latency for each number of entities. We had the client program transmit data continuously every 500 ms. Figure 13 illustrates a plot of the average data latency against the number of entities.

5   Related Work

The past fifteen years have witnessed numerous research contributions in the area of robotic middleware. In this section, a concise comparison and contrast between Axon and similar two state-of-the-art frameworks is provided. The comparison is done based on four main points: the overall architecture of the framework, the framework representation of robots, how the framework handles communication between multiple robots, and how the framework facilitates the development of robotic applications.

A notable work is the Player/Stage project by Gerkey *et al.* [5]. It started out as an ambitious multi-robot framework that grew to become a de facto standard in robotic frameworks and middleware. The project is composed of two parts: the Player robot device server and the Stage multi-robot simulator. Player is a client-server based framework where robotic components act as servers to client programs. Multiple clients can connect to the same component simultaneously. Similarly, Axon is based on a client-server architecture. However, unlike Player, in Axon each entity can only be assigned to one client at a time. On one hand, Player provides fine-grained control of robots. It models a robot as a collection of hardware devices such as sensors and actuators, and implements device drivers for them. On the other hand, Axon provides a much simpler high-level representation of robots. It models every robot as a single entity in the environment. Communication between clients and devices in Player is TCP-based; Player provides TCP socket-based interfaces to robotic devices and components. In Axon, we rely on a sophisticated COTS network middleware as the basis for communication between clients and robots. Player provides an extensible and flexible device model that allows for implementing sophisticated algorithms in drivers, as well as the exchange of data and information with other drivers. Axon provides a robust and simple interface for developers to rapidly prototype applications and algorithms. The second part of Player/Stage is the Stage simulator. Stage is a sophisticated multi-robot simulator that enables fast creation of multi-robot simulations that could be later realized in real environments. Currently, our middleware does not provide simulation capability.

One of the high-profile work in the robotic frameworks field is the Robot Operating System (ROS) developled by Quigley *et al.* [10]. It is an open-source robotic operating system that tackles the problem of designing and developing large-scale distributed robotic applications. ROS is a distributed framework that greatly promotes modularity as one of its cornerstone philosophies. It is based on four fundamental concepts: nodes, messages, topics, and services. Unlike Axon, ROS provides varying levels of control through nodes. In ROS, a node is the elementary building block of any application, analogous to a software module, and can implement various levels of control, from a low-level device driver up to a sophisticated robotic algorithm. Nodes on a single robot communicate through interprocess communication. On one hand, in ROS, modules residing on different robots and machines communicate through message passing in a publish-subscribe manner by publishing and/or subscribing to different topics. Communication is done in a peer-to-peer fashion through XML-RPC. ROS provides its own IDL for specifying communication interfaces between different modules. On the other hand, Axon handles communication in a centralized manner through ICE RPC and GPB. For application development, ROS provides a powerful build tool to group different nodes into packages. Nodes can be added and removed at runtime. Packaging enables simultaneous development and deployment by multiple developers. In Axon, robots can be added and removed at runtime. Moreover, the ability of different robots to collaborate implicitly groups them into systems.

6 Conclusion

In this paper, we tackled the design of robotic middleware from a novel perspective. Our robotic middleware Axon models a robot as a single high-level entity by providing a simple software abstraction layer atop its vendor-provided software libraries. We introduced the notion of structured and efficient data exchange as a substantial explicit issue in robotic middleware research.

Axon was evaluated in terms of its capability to handle heterogeneous robotic systems. In addition, performance results clearly demonstrate that the prototype is able to accommodate heterogeneous robots, and is robust enough to be used in prototype experiments.

The modular and well structured design of the system allows for easy expansion and upgrade. Given that the current version is only a prototype, this opens the door to a number of ideas for enhancing the system such as the addition of real-time capabilities, deployment in multiple laboratories environments, and developing a plugin framework that can support complex experiments and scenarios.

References

1. Bakken, D.E., Dasgupta, P., Urban, J.: Middleware. Encyclopedia of Distributed Computing (2001)
2. Mohamed, N., Al-Jaroodi, J., Jawhar, I.: Middleware for Robotics: A Survey. In: IEEE International Conference on Robotics, Automation, and Mechatronics (RAM 2008), pp. 736–742 (2008)
3. Bruyninckx, H.: Open robot control software: the OROCOS project. In: 2001 IEEE International Conference on Robotics and Automation (ICRA), vol. 3, pp. 2523–2528 (2001)
4. Utz, H., Sablatnög, S., Enderle, S., Kraetzschmar, G.: Miro - Middleware for Mobile Robot Applications. IEEE Transaction on Robotics and Automation 18(4), 493–497 (2002)
5. Gerkey, B.P., Vaughan, R.T., Howard, A.: The Player/Stage Project: Tools for Multi-Robot and Distributed Sensor Systems. In: Proceedings of the International Conference on Advanced Robotics (ICAR), pp. 317–323 (2003)
6. Ando, N., Suehiro, T., Kitagaki, K., Kotoku, T., Yoon, W.K.: RT-Middleware: Distributed Component Middleware for RT (Robot Technology). In: 2005 IEEE/RSJ International Conference on Intelligent Robots and Systems, pp. 3933–3938 (2005)
7. Côté, C., Brosseau, Y., Létourneau, D., Raevsky, C., Michaud, F.: Robotic Software Integration Using MARIE. International Journal of Advanced Robotic Systems 3(4), 55–60 (2006)
8. Makarenko, A., Brooks, A., Kaupp, T.: Orca: Components for Robotics. In: International Conference on Intelligent Robots and Systems, IROS (2006)

9. Song, B., Jung, S., Jang, C., Kim, S.: An Introduction to Robot Component Model for OPRoS (Open Platform for Robotic Services). In: Proceedings of the International Conference Simulation, Modeling Programming for Autonomous Robots Workshop, pp. 592–603 (2008)

10. Quigley, M., Gerkey, B., Conley, K., Faust, J., Foote, T., Leibs, J., Berger, E., Wheeler, R., Ng, A.: ROS: an open-source Robot Operating System. In: ICRA 2009 Workshop on Open Source Software in Robotics, pp. 1–6 (2009)

11. Santos, F., Almeida, L., Pedreiras, P., Lopes, L.S.: A real-time distributed software infrastructure for cooperating mobile autonomous robots. In: International Conference on Advanced Robotics (ICAR), pp. 1–6 (2009)

12. Chishiro, H., Fujita, Y., Takeda, A., Kojima, Y., Funaoka, K., Kato, S., Yamasaki, N.: Extended RT-Component Framework for RT-Middleware. In: IEEE International Symposium on Object/Component/Service-Oriented Real-Time Distributed Computing, ISORC 2009, pp. 161–168 (2009)

13. Jung, M.Y., Deguet, A., Kazanzides, P.: A Component-based Architecture for Flexible Integration of Robotic Systems. In: 2010 IEEE/RSJ International Conference on Intelligent Robots and Systems, pp. 6107–6112 (October 2010)

14. Smart, W.D.: Is a Common Middleware for Robotics Possible? In: IEEE/RSJ International Conference on Intelligent Robots and Systems, IROS 2007 (2007)

15. Martínez, J., Romero-Garcés, A., Vázquez-Martín, R., Bandera, A.: Recipes for Designing High-performance and Robust Software for Robots. In: 2010 IEEE Conference on Robotics Automation and Mechatronics (RAM), pp. 250–255 (2010)

16. Song, I., Guedea, F., Karray, F.: CONCORD: A Control Framework for Distributed Real-time Systems. IEEE Sensors Journal 7(7), 1078–1090 (2007)

17. Guorui, F., Jian, W.: Research of Heterogeneous Robots System Based on CORBA. In: 2011 International Conference on Consumer Electronics, Communications and Networks (CECNet), pp. 569–573 (2011)

18. Knoop, S., Vacek, S., Zollner, R., Au, C., Dillmann, R.: A CORBA-based distributed software architecture for control of service robots. In: 2004 IEEE/RSJ International Conference on Intelligent Robots and Systems, vol. 4, pp. 3656–3661 (2004)

19. Hongxing, W., Shiyi, L., Ying, Z., Liang, Y., Tianmiao, W.: A Middleware Based Control Architecture for Modular Robot Systems. In: 2008 IEEE/ASME International Conference on Mechtronic and Embedded Systems and Applications, pp. 327–332 (2008)

20. Corke, P., Hu, W., Dunbabin, M.: An RPC-based Service Framework for Robot and Sensor Network Integration. In: 2011 IEEE 73rd Vehicular Technology Conference (VTC Spring), pp. 1–6 (2011)

21. Rashid, J., Broxvall, M., Saffiotti, A.: A middleware to integrate robots, simple devices and everyday objects into an ambient ecology. Journal of Pervasive and Mobile Computing 8(4), 522–541 (2012)

22. Marin, R., León, G., Wirzm, R., Sales, J., Claver, J.M., Sanz, P.J., Fernández, J.: Remote Programming of Network Robots Within the UJI Industrial Robotics Telelaboratory: FPGA Vision and SNRP network Protocol. IEEE Transactions on Industrial Electronics 56(12), 4806–4816 (2009)

23. Mizukawa, M., Inukai, H.M.T.K.T., Nodad, A., Tezuka, H., Noguchi, Y., Otera, N.: ORiN: open robot interface for the network - the standard and unified network interface for industrial robot applications. In: 41st SICE Annual Conference, vol. 2, pp. 925–928 (2002)

24. de Rivera, G.G., Ribalda, R., Cols, J., Garrido, J.: A generic software platform for controlling collaborative robotic system using XML-RPC. In: 2005 International Conference on Advanced Intelligent Mechatronics (IEEE/ASME), pp. 1336–1341 (2005)
25. Protocol Buffers – Google Developers (2012), https://developers.google.com/protocol-buffers/
26. Kaur, G., Fuad, M.M.: An Evaluation of Protocol Buffer. In: Proceedings of the IEEE SoutheastCon 2010 (SoutheastCon), pp. 459–462 (2010)
27. Muller, J., Lorenz, M., Geller, F., Zeier, A., Plattner, H.: Assessment of Communication Protocols in the EPC Network - Replacing Textual SOAP and XML with Binary Google Protocol Buffers Encoding. In: 2010 IEEE 17th International Conference on Industrial Engineering and Engineering Management (IE&EM), pp. 404–409 (2010)
28. Wienke, J., Wrede, S.: A Middleware for Collaborative Research in Experimental Robotics. In: 2011 IEEE/SICE International Symposium on System Integration (SII), pp. 1183–1190 (2011)
29. ZeroC - the Internet Communications Engine (2012), http://zeroc.com/ice.html
30. Rise and fall of CORBA (2012), http://www.zeroc.com/documents/riseAndFallOfCorba.pdf
31. ZeroC - Our Customers (2012), http://www.zeroc.com/customers.html
32. Krizsán, Z., Kovács, S.: Some Structural Improvements of the OpenRTM Robot Middleware. In: 2011 IEEE 12th International Symposium on Computational Intelligence and Informatics (CINTI), pp. 345–350 (2011)
33. Gamma, E., Helm, R., Johnson, R., Vlissides, J.: Design Patterns: Elements of Reusable Object-Oriented Software. Addison-Wesley Professional, USA (1994)
34. Peoplebot Robot Makes Human-Robot Interaction Research Affordable (2012), http://www.mobilerobots.com/ResearchRobots/
35. Intelligent Mobile Robotic Platforms for Service pobots, Research and Rapid Prototyping (2012), http://www.mobilerobots.com/
36. Mobilerobots Research Development Software (2012), http://www.mobilerobots.com/
37. Robai - Powerful Affordable Robots (2012), http://www.robai.com/

Enhancing Wi-Fi Signal Strength of a Dynamic Heterogeneous System Using a Mobile Robot Provider

Esther Rolf[1], Matt Whitlock[2], Byung-Cheol Min[3], and Eric T. Matson[3]

[1] Princeton University, Princeton, NJ, USA
[2] The University of Alabama, Tuscaloosa, AL, USA
[3] Machine-to-Machine (M2M) Lab, Department of Computer and Information Technology,
Purdue University, West Lafayette, IN 47907 USA
erolf@princeton.edu, mswhitlock@crimson.ua.edu, {minb,ematson}@purdue.edu

Abstract. Heterogeneous networks of humans, robots, and agents are becoming increasingly common. Clients of wireless networks have continuously changing requirements for providers. In this project, a system to provide a sufficient signal for clients of a network as conditions change is proposed and validated. The system is comprised of hardware features such as a mobile access point and three heterogeneous client devices, and a movement algorithm. The mobile provider's autonomy is verified by the independence of initial position or orientation from success of the system. The system is designed for ease of reconfiguration; modularity in system design allows for advancements to be implemented simply and effectively.

1 Introduction

Wi-Fi enabled devices have become a ubiquitous aspect of society. Wi-Fi networks serve a variety of heterogeneous client systems, from laptops and smartphones connecting to and disconnecting from internet in a coffee shop to teams of heterogeneous mobile robots performing search and rescue missions. An inherent aspect of heterogeneous systems such as these is that some activities or devices will have different Wi-Fi requirements than others. Networks must provide high-quality signals regardless of variable client use conditions. Since needs for Wi-Fi propagation are dynamic, it follows that a Wi-Fi provider should have the ability to accommodate varying environmental conditions. This project seeks to achieve this by mobilizing a Wi-Fi provider.

Received signal strength indication (RSSI), measured in decibel-milliwatts (dBm), is commonly used because Ethernet infrastructure with off-the-shelf utility for measuring RSSI already exists in most indoor spaces [1][6]. However, due to fluctuations caused by phenomena such as multi-path fading and non-uniform propagation of radio signals, RSSI measurements can be unreliable. In [4], RSSI measurements are found too unreliable to be used as the sole source of indoor localization, but in [6] successful position estimation was obtained using wireless Ethernet signal strength. The algorithm proposed in this paper does not use RSSI measurements to determine distance or to localize. Therefore, while complications due to RSSI signal noise are present, they may not be as impeding in the physical realization of the algorithm.

J.-H. Kim et al. (eds.), *Robot Intelligence Technology and Applications 2*,
Advances in Intelligent Systems and Computing 274,
DOI: 10.1007/978-3-319-05582-4_81, © Springer International Publishing Switzerland 2014

A number of methods for exploration and optimization of sensor networks use mapping techniques to traverse an area, including mapping the gradient of RSSI values [13], and intentionally avoiding previously explored areas [14]. This project does not involve mapping the area of interest because RSSI values at specific points are expected to change over time given the changing needs of clients, therefore mapping would be ineffective. Our intent to enhance Wi-Fi signal of clients omni-directionally using robots without mapping capabilities is similar to that of [2], however we propose a more systematic algorithm than random movement.

This project proposes and validates an algorithm that enhances the Wi-Fi strength of a dynamic heterogeneous network such that the average RSSI value for all clients is within a predetermined allowable threshold. As in [5], this project provides a pattern-based algorithm which includes processing sensor information to dictate movements for traversing an area. This algorithm extends to many scenarios and conditions, such as mobile clients, different initial position and orientation of the mobile Wi-Fi provider, different weighting schemes for the clients, and different RSSI thresholds. The algorithm is simple enough to be implemented with a variety of mobile robots and easily adaptable for future improvements in RSSI collection and analysis as well as movement patterns.

The remainder of this paper is outlined as follows: Section 2 details the system configuration including hardware components, methods for RSSI collection and analyzation, and movement protocol. Section 3 includes a condition testing for the test environment and three test cases used to validate the algorithm. Section 4 provides insight for future works and applications. Section 5 concludes the paper with an overall evaluation of the algorithm.

2 Configuration

2.1 Provider and Clients

Provider. The mobile provider consists of a two-tiered shelf system on top of an iRobot Create (Fig. 1). Below the shelves is a 12V lead battery and a transformer to provide 24V of power. On the lower shelf is an Eee laptop, connected by an RJ-45 cable to the Power over Ethernet (PoE) network switch and by a serial connection to the iRobot Create. On the upper shelf is the PoE switch, given power by the battery and transformer, and providing the power to the PicoStation [9] access point (AP). The access point is positioned vertically above the center of the iRobot Create. The Ethernet connections are as such to set up the wired local area network (WLAN). The laptop accesses the PicoStation's TinyOS web interface through its connection to the switch. A bash script using the Linux "grep" command allows the laptop to read RSSI values from the access point using TinyOS. The laptop sends commands to the iRobot Create through an established serial connection. PicoStation RSSI value collection and iRobot Create movement are linked together by a C++ script executed by the laptop.

Clients. Three representative clients make up the heterogeneous team for this project. The DARwin-OP robot is a robot client with a wireless adapter. DARwin is a humanoid robot commonly used for research and education in robotics research [3]. The Nexus tablet is a human controlled device. The Edimax access point/range extender represents a packet-forwarding or routing device. The three clients, as well as the laptop and the PicoStation AP make up the WLAN.

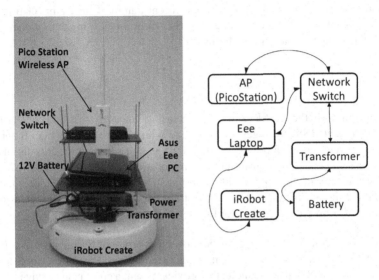

Fig. 1. Mobile Provider

2.2 Algorithm

Our algorithm consists of two procedures: current state analysis and a movement proto-col. During current state analysis, RSSI values are obtained and analyzed to determine whether the average value is within the acceptable threshold, and if not, whether RSSI values are increasing. The movement protocol determines which action to take based on these results. The algorithm is constructed in such a way that it reacts in real-time to client movement, without reliance on a particular initial position or orientation of the mobile provider. It is easily adaptable for different client needs, such as the weight-ing scheme for clients and the overall acceptable threshold limit. A limited number of commands in the movement procedure make the algorithm adaptable for use with different robots. The separation of data collection and integration from physical move-ment commands allows for improvements in either to be implemented without needing to reconfigure the entire system.

Current State Analysis. At each time step, RSSI values are measured by the PicoStation for each client connected to the network. Each client device is given a weight which reflects its priority in the system. Test Cases 1 and 2 weight clients equally, and Test Case 3 examines the effect of different weighting schemes on provider movement. From this step on, the algorithm deals with this weighted average as the singular value to minimize.

An exponentially weighted moving average (EWMA) originally proposed in [10] is used to smooth inherent noise in RSSI measurements while maintaining priority of the most recent data (Eq. (1)). S_t represents the EWMA at time t, Y_t represents the data, in this case RSSI measurement averaged over the three clients, at time t. A constant λ is used to control the weight of the most recent datum at time t in relation to the previously calculated EWMA at time $t - 1$. The selection of λ involves a compromise between detection delay and false alarms [7].

$$S_t = \lambda \cdot Y_t + (1 - \lambda) \cdot S_{t-1} \tag{1}$$

$$S_1 = Y_1 \tag{2}$$

To determine whether RSSI values are increasing, linear regression is performed on the 8 most recent RSSI averages. A trend line with a slope greater than 0 and standard error of regression slope greater than 0.05 is considered an increasing trend in RSSI measurements. These constraints vary by location and can be determined through preliminary testing in the experiment environment to recognize significant changes in RSSI. Validating optimization of these constraints for various environments is outside the intent of this project and is left for future research.

Movement Protocol. Figure 2 details the movement protocol for the provider. RSSI is measured continuously and noise is smoothed using the processes described above. If the RSSI value is within the threshold, the provider is sent a command to stay in place. If the RSSI value is above the threshold and not increasing, the provider is sent a command to move straight. If the RSSI value is above the threshold and increasing, the provider is sent a command to turn 90° counter-clockwise. Thus, with only three commands, the provider moves only when the average RSSI value is above the threshold, moves in directions of decreasing RSSI, and changes trajectory when RSSI values along are first found to be increasing.

RSSI Collection and Movement Algorithm

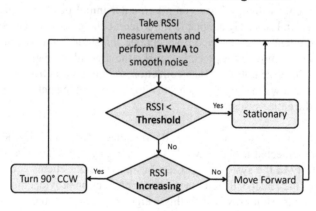

Fig. 2. RSSI Collection Analysis and Movement Algorithm

3   Experiment

The experimentation portion of this project involves condition testing to validate the choice of λ used in calculating the EWMA for the testing environment, followed by three test cases designed to isolate the effects of initial position, initial orientation, and weighting scheme of clients on the mobile provider's subsequent path. A trial is defined as successful if the provider reaches a location where threshold conditions are met (RSSI < 42 dBm in this environment), and remains stationary for more than 10 seconds. The provider's speed is set to 0.2 meters/second (m/s) throughout the tests. The testing location is the ROTC armory at Purdue University in a room which measures roughly the dimensions of a football field. Our testing grid consisted of a 40 meters (m) x 40 meters (m) square.

3.1   Conditions Testing: Optimizing λ

The first experiment tests for an optimal λ value. The provider is sent down a straight path which goes between the three clients, and collects RSSI values along its path. This process is repeated for λ values of 0.1, 0.2, 0.3, 0.4, 0.5, 0.7, and 0.9. Figure 3 displays the outcomes of trials 0.1, 0.3, and 0.5 in order to highlight the effect of changing λ on the EWMA measurement trends. A λ of 0.5 is too large to smooth local noise in the RSSI values caused by unwanted phenomena, while a λ of 0.1 is so small that it inhibits the display of significant changes in RSSI caused by the movement of the provider. A λ of 0.3 is optimal as it captures the desired, global trend of RSSI but smoothes unwanted, local noise.

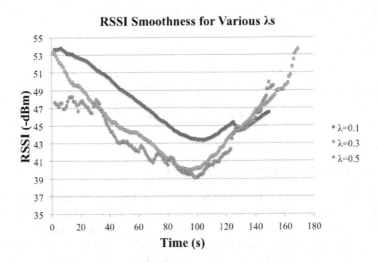

Fig. 3. Mobile Provider

3.2   Test Case 1: Varying Initial Position

The first test case assesses the effect of different initial positions of the provider. The mobile provider starts at 8 different points around the center of the configuration of clients as shown in Fig. 4. In all cases, the provider stops at a position within the threshold, validating that the success of the algorithm is largely independent of initial position. The average number of turns taken per path is 3.375, and the average time to reach the final location is 2 minutes, 50 seconds. The maximum number of turns taken is 8, and the maximum time taken is 6 minutes 30 seconds, both of which occur on the trial originating from the bottom left corner of the grid ($-16m$, $-7.5m$). In some trials, for example that originating at the top-right corner of the grid ($16m$, $27.5m$), the provider takes more than the minimum number of turns necessary to reach its final location, but in such cases the provider corrects quickly, and therefore the efficiency of the algorithm is not compromised.

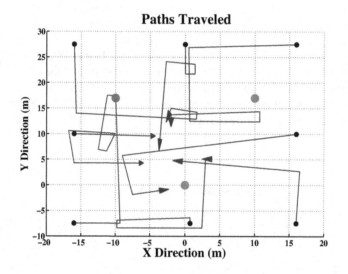

Fig. 4. Initial Positions

3.3   Test Case 2: Varying Initial Orientation

The second test case isolates the effect of initial orientation of the provider on the path taken. Three trials are performed with the provider starting facing each of the cardinal directions as shown in Fig. 5. Table 1 gives the number of turns taken for three trials facing each direction, the number of trials necessary to reach the final location of the provider, and the average time taken. Of the 12 trials, 11 are successful. In some trials,

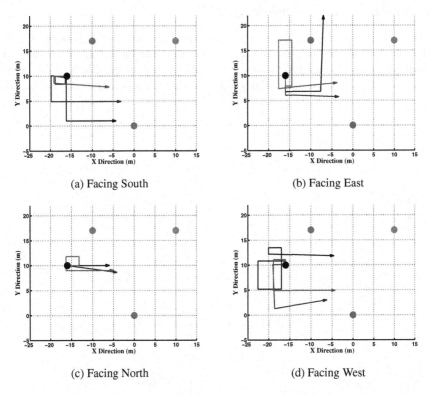

Fig. 5. Provider Paths for Different Initial Orientations

Table 1. Turns Taken and Necessary; Average Time for Different Initial Orientations

Orientation	East	North	West	South
Turns Taken	2 1 5 0 0 4	7 7 3	2 6 2	
Turns Necessary	2 1 3 0 0 3	3 3 3	2 3 2	
Average Time (min:sec)	2:32	1:17	2:56	2:14

the provider takes the most direct route, and in the others, mistakes are quickly corrected. In the first trial facing East, the provider moved off the grid without reaching a successful final location. The average time given for this orientation is the average of the second two trials. After examining the averaged RSSI values over time for this trial, the most likely cause of this is that RSSI values along the last leg of this path were in fact not increasing, and therefore the provider never changed direction. This serves as further proof that RSSI fields do not always follow expected gradient trends. Apart from this one trial, however, success was largely independent orientation.

3.4   Test Case 3: Varying Client Weights

The third case tested the the effect of different schemes for weighting clients (Fig. 6). Figures 6 (a)-(e) show two trials for each of the paths taken for five different weighting weighting schemes. Figure 6 (f) shows the variability of RSSI per client while the mobile provider remains stationary at its starting position for a period of about 3 minutes

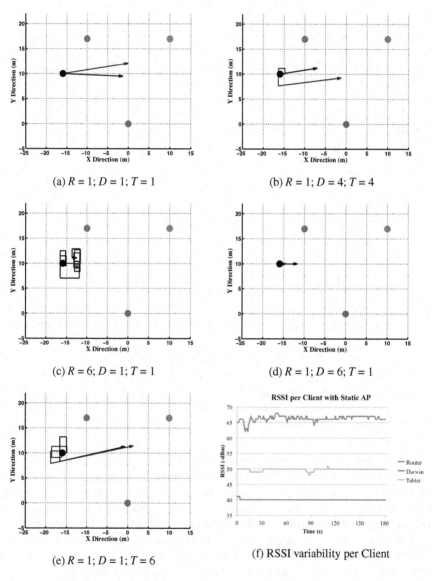

(a) $R = 1; D = 1; T = 1$

(b) $R = 1; D = 4; T = 4$

(c) $R = 6; D = 1; T = 1$

(d) $R = 1; D = 6; T = 1$

(e) $R = 1; D = 1; T = 6$

(f) RSSI variability per Client

Fig. 6. Provider Paths for Different Client Weights. (R = Router, D = DARwin, T = Tablet. Stars mark clients with the highest weights in each scheme.)

$(-16m, 10m)$. The standard deviations of RSSI values from the devices to the provider over this period are as follows: $\sigma_{DARwin} = 0.147$, $\sigma_{tablet} = 0.400$, and $\sigma_{router} = 0.924$.

The most efficient schemes were that in which all clients were weighted equally, (Fig. 6 (a)), that in which the router is weighted less than the other two clients (Fig. 6 (b)), and that in which DARwin is weighted more than the tablet and the router (Fig. 6 (d)). In each of these three schemes, success is reached in both trials, for a total of six successful trials. While both trials for the scheme in which the tablet is weighted more than DARwin and the router are ultimately successful, these trials are not as efficient, as paths double back on themselves (Fig. 6 (e)). In the scheme that weights the router more than DARwin and the tablet, success is not reached in either trial, and paths double back on themselves multiple times (Fig. 6 (c)). Consulting the variability of RSSI signals from each client (Fig. 6 (f)), it is evident that paths are most efficient when clients with lower signal variability are given higher weight (Figs. 6 (a), (b), (d)) and less efficient when clients with higher signal variability are given higher weight (Figs. 6 (c), (e)).

Each client had a different average RSSI value during static testing (Router 66.25 dBm; DARwin 40.02 dBm; Tablet 49.85 dBm). Since the router's average RSSI value of 66.25 was much higher than the testing threshold of 42 dBm, it is realistic that when weighted the router is weighted the most, the threshold condition would never be met. However, we would expect the provider to find a position more optimal for the signal strength of all clients, rather than double back multiple times near its starting location as in Fig. 6 (c). Therefore, while reaching a threshold of 42 dBm may not have been realizable in this trial, the motion taken by the provider suggests that the noise of individual clients has a significant influence on provider path.

4   Future Works and Applications

Adaptability is a key focus of this project. Modularity between data collection, data analysis, and movement protocol allows for improvements in a certain process to be implemented without reconfiguration of the algorithm at large. The algorithm could easily be adapted to measure parameters other than RSSI for connection strength, for example signal quality or throughput. Improvements in smoothing raw data values and interpreting them or in the movement protocol could increase efficiency and reduce the time and energy consumptions for reaching an allowable threshold. This system could also be applied to different environments, in which case constraints and thresholds should be re-evaluated. Multiple APs could be used to create a robotic mesh network in a larger environment as suggested in [8] and [12]. The algorithm could easily be adjusted to accept new clients and account for clients leaving, furthering possibilities for applications involving dynamic systems.

An unavoidable consequence of using mobile robot communication is the possibility of malicious destructive intrusion to the system. Prior to implementation, validating a secure connection between mobile robots of the same network is particularly important. Suggestions for secure implementation detailed in [11].

5 Conclusion

This project validates an algorithm for enhancing the strength of signal connections between a Wi-Fi provider and clients by mobilizing the provider. The algorithm involves current state collection and analysis of RSSI values, which relies on location-dependent constraints, and a simple movement protocol. Since the algorithm is a heuristic method involving pattern-based algorithms, the path taken by a mobile provider is often not the most efficient possible toward the allowable RSSI threshold, however the algorithm is efficient and timely enough to be effective in a physical real-time environment. The algorithm is validated by a high success rate for a variety of initial positions and initial orientations of the mobile provider. This resistance of the algorithm to initial conditions simplifies deployment for the user. The main contribution of this work is the presentation and physical realization of a system that successfully implements a modular design for enhancing Wi-Fi multi-directionally, which easily accommodates future improvements and implementations.

References

1. Chang, N., Rashidzadeh, R., Ahmadi, M.: Robust Indoor Positioning Using Differential Wi-Fi Access Points. IEEE Transactions on Consumer Electronics 56(3), 1860–1867 (2010)
2. Correll, N., Bachrach, J., Vickery, D., Rus, D.: Ad-hoc Wireless Network Coverage with Networked Robots that Cannot Localize. In: IEEE International Conference on Robotics and Automation, ICRA 2009, pp. 3878–3885. IEEE (2009)
3. DARwIn robot: Robotis (2013), http://robotis.com
4. Dong, Q., Dargie, W.: Evaluation of the Reliability of RSSI for Indoor Localization. In: 2012 International Conference on Wireless Communications in Unusual and Confined Areas (ICWCUCA), pp. 1–6. IEEE (2012)
5. Hayes, A.T., Martinoli, A., Goodman, R.M.: Distributed Odor Source Localization. IEEE Sensors Journal 2(3), 260–271 (2002)
6. Ladd, A.M., Bekris, K.E., Rudys, A., Kavraki, L.E., Wallach, D.S.: Robotics-based Location Sensing Using Wireless Ethernet. Wireless Networks 11, 189–204 (2005)
7. Lucas, J.M., Saccucci, M.S.: Exponentially Weighted Moving Average Control Schemes: Properties and Enhancements. Technometrics 32(1), 1–12 (1990)
8. Pabst, R., Walke, B.H., Schultz, D.C., Herhold, P., Yanikomeroglu, H., Mukherjee, S., Viswanathan, H., Lott, M., Zirwas, W., Dohler, M., et al.: Relay-based deployment concepts for wireless and mobile broadband radio. IEEE Communications Magazine 42(9), 80–89 (2004)
9. PicoStation: Ubiquiti Networks (2013)
10. Roberts, S.W.: Control chart tests based on geometric moving averages. Technometrics 1(3), 239–250 (1959)
11. Stoleru, R., Wu, H., Chenji, H.: Secure neighbor discovery and wormhole localization in mobile ad hoc networks. Ad Hoc Networks 10(7), 1179–1190 (2012)

12. Tekdas, O., Yang, W., Isler, V.: Robotic Routers: Algorithms and Implementation. The International Journal of Robotics Research 29(1), 110–126 (2010)
13. Twigg, J.N., Fink, J.R., Yu, P., Sadler, B.M.: RSS Gradient-Assisted Frontier Exploration and Radio Source Localization. In: 2012 IEEE International Conference on Robotics and Automation (ICRA), pp. 889–895. IEEE (2012)
14. Zhang, X., Sun, Y., Xiao, J., Cabrera-Mora, F.: Theseus gradient guide: An indoor transmitter searching approach using received signal strength. In: 2011 IEEE International Conference on Robotics and Automation (ICRA), pp. 2560–2565. IEEE (2011)

Adaptive Neuro-Fuzzy Control for Ionic Polymer Metal Composite Actuators

Nguyen Truong Thinh and Dang Tri Dung

Dept. of Mechatronics, Ho Chi Minh City University of Technical Education
1 Vo Van Ngan St., Thu Duc District, Ho Chi Minh City, Viet Nam
thinhnt@hcmute.edu.vn, dangtridung1991@gmail.com

Abstract. Electroactive polymers (EAPs) have many attractive characteristics for applications, especially for biomimetics robots and bio-medical devices. Among the electroactive polymers, the ionic polymer metal composite (IPMC) is the commonly used EAPs. The IPMC is new generation of smart materials with significant potential in producing biomimetic robots and smart structures, and for medical applications. Ionic polymer metal composites (IPMC) have attracted great attention in the past years due to its large strain. IPMC materials have quite an unpredictable behavior due to several critical aspects that produce a change in their dynamic response. For modeling of controlling the IPMC, it is required to find suitable algorithm. In order to avoid difficult problems in control, a controller based Adaptive Neuro-Fuzzy Inference System (ANFIS), which combines the merits of fuzzy logic and neural network, is used for tracking position of IPMC actuator. This paper describes the using of controller based on Neuro and Fuzzy for controlling an IPMC actuator under water to improve tracking ability for an IPMC actuator like as biomimetics and bio-medical devices. The results showed that ANFIS algorithm is reliable in controlling IPMC actuator. In addition, experimental results show that the ANFIS performed better than the pure fuzzy controller (PFC). Present results show that the current adaptive neuro-fuzzy controller can be successfully applied to the real-time control of the ionic polymer metal composite actuator for which the performance degrades under long-term actuation.

Keywords: IPMC, Neuro-Fuzzy control, ANFIS, ANFC, Position control.

1 Introduction

The IPMC is new generation of smart materials with significant potential in the development of microdevices, actuators, sensors, micromanipulators and microgrippers for multidisciplinary areas such as biomedical, entertainment, and space industries [1]. IPMCs have several advantages as they require low driving voltage (less than 5V) and can produce high displacement. Researches have demonstrated different applications of IPMCs in artificial muscles, distributed actuation devices and also robotics, some of these applications are summarized in [2]. The use of IPMC is being extended to several different applications due to their interesting capabilities such as large stroke with actuating voltage, low stiffness material, biocompatible properties, good operation in wet and underwater operation, etc.

IPMC (Ionic Polymer-Metal Composite) actuator is one of the most promising EAP actuators for replacement of traditional actuators. IPMC is consist a thin polymer membrane with metal electrodes plated on both faces. The bending configuration is made when voltage is applied to metal electrodes on both surface. The motion principle is following: when input voltage is applied to it, action ion which moves freely in membrane is drawn to electrode by electric field with water molecule. Then one side of electrodes is dilated, and another shrank, thus IPMC membrane bends. In fish-like robot, the controlled object is a IPMC membrane fin. The purpose of control system is controlling the angle of the tail fin base on controlling the amplitude and the frequency of caudal fin. IPMC fin is the only actuator for the operation of the fish and IPMC actuator curve displacement by change of input voltage and input frequency so that the control of the fish become the position control of the IPMC membrane. From the experiments results, it can be known that changing the voltage frequency can control the walking speed for the fish-like micro robot.

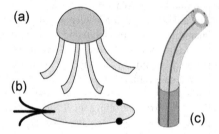

Fig. 1. Several bio-medical products based on IPMC: (a) robotic jellyfish, (b) robotic fish, (c) active catheter

From a bio-engineering viewpoint, this study aims to use IPMC to design several bio-medical products as active catheter, biomimetics, bio-robot... In recent years, the catheter type IPMC was researched and designed[2]. The IPMC is made up of a nafion core with a circular hole through the middle. This hole allows the IPMC to be used for the insertion of specialized tools(Fig. 1(c)). The nafion core is surrounded by two (or four or six) sectioned, chemically plated platinum electrodes, with one(or two or three) DOF respectively to get the specific locations. Besides, there has been growing interest in robotic research where robots have either been used to address specific biological questions or have been directly inspired by biological systems. A variety of biomimetic robots ranging from swinging to swimming have been designed to imitate jellyfishes, waterspiders or fishes. One more type of fish-like robot baed on IPMC concept was researched as in Fig. 1(a). It has a movement process as the jellyfish does when it is floating and sinking. Moreover, the legs based IPMC actuators are designed to replace the antennae of the jellyfish. Controlling the frequency of the actuator's shrinkage and the voltage between its two ends, the upward force can be changed. The IPMC strip can be used to generate the tail of robotic fish and the robotic fish can swim based on IPMC tail (Fig. 1(b)).

In above examples and the most IPMC based applications, the control problem is always position control problem. Base on controlling the IPMC actuator frequency and displacement, the bending of IPMC will make the disire motion. But the disadvantage of the IPMC actuator is the nonlinearity and the degradation in time. So we need a control method that can meet this problem.

Due to the increasing need for intelligent systems in biomimetics as well as bio-medical devices, the ANFIS, which combines the merits of fuzzy logic and neural network, has recently attracted the attention of researchers in various scientific and engineering areas. This control algorithm is a powerful tool to give a suitable solution for controlling the nonlinear plant. Artificial Neural Networks (ANNs) have been used as computational tools for data quality identification because of the belief that they have greater predictive power than signal analysis techniques. However, fuzzy set theory plays an important role in dealing with uncertainty when making decisions in control. Therefore, fuzzy sets have attracted the growing attention and interest in modern information technology, production technique, decision making, and data analysis, etc. ANFIS are fuzzy systems, which use ANNs theory in order to determine their properties (fuzzy sets and fuzzy rules) by processing data samples. Neuro-fuzzy systems harness the power of the two paradigms: fuzzy logic and ANNs, by utilizing the mathematical properties of ANNs in tuning rule-based fuzzy systems that approximate the complex unknown information. A specific approach in neuro-fuzzy development is the adaptive neuro-fuzzy inference system (ANFIS), which has shown significant results in modeling nonlinear functions. In ANFIS, the membership function parameters are extracted from a data set that describes the system behavior. The ANFIS learns features in the data set and adjusts the system parameters according to a given error criterion.

IPMC materials have quite an unpredictable behavior due to several critical aspects that produce a change in their dynamic response. In some applications this unpredictable behavior will produce unacceptable errors[3]. In order to avoid these problems, an ANFIS controller is used in the paper for tracking position of IPMC actuator. This paper presents a method that combines fuzzy logic and neural network to form an algorithm for the target tracking control.

2   Background Developed Controller Based Model of ANFIS

Artificial intelligence techniques have been used with success in different control applications. Main objective of this paper is to investigate the suitability of artificial intelligence system (Neural Networks and Fuzzy-logic) for controlling the IPMC actuator. However, we can consider the IPMC actuator as nonlinear system. Artificial Intelligent (AI) techniques (Neural Networks, Fuzzy logic, Genetic algorithm, Neural-Fuzzy…) give a suitable solution for controlling the nonlinear system, besides they can be used for modeling, prediction and optimization of complex systems. This paper deals with controlling of IPMC actuator using an Adaptive Neuro-Fuzzy Inference System (ANFIS). The Neuro-Fuzzy algorithm was developed by Wang [2] used the hybrid model developed by Takagi-Sugeno. The neuro-fuzzy system used here is the Adaptive Network-based Fuzzy Inference System or Adaptive Neuro-Fuzzy Inference System (ANFIS). The system is an adaptive network functionally equivalent to a first order Sugeno fuzzy inference system. The ANFIS uses a hybrid-learning rule combining back-propagation, gradient-descent and least-squares algorithm to identify and optimize the Segono system's signal.

The main objective of this paper is to use an algorithm of ANFIS for controlling the IPMC actuator in a cantilever configuration. The control inputs are defined as an error (e) and change in error (c) and control output (v) is defined as input voltage to actuate the IPMC. To obtain some data for the training process, the IPMC dynamic system should be considered. The training data, that characterizes a significant part of operating domain, is acquired. We can not use the system in an open loop mode since we can not control the system. So, to assure that training data contain representative data and attenuate the nonlinear characteristic effects, we used the IPMC actuator in a closed loop with gain for tip displacement control. The process of extracting the knowledge of a human operator, in the form of fuzzy control rule, is by no means trivial, nor is the process of deriving the rules based on heuristics and a good understanding of the plant and control theory. The fuzzy controller is designed to have two fuzzy state variables and one control variable for tracking the tip-displacement as reference. Before training can occur, an initial FIS is created based skills of expert in controlling the IPMC actuator. After training, the algorithm of ANFIS used contains rules based the membership functions being assigned to each input and output variable. After several trials, are found to attain best results. The parameters use in ANFIS shown in Table 1.

Table 1. Parameters for ANFIS

Architecture	
The number of layers	5
The number of inputs, rules and output	Input: 2, Rules number: 25, Output: 1
Type of input membership functions	triangle-shaped
Activation functions	Log-sigmoid
Training parameters	
Learning rule	Hybrid Learning Algorithm
Momentum constant	0.9
Number of training patterns	3000
Sum-squared error of training	0.0032
Sum-squared error of testing	0.0043

3   Experiments, Results and Discussions

To verify the adaptive neuro-fuzzy logic controller, a test setup of IPMC was built to perform closed-loop control experiments to evaluate the ANFIS controller. This experimental implementation is built on data acquisition card (DAQ card). The schematic shows the experimental setup for control and measurement of tip displacement of IPMC. The displacement at approximately the tip of IPMC is measured with laser sensor. The experiment was performed IPMC actuator under wet condition. The schematic test setup consists of a set of two electrodes placed at the fixed end of the actuator. The sample of IPMC strip is cantilevered from the electrodes. The IPMC is excited by a voltage supplied by a power amplifier that is controlled by the signal from a digital signal processor.

To demonstrate the capability of the IPMC strip to follow position trajectories under ANFIS controller, several responses were taken to various commanded trajectories. In these experiments, tip-displacement is controlled tracking along several predefined paths to test the performance of the ANFIS controller for different reference inputs: step with long time, sine wave, sawtooth wave, square wave. To validate the ANFC controller that was designed as described in the previous section, an experimental setup was built to execute the closed-loop control of the IPMC actuator. The closed-loop control system with a laser displacement sensor, the IPMC actuator and a current amplifier is shown in Fig. 2. The schematic diagram shows the experimental setup for the control and measurement of the tip displacement of the IPMC actuator. A laser displacement sensor with the resolution of 1 μm is used to measure the tip displacement of the IPMC actuator. In these domains, a fuzzy controller has also five triangular fuzzy sets for both inputs. The control voltage is saturated beyond ±4V and no control action occurs outside those limits.

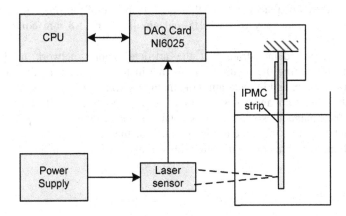

Fig. 2. Experiment setup for feedback control system of ipmc actuator

In the section, some experimental results for the ANFC of the IPMC actuators are addressed. To demonstrate the capability of the ANFC for highly nonlinear IPMC actuators, several responses were taken into account with various command trajectories. In these experiments, the tip-displacement of the IPMC actuator is regulated by following several predefined paths, including step input, sine wave, sawtooth wave and square wave. The first experiment deals with the ANFC control of the IPMC actuator according to the abrupt changes of the multiple step references, as shown in Fig. 3. The IPMC actuator was controlled over 9,000 seconds with the changes of the step inputs. The controlled displacement by the ANFC as shown with the solid line nicely tracks the desired displacement, as shown with the dashed line, and also the controlled signal dramatically adapts to the abrupt change of the target displacement with good performance of the ANFC. One of most important tasks in controller of IPMC actuator is to track the step response against the straightening-back. The step responses of the IPMC actuator reveal that the experimental results are almost near the desired ones.

We found that the ANFC works well as time goes beyond the initial settling time. Over the settling time, the response of the control system is stable and the difference between the commanded reference and the output response is remarkably reduced. Also, it can be seen that the response by the ANFC is faster than results of previous studied methods, which used in reference papers [4, 5, 6] being only valid for short-time control. With the ANFC controller, the time response of step-signals is faster with same desired signals and is controlled within long time. The ANFC controller can track the step response against the straightening-back and also keep the tip position for long time without the degradation of the actuation performance.

Next, other responses for harmonic reference commands with different frequencies and amplitudes are presented to evaluate the performance of the ANFC controller. Fig.4(a) shows the output and input responses for sinusoidal reference signals with frequency of 0.01 Hz and amplitude of 1.6 mm. Fig. 4(b) shows the actual time response signals by the ANFC controller and the command signals of the IPMC strip to a sawtooth profile, and the time history of the controlled voltage output. The tip-displacement followed the commanded signals quite well for long time. Fig. 4(c) shows the desired displacement and the actual response of the IPMC strip to a square waveform with frequency of 0.05 Hz and amplitude of 1.8 mm controlled by the ANFC controller.

Present results demonstrate that the artificial neural network model is quite successful in learning the relationship in time history between the displacement and the control voltage, and in generating the membership functions and rules. The trained membership function and rules are then used to control the IPMC actuator with different commanded signals. As shown in the figures corresponding to the experimental results, the general performance of the adaptive controller show stable and excellent responses in controlling highly nonlinear IPMC actuators.

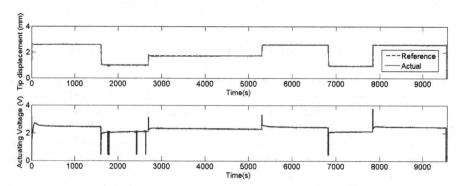

Fig. 3. Experimental results of step response of long time ANFC with abrupt change of desired displacement

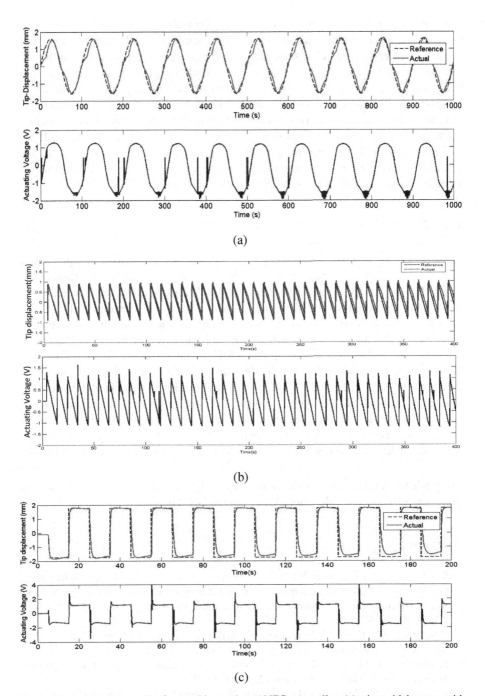

Fig. 4. Experimental results for tracking using ANFC controller (a) sinusoidal wave with frequency of 0.01 Hz and amplitude of 1.6mm, (b) saw wave with frequency of 0.1 Hz and amplitude of 0.9 mm, (c) square wave with frequency of 0.05 Hz and amplitude of 1.8 mm

It can be seen in present results that AFNC controller can operate successfully in the high frequency harmonic regulation.The controlled results of the very low frequency harmonic regulation show relatively large errors. It is well known that the harmonic distortion at the low frequency is much larger than that at the high frequencies because of the actuation mechanism by ion migration in the IPMC actuators. The open-loop control is   inefficient for change of system characteristics and lead to large error in the position control of the IPMC actuator because dynamic characteristics of the polymer actuators are highly nonlinear and very complex. If one requires precise and fast performance and long-time control, the control laws need to adapt the change of the environments and the degradation of actuation performance because classical models become inappropriate and limited.

4   Conclusion

In this study, an adaptive neuro-fuzzy controller has been described to regulate and track the desired tip displacements of the IPMC actuators. The main advantage of this approach is that the control performance of the ANFC controller can be improved through the learning process. The artificial neural network algorithm in the ANFC controller can be used to train the membership functions and the rules for the unpredictable IPMC actuators. For the long-time regulation of the tip displacement, the straightening-back   problem of the IPMC actuator can be directly solved by using an adaptive neuro-fuzzy control algorithm and the long-time activation can be controlled by recovering the degradation of the performance without large control input and changes of the system model. The present control algorithm can be successfully applied to the dynamic tracking of sinusoidal, sawtooth and square waves. The proposed control algorithm can be used to adaptively recover the performance degradation that is a critical problem of electro-active polymer actuators.

Acknowledgements. This study was financially supported Ho Chi Minh city University of Technical Education, Viet Nam (HCMUTE).

References

1. Shahinpoor, M., Bar-Cohen, Y., Xue, T., Simpson, J.O., Smith, J.: Ionic Polymer-Metal Composites (IPMC) as Biomimetic Sensors and Actuators. In: Proceedings of SPIE's 5th Annual International Symposium on Smart Structures and Materials, San Diego, CA, March 1-5, pp. 3324–3340 (1998)
2. Shahinpoor, M., Kim, K.J.: Ionic polymer-metal composites: IV. Industrial and medical applications, Smart Materials and Structures 14, 197–214 (2005)
3. Mallavarapu, K., Leo, D.: Feedback control of the bending response of ionic polymer actuators J. Intell. Mater. Syst. Struct. 12, 143–155 (2001)
4. Richardson, R.C., Levesley, M.C., Brown, M.D., Hawkes, J.A., Watterson, K., Walker, P.G.: Control of ionic polymer metal composites. IEEE/ASME Trans. Mechatron. 8, 245–253 (2003)

5. Brufau-Penella, J., Tsiakmakis, K., Laopoulos, T., Puig-Vidal, M.: Model reference adaptive control for an ionic polymer metal composite in underwater applications. Smart Mater. Struct. 17, 045020 (2008)
6. Oh, S.J., Kim, H.: A study on the control of an IPMC actuator using an adaptive fuzzy algorithm. KSME International Journal 18, 1–11 (2004)

Fall Detection Interface of Remote Excavator Control System

Sun Lim, Han Youn Jin, Jae Soon Park, and Bong-Seok Kim

Intelligent Robotics Research Center, Korea Electronics Technology Institute(KETI),
Bucheon Techno Park 401-402, Wonmi-gu, Bucheon-si, Korea
sunishot@keti.re.kr

Abstract. This paper describes the fall detection algorithm for wireless excavator control system. During the using the it, user's unintentional fall cause the serious and sick problem such as overturned excavator and excavator failure. The distinguish of fall and fall-like activities is very difficult on practice environment. The adaptive band pass filter mechanism is very useful to determine the fall detection and to distinguish the state of excavator control system. Our algorithm reduce both false detection rate while improving fall detection accuracy. In addition, our solution features low computational cost and real time response. Thus most system will be equipped easy.

Keywords: Fall detection, Gyro sensor, remote excavator control system.

1 Introduction

Unintentional falls are determinable events and a common issue, but they are difficult to define rigorously. Remote control of an excavator plays a significant role in real-life application, such as nuclear decommissioning, building demolition, military operations and rescue missions, etc. The advantage of remote control is that it allows the operator to control the mechanism in a remote safe region (zone) via the wired/wireless network. The control Station was developed by KETI[1]. This is vehicle platform for remote control of excavator. It is equipped the three server and chair similar with cabin of excavator. But this is not useful for simple work.

Fall detection work had been mainly studied in the medical devices. For fall detection, some devices are attached on the body. Numerous researchers have performed fall detection using body-worn accelerometers.

There is nothing of report and previous work for wireless excavator control system.

Several accelerometer-based fall detection reports use simple threshold value to differentiate between falls and normal state. These systems mostly rely on single threshold value to detect whether an activity is considered as a fall or simply an normal state. In addition, an accelerometer-only system cannot measure movements with constant velocities and is limited in its use for serious gait analysis for potential fall prevention for any individual patient. Similarly, research has been conducted where gyroscopes were used to detect falls[3-4]. This kind of system has been able to clearly distinguish between falls and normal state using simple threshold-based algorithms, as well as providing more info on gaits.

J.-H. Kim et al. (eds.), *Robot Intelligence Technology and Applications 2*,
Advances in Intelligent Systems and Computing 274,
DOI: 10.1007/978-3-319-05582-4_83, © Springer International Publishing Switzerland 2014

Our solution divides human activities into two categories: static state, and dynamic transition state between states. We define falling as an unintentional transition during operation of excavator. Our system can recognize five kinds of posture: standing, sitting, bending, walking and running. To determine whether a transition is intentional or unintentional, our system measures not only linear acceleration but also angular velocity with gyroscopes. By using both accelerometer and gyroscopes, our distinguish state and fall detection algorithm is more accurate than existing methods. Moreover, our solution has low computational cost and fast response.

The rest of paper is divided into four sections. Sec. 2 describes remote excavator control system with sensor. Sec. 3 proposes our fall detection solution. The evaluation is also presented this section. The last sec. 4 concludes the paper and gives directions for future work.

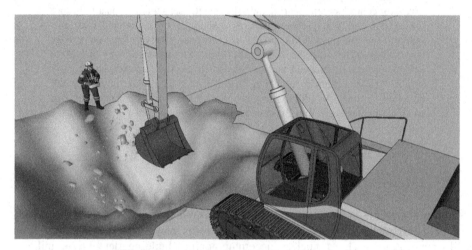

Fig. 1. The remote excavator control environment

2   Remote Excavator Control System

2.1   Gyro Sensor

The RECS(remote excavator control system) includes the accelerometers and gyroscopes. Finding a suitable location for the placement of both and accelerometer and gyroscope is important. Thus we prove and test the reasonable location of sensor unit. During our testing two locations are considered that is depicted figure 1. The accelerometer used is MicroStrain's 3DM-GX2 , which has a selectable range of +/-2g~10g(5g is standard). In this work the accelerometer is set to 5g. The gyroscope used are the +/- 75 degree/se, 150 degree/sec, 300 degree/sec and 1200degree.sec. The all sensors are controlled by an TI TMS320F28335-150Mhz DSP controller. The sampling rate is set to 1kHz, a bandwidth exceeding the characteristic response or human movement action.

2.2   Remote Control Excavator System

The objective is to develop a fall detection system that is based on expert knowledge and band pass filter.

Apparatus
1) It must be able to detect falls while user is standing, walking, running and acting.
2) It must be distinguish the using state.
3) It must be prevent the unintentional action on falling .
4) It must be small and simple to easy to add the other system.
5) It must be activated only during falls and not during daily activities.

Based on these considerations, our system consists of a gyro-sensor and monitoring embedded board. Fig.2 shows a block diagram of the device.

The gyro sensor(spec) were used to measure the subject's movement. The sensing data was converted into digital data with 12bit resolution. The received data was transferred to the CPU and then analyzed with the fall detection and distinguish algorithm.

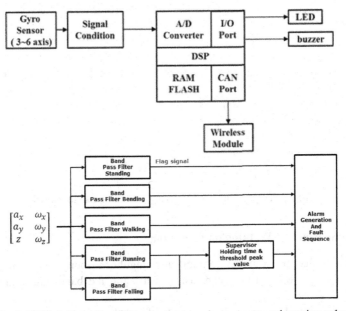

Fig. 2. Multiple Band Pass filter w.r.t state and gyro measured sensing value

3   Fall Detection Algorithm Strategy

The data collected are segmented into one 10 msecond intervals. If the change of sensor reading within interval falls into the matched model output, it is classified as a static state. Otherwise, a dynamic transition is assumed. The detail mathematical equation and proof is omitted. For static state, the measure data with the data acquisition is calculated band-pass filter. The band pass filter is step up for individual state such as stating bandwidth 0~1Hz and walking bandwidth 3~5 Hz. These individual band pass filter is updated by the operation type and parameter of physical dimension of RECS on real time.

If the operator is running, the measured data is very similar on the falling detection as shown in Fig. 3. If the transition was unintentional, it is flagged as a fall. The flagged are cleared only by the operator. The band pass filter configuration parameter is also tuned by the operator or automatic procedure. This configuration parameter is considered the sensitivity of state transition. An unintentional transition to static state is regarded as a fall, and it features large accelerometer and gyroscope reading. This condition has very large peak value of acceleration and angular rate from sensing value.

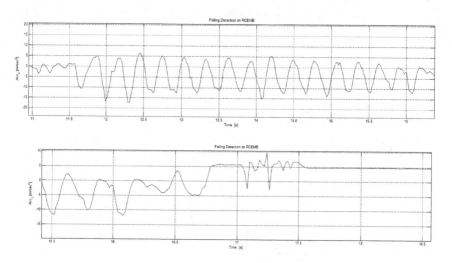

Fig. 3. The running operation *(top)* and falling state *(down)*

3.1   Comparison of Walking and Running

Fig. 4 shows that tri-axial acceleration and angular velocity graph of walking a running differ greatly and they can distinguish using the bandwidth and maximum difference of wave.

The below table show the predefined state of remote controller usage. The standing state is nominal reference value of overall situation. In this during time, the controller and supervisor calculation and initialization all band pass filter and falling detection model parameters. If this duration time is not guaranteed for long time, the distinguishing and falling detection algorithm's accuracy decreased.

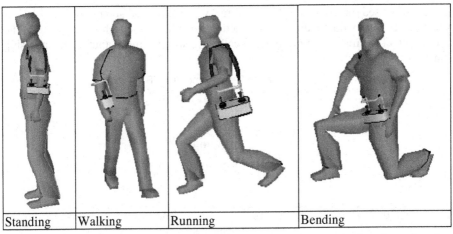

| Standing | Walking | Running | Bending |

The walking state has the very short interval and low magnitude of acceleration w.r.t. z axis. Thus sum vector of x,y and z axis acceleration is very low. Thus the comparison of walking and running is not difficult. The detail detection is described the bellow algorithm structure. The individual band pass filter is designed based on the active frequency and magnitude of state.

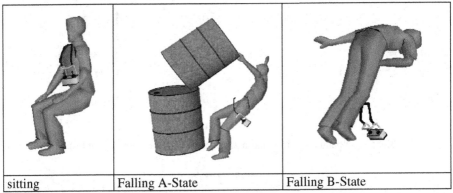

| sitting | Falling A-State | Falling B-State |

The scenario is defined on standing state(5'), walking(5'), running(5'), falling(5'), bending(5') and standing(5'). The z a axis raw value illustrated on Fig.4. The magnitude and frequency of signal is also capture on same signal. Thus the distinguish and detection mechanism is constructed by the band pass filter and peak magnitude value.

The band-pass filters have the range 0.2~20Hz and the resolution is determined and configured by the user on initial setting. If the cut-off frequency is detailed configured, the implementation board require the amount of memory size and calculation time. This is trade off relation between precise and memory space.

Figure 5 and 6 are depicted on the band pass filter of individual cut-off frequency and x,y,z axis sum vector, i.e., $\sqrt{a_x^2 + a_y^2 + a_z^2}$. This value is useful for detection change of state transition. However it is not enough to determine the state only to use sum vector because the value is very similar running and falling state.

The proposed band pass filter with fuzzy detection is good work on the all situation. The individual membership function generates the matching frequency. The supervisor of falling detection automatically decision by the output of membership function.

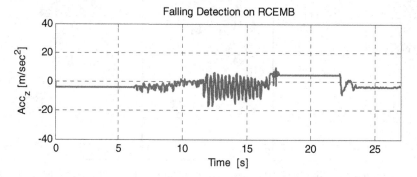

Fig. 4. The magnitude and frequency of z-axis signal

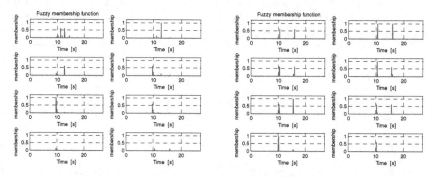

Fig. 5. The band pass filter and fuzzy membership function output result

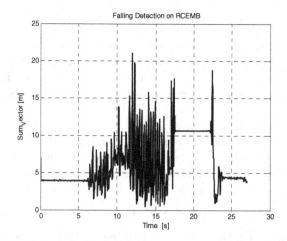

Fig. 6. The magnitude of all accelerometer vectors (x,y and z axis)

4   Conclusion and Future Work

The experiment results have demonstrated the proposed fall detection is effective and useful. The algorithm is typically based upon the value of acceleration and angular velocity. The tri-axial acceleration and angular velocity is used to distinguish the state of RECS and detect an unintentional falling. The band pass filter of individual state such as standing, bending, walking and running, is designed on measured data is matching the output of filter is flagged. The hold time and threshold can determine fall direction. These are configuration parameters of band pass filter. This filter is required also low- computation than other algorithm. The improving accuracy of this algorithm is the major work in the future.

Acknowledgements. This work was supported by the Industrial Strategic technology development program, 10041965, Development of a parallel link robot with 6kg payload and 0.43 sec cycling time, funded by the Ministry of Knowledge Economy (MKE), Korea.

References

1. Stoev, J., Cho, K., Shim, J.-S., Lee, H.S.: Free Fall Detection Algorithms for Hard Disk Drives. In: SICE-ICASE International Joint Conference, October 18-21, pp. 2760–2764 (2006)
2. Bourke, A.K., O'Brien, J.V., Lyons, G.M.: Evaluation of a threshold-based tri-axial accelerometer fall detection algorithm. Gait and Posture 26, 194–199 (2007)
3. Najafi, B., et al.: Ambulatory system for human motion analysis using a kinematic sensor: monitoring of daily physical activity in the elderly. IEEE Trans. Biomed. Eng. 50, 711–723 (2003)
4. Li, Q., Stankovic, J.A., Hanson, M.A., Barth, A.T., Lach, J., Zhou, G.: Accurate, Fast Fall Detection Using Gyroscopes and Accelerometer-Derived Posture Information. In: Proc. Sixth Int'l Workshop on Wearable and ImplantableBody Sensor Networks, June 03-05, pp. 138–143 (2009)
5. Bourke, A.K., Lyons, G.M.: A threshold-based fall-detection algorithm using a bi-axial gyroscope sensor. Med. Eng. Phys. 30, 84–90 (2009)

Solar Energy as an Alternative Energy Source to Power Mobile Robots

Abdusalam Sulaiman[*,**], Freddie Inambao, and Glen Bright

Mechanical Engineering, University of KwaZulu-Natal,
Howard College Campus, Glenwood, Durban, South Africa
200300000@stu.ukzn.ac.za, {Inambaof,bright}@ukzn.ac.za

Abstract. Solar energy can provide a viable alternative energy source to meet the special energy demands that are typically required to operate mobile robots. Conventional energy sources cannot fulfil these demands as satisfactorily as solar energy can, given the disfavour that conventional energy sources find in an eco-conscious world, and also given the practical limitations associated with conventional energy sources which cannot conveniently be accessed in places where mobile robots are normally put to use which are often inaccessible and beyond human reach. This study seeks to demonstrate that solar energy can be harnessed and stored using hydrogen as a medium to store an otherwise intermittent supply of energy that is characteristic of solar energy. In this study, an Industrial Mobile Robot Platform (IMRP) was designed to run on fuel cells using a low-pressure metal hydride hydrogen storage system which would store more energy on board than a rechargeable battery could.

Keywords: Energy, Solar, Hydrogen, Mobile Robot, Fuel cell.

1 Introduction

In an ever more eco-conscious world, the quest for cleaner and environmentally-friendly energy sources has become increasingly pressing. Conventional energy sources, with all the negative connotations associated with the toxic nature of its production, are being shunned in favour of renewable energy sources.

Solar energy, as one form of renewable energy, offers enormous potential as an alternative to conventional energy sources. It does not have all the negative draw-backs of traditional energy sources such as fossil-based electricity and petroleum oil which carries a huge threat of pollution to the environment [1, 2].

Apart from environmental concerns, there are other draw-backs to conventional energy sources which are key limiting factors for a mobile robot which requires versatility to be efficient. Electricity, for instance, may not be available in a disaster area, or it may not be practical for a robot to be constantly attached to a power source by

---

[*] The author is a candidate for a PhD degree in Engineering in the Mechanical Department, University of KwaZulu-Natal, South Africa.

[**] Corresponding author.

J.-H. Kim et al. (eds.), *Robot Intelligence Technology and Applications 2*,
Advances in Intelligent Systems and Computing 274,
DOI: 10.1007/978-3-319-05582-4_84, © Springer International Publishing Switzerland 2014

means of an electrical cord in situations where it is required to be manoeuvred in difficult areas of access. Petroleum also, has its own draw-backs as it would necessitate the storing of bulk fuel on-board, thereby hampering the operation of a mobile robot which, by its very nature is required to be light, compact and versatile enough to be able to manoeuvre in the tightest of spaces [1, 3].

Rechargeable batteries on the other hand, appear to offer a more practical alternative in that they are light, and therefore can be taken on-board without making the mobile robot cumbersome as would be the case if bulk fuel like petroleum had to be taken on-board. Also, since batteries are a portable source of power, they do not have the draw-backs of electricity which requires the mobile robot to be constantly attached to a power source, thereby limiting its range of movement and distance. But batteries too, have their draw-backs. Rechargeable batteries, by nature, have very low densities and high rates of discharge, thereby making them ineffective to sustain a mobile robot's energy demands in times of high peak demands and over long duration missions.

As for nuclear energy, there is very strong global political pressure against its use, and hence, it has become the central aim of world energy policy today to seek alternative sources of energy, preferably from renewable forms such as solar.

But solar energy presents certain challenges. It cannot provide a constant supply of energy, since it is dependent on available sunshine, which is not always present. This means that solar power is intermittent and unreliable. But if suitable means can be developed to store this energy, it could offer a viable alternative that could address all the draw-backs to be found in the energy sources already covered above [4, 5].

2   Research

The research presented here will demonstrate that solar energy can indeed provide a constant and reliable source of energy to meet the energy demands of a mobile robot. The challenges posed in finding a suitable method of storing that energy can be overcome by the use of hydrogen as a medium for storage, using metal hydride.

3   Methodology

As a first step, sunlight is converted to electricity by the use of photo-voltaic (PV) cells. Apart from the high initial cost in setting up the PV panels, once built, the system has the advantage of very low operating costs and the energy produced by the PV panels are cost-free, and also free of any waste products.

However, as the availability of sunlight is intermittent, there is a need to devise a back-up power system so as to be able to store the electricity produced by the PV panels and thereby have a readily available source of energy at all times.

Hydrogen offers the best energy storage solution. Hydrogen is an ideal energy carrier. To produce hydrogen, the process of electrolysis is used to split water or H_2O into its constituent elements of hydrogen and oxygen. The electricity that will be required for this process of electrolysis, would already be available from the electricity produced by the PV cells from sunlight, as described above.

So, during periods of high availability when there is abundant sunlight, the excess electricity that is produced from solar will be directed to the electrolyzer to produce hydrogen, thereby rendering the surplus electricity in a form that would be capable of storage [6].

Next, a medium must be found to store the hydrogen. Metal hydrides are the most compact way to store hydrogen. They can be used as a storage medium for the hydrogen, often reversibly so that when the renewable source isn't available, the hydrogen can be converted back to electricity to provide constant power. Reversible metal hydrides offer several benefits over other means of storing hydrogen. They operate at low pressure, especially when compared to compressed hydrogen, and do not need to be kept at the cryogenic temperatures required for liquid hydrogen storage. Reversible hydride storage typically requires less energy on a system basis, is compact, and can be conformable [7] to fit space available on an application such as a mobile robot.

The conversion of hydrogen back to electricity requires a fuel cell. Fuel cells are devices that produce electric power by direct conversion of a fuel's chemical energy. Fuel cell systems offer many potential benefits as a distributed generation system. They are small, and modular, and capital costs are relatively insensitive to scale. A hydrogen fuel cell does not generate any pollution. The only by-product is pure water, which is emitted as both liquid and vapour, depending on the operating conditions (temperature and pressure) and system configuration [8].

Fuel cells are being extensively studied in many research environments for the potential they offer in converting energy without the losses associated with thermal cycles, thereby having the potential to increase efficiency. Compared to other power sources, they operate silently, have no major moving parts, and can be assembled easily into larger stacks.

Fuel cells are essentially energy converters. They resemble batteries in many ways, but in contrast to batteries, they do not store the chemical energy: fuel has to be continuously provided to the cell to maintain the power output. Various designs for fuel cells have been proposed, the most popular being the proton-exchange-membrane (PEM) fuel cell, operating at temperatures up to about 100 °C, and the solid-oxide fuel cell (SOFC), operating at temperatures of about 800 °C or higher. Whereas the underlying principle is always to extract electricity without combustion, each design presents different problems and advantages, and has unique characteristics that make it more appropriate for different environments [9].

Hydrogen and fuel cells are now widely regarded as key energy solutions for the 21st century. These technologies will contribute significantly to a reduction in environmental impact, enhanced energy security (and diversity) and the creation of new energy industries [6].

One of the big advantages of fuel cells and a difference to batteries is that they decouple energy storage from power production. This makes it easy to provide more energy (in the form of fuel) as needed, and as long as fuel is supplied, the power available stays the same. In effect, a fuel cell gives similar benefits as an internal combustion engine, but it is also quieter, more efficient, non-polluting, and more easily scaled-down [2, 10].

4   Case Study : Powering a Mobile Robot with Renewable Energy Using Hydrogen Storage Solutions

4.1   The Proposed System for Hydrogen Storage

The electricity that is produced by the PV panels is direct current (DC). To meet the immediate energy requirements of the mobile robot, this DC voltage is firstly passed to the batteries. From there the DC voltage will need to be converted into alternative current (AC) by means of a DC/AC inverter, so that the current is in the required form to power the mobile robot.

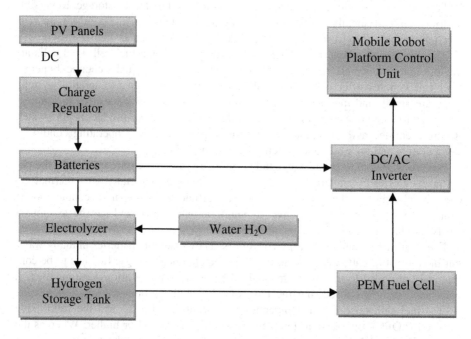

Fig. 1. Block diagram of the proposed power supply for the Mobile Robot

As for the storage of the electricity, the system will operate as follows : The electricity that is obtained from the PV panels, which is in DC voltage, will be used for the electrolysis unit to generate hydrogen. The hydrogen obtained by electrolysis is stored in hydrogen tanks containing the metal hydrides. Thereafter, a Proton Exchange Membrane (PEM) fuel cell is used to re-convert hydrogen back to electricity, as illustrated in Figure 1. This electricity is in DC voltage and will then be converted to AC voltage, using the DC/AC inverter, which can now be applied to meet the electrical demands of the mobile robot platform.

The PV system includes, in total, 6 PV modules and the total installed power is 1.2kWp in standard conditions. The specifications of the solar modules are determined as follows: maximum power is 200 W; Open circuit voltage (V_{oc}) is 59.5V;

Optimum operating voltage (V_{mp}) is 46.1V; Optimum operating current (I_{mp}) is 4.37A.

The DC energy obtained from the PV panel groups is stored in 4 units of solar batteries such that each unit has 12 V to 350 Ah.

The electricity required for electrolysis is supplied from the PV system. The electrolysis process will be facilitated by an electrolyser. The hydrogen that is produced from the electrolysis process will be stored in hydrogen tanks.

Current hydrogen storage methods include gas compression and liquefaction, and are not optimal because not only are they energy-intensive and expensive but present safety issues. A promising alternative is solid-state hydrogen storage, which utilizes metal hydrides to absorb/desorb hydrogen at relatively low pressure offering safety and cost advantages with potentially unparalleled hydrogen storage density.

Metal hydrides are used for hydrogen storage whereby the hydrogen is chemically bonded to one or more metals and released with a catalyzed reaction or heating. The hydrides can be used for storage in a solid form or in a water-based solution. When a hydride has released its hydrogen, a by-product remains in the fuel tank to be either replenished or disposed of. Hydrides may be reversible or irreversible [7]. Reversible hydrides act like sponges, soaking up the hydrogen. They are usually solids. These compounds release hydrogen at specific pressures and temperatures. They may be replenished by adding pure hydrogen. Irreversible hydrides are compounds that go through reactions with other reagents, including water, and produce a by-product. Metal hydrides can hold a large amount of hydrogen in a small volume. [7, 8].

This technique is very advantageous because of the safe handling of hydrogen and the convenience it provides for mobile applications [10, 11]. The hydrogen stored in metal hydride tanks is used to run the fuel cell system, which in turn provides the power to operate the Mobile Robot. The experiment set-up for this system is shown in figure 2. It consists of a metal hydride hydrogen tank, a hydrogen supply valve, a pressure regulator, a controller, a PEM fuel cell, power recording and analyzing instruments.

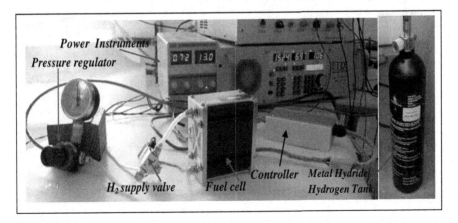

Fig. 2. Experiment set-up showing the hydrogen fuel cell power supply system

4.2   Hydrogen Fuel Cell System Simulation Model

To analyse the performance of this system, a simulation tool is used in the form of the TRNSYS (Transient Energy System Simulation Tool) software. TRNSYS has a wide range of use in renewable energy systems, and has been used extensively to simulate solar energy applications. The TRNSYS model simulates the performance of the entire energy-system by breaking it down into individual components. To create a model, the end user is able to create custom components or choose from the TRNSYS standard library of components. It is a very flexible tool that allows any user with a FORTRAN compiler to define their own elements into the software if necessary. Each component in the software is a FORTRAN sub-routine with input, output and calculation parameters. Every component can be linked to each other with output/input relations. A fuel cell component reads its input such as inlet pressures, physical properties, cell current, number of cells, cooling data and membrane properties, and then runs the sub-routine and calculates the output data such as cell voltage, power and temperature, hydrogen consumption or energy efficiency. By linking the hydrogen consumption output of the fuel cell, and hydrogen production output of the electrolyzer to the hydrogen outflow and hydrogen inflow inputs of the hydrogen tank respectively, the hydrogen tank sub-routine can calculate the hydrogen level in the tank. Upon linking the hydrogen tank output, user power demand and electricity production output of the PV panels to the system controller, the controller can decide how the system should work. The hydrogen fuel cell system TRNSYS simulation model is shown in figure 3.

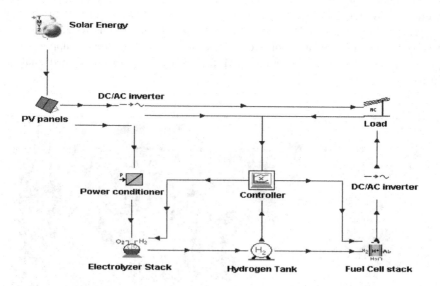

Fig. 3. Hydrogen fuel cell system simulation model

TRNSYS calculates the state of each component at every step. The system consists of several inter-connected components. The components used in the model are: a photovoltaic array module, a fuel cell module, and a hydrogen storage module [12]. The fuel cell converts chemical energy to electricity, in much the same way as a battery. The difference between a battery and a fuel cell is that the fuel cell does not have any internal storage of chemical energy, but is supplied externally by the fuel. The fuel is pure hydrogen supplied from the hydrogen storage tank. It is possible to use the excess heat from the fuel cell. The cell voltage takes the form:

$$U_{cell} = E + \eta_{act} + \eta_{ohmic} \tag{1}$$

$$U_{cell} = U_{low} + (U_{high} - U_{low}) \cdot \frac{T_{fc} - T_{low}}{T_{high} - T_{low}} \tag{2}$$

Where:

η_{act}	Activation voltage loss
η_{ohmic}	Voltage loss due to resistance
U_{cell}	Cell voltage at the given temperature
U_{low}	Maximum voltage for low temp I-V curve
U_{high}	Maximum voltage for high temp I-V curve
T_{fc}	Temperature of fuel cell
T_{high}	Temperature for high temp I-V curve
T_{low}	Temperature for low temp I-V curve

The current for the high (I_{high}) and the low temperature (I_{low}) curve is calculated from this equation 3. To find the current at the working temperature of the fuel cell, the current is calculated by linear interpolation [13-15]:

$$I_{temp} = I_{low} + (I_{high} - I_{low}) \cdot \frac{T_{fc} - T_{low}}{T_{high} - T_{low}} \tag{3}$$

Where:

E	Thermodynamic potential
I_{high}	Maximum current for high temp I-V curve
I_{low}	Maximum current for low temp I-V curve
I_{temp}	Maximum current at the given temperature
U_{low}	Maximum voltage for low temp I-V curve
U_{high}	Maximum voltage for high temp I-V curve
U_{temp}	Maximum voltage at the given temperature
T_{fc}	Temperature of fuel cell
T_{high}	Temperature for high temp I-V curve
T_{low}	Temperature for low temp I-V curve

Two main efficiencies are calculated, the electric efficiency, η_{el} and the total efficiency η_{eff}. The reason for calculating two efficiencies is that it is only the electric efficiency that will heat up the cells. The total efficiency also includes the loss of hydrogen that will not heat up the fuel cell (the electric efficiency and the total efficiency will be very close at normal or high production, but will differ at a very low production rate). Thus:

$$\eta_{el} = \frac{V_{fc} \cdot I_{fc}}{V_{fc} \cdot (I_{fc} + k_{kurloss} \cdot V_{f_c})}_{...} \qquad (4)$$

$$\eta_{eff} = \frac{V_{fc} \cdot I_{fc}}{V_{ref} \cdot (I_{fc} + k_{kurloss} \cdot V_{f_c} + k_{hydloss} \cdot I_{min})}_{...} \qquad (5)$$

where:

I_{fc}	Current for fuel cell
K	Constant for Temperature-Voltage equation
V_{fc}	Voltage over fuel cell
V_{ref}	Reference voltage
η_{el}	electric efficiency
η_{eff}	total efficiency

The hydrogen energy storage sub-system, comprising an electrolyzer, hydrogen storage tank, and a fuel cell, is an integral part of a solar-hydrogen power supply system for supplying power to the Mobile Robot. This energy storage is required due to variation of the intermittent and variable primary energy source [15].

4.3   Mobile Robot System Hardware Architecture

An Industrial Mobile Robot Platform (IMRP) is used for this case study, and consists of an Automatic Guided Vehicle (AGV), Mobile Robot Frame, motors and mecanum wheels, as shown in figures 4 and 5. The development for this case study can be divided into the following major processes: the mechanical design for the mecanum wheels and mobile robot chassis; the electronics design for the 4-channel motor driver; interfacing with a Basic Stamp micro-controller board; and software development for motion control. The AGV has the unique ability to move laterally. It is unlike conventional vehicles that are limited to forward and reverse motion.

This is advantageous at decision points in manufacturing environments where branching routes are employed for routing purposes. Lateral movement was made possible by the AGV's mecanum wheels. Each wheel consists of a steel rim with eight nylon rollers. The rollers were positioned at 45 degrees to the rotation direction and rotated freely about their own longitudinal axis. A small area of contact was made between a single roller and the floor when the wheel was stationary. As the wheel rotated, during forward or reverse motion, a particular roller left the floor and the contact area was picked up by the next roller and so on. When the AGV moved laterally or rotated about its own axis, the forces from the counter-wheel-rotation caused the contact area to rotate the particular roller. When the roller left the floor the area was picked up by the next roller and it starts to rotate [16].

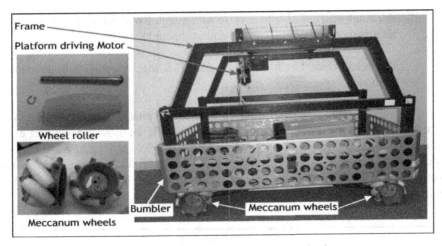

Fig. 4. Mobile Robot system mecanum wheels

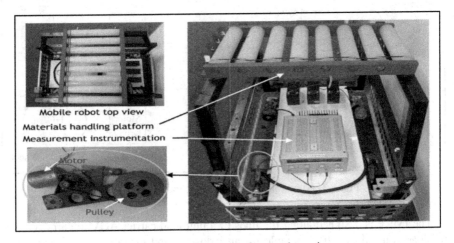

Fig. 5. Mobile robot system front and top view

The Electronic design of the Mobile Robot, which consists of a four-channel bi-directional motor driver, is designed to drive all four mecanum wheels, as illustrated in figure 6. The slave devices consist of five Motor Boards, two light-sensor boards, four ultra-sonic sensor boards, and a Limit Switch Board. An LCD module acts as a diagnostic device and verifies signal transfer to the robot, as illustrated in figure 7. Digital signals are sent back to the main computer via Radio Frequency (RF). The inputs are interpreted and used accordingly in the main program.

The specifications developed for the necessary Driver Board are as follows:

a)   A circuit which is compatible with a single logic-level PWM input signal for speed control of each wheel, and a single logic-level input line for the direction of motor rotation for each wheel.

b) A circuit which is able to operate with a high PWM carrier frequency from the microcontroller to provide inaudible operation.

c) A circuit which has four independent H-Bridge drivers for bi-directional motion.

Each H-Bridge driver circuit is capable of providing suitable continuous current at 12V DC. The motor controller board is illustrated in figure 6. A maximum of eight motor boards are addressable by the robot CPU. The AGV has five 12V DC motors. Four motors drive the four wheels, one motor per wheel, and one motor drives the materials handling platform. Five motor controller boards are used. Each board is individually addressed, and has a micro-controller [17, 18].

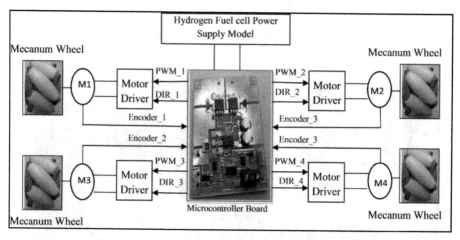

Fig. 6. Hardware architecture of the Mobile Robot system

The IMRP is designed along the lines of a unit load carrier AGV to carry individual loads on a deck. It is suitable for an industrial and manufacturing environment where it can move in confined spaces and travel over short distances. It is designed to move in all directions and carry single loads on its deck. The deck is a moving platform with payloads not exceeding 40 Kg. The loads could be picked up at a predetermined station and deposited at another. To achieve mobility, an electronic drive controller unit has been designed to provide the vital link between the motors and the computer system [17].

A micro-computer is used to control this unit with the aid of appropriate software via a computer interface card. This allows the IMRP to move in any desired planer direction. The IMRP is able to follow pre-determined paths to transport the incoming raw materials and outgoing machine parts between work stations. It is envisaged that the IMRP would be able to carry a pay load of 40 Kg. Thus a gross weight of 120 Kg was chosen. The frame was made from 38mm x 38mm square tubing and steel joints were arc-welded.

4.4   Mobile Robot Telemetry System Components

The system has been custom-made. It was made according to specifications of the guidance and navigation requirements of a mobile Robot [17, 19]. As illustrated in figure 7, the telemetry system consists of two main components: a USB-Transceiver unit and a robot CPU unit that communicates with each other via RF. The robot CPU unit transfers data to the respective slave modules via the RS485BUS. The robot CPU unit acts as a data acquisition device that one can read and write from. Higher level programming to control the Mobile Robot has to be done by the user.

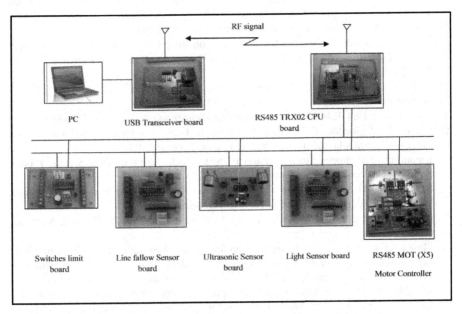

Fig. 7. Mobile Robot telemetry system components

5   Conclusion

The use of renewable energy from solar power offers a viable alternative to traditional energy sources such as petroleum and fossil-based electricity, and is one of the potential options for overcoming current environmental and sustainability issues. Solar energy can efficiently meet the energy demands of modern technologies, as demonstrated in this case study, where it has been shown as an efficient way of meeting the energy demands of an Industrial Mobile Robot Platform. The various challenges associated with developing solar energy in a viable way, can be addressed by the use of efficient storage methods as dealt with in this study. There are, however, some challenges in improving hydrogen storage technologies with regard to their efficiency, size, weight, capacity and, ultimately their cost. Durability remains an issue, as does the development of unified international codes and safety standards to facilitate safe deployment of commercial technologies. Energy efficiency is also a challenge for all hydrogen storage approaches. And although the cost of on-board hydrogen storage systems may be currently too high, particularly in comparison to conventional storage systems for petroleum fuels, world energy policy is gradually driving demand in this area. Increasing demand and mass production is certain to drive down costs and make this source of energy more cost-effective.

References

1. Zuttel, A., Borgschulte, A., Schlapbach, L.: Hydrogen as a Future Energy Carrier. WILEY-VCH Verlag GmbH & Co. KGaA (2008)
2. Hordeski, M.F.: Alternative Fuels: The Future of Hydrogen, 2nd edn. CRC Press (2008)
3. Yilanci, A., Dincer, I., Ozturk, H.K.: A review on solar-hydrogen/fuel cell hybrid energy systems for stationary applications. Progress in Energy and Combustion Science 35, 231–244 (2009)
4. Dincer, I., Rosen, M.A.: Sustainability aspects of hydrogen and fuel cell systems. Energy for Sustainable Development 15, 137–146 (2011)
5. Rand, D.A.J., Dell, R.M.: Hydrogen Energy Challenges and Prospects. RSC (2008)
6. Corbo, P., Migliardini, F., Veneri, O.: Hydrogen Fuel Cells for Road Vehicles (Green Energy and Technology), Springer-Verlag London Limited (2011)
7. Broom, D.P.: Hydrogen Storage Materials the Characterization of Their Storage Properties. Springer-Verlag London Limited (2011)
8. Hirscher, M.: Handbook of Hydrogen Storage New Materials for Future Energy Storage. WILEY-VCH Verlag GmbH & Co. KGaA, Weinheim (2010)
9. Larminie, J., Dicks, A.: Fuel Cell Systems Explained, 2nd edn. John Wiley & Sons Ltd. (2003)
10. Masten, D.A., Bosco, A.D.: System Design for Vehicle Applications: GM/Opel. In: Vielstich, W., Gasteiger, H.A., Lamm, A. (eds.) Handbook of Fuel Cells Fundamentals, Technology and Applications, vol. 4, pp. 714–772. John Wiley, New York (2003)
11. Ali, S.M., Andrews, J.: Low-cost storage options for solar hydrogen systems for remote area power supply. In: 16th World Hydrogen Energy Conference, Lyon, France (2005)
12. Gibril, S.: Solar hydrogen energy House for Libya. In: World Hydrogen Energy Conference, Brisbane, Australia (2008)
13. Lee, S.: Development of a 600 W Proton Exchange Membrane Fuel Cell Power System for the Hazardous Mission Robot. Journal of Fuel Cell Science and Technology (2010)
14. Rubio, M.A., Urquia, A., Dormida, S.: Diagnosis of Performance Degradation Phenomena in PEM Fuel Cells. International Journal of Hydrogen Energy, 1–5 (2009)
15. Ulleberg, O., Glockner, R.: HYDROGEMS, Hydrogen Energy Models. In: WHEC. 14th World Hydrogen Energy Conference, Montreal (2002)
16. Pancham, A.: Variable Sensor System for Guidance and Navigation of AGVs. MSc Thesis, University of KwaZulu Natal (2008)
17. Mccomb, G.: Robot Builder's Sourcebook. McGraw-Hill (2003)
18. Xu, W.L., Bright, G., Cooney, J.A.: Visual dead-reckoning for motion control of a Mecanum-wheeled mobile robot. Mechatronics International Journal 14, 623–637 (2003)
19. Bright, G.: M.Sc. Thesis: Automated guided Vehicle System for manufacturing operations, University of KwaZulu Natal (1989)

Appendix A: Mobile Robot Power Supply

The electrical power required by the five DC motors, the motor controller unit (RS485MOT and Encoder), the Central Processing Unit (CPU- RS485TRX02), the USB unit, the ultrasonic sensor, the light sensor, the line follow sensor, the limit switch sensor, and LCD (RS485 BUS Monitor) on the IMRP, is 195.72 watt as shown in Table 1.

Table 1. Power consumption

	Component	Description	V	I	Q	Current	Power (w)
1	DC Motor		12V	3 A	5	15A	180
2	RS485MOT & Encoder	Motor controller	12V	0.2A	5	1A	12
3	RS485TRX02	Robot CPU	12V	0.1A	1	0.1A	1.2
4	USBTRX02	USB unit	12V	0.05A	1	0.05A	0.6
5	Add-on Sensors	Ultrasonic Sensor	12V	0.02 A	4	0.08A	0.96
6		Light Sensor	12V	0.02 A	1	0.02 A	0.24
7		Line follow Sensor	12V	0.02 A	1	0.02 A	0.24
8		Limit switch Sensor	12V	0.02 A	1	0.02 A	0.24
9	RS485 BUS Monitor	LCD	12V	0.02 A	1	0.02 A	0.24
	Total power required onboard for Mobile Robot						195.72
10	PC (laptop)		19.5	3.34A	1	3.34A	65.13

Author Index

Printed in the United States
By Bookmasters